# Matrix Computations

Johns Hopkins Series in the Mathematical Sciences

1. *Nonparametric Probability Density Estimation*
   Richard A. Tapia and James R. Thompson
2. *Matrix-Geometric Solutions in Stochastic Models: An Algorithmic Approach*
   Marcel F. Neuts
3. *Matrix Computations*
   Gene H. Golub and Charles F. Van Loan
4. *Traffic Processing in Queueing Networks*
   Ralph L. Disney and Peter C. Kiessler
5. *Nonlinear Preference and Utility Theory*
   Peter C. Fishburn

# Matrix Computations

SECOND EDITION

Gene H. Golub
*Department of Computer Science*
*Stanford University*

Charles F. Van Loan
*Department of Computer Science*
*Cornell University*

The Johns Hopkins University Press
Baltimore and London

Originally published, hardcover and paperback, 1989
Third printing, 1991

The Johns Hopkins University Press
701 West 40th Street
Baltimore, Maryland 21211
The Johns Hopkins Press Ltd., London

The paper used in this publication meets the
minimum requirements of American National
Standard for Information Sciences—
Permanence of Paper for Printed Library
Materials, ANSI Z39.48-1984.

ISBN 0-8018-3772-3
ISBN 0-8018-3739-1 (pbk.)

Library of Congress Catalog Number 88-45404

Dedicated to

ALSTON S. HOUSEHOLDER

AND

JAMES H. WILKINSON

# Contents

# Preface to the First Edition

It can be argued that the "mission" of numerical analysis is to provide the scientific community with effective software tools. What makes this enterprise so interesting is that its participants require skills from both mathematics and computer science. Indeed, good software development demands a mathematical understanding of the problem to be solved, a flair for algorithmic expression, and an appreciation for finite precision arithmetic. The aim of this book is to provide the reader with these skills as they pertain to matrix computations.

Great progress has been made in this area since the mid-1950's. This is borne out by the existence of quality programs for many linear equations, least squares, and eigenvalue problems. Typical are the routines in Eispack and Linpack, whose widespread use has had the effect of elevating the level of algorithmic thought in various applied areas. By using the programs in these packages as building blocks, scientists and engineers can piece together more complicated software tools that are tailored specifically for their needs. This development encourages the writing of well-structured programs, a welcome trend since more and more scientific research is manifested in software.

The impact of numerical linear algebra is felt in other ways. The habits of the field—our reliance on orthogonal matrices, our appreciation of problem sensitivity, our careful consideration of roundoff—have spilled over into many areas of research. A prime example of this is the increased use of the singular value decomposition (SVD) as an analytical tool by many statisticians and control engineers. People in these areas are reformulating numerous theoretical concepts in the "language" of the SVD and as a result are finding it much easier to implement their ideas in the presence of roundoff error and inexact data.

Further evidence of the growing impact of numerical linear algebra is in the area of hardware design. Recent developments in floating point arithmetic and parallel processing have in no small measure been provoked by activity in the matrix computation field.

We have written this book in order to impart a sense of unity to this expanding and exciting field. Much has been accomplished since the publication in 1965 of Wilkinson's monumental treatise *The Algebraic Eigen-*

*value Problem.* Many of these modern developments have been discussed in survey articles and in specialized volumes such as *Solving Least Squares Problems* by Lawson and Hanson and *The Symmetric Eigenvalue Problem* by Parlett. We feel that the time is appropriate for a synthesis of this material. In this regard we see *Matrix Computations* as a comprehensive, somewhat more advanced version of Stewart's *Introduction to Matrix Computations.*

We anticipate three categories of readers: graduate students in technical areas, computational scientists and engineers, and our colleagues in numerical analysis. We have included special features addressed to each group.

For students (and their instructors), we have included a large number of problems. Many of these are computational and can form the nucleus of a programming project. Our experience in teaching with the book is that Eispack and Linpack assignments greatly enliven its contents. Matlab, an easy-to-use system for performing matrix calculations, has also been used successfully in conjunction with this text.

For practicing engineers and scientists who wish to use the volume as a reference book, we have tried to minimize the interconnections among chapters. We have also included an annotated bibliography at the end of almost every section in order to hasten the search for reference material on any given topic.

For our colleagues in numerical analysis, we have sprinkled our algorithmic discussions with a generous amount of perturbation theory and error analysis. It is our intention to provide these readers with enough detail so that they can understand and solve the matrix problems that arise in their own work. Research in numerical linear algebra is frequently instigated by ongoing work in other areas of numerical analysis. For example, some of the best methods for solving sparse linear systems have been developed by researchers concerned with the numerical solution of partial differential equations. Similarly, it was the activity in the quasi-Newton area that prompted the development of techniques for updating various matrix factorizations.

This book took six years to write and underwent several title changes: (1) *A Last Course in Matrix Computations*, (2) *Applied Matrix Computations*, and (3) *Advanced Matrix Computations.* The first title, aside form being "cutesy," is misleading. We do not pretend that our treatment of matrix computations is complete. In particular many topics in the vibrant area of sparse matrix computation were excluded from the text simply because we did not wish to delve into graph theory and data structures. The second title was dismissed because we do not dwell at great length on applications. For the most part, the matrix problems we consider are treated as given—their origins are not chased down. We recognize this as a pedagogic shortcoming but one which can be offset by an experienced teacher. Moreover, we suspect that many of our readers will be experienced themselves, thereby obviating the need for excessive motivation. Finally, the last title was dispensed with because the book does contain introductory material. We chose to include elementary topics for the sake of completeness and because we think that our approach to the rudiments of the subject will interest teachers of undergraduate numerical analysis.

What, then, is the book about if it is incomplete, less than applied, and not entirely advanced? A brief synopsis of its contents should answer this question.

The first three chapters contain the necessary background material. Matrix algebra is reviewed, and some key algorithms are established. The pace is rather brisk. Readers who struggle with the problems in these early chapters will no doubt struggle with the remainder of the book.

Much current research in numerical linear algebra focuses on problems in which the matrices have special structure, e.g., are large and sparse. The art of exploiting structure is the central theme of Chapter 5, where various special-purpose linear equation solvers are described.

Chapter 6 picks up on another trend in the field—the increased reliance on orthogonal matrices. We discuss several orthogonalization methods and show how they can be applied to the least squares problem. Special attention is paid to the handling of rank deficiency.

The all-powerful QR algorithm for the unsymmetric eigenvalue problem is the centerpiece of Chapter 7. Our pedagogic derivation should help to demystify this important technique. We also comment on invariant subspace calculation and the generalized eigenvalue problem.

In Chapter 8, we continue our discussion of the eigenvalue problem by focusing on the important symmetric case. We first describe the symmetric QR algorithm and then proceed to show how symmetry permits several alternative computational procedures.

Up to this point in the book, our treatment of sparsity is rather scattered. Banded linear system solvers are discussed in Chapter 5, simultaneous iteration is described in Chapter 7, Rayleigh quotient iteration in Chapter 8, and so on. Chapters 9 and 10, however, are entirely devoted to the solving of sparse matrix problems. The discussion revolves around the Lanczos method and its country cousin, the method of conjugate gradients. We show how various sparse eigenvalue, least squares, and linear equation problems can be solved using these important algorithms.

The purpose of the last two chapters in the book is to illustrate the wide applicability of the algorithms presented in earlier chapters. Chapter 11 deals with the problem of computing a function of a matrix, something that is frequently required in applications of control theory. Chapter 12 describes a selection of matrix problems, several of which highlight the power of the singular value decomposition.

Indeed, perhaps the most recurring theme in the book is the practical and theoretical value of this matrix decomposition. Its algorithmic and mathematical properties have a key role to play in nearly every chapter. In many respects *Matrix Computations* is an embellishment of the manuscript "Everything You Wanted to Know About the Singular Value Decomposition (But Were Afraid to Ask)" authored by our colleague Alan Cline.

A word is in order about references to available software. We have concentrated on Eispack and Linpack, and just about every subroutine in these packages is alluded to in the text. In addition, numerous "tech report" references that we cite in the annotated bibliographies are in fact references to software. It should be stressed, however, that we have no direct experience with much of the referenced software aside from Eispack and Linpack. Caveat emptor.

Many people assisted in the production of the book. Richard Bartels helped to gather references and to revise the first draft of some early chapters. The writing of this book was prompted by his organization of the Workshop in Matrix Computations held at Johns Hopkins University in August 1977.

Bob Plemmons, John Dennis, Alan Laub, and Don Heller taught from various portions of the text and provided numerous constructive criticisms. George Cybenko generously helped us with the section on Toeplitz matrices, while Bo Kågström offered many intelligent comments on our treatment of invariant subspace computation. Per-Åke Wedin diligently read early versions of Chapter 5 and expertly guided our revisions. Uri Ascher and Roger Horn have our gratitude for independently suggesting the book's final title and so does an anonymous reviewer who made many valuable suggestions.

Last, but not least, we happily acknowledge the influence of our colleagues Cleve Moler and Pete Stewart. Their own work, which so perfectly captures the spirit of the field, has strongly affected the balance and style of our own presentation.

# Preface to the Second Edition

We were motivated to write a revised and expanded version of *Matrix Computations* for two reasons. First, after five years of teaching from the old edition and hearing from colleagues, it became clear that much of what we wrote could be written better. Many of the proofs and derivations have been clarified and we have adopted a stylized Matlab notation that facilitates matrix/vector thinking. Another new stylistic feature of the new edition is the classification of material down to the subsection level. This should make the volume easier to use as a textbook and as a reference.

The second reason for the new edition has to do with the blossoming of the parallel matrix computation area. The multiprocessor is revolutionizing the role of computation in science and engineering and it is crucial that we document the contributions of numerical linear algebra to this process. We do so in a machine-independent way that stresses high-level algorithmic ideas in preference to specific implementations. As we say in the new chapter on parallel matrix computations, the field is very fluid and the literature is overwhelmed with case studies. Nevertheless, there are a handful of key algorithmic developments that are likely to be with us for a long time and these are most definitely ready for textbook level discussion.

Here is a chapter-by-chapter tour of what is new in the Second Edition:

In Chapter 1 (Matrix Multiplication Problems) we introduce notation and fundamental concepts using matrix multiplication as an example. Block matrix manipulation is given a very high profile and we have added a new section on vector pipeline computing.

In Chapter 2 (Matrix Analysis) we have concentrated all the mathematical background required for the derivation and analysis of least squares and linear system algorithms.

In Chapter 3 (General Linear Systems) and Chapter 4 (Special Linear Systems) there is new material on how to organize for "high performance" the Gaussian elimination and Cholesky procedures. Emphasis is placed on implementations that are rich in matrix-vector and matrix-matrix multiplication. A subsection on positive semi-definite matrices has been added.

Chapter 5 (Orthogonalization and Least Squares) has new material on block Householder computations. We have also adjusted the order of presentation so as to decouple the discussion of orthogonal factorizations from the discussion linear least squares fitting.

In Chapter 6 (Parallel Matrix Computations) we use the gaxpy operation to introduce distributed memory and shared memory computation. The organization of matrix multiplication and various matrix factorizations in these two multiprocessing environments is also discussed.

In Chapter 7 (The Unsymmetric Eigenvalue Problem) we added a subsection on block Hessenberg reductions. In Chapter 8 (The Symmetric Eigenvalue Problem) has two more fundamental alterations. The Jacobi method section was completely redone in light of recent developments in parallel computing. We also added a section on a new highly parallel divide and conquer algorithm for the tridiagonal problem.

In Chapter 9 (The Lanczos Method) we have added a subsection on the Arnoldi method. Chapter 10 (Iterative Methods for Linear Systems) includes an expanded discussion of the preconditioned conjugate gradient algorithm. The only significant change in Chapter 11 (Functions of Matrices) and Chapter 12 (Special Topics) is a new subsection in §12.6 on hyperbolic downdating.

We are deeply indebted to the many individuals who have pointed out the typos, mistakes, and expository shortcomings of the First Edition. It has been a pleasure to deal with such an interested and friendly readership. From this large group of correspondents we would like to thank Å. Björck, J. Bunch, J. Dennis, T. Ericsson, O. Hald, N. Higham, A. Laub, R. Le Veque, M. Overton, B.N. Parlett, R. Plemmons, R. Skeel, E. Stickel, S. Van Huffel, and R.S. Varga for their particularly detailed and cogent remarks.

We also wish to acknowledge the contributions of individuals and organizations that have had a critical role to play in the production and shaping of this second edition. At the LaTeX level we were assisted by Alex Aiken, Chris Bischof, Gil Neiger, and Hal Perkins at Cornell and by Mark Kent at Stanford. Cindy Robinson-Hubbell at Cornell machine-coded the first edition and was absolutely indispensible during all subsequent phases of production. The research facilities made available by Iain Duff at Harwell Laboratory in the United Kingdom and Bill Morton at Oxford University were essential to the revision process.

Iain Duff, Bo Kågström, Chris Paige, and Nick Trefethen each contributed to the work in profound philosophical and technical ways. Our sincere gratitude goes to these friends and colleagues.

Nick Higham carefully read the entire manuscript and offered countless suggestions. The publication of the Second Edition would have been much delayed and much inferior without Nick's help.

Finally, we wish to acknowledge the many contributions of Jim Wilkinson, our very dear colleague who passed away in 1986. Jim continues to be a great inspiration and it is with pleasure that we dedicate the new volume to both him and Alston Householder.

# Using the Book

### Abbreviations

The following references are frequently cited in the text:

**SLE** G.E. Forsythe and C. Moler (1967). *Computer Solution of Linear Algebraic Systems,* Prentice-Hall, Englewood Cliffs, NJ.

**SLS** C.L. Lawson and R.J. Hanson (1974). *Solving Least Squares Problems*, Prentice-Hall, Englewood Cliffs, NJ.

**SEP** B.N. Parlett (1980). *The Symmetric Eigenvalue Problem,* Prentice-Hall, Englewood Cliffs, NJ.

**IMC** G.W. Stewart (1973). *Introduction to Matrix Computations,* Academic Press, New York.

**AEP** J.H. Wilkinson (1965). *The Algebraic Eigenvalue Problem,* Clarendon Press, Oxford, England.

Mnemonics are used to identify these "global" references, e.g., "Wilkinson (AEP, chapter 5)." Bibliographic information associated with a given section appears locally. A master bibliography is also included.

References to the software packages Linpack and Eispack are tacit references to the corresponding manuals:

B.T. Smith, J.M. Boyle, Y. Ikebe, V.C. Klema, and C.B. Moler (1970). *Matrix Eigensystem Routines: EISPACK Guide,* 2nd ed., Springer-Verlag, New York.

B.S. Garbow, J.M. Boyle, J.J. Dongarra, and C.B. Moler (1972). *Matrix Eigensystem Routines: EISPACK Guide Extension,* Springer-Verlag, New York.

J. Dongarra, J.R. Bunch, C.B. Moler, and G.W. Stewart (1978). *LINPACK Users Guide,* SIAM Publications, Philadelphia.

Thus, "subroutine xyz may be found in Linpack (Chapter 2)" means that subroutine xyz is described in the second chapter of the Linpack manual.

Algol versions of many Eispack and Linpack subroutines are collected in

**HACLA** J.H. Wilkinson and C. Reinsch, eds. (1971). *Handbook for Automatic Computation, Vol. 2, Linear Algebra,* Springer-Verlag, New York.

At the time of writing (1989), it was possible to obtain software for many of the algorithms in the book by sending electronic mail to any of the following three addresses:

netlib@anl-mcs.arpa
netlib@research.att.com
research!netlib

Typical messages include

| | |
|---|---|
| send index | {For a list of available libraries} |
| send index for linpack | {For a list of Linpack codes.} |
| send svd from eispack | {For the Eispack svd code.} |

This electronic distibution service is described in

J.J. Dongarra and E. Grosse (1987). "Distribution of Mathematical Software via Electronic Mail," *Comm. ACM 30,* 403-407.

We also mention that a master bibliography of *Matrix Computations* (in LaTeX format) is available from netlib.

### A Note to Instructors

The book can be used in several different types of courses. Here are some samples:

**Title** Introduction to Matrix Computations (1 semester)
**Syllabus** Chapters 1-4 and 5.1-5.3, 7.1-7.6, 8.1-8.2

**Title** Matrix Computations (2 semesters)
**Syllabus** Chapters 1-12

**Title** Linear Equation and Least Square Problems (1 semester)
**Syllabus** Chapters 3-5, and 12.1-12.4, 12.6

**Title** Eigenvalue Problems (1 semester)
**Syllabus** Chapters 7-9, 11, and 12.5

**Title** Parallel Matrix Computations (1 semester)
**Syllabus** Chapters 1, 6, 8.5-8.6.

In each case, we strongly recommend the inclusion of computing assignments that involve the use of state-of-the-art software packages like Eispack and Linpack. The subject can be further enlivened by using Matlab, an easy-to-use-system for performing matrix computations. In this regard we recommend the companion volume

T.F. Coleman and C.F. Van Loan (1988). *Handbook for Matrix Computations*, SIAM Publications, Philadelphia, PA.

It has chapters on Fortran 77, the basic linear algebra subprograms (the Blas), Linpack, and Matlab.

Finally, we mention the new text

W.W. Hager (1988). *Applied Numerical Linear Algebra,* Prentice Hall, Englewood Cliffs, NJ.

which may be more suitable than *Matrix Computations* for the beginning student.

# Matrix Computations

# Chapter 1

# Matrix Multiplication Problems



The proper study of matrix computations begins with the study of the matrix-matrix multiplication problem. Although this problem is simple mathematically it is very rich from the computational point of view. We begin in §1.1 by looking at the several ways that the matrix multiplication problem can be organized. The "language" of partitioned matrices is established early on and is used to characterize several linear algebraic "levels" of computation.

If a matrix has structure then it is usually possible to exploit it. For example, a symmetric matrix can be stored in half the space as a general matrix. A matrix-vector product that involves a matrix with many zero entries may require much less time to execute than a full matrix times a vector. These matters are discussed in §1.2. In §1.3 block matrix notation is established. A block matrix is a matrix with matrix entries. This concept is very important from the standpoint of both theory and practice. On the theoretical side, block matrix notation allows us to prove important matrix factorizations very succinctly. These factorizations are the cornerstone of numerical linear algebra. From the computational point of view, block algorithms are important because they are rich in matrix multiplication, the operation of choice for many new high performance computer architectures.

These new architectures require the algorithm designer to pay as much attention to memory traffic as to the actual amount of arithmetic. This new dimension of scientific computation is illustrated in §1.4 where the critical aspects of vector pipeline computing are discussed. These aspects include vector stride, vector length, the number of vector loads and stores, and the level of vector re-use.

## 1.1 Basic Algorithms and Notation

Matrix computations are built upon a hierarchy of linear algebraic operations. Dot products involve the scalar operations of addition and multiplication. Matrix-vector multiplication is made up of dot products. Matrix-matrix multiplication amounts to a collection of matrix-vector products. All of these operations can be described in algorithmic form or in the language of linear algebra. Our primary objective in this section is to show how these two styles of expression complement one another. Along the way we pick up notation and acquaint the reader with the kind of thinking that underpins the matrix computation area. The discussion revolves around the matrix multiplication problem, which we show can be organized in several ways.

### 1.1.1 Matrix Notation

Let $\mathbb{R}$ denote the set of real numbers. We denote the vector space of all $m$-by-$n$ real matrices by $\mathbb{R}^{m \times n}$:

$$A \in \mathbb{R}^{m \times n} \quad \Longleftrightarrow \quad A = (a_{ij}) = \begin{bmatrix} a_{11} & \cdots & a_{1n} \\ \vdots & & \vdots \\ a_{m1} & \cdots & a_{mn} \end{bmatrix} \quad a_{ij} \in \mathbb{R}\,.$$

When a capital letter is used to denote a matrix (e.g. $A$, $B$, $\Delta$) the corresponding lower case letter with subscript $ij$ refers to the $(i,j)$ entry (e.g., $a_{ij}$ , $b_{ij}$, $\delta_{ij}$). We also use the notation $[\,A\,]_{ij}$ to designate the $(i,j)$ entry of a matrix $A$. When presenting a very detailed algorithm, matrix entries are usually specified in "programming language" style: $A(i,j)$.

### 1.1.2 Matrix Operations

The basic manipulations with matrices include transposition ($\mathbb{R}^{m \times n} \rightarrow \mathbb{R}^{n \times m}$),

$$C = A^T \quad \Longrightarrow \quad c_{ij} = a_{ji},$$

addition ($\mathbb{R}^{m\times n} \times \mathbb{R}^{m\times n} \to \mathbb{R}^{m\times n}$),

$$C = A + B \quad \Longrightarrow \quad c_{ij} = a_{ij} + b_{ij},$$

scalar-matrix multiplication, ($\mathbb{R} \times \mathbb{R}^{m\times n} \to \mathbb{R}^{m\times n}$),

$$C = \alpha A \quad \Longrightarrow \quad c_{ij} = \alpha a_{ij},$$

and matrix-matrix multiplication ($\mathbb{R}^{m\times r} \times \mathbb{R}^{r\times n} \to \mathbb{R}^{m\times n}$),

$$C = AB \quad \Longrightarrow \quad c_{ij} = \sum_{k=1}^{r} a_{ik} b_{kj}.$$

These are the building blocks of matrix computations.

### 1.1.3 Vector Notation

Let $\mathbb{R}^n$ denote the vector space of real $n$-vectors:

$$x \in \mathbb{R}^n \quad \Longleftrightarrow \quad x = \begin{bmatrix} x_1 \\ \vdots \\ x_n \end{bmatrix} \quad x_i \in \mathbb{R}.$$

We refer to $x_i$ as the $i$th component of $x$. We also use $x(i)$ in algorithmic contexts when referring to the $i$th component of a vector $x$.

Notice that we are identifying $\mathbb{R}^n$ with $\mathbb{R}^{n\times 1}$ and so the members of $\mathbb{R}^n$ are *column* vectors. On the other hand, $\mathbb{R}^{1\times n}$ is composed of *row* vectors:

$$x \in \mathbb{R}^{1\times n} \quad \Longleftrightarrow \quad x = (x_1, \ldots, x_n).$$

If $x$ is a column vector then $y = x^T$ is a row vector.

An operation between vectors that "looks funny" is the *outer product* :

$$C = xy^T \qquad x \in \mathbb{R}^m,\ y \in \mathbb{R}^n.$$

This is a perfectly legal matrix multiplication because the number of columns in $x$ matches the number of rows in $y^T$. If $C = xy^T$ then $c_{ij} = x_i y_j$ and so, for example,

$$\begin{bmatrix} 1 \\ 2 \\ 3 \end{bmatrix} \begin{bmatrix} 4 & 5 \end{bmatrix} = \begin{bmatrix} 4 & 5 \\ 8 & 10 \\ 12 & 15 \end{bmatrix}.$$

### 1.1.4 Vector Operations

There are four basic vector operations. If $\alpha \in \mathbb{R}$, $x \in \mathbb{R}^n$ and $y \in \mathbb{R}^n$ then we have scalar-vector multiplication $z = \alpha x$ ($z_i = \alpha x_i$), vector addition $z = x + y$ ($z_i = x_i + y_i$), dot product $c = x^T y$ ($c = \sum_{i=1}^n x_i y_i$), and vector multiply $z = x.*y$ ($z_i = x_i y_i$). The *saxpy* operation is a fifth operation that is so important in matrix computations that we add it to our list of basic vector operations even though it makes that list redundant. The saxpy operation is defined as follows:

$$z = \alpha x + y \quad \Longrightarrow \quad z_i = \alpha x_i + y_i$$

The name "saxpy" is used in Linpack, a software package that implements many of the algorithms in this book. One can think of "saxpy" as a mnemonic for "scalar alpha $x$ plus $y$."

### 1.1.5 The Computation of Dot Products and Saxpys

We have chosen to express algorithms in a stylized version of the Matlab language. Matlab is an elegant interactive system ideal for matrix computation work. See Coleman and Van Loan (1988, Chapter 4). We gradually introduce our stylized Matlab notation in this chapter beginning here with a function for computing dot products.

**Algorithm 1.1.1 (Dot Product)** Given $n$-vectors $x$ and $y$, the following algorithm computes their dot product $c = x^T y$.

```
function: c = dot(x, y)
    c = 0
    n = length(x)
    for i = 1:n
        c = c + x(i)y(i)
    end
end dot
```

In this procedure, the dimension of the vectors is obtained by referencing **length**. The **for** statement indicates that the count variable $i$ takes on the values $1, 2, \ldots, n$. The value of the dot product is returned in $c$. The dot product of two $n$ vectors involves $n$ multiplications and $n$ additions. More informally, it is an $O(n)$ operation, meaning that the amount of work is linear in the dimension.

The saxpy computation is also an $O(n)$ operation but it returns a vector instead of a scalar:

**Algorithm 1.1.2 (Saxpy)** Given $n$-vectors $x$ and $y$ and a scalar $\alpha$, the following algorithm computes $z = \alpha x + y$.

```
function: z = saxpy(α, x, y)
    n = length(x)
    for i = 1:n
        z(i) = αx(i) + y(i)
    end
end saxpy
```

### 1.1.6 A Note on Formality and Style

It should be stressed that the algorithms in this book like **dot** and **saxpy** are encapsulations of critical computational ideas and not "production codes." Moreover, these encapsulations may be written quite informally if it suits the exposition. It so happens that Matlab functions are a handy way of describing the computations in this and the next few sections. But as we progress through the book greater informality is called for because the algorithms get too complicated to detail at the "$ij$" level. This should pose no difficulty to the reader who acquires an intuitive sense about matrix computations from the early chapters.

### 1.1.7 Matrix-Vector Multiplication

Suppose $A \in \mathbb{R}^{m\times n}$ and we wish to compute the matrix-vector product $z = Ax$ where $x \in \mathbb{R}^n$. The usual way this computation proceeds is to compute the dot products

$$z_i = \sum_{j=1}^{n} a_{ij}x_j$$

one at a time. This gives the following algorithm.

**Algorithm 1.1.3 (Matrix-Vector Multiplication: Row Version)** If $A \in \mathbb{R}^{m\times n}$ and $x \in \mathbb{R}^n$, then the following algorithm computes $z = Ax$.

```
function: z = matvec.ij(A,x)
    m = rows(A) ; n = cols(A)
    z(1:m) = 0
    for i = 1:m
        for j = 1:n
            z(i) = z(i) + A(i,j)x(j)
        end
    end
end matvec.ij
```

The functions **rows** and **cols** have matrix arguments and return row and column dimensions. The statement $z(1{:}m) = 0$ initializes $z$ to be the $m$-by-1 zero vector.

An alternative algorithm results if we regard $z = Ax$ as a linear combination of $A$'s columns, e.g.,

$$\begin{bmatrix} 1 & 2 \\ 3 & 4 \\ 5 & 6 \end{bmatrix} \begin{bmatrix} 7 \\ 8 \end{bmatrix} = \begin{bmatrix} 1\cdot 7 + 2\cdot 8 \\ 3\cdot 7 + 4\cdot 8 \\ 5\cdot 7 + 6\cdot 8 \end{bmatrix} = 7\begin{bmatrix} 1 \\ 3 \\ 5 \end{bmatrix} + 8\begin{bmatrix} 2 \\ 4 \\ 6 \end{bmatrix} = \begin{bmatrix} 23 \\ 53 \\ 83 \end{bmatrix}.$$

**Algorithm 1.1.4 (Matrix-Vector Multiplication: Column Version)** If $A \in \mathbb{R}^{m\times n}$ and $x \in \mathbb{R}^n$, then the following algorithm computes $z = Ax$.

```
function: z = matvec.ji(A, x)
    m = rows(A); n = cols(A); z(1:m) = 0
    for j = 1:n
        for i = 1:m
            z(i) = z(i) + x(j)A(i, j)
        end
    end
end matvec.ji
```

Note that the $i$-loop carries out a saxpy operation. We derived the column algorithm by rethinking what matrix-vector multiplication "means" at the vector level. Alternatively, we could derive the column algorithm by interchanging the order of the loops in the row algorithm. In matrix computations, it is important to understand program alterations like loop interchanges in terms of the underlying linear algebra.

### 1.1.8 Partitioning a Matrix into Rows and Columns

By looking at the inner loops we see that Algorithm 1.1.3 accesses $A$ by row while Algorithm 1.1.4 accesses $A$ by column. To characterize these methods of access more clearly we need the language of *partitioned matrices.* From the row point of view, a matrix is a stack of row vectors:

$$A \in \mathbb{R}^{m\times n} \quad \Longleftrightarrow \quad A = \begin{bmatrix} a_1^T \\ \vdots \\ a_m^T \end{bmatrix} \quad a_k \in \mathbb{R}^n . \qquad (1.1.1)$$

This is called a *row partition* of $A$. Thus, if we row partition

$$\begin{bmatrix} 1 & 2 \\ 3 & 4 \\ 5 & 6 \end{bmatrix}$$

then we are choosing to look at $A$ as a collection of rows with

$$a_1^T = [\,1 \quad 2\,], \qquad a_2^T = [\,3 \quad 4\,], \quad \text{and} \quad a_3^T = [\,5 \quad 6\,].$$

With the row partitioning (1.1.1) we see that **matvec.ij** is essentially organized as follows:

> **for** $i = 1{:}m$
>     $z_i = a_i^T x$
> **end**

Alternatively, a matrix is a collection of column vectors:

$$A \in \mathbb{R}^{m\times n} \quad \Longleftrightarrow \quad A = [\,a_1, \ldots, a_n\,], \quad a_k \in \mathbb{R}^m . \qquad (1.1.2)$$

We refer to this as a *column partition* of $A$. In the 3-by-2 example above, we thus would set $a_1$ and $a_2$ to be the first and second columns of $A$ respectively:

$$a_1 = \begin{bmatrix} 1 \\ 3 \\ 5 \end{bmatrix} \qquad a_2 = \begin{bmatrix} 2 \\ 4 \\ 6 \end{bmatrix}.$$

With (1.1.2) we see that **matvec.ji** is a saxpy procedure that accesses $A$ by columns:

> $z(1{:}m) = 0$
> **for** $j = 1{:}n$
>     $z = z + x_j a_j$
> **end**

Observe that $z$ is a running vector sum that is updated via saxpy operations.

### 1.1.9 The Colon Notation

A handy way to specify a column or a row of a matrix is with the "colon" notation. If $A \in \mathbb{R}^{m\times n}$ then we designate the $k$th row of $A$ with the notation $A(k,:)$, i.e.,

$$A(k,:) = [a_{k1}, \ldots, a_{kn}] \; .$$

Likewise, we designate the $k$th column of $A$ by $A(:,k)$,

$$A(:,k) = \begin{bmatrix} a_{1k} \\ \vdots \\ a_{mk} \end{bmatrix}.$$

With these conventions we can rewrite **matvec.ij** as

```
for i = 1:m                         for i = 1:m
    z(i) = A(i,:)x          or          z(i) = dot(A(i,:), x)
end                                 end
```

The saxpy version **matvec.ji** has the form

```
z(1:m) = 0                          z(1:m) = 0
for j = 1:n                         for j = 1:n
    z = z + x(j)A(:,j)      or          z = saxpy(x(j), A(:,j), z)
end                                 end
```

When describing a matrix algorithm, it is important to choose carefully the level and style of the description. With the colon notation we are able to suppress inner loops and thereby focus attention on the "higher level" operations. With the function notation, we are able to identify key computational kernels that may reflect good software design, e.g., **dot** and **saxpy**.

### 1.1.10 Outer Product Updates

As we have seen, an important attribute of a matrix algorithm is its mode of access. For reasons that we give later, algorithms that access arrays by column turn out to be preferable. Thus, **matvec.ji** is generally preferred to **matvec.ij**. Recall that the only difference between these two algorithms is the order of the two loops. The connection between loop order and data access is important to appreciate. To illustrate this further we consider the computation of the outer product update

$$A \longleftarrow A + xy^T, \qquad A \in \mathbb{R}^{m\times n},\ x \in \mathbb{R}^m,\ y \in \mathbb{R}^n\ .$$

Here, the $\longleftarrow$ denotes assignment. Recall that $xy^T$ is just a special matrix-matrix product and that the entries of $A$ are prescribed by

$$a_{ij} = a_{ij} + x_i y_j, \qquad i = 1{:}m,\ j = 1{:}n\ .$$

The "$ij$" version of this computation "says" add a multiple of $y^T$ to each row of $A$:

```
for i = 1:m
    A(i,:) = A(i,:) + x(i)y^T
end
```

On the other hand, the "$ji$" version

```
for j = 1:n
    A(:,j) = A(:,j) + y(j)x
end
```

accesses $A$ by column. Note that both outer product procedures are saxpy-based.

### 1.1.11 The Gaxpy Computation

Outer product updates figure heavily in the traditional formulation of many important matrix algorithms. It turns out that most of these algorithms can be rearranged so that the dominant operation is a *gaxpy* operation. A gaxpy is merely a computation of the form

$$z = y + Ax \qquad x \in \mathbb{R}^n,\ y \in \mathbb{R}^m,\ A \in \mathbb{R}^{m\times n}.$$

Like "saxpy", the term "gaxpy" has its origins in software. We can think of it as a mnemonic for "general $A$ $x$ plus $y$ . As we explain in §1.4, gaxpy formulations (properly implemented) are usually to be preferred to outer product update formulations. This is a very important computational development and so the gaxpy kernel is worthy of formal encapsulation.

**Algorithm 1.1.5 (Gaxpy)** Given $x \in \mathbb{R}^n$, $y \in \mathbb{R}^m$, and $A \in \mathbb{R}^{m\times n}$, the following algorithm computes $z = y + Ax$.

```
function: z = gaxpy(A, x, y)
    n = cols(A); z = y
    for j = 1:n
        z = z + x(j)A(:,j)
    end
end gaxpy
```

As we have arranged it, $y$ is a running vector sum that is updated by a sequence of saxpy operations. With this organization, we see that a gaxpy is a generalized saxpy.

### 1.1.12 The Notion of "Level"

The dot product and saxpy operations are examples of "level-1" operations. Level-1 operations involve an amount of data and an amount of arithmetic that is linear in the dimension of the operation. An $m$-by-$n$ outer product update or gaxpy operation involve a quadratic amount of data ($O(mn)$) and a quadratic amount of work ($O(mn)$). They are examples of "level-2" operations.

The design of matrix algorithms that are rich in high-level linear algebra operations is an area of intensive research and a recurring theme in the book. For example, a high performance linear equation solver may require a level-2 gaxpy organization of Gaussian elimination. This requires some algorithmic rethinking as the usual specification of Gaussian elimination is at level-1, e.g., "multiply row 1 by a constant and add the result to row 2."

It is also possible to arrange certain algorithms so that they are rich in matrix-matrix multiplication, which is regarded as a level-3 operation. Level-3 operations involve a quadratic amount of data and a cubic amount of work. If $A$, $B$, and $C$ are matrices then the computations $C = C + AB$ and $C = C + AB^T$ are sample level-3 operations. Computations that are this extensive can be arranged in numerous ways as we now show with respect to ordinary matrix-matrix multiplication $C = AB$.

### 1.1.13 Matrix-Matrix Multiplication

Consider the 2-by-2 matrix-matrix multiplication problem $C = AB$. In the dot product formulation each entry in $C$ is computed as a dot product:

$$\begin{bmatrix} 1 & 2 \\ 3 & 4 \end{bmatrix} \begin{bmatrix} 5 & 6 \\ 7 & 8 \end{bmatrix} = \begin{bmatrix} 1\cdot 5 + 2\cdot 7 & 1\cdot 6 + 2\cdot 8 \\ 3\cdot 5 + 4\cdot 7 & 3\cdot 6 + 4\cdot 8 \end{bmatrix}.$$

In the saxpy version each column of $C$ is regarded as a linear combination of columns of $A$:

$$\begin{bmatrix} 1 & 2 \\ 3 & 4 \end{bmatrix} \begin{bmatrix} 5 & 6 \\ 7 & 8 \end{bmatrix} = \left[\; 5\begin{bmatrix} 1 \\ 3 \end{bmatrix} + 7\begin{bmatrix} 2 \\ 4 \end{bmatrix} \quad 6\begin{bmatrix} 1 \\ 3 \end{bmatrix} + 8\begin{bmatrix} 2 \\ 4 \end{bmatrix} \;\right].$$

Finally, in the outer product version, $C$ is regarded as the sum of outer products:

$$\begin{bmatrix} 1 & 2 \\ 3 & 4 \end{bmatrix} \begin{bmatrix} 5 & 6 \\ 7 & 8 \end{bmatrix} = \begin{bmatrix} 1 \\ 3 \end{bmatrix} \begin{bmatrix} 5 & 6 \end{bmatrix} + \begin{bmatrix} 2 \\ 4 \end{bmatrix} \begin{bmatrix} 7 & 8 \end{bmatrix}.$$

Although equivalent mathematically, it turns out that the different versions of matrix multiplication can have very different levels of performance because of the different ways that they access memory. This matter is pursued in §1.4. For now, it is worth detailing the above three approaches to matrix multiplication as it gives us a chance to use our notation and to practice algorithmic redesign.

### 1.1.14 Dot Product Matrix Multiply

Suppose $A \in \mathbb{R}^{m\times r}$ and $B \in \mathbb{R}^{r\times n}$ and that we wish to compute $C = AB$. The usual matrix multiplication procedure regards $C$ as an array of dot products to be computed one at a time in left to right, top to bottom order.

**Algorithm 1.1.6 (Matrix Multiplication: Dot Version)** If $A \in \mathbb{R}^{m\times r}$ and $B \in \mathbb{R}^{r\times n}$ then the following algorithm computes $C = AB$.

```
function: C = matmat.ijk(A, B)
    m = rows(A); r = cols(A); n = cols(B)
    C(1:m, 1:n) = 0
    for i = 1:m
        for j = 1:n
            for k = 1:r
                C(i, j) = C(i, j) + A(i, k)B(k, j)
            end
        end
    end
end matmat.ijk
```

In this procedure we compute $c_{ij}$ as the dot product of $A$'s $i$th row and $B$'s $j$th column. In the language of partitioned matrices, if we have the partitionings

$$A = \begin{bmatrix} a_1^T \\ \vdots \\ a_m^T \end{bmatrix} \qquad a_k \in \mathbb{R}^r$$

and

$$B = [\, b_1, \ldots, b_n \,] \qquad b_k \in \mathbb{R}^r$$

then Algorithm 1.1.6 has the following level-1 description:

```
for i = 1:m
    for j = 1:n
        c_ij = a_i^T b_j
    end
end
```

Note that the "mission" of the $j$-loop is to compute the $i$th row of $C$. To emphasize this we could write

```
for i = 1:m
    c_i^T = a_i^T B
end
```

where $c_i^T$ is the $i$th row of $C$ . This matrix-vector product characterization of Algorithm 1.1.6 may be regarded as level-2. Of course, at the highest level we have the notation $C = AB$ which suppresses all the looping and subscripts.

Choosing the right level of description is important both in exposition and in software design.

### 1.1.15 Gaxpy Matrix Multiply

Suppose $A$, $B$, and $C$ are column-partitioned as follows

$$\begin{aligned} A &= [\, a_1, \ldots, a_r \,] & a_j &\in \mathbb{R}^m \\ B &= [\, b_1, \ldots, b_n \,] & b_j &\in \mathbb{R}^r \\ C &= [\, c_1, \ldots, c_n \,] & c_j &\in \mathbb{R}^m \end{aligned}$$

and that we want to compute $C = AB$. Since

$$c_j = \sum_{k=1}^{r} b_{kj} a_k, \qquad j = 1{:}n$$

we see that each column of $C$ is a linear combination of columns of $A$. These vector sums can be built up with a sequence of saxpy operations giving:

**Algorithm 1.1.7 (Matrix Multiplication: Gaxpy Version)** If the matrices $A \in \mathbb{R}^{m \times r}$ and $B \in \mathbb{R}^{r \times n}$ are given then the following algorithm computes $C = AB$.

```
function: C = matmat.jki(A, B)
    m = rows(A); n = cols(B); r = cols(A)
    C(1:m, 1:n) = 0
    for j = 1:n
        for k = 1:r
            for i = 1:m
                C(i, j) = C(i, j) + A(i, k)B(k, j)
            end
        end
    end
end matmat.jki
```

Note that although we derived **matmat.jki** by considering a column partitioning of the matrix equation $C = AB$, it can also be obtained from **matmat.ijk** by rearranging the order of the loops. The $jki$ algorithm is clearly a gaxpy-based procedure as it can be expressed in the following form:

```
C(1:m, 1:n) = 0
for j = 1:n
    C(:, j) = gaxpy(A, B(:, j), C(:, j))
end
```

### 1.1.16 Outer Product Matrix Multiply

Let us play the loop reordering game one more time and analyze afterwards what it means in linear algebraic terms.

**Algorithm 1.1.8 (Matrix Multiplication: Outer Product Version)** If $A \in \mathbb{R}^{m\times r}$ and $B \in \mathbb{R}^{r\times n}$ then the following algorithm computes $C = AB$.

```
function: C = matmat.kji(A, B)
    m = rows(A); n = cols(B); r = cols(A)
    C(1:m, 1:n) = 0
    for k = 1:r
        for j = 1:n
            for i = 1:m
                C(i, j) = C(i, j) + A(i, k)B(k, j)
            end
        end
    end
end matmat.kji
```

In this *kji* formulation, the $c_{ij}$ are "built up" simultaneously. For a given $k$, the inner two loops handle the outer product update $C \leftarrow C + a_k b_k^T$ where

$$A = [\, a_1, \ldots, a_r \,], \qquad a_k \in \mathbb{R}^m \tag{1.1.3}$$

and

$$B = \begin{bmatrix} b_1^T \\ \vdots \\ b_r^T \end{bmatrix} \qquad b_k \in \mathbb{R}^n \,. \tag{1.1.4}$$

The inner loop in **matmat.kji** carries out a saxpy operation. In particular, a multiple of $a_k$ is added to the $j$th column of $C$.

### 1.1.17 Loop Reordering

Just as "double loop" matrix-vector multiplication can be arranged in 2! = 2 ways so we find that "triple loop" matrix-matrix multiplication can be rearranged in 3! = 6 ways. Three of these arrangements have been detailed above: *ijk*, *jki*, and *kji*. Each of the six possibilities has its own featured operation (dot product, saxpy) and its own mode of access. These are summarized in Table 1.1.1. Which of the possibilities is to be preferred depends upon the underlying computer architecture. This is discussed in §1.4.

| Loop Order | Inner Loop | Middle Loop | Inner Loop Data Access |
|---|---|---|---|
| $ijk$ | dot | vector × matrix | $A$ by row, $B$ by column |
| $jik$ | dot | matrix × vector | $A$ by row, $B$ by column |
| $ikj$ | saxpy | row gaxpy | $B$ by row |
| $jki$ | saxpy | column gaxpy | $A$ by column |
| $kij$ | saxpy | row outer product | $B$ by row |
| $kji$ | saxpy | column outer product | $A$ by column |

Table 1.1.1. Matrix Multiplication: Orderings and Properties

### 1.1.18 A Note on Matrix Equations

In striving to understand matrix multiplication via outer products, we essentially established the matrix equation

$$AB = \sum_{k=1}^{r} a_k b_k^T$$

where the $a_k$ and $b_k$ are defined by the partitionings (1.1.3) and (1.1.4).

Numerous matrix equations are developed in subsequent chapters. Sometimes they are established algorithmically like the above outer product expansion and other times they are proved at the $ij$-component level. As an example of the latter style of development we prove an important result that characterizes transposes of products.

**Theorem 1.1.1** *If $A \in \mathbb{R}^{m \times r}$ and $B \in \mathbb{R}^{r \times n}$, then $(AB)^T = B^T A^T$.*

**Proof.** If $C = (AB)^T$ then

$$c_{ij} = [(AB)^T]_{ij} = (AB)_{ji} = \sum_{k=1}^{r} a_{jk} b_{ki} \,.$$

On the other hand, if $D = B^T A^T$ then

$$d_{ij} = [B^T A^T]_{ij} = \sum_{k=1}^{r} (B^T)_{ik} (A^T)_{kj} = \sum_{k=1}^{r} b_{ki} a_{jk}$$

and so $C = D$. □

As evidenced by this proof, certain $ijk$ developments are not very insightful. However, they are sometimes the only way to develop a high level linear algebra result.

### 1.1.19 Specifying Algorithms

Throughout the book we require various levels of detail when specifying algorithms. The algorithms in this section have all been written as formal functions. As an example of a looser style of algorithmic expression, we give a procedure for computing the matrix multiplication $C = A^T B$.

**Algorithm 1.1.9** Given $m$-by-$n$ matrices $A$ and $B$, the following algorithm computes the product $C = A^T B$.

```
for i = 1:n
    for j = 1:n
        C(i,j) = A(1:m,i)^T B(1:m,j)
    end
end
```

### 1.1.20 Complex Matrices

This book focusses on real matrix computations for notational convenience and because most practical problems involve real data. However, whenever appropriate we shall include problems about complex matrix manipulation in subsequent sections.

We must establish some notation here at the outset. The vector space of $m$-by-$n$ complex matrices is designated by $\mathbb{C}^{m \times n}$. The scaling, addition, and multiplication of complex matrices corresponds exactly to the real case. However, transposition becomes *conjugate transposition*:

$$C = A^H \quad \Longrightarrow \quad c_{ij} = \bar{a}_{ji} \, .$$

The vector space of complex $n$-vectors is designated by $\mathbb{C}^n$. The dot product of complex $n$-vectors $x$ and $y$ is prescribed by

$$s = x^H y = \sum_{i=1}^{n} \bar{x}_i y_i \, .$$

Finally, if $A = B + iC \in \mathbb{C}^{m \times n}$ then we designate the real and imaginary parts of $A$ by $\text{Re}(A) = B$ and $\text{Im}(A) = C$ respectively.

#### Problems

**P1.1.1** Suppose $A \in \mathbb{R}^{n \times n}$ and $x \in \mathbb{R}^r$ are given. Give a saxpy algorithm for computing the first column of $M = (A - x_1 I) \cdots (A - x_r I)$

**P1.1.2** In the conventional 2-by-2 matrix multiplication $C = AB$, there are eight multiplications: $a_{11}b_{11}$, $a_{11}b_{12}$, $a_{21}b_{11}$, $a_{21}b_{12}$, $a_{12}b_{21}$, $a_{12}b_{22}$, $a_{22}b_{21}$ and $a_{22}b_{22}$. Make a table that indicates the order that these multiplications are performed for the $ijk$, $jik$, $kij$, $ikj$, $jki$, and $kji$ matrix multiply algorithms.

**P1.1.3** Give an algorithm for computing $C = (xy^T)^k$ where $x$ and $y$ are $n$-vectors. Use **dot** and **saxpy**.

**P1.1.4** Suppose we have real $n$-by-$n$ matrices $C$, $D$, $E$, and $F$. Show how to compute real $n$-by-$n$ matrices $A$ and $B$ with just three real $n$-by-$n$ calls to **matmat.jki** so that $(A + iB) = (C + iD)(E + iF)$. Hint: Compute $W = (C + D)(E - F)$.

**Notes and References for Sec. 1.1**

For the specification of algorithms we use a corrupted version of the Matlab language which is formally described in

T.F. Coleman and C. Van Loan (1988). *Handbook for Matrix Computations,* SIAM Publications, Philadelphia, PA.

C.B. Moler, J.N. Little, and S. Bangert (1987). *PC-Matlab User's Guide*, The Math Works Inc., 20 N. Main St., Sherborn, Mass.

We wish to stress that the algorithms in this text are not production codes. The development of quality software from any of our algorithmic specifications is a long and arduous task if done properly. Even the implementation of low level procedures like the saxpy and gaxpy require time and care. To see what is involved, read the following papers:

J. Dongarra, J. DuCroz, S. Hammarling, and R.J. Hanson (1988). "An Extended Set of Fortran Basic Linear Algebra Subprograms," *ACM Trans. Math. Soft. 14,* 1–17.

J. Dongarra, J. DuCroz, S. Hammarling, and R.J. Hanson (1988). "Algorithm 656 An Extended Set of Fortran Basic Linear Algebra Subprograms: Model Implementation and Test Programs," *ACM Trans. Math. Soft. 14,* 18–32.

J. Dongarra, J. Du Croz, I.S. Duff, and S. Hammarling (1988). "A Set of Level 3 Basic Linear Algebra Subprograms," Argonne National Laboratory Report, ANL-MCS-TM-88.

C.L. Lawson, R.J. Hanson, , D.R. Kincaid, and F.T. Krogh (1979a). "Basic Linear Algebra Subprograms for FORTRAN Usage," *ACM Trans. Math. Soft. 5,* 308–23.

C.L. Lawson, R.J. Hanson, D.R. Kincaid, and F.T. Krogh (1979b). "Algorithm 539, Basic Linear Algebra Subprograms for FORTRAN Usage," *ACM Trans. Math. Soft. 5,* 324–25.

We also recommend

J.R. Rice (1981). *Matrix Computations and Mathematical Software*, Academic Press, New York.

The effect of loop orderings on performance is detailed in the following excellent survey article:

J.J. Dongarra, F.G. Gustavson, and A. Karp (1984). "Implementing Linear Algebra Algorithms for Dense Matrices on a Vector Pipeline Machine," *SIAM Review 26,* 91–112.

## 1.2 Exploiting Structure

The efficiency of a given matrix algorithm depends on many things. Most obvious and what we treat in this section is the amount of arithmetic and storage required by a procedure. Less obvious but often crucial to performance are memory accesses, stride, and a host of other data movement overheads. These issues are treated in §1.4 and elsewhere.

We continue to use matrix-vector and matrix-matrix multiplication as a vehicle for introducing the key ideas. As examples of exploitable structure we have chosen the properties of bandedness and symmetry. Band matrices have many zero entries and so it is no surprise that band matrix manipulation allows for many arithmetic and storage shortcuts. "Flops" and data structures are discussed in this context.

Symmetric matrices provide another set of examples that can be used to illustrate structure exploitation. Symmetric linear systems and eigenvalue problems have a very prominent role in matrix computations and so it is important to be familiar with their manipulation.

We conclude with some remarks about overwriting, another technique for controlling the amount of required storage.

### 1.2.1 Band Matrices and the x-0 Notation

We say that $A \in \mathbb{R}^{m \times n}$ has *lower bandwidth* $p$ if $a_{ij} = 0$ whenever $i > j+p$ and *upper bandwidth* $q$ if $j > i + q$ implies $a_{ij} = 0$ . Here is an example of an 8-by-5 matrix that has lower bandwidth 1 and upper bandwidth 2:

$$\begin{bmatrix} \times & \times & \times & 0 & 0 \\ \times & \times & \times & \times & 0 \\ 0 & \times & \times & \times & \times \\ 0 & 0 & \times & \times & \times \\ 0 & 0 & 0 & \times & \times \\ 0 & 0 & 0 & 0 & \times \\ 0 & 0 & 0 & 0 & 0 \\ 0 & 0 & 0 & 0 & 0 \end{bmatrix}$$

The $\times$'s designates arbitrary nonzero entries. This notation is handy to indicate the zero-nonzero structure of a matrix and we use it extensively. Certain band structures occur frequently and we tabulate them in Table 1.2.1.

### 1.2.2 Diagonal Matrix Manipulation

We have a special notation for diagonal matrices. If $D \in \mathbb{R}^{m \times n}$ then

$$D = \text{diag}(d_1, \ldots, d_q), \quad q = \min\{m, n\} \quad \Longleftrightarrow \quad d_i = d_{ii}$$

If $D$ is diagonal and $A$ is a matrix, then $DA$ is a *row scaling* of $A$ and $AD$ is a *column scaling* of $A$.

Suppose $y = Dx$ where $x \in \mathbb{R}^n$ and $y \in \mathbb{R}^n$. It follows that $y$ is the consequence of a vector multiply: $y = d. * x$. Here, $d \in \mathbb{R}^n$ is made up of $D$'s diagonal entries.

| Type of Matrix | Lower Bandwidth | Upper Bandwidth |
|---|---|---|
| diagonal | 0 | 0 |
| upper triangular | 0 | $n-1$ |
| lower triangular | $m-1$ | 0 |
| tridiagonal | 1 | 1 |
| upper bidiagonal | 0 | 1 |
| lower bidiagonal | 1 | 0 |
| upper Hessenberg | 1 | $n-1$ |
| lower Hessenberg | $m-1$ | 1 |

Table 1.2.1. Band Terminology for $m$-by-$n$ Matrices.

### 1.2.3 Triangular Matrix Multiplication

To introduce band matrix "thinking" we look at the matrix multiplication problem $C = AB$ when $A$ and $B$ are both $n$-by-$n$ and upper triangular. The 3-by-3 case is illuminating:

$$C = \begin{bmatrix} a_{11}b_{11} & a_{11}b_{12} + a_{12}b_{22} & a_{11}b_{13} + a_{12}b_{23} + a_{13}b_{33} \\ 0 & a_{22}b_{22} & a_{22}b_{23} + a_{23}b_{33} \\ 0 & 0 & a_{33}b_{33} \end{bmatrix}$$

It suggests that the product $C$ is upper triangular and that the upper triangular entries in $C$ are the result of abbreviated inner products. Indeed, since $a_{ik}b_{kj} = 0$ whenever $k < i$ or $j < k$ we see that $c_{ij} = \sum_{k=i}^{j} a_{ik}b_{kj}$ and so we obtain:

**Algorithm 1.2.1** Given $n$-by-$n$ upper triangular matrices $A$ and $B$, the following algorithm computes $C = AB$.

```
C(1:n, 1:n) = 0
for i = 1:n
    for j = i:n
        for k = i:j
            C(i,j) = C(i,j) + A(i,k)B(k,j)
        end
    end
end
```

To quantify the savings in this algorithm we need some tools for measuring the amount of work.

### 1.2.4 Flops

Obviously, upper triangular matrix multiplication involves less arithmetic than when the matrices are full. One way to quantify this is with the notion of a *flop*. A flop[1] is a floating point operation. A dot product or saxpy operation of length $n$ involves $2n$ flops because there are $n$ multiplications and $n$ adds in either of these vector operations.

The matrix-vector multiplication $z = Ax$ where $A \in \mathbb{R}^{m\times n}$ involves $2mn$ flops as does an $m$-by-$n$ outer product update of the form $C = A + xy^T$. The matrix-matrix multiplication $C = AB$ where $A \in \mathbb{R}^{m\times r}$ and $B \in \mathbb{R}^{r\times n}$ involves $2mnr$ flops.

Flop counts are usually obtained by summing the amount of arithmetic associated with the most deeply nested statements in an algorithm. For matrix-matrix multiplication, this is the statement,

$$C(i,j) = C(i,j) + A(i,k)B(k,j)$$

which involves two flops and is executed $mnr$ times as a simple loop accounting indicates. Hence the conclusion that general matrix multiplication requires $2mnr$ flops.

Now let us investigate the amount of work involved in Algorithm (1.2.1). Note that $c_{ij}$, $(i \le j)$ requires $2(j-i+1)$ flops. Using the heuristics

$$\sum_{p=1}^{q} p = \frac{q(q+1)}{2} \approx \frac{q^2}{2}$$

and

$$\sum_{p=1}^{q} p^2 = \frac{q^3}{3} + \frac{q^2}{2} + \frac{q}{6} \approx \frac{q^3}{3}$$

we find that triangular matrix multiplication requires approximately $n^3/3$ flops:

$$\sum_{i=1}^{n}\sum_{j=i}^{n} 2(j-i+1) = \sum_{i=1}^{n}\sum_{j=1}^{n-i+1} 2j \approx \sum_{i=1}^{n} \frac{2(n-i+1)^2}{2} = \sum_{i=1}^{n} i^2 \approx \frac{n^3}{3}.$$

[1] In the first edition of this book we defined a flop to be the amount of work associated with an operation of the form $a_{ij} = a_{ij} + a_{ik}a_{kj}$, i.e., a floating point add, a floating point multiply, and some subscripting. Thus, an "old flop" involves two "new flops." In defining a flop to be a single floating point operation we are opting for a more precise measure of arithmetic complexity. Moreover, new flops have replaced old flops in the supercomputing field to the delight of the manufacturers whose machines suddenly doubled in speed with the new terminology! After some agonizing, we have decided to go along with the change.

We throw away the low order terms since their inclusion does not contribute to what the flop count "says." For example, an exact flop count of Algorithm 1.2.1 reveals that precisely $n^3/3 + n^2 + 2n/3$ flops are involved. For large $n$ (the typical situation of interest) we see that the exact flop count offers no insight beyond our $n^3/3$ estimate.

Flop counting is a necessarily crude approach to the measuring of program efficiency since it ignores subscripting, memory traffic, and the countless other overheads associated with program execution. We must not infer too much from a comparison of flops counts. We cannot conclude, for example, that triangular matrix multiplication is six times faster than square matrix multiplication. Flop counting is just a "quick and dirty" accounting method that captures only one of the several dimensions of the efficiency issue. As we will see, in various "high performance" architectures the efficiency of an algorithm such as matrix multiplication is much more dependent on communication/memory traffic overheads and a more accurate efficiency measure might involve looking at the number of **saxpy** calls and the length of the vector operands.

### 1.2.5 The Colon Notation–Again

The dot product that the $k$-loop performs in Algorithm 1.2.1 can be succinctly stated if we extend the colon notation introduced in §1.1.9. Suppose $A \in \mathbb{R}^{m\times n}$ and the integers $p$, $q$, and $r$ satisfy $1 \le p \le q \le n$ and $1 \le r \le m$. We then define

$$A(r, p{:}q) = [\, a_{rp}, \ldots, a_{rq} \,] \in \mathbb{R}^{1\times(q-p+1)} .$$

Likewise, if $1 \le p \le q \le m$ and $1 \le c \le n$ then

$$A(p{:}q, c) = \begin{bmatrix} a_{pc} \\ \vdots \\ a_{qc} \end{bmatrix} \in \mathbb{R}^{q-p+1}$$

With this notation we can rewrite Algorithm (1.2.1) as

$$
\begin{array}{l}
C(1{:}n, 1{:}n) = 0 \\
\textbf{for } i = 1{:}n \\
\quad \textbf{for } j = i{:}n \\
\qquad C(i,j) = C(i,j) + A(i, i{:}j)B(i{:}j, j) \\
\quad \textbf{end} \\
\textbf{end}
\end{array}
$$

We mention one additional feature of the colon notation. Negative increments are allowed. Thus, if $x$ and $y$ are $n$-vectors then $s = x^T y(n{:}-1{:}1)$ is the summation

$$s \;=\; \sum_{i=1}^{n} x_i y_{n-i+1} \,,$$

i.e., the convolution of $x$ and $y$.

### 1.2.6 Band Storage

Suppose $A \in \mathbb{R}^{n \times n}$ has lower bandwidth $p$ and upper bandwidth $q$ and assume that $p$ and $q$ are much smaller than $n$. Such a matrix can be stored in a $(p+q+1)$-by-$n$ array $A.band$ with the convention that

$$a_{ij} = A.band(i-j+q+1, j) \qquad (1.2.1)$$

for all $(i, j)$ that fall inside the band. Thus, if $n = 5$, $p = 1$, and $q = 2$ as in §1.2.1 we have

$$A.band = \begin{bmatrix} 0 & 0 & a_{13} & a_{24} & a_{35} \\ 0 & a_{12} & a_{23} & a_{34} & a_{45} \\ a_{11} & a_{22} & a_{33} & a_{44} & a_{55} \\ a_{21} & a_{32} & a_{43} & a_{54} & 0 \end{bmatrix}.$$

Here, the "0" entries are unused. With this data structure, a column access matrix-vector multiply algorithm could be implemented as follows:

**Algorithm 1.2.2** Suppose $A \in \mathbb{R}^{n \times n}$ has lower bandwidth $p$ and upper bandwidth $q$ and is stored in the $A.band$ format (1.2.1). If $x \in \mathbb{R}^n$ then the following algorithm computes $z = Ax$.

$z(1{:}n) = 0$
**for** $j = 1{:}n$
  $z_{top} = \max(1, j-q)$; $z_{bot} = \min(n, j+p)$
  $a_{top} = \max(1, q+2-j)$; $a_{bot} = a_{top} + z_{bot} - z_{top}$
  $z(z_{top}{:}z_{bot}) = z(z_{top}{:}z_{bot}) + x(j)A.band(a_{top}{:}a_{bot}, j)$
**end**

Notice that by storing $A$ by columns in $A.band$, we obtain a saxpy, column access procedure. Indeed, Algorithm (1.2.2) is obtained from **matvec.ji** by recognizing that each saxpy involves a vector with a small number of nonzeros. The integer arithmetic is used to identify the location of these nonzeros. As a result of this care the algorithm involves about $2n(p+q+1)$ flops with the assumption that $p$ and $q$ are much smaller than $n$.

### 1.2.7 Symmetry

We say that $A \in \mathbb{R}^{n \times n}$ is *symmetric* if $A^T = A$. Thus,

$$A = \begin{bmatrix} 1 & 2 & 3 \\ 2 & 4 & 5 \\ 3 & 5 & 6 \end{bmatrix}$$

is symmetric. Storage requirements can be halved if we just store the lower triangle of elements, e.g., $A.vec = [\ 1\ \ 2\ \ 3\ \ 4\ \ 5\ \ 6\ ]$. In general with this data structure we agree to store the $a_{ij}$ as follows:

$$a_{ij} \;=\; A.vec((j-1)n - j(j-1)/2 + i) \qquad (i \geq j) \tag{1.2.2}$$

Let us look at matrix-vector multiplication with the matrix $A$ represented in $A.vec$.

**Algorithm 1.2.3** Suppose $A \in \mathbb{R}^{n \times n}$ is symmetric and stored in the $A.vec$ style (1.2.2). If $x \in \mathbb{R}^n$ then the following algorithm computes $z = Ax$.

```
z(1:n) = 0
for j = 1:n
    for i = 1:j - 1
        z(i) = z(i) + A.vec((i - 1)n - i(i - 1)/2 + j)x(j)
    end
    for i = j:n
        z(i) = z(i) + A.vec((j - 1)n - j(j - 1)/2 + i)x(j)
    end
end
```

This algorithm requires the same $2n^2$ flops that general matrix-vector multiplication requires. Notice how awkward the data structure makes the multiplication.

### 1.2.8 Store by Diagonal

In algorithms that access arrays we should try to organize the computations so that the inner loops access contiguous data. In the two-dimensional Fortran setting (for example) this amounts to designing procedures that access matrix data by columns as we discussed in §1.1. A problem with Algorithm 1.2.3 is that the first $i$-loop does not access contiguous array entries. This is the fault of the chosen data structure.

To get around this problem we consider a compact storage scheme for symmetric matrices in which the matrix is stored by diagonal. If

$$A \;=\; \begin{bmatrix} 1 & 2 & 3 \\ 2 & 4 & 5 \\ 3 & 5 & 6 \end{bmatrix}$$

then in store-by-diagonal we represent $A$ with the vector

$$A.diag = [\ 1\ \ 4\ \ 6\ \ 2\ \ 5\ \ 3\ ]\ .$$

In general, if $i \geq j$, then

$$a_{i+k,i} \;=\; A.diag(i + nk - k(k-1)/2) \qquad (k \geq 0) \tag{1.2.3}$$

Some notation simplifies the discussion of how to use this data structure in a matrix-vector multiplication.

If $A \in \mathbb{R}^{m \times n}$ then let $D(A,k) \in \mathbb{R}^{m \times n}$ designate the $k$th diagonal part of $A$ as follows:

$$[D(A,k)]_{ij} = \begin{cases} a_{ij} & j = i + k, \quad 1 \le i \le m, \quad 1 \le j \le n \\ 0 & \text{otherwise} \end{cases}$$

Thus,

$$A = \begin{bmatrix} 1 & 2 & 3 \\ 2 & 4 & 5 \\ 3 & 5 & 6 \end{bmatrix} = \underbrace{\begin{bmatrix} 0 & 0 & 3 \\ 0 & 0 & 0 \\ 0 & 0 & 0 \end{bmatrix}}_{D(A,2)} + \underbrace{\begin{bmatrix} 0 & 2 & 0 \\ 0 & 0 & 5 \\ 0 & 0 & 0 \end{bmatrix}}_{D(A,1)}$$

$$+ \underbrace{\begin{bmatrix} 1 & 0 & 0 \\ 0 & 4 & 0 \\ 0 & 0 & 6 \end{bmatrix}}_{D(A,0)} + \underbrace{\begin{bmatrix} 0 & 0 & 0 \\ 2 & 0 & 0 \\ 0 & 5 & 0 \end{bmatrix}}_{D(A,-1)} + \underbrace{\begin{bmatrix} 0 & 0 & 0 \\ 0 & 0 & 0 \\ 3 & 0 & 0 \end{bmatrix}}_{D(A,-2)}$$

Returning to our store-by-diagonal data structure, we see that the nonzero parts of $D(A,0)$, $D(A,1)$,..., $D(A,n-1)$ are sequentially stored in the *A.diag* scheme (1.2.3). The matrix-vector multiplication $z = Ax$ can then be organized as follows:

$$z = D(A,0)x + \sum_{k=1}^{n-1} (D(A,k) + D(A,k)^T)x \,.$$

Multiplications of the form $D(A,k)x$ and $D(A,k)^T x$ access contiguous data. Indeed, if we work out the details we obtain:

**Algorithm 1.2.4** Suppose $A \in \mathbb{R}^{n \times n}$ is symmetric and stored in the *A.diag* style (1.2.3). If $x \in \mathbb{R}^n$ then the following algorithm computes $z = Ax$.

```
for i = 1:n
    z(i) = A.diag(i)x(i)
end
for k = 1:n - 1
    t = nk - k(k - 1)/2
    {z = z + D(A,k)x}
    for i = 1:n - k
        z(i) = z(i) + A.diag(i + t)x(i + k)
    end
    {z = z + D(A,k)^T x}
```

```
        for i = 1:n − k
            z(i + k) = z(i + k) + A.diag(i + t)x(i)
        end
    end
```

The inner loops do not handle a saxpy operation. Instead, they perform a *vector multiply* operation. See §1.1.4. Note that inner loops access contiguous data.

We have inserted "comments" in the above algorithm for clarity. Comments are enclosed within curly brackets.

### 1.2.9 Overwriting

An undercurrent in the above discussion has been the economical use of storage. Overwriting input data is another way to control the amount of memory that a procedure requires. Consider the $n$-by-$n$ matrix multiplication problem $C = AB$ with the proviso that the "input matrix" $B$ is to be overwritten by the "output matrix" $C$ . Suppose we start with the non-overwriting $jki$ algorithm developed in §1.1.15:

```
C(1:n, 1:n) = 0
for j = 1:n
    for k = 1:n
        for i = 1:n
            C(i, j) = C(i, j) + A(i, k)B(k, j)
        end
    end
end
```

A correct overwriting version does *not* result by dropping the $C = 0$ initialization and merely substituting "$B$" for "$C$." This is because $B(:, j)$ is needed throughout the entire $k$-loop. A linear workspace is needed to hold the $j$th column of $C$ until it is "safe" to overwrite $B(:, j)$:

```
for j = 1:n
    w(1:n) = 0
    for k = 1:n
        for i = 1:n
            w(i) = w(i) + A(i, k)B(k, j)
        end
    end
    B(:, j) = w(:)
end
```

In this example, a linear workspace is required. Usually, this is of no consequence in a procedure that has a 2-dimensional array of the same order.

**Problems**

**P1.2.1** Give an algorithm that overwrites $A$ with $A^2$ where $A \in \mathbb{R}^{n \times n}$ is (a) upper triangular and (b) square. Strive for a minimum workspace in each case.

**P1.2.2** Suppose $A \in \mathbb{R}^{n \times n}$ is upper Hessenberg and that scalars $\lambda_1, \ldots, \lambda_r$ are given. Give a saxpy algorithm for computing the first column of $M = (A - \lambda_1 I) \cdots (A - \lambda_r I)$.

**P1.2.3** Give a column saxpy algorithm for the $n$-by-$n$ matrix multiplication problem $C = AB$ where $A$ is upper triangular and $B$ is lower triangular.

**P1.2.4** Extend Algorithm 1.2.2 so that it can handle rectangular band matrices. Be sure to describe the underlying data structure.

**P1.2.5** $A \in \mathbb{C}^{n \times n}$ is *Hermitian* if $A^H = A$. If $A = B + iC$ then it is easy to show that $B^T = B$ and $C^T = -C$. Suppose we represent $A$ in an array $A.herm$ with the property that $A.herm(i,j)$ houses $b_{ij}$ if $i \geq j$ and $c_{ij}$ if $j > i$. Using this data structure write a matrix-vector multiply function that computes $\text{Re}(z)$ and $\text{Im}(z)$ from $\text{Re}(x)$ and $\text{Im}(x)$ so that $z = Ax$.

**P1.2.6** Suppose $X \in \mathbb{R}^{n \times p}$ and $A \in \mathbb{R}^{n \times n}$, with $A$ symmetric and stored by diagonal. Give an algorithm that computes $Y = X^T A X$ and stores the result by diagonal. Use separate arrays for $A$ and $Y$.

**Notes and References for Sec. 1.2**

See the Linpack manual for details on appropriate data structures when symmetry and/or bandedness is present.

# 1.3 Block Matrices and Algorithms

A facility with block matrix notation is crucial in matrix computations. It simplifies the derivation of many central algorithms. Moreover, "block algorithms" are increasingly important in high performance computing. By a block algorithm we essentially mean an algorithm that is rich in matrix-matrix multiplication. Algorithms of this type are likely to be more efficient than those that just manipulate scalars because more computation is associated with a given data movement. For example, $k$-by-$k$ matrix multiplication involves $2k^3$ flops but only $2k^2$ data. We will see in §1.4 that when $k$ is large the data movement overhead is relatively less costly.

## 1.3.1 Block Matrix Notation

Column and row partitionings are special cases of matrix blocking. In general we can partition both the rows and columns of an $m$-by-$n$ matrix

$A$ to obtain

$$A = \begin{bmatrix} A_{11} & \dots & A_{1q} \\ \vdots & & \vdots \\ A_{p1} & \cdots & A_{pq} \end{bmatrix} \begin{matrix} m_1 \\ \\ m_p \end{matrix}$$
$$\qquad n_1 \qquad n_q$$

Here $m_1 + \cdots + m_p = m$, $n_1 + \cdots + n_q = n$, and $A_{ij}$ designates the $(i, j)$ block or submatrix. Block $A_{ij}$ has dimension $m_i$-by-$n_j$ and we say that $A = (A_{ij})$ is a $p$-by-$q$ block matrix.

## 1.3.2 Block Matrix Manipulation

Block matrices combine just as matrices with scalar entries so long as certain dimension requirements are met. For example, if

$$B = \begin{bmatrix} B_{11} & \dots & B_{1q} \\ \vdots & & \vdots \\ B_{p1} & \cdots & B_{pq} \end{bmatrix} \begin{matrix} m_1 \\ \\ m_p \end{matrix}$$
$$\qquad n_1 \qquad n_q$$

then we say that $B$ is partitioned *conformably* with $A$ and their sum $C = A + B$ is a $p$-by-$q$ block matrix $C = (C_{ij})$ with $C_{ij} = A_{ij} + B_{ij}$.

The multiplication of block matrices is a little trickier. Here is the key result.

**Lemma 1.3.1** *Suppose $A \in \mathbb{R}^{m \times r}$ and $B \in \mathbb{R}^{r \times n}$ are partitioned as follows:*

$$A = \begin{bmatrix} A_{11} & A_{12} \\ A_{21} & A_{22} \end{bmatrix} \begin{matrix} m_1 \\ m_2 \end{matrix} \qquad B = \begin{bmatrix} B_{11} & B_{12} \\ B_{21} & B_{22} \end{bmatrix} \begin{matrix} r_1 \\ r_2 \end{matrix}$$
$$\qquad r_1 \quad r_2 \qquad\qquad\qquad n_1 \quad n_2$$

*It then follows that*

$$AB = C = \begin{bmatrix} C_{11} & C_{12} \\ C_{21} & C_{22} \end{bmatrix} \begin{matrix} m_1 \\ m_2 \end{matrix}$$
$$\qquad n_1 \quad n_2$$

*where $C_{ij} = A_{i1}B_{1j} + A_{i2}B_{2j}$ for $i = 1{:}2$ and $j = 1{:}2$.*

**Proof.** The proof is an exercise in subscripts:

$$[C_{ij}]_{pq} = \sum_{k=1}^{r} a_{(i-1)m_1+p,k} b_{k,(j-1)n_1+q}$$

$$= \sum_{k=1}^{r_1} a_{(i-1)m_1+p,k} b_{k,(j-1)n_1+q}$$

$$+ \sum_{k=r_1+1}^{r} a_{(i-1)m_1+p,k} b_{k,(j-1)n_1+q}$$

$$= \sum_{k=1}^{r_1} [A_{i1}]_{pk} [B_{1j}]_{kq} + \sum_{k=1}^{r_2} [A_{i2}]_{pk} [B_{2j}]_{kq}$$

$$= [A_{i1}B_{1j}]_{pq} + [A_{i2}B_{2j}]_{pq} = [A_{i1}B_{1j} + A_{i2}B_{2j}]_{pq} \,. \square$$

Lemma 1.3.1 tells us everything about conventional 2-by-2 block matrix multiplication. For general block matrices we have the following:

**Theorem 1.3.2** *If*

$$A = \begin{bmatrix} A_{11} & \dots & A_{1q} \\ \vdots & & \vdots \\ A_{p1} & \cdots & A_{pq} \end{bmatrix} \begin{matrix} m_1 \\ \\ m_p \end{matrix}, \qquad B = \begin{bmatrix} B_{11} & \dots & B_{1t} \\ \vdots & & \vdots \\ B_{q1} & \cdots & B_{qt} \end{bmatrix} \begin{matrix} r_1 \\ \\ r_q \end{matrix}$$
$$\qquad \begin{matrix} r_1 & \qquad & r_q \end{matrix} \qquad\qquad\qquad\qquad \begin{matrix} n_1 & \qquad & n_t \end{matrix}$$

*and we partition the product $C = AB$ as follows,*

$$C = \begin{bmatrix} C_{11} & \dots & C_{1t} \\ \vdots & & \vdots \\ C_{p1} & \cdots & C_{pt} \end{bmatrix} \begin{matrix} m_1 \\ \\ m_p \end{matrix}$$
$$\begin{matrix} n_1 & \qquad & n_t \end{matrix}$$

*then*

$$C_{ij} = \sum_{k=1}^{q} A_{ik} B_{kj} \qquad i = 1{:}p, \quad j = 1{:}t.$$

**Proof.** Use induction with Lemma 1.3.1. The details are not instructive and so are omitted. □

As with "ordinary" matrix multiplication, block matrix multiplication can be organized in several ways. To specify the computations precisely, we need some notation.

## 1.3.3 Submatrix Designation

Consider the identification of specific blocks in a scalar matrix. Suppose $A \in \mathbb{R}^{m \times n}$ and that $i = (i_1, \dots, i_r)$ and $j = (j_1, \dots, j_c)$ are integer vectors with the property that $i_1, \ldots, i_r \in \{1, 2, \ldots, m\}$ and $j_1, \ldots, j_c \in \{1, 2, \ldots, n\}$. We let $A(i, j)$ denote the $r$-by-$c$ submatrix

$$A(i,j) \quad = \quad \begin{bmatrix} A(i_1, j_1) & \cdots & A(i_1, j_c) \\ \vdots & & \vdots \\ A(i_r, j_1) & \cdots & A(i_r, j_c) \end{bmatrix}.$$

Often the entries in $i$ and $j$ are contiguous and on these occasions we use the "colon" notation. In particular, if $1 \le i \le j \le m$ and $1 \le p \le q \le n$ then $A(i{:}j, p{:}q)$ is the submatrix that we obtain by extracting rows $i$ through $j$ and columns $p$ through $q$. Thus,

$$A(3{:}5, 1{:}2) = \begin{bmatrix} a_{31} & a_{32} \\ a_{41} & a_{42} \\ a_{51} & a_{52} \end{bmatrix}.$$

While on the subject of submatrices, recall from §1.2.5 that if $i$ and $j$ are scalars then $A(i,:)$ designates the $i$th row of $A$ and $A(:,j)$ designates the $j$th column of $A$.

### 1.3.4 Block Matrix Times Vector

An important situation covered by Theorem 1.3.2 is the case of a block matrix times a vector. Suppose we partition $A \in \mathbb{R}^{m \times n}$ and $z \in \mathbb{R}^m$ as follows:

$$A \ = \begin{bmatrix} A_1 \\ \vdots \\ A_p \end{bmatrix} \begin{matrix} m_1 \\ \\ m_p \end{matrix} \qquad z \ = \begin{bmatrix} z_1 \\ \vdots \\ z_p \end{bmatrix} \begin{matrix} m_1 \\ \\ m_p \end{matrix}.$$

Here $m_1 + \cdots + m_p = m$. We refer to $A_i$ as the $i$th block row. Note that if $x \in \mathbb{R}^n$ and $z = Ax$ then $z_i = A_i x$, $i = 1{:}p$. Thus, if $m.vec = (m_1, ..., m_p)$ then we obtain the following block form of matrix-vector multiplication:

$$\begin{array}{l} last = 0 \\ \textbf{for } i = 1{:}p \\ \quad first = last + 1;\ last = first + m.vec(i) - 1 \\ \quad z(first{:}last) = A(first{:}last, :)x \\ \textbf{end} \end{array} \tag{1.3.1}$$

Of course, the matrix-vector products $z_i = A_i x$ can be found using either **matvec.ij** or **matvec.ji**.

Another way to block the matrix-vector multiplication problem is to partition $A$ and $x$ as follows:

$$A \ = \ \begin{matrix} [\ A_1 & ,\ldots, & A_q\ ] \\ n_1 & & n_q \end{matrix} \qquad x \ = \begin{bmatrix} x_1 \\ \vdots \\ x_q \end{bmatrix} \begin{matrix} n_1 \\ \\ n_q \end{matrix}.$$

In this case we refer to the $A_j$ as the $j$th block column and the product $z = Ax$ has the form $z = A_1x_1 + \cdots + A_qx_q$. If $n.vec = (n_1, \ldots, n_q)$ is the vector of block column dimensions, then here is a block column version of (1.3.1):

$$\begin{array}{l} z(1{:}m) = 0;\ last = 0 \\ \textbf{for } j = 1{:}q \\ \quad first = last + 1;\ last = first + n.vec(j) - 1 \\ \quad w = A(:, first{:}last)x(first{:}last) \\ \quad z = z + w \\ \textbf{end} \end{array} \tag{1.3.2}$$

Again, the individual matrix-vector products (the $w$ computations) can be carried out with either **matvec.ij** or **matvec.ji**.

### 1.3.5 Block Matrix Multiplication

Just as ordinary, scalar level matrix multiplication can be arranged in several possible ways, so can the multiplication of block matrices. For simplicity we assume that $A = (A_{ij})$ and $B = (B_{ij})$ are each $N$-by-$N$ block matrices with $\alpha$-by-$\alpha$ blocks, i.e., $n = N\alpha$. With this uniform blocking Theorem 1.3.2 says that if $C = (C_{ij}) = AB$ then

$$C_{ij} = \sum_{k_b=1}^{N} A_{ik_b}B_{k_bj} \qquad i = 1{:}N, \quad j = 1{:}N.$$

If we organize a matrix multiplication procedure around this block dot product equation we obtain a block dot product procedure:

$$\begin{array}{l} C(1{:}n, 1{:}n) = 0 \\ \textbf{for } i_b = 1{:}N \\ \quad i = (i_b - 1)\alpha + 1{:}i_b\alpha \\ \quad \textbf{for } j_b = 1{:}N \\ \quad\quad j = (j_b - 1)\alpha + 1{:}j_b\alpha \\ \quad\quad \textbf{for } k_b = 1{:}N \\ \quad\quad\quad k = (k_b - 1)\alpha + 1{:}k_b\alpha \\ \quad\quad\quad C(i,j) = C(i,j) + A(i,k)B(k,j) \\ \quad\quad \textbf{end} \\ \quad \textbf{end} \\ \textbf{end} \end{array} \tag{1.3.3}$$

By virtue of our notation which permits $i$, $j$, and $k$ to be vector subscripts, the assignment $C(i,j) = C(i,j) + A(i,k)B(k,j)$ looks just as it did in our scalar level formulations.

Other block matrix multiplication procedures result if we play with the loop orderings in (1.3.3). For example,

$$
\begin{array}{l}
C(1{:}n, 1{:}n) = 0 \\
\textbf{for } j_b = 1{:}N \\
\quad j = (j_b - 1)\alpha + 1{:}j_b\alpha \\
\quad \textbf{for } k_b = 1{:}N \\
\quad\quad k = (k_b - 1)\alpha + 1{:}k_b\alpha \\
\quad\quad \textbf{for } i_b = 1{:}N \\
\quad\quad\quad i = (i_b - 1)\alpha + 1{:}i_b\alpha \\
\quad\quad\quad C(i,j) = C(i,j) + A(i,k)B(k,j) \\
\quad\quad \textbf{end} \\
\quad \textbf{end} \\
\textbf{end}
\end{array}
\tag{1.3.4}
$$

The best way to think about this arrangement is to partition $A$ into block columns

$$
A = [\ \underset{\alpha}{A_1}\ ,\ldots,\ \underset{\alpha}{A_N}\ ]
$$

If $B = (B_{ij})$ with $B_{ij} \in \mathbb{R}^{\alpha\times\alpha}$, then

$$
AB = C = [\ \underset{\alpha}{C_1}\ ,\ldots,\ \underset{\alpha}{C_N}\ ] \qquad\qquad C_j = \sum_{k=1}^{N} A_k B_{kj}\ .
$$

Thus, $C_j$ is a *block* linear combination of $A$'s block columns. The $k_b$ loop carries out a *block gaxpy* operation.

A *block outer product* scheme results if we rearrange the loops in (1.3.3) as follows:

$$
\begin{array}{l}
C(1{:}n, 1{:}n) = 0 \\
\textbf{for } k_b = 1{:}N \\
\quad k = (k_b - 1)\alpha + 1{:}k_b\alpha \\
\quad \textbf{for } j_b = 1{:}N \\
\quad\quad j = (j_b - 1)\alpha + 1{:}j_b\alpha \\
\quad\quad \textbf{for } i_b = 1{:}N \\
\quad\quad\quad i = (i_b - 1)\alpha + 1{:}i_b\alpha \\
\quad\quad\quad C(i,j) = C(i,j) + A(i,k)B(k,j) \\
\quad\quad \textbf{end} \\
\quad \textbf{end} \\
\textbf{end}
\end{array}
\tag{1.3.5}
$$

The best way to think about this computation is through the partitionings

$$
A = [\ \underset{\alpha}{A_1}\ ,\ldots,\ \underset{\alpha}{A_N}\ ] \qquad\qquad B = \begin{bmatrix} B_1 \\ \vdots \\ B_N \end{bmatrix} \begin{array}{l} \alpha \\ \\ \alpha \end{array}
$$

for then we see that the $k_b$ loop performs the summation

$$C = AB = \sum_{k_b=1}^{N} A_{k_b} B_{k_b} .$$

In this formulation of matrix multiplication we regard $C$ as the sum of block outer products.

## 1.3.6 An Important Special Case

In §1.3.4 we presented a block row and a block column approach to the matrix-vector product problem $z = Ax$. Another important way to look at this computation is to regard $A$ as a 2-by-2 block matrix and $x$ as a 2-dimensional block vector:

$$\begin{bmatrix} z_1 \\ z_2 \end{bmatrix} = \begin{bmatrix} A_{11} & A_{12} \\ A_{21} & A_{22} \end{bmatrix} \begin{bmatrix} x_1 \\ x_2 \end{bmatrix} = \begin{bmatrix} A_{11}x_1 + A_{12}x_2 \\ A_{21}x_1 + A_{22}x_2 \end{bmatrix}$$

This particular partitioning of $z = Ax$ is used over and over again in subsequent chapters.

## 1.3.7 Block Matrix Data Structures

Arrays in Fortran are stored in column major order. This means that the entries within a column are contiguous in memory. Thus, if 24 storage locations are allocated for $A \in \mathbb{R}^{4\times 6}$ then in traditional store-by-column format the matrix entries are "lined up" in memory as depicted in Figure 1.3.1.

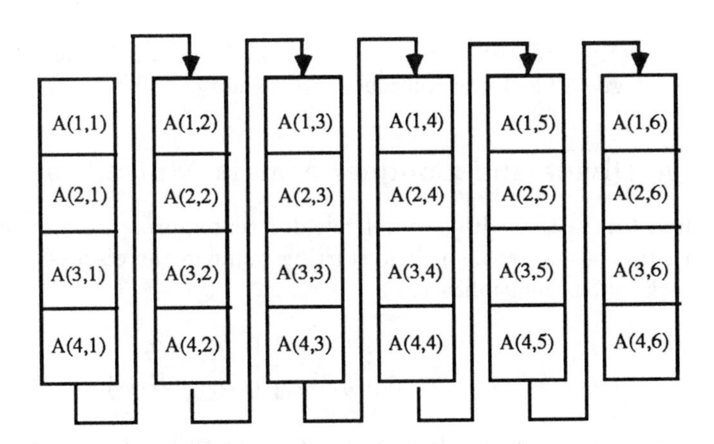


Figure 1.3.1. Store by Column (4-by-6 case).

In other words, if $A \in \mathbb{R}^{m \times n}$ is stored in $v(1{:}mn)$ then we identify $A(i,j)$ with $v((j-1)m+i)$. For algorithms that access matrix data by column this is a good arrangement since the column entries are contiguous in memory.

In certain block matrix algorithms it is sometimes useful to store matrices by blocks rather than by column. Suppose, for example, that the matrix $A$ above is a 2-by-3 block matrix with 2-by-2 blocks. In a store-by-column block scheme with store-by-column within each block, the 24 entries are arranged in memory as shown in Figure 1.3.2.

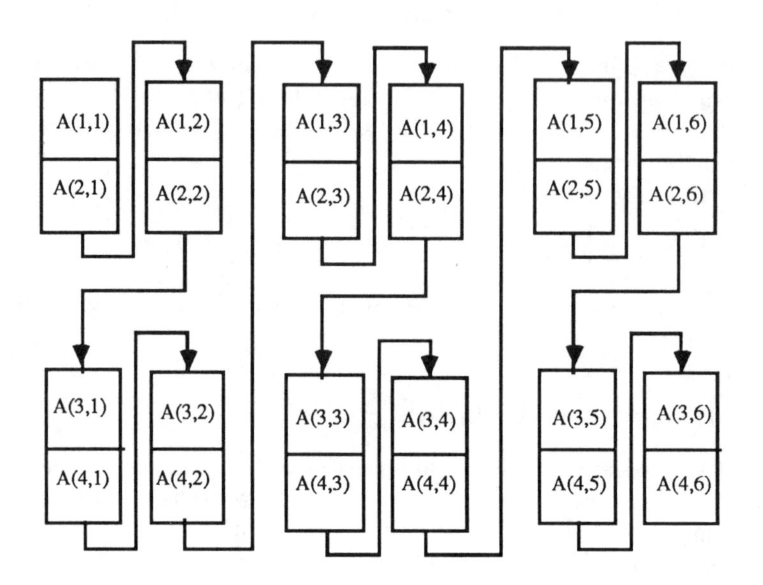


Figure 1.3.2. Store-by-Blocks (4-by-6 case with 2-by-2 Blocks).

This data structure can be attractive for block algorithms because the entries within a given block are contiguous in memory.

## 1.3.8 A Divide and Conquer Matrix Multiplication

We conclude this section with a completely different approach to the matrix-matrix multiplication problem. The starting point in the discussion is the 2-by-2 block matrix multiplication

$$\begin{bmatrix} C_{11} & C_{12} \\ C_{21} & C_{22} \end{bmatrix} = \begin{bmatrix} A_{11} & A_{12} \\ A_{21} & A_{22} \end{bmatrix} \begin{bmatrix} B_{11} & B_{12} \\ B_{21} & B_{22} \end{bmatrix}$$

where each block is square. In the ordinary algorithm, $C_{ij} = A_{i1}B_{1j} + A_{i2}B_{2j}$. There are 8 multiplies and 4 adds. Strassen(1969) has shown how to compute $C$ with just 7 multiplies and 18 adds:

$$
\begin{aligned}
P_1 &= (A_{11} + A_{22})(B_{11} + B_{22}) \\
P_2 &= (A_{21} + A_{22})B_{11} \\
P_3 &= A_{11}(B_{12} - B_{22}) \\
P_4 &= A_{22}(B_{21} - B_{11}) \\
P_5 &= (A_{11} + A_{12})B_{22} \\
P_6 &= (A_{21} - A_{11})(B_{11} + B_{12}) \\
P_7 &= (A_{12} - A_{22})(B_{21} + B_{22}) \\
C_{11} &= P_1 + P_4 - P_5 + P_7 \\
C_{12} &= P_3 + P_5 \\
C_{21} &= P_2 + P_4 \\
C_{22} &= P_1 + P_3 - P_2 + P_6
\end{aligned}
$$

These equations are easily confirmed by substitution. Suppose $n = 2m$ so that the blocks are $m$-by-$m$. Counting adds and multiplies in the computation $C = AB$ we find that conventional matrix multiplication involves $(2m)^3$ multiplies and $(2m)^3 - (2m)^2$ adds. In contrast, if Strassen's algorithm is applied *with conventional multiplication at the block level,* then $7m^3$ multiplies and $7m^3 + 11m^2$ adds are required. If $m \gg 1$ then the Strassen method involves about 7/8ths the arithmetic of the fully conventional algorithm.

Now recognize that we can recur on the Strassen idea. In particular, we can apply the Strassen algorithm to each of the half-sized block multiplications associated with the $P_i$. Thus, if the original $A$ and $B$ are $n$-by-$n$ and $n = 2^q$ then we can repeatedly apply the Strassen multiplication algorithm. At the bottom "level," the blocks are 1-by-1. Of course, there is no need to recur down to the $n = 1$ level. When the block size gets sufficiently small, $(n \leq n_{min})$, it may be sensible to use conventional matrix multiplication when finding the $P_i$ . Here is the overall procedure:

**Algorithm 1.3.1 (Strassen Multiplication)** Suppose $n = 2^q$ and that $A \in \mathbb{R}^{n\times n}$ and $B \in \mathbb{R}^{n\times n}$. If $n_{min} = 2^d$ with $d \leq q$, then the following algorithm computes $C = AB$ by applying Strassen procedure recursively $q - d$ times.

**function:** $C =$ **strass**$(A, B, n_{min})$
  $n = rows(A)$
  **if** $n \leq n_{min}$
    $C = AB$
  **else**
    $m = n/2; u = 1{:}m; v = m + 1{:}n;$
    $P_1 =$ **strass**$(A(u, u) + A(v, v), B(u, u) + B(v, v), n_{min})$
    $P_2 =$ **strass**$(A(v, u) + A(v, v), B(u, u), n_{min})$

$P_3 = \textbf{strass}(A(u,u), B(u,v) - B(v,v), n_{min})$
$P_4 = \textbf{strass}(A(v,v), B(v,u) - B(u,u), n_{min})$
$P_5 = \textbf{strass}(A(u,u) + A(u,v), B(v,v), n_{min})$
$P_6 = \textbf{strass}(A(v,u) - A(u,u), B(u,u) + B(u,v), n_{min})$
$P_7 = \textbf{strass}(A(u,v) - A(v,v), B(v,u) + B(v,v), n_{min})$
$C(u,u) = P_1 + P_4 - P_5 + P_7$
$C(u,v) = P_3 + P_5$
$C(v,u) = P_2 + P_4$
$C(v,v) = P_1 + P_3 - P_2 + P_6$
**end**
**end strass**

Unlike any of our previous algorithms **strass** is recursive, meaning that it calls itself. Divide and conquer algorithms are often best described in this manner.

The amount of arithmetic associated with **strass** is a complicated function of $n$ and $n_{min}$. If $n_{min} \gg 1$ then it suffices to count multiplications as the number of additions is roughly the same. If we just count the multiplications then it suffices to examine the deepest level of the recursion as that is where all the multiplications occur. In **strass** there are $q - d$ subdivisions and thus, $7^{q-d}$ conventional matrix-matrix multiplications to perform. These multiplications have size $n_{min}$ and thus **strass** involves about $s = (2^d)^3 7^{q-d}$ multiplications compared to $c = (2^q)^3$, the number of multiplications in the conventional approach. Notice that

$$\frac{s}{c} = \left(\frac{2^d}{2^q}\right)^3 7^{q-d} = \left(\frac{7}{8}\right)^{q-d}.$$

If $d = 0$ , i.e., we recur on down to the 1-by-1 level, then

$$s = \left(\frac{7}{8}\right)^q c = 7^q = n^{\log_2 7} \approx n^{2.807}.$$

Thus, asymptotically, the number of multiplications in the Strassen procedure is $O(n^{2.807})$. However, the number of additions (relative to the number of multiplications) becomes significant as $n_{min}$ gets small.

**Example 1.3.1** If $n = 1024$ and $n_{min} = 64$, then **strass** involves $(7/8)^{10-6} \approx .6$ the arithmetic of the conventional algorithm.

**Problems**

**P1.3.1** Suppose $A \in \mathbb{R}^{m\times n}$ is stored by column in $A.col(1{:}mn)$. Assume that $n = \alpha N$ and $m = \beta M$ and that we regard $A$ as an $M$-by-$N$ block matrix with $\beta$-by-$\alpha$ blocks. Given $i$, $j$, $r$, and $s$ that satisfy $1 \leq i \leq \beta$, $1 \leq j \leq \alpha$, $1 \leq r \leq M$, and $1 \leq s \leq N$ determine $k$ so that $A.col(k)$ houses the $(i,j)$ entry of the $(r,s)$ block. Give an algorithm that overwrites $A.col$ with $A$ stored by block as in Figure 1.3.2. How big must the work array be?

**P1.3.2** Generalize (1.3.4) so that it can handle the general block-size problem covered by Theorem 1.3.2.

**P1.3.3** Adapt **strass** so that it can handle square matrix multiplication of any order. Hint: If the "current" $A$ has odd dimension, append a zero row and column.

**P1.3.4** Suppose $A = A_1 + iA_2 \in \mathbb{C}^{m\times n}$ and $x = x_1 + ix_2 \in \mathbb{C}^n$. Show how an appropriate real call to **matvec.ji** can be used to compute $y = y_1 + iy_2 = Ax$.

**P1.3.5** Suppose $n$ is even and define the following function from $\mathbf{R}^n$ to $\mathbf{R}$:

$$f(x) \;=\; x(1{:}2{:}n)^T x(2{:}n) \;=\; \sum_{i=1}^{n/2} x_{2i-1}x_{2i}$$

(a) Show that if $x, y \in \mathbf{R}^n$ then

$$x^T y \;=\; \sum_{i=1}^{n/2}(x_{2i-1} + y_{2i})(x_{2i} + y_{2i-1}) - f(x) - f(y)$$

(b) Now consider the $n$-by-$n$ matrix multiplication $C = AB$. Give an algorithm for computing this product that requires $n^3/2$ multiplies once $f$ is applied to the rows of $A$ and the columns of $B$. See Winograd (1968) for details.

### Notes and References for Sec. 1.3

For quite some time fast methods for matrix multiplication have attracted a lot of attention within computer science. See

V. Pan (1984). "How Can We Speed Up Matrix Multiplication?," *SIAM Review 26*, 393–416.

V. Strassen (1969). "Gaussian Elimination is Not Optimal," *Numer. Math. 13*, 354–356.

S. Winograd (1968). "A New Algorithm for Inner Product," *IEEE Trans. Comp. C-17*, 693–694.

Many of these methods have dubious practical value. However, with the publication of

D. Bailey (1988). "Extra High Speed Matrix Multiplication on the Cray-2," *SIAM J. Sci. and Stat. Comp. 9*, 603–607.

it is clear that the blanket dismissal of these fast procedures is unwise. Bailey essentially implemented Algorithm 1.3.1 on a Cray-2 with $n_{min} = 128$. His Strassen approach requires about 60% of the time needed by the conventional matrix multiply routine. We discuss the "stability" of the Strassen algorithm in §2.7.7.

Finally we mention that the matrices that arise in practice frequently have a block structure of their own. For example, many discretized differential operators give rise to block matrices. Or, in a Markov process clusters of "states" may have special interactions that give rise to a block structured matrix problem. See

L. Kaufman (1983). "Matrix Methods for Queueing Problems," *SIAM J. Sci. and Stat. Comp. 4*, 525–552.

## 1.4 Aspects of Vector Pipeline Computing

Matrix manipulation is mostly built upon dot products and saxpy operations. *Vector pipeline computers* are able to perform vector operations such as these very fast because of special hardware that is able to exploit the fact that a vector operation is a very regular sequence of scalar operations. Whether or not high performance is extracted from such a computer depends upon the length of the vector operands and a number of other factors that pertain to the movement of data such as vector stride, the number of vector loads and stores, and the level of vector re-use. Our goal is to build a useful awareness of these issues. We are *not* trying to build a comprehensive model of vector pipeline computing that might be used to predict performance. Our goal is simply to identify the kind of *thinking* that goes into the design of an effective vector pipeline code. We do not mention any particular machine. The literature is filled with case studies for readers who want them.

## 1.4.1 Pipelining Arithmetic Operations

The primary reason why vector computers are fast has to do with *pipelining*. The concept of pipelining is best understood by making an analogy to assembly line production. Suppose the assembly of an individual automobile requires one minute at each of sixty workstations along an assembly line. If the line is well staffed and able to initiate the assembly of a new car every minute, then 1000 cars can be produced from scratch in about 1000 + 60 = 1060 minutes. For a work order of this size the line has an effective "vector speed" of 1000/1060 automobiles per minute. On the other hand, if the assembly line is understaffed and a new assembly can be initiated just once an hour, then 1000 hours are required to produce a 1000 cars. In this case the line has an effective "scalar speed" of 1/60th automobile per minute.

So it is with a pipelined vector operation such as the vector add $z = x+y$. The scalar operations $z_i = x_i + y_i$ are the cars. The number of elements is the size of the work order. If the start-to-finish time required for each $z_i$ is $\tau$ then a pipelined, length $n$ vector add could be completed in time much less than $n\tau$. This gives vector speed. Without the pipelining, the vector computation would proceed at a scalar rate and would approximately require time $n\tau$ for completion.

Let us see how a sequence of floating point operations can be pipelined. Floating point operations usually require several cycles to complete. For example, a 3-cycle addition of two scalars $x$ and $y$ may proceed as in Figure 1.4.1.

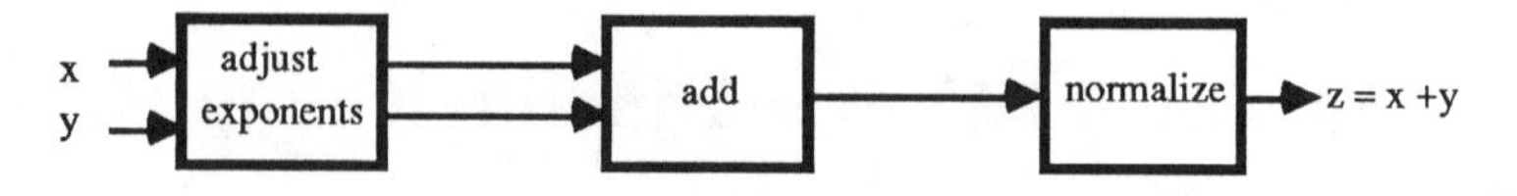


Figure 1.4.1. A 3-Cycle Adder

To visualize the operation, think of the addition unit as an assembly line with three "work stations" as indicated. The input scalars $x$ and $y$ proceed along the assembly line spending one cycle at each of three stations. The sum $z$ emerges after three cycles. Note that when a single, "free standing" addition is performed, only one of the three stations is active during the computation.

Now consider a vector addition $z = x + y$ . With pipelining, the $x$ and $y$ vectors are streamed through the addition unit. Once the pipeline is filled and steady state reached, a $z_i$ is produced every cycle. In Figure 1.4.2 we depict what the pipeline might look like once this steady state is achieved:

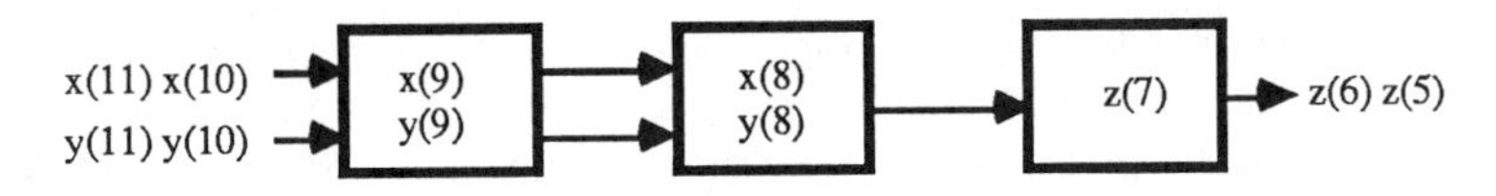


Figure 1.4.2. Pipelined Addition

In this case, vector speed is about three times scalar speed because the time for an individual add is three cycles.

Vector operations such as dot product and saxpy involve both additions and multiplications. In our model vector computer we assume that the addition and multiplication units can be linked together to form a single pipeline. For example, in a saxpy operation $z = \alpha x + y$ the pipelining would organized as in Figure 1.4.3.

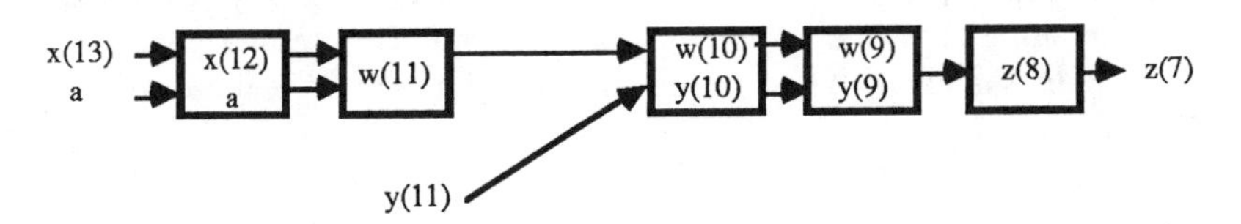


Figure 1.4.3 Pipelined Saxpy

Here $w = \alpha x$. Thus, once the pipeline is full, a component of $z$ is produced each cycle. Note that once "steady state" is achieved, a floating point add and multiply are completed each cycle. This is why the "peak performance" of a typical vector processor is given by $2/\mu$ floating point operations per second where $\mu$ is the cycle time.

## 1.4.2 Vector Operations

A vector pipeline computer comes with a repertoire of *vector instructions*, such as vector add ($z_i = x_i + y_i$), vector multiply ($z_i = x_i y_i$), vector scale ($z_i = \alpha x_i$), dot product, and saxpy. We assume for clarity that these operations take place in *vector registers.* Vectors travel between the registers and memory by means of *vector load* and *vector store* instructions.

An important attribute of a vector processor is the length of its vector registers which we designate by $v_L$. A length $n$ vector operation must be broken down into subvector operations of length $v_L$ or less. Here is how such a partitioning might be managed in the case of a vector addition $z = x + y$ where $x$ and $y$ are $n$-vectors:

$$
\begin{array}{l}
first = 1 \\
\textbf{while } first \le n \\
\quad last = \min\{n, first + v_L - 1\} \\
\quad \text{Vector load } x(first{:}last). \\
\quad \text{Vector load } y(first{:}last). \\
\quad \text{Vector add: } z(first{:}last) = x(first{:}last) + y(first{:}last). \\
\quad \text{Vector store } z(first{:}last). \\
\quad first = last + 1 \\
\textbf{end}
\end{array}
\qquad (1.4.1)
$$

Notice that all the operations (load, store, add) are vector operations. A reasonable compiler for a vector computer would automatically generate (1.4.1) from a programmer specified $z = x + y$ instruction. Of course, the programmer may have to specify the vector add in the form of a Fortran loop.

### 1.4.3 The Vector Length Issue

Suppose the pipeline for the vector operation op takes $\tau_{\text{op}}$ cycles to "set up." Assume that one component of the result is obtained per cycle once the pipeline is filled. The time required to perform an $n$-dimensional op is then given by

$$T_{\text{op}}(n) = (\tau_{\text{op}} + n)\mu \qquad n \le v_L$$

where $\mu$ is the cycle time and $v_L$ is the length of the vector hardware.

If the vectors to be combined are longer than the vector hardware length, then as we have seen the overall vector operation must be broken down into hardware-manageable chunks. Thus, if

$$n = n_1 v_L + n_0 \qquad 0 \le n_0 < v_L$$

then we assume that

$$T_{\text{op}}(n) = \begin{cases} n_1(\tau_{\text{op}} + v_L)\mu & n_0 = 0 \\ (n_1(\tau_{\text{op}} + v_L) + \tau_{\text{op}} + n_0)\mu & n_0 \ne 0 \end{cases}$$

specifies the overall time required to perform a length $n$ op. This simplifies to

$$T_{\text{op}}(n) = \left(n + \tau_{\text{op}}\text{ceil}(n/v_L)\right)\mu$$

where ceil($\alpha$) is the smallest integer such that $\alpha \leq$ ceil($\alpha$). If $\rho$ flops per component are involved then the effective rate of computation for general $n$ is given by

$$R_{\text{op}}(n) = \frac{\rho n}{T_{op}(n)} = \frac{\rho}{\mu} \frac{1}{1 + \frac{\tau_{\text{op}}}{n} \text{ceil}\left(\frac{n}{v_L}\right)}.$$

(If $\mu$ is in seconds then $R_{\text{op}}$ is in flops per second.) The asymptotic rate of performance is given by

$$\lim_{n \to \infty} R_{\text{op}}(n) = \frac{1}{1 + \frac{\tau_{\text{op}}}{v_L}} \frac{\rho}{\mu}.$$

As a way of assessing how serious the start-up overhead is for a vector operation Hockney and Jesshope (1988) define the quantity $n_{1/2}$ to be the smallest $n$ for which half of peak performance is achieved, i.e.,

$$\frac{\rho n_{1/2}}{T_{op}(n_{1/2})} = \frac{1}{2} \frac{\rho}{\mu}.$$

Machines that have big $n_{1/2}$ factors do not perform well on short vector operations.

### 1.4.4 Optimizing Code for Vector Length

We have outlined a simplified model of vector computing where performance is a function of vector length and the start-up factor $\tau_{\text{op}}$. Let us see how the model advises us in the design of a matrix-matrix multiply procedure ignoring all other factors such as mode of access.

Suppose that we want to compute the product $C = AB$ where $A \in \mathbb{R}^{m \times r}$ and $B \in \mathbb{R}^{r \times n}$. Recall from §1.1 that there are six possible versions of the conventional algorithm and that the innermost loop performs either a dot product or a saxpy. Consider the $ijk$ version:

```
for i = 1:m
    for j = 1:n
        for k = 1:r
            C(i,j) = C(i,j) + A(i,k)B(k,j)
        end
    end
end
```

The inner loop is a length $r$ dot product. Thus, according to our simple model above, if we ignore everything except the dot product executions, then the algorithm requires time

$$T_{ijk} = mn[r + \text{ceil}(r/v_L)\tau_{dot}]\mu$$

to complete. The *jki* algorithm has a saxpy innermost loop of length $m$ while the *kij* version also has a saxpy innermost loop but of length $n$. Thus the predicted execution times for these two algorithms are given by

$$T_{jki} = nr\left[m + \text{ceil}(m/v_L)\tau_{sax}\right]\mu$$

and

$$T_{kij} = mr\left[n + \text{ceil}(n/v_L)\tau_{sax}\right]\mu\,.$$

(We need not consider the *jik*, *kji*, or *ikj* versions because in our model, $T_{jik} = T_{ijk}$ , $T_{kji} = T_{jki}$, and $T_{ikj} = T_{kij}$.) It is apparent from these three formulae that the most efficient version of matrix multiplication is a function of the dimensions $m$, $n$, and $r$ and the start-up constants $\tau_{sax}$ and $\tau_{dot}$. An optimized, machine dependent matrix multiply procedure could thus choose the most efficient of the possibilities:

**function:** $C =$ **matmat.opt**$(A, B, v_L, \tau_{dot}, \tau_{sax})$
  $m =$ **rows**$(A)$; $n =$ **cols**$(B)$; $r =$ **rows**$(B)$
  $t_i = nr\tau_{sax}\text{ceil}(m/v_L)$
  $t_j = mr\tau_{sax}\text{ceil}(n/v_L)$
  $t_k = mn\tau_{dot}\text{ceil}(r/v_L)$
  **if** $t_i = \min(t_i, t_j, t_k)$
    $C =$ **matmat.jki**$(A, B)$
  **elseif** $t_j = \min(t_i, t_j, t_k)$
    $C =$ **matmat.kij**$(A, B)$
  **elseif** $t_k = \min(t_i, t_j, t_k)$
    $C =$ **matmat.ijk**$(A, B)$
  **end**
**end matmat.opt**

We make a few observations based upon some elementary integer arithmetic manipulation. Assume that $\tau_{sax}$ and $\tau_{dot}$ are roughly the same size. If $m$, $n$, and $r$ are all less than $v_L$, then **matmat.opt** chooses the version with roughly the longest innnermost loop. If $m$, $n$, and $r$ are much bigger than $v_L$then the distinction between the three options is small.

Remember that these observations and **matmat.opt** itself, are premised on a "one-dimensional" view of vector pipeline computation that is concerned only with vector length. To refine our thinking, we must also pay attention to the movement of data between memory and the registers that perform the actual vector computation.

### 1.4.5 Hierarchical Memory

Before a vector operation can be performed in a vector computer, the operands must be in the "right place" and getting to the right place entails a cost. The problem is that memory is usually arranged in a *hierarchy*. A relatively small high speed memory called a *cache* may be situated between the functional units that perform the arithmetic and the "main memory" that houses the vector data.

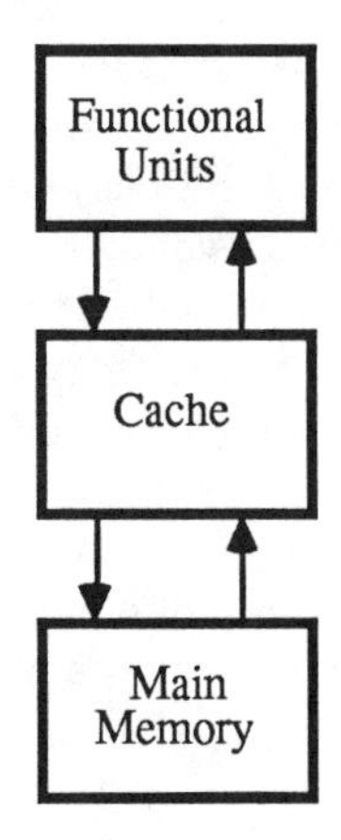


Figure 1.4.4. Memory Hierarchy

Details vary from machine to machine, but there are essentially two issues to bear in mind when working with a memory hierarchy:

- Each level in the hierarchy has a limited capacity and this capacity, for economic reasons, is usually smaller as we ascend the hierarchy.

- There is a cost, sometimes relatively great, associated with the moving of data between two levels in the hierarchy.

The implication here is obvious. *If we are to design efficient matrix algorithms for the typical vector computer, then we must pay strict attention to the flow of data within the memory hierarchy.* This is even the case for computations that involve relatively small matrices because present day compilers are just not smart enough to map basic matrix computations onto a hierarchical system.

Of course, monitoring memory traffic is not a new activity in scientific computation. The solution of matrix problems on the earliest electronic computers involved the intelligent flow of information to and from paper tapes and other primitive secondary storage devices. In sparse matrix computations the design of efficient code has always required strict attention to memory traffic issues.

In the remainder of this section we illustrate the kind of algorithmic thinking that memory hierarchies impose upon the designer of efficient matrix code.

### 1.4.6 The Unit Stride Issue

Whenever data is moved between the levels in a hierarchical memory system, it is usually important that the involved data be contiguous. For vectors this means *unit stride.* The *stride* of a stored floating point vector is the distance (in memory locations) between the vector's components. Accessing a row in a two-dimensional Fortran array is not a unit stride operation because arrays are stored by column.

Non-unit stride vector operations may interfere with the pipelining capability of a computer. For example, the loading of a non-unit stride vector from an interleaved memory may require an excess number of cycles due to a phenomena known as *memory bank conflict.*

Now consider the performance of our optimized matrix multiply procedure **matmat.opt** when stride becomes a concern. Of the three possible innermost loops,

**for** $i = 1{:}m$
  $C(i,j) = C(i,j) + A(i,k)B(k,j)$
**end**

**for** $j = 1{:}n$
  $C(i,j) = C(i,j) + A(i,k)B(k,j)$
**end**

**for** $k = 1{:}r$
  $C(i,j) = C(i,j) + A(i,k)B(k,j)$
**end**

only the first accesses the arrays $A$, $B$, and $C$ in unit stride fashion. Clearly, a Fortran implementation of **matmat.opt** should be revised so that the stride issue is addressed. How much the refined procedure should gravitate toward the selection of **matmat.jki** depends on the penalty associated with non-unit stride vector operations. We do not pursue this point as it is a particularly machine dependent issue. For us the key point has already been made: stride and vector length requirements may be difficult to reconcile.

Dilemmas of this type are typical in high performance computing. One goal (maximize vector length) can conflict with another (impose unit stride). Sometimes the best compromise can be deduced from a detailed timing study.

### 1.4.7 Effect of Data Structure

Vector stride/vector length tensions can sometimes be resolved through the intelligent choice of data structures. Consider again the matrix-vector multiplication $z = Ax$ where $A$ is symmetric. Assume that $A$ is $n$-by-$n$ and for simplicity that $n \leq v_L$. If $A$ is stored conventionally and **matvec.ji** is used then the central computation entails $n$ unit stride saxpy's each having length $n$:

$z(1{:}n) = 0$
**for** $j = 1{:}n$
    $z = z + A(:, j)x(j)$
**end**

If we only account for the saxpy operations then the predicted execution time is given by

$$T_1 = n(\tau_{sax} + n)\mu\ .$$

Let us see how this level of performance can be degraded when a symmetry-exploiting data structure is involved. If we store the lower triangular elements of $A$ by subdiagonal column in an array $A.vec(:)$ as discussed in §1.2.7, then the corresponding version of **matvec.ji** is as follows:

$z(1{:}n) = 0$
**for** $j = 1{:}n$
    **for** $i = 1{:}j-1$
        $z(i) = z(i) + A.vec((i-1)n - i(i-1)/2 + j)x(j)$
    **end**
    **for** $i = j{:}n$
        $z(i) = z(i) + A.vec((j-1)n - j(j-1)/2 + i)x(j)$
    **end**
**end**

Although the second $i$-loop, carries out a unit stride saxpy of length $n-j+1$, the first $i$-loop performs a nonunit stride, length $j-1$ saxpy.

To quantify the damage of the nonunit stride computation we must make some assumptions. Suppose a non-unit stride vector computation implies processing at scalar rate $R_s$ flops per second. Thus, time $n^2/R_s$ is spent in the second $i$-loop which involves about half of the total $2n^2$ flops. On the other hand, even though the first $i$-loop involves ever shorter vectors, it can (on average) proceed at a much faster rate, say $R_v$ . Thus, the overall time of execution is given by

$$T_2 = n^2\left(\frac{1}{R_s} + \frac{1}{R_v}\right) = \frac{n^2}{R_s}\left(1 + \frac{R_s}{R_v}\right)\ .$$

Note that even if $R_v = \infty$, the speed-up over a pure scalar rate execution is no more than a factor of 2. This is an illustration of *Amdahl's rule.* If $f \in [0,1]$ is the fraction of scalar code in a procedure, then

$$\frac{\text{Time with complete scalar execution}}{\text{Time with infinitely fast vector execution}} = \frac{1}{f}\,.$$

We may conclude that in an environment where stride is important, the procedure **matvec.ji** with the conventional storage of $A$ is preferable to **matvec.ji** with lower triangular storage.

Now suppose that $A$ is stored by diagonal in $A.diag(\cdot)$ as described in §1.2.8 and that Algorithm 1.2.4 is used to compute $z = Ax$. This algorithm consists entirely of unit stride vector multiplies. Thus, the anticipated execution time depends upon the vector multiply start-up overhead $\tau_{vm}$ :

$$\begin{aligned} T_3 &= (\tau_{vm} + n)\mu + \sum_{k=1}^{n-1}[(\tau_{vm} + (n-k)) + (\tau_{vm} + (n-k))]\mu \\ &= \left((2n-1)\tau_{vm} + n^2\right)\mu \qquad (1.4.2) \end{aligned}$$

Hence, it is the size of $T_3 - T_1 = ((2n-1)\tau_{vm} - n\tau_{sax})\mu$ and the importance of having a compact storage scheme that determines whether store-by-diagonal is attractive. *But it is clear from the discussion that the value of compact storage schemes for symmetric matrices is somewhat diminished in the vector/pipeline environment.*

### 1.4.8 Gaxpy's, Outer Products, and Vector Touches

In many environments, minimizing the number of vector loads and stores goes a long way toward producing good vector code. For convenience, we refer to vector loads and stores as *vector touches.*

Many of the algorithms in numerical linear algebra can be arranged so that the dominant computation is either an outer product update or a gaxpy operation. (See §1.1.15 and §1.1.16.) The $m$-by-$n$ versions of these operations each involve $2mn$ flops but the gaxpy involves half the number of vector touches.

To see this we assume for clarity that $m = m_1 v_L$ and $n = n_1 v_L$ . In a vector computing environment the outer product $A \leftarrow A + uv^T$ would be arranged as follows:

**for** $i_v = 1{:}m_1$
  $i = (i_v - 1)v_L + 1{:}i_v v_L;\ w = u(i)$
  **for** $j_v = 1{:}n_1$
    $j = (j_v - 1)v_L + 1{:}j_v v_L;\ A(i,j) = A(i,j) + wv(j)^T$
  **end**
**end**

Each column of the submatrix $A(i, j)$ must be loaded, updated, and then stored. Accounting for the $u$ and $v$ touches as well we see that a total of

$$\sum_{i_v=1}^{m_1} \left(1 + \sum_{j_v=1}^{n_1} (1 + 2v_L)\right) \approx 2m_1 n$$

vector touches are involved. (Notice how we disregard low order terms which do not contribute to our understanding.)

Now consider the $m$-by-$n$ gaxpy update $u = u + Av$ where $u \in \mathbb{R}^m$, $v \in \mathbb{R}^n$ and $A \in \mathbb{R}^{m \times n}$. Breaking this computation down into segments of length $v_L$ gives

**for** $i_v = 1{:}m_1$
    $i = (i_v - 1)v_L + 1{:}i_v v_L$; $w = u(i)$
    **for** $j_v = 1{:}n_1$
        $j = (j_v - 1)v_L + 1{:}j_v v_L$; $w = w + A(i, j)v(j)$
    **end**
    $u(i) = w$
**end**

Again, each column of submatrix $A(i, j)$ must be read but the only writing to memory involves the vector $w$. Thus, the number of vector touches involved is

$$\sum_{i_v=1}^{m_1} \left(2 + \sum_{j_v=1}^{n_1} (1 + v_L)\right) \approx m_1 n ,$$

about half the number required by the outer product.

*In practical vector computing, it is preferable to arrange matrix codes so that they are rich in gaxpy's rather than outer products.*

### 1.4.9 Block Algorithms, Cache, and Data Re-Use

The careful monitoring of memory traffic is central to good vector computing. The vector touch discussion above illustrates this point. Related observations can be made about the general movement of data up and down the memory hierarchy as a computation progresses. As a concrete example we consider how a cache can effect the design of a matrix multiplication algorithm. Recall from §1.4.5 that a memory cache can be thought of as a local memory situated between the main memory and the processing unit. See Figure 1.4.4. In this fairly typical situation, we assume that the operands for a vector operation must reside in the cache and it is relatively expensive to move data between cache and main memory. Thus, when data moves from main memory to the cache, it is important to use it as much as possible.

Consider the $n$-by-$n$ matrix multiply $C = AB$. Assume that the cache can hold $M$ floating point numbers and that $M << n^2$. Partition $B$ and $C$ into block columns,

$$B = [\underbrace{B_1}_{\alpha}, \ldots, \underbrace{B_N}_{\alpha}] \qquad C = [\underbrace{C_1}_{\alpha}, \ldots, \underbrace{C_N}_{\alpha}]$$

where $n = \alpha N$. Assume that $\alpha$ has been chosen such that $M \approx n(2\alpha+1) \approx 2n^2/N$ implying that a block column of $B$, a block column of $C$, and a column of $A$ can just fit in cache. Here is a matrix multiply procedure that gets a fair amount of re-use for each $B_j$ brought into cache:

> **for** $j = 1{:}N$
>     Load $B_j$ and $C_j = 0$ into cache.
>     **for** $k = 1{:}n$
>         Load $A(:,k)$ into cache and update $C_j$
>     **end**
>     Store completed $C_j$ in main memory.
> **end**

In effect, the $k$-loop performs a saxpy oriented matrix product $C_j = AB_j$. If we count the number $\Gamma_1$ of floating point numbers that move (in either direction) through the cache/main memory "door" during execution our block saxpy algorithm, then we find

$$\Gamma_1 = \sum_{j=1}^{N}[n(n/N) + n^2 + n(n/N)] = (2+N)n^2 = 2n^2(1+n^2/M)\,.$$

The formula clearly indicates the advantage of a large cache.

Now we show that a more effective cache utilization results if we use a block dot product approach to the multiplication. Regard $A = (A_{ij})$, $B = (B_{ij})$, and $C = (C_{ij})$ as $N$-by-$N$ block matrices with uniform block size $\alpha = n/N$. Assume that $N$is chosen so that $3\alpha^2 \approx M$ implying that 3 blocks can fit in the cache. With $C_L$ being a cache workspace, we proceed as follows:

> **for** $i = 1{:}N$
>     **for** $j = 1{:}N$
>         $C_L(1{:}\alpha, 1{:}\alpha) = 0$
>         **for** $k = 1{:}N$
>             Load $A_{ik}$ and $B_{kj}$ into cache.
>             $C_L = C_L + A_{ik}B_{kj}$
>         **end**
>         Store $C_L$ in main memory location for $C_{ij}$
>     **end**
> **end**

With this organization the main memory/cache traffic sums as follows:

$$\Gamma_2 = \sum_{i=1}^{N}\sum_{j=1}^{N}\left(\alpha^2 + \sum_{k=1}^{N} 2\alpha^2\right) = n^2(1+2N) \approx \frac{2n^3\sqrt{3}}{\sqrt{M}} .$$

Since $\Gamma_1/\Gamma_2 \approx n/\sqrt{3M}$ we see that there is a decided advantage to the block dot product approach. For example, if $n = 2^{10}$ and $M = 2^{14}$ then the memory-cache traffic for the two methods differs by a factor of about 4.5.

*From the example we see that computers having a cache tend to perform better on block algorithms.*

### 1.4.10 Summary

We have shown that the efficiency of a vector-pipeline matrix computation depends upon the vector length, the vector stride, the vector touch, and the data re-use properties of the algorithm. Optimizing with respect to all these attributes is very complicated and something of an art. All we have done here is to identify the key things to think about when working in a vector/pipeline environment. A good compiler can of course do some of the thinking for us, but do not count on it!

#### Problems

**P1.4.1** Consider the matrix product $D = ABC$ where $A \in \mathbb{R}^{m\times r}$, $B \in \mathbb{R}^{r\times n}$ and $C \in \mathbb{R}^{n\times q}$. Assume that all the matrices are stored by column and that the time required to execute a unit-stride saxpy operation of length $k$ is of the form $t(k) = (L+k)\mu$ where $L$ is a constant and $\mu$ is the cycle time. Based on this model, when is it more economical to compute $D$ as $D = (AB)C$ instead of as $D = A(BC)$? Assume that all matrix multiplies are done using the $jki$, gaxpy algorithm.

**P1.4.2** What is the total time spent in **matmat.jki** on the saxpy operations assuming that all the matrices are stored by column and that the time required to execute a unit-stride saxpy operation of length $k$ is of the form $t(k) = (L+k)\mu$ where $L$ is a constant and $\mu$ is the cycle time? Specialize **matmat.jki** so that it efficiently handles the case when $A$ and $B$ are $n$-by-$n$ and upper triangular. Does it follow that the triangular algorithm is six times faster as the flop count suggests?

**P1.4.3** Give an algorithm for computing $C = A^TBA$ where $A$ and $B$ are $n$–by–$n$ and $B$ is symmetric. Arrays should be accessed in unit stride fashion within all innermost loops.

#### Notes and References for Sec. 1.4

Two excellent expositions about vector computation are

J.J. Dongarra, F.G. Gustavson, and A. Karp (1984). "Implementing Linear Algebra Algorithms for Dense Matrices on a Vector Pipeline Machine," *SIAM Review 26*, 91-112.

J.M. Ortega and R.G. Voigt (1985). "Solution of Partial Differential Equations on Vector and Parallel Computers," *SIAM Review 27*, 149-240.

A very detailed look at matrix computations in hierarchical memory systems can be found in

K. Gallivan, W. Jalby, U. Meier, and A.H. Sameh (1988). "Impact of Hierarchical Memory Systems on Linear Algebra Algorithm Design," *Int'l J. Supercomputer Applic. 2*, 12-48.

See also

R.W. Hockney and C.R. Jesshope (1988). *Parallel Computers 2*, Adam Hilger, Bristol and Philadelphia.

W. Schönauer (1987). *Scientific Computing on Vector Computers*, North Holland, Amsterdam.

where various models of vector processor performance are set forth.

Some of the many papers on the practical aspects of vector computing include

B.L. Buzbee (1986) "A Strategy for Vectorization," *Parallel Computing 3*, 187-192.

J. Dongarra and S. Eisenstat (1984). "Squeezing the Most Out of an Algorithm in Cray Fortran," *ACM Trans. Math. Soft. 10*, 221-230.

J. Dongarra and A. Hinds (1979). "Unrolling Loops in Fortran," *Software Practice and Experience 9*, 219-229.

K. Gallivan, W. Jalby, and U. Meier (1987). "The Use of BLAS3 in Linear Algebra on a Parallel Processor with a Hierarchical Memory," *SIAM J. Sci. and Stat. Comp. 8*, 1079-1084.

B. Kågström and P. Ling (1988). "Level 2 and 3 BLAS Routines for the IBM 3090 VF/400: Implementation and Experiences," Report UMINF–154.88, Inst. of Inf. Proc., University of Umeå, S-901 87 Umeå, Sweden.

N. Madsen, G. Roderigue, and J. Karush (1976). "Matrix Multiplication by Diagonals on a Vector Parallel Processor," *Infomation Processing Letters 5*, 41-45.

# Chapter 2

# Matrix Analysis



The analysis and derivation of algorithms in the matrix computation area requires a facility with certain aspects of linear algebra. Some of the basics are reviewed in §2.1. Norms and their manipulation are covered in §2.2 and §2.3. In §2.4 we develop a model of finite precision arithmetic and then illustrate the kind of matrix analysis that is required when we quantify the effect of roundoff error.

The next two sections deal with orthogonality, which has a prominent role to play in matrix computations. The singular value decomposition and the CS decomposition are a pair of orthogonal reductions that provide critical insight into the important notions of rank and distance between subspaces. These issues are treated in §2.5 and §2.6.

Finally, in the last section we examine how the solution of a linear system $Ax = b$ changes if $A$ and $b$ are perturbed. The important concept of matrix condition is introduced.

## 2.1 Basic Ideas From Linear Algebra

This section is a quick review of linear algebra. Readers who wish a more detailed coverage should consult the references at the end of the section.

### 2.1.1 Independence, Subspace, Basis, and Dimension

A set of vectors $\{a_1, \ldots, a_n\}$ in $\mathbb{R}^m$ is *linearly independent* if $\sum_{j=1}^n \alpha_j a_j = 0$ implies $\alpha(1{:}n) = 0$. Otherwise, a nontrivial combination of the $a_i$ is zero and $\{a_1, \ldots, a_n\}$ is said to be *linearly dependent* .

A *subspace* of $\mathbb{R}^m$ is a subset that is also a vector space. Given a collection of vectors $a_1, \ldots, a_n \in \mathbb{R}^m$, the set of all linear combinations of these vectors is a subspace referred to as the *span* of $\{a_1, \ldots, a_n\}$:

$$\text{span}\{a_1, \ldots, a_n\} = \left\{ \sum_{j=1}^n \beta_j a_j : \beta_j \in \mathbb{R} \right\}.$$

If $\{a_1, \ldots, a_n\}$ is independent and $b \in \text{span}\{a_1, \ldots, a_n\}$ then $b$ is a unique linear combination of the $a_j$

If $S_1, \ldots, S_k$ are subspaces of $\mathbb{R}^m$, then their sum is the subspace defined by $S = \{ a_1 + a_2 + \cdots + a_k : a_i \in S_i,\ i = 1{:}k \}$. $S$ is said to be a *direct sum* if each $v \in S$ has a unique representation $v = a_1 + \cdots + a_k$ with $a_i \in S_i$. In this case we write $S = S_1 \oplus \cdots \oplus S_k$. The intersection of the $S_i$ is also a subspace, $S = S_1 \cap S_2 \cap \cdots \cap S_k$.

The subset $\{a_{i_1}, \ldots, a_{i_k}\}$ is a *maximal linearly independent subset* of $\{a_1, \ldots, a_n\}$ if it is linearly independent and is not properly contained in any linearly independent subset of $\{a_1, \ldots, a_n\}$. If $\{a_{i_1}, \ldots, a_{i_k}\}$ is maximal, then $\text{span}\{a_1, \ldots, a_n\} = \text{span}\{a_{i_1}, \ldots, a_{i_k}\}$ and $\{a_{i_1}, \ldots, a_{i_k}\}$ is a *basis* for $\text{span}\{a_1, \ldots, a_n\}$ . If $S \subseteq \mathbb{R}^m$ is a subspace, then it is possible to find independent basic vectors $a_1, \ldots, a_k \in S$ such that $S = \text{span}\{a_1, \ldots, a_k\}$ . All bases for a subspace $S$ have the same number of elements. This number is the *dimension* of $S$ and is denoted by $\dim(S)$.

### 2.1.2 Range, Null Space, and Rank

There are two important subspaces associated with an $m$-by-$n$ matrix $A$. The *range* of $A$ is defined by

$$\text{range}(A) = \{y \in \mathbb{R}^m : y = Ax \text{ for some } x \in \mathbb{R}^n\},$$

and the *null space* of $A$ is defined by

$$\text{null}(A) = \{x \in \mathbb{R}^n : Ax = 0\}.$$

If $A = [\, a_1, \ldots, a_n \,]$ is a column partitioning then

$$\text{range}(A) = \text{span}\{a_1, \ldots, a_n\} .$$

The *rank* of a matrix $A$ is defined by

$$\text{rank}(A) = \dim(\text{range}(A)) .$$

It can be shown that $\text{rank}(A) = \text{rank}(A^T)$ , and thus, the rank of a matrix equals the maximal number of independent rows or columns. For any $A \in \mathbb{R}^{m \times n}$ we have $\dim(\text{null}(A)) + \text{rank}(A) = n$.

### 2.1.3 Matrix Inverse

The $n$-by-$n$ *identity matrix* $I_n$ is defined by the column partitioning

$$I_n = [\, e_1, \ldots, e_n \,]$$

where $e_k$ is the $k$th "canonical" vector:

$$e_k = (\, \underbrace{0, \ldots, 0}_{k-1}, 1, \underbrace{0, \ldots, 0}_{n-k} \,)^T.$$

The canonical vectors arise frequently in matrix analysis and if their dimension is ever ambiguous, we use superscripts, i.e., $e_k^{(n)} \in \mathbb{R}^n$.

If $A$ and $X$ are in $\mathbb{R}^{n \times n}$ and satisfy $AX = I$, then $X$ is the *inverse* of $A$ and is denoted by $A^{-1}$. If $A^{-1}$ exists, then $A$ is said to be *nonsingular.* Otherwise $A$ is *singular.*

Several matrix inverse properties have an important role to play in matrix computations. The inverse of a product is the reverse product of the inverses:

$$(AB)^{-1} = B^{-1}A^{-1}. \tag{2.1.1}$$

The transpose of the inverse is the inverse of the transpose:

$$(A^{-1})^T = (A^T)^{-1} \equiv A^{-T}. \tag{2.1.2}$$

The identity

$$B^{-1} \;=\; A^{-1} - B^{-1}(B - A)A^{-1} \tag{2.1.3}$$

shows how the inverse changes if the matrix changes.

The *Sherman-Morrison-Woodbury formula* gives a convenient expression for the inverse of $(A + UV^T)$ where $A \in \mathbb{R}^{n \times n}$ and $U$ and $V$ are $n$-by-$k$:

$$(A + UV^T)^{-1} = A^{-1} - A^{-1}U(I + V^TA^{-1}U)^{-1}V^TA^{-1}. \tag{2.1.4}$$

A rank $k$ correction to a matrix results in a rank $k$ correction of the inverse. In (2.1.4) we assume that both $A$ and $(I + V^TA^{-1}U)$ are nonsingular.

Any of these facts can be verified by just showing that the "proposed" inverse does the job. For example, here is how to confirm (2.1.3):

$$B\,(A^{-1} - B^{-1}(B - A)A^{-1}) = BA^{-1} - (B - A)A^{-1} = I.$$

### 2.1.4 The Determinant

If $A = (a) \in \mathbb{R}^{1 \times 1}$, then its *determinant* is given by $\det(A) = a$. The determinant of an $n$-by-$n$ matrix is defined in terms of the determinant of an $(n-1)$-by-$(n-1)$ matrix. In particular, if $A \in \mathbb{R}^{n \times n}$ we have

$$\det(A) = \sum_{j=1}^{n} (-1)^{j+1} a_{1j} \det(A_{1j})$$

where $A_{1j}$ is an $(n-1)$-by-$(n-1)$ matrix obtained by deleting the first row and $j$th column of $A$. Useful properties of the determinant include

$$\begin{array}{lcll} \det(AB) & = & \det(A)\det(B) & A, B \in \mathbb{R}^{n\times n} \\ \det(A^T) & = & \det(A) & A \in \mathbb{R}^{n\times n} \\ \det(cA) & = & c^n\det(A) & c \in \mathbb{R}, A \in \mathbb{R}^{n\times n} \\ \det(A) \neq 0 & \Leftrightarrow & A \text{ is nonsingular} & A \in \mathbb{R}^{n\times n} \end{array}$$

### 2.1.5 Differentiation

Suppose $\alpha$ is a scalar and that $A(\alpha)$ is an $m$-by-$n$ matrix with entries $a_{ij}(\alpha)$. If $a_{ij}(\alpha)$ is a differentiable function of $\alpha$ for all $i$ and $j$, then by $\dot{A}(\alpha)$ we mean the matrix

$$\dot{A}(\alpha) = \frac{d}{d\alpha}A(\alpha) = \left(\frac{d}{d\alpha}a_{ij}(\alpha)\right) = \left(\dot{a}_{ij}(\alpha)\right).$$

The differentiation of a parameterized matrix turns out to be a handy way to examine the sensitivity of various matrix problems.

#### Problems

**P2.1.1** Show that if $A \in \mathbb{R}^{m\times n}$ has rank $p$, then there exists an $X \in \mathbb{R}^{m\times p}$ and a $Y \in \mathbb{R}^{n\times p}$ such that $A = XY^T$, where $\text{rank}(X) = \text{rank}(Y) = p$.

**P2.1.2** Suppose $A(\alpha) \in \mathbb{R}^{m\times r}$ and $B(\alpha) \in \mathbb{R}^{r\times n}$ are matrices whose entries are differentiable functions of the scalar $\alpha$. Show

$$\frac{d}{d\alpha}[A(\alpha)B(\alpha)] = \left[\frac{d}{d\alpha}A(\alpha)\right]B(\alpha) + A(\alpha)\left[\frac{d}{d\alpha}B(\alpha)\right].$$

**P2.1.3** Suppose $A(\alpha) \in \mathbb{R}^{n\times n}$ has entries that are differentiable functions of the scalar $\alpha$. Assuming $A(\alpha)$ is always nonsingular, show

$$\frac{d}{d\alpha}\left[A(\alpha)^{-1}\right] = -A(\alpha)^{-1}\left[\frac{d}{d\alpha}A(\alpha)\right]A(\alpha)^{-1}.$$

**P2.1.4** Suppose $A \in \mathbb{R}^{n\times n}$, $b \in \mathbb{R}^n$, and that $\phi{:}\mathbb{R}^n \to \mathbb{R}^n$ is defined by $\phi(x) = \frac{1}{2}x^TAx - x^Tb$. Show that the gradient of $\phi$ is given by $\nabla\phi(x) = \frac{1}{2}(A^T + A)x - b$.

#### Notes and References for Sec. 2.1

There are many introductory linear algebra texts but only a few that provide the necessary background for the beginning student of matrix computations. Among them, we have found the following very useful:

P.R. Halmos (1958). *Finite Dimensional Vector Spaces*, 2nd ed., Van Nostrand-Reinhold, Princeton.

S.J. Leon (1980). *Linear Algebra with Applications*. Macmillan, New York.

B. Noble and J.W. Daniel (1977). *Applied Linear Algebra*, Prentice-Hall, Englewood Cliffs.

G. Strang (1988). *Linear Algebra and Its Applications*, 3rd edition, Harcourt, Brace, Jovanovich, San Diego.

More encyclopedic treatments include

R. Bellman (1970). *Introduction to Matrix Analysis*, 2nd ed., McGraw-Hill, New York.
F.R. Gantmacher (1959). *The Theory of Matrices, vols. 1 and 2*, Chelsea, New York.
A.S. Householder (1964). *The Theory of Matrices in Numerical Analysis*, Ginn (Blaisdell), Boston.

Stewart (IMC, Chapter 1) also provides an excellent review of matrix algebra.

## 2.2 Vector Norms

Norms serve the same purpose on vector spaces that absolute value does on the real line: they furnish a measure of distance. More precisely, $\mathbb{R}^n$ together with a norm on $\mathbb{R}^n$ defines a metric space. Therefore, we have the familiar notions of neighborhood, open sets, convergence, and continuity when working with vectors and vector-valued functions.

### 2.2.1 Definitions

A *vector norm* on $\mathbb{R}^n$ is a function $f:\mathbb{R}^n \rightarrow \mathbb{R}$ that satisfies the following properties:

$$\begin{array}{lll} f(x) \geq 0 & x \in \mathbb{R}^n, & (f(x) = 0 \text{ iff } x = 0) \\ f(x+y) \leq f(x) + f(y) & x, y \in \mathbb{R}^n & \\ f(\alpha x) = |\alpha| f(x) & \alpha \in \mathbb{R}, x \in \mathbb{R}^n & \end{array}$$

We denote such a function with a double bar notation: $f(x) = \| x \|$. Subscripts on the double bar are used to distinguish between various norms.

A useful class of vector norms are the *p-norms* defined by

$$\| x \|_p = (|x_1|^p + \cdots + |x_n|^p)^{\frac{1}{p}} \qquad p \geq 1 . \tag{2.2.1}$$

Of these the 1, 2, and $\infty$ norms are the most important:

$$\begin{aligned} \| x \|_1 &= |x_1| + \cdots + |x_n| \\ \| x \|_2 &= \left(|x_1|^2 + \cdots + |x_n|^2\right)^{\frac{1}{2}} = \left(x^T x\right)^{\frac{1}{2}} \\ \| x \|_\infty &= \max_{1 \leq p \leq n} |x_i| \end{aligned}$$

A *unit vector* with respect to the norm $\| \cdot \|$ is a vector $x$ that satisfies $\| x \| = 1$.

### 2.2.2 Some Vector Norm Properties

A classic result concerning $p$-norms is the *Holder inequality*:

$$|x^T y| \leq \| x \|_p \| y \|_q \qquad \frac{1}{p} + \frac{1}{q} = 1. \tag{2.2.2}$$

A very important special case of this is the *Cauchy-Schwartz inequality*:

$$|x^T y| \leq \| x \|_2 \| y \|_2 \tag{2.2.3}$$

All norms on $\mathbb{R}^n$ are *equivalent* , i.e., if $\| \cdot \|_\alpha$ and $\| \cdot \|_\beta$ are norms on $\mathbb{R}^n$, then there exist positive constants, $c_1$ and $c_2$ such that

$$c_1 \| x \|_\alpha \leq \| x \|_\beta \leq c_2 \| x \|_\alpha \tag{2.2.4}$$

for all $x \in \mathbb{R}^n$. For example, if $x \in \mathbb{R}^n$ then

$$\| x \|_2 \leq \| x \|_1 \leq \sqrt{n} \| x \|_2 \tag{2.2.5}$$
$$\| x \|_\infty \leq \| x \|_2 \leq \sqrt{n} \| x \|_\infty \tag{2.2.6}$$
$$\| x \|_\infty \leq \| x \|_1 \leq n \| x \|_\infty . \tag{2.2.7}$$

### 2.2.3 Absolute and Relative Error

Suppose $\hat{x} \in \mathbb{R}^n$ is an approximation to $x \in \mathbb{R}^n$. For a given vector norm $\| \cdot \|$ we say that

$$\epsilon_{abs} = \| \hat{x} - x \|$$

is the *absolute error* in $\hat{x}$ while if $x \neq 0$ then

$$\epsilon_{rel} = \frac{\| \hat{x} - x \|}{\| x \|}$$

prescribes the *relative error* in $\hat{x}$. Relative error in the $\infty$-norm can be translated into a statement about the number of correct significant digits in $\hat{x}$. In particular, if

$$\frac{\| \hat{x} - x \|_\infty}{\| x \|_\infty} \approx 10^{-p}$$

then the largest component of $\hat{x}$ has approximately $p$ correct significant digits.

**Example 2.2.1** If $x = (1.234 \;\; .05674)^T$ and $\hat{x} = (1.235 \;\; .05128)^T$, then $\| \hat{x} - x \|_\infty / \| x \|_\infty \approx .0043 \approx 10^{-3}$. Note than $\hat{x}_1$ has about three significant digits that are correct while only one significant digit in $\hat{x}_2$ is right.

### 2.2.4 Convergence

We say that a sequence $\{x^{(k)}\}$ of $n$-vectors converges to $x$ if

$$\lim_{k\to\infty} \| x^{(k)} - x \| = 0 .$$

Note that because of (2.2.4), convergence in the $\alpha$-norm implies convergence in the $\beta$-norm and vice versa.

**Problems**

**P2.2.1** Show that if $x \in \mathbf{R}^n$ then $\lim_{p\to\infty} \| x \|_p = \| x \|_\infty$.

**P2.2.2** Prove the Cauchy-Schwartz inequality (2.2.3) by considering the inequality $0 \le (ax + by)^T(ax + by)$ for suitable scalars $a$ and $b$ .

**P2.2.3** Verify that $\| \cdot \|_1$, $\| \cdot \|_2$, and $\| \cdot \|_\infty$ are vector norms.

**P2.2.4** Verify (2.2.5)-(2.2.7). When is equality achieved in each result?

**P2.2.5** Show that in $\mathbf{R}^n$, $x^{(i)} \to x$ if and only if $x_k^{(i)} \to x_k$ for $k = 1{:}n$.

**P2.2.6** Show that any vector norm on $\mathbf{R}^n$ is uniformly continuous by verifying the inequality $|\, \| x \| - \| y \| \,| \le \| x - y \|$.

**P2.2.7** Let $\| \cdot \|$ be a vector norm on $\mathbf{R}^m$ and assume $A \in \mathbf{R}^{m\times n}$ . Show that if $\text{rank}(A) = n$ then $\| x \|_A = \| Ax \|$ is a vector norm on $\mathbf{R}^n$.

**P2.2.8** Let $x$ and $y$ be in $\mathbf{R}^n$ and define $\psi{:}\mathbf{R} \to \mathbf{R}$ by $\psi(\alpha) = \| x - \alpha y \|_2$. Show that $\psi$ is minimized when $\alpha = x^Ty/y^Ty$.

**P2.2.9** (a) Verify that $\| x \|_p = (|x_1|^p + \cdots + |x_n|^p)^{\frac{1}{p}}$ is a vector norm on $\mathbb{C}^n$. (b) Show that if $x \in \mathbb{C}^n$ then $\| x \|_p \le c\left(\| \text{Re}(x) \|_p + \| \text{Im}(x) \|_p\right)$. (c) Find a constant $c_n$ such that $c_n\left(\| \text{Re}(x) \|_2 + \| \text{Im}(x) \|_2\right) \le \| x \|_2$ for all $x \in \mathbb{C}^n$.

**Notes and References for Sec. 2.2**

Stewart(IMC) has an excellent discussion of vector norms. See also

A.S. Householder (1974). *The Theory of Matrices in Numerical Analysis,* Dover Publications, New York.

J.D. Pryce (1984). "A New Measure of Relative Error for Vectors," *SIAM J. Numer. Anal. 21,* 202-21.

## 2.3 Matrix Norms

The analysis of matrix algorithms frequently requires use of matrix norms. For example, the quality of a linear system solver may be poor if the matrix of coefficients is "nearly singular." To quantify the notion of near-singularity we need a measure of distance on the space of matrices. Matrix norms provide that measure.

### 2.3.1 Definitions

Since $\mathbb{R}^{m\times n}$ is isomorphic to $\mathbb{R}^{mn}$, the definition of a matrix norm should be equivalent to the definition of a vector norm. In particular, $f:\mathbb{R}^{m\times n} \rightarrow \mathbb{R}$ is a matrix norm if

$$\begin{array}{lll} f(A) \geq 0 & A \in \mathbb{R}^{m\times n}, & (f(A) = 0 \text{ iff } A = 0) \\ f(A+B) \leq f(A) + f(B) & A, B \in \mathbb{R}^{m\times n} & \\ f(\alpha A) = |\alpha| f(A) & \alpha \in \mathbb{R}, A \in \mathbb{R}^{m\times n} & \end{array}$$

As with vector norms, we use a double bar notation with subscripts to designate matrix norms, i.e., $\| A \| = f(A)$.

The most frequently used matrix norms in numerical linear algebra are the Frobenius norm,

$$\| A \|_F = \sqrt{\sum_{i=1}^{m} \sum_{j=1}^{n} |a_{ij}|^2} \tag{2.3.1}$$

and the $p$-norms

$$\| A \|_p = \sup_{x \neq 0} \frac{\| Ax \|_p}{\| x \|_p}. \tag{2.3.2}$$

Note that the matrix $p$-norms are defined in terms of the vector $p$-norms that we discussed in the previous section. The verification that (2.3.1) and (2.3.2) are matrix norms is left as an exercise. It is clear that $\| A \|_p$ is the $p$-norm of the largest vector obtained by applying $A$ to a unit $p$-norm vector:

$$\| A \|_p = \sup_{x \neq 0} \left\| A \left( \frac{x}{\| x \|_p} \right) \right\|_p = \max_{\| x \|_p = 1} \| Ax \|_p .$$

It is important to understand that (2.3.1) and (2.3.2) define families of norms—the 2-norm on $\mathbb{R}^{3\times 2}$ is a different function from the 2-norm on $\mathbb{R}^{5\times 6}$. Thus, the easily verified inequality

$$\| AB \|_p \leq \| A \|_p \| B \|_p \qquad A \in \mathbb{R}^{m\times n}, B \in \mathbb{R}^{n\times q} \tag{2.3.3}$$

is really an observation about the relationship between three different norms. Formally, we say that norms $f_1$, $f_2$ , and $f_3$ on $\mathbb{R}^{m\times q}$ ,$\mathbb{R}^{m\times n}$, $\mathbb{R}^{n\times q}$ are *mutually consistent* if for all $A \in \mathbb{R}^{m\times n}$ and $B \in \mathbb{R}^{n\times q}$ we have $f_1(AB) \leq f_2(A) f_3(B)$.

Not all matrix norms satisfy the submultiplicative property

$$\| AB \| \leq \| A \| \| B \| . \tag{2.3.4}$$

For example, if $\| A \|_\Delta = \max |a_{ij}|$ and

$$A = B = \begin{bmatrix} 1 & 1 \\ 1 & 1 \end{bmatrix},$$

then $\| AB \|_\Delta > \| A \|_\Delta \| B \|_\Delta$. For the most part we work with norms that satisfy (2.3.4).

The $p$-norms have the important property that for every $A \in \mathbb{R}^{m\times n}$ and $x \in \mathbb{R}^n$ we have $\| Ax \|_p \leq \| A \|_p \| x \|_p$. More generally, for any vector norm $\| \cdot \|_\alpha$ on $\mathbb{R}^n$ and $\| \cdot \|_\beta$ on $\mathbb{R}^m$ we have $\| Ax \|_\beta \leq \| A \|_{\alpha,\beta} \| x \|_\alpha$ where $\| A \|_{\alpha,\beta}$ is a matrix norm defined by

$$\| A \|_{\alpha,\beta} = \sup_{x\neq 0} \frac{\| Ax \|_\beta}{\| x \|_\alpha}. \tag{2.3.5}$$

We say that $\| \cdot \|_{\alpha,\beta}$ is *subordinate* to the vector norms $\| \cdot \|_\alpha$ and $\| \cdot \|_\beta$. Since the set $\{x \in \mathbb{R}^n : \| x \|_\alpha = 1\}$ is compact and closed and since $\| \cdot \|_\beta$ is continuous, it follows that

$$\| A \|_{\alpha,\beta} = \max_{\|x\|_\alpha=1} \| Ax \|_\beta = \| Ax^* \|_\beta \tag{2.3.6}$$

for some $x^* \in \mathbb{R}^n$ having unit $\alpha$-norm.

### 2.3.2 Some Matrix Norm Properties

The Frobenius and $p$-norms (especially $p = 1, 2, \infty$) satisfy certain inequalities that are frequently used in the analysis of matrix computations. For $A \in \mathbb{R}^{m\times n}$we have

$$\| A \|_2 \leq \| A \|_F \leq \sqrt{n} \| A \|_2 \tag{2.3.7}$$

$$\max_{i,j} |a_{ij}| \leq \| A \|_2 \leq \sqrt{mn} \max_{i,j} |a_{ij}| \tag{2.3.8}$$

$$\| A \|_1 = \max_{1\leq j\leq n} \sum_{i=1}^{m} |a_{ij}| \tag{2.3.9}$$

$$\| A \|_\infty = \max_{1\leq i\leq m} \sum_{j=1}^{n} |a_{ij}| \tag{2.3.10}$$

$$\frac{1}{\sqrt{n}} \| A \|_\infty \leq \| A \|_2 \leq \sqrt{m} \| A \|_\infty \tag{2.3.11}$$

$$\frac{1}{\sqrt{m}} \| A \|_1 \leq \| A \|_2 \leq \sqrt{n} \| A \|_1 \tag{2.3.12}$$

The proofs of these relations are not hard and are left as exercises.

A sequence $\{A^{(k)}\} \in \mathbb{R}^{m \times n}$ converges if $\lim_{k\to\infty} \| A^{(k)} - A \| = 0$. Choice of norm is irrelevant since all norms on $\mathbb{R}^{m \times n}$ are equivalent.

### 2.3.3 The Matrix 2-Norm

A nice feature of the matrix 1-norm and the matrix $\infty$-norm is that they are easily computed from (2.3.9) and (2.3.10). A characterization of the 2-norm is considerably more complicated.

**Theorem 2.3.1** *If $A \in \mathbb{R}^{m \times n}$ then there exists a unit 2-norm n-vector $z$ such that $A^TAz = \mu^2 z$ where $\mu = \| A \|_2$.*

**Proof.** Suppose $z \in \mathbb{R}^n$ is a unit vector such that $\| Az \|_2 = \| A \|_2$. Since $z$ maximizes the function

$$g(x) = \frac{1}{2}\frac{\| Ax \|_2^2}{\| x \|_2^2} = \frac{1}{2}\frac{x^TA^TAx}{x^Tx}$$

it follows that it satisfies $\nabla g(z) = 0$ where $\nabla g$ is the gradient of $g$. But a tedious differentiation shows that for $i = 1{:}n$

$$\frac{\partial g(z)}{\partial z_i} = \left[ (z^Tz)\sum_{j=1}^{n}(A^TA)_{ij}z_j - (z^TA^TAz)z \right] \Big/ (z^Tz)^2 .$$

In vector notation this says $A^TAz = (z^TA^TAz)z$. The theorem follows by setting $\mu = \| Az \|_2$. $\square$

The theorem implies that $\| A \|_2^2$ is a zero of the polynomial $p(\lambda) = \det(A^TA - \lambda I)$. In particular, the 2-norm of $A$ is the square root of the largest eigenvalue of $A^TA$. We have much more to say about eigenvalues in Chapters 7 and 8 but now we merely observe that 2-norm computation is iterative and decidedly more complicated than the computation of the matrix 1-norm or $\infty$-norm. Fortunately, if the object is to obtain an order-of-magnitude estimate of $\| A \|_2$, then (2.3.7), (2.3.11), or (2.3.12) can be used.

As another example of "norm analysis," here is a handy result for 2-norm estimation.

**Corollary 2.3.2** *If $A \in \mathbb{R}^{m \times n}$, then $\| A \|_2 \leq \sqrt{\| A \|_1 \| A \|_\infty}$ .*

**Proof.** If $z \neq 0$ is such that $A^TAz = \mu^2 z$ with $\mu = \| A \|_2$, then $\mu^2 \| z \|_1 = \| A^TAz \|_1 \leq \| A^T \|_1 \| A \|_1 \| z \|_1 = \| A \|_\infty \| A \|_1 \| z \|_1$. $\square$

### 2.3.4 Perturbations and the Inverse

We frequently use norms to quantify the effect of perturbations or to prove that a sequence of matrices converges to a specified limit. As an illustration of these norm applications, let us quantify the change in $A^{-1}$ as a function of change in $A$.

**Lemma 2.3.3** *Suppose $F \in \mathbb{R}^{n\times n}$ and that $\|\cdot\|$ is a norm satisfying the submultiplicative property (2.3.4). If $\| F \| < 1$, then $I - F$ is nonsingular and*

$$(I-F)^{-1} = \sum_{k=0}^{\infty} F^k$$

*with*

$$\| (I-F)^{-1} \| \le \frac{1}{1 - \| F \|}.$$

**Proof.** Suppose $I - F$ is singular. It follows that $(I-F)x = 0$ for some nonzero $x$. But then $\| x \| = \| Fx \|$ implies $\| F \| \ge 1$, a contradiction. Thus, $I - F$ is nonsingular. To obtain an expression for its inverse consider the identity

$$\left( \sum_{k=0}^{N} F^k \right) (I - F) = I - F^{N+1}.$$

Since $\| F \| < 1$ it follows that $\lim_{k\to\infty} F^k = 0$ because $\| F^k \| \le \| F \|^k$. Thus,

$$\left( \lim_{N\to\infty} \sum_{k=0}^{N} F^k \right) (I - F) = I .$$

It follows that $(I-F)^{-1} = \lim_{N\to\infty} \sum_{k=0}^{N} F^k$. From this it is easy to show that

$$\| (I-F)^{-1} \| \le \sum_{k=0}^{\infty} \| F \|^k = \frac{1}{1 - \| F \|}. \quad \square$$

Note that $\| (I-F)^{-1} - I \| \le \| F \| / (1 - \| F \|)$ as a consequence of the lemma. Thus, if $\epsilon \ll 1$ then $O(\epsilon)$ perturbations in $I$ induce $O(\epsilon)$ perturbations in the inverse. We next extend this result to general matrices.

**Theorem 2.3.4** *If $A$ is nonsingular and $\| A^{-1}E \| = r < 1$, then $A + E$ is nonsingular and $\| (A+E)^{-1} - A^{-1} \| \le \| E \| \, \| A^{-1} \|^2 / (1 - r)$.*

**Proof.** Since $A$ is nonsingular $A + E = A(I - F)$ where $F = -A^{-1}E$. Since $\| F \| = r < 1$ it follows from Lemma 2.3.3 that $I - F$ is nonsingular and $\| (I-F)^{-1} \| < 1/(1-r)$. Now $(A+E)^{-1} = (I-F)^{-1}A^{-1}$ and so

$$\| (A+E)^{-1} \| \leq \frac{\| A^{-1} \|}{1-r}.$$

Now (2.1.3) says that $(A+E)^{-1} - A^{-1} = -A^{-1}E(A+E)^{-1}$ and so by taking norms we find

$$\| (A+E)^{-1} - A^{-1} \| \leq \| A^{-1} \| \| E \| \| (A+E)^{-1} \| \leq \frac{\| A^{-1} \|^2 \| E \|}{1-r}. \qquad \square$$

**Problems**

**P2.3.1** Show $\| AB \|_p \leq \| A \|_p \| B \|_p$ where $1 \leq p \leq \infty$.

**P2.3.2** Let $B$ be any submatrix of $A$. Show that $\| B \|_p \leq \| A \|_p$.

**P2.3.3** Show that if $D = \text{diag}(\mu_1, \ldots, \mu_k) \in \mathbf{R}^{m \times n}$ with $k = \min\{m, n\}$, then $\| D \|_p = \max |\mu_i|$.

**P2.3.4** Verify (2.3.7) and (2.3.8).

**P2.3.5** Verify (2.3.9) and (2.3.10).

**P2.3.6** Verify (2.3.11) and (2.3.12).

**P2.3.7** Show that if $0 \neq s \in \mathbf{R}^n$ and $E \in \mathbf{R}^{n \times n}$, then

$$\left\| E\left(I - \frac{ss^T}{s^Ts}\right) \right\|_F^2 = \| E \|_F^2 - \frac{\| Es \|_2^2}{s^Ts}.$$

**P2.3.8** Suppose $u \in \mathbf{R}^m$ and $v \in \mathbf{R}^n$. Show that if $E = uv^T$ then $\| E \|_F = \| E \|_2 = \| u \|_2 \| v \|_2$ and that $\| E \|_\infty \leq \| u \|_\infty \| v \|_1$.

**P2.3.9** Suppose $A \in \mathbf{R}^{m \times n}$, $y \in \mathbf{R}^m$, and $0 \neq s \in \mathbf{R}^n$. Show that $E = (y - As)s^T/s^Ts$ has the smallest 2-norm of all $m$-by-$n$ matrices $E$ that satisfy $(A+E)s = y$.

**Notes and References for Sec. 2.3**

Stewart (IMC, pp. 160-84) has a complete review of norms. Further references include

F.L. Bauer and C.T. Fike (1960). "Norms and Exclusion Theorems," *Numer. Math. 2*, 137-44.

A.S. Householder (1964). *The Theory of Matrices in Numerical Analysis* , Dover Publications, New York.

L. Mirsky (1960). "Symmetric Gauge Functions and Unitarily Invariant Norms," *Quart. J. Math. 11*, 50-59.

J.M. Ortega (1972). *Numerical Analysis: A Second Course*, Academic Press, New York.

## 2.4 Finite Precision Matrix Computations

Rounding errors are in part what makes the matrix computation area so nontrivial and interesting. In this section we set up a model of floating point arithmetic and then use it to develop error bounds for floating point dot products, saxpy's, matrix-vector products and matrix-matrix products.

## 2.4.1 The Floating Point Numbers

When calculations are performed on a computer, each arithmetic operation is generally affected by *roundoff error.* This error arises because the machine hardware can only represent a subset of the real numbers. We denote this subset by **F** and refer to its elements as *floating point numbers.* Following conventions set forth in Forsythe, Malcolm, and Moler (1977, pp. 10-29), the floating point number system on a particular computer is characterized by four integers: the *base* $\beta$, the *precision* $t$, and the *exponent range* $[L, U]$. In particular, **F** consists of all numbers $f$ of the form

$$f = \pm .d_1 d_2 \ldots d_t \times \beta^e \qquad 0 \le d_i < \beta, \quad d_1 \ne 0, \quad L \le e \le U$$

together with zero. Notice that for a nonzero $f \in \mathbf{F}$ we have $m \le |f| \le M$ where

$$m = \beta^{L-1} \quad \text{and} \quad M = \beta^U(1 - \beta^{-t}). \tag{2.4.1}$$

As an example, if $\beta = 2$, $t = 3$, $L = 0$, and $U = 2$, then the non-negative elements of **F** are represented by hash marks on the axis of Figure 2.4.1.

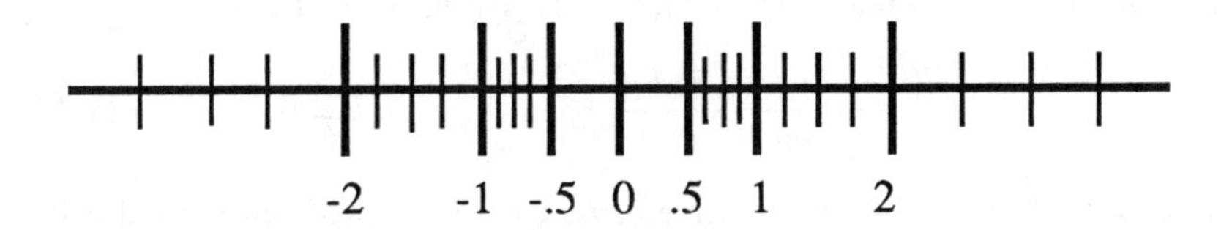


Figure 2.4.1. Sample Floating Point Number System

Notice that the floating point numbers are not equally spaced. A typical value of $(\beta, t, L, U)$ might be (2, 56, -64, 64).

## 2.4.2 A Model of Floating Point Arithmetic

To make general pronouncements about the effect of rounding errors on a given algorithm, it is necessary to have a model of computer arithmetic on **F**. To this end define the set $G$ by

$$G = \{\, x \in \mathbb{R} : m \le |x| \le M \,\} \cup \{0\} \tag{2.4.2}$$

and the operator $fl$: $G \to \mathbf{F}$ by

$$fl(x) = \begin{cases} \text{nearest } c \in \mathbf{F} \text{ to } x \text{ if } \textit{rounded arithmetic} \text{ is used. If tie then round away from zero.} \\ \\ \text{nearest } c \in \mathbf{F} \text{ satisfying } |c| \le |x| \text{ if } \textit{chopped arithmetic} \text{ is used.} \end{cases}$$

Let $a$ and $b$ be any two floating point numbers and let "op" denote any of the four arithmetic operations $+$, $-$, $\times$, $\div$. If $|a \text{ op } b| \notin G$, then an *arithmetic fault* occurs implying either overflow ($|a \text{ op } b| > M$) or underflow ($0 < |a \text{ op } b| < m$). If $|a \text{ op } b| \in G$, then in our model of floating point arithmetic we assume that the computer version of $(a \text{ op } b)$ is given by $fl(a \text{ op } b)$.

An arithmetic fault usually implies program termination, although with some machines and compilers, underflows are set to zero. The handling of arithmetic faults is an interesting and important issue. See Demmel(1984).

If an arithmetic fault does not occur, then an important question to be answered concerns the accuracy of $fl(a \text{ op } b)$. The $fl$ operator can be shown to satisfy

$$fl(x) = x(1+\epsilon) \qquad |\epsilon| \le \mathbf{u} \tag{2.4.3}$$

where $\mathbf{u}$ is the *unit roundoff* defined by

$$\mathbf{u} = \begin{cases} \frac{1}{2}\beta^{1-t} & \text{for rounded arithmetic} \\ \\ \beta^{1-t} & \text{for chopped arithmetic} \end{cases} \tag{2.4.4}$$

It follows that $fl(a \text{ op } b) = (a \text{ op } b)(1+\epsilon)$ with $|\epsilon| \le \mathbf{u}$. Thus,

$$\frac{|fl(a \text{ op } b) - (a \text{ op } b)|}{|a \text{ op } b|} \le \mathbf{u} \qquad a \text{ op } b \ne 0 \tag{2.4.5}$$

showing that there is small relative error associated with individual arithmetic operations. It is important to realize, however, that this is not necessarily the case when a sequence of operations is involved.

**Example 2.4.1** If $\beta = 10$, $t = 3$ floating point chopped arithmetic is used, then it can be shown that $fl[fl(10^{-3}+1)-1] = 0$ implying a relative error of 1. On the other hand the exact answer is given by $fl[fl(10^{-3} + fl(1-1)] = 10^{-3}$. This shows that floating point arithmetic is not always associative.

## 2.4.3 Cancellation

Another important aspect of finite precision arithmetic is the phenomenon of *catastrophic cancellation*. Roughly speaking, this term refers to the extreme loss of correct significant digits when small numbers are additively computed from large numbers. A well-known example taken from Forsythe, Malcolm and Moler (1977, pp. 14-16) is the computation of $e^{-a}$ with $a > 0$ via Taylor series. The roundoff error associated with this method is approximately $\mathbf{u}$ times the largest partial sum. For large $a$, this error can actually be larger than the correct exponential and there will be no correct digits in the answer no matter how many terms in the series are summed. On the other hand, if enough terms in the Taylor series for $e^a$ are added and the result reciprocated, then an estimate of $e^{-a}$ to full precision is attained.

### 2.4.4 The Absolute Value Notation

Before we proceed with the roundoff analysis of some basic matrix calculations, we acquire some useful notation. Suppose $A \in \mathbb{R}^{m \times n}$ and that we wish to quantify the errors associated with its floating point representation. Denoting the stored version of $A$ by $fl(A)$, we see that

$$[fl(A)]_{ij} = fl(a_{ij}) = a_{ij}(1+\epsilon_{ij}) \qquad |\epsilon_{ij}| \le \mathbf{u} \tag{2.4.6}$$

for all $i$ and $j$. A better way to say the same thing results if we adopt two conventions. If $A$ and $B$ are in $\mathbb{R}^{m \times n}$ then

$$B = |A| \quad \Rightarrow \quad b_{ij} = |a_{ij}|,\ i = 1{:}m,\ j = 1{:}n$$

$$B \le A \quad \Rightarrow \quad b_{ij} \le a_{ij},\ i = 1{:}m,\ j = 1{:}n\,.$$

With this notation we see that (2.4.6) has the form

$$|fl(A) - A| \le \mathbf{u}|A|\,.$$

A relation such as this can be easily turned into a norm inequality, e.g., $\|\, fl(A) - A\,\|_1 \le \mathbf{u}\|\,A\,\|_1$. However, when quantifying the rounding errors in a matrix manipulation, the absolute value notation can be a lot more informative because it provides a comment on each $(i,j)$ entry.

### 2.4.5 Roundoff in Dot Products

We begin our study of finite precision matrix computations by considering the rounding errors that result in the standard dot product algorithm:

$$\begin{array}{l} s = 0 \\ \textbf{for } k = 1{:}n \\ \qquad s = s + x_k y_k \\ \textbf{end} \end{array} \tag{2.4.7}$$

Here, $x$ and $y$ are $n$-by-1 floating point vectors.

In trying to quantify the rounding errors in this algorithm, we are immediately confronted with a notational problem: the distinction between computed and exact quantities. When the underlying computations are clear, we shall use the $fl(\cdot)$ operator to signify computed quantities. Thus, $fl(x^Ty)$ denotes the computed output of (2.4.7). Let us bound $|fl(x^Ty) - x^Ty|$. If

$$s_p = fl\left(\sum_{k=1}^{p} x_k y_k\right)$$

then $s_1 = x_1y_1(1+\delta_1)$ with $|\delta_1| \le \mathbf{u}$ and for $p = 2{:}n$

$$\begin{aligned} s_p &= fl(s_{p-1} + fl(x_p y_p)) \\ &= (s_{p-1} + x_p y_p(1+\delta_p))(1+\epsilon_p) \qquad |\delta_p|, |\epsilon_p| \le \mathbf{u}\,. \end{aligned} \tag{2.4.8}$$

A little algebra shows that

$$fl(x^T y) = s_n = \sum_{k=1}^{n} x_k y_k (1+\gamma_k)$$

where

$$(1+\gamma_k) = (1+\delta_k)\prod_{j=k}^{n}(1+\epsilon_j)$$

with the convention that $\epsilon_1 = 0$. Thus,

$$|fl(x^T y) - x^T y| \;\le\; \sum_{k=1}^{n} |x_k y_k||\gamma_k|. \tag{2.4.9}$$

To proceed further, we must be able to bound the quantities $|\gamma_k|$ in terms of $\mathbf{u}$. The following result is useful for this purpose.

**Lemma 2.4.1** *If* $(1+\alpha) = \prod_{k=1}^{n}(1+\alpha_k)$ *where* $|\alpha_k| \le \mathbf{u}$ *and* $n\mathbf{u} \le .01$, *then* $|\alpha| \le 1.01n\mathbf{u}$.

**Proof.** See Forsythe and Moler (SLE, p. 92). □

Applying this result to (2.4.9) under the "reasonable" assumption $n\mathbf{u} \le .01$ gives

$$|fl(x^T y) - x^T y| \;\le\; 1.01n\mathbf{u}|x|^T|y|. \tag{2.4.10}$$

Notice that if $|x^T y| \ll |x|^T|y|$ then the relative error in $fl(x^T y)$ may not be small.

## 2.4.6 Alternative Ways to Quantify Roundoff Error

An easier but less rigorous way of bounding $\alpha$ in Lemma 2.4.1 is to say $|\alpha| \le n\mathbf{u} + O(\mathbf{u}^2)$. With this convention we have

$$|fl(x^T y) - x^T y| \le n\mathbf{u}|x|^T|y| + O(\mathbf{u}^2). \tag{2.4.11}$$

Other ways of expressing the same result include

$$|fl(x^T y) - x^T y| \le \phi(n)\mathbf{u}|x|^T|y| \tag{2.4.12}$$

and

$$|fl(x^T y) - x^T y| \leq cn\mathbf{u}|x|^T|y|, \tag{2.4.13}$$

where in (2.4.12) $\phi(n)$ is a "modest" function of $n$ and in (2.4.13) $c$ is a constant of order unity.

We shall not express a preference for any of the error bounding styles shown in (2.4.10)-(2.4.13). This spares us the necessity of translating the roundoff results that appear in the literature into a fixed format. Moreover, paying overly close attention to the details of an error bound is inconsistent with the "philosophy" of roundoff analysis. As Wilkinson (1971, p. 567) says,

> There is still a tendency to attach too much importance to the precise error bounds obtained by an à priori error analysis. In my opinion, the bound itself is usually the least important part of it. The main object of such an analysis is to expose the potential instabilities, if any, of an algorithm so that hopefully from the insight thus obtained one might be led to improved algorithms. Usually the bound itself is weaker than it might have been because of the necessity of restricting the mass of detail to a reasonable level and because of the limitations imposed by expressing the errors in terms of matrix norms. À priori bounds are not, in general, quantities that should be used in practice. Practical error bounds should usually be determined by some form of à posteriori error analysis, since this takes full advantage of the statistical distribution of rounding errors and of any special features, such as sparseness, in the matrix.

It is for these reasons that roundoff error analysis has a different flavor than "pure math" analysis.

### 2.4.7 Dot Product Accumulation

Some computers have provision for accumulating dot products in *double precision*. This means that if $x$ and $y$ are floating point vectors with length $t$ mantissas, then the running sum $s$ in (2.4.7) is built up in a register with a $2t$ digit mantissa. Since the multiplication of two $t$-digit floating point numbers can be stored exactly in a double precision variable, it is only when $s$ is written to single precision memory that any roundoff occurs. In this situation one can usually assert that a computed dot product has good *relative* error, i.e., $fl(x^T y) = x^T y(1 + \delta)$ where $|\delta| \approx \mathbf{u}$. Thus, the ability to accumulate dot products is very attractive in matrix computation work.

### 2.4.8 Roundoff in Other Basic Matrix Computations

It is easy to show that if $A$ and $B$ are floating point matrices and $\alpha$ is a floating point number, then

$$fl(\alpha A) = \alpha A + E \qquad |E| \le \mathbf{u}|\alpha A| \tag{2.4.14}$$

and

$$fl(A+B) = (A+B) + E \qquad |E| \le \mathbf{u}|A+B|. \tag{2.4.15}$$

As a consequence of these last two results it is easy to verify that computed saxpy's and outer product updates satisfy

$$fl(\alpha x + y) = \alpha x + y + z \qquad |z| \le \mathbf{u}\,(2|\alpha x| + |y|) + O(\mathbf{u}^2) \tag{2.4.16}$$

$$fl(C + uv^T) = C + uv^T + E \qquad |E| \le \mathbf{u}\left(|C| + 2|uv^T|\right) + O(\mathbf{u}^2). \tag{2.4.17}$$

Using (2.4.10) it is easy to show that a dot product based multiplication of two floating point matrices $A$ and $B$ satisfies

$$fl(AB) = AB + E \qquad |E| \le n\mathbf{u}|A||B| + O(\mathbf{u}^2). \tag{2.4.18}$$

The same result applies if a gaxpy or outer product based procedure is used. Notice that matrix multiplication does not necessarily give small relative error since $|AB|$ may be much smaller than $|A||B|$, e.g.,

$$\begin{bmatrix} 1 & 1 \\ 0 & 0 \end{bmatrix} \begin{bmatrix} 1 & 0 \\ -.99 & 0 \end{bmatrix} = \begin{bmatrix} .01 & 0 \\ 0 & 0 \end{bmatrix}.$$

It is easy to obtain norm bounds from the roundoff results developed thus far. If we look at the 1-norm error in floating point matrix multiplication then it is easy to show from (2.4.18) that

$$\| fl(AB) - AB \|_1 \;\le\; n\mathbf{u}\| A \|_1 \| B \|_1 + O(\mathbf{u}^2). \tag{2.4.19}$$

### 2.4.9 Forward and Backward Error Analyses

Each roundoff bound given above is the consequence of a *forward error analysis.* An alternative style of characterizing the roundoff errors in an algorithm is accomplished through a technique known as *backward error analysis.* Here, the rounding errors are related to the data of the problem rather than to its solution. By way of illustration, consider the $n = 2$ version of triangular matrix multiplication. It can be shown that:

$$fl(AB) = \begin{bmatrix} a_{11}b_{11}(1+\epsilon_1) & (a_{11}b_{12}(1+\epsilon_2) + a_{12}b_{22}(1+\epsilon_3))(1+\epsilon_4) \\ 0 & a_{22}b_{22}(1+\epsilon_5) \end{bmatrix}$$

where $|\epsilon_i| \leq u$, for $i = 1{:}5$. However, if we define

$$\hat{A} = \left[ \begin{array}{cc} a_{11} & a_{12}(1+\epsilon_3)(1+\epsilon_4) \\ \\ 0 & a_{22}(1+\epsilon_5) \end{array} \right]$$

and

$$\hat{B} = \left[ \begin{array}{cc} b_{11}(1+\epsilon_1) & b_{12}(1+\epsilon_2)(1+\epsilon_4) \\ \\ 0 & b_{22} \end{array} \right]$$

then it is easily verified that $fl(AB) = \hat{A}\hat{B}$. Moreover,

$$\hat{A} = A + E \qquad |E| \leq 2\mathbf{u}|A| + O(\mathbf{u}^2)$$

$$\hat{B} = B + F \qquad |F| \leq 2\mathbf{u}|B| + O(\mathbf{u}^2).$$

In other words, the computed product is the exact product of slightly perturbed $A$ and $B$.

### 2.4.10 Error in Strassen Multiplication

In §1.3.8 we outlined an unconventional matrix multiplication procedure due to Strassen(1969). It is instructive to compare the effect of roundoff in this method with the effect of roundoff in any of the conventional matrix multiplication methods of §1.1.

It can be shown that the Strassen approach (Algorithm 1.3.1) produces a $\hat{C} = fl(AB)$ that satisfies an inequality of the form (2.4.19). This is perfectly satisfactory in many applications. However, the $\hat{C}$ that Strassen's method produces does *not* always satisfy an inequality of the form (2.4.18). To see this, suppose

$$A = B = \left[ \begin{array}{cc} .99 & .0010 \\ .0010 & .99 \end{array} \right]$$

and that we execute Algorithm 1.3.1 using 2-digit rounded arithmetic. Among other things the following quantities are computed:

$$\begin{aligned} \hat{P}_3 &= fl(.99(.001 - .99)) = -.98 \\ \hat{P}_5 &= fl((.99 + .001).99) = .98 \\ \hat{c}_{12} &= fl(\hat{P}_3 + \hat{P}_5) = 0.0 \end{aligned}$$

Now in exact arithmetic $c_{12} = 2(.001)(.99) = .00198$ and thus Algorithm 1.3.1 produces a $\hat{c}_{12}$ with no correct significant digits. The Strassen approach gets into trouble in this example because small off-diagonal entries are combined with large diagonal entries. Note that in conventional matrix multiplication neither $b_{12}$ and $b_{22}$ nor $a_{11}$ and $a_{12}$ are summed. Thus the contribution of the small off-diagonal elements is not lost. Indeed, for the above $A$ and $B$ a conventional matrix multiply gives $\hat{c}_{12} = .0020$.

Failure to produce a componentwise accurate $\hat{C}$ can be a serious shortcoming in some applications. For example, in Markov processes the $a_{ij}$, $b_{ij}$, and $c_{ij}$ are transition probabilities and are therefore nonnegative. It may be critical to compute $c_{ij}$ accurately if it reflects a particularly important probability in the modeled phenomena. Note that if $A \geq 0$ and $B \geq 0$ then conventional matrix multiplication produces a $\hat{C}$ that has small componentwise relative error:

$$|\hat{C} - C| \;\leq\; n\mathbf{u}|A|\,|B| + O(\mathbf{u}^2) \;=\; n\mathbf{u}|C| + O(\mathbf{u}^2)\,.$$

This follows from (2.4.18). Because we cannot say the same for the Strassen approach, we conclude that Algorithm 1.3.1 is not attractive for *certain* nonnegative matrix multiplication problems *if* relatively accurate $\hat{c}_{ij}$ are required.

Extrapolating from this discussion we reach two fairly obvious but important conclusions:

- Different methods for computing the same quantity can produce substantially different results.
- Whether or not an algorithm produces satisfactory results depends upon the type of problem solved and the goals of the user.

These observations are clarified in subsequent chapters and are intimately related to the concepts of algorithm stability and problem condition.

### Problems

**P2.4.1** Show that if (2.4.7) is applied with $y = x$, then $fl(x^Tx) = x^Tx(1+\alpha)$ where $|\alpha| \leq n\mathbf{u} + O(\mathbf{u}^2)$.

**P2.4.2** Prove (2.4.18).

**P2.4.3** Show that if $E \in \mathbb{R}^{m\times n}$ with $m \geq n$, then $\|\,|E|\,\|_2 \leq \sqrt{n}\|\,E\,\|_2$. This result is useful when deriving norm bounds from absolute value bounds.

**P2.4.4** Assume the existence of a square root function satisfying $fl(\sqrt{x}) = \sqrt{x}(1+\epsilon)$ with $|\epsilon| \leq \mathbf{u}$. Give an algorithm for computing $\|\,x\,\|_2$ and bound the rounding errors.

**P2.4.5** Suppose $A$ and $B$ are $n$-by-$n$ upper triangular floating point matrices. If $\hat{C} = fl(AB)$ is computed using one of the conventional §1.1 algorithms, does it follow that $\hat{C} = \hat{A}\hat{B}$ where $\hat{A}$ and $\hat{B}$ are close to $A$ and $B$?

**P2.4.6** Suppose $A$ and $B$ are $n$-by-$n$ floating point matrices and that $A$ is nonsingular with $\|\,|A^{-1}|\,|A|\,\|_\infty = \tau$. Show that if $\hat{C} = fl(AB)$ is obtained using any of the algorithms in §1.1, then there exists a $\hat{B}$ so $\hat{C} = A\hat{B}$ and $\|\,\hat{B} - B\,\|_\infty \leq n\mathbf{u}\tau\|\,B\,\|_\infty + O(\mathbf{u}^2)$.

### Notes and References for Sec. 2.4

The most complete treatment of roundoff error analysis is Wilkinson (AEP, chapter 3).

The treatments in Forsythe and Moler (SLE, pp. 87-97) and Stewart (IMC, pp. 69-82) are also excellent. For a general introduction to the effects of roundoff error, we recommend chapter 2 of

G.E. Forsythe, M.A. Malcolm, and C.B. Moler (1977). *Computer Methods for Mathematical Computations*, Prentice-Hall, Englewood Cliffs, NJ.

and its revision

D. Kahaner, C.B. Moler, and S. Nash (1988). *Numerical Methods and Software*, Prentice-Hall, Englewood Cliffs, NJ.

We also highly recommend the classical reference

J.H. Wilkinson (1963). *Rounding Errors in Algebraic Processes*, Prentice-Hall, Englewood Cliffs, NJ.

A philosophical and historical overview of roundoff analysis was given in the 1970 John von Neumann Lecture delivered by J.H. Wilkinson and appears in

J.H. Wilkinson (1971). "Modern Error Analysis," *SIAM Review 13*, 548–68.

This paper gives a critique of the early papers in error analysis authored by Von Neumann and Goldstine, Turing and Givens.

More recent developments in error analysis involve interval analysis, the building of statistical models of roundoff error, and the automating of the analysis itself. See

T.E. Hull and J.R. Swensen (1966). "Tests of Probabilistic Models for Propagation of Roundoff Errors," *Comm. ACM. 9*, 108–13.

J. Larson and A. Sameh (1978). "Efficient Calculation of the Effects of Roundoff Errors," *ACM Trans. Math. Soft. 4*, 228–36.

W. Miller and D. Spooner (1978). "Software for Roundoff Analysis, II," *ACM Trans. Math. Soft. 4*, 369–90.

J.M. Yohe (1979). "Software for Interval Arithmetic: A Reasonable Portable Package," *ACM Trans. Math. Soft. 5*, 50–63.

Anyone engaged in serious software development needs a thorough understanding of floating point arithmetic. A good way to begin acquiring knowledge in this direction is to read about the IEEE floating point standard in

D. Stevenson (1981). "A Proposed Standard for Binary Floating Point Arithmetic," *Computer 14 (March)*, 51–62.

The properties that a floating point system should have is still an area of intense interest. See

J.W. Demmel (1984). "Underflow and the Reliability of Numerical Software," *SIAM J. Sci. and Stat. Comp. 5*, 887–919.

U.W. Kulisch and W.L. Miranker (1986). "The Arithmetic of the Digital Computer," *SIAM Review 28*, 1–40.

The subtleties associated with the development of high-quality software, even for "simple" problems, are immense. A good example is the design of a subroutine to compute 2-norms

J.M. Blue (1978). "A Portable FORTRAN Program to Find the Euclidean Norm of a Vector," *ACM Trans. Math. Soft. 4*, 15–23.

For an analysis of the Strassen algorithm and other "fast" linear algebra procedures see

R.P. Brent (1970). "Error Analysis of Algorithms for Matrix Multiplication and Triangular Decomposition Using Winograd's Identity," *Numer. Math. 16,* 145–156.
W. Miller (1975). "Computational Complexity and Numerical Stability," *SIAM J. Computing 4,* 97–107.

## 2.5 Orthogonality and the SVD

Orthogonality has a very prominent role to play in matrix computations. After establishing a few definitions we prove the extremely useful singular value decomposition (SVD). Among other things, the SVD enables us to intelligently handle the matrix rank problem. The concept of rank, though perfectly clear in the exact arithmetic context, is tricky in the presence of roundoff error and fuzzy data. With the SVD we can introduce the practical notion of numerical rank.

### 2.5.1 Orthogonality

A set of vectors $\{x_1, \ldots, x_p\}$ in $\mathbb{R}^m$ is *orthogonal* if $x_i^T x_j = 0$ whenever $i \neq j$ and *orthonormal* if $x_i^T x_j = \delta_{ij}$. Intuitively, orthogonal vectors are maximally independent for they point in totally different directions.

A collection of subspaces $S_1, \ldots, S_p$ in $\mathbb{R}^m$ is *mutually orthogonal* if $x^T y = 0$ whenever $x \in S_i$ and $y \in S_j$ for $i \neq j$. The *orthogonal complement* of a subspace $S \subseteq \mathbb{R}^m$ is defined by

$$S^\perp = \{y \in \mathbb{R}^m : y^T x = 0 \text{ for all } x \in S\}.$$

It can be shown that $\text{range}(A)^\perp = \text{null}(A^T)$.

The vectors $v_1, \ldots, v_k$ form an *orthonormal* basis for a subspace $S \subseteq \mathbb{R}^m$ if they are orthonormal and span $S$. It is always possible to extend such a basis to a full orthonormal basis $\{v_1, \ldots, v_m\}$ for $\mathbb{R}^m$. Note that in this case, $S^\perp = \text{span}\{v_{k+1}, \ldots, v_m\}$ .

A matrix $Q \in \mathbb{R}^{m \times m}$ is said to be *orthogonal* if $Q^T Q = I$. If $Q = [\, q_1, \ldots, q_m \,]$ is orthogonal, then the $q_i$ form an orthonormal basis for $\mathbb{R}^m$.

### 2.5.2 Norms and Orthogonal Transformations

The 2-norm is invariant under orthogonal transformation, for if $Q^T Q = I$, then $\| Qx \|_2^2 = x^T Q^T Q x = x^T x = \| x \|_2^2$ . The matrix 2-norm and the Frobenius norm are also invariant with respect to orthogonal transformations. In particular, it is easy to show that for all orthogonal $Q$ and $Z$ of appropriate dimensions we have

$$\| QAZ \|_F = \| A \|_F \tag{2.5.1}$$

and

$$\| QAZ \|_2 = \| A \|_2 \,. \tag{2.5.2}$$

### 2.5.3 The Singular Value Decomposition

The theory of norms developed in the previous two sections can be used to prove the extremely useful singular value decomposition.

**Theorem 2.5.1 (Singular Value Decomposition (SVD))** *If $A$ is a real $m$-by-$n$ matrix then there exist orthogonal matrices*

$$U = [\, u_1, \ldots, u_m \,] \in \mathbb{R}^{m \times m} \quad \textit{and} \quad V = [\, v_1, \ldots, v_n \,] \in \mathbb{R}^{n \times n}$$

*such that*

$$U^T A V = \text{diag}(\sigma_1, \ldots, \sigma_p) \in \mathbb{R}^{m \times n} \qquad p = \min\{m, n\}$$

*where $\sigma_1 \geq \sigma_2 \geq \ldots \geq \sigma_p \geq 0$.*

**Proof.** Let $x \in \mathbb{R}^n$ and $y \in \mathbb{R}^m$ be unit 2-norm vectors that satisfy $Ax = \sigma y$ with $\sigma = \| A \|_2$. Because any orthonormal set can be extended to form an orthonormal basis for the whole space, it is possible to find $V_1 \in \mathbb{R}^{n \times (n-1)}$ and $U_1 \in \mathbb{R}^{m \times (m-1)}$ so $V = [\, x \;\; V_1 \,] \in \mathbb{R}^{n \times n}$ and $U = [\, y \;\; U_1 \,] \in \mathbb{R}^{m \times m}$ are orthogonal. It is not hard to show that $U^T A V$ has the following structure:

$$U^T A V \;=\; \begin{bmatrix} \sigma & w^T \\ 0 & B \end{bmatrix} \;\equiv\; A_1 .$$

Since

$$\left\| A_1 \left( \begin{bmatrix} \sigma \\ w \end{bmatrix} \right) \right\|_2^2 \;\geq\; (\sigma^2 + w^T w)^2$$

we have $\| A_1 \|_2^2 \geq (\sigma^2 + w^T w)$. But $\sigma^2 = \| A \|_2^2 = \| A_1 \|_2^2$, and so we must have $w = 0$. An obvious induction argument completes the proof of the theorem. □

The $\sigma_i$ are the *singular values* of $A$ and the vectors $u_i$ and $v_i$ are the $i$th *left singular vector* and the $i$th *right singular vector* respectively. It is easy to verify by comparing columns in the equations $AV = \Sigma U$ and $A^T U = \Sigma^T V$ that

$$\left.\begin{aligned} Av_i &= \sigma_i u_i \\ A^T u_i &= \sigma_i v_i \end{aligned}\right\} \; i = 1{:}\min\{m, n\}$$

It is convenient to have the following notation for designating singular values:

$$\begin{aligned} \sigma_i(A) &= \text{ the } i\text{th largest singular value of } A, \\ \sigma_{max}(A) &= \text{ the largest singular value of } A, \\ \sigma_{min}(A) &= \text{ the smallest singular value of } A. \end{aligned}$$

The singular values of a matrix $A$ are precisely the lengths of the semi-axes of the hyperellipsoid $E$ defined by $E = \{ Ax : \| x \|_2 = 1 \}$.

**Example 2.5.1**

$$A = \begin{bmatrix} .96 & 1.72 \\ 2.28 & .96 \end{bmatrix} = U\Sigma V^T = \begin{bmatrix} .6 & -.8 \\ .8 & .6 \end{bmatrix} \begin{bmatrix} 3 & 0 \\ 0 & 1 \end{bmatrix} \begin{bmatrix} .8 & .6 \\ .6 & -.8 \end{bmatrix}^T .$$

The SVD reveals a great deal about the structure of a matrix. If the SVD of $A$ is given by Theorem 2.5.1, and we define $r$ by

$$\sigma_1 \geq \cdots \geq \sigma_r > \sigma_{r+1} = \cdots = \sigma_p = 0,$$

then

$$\begin{aligned} \text{rank}(A) &= r & (2.5.3) \\ \text{null}(A) &= \text{span}\{v_{r+1}, \ldots, v_n\} & (2.5.4) \\ \text{range}(A) &= \text{span}\{u_1, \ldots, u_r\} . & (2.5.5) \end{aligned}$$

Moreover, if $U_r = U(:, 1:r)$, $\Sigma_r = \Sigma(1:r, 1:r)$, and $V_r = V(:, 1:r)$, then we have the *SVD expansion*

$$A = U_r \Sigma_r V_r^T = \sum_{i=1}^{r} \sigma_i u_i v_i^T . \qquad (2.5.6)$$

Finally, both the 2-norm and the Frobenius norm are neatly characterized in terms of the SVD:

$$\begin{aligned} \| A \|_F^2 &= \sigma_1^2 + \cdots + \sigma_p^2 \qquad p = \min\{m, n\} & (2.5.7) \\ \| A \|_2 &= \sigma_1 . & (2.5.8) \end{aligned}$$

## 2.5.4 Rank Deficiency and the SVD

One of the most valuable aspects of the SVD is that it enables us to deal sensibly with the concept of matrix rank. Numerous theorems in linear algebra have the form "if such-and-such a matrix has full rank, then such-and-such a property holds." While neat and aesthetic, results of this flavor do not help us address the numerical difficulties frequently encountered in situations where near rank deficiency prevails. Rounding errors and fuzzy data make rank determination a nontrivial exercise. Indeed, for some small $\epsilon$ we may be interested in the $\epsilon$-rank of a matrix which we define by

$$\text{rank}(A, \epsilon) = \min_{\|A-B\|_2 \leq \epsilon} \text{rank}(B).$$

Thus, if $A$ is obtained in a laboratory with each $a_{ij}$ correct to within $\pm .001$, then it might make sense to look at rank$(A, .001)$. Along the same lines, if $A$ is an $m$-by-$n$ floating point matrix then it is reasonable to regard $A$ as *numerically rank deficient* if $\text{rank}(A,\epsilon) < \min\{m,n\}$ with $\epsilon = \mathbf{u}\| A \|_2$.

Numerical rank deficiency and $\epsilon$-rank are nicely characterized in terms of the SVD because the singular values indicate how near a given matrix is to a matrix of lower rank.

**Theorem 2.5.2** *Let the SVD of $A \in \mathbb{R}^{m\times n}$ be given by Theorem 2.5.1. If $k < r = \text{rank}(A)$ and*

$$A_k = \sum_{i=1}^{k} \sigma_i u_i v_i^T \tag{2.5.9}$$

*then*

$$\min_{\text{rank}(B)=k} \| A - B \|_2 = \| A - A_k \|_2 = \sigma_{k+1} . \tag{2.5.10}$$

**Proof.** Since $U^T A_k V = \text{diag}(\sigma_1, \ldots, \sigma_k, 0, \ldots, 0)$ it follows that $\text{rank}(A_k) = k$ and that $U^T(A - A_k)V = \text{diag}(0, \ldots, 0, \sigma_{k+1}, \ldots, \sigma_p)$ and so $\| A - A_k \|_2 = \sigma_{k+1}$.

Now suppose $\text{rank}(B) = k$ for some $B \in \mathbb{R}^{m\times n}$. It follows that we can find orthonormal vectors $x_1, \ldots, x_{n-k}$ so $\text{null}(B) = \text{span}\{x_1, \ldots, x_{n-k}\}$. A dimension argument shows that

$$\text{span}\{x_1, \ldots, x_{n-k}\} \cap \text{span}\{v_1, \ldots, v_{k+1}\} \neq \{0\}.$$

Let $z$ be a unit 2-norm vector in this intersection. Since $Bz = 0$ and

$$Az = \sum_{i=1}^{k+1} \sigma_i (v_i^T z) u_i$$

we have

$$\| A - B \|_2^2 \geq \| (A - B)z \|_2^2 = \| Az \|_2^2 = \sum_{i=1}^{k+1} \sigma_i^2 (v_i^T z)^2 \geq \sigma_{k+1}^2$$

completing the proof of the theorem. □

Theorem 2.5.2 says that the smallest singular value of $A$ is the 2-norm distance of $A$ to the set of all rank-deficient matrices. It also follows that the set of full rank matrices in $\mathbb{R}^{m\times n}$ is both open and dense.

Finally, if $r_\epsilon = \text{rank}(A, \epsilon)$ then

$$\sigma_1 \geq \cdots \geq \sigma_{r_\epsilon} > \epsilon \geq \sigma_{r_\epsilon + 1} \geq \cdots \geq \sigma_p \qquad p = \min\{m, n\}.$$

We have more to say about the numerical rank issue in §5.5 and §12.2.

### Problems

**P2.5.1** Show that if $S^T = -S$ then $I - S$ is nonsingular and the matrix $(I-S)^{-1}(I+S)$ is orthogonal. This is known as the *Cayley transform* of $S$.

**P2.5.2** Show that a triangular orthogonal matrix is diagonal.

**P2.5.3** Show that if $Q = Q_1 + iQ_2$ is unitary with $Q_1, Q_2 \in \mathbf{R}^{n\times n}$, then the $2n$-by-$2n$ real matrix

$$Z = \begin{bmatrix} Q_1 & -Q_2 \\ Q_2 & Q_1 \end{bmatrix}$$

is orthogonal.

**P2.5.4** Establish properties (2.5.3)-(2.5.8).

**P2.5.5** Prove that

$$\sigma_{max}(A) = \max_{y \in \mathbf{R}^m, x \in \mathbf{R}^n} \frac{y^T A x}{\| x \|_2 \| y \|_2}$$

**P2.5.6** For the 2-by-2 matrix $A = \begin{bmatrix} w & x \\ y & z \end{bmatrix}$, derive expressions for $\sigma_{max}(A)$ and $\sigma_{min}(A)$ that are functions of $w$, $x$, $y$, and $z$.

**P2.5.7** Show that any matrix in $\mathbf{R}^m n$ is the limit of a sequence of full rank matrices.

**P2.5.8** Show that if $A \in \mathbf{R}^{m\times n}$ has rank $n$, then $\| A(A^TA)^{-1}A^T \|_2 = 1$.

**P2.5.9** What is the nearest rank-one matrix to $A = \begin{bmatrix} 1 & M \\ 0 & 1 \end{bmatrix}$ in the Frobenius norm?

**P2.5.10** Show that if $A \in \mathbf{R}^{m\times n}$ then $\| A \|_F \le \sqrt{\text{rank}(A)}\, \| A \|_2$, thereby sharpening (2.3.7).

### Notes and References for Sec. 2.5

Forsythe and Moler (SLE) offer a good account of the SVD's role in the analysis of the $Ax = b$ problem. Their proof of the decomposition is more traditional than ours in that it makes use of the eigenvalue theory for symmetric matrices. Some early SVD references include

E. Beltrami (1873). "Sulle Funzioni Bilineari," *Gionale di Mathematiche 11,* 98–106.

C. Eckart and G. Young (1939). "A Principal Axis Transformation for Non-Hermitian Matrices," *Bull. Amer. Math. Soc. 45,* 118–21.

One of the most significant developments in scientific computation has been the increased use of the SVD in application areas that require the intelligent handling of matrix rank. The range of applications is impressive. One of the most interesting is

C.B. Moler and D. Morrison (1983). "Singular Value Analysis of Cryptograms," *Amer. Math. Monthly 90,* 78–87.

For generalizations of the SVD to infinite dimensional Hilbert space, see

I.C. Gohberg and M.G. Krein (1969). *Introduction to the Theory of Linear Non-Self Adjoint Operators* , Amer. Math. Soc., Providence, R.I.

F. Smithies (1970). *Integral Equations,* Cambridge University Press, Cambridge.

Reducing the rank of a matrix as in Theorem 2.5.2 when the perturbing matrix is constrained is discussed in

J.W. Demmel (1987). "The smallest perturbation of a submatrix which lowers the rank and constrained total least squares problems, *SIAM J. Numer. Anal. 24,* 199–206.
G.H. Golub, A. Hoffman, and G.W. Stewart (1988). "A Generalization of the Eckart-Young-Mirsky Approximation Theorem." *Lin. Alg. and Its Applic. 88/89,* 317–328.
G.A. Watson (1988). "The Smallest Perturbation of a Submatrix which Lowers the Rank of the Matrix," *IMA J. Numer. Anal. 8,* 295–304.

## 2.6 Projections and the CS Decomposition

If the object of a computation is to compute a matrix or a vector, then norms are useful for assessing the accuracy of the answer or for measuring progress during an iteration. If the object of a computation is to compute a subspace, then to make similar comments we need to be able to quantify the distance between two subspaces. Orthogonal projections are critical in this regard. After the elementary concepts are established we discuss the CS decomposition. This is an SVD-like decomposition that is handy when having to compare a pair of subspaces. We begin with the notion of an orthogonal projection.

### 2.6.1 Orthogonal Projections

Let $S \subseteq \mathbb{R}^n$ be a subspace. $P \in \mathbb{R}^{n\times n}$ is the *orthogonal projection* onto $S$ if range$(P) = S$, $P^2 = P$, and $P^T = P$. From this definition it is easy to show that if $x \in \mathbb{R}^n$ then $Px \in S$ and $(I - P)x \in S^\perp$.

If $P_1$ and $P_2$ are each orthogonal projections, then for any $z \in \mathbb{R}^n$ we have

$$\| (P_1 - P_2)z \|_2^2 \;=\; (P_1 z)^T(I - P_2)z + (P_2 z)^T(I - P_1)z$$

If range$(P_1)$ = range$(P_2) = S$, then the right-hand side of this expression is zero showing that the orthogonal projection for a subspace is unique.

If the columns of $V = [\, v_1, \ldots, v_k \,]$ are an orthonormal basis for a subspace $S$, then it follows that $P = VV^T$ is the unique orthogonal projection onto $S$. Note that if If $v \in \mathbb{R}^n$ then $P = vv^T/v^Tv$ is the orthogonal projection onto $S = \text{span}\{v\}$.

### 2.6.2 SVD-Related Projections

There are several important orthogonal projections associated with the singular value decomposition. Suppose $A = U\Sigma V^T \in \mathbb{R}^{m\times n}$ is the SVD of $A$ and that $r = \text{rank}(A)$. If we have the $U$ and $V$ partitionings

$$U = [\, \underset{r}{U_r} \quad \underset{m-r}{\tilde{U}_r} \,] \qquad\qquad V = [\, \underset{r}{V_r} \quad \underset{n-r}{\tilde{V}_r} \,]$$

then

$$\begin{aligned}
V_r V_r^T &= \text{projection on to null}(A)^\perp = \text{range}(A^T) \\
\tilde{V}_r \tilde{V}_r^T &= \text{projection on to null}(A) \\
U_r U_r^T &= \text{projection on to range}(A) \\
\tilde{U}_r \tilde{U}_r^T &= \text{projection on to range}(A)^\perp = \text{null}(A^T)
\end{aligned}$$

### 2.6.3 Distance Between Subspaces

The one-to-one correspondence between subspaces and orthogonal projections enables us to devise a notion of distance between subspaces. Suppose $S_1$ and $S_2$ are subspaces of $\mathbb{R}^n$ and that $\dim(S_1) = \dim(S_2)$. We define the *distance* between these two spaces by

$$\text{dist}(S_1, S_2) = \| P_1 - P_2 \|_2 \tag{2.6.1}$$

where $P_i$ is the orthogonal projection onto $S_i$.

By considering the case of one-dimensional subspaces in $\mathbb{R}^2$, we can obtain a geometrical interpretation of $\text{dist}(\cdot,\cdot)$. Suppose $S_1 = \text{span}\{x\}$ and $S_2 = \text{span}\{y\}$, where $x$ and $y$ are unit 2-norm vectors in $\mathbb{R}^2$. Because they have unit 2-norm, we may write

$$x = \begin{bmatrix} \cos(\theta_1) \\ \sin(\theta_1) \end{bmatrix} \qquad y = \begin{bmatrix} \cos(\theta_2) \\ \sin(\theta_2) \end{bmatrix}$$

for some $\theta_1$ and $\theta_2$ taken from the interval $[0, 2\pi]$. If we define the orthogonal matrices

$$W = \begin{bmatrix} \cos(\theta_1) & -\sin(\theta_1) \\ \sin(\theta_1) & \cos(\theta_1) \end{bmatrix} \qquad Z = \begin{bmatrix} \cos(\theta_2) & -\sin(\theta_2) \\ \sin(\theta_2) & \cos(\theta_2) \end{bmatrix}$$

then

$$W^T y = \begin{bmatrix} \cos(\theta_2 - \theta_1) \\ \sin(\theta_2 - \theta_1) \end{bmatrix} \qquad Z^T x = \begin{bmatrix} \cos(\theta_1 - \theta_2) \\ \sin(\theta_1 - \theta_2) \end{bmatrix}.$$

A calculation shows that

$$W^T \left( xx^T - yy^T \right) Z = \begin{bmatrix} 0 & \sin(\theta_1 - \theta_2) \\ \sin(\theta_1 - \theta_2) & 0 \end{bmatrix}$$

and so $\| xx^T - yy^T \|_2 = \| W^T \left( xx^T - yy^T \right) Z \|_2 = |\sin(\theta_1 - \theta_2)|$. Consequently, $\text{dist}(S_1, S_2) = |\sin(\theta_1 - \theta_2)|$, the magnitude of the sine of the angle between the two subspaces.

This nice geometrical interpretation of the distance function also applies in higher dimensions, but we need to extend the notion of angles between subspaces. The vehicle for doing this is the following theorem:

**Theorem 2.6.1 (CS Decomposition)** *If*

$$Q = \begin{bmatrix} Q_{11} & Q_{12} \\ Q_{21} & Q_{22} \end{bmatrix} \begin{matrix} k \\ j \end{matrix}$$
$$\qquad\quad k \qquad j$$

*is orthogonal with* $k \geq j$, *then there exist orthogonal matrices* $U_1, V_1 \in \mathbb{R}^{k\times k}$ *and orthogonal matrices* $U_2, V_2 \in \mathbb{R}^{j\times j}$ *such that*

$$\begin{bmatrix} U_1 & 0 \\ 0 & U_2 \end{bmatrix}^T \begin{bmatrix} Q_{11} & Q_{12} \\ Q_{21} & Q_{22} \end{bmatrix} \begin{bmatrix} V_1 & 0 \\ 0 & V_2 \end{bmatrix} = \begin{bmatrix} I_{k-j} & 0 & 0 \\ 0 & C & S \\ 0 & -S & C \end{bmatrix} \qquad (2.6.2)$$

*where*

$$\begin{aligned} C &= \text{diag}(c_1, \ldots, c_j) \in \mathbb{R}^{j\times j} \qquad c_i = \cos(\theta_i) \\ S &= \text{diag}(s_1, \ldots, s_j) \in \mathbb{R}^{j\times j} \qquad s_i = \sin(\theta_i) \end{aligned}$$

*and* $0 \leq \theta_1 \leq \theta_2 \leq \cdots \leq \theta_j \leq \pi/2$.

**Proof.** See Davis and Kahan (1970) or Stewart (1977). The assumption $k \geq j$ is made for notational convenience only. □

Roughly speaking, the CS decomposition amounts to a simultaneous diagonalization of the blocks of an orthogonal matrix.

**Example 2.6.1** The matrices

$$Q = \begin{bmatrix} -.716 & .548 & .433 \\ -.698 & -.555 & -.451 \\ -.006 & -.626 & .780 \end{bmatrix}, \qquad U = \begin{bmatrix} -.721 & -.692 & .000 \\ -.692 & .721 & .000 \\ .000 & .000 & 1 \end{bmatrix}$$

and

$$V = \begin{bmatrix} .999 & -.010 & .000 \\ -.010 & -.999 & .000 \\ .000 & .000 & 1 \end{bmatrix}$$

are orthogonal to three decimals and satisfy

$$U^TQV = \begin{bmatrix} 1.000 & .000 & .000 \\ .000 & .780 & .625 \\ .000 & -.625 & .780 \end{bmatrix}.$$

Thus, in the language of Theorem 2.6.1, $k = 2$, $j = 1$, $c_1 = .780$ and $s_1 = .625$.

We now show how an appropriately chosen CS decomposition reveals much about the "closeness" of a pair of equidimensional subspaces.

**Corollary 2.6.2** *Let* $W = [W_1, W_2]$ *and* $Z = [Z_1, Z_2]$ *be orthogonal matrices where* $W_1, Z_1 \in \mathbb{R}^{n\times k}$ *and* $W_2, Z_2 \in \mathbb{R}^{n\times(n-k)}$. *If* $S_1 = \text{range}(W_1)$ *and* $S_2 = \text{range}(Z_1)$ *then* $\text{dist}(S_1, S_2) = \sqrt{1 - \sigma^2_{min}(W_1^T Z_1)}$.

**Proof.** Let $Q = W^T Z$ and assume that $k \geq j = n - k$. Let the CS decomposition of $Q$ be given by (2.6.2) with $Q_{ip} = W_i^T Z_p$. It follows that

$$\| W_1^T Z_2 \|_2 = \| W_2^T Z_1 \|_2 = s_j = \sqrt{1 - c_j^2} = \sqrt{1 - \sigma_{min}^2(W_1^T Z_1)}\,.$$

Since $W_1 W_1^T$ and $Z_1 Z_1^T$ are the orthogonal projections onto $S_1$ and $S_2$, respectively, we have

$$\begin{aligned} \text{dist}(S_1, S_2) &= \| W_1 W_1^T - Z_1 Z_1^T \|_2 = \| (W^T (W_1 W_1^T - Z_1 Z_1^T) Z \|_2 \\ &= \left\| \begin{bmatrix} 0 & W_1^T Z_2 \\ -W_2^T Z_1 & 0 \end{bmatrix} \right\|_2 = s_j\,. \end{aligned}$$

If $k < j$, then the above argument goes through merely by setting

$$Q = [W_2, W_1]^T \, [Z_2, Z_1]$$

and noting that $\sigma_{max}(W_2^T Z_1) = \sigma_{max}(W_1^T Z_2) = s_j$. □

**Example 2.6.2** If $S_1 = \text{span}\{e_1, e_2\}$ and $S_2 = \text{span}\{e_2, e_3\}$ are subspaces of $\mathbf{R}^n$ then the matrix $Q$ in the corollary is given by

$$Q = [e_1, e_2, e_3, e_4, \ldots, e_n]^T \, [e_2, e_3, e_1, e_4, \ldots, e_n] = [e_2, e_3, e_1, e_4, \ldots, e_n]\,.$$

Thus, $\text{dist}(S_1, S_2) = 1$. The example shows that subspace intersection is not equivalent to zero distance.

### Problems

**P2.6.1** Show that if $P$ is an orthogonal projection, then $Q = I - 2P$ is orthogonal.

**P2.6.2** What are the singular values of an orthogonal projection?

**P2.6.3** Suppose $k = j$ in Theorem 2.6.1. (a) Show that

$$\| Q_{11} v \|_2 = \sigma_{max}(Q_{11}) \| v \|_2 \quad \Rightarrow \quad \| Q_{21} v \|_2 = \sigma_{min}(Q_{21}) \| v \|_2.$$

(b) Show that if $Q_{11}$ is singular then $Q_{22}$ is singular. Prove these results *without* using the CS decomposition.

### Notes and References for Sec. 2.6

A proof of the CS decomposition appears in

G.W. Stewart (1977). "On the Perturbation of Pseudo-Inverses, Projections and Linear Least Squares Problems," *SIAM Review 19*, 634–62.

The following papers discuss other aspects of this important decomposition:

C. Davis and W. Kahan (1970). "The Rotation of Eigenvectors by a Perturbation III," *SIAM J. Num. Anal. 7*, 1–46.

C.C. Paige and M. Saunders (1981). "Toward a Generalized Singular Value Decomposition," *SIAM J. Num. Anal. 18*, 398–405.

See §8.7 for computational details.

## 2.7 The Sensitivity of Square Systems

We now use some of the tools developed in previous sections to analyze the linear system problem $Ax = b$ where $A \in \mathbb{R}^{n \times n}$ is nonsingular and $b \in \mathbb{R}^n$. Our aim is to examine how perturbations in $A$ and $b$ affect the solution $x$.

### 2.7.1 An SVD Analysis

If

$$A = \sum_{i=1}^{n} \sigma_i u_i v_i^T = U\Sigma V^T$$

is the SVD of $A$ then

$$x = A^{-1}b = (U\Sigma V^T)^{-1}b = \sum_{i=1}^{n} \frac{u_i^T b}{\sigma_i} v_i. \tag{2.7.1}$$

This expansion shows that small changes in $A$ or $b$ can induce relatively large changes in $x$ if $\sigma_n$ is small. Indeed, if $\cos(\theta) = |u_n^T b| / \| b \|_2$ and

$$(A - \epsilon u_n v_n^T)y = b + \epsilon \ \mathrm{sign}(u_n^T b) u_n \qquad \sigma_n > \epsilon \geq 0$$

then it can be shown that $\| y - x \|_2 \geq (\epsilon/\sigma_n) \| x \|_2 \cos(\theta)$. Thus, $O(\epsilon)$ perturbations can alter the solution by an amount $\epsilon/\sigma_n$.

It should come as no surprise that the magnitude of $\sigma_n$ should have a bearing on the sensitivity of the $Ax = b$ problem when we recall from Theorem 2.5.2 that $\sigma_n$ is the distance from $A$ to the set of singular matrices. As we approach this set, it is intuitively clear that the solution $x$ should be increasingly sensitive to perturbations.

### 2.7.2 Condition

A precise measure of linear system sensitivity can be obtained by considering the parameterized system

$$(A + \epsilon F)x(\epsilon) = b + \epsilon f \qquad x(0) = x$$

where $F \in \mathbb{R}^{n \times n}$ and $f \in \mathbb{R}^n$. If $A$ is nonsingular, then it is clear that $x(\epsilon)$ is differentiable in a neighborhood of zero. Moreover, $\dot{x}(0) = A^{-1}(f - Fx)$ and thus, the Taylor series expansion for $x(\epsilon)$ has the form

$$x(\epsilon) = x + \epsilon \dot{x}(0) + O(\epsilon^2).$$

Using any vector norm and consistent matrix norm we obtain

$$\frac{\| x(\epsilon) - x \|}{\| x \|} \leq \epsilon \| A^{-1} \| \left\{ \frac{\| f \|}{\| x \|} + \| F \| \right\} + O(\epsilon^2). \tag{2.7.2}$$

For square matrices $A$ define the *condition number* $\kappa(A)$ by

$$\kappa(A) = \| A \| \| A^{-1} \| \tag{2.7.3}$$

with the convention that $\kappa(A) = \infty$ for singular $A$. Using the inequality $\| b \| \leq \| A \| \| x \|$ it follows from (2.7.2) that

$$\frac{\| x(\epsilon) - x \|}{\| x \|} \leq \kappa(A)(\rho_A + \rho_b) + O(\epsilon^2) \tag{2.7.4}$$

where

$$\rho_A = \epsilon \frac{\| F \|}{\| A \|} \quad \text{and} \quad \rho_b = \epsilon \frac{\| f \|}{\| b \|}$$

represent the relative errors in $A$ and $b$, respectively. Thus, the relative error in $x$ can be $\kappa(A)$ times the relative error in $A$ and $b$. In this sense, the condition number $\kappa(A)$ quantifies the sensitivity of the $Ax = b$ problem.

Note that $\kappa(\cdot)$ depends on the underlying norm. When this norm is to be stressed, we use subscripts, e.g.,

$$\kappa_2(A) = \| A \|_2 \| A^{-1} \|_2 = \frac{\sigma_1(A)}{\sigma_n(A)}. \tag{2.7.5}$$

Thus, the 2-norm condition of a matrix $A$ measures the elongation of the hyperellipsoid $\{ Ax : \| x \|_2 = 1 \}$.

We mention two other characterizations of the condition number. For $p$-norm condition numbers, we have

$$\frac{1}{\kappa_p(A)} = \min_{A+E \text{ singular}} \frac{\| E \|_p}{\| A \|_p}. \tag{2.7.6}$$

This result may be found in Kahan (1966) and shows that $\kappa_p(A)$ measures the relative $p$-norm distance from $A$ to the set of singular matrices.

For any norm, we also have

$$\kappa(A) = \lim_{\delta \to 0} \sup_{\|E\| \leq \delta \|A\|} \frac{\| (A+E)^{-1} - A^{-1} \|}{\delta} \frac{1}{\| A^{-1} \|}. \tag{2.7.7}$$

This imposing result merely says that the condition number is a normalized Frechet derivative of the map $A \rightarrow A^{-1}$. Further details may be found in Rice (1966). Recall that we were initially led to $\kappa(A)$ through differentiation.

If $\kappa(A)$ is large, then $A$ is said to be an *ill-conditioned* matrix Note that this is a norm-dependent property. However, any two condition numbers $\kappa_\alpha(\cdot)$ and $\kappa_\beta(\cdot)$ on $\mathbb{R}^{n \times n}$ are equivalent in that constants $c_1$ and $c_2$ can be found for which

$$c_1 \kappa_\alpha(A) \leq \kappa_\beta(A) \leq c_2 \kappa_\alpha(A) \qquad A \in \mathbb{R}^{n \times n}.$$

For example, on $\mathbb{R}^{n\times n}$ we have

$$\begin{array}{rcl} \frac{1}{n}\,\kappa_2(A) & \le & \kappa_1(A) \;\le\; n\kappa_2(A) \\ \frac{1}{n}\kappa_\infty(A) & \le & \kappa_2(A) \;\le\; n\kappa_\infty(A) \\ \frac{1}{n^2}\kappa_1(A) & \le & \kappa_\infty(A) \le n^2\kappa_1(A)\,. \end{array} \qquad (2.7.8)$$

Thus, if a matrix is ill-conditioned in the $\alpha$-norm, it is ill-conditioned in the $\beta$-norm modulo the constants $c_1$ and $c_2$ above.

For any of the $p$-norms, we have $\kappa_p(A) \ge 1$. Matrices with small condition numbers are said to be *well-conditioned* . In the 2-norm, orthogonal matrices are perfectly conditioned in that $\kappa_2(Q) = 1$ if $Q$ is orthogonal.

### 2.7.3 Determinants and Nearness to Singularity

It is natural to consider how well determinant size measures ill-conditioning. If $\det(A) = 0$ is equivalent to singularity, is $\det(A) \approx 0$ equivalent to near singularity? Unfortunately, there is little correlation between $\det(A)$ and the condition of $Ax = b$. For example, the matrix $B_n$ defined by

$$B_n \;=\; \begin{bmatrix} 1 & -1 & \cdots & -1 \\ 0 & 1 & \cdots & -1 \\ \vdots & \vdots & \ddots & \vdots \\ 0 & 0 & \cdots & 1 \end{bmatrix} \in \mathbb{R}^{n\times n} \qquad (2.7.9)$$

has determinant 1, but $\kappa_\infty(B_n) = n2^{n-1}$. On the other hand, a very well conditioned matrix can have a very small determinant. For example,

$$D_n = \text{diag}(10^{-1}, \ldots, 10^{-1}) \in \mathbb{R}^{n\times n}$$

satisfies $\kappa_p(D_n) = 1$ although $\det(D_n) = 10^{-n}$.

### 2.7.4 A Rigorous Norm Bound

Recall that the derivation of (2.7.4) was valuable because it highlighted the connection between $\kappa(A)$ and the rate of change of $x(\epsilon)$ at $\epsilon = 0$. However, it is a little dissatisfying because it is contingent on $\epsilon$ being "small enough" and because it sheds no light on the size of the $O(\epsilon^2)$ term. In this and the next subsection we develop some additional $Ax = b$ perturbation theorems that are completely rigorous.

We first establish a useful lemma that indicates in terms of $\kappa(A)$ when we can expect a perturbed system to be nonsingular.

**Lemma 2.7.1** *Suppose*

$$\begin{aligned} Ax &= b & A \in \mathbb{R}^{n\times n},\ 0 \neq b \in \mathbb{R}^n \\ (A+\Delta A)y &= b + \Delta b & \Delta A \in \mathbb{R}^{n\times n},\ \Delta b \in \mathbb{R}^n \end{aligned}$$

*with* $\| \Delta A \| \leq \delta \| A \|$ *and* $\| \Delta b \| \leq \delta \| b \|$. *If* $\delta\kappa(A) = r < 1$ *then* $A + \Delta A$ *is nonsingular and*

$$\frac{\| y \|}{\| x \|} \leq \frac{1+r}{1-r}.$$

**Proof.** Since $\| A^{-1}\Delta A \| < \delta \| A^{-1} \| \, \| A \| = r < 1$ it follows from Theorem 2.3.4 that $(A + \Delta A)$ is nonsingular.

Now $(I + A^{-1}\Delta A)y = x + A^{-1}\Delta b$ and so using Lemma 2.3.3 we find

$$\begin{aligned} \| y \| &\leq \| (I + A^{-1}\Delta A)^{-1} \| \left( \| x \| + \delta \| A^{-1} \| \, \| b \| \right) \\ &\leq \frac{1}{1-r} \left( \| x \| + \delta \| A^{-1} \| \, \| b \| \right) \\ &= \frac{1}{1-r} \left( \| x \| + r\frac{\| b \|}{\| A \|} \right). \end{aligned}$$

Since $\| b \| = \| Ax \| \leq \| A \| \, \| x \|$ it follows that

$$\| y \| \leq \frac{1}{1-r} \left( \| x \| + r \| x \| \right). \ \square$$

We are now set to establish a rigorous $Ax = b$ perturbation bound.

**Theorem 2.7.2** *If the conditions of Lemma 2.7.1 hold then*

$$\frac{\| x - y \|}{\| x \|} \leq \frac{2\delta}{1-r}\kappa(A) \tag{2.7.10}$$

**Proof.** Since

$$y - x = A^{-1}\Delta b - A^{-1}\Delta A y \tag{2.7.11}$$

we have $\| y - x \| \leq \delta \| A^{-1} \| \, \| b \| + \delta \| A^{-1} \| \, \| A \| \, \| y \|$ and so

$$\begin{aligned} \frac{\| y - x \|}{\| x \|} &\leq \delta\kappa(A)\frac{\| b \|}{\| A \| \, \| x \|} + \delta\kappa(A)\frac{\| y \|}{\| x \|} \\ &\leq \delta\kappa(A)\left(1 + \frac{1+r}{1-r}\right) = \frac{2\delta}{1-r}\kappa(A). \ \square \end{aligned}$$

**Example 2.7.1** The $Ax = b$ problem

$$\begin{bmatrix} 1 & 0 \\ 0 & 10^{-6} \end{bmatrix} \begin{bmatrix} x_1 \\ x_2 \end{bmatrix} = \begin{bmatrix} 1 \\ 10^{-6} \end{bmatrix}$$

has solution $x = (1\ 1)^T$ and condition $\kappa_\infty(A) = 10^6$. If $\Delta b = (10^{-6}\ 0)^T$, $\Delta A = 0$, and $(A + \Delta A)y = b + \Delta b$, then the inequality (2.7.10) in the $\infty$-norm has the form

$$\frac{10^{-6}}{10^0} \le 10^{-6}10^6 .$$

Thus, the upper bound can be a gross overestimate of the error induced by the perturbation. On the other hand, if $\Delta b = (0\ 10^{-6})^T$, $\Delta A = 0$, and $(A + \Delta A)y = b + \Delta b$, then the inequality says

$$\frac{10^0}{10^0} \le 2 \times 10^{-6}10^6 .$$

Thus, there are perturbations for which the bound in (2.7.10) is essentially attained.

### 2.7.5 Some Rigorous Componentwise Bounds

We conclude this section by showing that a more refined perturbation theory is possible if componentwise perturbation bounds are in effect and if we make use of the absolute value notation.

**Theorem 2.7.3** *Suppose*

$$\begin{aligned} Ax &= b & A \in \mathbb{R}^{n \times n},\ 0 \ne b \in \mathbb{R}^n \\ (A + \Delta A)y &= b + \Delta b & \Delta A \in \mathbb{R}^{n \times n},\ \Delta b \in \mathbb{R}^n \end{aligned}$$

*and that* $|\Delta A| \le \delta |A|$ *and* $|\Delta b| \le \delta |b|$. *If* $\delta \kappa_\infty(A) = r < 1$ *then* $(A + \Delta A)$ *is nonsingular and*

$$\frac{\| y - x \|_\infty}{\| x \|_\infty} \le \frac{2\delta}{1 - r} \| \, |A^{-1}|\,|A| \, \|_\infty .$$

**Proof.** Since $\| \Delta A \|_\infty \le \delta \| A \|_\infty$ and $\| \Delta b \|_\infty \le \delta \| b \|_\infty$ the conditions of Lemma 2.7.1 are satisfied in the infinity norm. This implies that $A + \Delta A$ is nonsingular and

$$\frac{\| y \|_\infty}{\| x \|_\infty} \le \frac{1 + r}{1 - r} .$$

Now using (2.7.11) we find

$$\begin{aligned} |y - x| &\le |A^{-1}|\,|\Delta b| + |A^{-1}|\,|\Delta A|\,|y| \\ &\le \delta |A^{-1}|\,|b| + \delta |A^{-1}|\,|A|\,|y| \le \delta |A^{-1}|\,|A| \left( |x| + |y| \right) . \end{aligned}$$

If we take norms then

$$\| y - x \|_\infty \le \delta \| \, |A^{-1}|\,|A| \, \|_\infty \left( \| x \|_\infty + \frac{1 + r}{1 - r} \| x \|_\infty \right) .$$

The theorem follows upon division by $\| x \|_\infty$. $\square$

We refer to the quantity $\| \, |A| \, |A^{-1}| \, \|_\infty$ as the *Skeel condition number.* It has been effectively used in the analysis of several important linear system computations. See §3.5.

Lastly, we report on the results of Oettli and Prager (1964) that indicate when an approximate solution $\hat{x} \in \mathbb{R}^n$ to the $n$-by-$n$ system $Ax = b$ satisfies a perturbed system with prescribed structure. In particular, suppose $E \in \mathbb{R}^{n \times n}$ and $f \in \mathbb{R}^n$ are given and have nonnegative entries. We seek $\delta A \in \mathbb{R}^{n \times n}$, $\delta b \in \mathbb{R}^n$, and $\omega \geq 0$ such that

$$(A + \delta A)\hat{x} = b + \delta b \qquad |\delta A| \leq \omega E \, , \; |\delta b| \leq \omega f \, . \tag{2.7.12}$$

Note that by properly choosing $E$ and $f$ the perturbed system can take on certain qualities. For example, if $E = |A|$ and $f = |b|$ and $\omega$ is small, then $\hat{x}$ satisfies a nearby system in the componentwise sense. Oettli and Prager (1964) show that for a given $A$, $b$, $\hat{x}$, $E$, and $f$ the smallest $\omega$ possible in (2.7.12) is given by

$$\omega_{min} = \max_{1 \leq i \leq n} \frac{|A\hat{x} - b|_i}{(E|\hat{x}| + f)_i} \, .$$

If $A\hat{x} = b$ then $\omega_{min} = 0$. On the other hand, if $\omega_{min} = \infty$ then $\hat{x}$ does not satisfy any system of the prescribed perturbation structure.

### Problems

**P2.7.1** Show that if $\| I \| \geq 1$ then $\kappa(A) \geq 1$.

**P2.7.2** Show that for a given norm, $\kappa(AB) \leq \kappa(A)\kappa(B)$ and that $\kappa(\alpha A) = \kappa(A)$ for all nonzero $\alpha$.

### Notes and References for Sec. 2.7

The condition concept is thoroughly investigated in

J. Rice (1966). "A Theory of Condition," *SIAM J. Num. Anal. 3,* 287–310.
W. Kahan (1966). "Numerical Linear Algebra," *Canadian Math. Bull. 9,* 757–801.

In some applications, it is necessary to examine the sensitivity of the solution to $Ax = b$ subject to structured perturbations. For example, we may wish to know $x$ varies as each element of $b$ varies over some known interval. A paper concerned with this type of more refined sensitivity analysis is

J.E. Cope and B.W. Rust (1979). "Bounds on solutions of systems with accurate data," *SIAM J. Num. Anal. 16,* 950–63.

Basic references for componentwise perturbation theory include

W. Oettli and W. Prager (1964). "Compatability of Approximate Solutions of Linear Equations with Given Error Bounds for Coefficients and Right Hand Sides," *Numer. Math. 6,* 405–09.

R.D. Skeel (1979). "Scaling for numerical stability in Gaussian Elimination," *J. ACM 26*, 494–526

The reciprocal of the condition number measures how near a given $Ax = b$ problem is to singularity. The importance of knowing how near a given problem is to a difficult or insoluble problem has come to be appreciated in many computational settings. See

J. L. Barlow (1986). "On the Smallest Positive Singular Value of an $M$-Matrix with Applications to Ergodic Markov Chains," *SIAM J. Alg. and Disc. Struct. 7*, 414–424.

J.W. Demmel (1987a). "On the Distance to the Nearest Ill-Posed Problem," *Numer. Math. 51*, 251–289.

J.W. Demmel (1987b). "A Counterexample for two Conjectures About Stability," *IEEE Trans. Auto. Cont. AC-32*, 340–342.

J.W. Demmel (1988). "The Probability that a Numerical Analysis Problem is Difficult," *Math. Comp. 50*, 449–480.

N.J. Higham (1985). "Nearness Problems in Numerical Linear Algebra," PhD Thesis, University of Manchester, England.

N.J. Higham (1988b). "Computing a Nearest Symmetric Positive Semidefinite Matrix," *Lin. Alg. and Its Applic. 103*, 103–118.

N.J. Higham (1988e). "Matrix Nearness Problems and Applications," Numerical Analysis Report 161, University of Manchester, Manchester, England. To appear in *The Proceedings of the JIMA Conference on Applications of Matrix Theory*, S. Barnett and M.J.C. Gover (eds), Oxford University Press.

A. Laub(1985). "Numerical Linear Algebra Aspects of Control Design Computations," *IEEE Trans. Auto. Cont. AC-30*, 97–108.

A. Ruhe (1987). "Closest Normal Matrix Found!," *BIT 27*, 585–598.

C. Van Loan (1985). "How Near is a Stable Matrix to an Unstable Matrix?," Contemporary Mathematics, Vol. 47.,465–477.

# Chapter 3

# General Linear Systems



The problem of solving a linear system $Ax = b$ is central in scientific computation. In this chapter we focus on the method of Gaussian elimination, the algorithm of choice when $A$ is square, dense, and unstructured. When $A$ does not fall into this category, then the algorithms of Chapters 4, 5, and 10 are of interest. Parallel $Ax = b$ solvers are discussed in Chapter 6.

We motivate the method of Gaussian elimination in §3.1 by discussing the ease with which triangular systems can be solved. The conversion of a general system to triangular form via Gauss transformations is then presented in §3.2 where the "language" of matrix factorizations is introduced. Unfortunately, the derived method behaves very poorly on a nontrivial class of problems. Our error analysis in §3.3 pinpoints the difficulty and motivates §3.4, where the concept of pivoting is introduced. In the final section we offer some remarks on the important practical issues associated with scaling, iterative improvement, and condition estimation.

## 3.1 Triangular Systems

Traditional factorization methods for linear systems involve the conversion of the given square system to a triangular system that has the same solution.

This section is about the solution of triangular systems.

### 3.1.1 Forward Substitution

Consider the following 2-by-2 lower triangular system:

$$\begin{bmatrix} \ell_{11} & 0 \\ \ell_{21} & \ell_{22} \end{bmatrix} \begin{bmatrix} x_1 \\ x_2 \end{bmatrix} = \begin{bmatrix} b_1 \\ b_2 \end{bmatrix}$$

If $\ell_{11}\ell_{22} \neq 0$, then the unknowns can be determined sequentially:

$$\begin{aligned} x_1 &= b_1/\ell_{11} \\ x_2 &= (b_2 - \ell_{21}x_1)/\ell_{22}. \end{aligned}$$

This is the 2-by-2 version of an algorithm known as *forward substitution.* The general procedure is obtained by solving the $i$th equation in $Lx = b$ for $x_i$:

$$x_i = \left( b_i - \sum_{j=1}^{i-1} \ell_{ij}x_j \right) \Bigg/ \ell_{ii}$$

If this is evaluated for $i = 1{:}n$ a complete specification of $x$ is obtained. Note that at the $i$th stage the dot product of $L(i, 1{:}i-1)$ and $x(1{:}i-1)$ is required. Since $b_i$ only is involved in the formula for $x_i$, the former may be overwritten by the latter:

**Algorithm 3.1.1 (Forward Substitution: Row Version)** Suppose $L \in \mathbb{R}^{n\times n}$ is lower triangular and $b \in \mathbb{R}^n$. This algorithm overwrites $b$ with the solution to $Lx = b$. $L$ is assumed to be nonsingular.

$b(1) = b(1)/L(1,1)$
**for** $i = 2{:}n$
  $b(i) = (b_(i) - L(i, 1{:}i-1)b(1{:}i-1))/L(i,i)$
**end**

This algorithm requires $n^2$ flops. Note that $L$ is accessed by row. The computed solution $\hat{x}$ satisfies:

$$(L + F)\hat{x} = b \qquad |F| \leq n\mathbf{u}|L| + O(\mathbf{u}^2) \tag{3.1.1}$$

This result is established in Forsythe and Moler (SLE, pp. 104-5). It says that the computed solution exactly satisfies a slightly perturbed system. Moreover, each entry in the perturbing matrix $F$ is small relative to the corresponding element of $L$.

### 3.1.2 Back Substitution

The analogous algorithm for upper triangular systems $Ux = b$ is called *back-substitution*. The recipe for $x_i$ is prescribed by

$$x_i = \left( b_i - \sum_{j=i+1}^{n} u_{ij} x_j \right) \Big/ u_{ii}$$

and once again $b_i$ can be overwritten by $x_i$.

**Algorithm 3.1.2 (Back Substitution: Row Version)** If $U \in \mathbb{R}^{n \times n}$ is upper triangular and $b \in \mathbb{R}^n$ then the following algorithm overwrites $b$ with the solution to $Ux = b$. $U$ is assumed to be nonsingular.

$b(n) = b(n)/U(n,n)$
**for** $i = n-1{:}-1{:}1$
  $b(i) = (b(i) - U(i, i+1{:}n)b(i+1{:}n))/U(i,i)$
**end**

This algorithm requires $n^2$ flops and accesses $U$ by row. The computed solution $\hat{x}$ obtained by the algorithm can be shown to satisfy

$$(U + F)\hat{x} = b \qquad |F| \le n\mathbf{u}|U| + O(\mathbf{u}^2). \tag{3.1.2}$$

### 3.1.3 Column Oriented Versions

Column oriented versions of the above procedures can be obtained by reversing loop orders. To understand what this means from the algebraic point of view, consider forward substitution. Once $x_1$ is resolved, it can be removed from equations 2 through $n$ and we proceed with the reduced system $L(2{:}n, 2{:}n)x(2{:}n) = b(2{:}n) - x(1)L(2{:}n, 1)$. We then compute $x_2$ and remove it from equations 3 through $n$, etc. Thus, if this approach is applied to

$$\begin{bmatrix} 2 & 0 & 0 \\ 1 & 5 & 0 \\ 7 & 9 & 8 \end{bmatrix} \begin{bmatrix} x_1 \\ x_2 \\ x_3 \end{bmatrix} = \begin{bmatrix} 6 \\ 2 \\ 5 \end{bmatrix}$$

we find $x_1 = 3$ and then deal with the 2-by-2 system

$$\begin{bmatrix} 5 & 0 \\ 9 & 8 \end{bmatrix} \begin{bmatrix} x_2 \\ x_3 \end{bmatrix} = \begin{bmatrix} 2 \\ 5 \end{bmatrix} - 3 \begin{bmatrix} 1 \\ 7 \end{bmatrix} = \begin{bmatrix} -1 \\ -16 \end{bmatrix}.$$

Here is the complete procedure with overwriting.

**Algorithm 3.1.3 (Forward Substitution: Column Version)** Suppose $L \in \mathbb{R}^{n \times n}$ is lower triangular and $b \in \mathbb{R}^n$. This algorithm overwrites $b$ with the solution to $Lx = b$. $L$ is assumed to be nonsingular.

```
for j = 1:n − 1
    b(j) = b(j)/L(j, j)
    b(j + 1:n) = b(j + 1:n) − b(j)L(j + 1:n, j)
end
b(n) = b(n)/L(n, n)
```

It is also possible to obtain a column-oriented saxpy procedure for back-substitution.

**Algorithm 3.1.4 (Back Substitution: Column Version)** Suppose $U \in \mathbb{R}^{n \times n}$ is upper triangular and $b \in \mathbb{R}^n$. This algorithm overwrites $b$ with the solution to $Ux = b$. $U$ is assumed to be nonsingular.

```
for j = n: − 1:2
    b(j) = b(j)/U(j, j)
    b(1:j − 1) = b(1:j − 1) − b(j)U(1:j − 1, j)
end
b(1) = b(1)/U(1, 1)
```

Note that the dominant operation in both Algorithms 3.1.3 and 3.1.4 is the saxpy operation. The roundoff behavior of these saxpy implementations is essentially the same as for the dot product versions.

The accuracy of a computed solution to a triangular system is often surprisingly good. See Higham (1988d).

### 3.1.4 Multiple Right Hand Sides

Consider the problem of computing a solution $X \in \mathbb{R}^{n \times q}$ to $LX = B$ where $L \in \mathbb{R}^{n \times n}$ is lower triangular and $B \in \mathbb{R}^{n \times q}$. This is the *multiple right hand side* forward substitution problem. We show that such a problem can be solved by a block algorithm that is rich in matrix multiplication assuming that $q$ and $n$ are large enough. This turns out to be important in subsequent sections where various block factorization schemes are discussed. We mention that although we are considering here just the lower triangular problem, everything we say applies to the upper triangular case as well.

To develop a block forward substitution algorithm we partition the equation $LX = B$ as follows:

$$\begin{bmatrix} L_{11} & 0 & \cdots & 0 \\ L_{21} & L_{22} & \cdots & 0 \\ \vdots & \vdots & \ddots & \vdots \\ L_{N1} & L_{N2} & \cdots & L_{NN} \end{bmatrix} \begin{bmatrix} X_1 \\ X_2 \\ \vdots \\ X_N \end{bmatrix} = \begin{bmatrix} B_1 \\ B_2 \\ \vdots \\ B_N \end{bmatrix}. \qquad (3.1.3)$$

Assume that the diagonal blocks are square. Paralleling the development of Algorithm 3.1.3, we solve the system $L_{11}X_1 = B_1$ for $X_1$ and then remove

$X_1$ from block equations 2 through $N$:

$$\begin{bmatrix} L_{22} & 0 & \cdots & 0 \\ L_{32} & L_{33} & \cdots & 0 \\ \vdots & \vdots & \ddots & \vdots \\ L_{N2} & L_{N3} & \cdots & L_{NN} \end{bmatrix} \begin{bmatrix} X_2 \\ X_3 \\ \vdots \\ X_N \end{bmatrix} = \begin{bmatrix} B_2 - L_{21}X_1 \\ B_3 - L_{31}X_1 \\ \vdots \\ B_N - L_{N1}X_1 \end{bmatrix}$$

Continuing in this way we obtain the following block saxpy forward elimination scheme:

```
for j = 1:N
    Solve L_jj X_j = B_j
    for i = j + 1:N                                  (3.1.4)
        B_i = B_i - L_ij X_j
    end
end
```

Notice that the $i$-loop oversees a single block saxpy update of the form

$$\begin{bmatrix} B_{j+1} \\ \vdots \\ B_N \end{bmatrix} = \begin{bmatrix} B_{j+1} \\ \vdots \\ B_N \end{bmatrix} - \begin{bmatrix} L_{j+1,j} \\ \vdots \\ L_{N,j} \end{bmatrix} X_j \,.$$

For this to be handled as a matrix multiplication in a given architecture it is clear that the blocking in (3.1.3) must give sufficiently "big" $X_j$. Let us assume that this is the case if each $X_j$ has at least $r$ rows. This can be accomplished if $N = \text{ceil}(n/r)$ and $X_1, \ldots, X_{N-1} \in \mathbb{R}^{r \times q}$ and $X_N \in \mathbb{R}^{(n-(N-1)r) \times q}$.

### 3.1.5 The Level-3 Fraction

It is handy to adopt a measure that quantifies the amount of matrix multiplication in a given algorithm. To this end we define the *level-3 fraction* of an algorithm to be the fraction of flops that occur in the context of matrix multiplication. We call such flops *level-3 flops.*

Let us determine the the level-3 fraction for (3.1.4) with the simplifying assumption that $n = rN$. (The same conclusions hold with the unequal blocking described above.) Because there are $N$ applications of $r$-by-$r$ forward elimination (the level-2 portion of the computation) and $n^2$ flops overall, the level-3 fraction is approximately given by

$$1 - \frac{Nr^2}{n^2} = 1 - \frac{1}{N}$$

Thus, for large $N$ almost all flops are level-3 flops and it makes sense to choose $N$ as large as possible subject to the constraint that the underlying architecture can achieve a high level of performance when processing block saxpy's of width at least $r = n/N$.

### 3.1.6 Non-square Triangular System Solving

The problem of solving nonsquare, $m$-by-$n$ triangular systems deserves some mention. Consider first the lower triangular case when $m \geq n$, i.e.,

$$\begin{bmatrix} L_{11} \\ L_{21} \end{bmatrix} x = \begin{bmatrix} b_1 \\ b_2 \end{bmatrix} \qquad \begin{array}{ll} L_{11} \in \mathbb{R}^{n \times n} & b_1 \in \mathbb{R}^n \\ L_{21} \in \mathbb{R}^{(m-n) \times n} & b_2 \in \mathbb{R}^{m-n} \end{array}$$

Assume that $L_{11}$ is lower triangular, and nonsingular. If we apply forward elimination to $L_{11}x = b_1$ then $x$ solves the system provided $L_{21}(L_{11}^{-1}b_1) = b_2$. Otherwise, there is no solution to the overall system. In such a case least squares minimization may be appropriate. See Chapter 5.

Now consider the lower triangular system $Lx = b$ when the number of columns $n$ exceeds the number of rows $m$. In this case apply forward substitution to the square system $L(1{:}m, 1{:}m)x(1{:}m, 1{:}m) = b$ and prescribe an arbitrary value for $x(m + 1{:}n)$. See §5.7 for additional comments on systems that have more unknowns than equations.

The handling of nonsquare upper triangular systems is similar. Details are left to the reader.

### 3.1.7 Unit Triangular Systems

A *unit* triangular matrix is a triangular matrix with ones on the diagonal. Many of the triangular matrix computations that follow have this added bit of structure. It clearly poses no difficulty in the above procedures.

### 3.1.8 The Algebra of Triangular Matrices

For future reference we list a few properties about products and inverses of triangular and unit triangular matrices.

- The inverse of an upper (lower) triangular matrix is upper (lower) triangular.
- The product of two upper (lower) triangular matrices is upper (lower) triangular.
- The inverse of a unit upper (lower) triangular matrix is unit upper (lower) triangular.
- The product of two unit upper (lower) triangular matrices is unit upper (lower) triangular.

**Problems**

**P3.1.1** Give an algorithm for computing a nonzero $z \in \mathbb{R}^n$ such that $Uz = 0$ where $U \in \mathbb{R}^{n \times n}$ is upper triangular with $u_{nn} = 0$ and $u_{11} \cdots u_{n-1,n-1} \neq 0$.

**P3.1.2** Discuss how the determinant of a square triangular matrix could be computed with minimum risk of overflow and underflow.

**P3.1.3** Rewrite Algorithm 3.1.4 given that $U$ is stored by column in a length $n(n+1)/2$ array $u.vec$.

**P3.1.4** Write a detailed version of (3.1.4). Do not assume that $N$ divides $n$.

**P3.1.5** Prove all the facts about triangular matrices that are listed at the end of the section

**Notes and References for Sec. 3.1**

Fortran codes for solving triangular linear systems may be found in Linpack (chapter 6). The accuracy of triangular system solvers is analyzed in

N.J. Higham (1988d). "The Accuracy of Solutions to Triangular Systems," Report 158, University of Manchester, Department of Mathematics. (To appear, *SIAM J. Numer. Anal.*.)

## 3.2 The LU Factorization

As we have just seen, triangular systems are "easy" to solve. The idea behind Gaussian elimination is to convert a given system $Ax = b$ to an equivalent triangular system. The conversion is achieved by taking appropriate linear combinations of the equations. For example, in the system

$$\begin{aligned} 3x_1 + 5x_2 &= 9 \\ 6x_1 + 7x_2 &= 4 \end{aligned}$$

if we multiply the first equation by 2 and subtract it from the second we obtain

$$\begin{aligned} 3x_1 + 5x_2 &= 9 \\ -3x_2 &= -14 \end{aligned}$$

This is $n = 2$ Gaussian elimination. Our objective in this section is to give a complete specification of this central procedure and to describe what it does in the language matrix factorizations. This means showing that the algorithm computes a unit lower triangular matrix $L$ and an upper triangular matrix $U$ so that $A = LU$, e.g.,

$$\begin{bmatrix} 3 & 5 \\ 6 & 7 \end{bmatrix} = \begin{bmatrix} 1 & 0 \\ 2 & 1 \end{bmatrix} \begin{bmatrix} 3 & 5 \\ 0 & -3 \end{bmatrix}.$$

The solution to the original $Ax = b$ problem is then found by a two step triangular solve process:

$$Ly = b, \quad Ux = y \qquad \Longrightarrow \qquad Ax = LUx = Ly = b\,.$$

The LU factorization is a "high-level" algebraic description of Gaussian elimination. Expressing the outcome of a matrix algorithm in the "language" of matrix factorizations is a worthwhile activity. It facilitates generalization and highlights connections between algorithms that may appear very different at the scalar level.

### 3.2.1 Gauss Transformations

To obtain a factorization description of Gaussian elimination we need a matrix description of the zeroing process. At the $n = 2$ level if $x_1 \neq 0$ and $\tau = x_2/x_1$, then

$$\begin{bmatrix} 1 & 0 \\ -\tau & 1 \end{bmatrix} \begin{bmatrix} x_1 \\ x_2 \end{bmatrix} = \begin{bmatrix} x_1 \\ 0 \end{bmatrix}$$

More generally, suppose $x \in \mathbb{R}^n$ with $x_k \neq 0$. If

$$\tau^T = (\underbrace{0, \ldots, 0}_{k}, \tau_{k+1}, \ldots, \tau_n) \qquad \tau_i = \frac{x_i}{x_k} \qquad i = k+1{:}n$$

and we define

$$M_k = I - \tau e_k^T \tag{3.2.1}$$

then

$$M_k x = \begin{bmatrix} 1 & \cdots & 0 & 0 & \cdots & 0 \\ \vdots & \ddots & \vdots & \vdots & & \vdots \\ 0 & & 1 & 0 & & 0 \\ 0 & & -\tau_{k+1} & 1 & & 0 \\ \vdots & \vdots & \vdots & \vdots & \ddots & \vdots \\ 0 & \cdots & -\tau_n & 0 & \cdots & 1 \end{bmatrix} \begin{bmatrix} x_1 \\ \vdots \\ x_k \\ x_{k+1} \\ \vdots \\ x_n \end{bmatrix} = \begin{bmatrix} x_1 \\ \vdots \\ x_k \\ 0 \\ \vdots \\ 0 \end{bmatrix}.$$

The matrix $M_k$ is a *Gauss transformation.* It is unit lower triangular. The components of $\tau(k+1{:}n)$ are the *multipliers.* The vector $\tau$ is called the *Gauss vector.*

To encapsulate these ideas, here is a function that computes the vector of multipliers:

**Algorithm 3.2.1** If $x \in \mathbb{R}^n$ and $x_1$ is nonzero then this function computes an $n-1$ vector $t$ such that if $M$ is a Gauss transform with $M(2{:}n, 1) = -t$ and $y = Mx$, then $y(2{:}n) = 0$.

```
function: t = gauss(x)
    n = length(x)
    t = x(2:n)/x(1)
end gauss
```

This algorithm involves $n - 1$ flops.

### 3.2.2 Applying Gauss Transformations

Multiplication by a Gauss transformation is particularly simple. If $C \in \mathbb{R}^{n \times r}$ and $M_k = I - \tau e_k^T$, then

$$M_k C \;=\; (I - \tau e_k^T)C \;=\; C - \tau(e_k^T C)$$

is an outer product update. Moreover, because $\tau(1{:}k) = 0$ only $C(k+1{:}n, :)$ is affected. We therefore obtain

**Algorithm 3.2.2** If $C \in \mathbb{R}^{n \times r}$ and $M$ is an $n$-by-$n$ Gauss transform with $M(2{:}n, 1) = -t$, then this function overwrites $C$ with $MC$.

```
function: C = gauss.app(C, t)
    n = rows(C)
    C(2:n, :) = C(2:n, :) − tC(1, :)
end gauss.app
```

This algorithm requires $2(n-1)r$ flops. Note that if $M(k+1{:}n, k) = -t$ then $C = \textbf{gauss.app}(C(k{:}n, :), t)$ overwrites $C$ with $M_k C$.

**Example 3.2.1**

$$C \;=\; \begin{bmatrix} 1 & 4 & 7 \\ 2 & 5 & 8 \\ 3 & 6 & 10 \end{bmatrix}, \; t \;=\; \begin{bmatrix} 1 \\ -1 \end{bmatrix} \;\Rightarrow\; MC = \begin{bmatrix} 1 & 4 & 7 \\ 1 & 1 & 1 \\ 4 & 10 & 17 \end{bmatrix}$$

### 3.2.3 Roundoff Properties of Gauss Transforms

If $\hat{t}$ denotes the computed version of $t$ in **gauss**, then it is easy to verify that

$$\hat{t} = t + e \qquad |e| \le \mathbf{u}|t|.$$

If $\hat{t}$ is used in **gauss.app** and $fl((I - \hat{t}e_1^T)C)$ denotes the resulting update, then

$$fl\left((I - \hat{t}e_1^T)C\right) \;=\; (I - te_1^T)C + E,$$

where

$$|E| \;\le\; 3\mathbf{u}\left(|C| + |t||C(1,:)|\right) + O(\mathbf{u}^2).$$

Clearly, if $t$ is big then the errors in the update may be large in comparison to $|C|$. For this reason, care must be exercised when Gauss transformations are employed, a matter that is pursued in §3.4.

### 3.2.4 Upper Triangularizing

Assume that $A \in \mathbb{R}^{n \times n}$. Gauss transformations $M_1, \ldots, M_{n-1}$ can usually be found such that $M_{n-1} \cdots M_2 M_1 A = U$ is upper triangular. To see this we first look at the $n = 3$ case. Suppose

$$A = \begin{bmatrix} 1 & 4 & 7 \\ 2 & 5 & 8 \\ 3 & 6 & 10 \end{bmatrix}.$$

If

$$M_1 = \begin{bmatrix} 1 & 0 & 0 \\ -2 & 1 & 0 \\ -3 & 0 & 1 \end{bmatrix},$$

then

$$M_1 A = \begin{bmatrix} 1 & 4 & 7 \\ 0 & -3 & -6 \\ 0 & -6 & -11 \end{bmatrix}.$$

Likewise,

$$M_2 = \begin{bmatrix} 1 & 0 & 0 \\ 0 & 1 & 0 \\ 0 & -2 & 1 \end{bmatrix} \quad \Rightarrow \quad M_2(M_1 A) = \begin{bmatrix} 1 & 4 & 7 \\ 0 & -3 & -6 \\ 0 & 0 & 1 \end{bmatrix}$$

Extrapolating from this example observe that during the $k$th step

- We are confronted with a matrix $A^{(k-1)} = M_{k-1} \cdots M_1 A$ that is upper triangular in columns 1 to $k-1$.
- The multipliers in $M_k$ are based on $A^{(k-1)}(k+1{:}n, k)$. In particular, we need $a_{kk}^{(k-1)} \neq 0$ to proceed.

Noting that complete upper triangularization is achieved after $n-1$ steps we therefore obtain

$$\begin{array}{l} k = 1 \\ \textbf{while } A(k,k) \neq 0 \ \wedge \ k \leq n-1 \\ \qquad t = \textbf{gauss}(A(k{:}n, k)) \\ \qquad A(k{:}n, :) = \textbf{gauss.app}(A(k{:}n, :), t) \\ \qquad k = k + 1 \\ \textbf{end} \end{array} \tag{3.2.2}$$

The entry $A(k,k)$ must be checked to avoid a zero divide in **gauss** where the multipliers $A(i,k)/A(k,k)$ are computed. The $A(k,k)$ are referred to as the *pivots* and their relative magnitude turns out to be critically important.

### 3.2.5 The LU Factorization

In matrix language, if (3.2.2) terminates with $k = n$ then it computes Gauss transforms $M_1, \ldots, M_{n-1}$ such that $M_{n-1} \cdots M_1 A = U$ is upper triangular. It is easy to check that if $M_k = I - \tau^{(k)} e_k^T$ then its inverse is prescribed by $M_k^{-1} = I + \tau^{(k)} e_k^T$ and so

$$A = LU \tag{3.2.3}$$

where

$$L = M_1^{-1} \cdots M_{n-1}^{-1}. \tag{3.2.4}$$

It is clear that $L$ is a unit lower triangular matrix because each $M_k^{-1}$ is unit lower triangular. The factorization (3.2.4) is called the *LU factorization* of $A$.

As evidenced by the need to check for zero pivots in (3.2.2), the LU factorization need not exist. For example, it is impossible to find $l_{ij}$ and $u_{ij}$ so

$$\begin{bmatrix} 1 & 2 & 3 \\ 2 & 4 & 7 \\ 3 & 6 & 3 \end{bmatrix} = \begin{bmatrix} 1 & 0 & 0 \\ \ell_{21} & 1 & 0 \\ \ell_{31} & \ell_{32} & 1 \end{bmatrix} \begin{bmatrix} u_{11} & u_{12} & u_{13} \\ 0 & u_{22} & u_{23} \\ 0 & 0 & u_{33} \end{bmatrix}.$$

To see this equate entries and observe that we must have $u_{11} = 1$, $u_{12} = 2$, $\ell_{21} = 2$, $u_{22} = 0$, and $\ell_{31} = 1$. But when we then look at the (3,2) entry we obtain the contradictory equation $6 = \ell_{31} u_{12} + \ell_{32} u_{22} = 4$.

As we now show, a zero pivot in (3.2.2) can be identified with a singular leading principal submatrix.

**Theorem 3.2.1** *$A \in \mathbb{R}^{n \times n}$ has an LU factorization if* $\det(A(1{:}k, 1{:}k)) \neq 0$ *for $k = 1{:}n-1$. If the LU factorization exists and $A$ is nonsingular, then the LU factorization is unique and* $\det(A) = u_{11} \cdots u_{nn}$.

**Proof.** Suppose $k-1$ steps in (3.2.2) have been executed. At the beginning of step $k$ the matrix $A$ has been overwritten by $M_{k-1} \cdots M_1 A = A^{(k-1)}$. Note that $a_{kk}^{(k-1)}$ is the $k$th pivot. Since the Gauss transformations are unit lower triangular it follows by looking at the leading $k$-by-$k$ portion of this equation that $\det(A(1{:}k, 1{:}k)) = a_{11}^{(k-1)} \cdots a_{kk}^{(k-1)}$. Thus, if $A(1{:}k, 1{:}k)$ is nonsingular then the $k$th pivot is nonzero.

As for uniqueness, if $A = L_1 U_1$ and $A = L_2 U_2$ are two LU factorizations of a nonsingular $A$, then $L_2^{-1} L_1 = U_2 U_1^{-1}$. Since $L_2^{-1} L_1$ is unit lower triangular and $U_2 U_1^{-1}$ is upper triangular, it follows that both of these matrices must equal the identity. Hence, $L_1 = L_2$ and $U_1 = U_2$.

Finally, if $A = LU$ then $\det(A) = \det(LU) = \det(L)\det(U) = \det(U) = u_{11} \cdots u_{nn}$. □

### 3.2.6 Some Practical Details

From the practical point of view there are several improvements that can be made to (3.2.2). First, because zeros have already been introduced in columns 1 through $k-1$, the Gauss transform update need only be applied to columns $k$ through $n$. Of course, we need not even apply the $k$th Gauss transform to $A(:,k)$ since we know the result. So the efficient thing to do is to invoke **gauss.app** as follows:

$$A(k{:}n, k+1{:}n) = \textbf{gauss.app}(A(k{:}n, k+1{:}n), t)$$

Another worthwhile observation is that the multipliers associated with $M_k$ can be stored in the locations that they zero, i.e., $A(k+1{:}n, k)$. With these changes we obtain the following version of (3.2.2):

**Algorithm 3.2.3 (Outer Product Gaussian Elimination)** Suppose $A \in \mathbb{R}^{n\times n}$ has the property that $A(1{:}k, 1{:}k)$ is nonsingular for $k = 1{:}n-1$. This algorithm computes the factorization $M_{n-1}\cdots M_1 A = U$ where $U$ is upper triangular and each $M_k$ is a Gauss transform. $U$ is stored in the upper triangle of $A$. The multipliers associated with $M_k$ are stored in $A(k+1{:}n, k)$, i.e., $A(k+1{:}n, k) = -M_k(k+1{:}n, k)$.

**for** $k = 1{:}n-1$
  $t = \textbf{gauss}(A(k{:}n, k))$
  $A(k+1{:}n, k) = t$
  $A(k{:}n, k+1{:}n) = \textbf{gauss.app}(A(k{:}n, k+1{:}n), t)$
**end**

This algorithm involves $2n^3/3$ flops and it is the classical formulation of *Gaussian Elimination.* Note that each pass through the $k$-loop involves an outer product.

### 3.2.7 Where is L?

Algorithm 3.2.3 represents $L$ in terms of the multipliers. In particular, if $t^{(k)}$ is the vector of multipliers associated with $M_k$ then upon termination, $A(k+1{:}n, k) = t^{(k)}$. One of the more happy "coincidences" in matrix computations is that if $L = M_1^{-1}\cdots M_{n-1}^{-1}$ then $L(k+1{:}k, k) = t^{(k)}$. This follows from a careful look at the product that defines $L$. Indeed,

$$L = \left(I + \tau^{(1)}e_1^T\right)\cdots\left(I + \tau^{(n-1)}e_{n-1}^T\right) \;=\; I + \sum_{k=1}^{n-1}\tau^{(k)}e_k^T\,.$$

Since $A(k+1{:}n, k)$ houses the $k$th vector of multipliers $t^{(k)}$, it follows that $A(i,k)$ houses $\ell_{ik}$ for all $i > k$.

### 3.2.8 Solving a Linear System

Once $A$ has been factored via Algorithm 3.2.3 then $L$ and $U$ are represented in the array $A$. We can then solve the system $Ax = b$ via the triangular systems $Ly = b$ and $Ux = y$ by using the methods of §3.1.

**Example 3.2.2** If Algorithm 3.2.3 is applied to

$$A = \begin{bmatrix} 1 & 4 & 7 \\ 2 & 5 & 8 \\ 3 & 6 & 10 \end{bmatrix} = \begin{bmatrix} 1 & 0 & 0 \\ 2 & 1 & 0 \\ 3 & 2 & 1 \end{bmatrix} \begin{bmatrix} 1 & 4 & 7 \\ 0 & -3 & -6 \\ 0 & 0 & 1 \end{bmatrix}$$

then upon completion,

$$A = \begin{bmatrix} 1 & 4 & 7 \\ 2 & -3 & -6 \\ 3 & 2 & 1 \end{bmatrix}$$

If $b = (1,1,1)^T$ then $y = (1,-1,0)^T$ solves $Ly = b$ and $x = (-1/3, 1/3, 0)^T$ solves $Ux = y$.

### 3.2.9 Gaxpy LU

Gaussian elimination, like matrix multiplication, is a triple loop procedure that can be arranged in several ways. Algorithm 3.2.3 corresponds to the "$kij$" version of Gaussian elimination for if we expand the function calls and do a bit of rearranging we obtain

```
for k = 1:n - 1
    A(k + 1:n, k) = A(k + 1:n, k)/A(k, k)
    for j = k + 1:n
        for i = k + 1:n
            A(i, j) = A(i, j) - A(i, k)A(k, j)
        end
    end
end
```

Notice that the computation is dominated by outer product updates.

We next derive a $jki$ version of Gaussian elimination that is rich in gaxpy's and forward elimination. In this new formulation the Gauss transformations are not immediately applied to $A$ as they are in the outer product version. Instead, their application is delayed. The original $A(:,j)$ is untouched until step $j$. At that point in the algorithm $A(:,j)$ is overwritten by $M_{j-1} \cdots M_1 A(:,j)$. The $j$th Gauss transformation is then computed.

To be precise, suppose $1 \le j \le n-1$ and assume that $L(:,1{:}j-1)$ and $U(1{:}j-1, 1{:}j-1)$ are known, i.e., we know the first $j-1$ columns of $L$ and $U$. To get the $j$th columns of $L$ and $U$ we equate $j$th columns in the equation $A = LU$: $A(:,j) = LU(:,j)$. From this we conclude that

$$A(1{:}j-1, j) \quad = \quad L(1{:}j-1, 1{:}j-1)U(1{:}j-1, j)$$

$$A(j{:}n, j) = \sum_{k=1}^{j} L(j{:}n, k)U(k, j).$$

The first equation is a lower triangular system that can be solved for the vector $U(1{:}j-1, j)$. Once this is accomplished, the second equation can be rearranged to produce recipes for $U(j,j)$ and $L(j+1{:}n, j)$ for if we set

$$\begin{aligned} v(j{:}n, j) &= A(j{:}n, j) - \sum_{k=1}^{j-1} L(j{:}n, k)U(k, j) \\ &= A(j{:}n, j) - L(j{:}n, 1{:}j-1)U(1{:}j-1, j) \end{aligned}$$

then $U(j,j) = v(j)$ and $L(j+1{:}n, j) = v(j+1{:}n)/U(j,j)$. Thus, $L(j+1{:}n, j)$ is a scaled gaxpy and we obtain

$$\begin{aligned} &\textbf{for } j = 1{:}n \\ &\qquad \text{Solve } L(1{:}j-1, 1{:}j-1)U(1{:}j-1, j) = A(1{:}j-1, j). \\ &\qquad v(j{:}n) = A(j{:}n, j) - L(j{:}n, 1{:}j-1)U(1{:}j-1, j) \\ &\qquad U(j,j) = v(j);\ \ L(j+1{:}n, j) = v(j+1{:}n)/U(j,j) \\ &\textbf{end} \end{aligned} \tag{3.2.5}$$

If we expand this and overwrite $A$ with $L$ and $U$ we get

**Algorithm 3.2.4 (Gaxpy Gaussian Elimination)** Let $A \in \mathbb{R}^{n\times n}$ have the property that $A(1{:}k, 1{:}k)$ is nonsingular for $k = 1{:}n-1$. This algorithm computes the factorization $A = LU$ where $L$ is unit lower triangular and $U$ is upper triangular. If $i > j$ then $A(i,j)$ houses $L(i,j)$. If $i \leq j$ then $A(i,j)$ houses $U(i,j)$.

$$\begin{aligned} &\textbf{for } j = 1{:}n \\ &\qquad \{ \text{ Solve } L(1{:}j-1, 1{:}j-1)U(1{:}j-1, j) = A(1{:}j-1, j) \ \} \\ &\qquad \textbf{for } k = 1{:}j-1 \\ &\qquad\qquad \textbf{for } i = k+1{:}j-1 \\ &\qquad\qquad\qquad A(i,j) = A(i,j) - A(i,k)A(k,j) \\ &\qquad\qquad \textbf{end} \\ &\qquad \textbf{end} \\ &\qquad \{ \ v(j{:}n) = A(j{:}n, j) - L(j{:}n, 1{:}j-1)U(1{:}j-1, j) \ \} \\ &\qquad \textbf{for } k = 1{:}j-1 \\ &\qquad\qquad \textbf{for } i = j{:}n \\ &\qquad\qquad\qquad A(i,j) = A(i,j) - A(i,k)A(k,j) \\ &\qquad\qquad \textbf{end} \\ &\qquad \textbf{end} \\ &\qquad \{ \ L(j+1{:}n, j) = v(j+1{:}n)/v(j) \ \} \\ &\qquad A(j+1{:}n, j) = A(j+1{:}n, j)/A(j,j) \\ &\textbf{end} \end{aligned}$$

This algorithm involves $2n^3/3$ flops.

### 3.2.10 Crout and Doolittle

We mention a *jik* formulation of the above procedures results in a dot product method for the LU factorization. This more or less corresponds to what are known as the Crout and Doolittle variants of Gaussian elimination.

### 3.2.11 Block LU

It is possible to organize Gaussian elimination so that matrix multiplication becomes the dominant operation. The key to the derivation of this block procedure is to partition $A \in \mathbb{R}^{n\times n}$ as follows

$$A = \begin{bmatrix} A_{11} & A_{12} \\ A_{21} & A_{22} \end{bmatrix} \begin{matrix} r \\ n-r \end{matrix}$$
$$\qquad\quad r \quad\; n-r$$

where $r$ is a blocking parameter. Suppose we compute the LU factorization $L_{11}U_{11} = A_{11}$ and then solve the multiple right hand side triangular systems $L_{11}U_{12} = A_{12}$ and $L_{21}U_{11} = A_{21}$ for $U_{12}$ and $L_{21}$ respectively. It follows that

$$\begin{bmatrix} A_{11} & A_{12} \\ A_{21} & A_{22} \end{bmatrix} = \begin{bmatrix} L_{11} & 0 \\ L_{21} & I_{n-r} \end{bmatrix} \begin{bmatrix} I_r & 0 \\ 0 & \tilde{A} \end{bmatrix} \begin{bmatrix} U_{11} & U_{12} \\ 0 & I_{n-r} \end{bmatrix} \qquad (3.2.6)$$

where $\tilde{A} = A_{22} - L_{21}U_{12}$. Notice that if $r \ll n$ then most flops occur in the level-3 update of $A_{22}$. We can repeat the process on the reduced matrix $\tilde{A}$. Continuing in this way we obtain

**Algorithm 3.2.5 (Block Outer Product LU)** Suppose $A \in \mathbb{R}^{n\times n}$ and that $\det(A(1{:}k, 1{:}k)$ is nonzero for $k = 1{:}n-1$. Assume that $r$ satisfies $1 \le r \le n$. The following algorithm computes $A = LU$ via rank $r$ updates. Upon completion, $A(i,j)$ is overwritten with $L(i,j)$ for $i > j$ and $A(i,j)$ is overwritten with $U(i,j)$ if $j \ge i$.

$\lambda = 1$
**while** $\lambda \le n$
  $\mu = \min(n, \lambda + r - 1)$
  Use Algorithm 3.2.3 to overwrite $A(\lambda{:}\mu, \lambda{:}\mu) = \tilde{L}\tilde{U}$ with $\tilde{L}$ and $\tilde{U}$.
  Solve: $\tilde{L}Z = A(\lambda{:}\mu, \mu+1{:}n)$. Overwrite: $A(\lambda{:}\mu, \mu+1{:}n) \leftarrow Z$.
  Solve: $W\tilde{U} = A(\mu+1{:}n, \lambda{:}\mu)$. Overwrite: $A(\mu+1{:}n, \lambda{:}\mu) \leftarrow W$
  $A(\mu+1{:}n, \mu+1{:}n) = A(\mu+1{:}n, \mu+1{:}n) - WZ$
  $\lambda = \mu + 1$
**end**

This algorithm involves $2n^3/3$ flops.

Recalling the discussion in §3.1.5, let us consider the level-3 fraction for this procedure. If $r \ll n$ (as is reasonable) then we may assume for clarity that $n = rN$. The only flops that are not level-3 flops occur in the context of the $r$-by-$r$ LU factorizations $A(\lambda{:}\mu, \lambda{:}\mu) = \tilde{L}\tilde{U}$. Since there are $N$ such systems solved in the overall computation we see that the level-3 fraction is given by

$$1 - \frac{N(2r^3/3)}{2n^3/3} = 1 - \frac{1}{N^2}.$$

It is also possible to develop a gaxpy-based block LU factorization. Note that after $r$ steps of Algorithm 3.2.4 we have the matrices $L_{11}$, $L_{21}$, and $U_{11}$ in (3.2.6). We then solve $L_{11}U_{12} = A_{12}$ for $U_{12}$ and compute $\tilde{A} = A_{22} - L_{21}U_{12}$. Repeating this process we obtain

**Algorithm 3.2.6 (Block Gaxpy LU)** Suppose $A \in \mathbb{R}^{n\times n}$ and that $\det(A(1{:}k, 1{:}k))$ is nonzero for $k = 1{:}n-1$. Assume that the integer $r$ satisfies $1 \le r \le n$. The following algorithm computes $A = LU$ via rank $r$ updates. Upon completion, $A(i,j)$ is overwritten with $L(i,j)$ for $i > j$ and $A(i,j)$ is overwritten with $U(i,j)$ if $j \ge i$.

```
λ = 1
while λ ≤ n
      μ = min(n, λ + r − 1)
      r₁ = μ − λ + 1
      Perform r₁ steps of Algorithm 3.2.4 on A(λ:n, λ:n).
      Overwrite A(λ:μ, μ + 1:n) with solution to
            A(λ:μ, λ:μ)Z = A(λ:μ, μ + 1:n).
      A(μ + 1:n, μ + 1:n) = A(μ + 1:n, μ + 1:n)
            −A(μ + 1:n, λ:μ)A(λ:μ, μ + 1:n)
      λ = μ + 1
end
```

This algorithm requires $2n^3/3$ flops.

The only flops in Algorithm 3.2.6 that are not level-3 flops are those that are associated with the calls to Algorithm 3.2.4. It is not hard to show as a consequence that the level-3 fraction for Algorithm 3.2.6 is given by $1 - 3/(2N)$ where again we assume $n = rN$. Thus, the level-3 fraction for this procedure is less than the level-3 fraction for Algorithm 3.2.5. However, if $N$ is large there is not a significant difference between the two procedures from the level-3 point point of view.

### 3.2.12 The LU Factorization of a Rectangular Matrix

The LU factorization of a rectangular matrix $A \in \mathbb{R}^{m\times n}$ can also be performed. The $m > n$ case is illustrated by

$$\begin{bmatrix} 1 & 2 \\ 3 & 4 \\ 5 & 6 \end{bmatrix} = \begin{bmatrix} 1 & 0 \\ 3 & 1 \\ 5 & 2 \end{bmatrix} \begin{bmatrix} 1 & 2 \\ 0 & -2 \end{bmatrix}$$

while

$$\begin{bmatrix} 1 & 2 & 3 \\ 4 & 5 & 6 \end{bmatrix} = \begin{bmatrix} 1 & 0 \\ 4 & 1 \end{bmatrix} \begin{bmatrix} 1 & 2 & 3 \\ 0 & -3 & -6 \end{bmatrix}$$

depicts the $m < n$ situation. The LU factorization of $A \in \mathbb{R}^{m\times n}$ is guaranteed to exist if $A(1:k, 1:k)$ is nonsingular for $k = 1:\min(m, n)$.

The square LU factorization algorithms above need only minor modification to handle the rectangular case. For example, to handle the $m \geq n$ case we modify (3.2.5) as follows:

**for** $j = 1:\min(m-1, n)$
    Solve $L(1:j, 1:j)U(1:j, j) = A(1:j, j)$ for $U(1:j, j)$
    $v(j+1:m) = (A(j+1:m, j) - L(j+1:m, 1:j-1)U(1:j, j))$
    $L(j+1:m, j) = v(j+1:m)/v(j)$
**end**

This algorithm requires $mn^2 - n^3/3$ flops.

### 3.2.13 A Note on Failure

As we know, Gaussian elimination fails unless the first $n-1$ principal submatrices are nonsingular. This rules out some very simple matrices, e.g.,

$$A = \begin{bmatrix} 0 & 1 \\ 1 & 0 \end{bmatrix}.$$

While $A$ has perfect 2-norm condition, it fails to have an LU factorization because it has a singular leading principal submatrix.

Clearly, modifications are necessary if Gaussian elimination is to be effectively used in general linear system solving. The error analysis in the following section suggests the needed modifications.

#### Problems

**P3.2.1** Suppose the entries of $A(\epsilon) \in \mathbb{R}^{n\times n}$ are continuously differentiable functions of the scalar $\epsilon$. Assume that $A \equiv A(0)$ and all its principal submatrices are nonsingular. Show that for sufficiently small $\epsilon$, the matrix $A(\epsilon)$ has an LU factorization $A(\epsilon) = L(\epsilon)U(\epsilon)$ and that $L(\epsilon)$ and $U(\epsilon)$ are both continuously differentiable.

**P3.2.2** Suppose we partition $A \in \mathbb{R}^{n \times n}$

$$A = \begin{bmatrix} A_{11} & A_{12} \\ A_{21} & A_{22} \end{bmatrix}$$

where $A_{11}$ is $r$-by-$r$. Assume that $A_{11}$ is nonsingular. The matrix $S = A_{22} - A_{21}A_{11}^{-1}A_{12}$ is called the *Schur complement* of $A_{11}$ in $A$. Show that if $A_{11}$ has an LU factorization, then after $r$ steps of Algorithm(3.2.3), $A(r+1{:}n, r+1{:}n)$ houses $S$. How could $S$ be obtained after $r$ steps of Algorithm(3.2.4)?

**P3.2.3** Suppose $A \in \mathbb{R}^{n \times n}$ has an LU factorization. Show how $Ax = b$ can be solved without storing the multipliers by computing the LU factorization of the $n$-by-$(n+1)$ matrix $[A\ b]$.

**P3.2.4** Describe a variant of Gaussian elimination that introduces zeros into the columns of $A$ in the order, $n{:}-1{:}2$ and which produces the factorization $A = UL$ where $U$ is unit upper triangular and $L$ is lower triangular.

**P3.2.5** Matrices in $\mathbb{R}^{n \times n}$ of the form $N(y,k) = I - ye_k^T$ where $y \in \mathbb{R}^n$ are said to be *Gauss-Jordan transformations.* (a) Give a formula for $N(y,k)^{-1}$ assuming it exists. (b) Given $x \in \mathbb{R}^n$, under what conditions can $y$ be found so $N(y,k)x = e_k$? (c) Give an algorithm using Gauss-Jordan transformations that overwrites $A$ with $A^{-1}$. What conditions on $A$ ensure the success of your algorithm?

**P3.2.6** Develop a fully detailed version of Algorithm 3.2.4 that handles arbitrary $m$-by-$n$ matrices.

### Notes and References for Sec. 3.2

Schur complements (P3.2.2) arise in many applications. For a survey of both practical and theoretical interest, see

R.W. Cottle (1974). "Manifestations of the Schur Complement," *Lin. Alg. and Its Applic. 8,* 189-211.

Schur complements are known as "Gauss transforms" in some application areas. The use of Gauss-Jordan transformations (P3.2.5) is detailed in:

L. Fox (1964). *An Introduction to Numerical Linear Algebra,* Oxford University Press, Oxford.

As we mentioned, inner product versions of Gaussian elimination have been known and used for some time. The names of Crout and Doolittle are associated with these *ijk* techniques. They were popular during the days of desk calculators because there are far fewer intermediate results than in Gaussian elimination. These methods still have attraction because they can be implemented with accumulated inner products. For remarks along these lines see Chapter 4 of the Fox reference above as well as Stewart (IMC, pp. 131–39). An ALGOL procedure may be found in

H.J. Bowdler, R.S. Martin, G. Peters, and J.H. Wilkinson (1966). "Solution of Real and Complex Systems of Linear Equations," *Numer. Math. 8,* 217–34. See also *HACLA,* pp. 93–110.

See also:

G.E. Forsythe (1960). "Crout with Pivoting," *Comm. ACM 3,* 507–8.
W.M. McKeeman (1962). "Crout with Equilibration and Iteration," *Comm. ACM. 5,* 553–55.

Loop orderings in LU computations are discussed in

M.J. Dayde and I.S. Duff (1988). "Use of Level-3 BLAS in LU Factorization on the Cray-2, the ETA-10P, and the IBM 3090-200/VF," Report CSS-229, Computer Science and Systems Division, Harwell Laboratory, Oxon OX11 ORA, England.

J.J. Dongarra, F.G. Gustavson, and A. Karp (1984). "Implementing Linear Algebra Algorithms for Dense Matrices on a Vector Pipeline Machine," *SIAM Review 26*, 91–112.

J.M. Ortega (1988). "The *ijk* Forms of Factorization Methods I: Vector Computers," *Parallel Computers 7*, 135–147.

The first paper investigates block dot product, block Gaxpy, and block outer product versions of Gaussian Elimination.

The computation of the LU factorization of large sparse matrices is a much-studied topic. See

Å. Björck, R.J. Plemmons, and H. Schneider (1981). *Large-Scale Matrix Problems*, North-Holland, New York.

I.S. Duff, A.M. Erisman, and J.K. Reid (1986). *Direct Methods for Sparse Matrices*, Oxford University Press.

I.S. Duff and G.W. Stewart, eds. (1979). *Sparse Matrix Proceedings*, 1978, SIAM Publications, Philadelphia, PA.

## 3.3 Roundoff Analysis of Gaussian Elimination

We now assess the effect of rounding errrors when the algorithms in the previous two sections are used to solve the linear system $Ax = b$. Before we proceed with the analysis, it is useful to consider the nearly ideal situation in which no roundoff occurs during the entire solution process except when $A$ and $b$ are stored. Thus, if $fl(b) = b+e$ and the stored matrix $fl(A) = A+E$ is nonsingular, then we are assuming that the computed solution $\hat{x}$ satisfies

$$(A+E)\hat{x} = (b+e) \qquad \| E \|_\infty \le \mathbf{u}\| A \|_\infty, \quad \| e \|_\infty \le \mathbf{u}\| b \|_\infty . \tag{3.3.1}$$

That is, $\hat{x}$ solves a "nearby" system exactly. Moreover, if $\mathbf{u}\kappa_\infty(A) \le \frac{1}{2}$ (say), then by using Theorem 2.7.1, it can be shown that

$$\frac{\| x - \hat{x} \|_\infty}{\| x \|_\infty} \le 4\mathbf{u}\kappa_\infty(A) . \tag{3.3.2}$$

The bounds (3.3.1) and (3.3.2) are "best possible" norm bounds. No general $\infty$-norm error analysis of a linear equation solver that requires the storage of $A$ and $b$ can render sharper bounds. As as consequence, we cannot justifiably criticize an algorithm for returning an inaccurate $\hat{x}$ if $A$ is ill-conditioned relative to the machine precision, e.g., $\mathbf{u}\kappa_\infty(A) \approx 1$.

### 3.3.1 Errors in the LU Factorization

Let us see how the error bounds for Gaussian elimination compare with the ideal bounds above. We work with the infinity norm for convenience and focus our attention on Algorithm 3.2.3, the outer product version. The error bounds that we derive also apply to Algorithm 3.2.4, the gaxpy formulation.

Our first task is to quantify the roundoff errors associated with the computed triangular factors.

**Theorem 3.3.1** *Assume that $A$ is an $n$-by-$n$ matrix of floating point numbers. If no zero pivots are encountered during the execution of Algorithm 3.2.3, then the computed triangular matrices $\hat{L}$ and $\hat{U}$ satisfy*

$$\hat{L}\hat{U} = A + H \tag{3.3.3}$$

$$|H| \le 3(n-1)\mathbf{u}\left(|A| + |\hat{L}||\hat{U}|\right) + O(\mathbf{u}^2). \tag{3.3.4}$$

**Proof.** The proof is by induction on $n$. The lemma obviously holds for $n = 1$. Assume it holds for all $(n-1)$-by-$(n-1)$ floating point matrices. If

$$A = \begin{bmatrix} \alpha & w^T \\ v & B \end{bmatrix} \begin{matrix} 1 \\ n-1 \end{matrix}$$
$$\quad\; 1 \quad n-1$$

then $\hat{z} = fl(v/\alpha)$ and $\hat{A}_1 = fl(B - \hat{z}w^T)$ are computed in the first step of the algorithm. We therefore have

$$\hat{z} = \frac{1}{\alpha}v + f \qquad |f| \le \mathbf{u}\frac{|v|}{\alpha} \tag{3.3.5}$$

and

$$\hat{A}_1 = B - \hat{z}w^T + F \qquad |F| \le 2\mathbf{u}\left(|B| + |\hat{z}||w|^T\right) + O(\mathbf{u}^2). \tag{3.3.6}$$

The algorithm now proceeds to calculate the LU factorization of $\hat{A}_1$. By induction $\hat{A}_1 = \hat{L}_1\hat{U}_1$ and the resulting triangular matrices $\hat{L}_1$ and $\hat{U}_1$ satisfy

$$\hat{L}_1\hat{U}_1 = \hat{A}_1 + H_1 \tag{3.3.7}$$

$$|H_1| \le 3(n-2)\mathbf{u}\left(|\hat{A}_1| + |\hat{L}_1||\hat{U}_1|\right) + O(\mathbf{u}^2). \tag{3.3.8}$$

Thus,

$$\begin{aligned} \hat{L}\hat{U} &\equiv \begin{bmatrix} 1 & 0 \\ \hat{z} & \hat{L}_1 \end{bmatrix} \begin{bmatrix} \alpha & w^T \\ 0 & \hat{U}_1 \end{bmatrix} \\ &= A + \begin{bmatrix} 0 & 0 \\ \alpha f & H_1 + F \end{bmatrix} \equiv A + H. \end{aligned}$$

From (3.3.6) it follows that

$$|\hat{A}_1| \le (1 + 2\mathbf{u})\left(|B| + |\hat{z}||w|^T\right) + O(\mathbf{u}^2),$$

and therefore by using (3.3.7) and (3.3.8) we have

$$|H_1 + F| \le 3(n-1)\mathbf{u}\left(|B| + |\hat{z}||w|^T + |\hat{L}_1||\hat{U}_1|\right) \quad + \quad O(\mathbf{u}^2).$$

Since $|\alpha f| \le \mathbf{u}|v|$ it is easy to verify that

$$|H| \le 3(n-1)\mathbf{u}\left\{\left[\begin{array}{cc} |\alpha| & |w|^T \\ |v| & |B| \end{array}\right] + \left[\begin{array}{cc} 1 & 0 \\ |\hat{z}| & |\hat{L}_1| \end{array}\right]\left[\begin{array}{cc} |\alpha| & |w|^T \\ 0 & |\hat{U}_1| \end{array}\right]\right\} + O(\mathbf{u}^2)$$

thereby proving the theorem. □

We mention that if $A$ is $m$-by-$n$, then the theorem applies with $n$ in (3.3.4) replaced by the smaller of $n$ and $m$ .

### 3.3.2 Triangular Solving with Inexact Triangles

We next examine the effect of roundoff error when $\hat{L}$ and $\hat{U}$ are used by the triangular system solvers of §3.1.

**Theorem 3.3.2** *Let $\hat{L}$ and $\hat{U}$ be the computed LU factors of the $n$-by-$n$ floating point matrix $A$ obtained by either Algorithm 3.2.3 or 3.2.4. Suppose the methods of §3.1 are used to produce the computed solution $\hat{y}$ to $\hat{L}y = b$ and the computed solution $\hat{x}$ to $\hat{U}x = \hat{y}$. Then $(A+E)\hat{x} = b$ with*

$$|E| \le n\mathbf{u}\left(3|A| + 5|\hat{L}||\hat{U}|\right) \; + \; O(\mathbf{u}^2)\,. \tag{3.3.9}$$

**Proof.** From (3.1.1) and (3.1.2) we have

$$\begin{array}{rclcrcl} (\hat{L}+F)\hat{y} & = & b & \qquad & |F| & \le & n\mathbf{u}|\hat{L}| + O(\mathbf{u}^2) \\ (\hat{U}+G)\hat{x} & = & \hat{y} & & |G| & \le & n\mathbf{u}|\hat{U}| + O(\mathbf{u}^2) \end{array}$$

and thus

$$(\hat{L}+F)(\hat{U}+G)\hat{x} \;=\; (\hat{L}\hat{U} + F\hat{U} + \hat{L}G + FG)\hat{x} = b.$$

From Theorem 3.3.1

$$\hat{L}\hat{U} = A + H,$$

with $|H| \le 3(n-1)\mathbf{u}(|A| + |\hat{L}||\hat{U}|) + O(\mathbf{u}^2)$, and so by defining

$$E \;=\; H + F\hat{U} + \hat{L}G + FG$$

we find $(A+E)\hat{x} = b$. Moreover,

$$\begin{array}{rcl} |E| & \le & |H| + |F|\,|\hat{U}| \,+\, |\hat{L}|\,|G| \,+\, O(\mathbf{u}^2) \\ & \le & 3n\mathbf{u}\left(|A| + |\hat{L}||\hat{U}|\right) + 2n\mathbf{u}\left(|\hat{L}||\hat{U}|\right) + O(\mathbf{u}^2). \quad \square \end{array}$$

Were it not for the possibility of a large $|\hat{L}||\hat{U}|$ term, (3.3.9) would compare favorably with the ideal bound in (3.3.1). (The factor $n$ is of no consequence, cf. the Wilkinson quotation in §2.4.6.) Such a possibility exists, for there is nothing in Gaussian elimination to rule out the appearance of small pivots. If a small pivot is encountered, then we can expect large numbers to be present in $\hat{L}$ and $\hat{U}$.

We stress that small pivots are not necessarily due to ill-conditioning as the example $A = \begin{bmatrix} \epsilon & 1 \\ 1 & 0 \end{bmatrix}$ bears out. Thus, Gaussian elimination can give arbitrarily poor results, even for well-conditioned problems. The method is unstable.

In order to repair this shortcoming of the algorithm, it is necessary to introduce row and/or column interchanges during the elimination process with the intention of keeping the numbers that arise during the calculation suitably bounded. This idea is pursued in the next section.

**Example 3.3.1** Suppose $\beta = 10, t = 3$, chopped arithmetic is used to solve:

$$\begin{bmatrix} .001 & 1.00 \\ 1.00 & 2.00 \end{bmatrix} \begin{bmatrix} x_1 \\ x_2 \end{bmatrix} = \begin{bmatrix} 1.00 \\ 3.00 \end{bmatrix}.$$

Applying Gaussian elimination we get

$$\hat{L} = \begin{bmatrix} 1 & 0 \\ 1000 & 1 \end{bmatrix} \qquad \hat{U} = \begin{bmatrix} .001 & 1 \\ 0 & -1000 \end{bmatrix}$$

and a calculation shows

$$\hat{L}\hat{U} = \begin{bmatrix} .001 & 1 \\ 1 & 2 \end{bmatrix} + \begin{bmatrix} 0 & 0 \\ 0 & -2 \end{bmatrix} \equiv A + H.$$

Moreover, $6\begin{bmatrix} 10^{-6} & .001 \\ 10^{-3} & 1.0001 \end{bmatrix}$ is the bounding matrix in (3.3.4), not a severe overestimate of $|H|$. If we go on to solve the problem using the triangular system solvers of §3.1, then using the same precision arithmetic we obtain a computed solution $\hat{x} = (0, 1)^T$. This is in contrast to the exact solution $x = (1.002..., .998...)^T$.

### Problems

**P3.3.1** Show that if we drop the assumption that $A$ is a floating point matrix in in Theorem 3.3.1, then (3.3.3) holds with the coefficient "3"replaced by "4."

**P3.3.2** Suppose $A$ is an $n$-by-$n$ matrix and that $\hat{L}$ and $\hat{U}$ are produced by Algorithm 3.2.3. (a) How many flops are required to compute $\| \, |\hat{L}| \, |\hat{U}| \, \|_\infty$? (b) Show $fl(|\hat{L}||\hat{U}|) \leq (1 + 2n\mathbf{u})|\hat{L}||\hat{U}| + O(\mathbf{u}^2)$.

**P3.3.3** Suppose $x = A^{-1}b$. Show that if $e = x - \hat{x}$ (the error) and $r = b - A\hat{x}$ (the residual), then

$$\frac{\| r \|}{\| A \|} \leq \| e \| \leq \| A^{-1} \| \| r \|.$$

Assume consistency between the matrix and vector norm.

**P3.3.4** Using 2-digit, base 10, rounded floating point arithmetic, compute the LU factorization of

$$A = \begin{bmatrix} 7 & 6 \\ 9 & 8 \end{bmatrix}.$$

For this example, what is the matrix $H$ in (3.3.3)?

**Notes and References for Sec. 3.3**

Our style of inverse error analysis is patterned after

C. de Boor and A. Pinkus (1977). "A Backward Error Analysis for Totally Positive Linear Systems," *Numer. Math. 27*, 485-90.

The original roundoff analysis of Gaussian elimination appears in

J.H. Wilkinson (1961). "Error Analysis of Direct Methods of Matrix Inversion," *J. ACM 8*, 281-330.

Various improvements in the bounds and simplifications in the analysis have occurred over the years. See

C.C. Paige (1973). "An Error Analysis of a Method for Solving Matrix Equations," *Math. Comp. 27*, 355–59.
J.K. Reid (1971). "A Note on the Stability of Gaussian Elimination," *J. Inst. Math. Applic. 8*, 374–75.
H.H. Robertson (1977). "The Accuracy of Error Estimates for Systems of Linear Algebraic Equations," *J. Inst. Math. Applic. 20*, 409–14.

## 3.4 Pivoting

The analysis in the previous section shows that we must take steps to ensure that no large entries appear in the computed triangular factors $\hat{L}$ and $\hat{U}$. The example

$$A = \begin{bmatrix} .0001 & 1 \\ 1 & 1 \end{bmatrix} = \begin{bmatrix} 1 & 0 \\ 10,000 & 1 \end{bmatrix} \begin{bmatrix} .0001 & 1 \\ 0 & -9999 \end{bmatrix} = LU$$

correctly identifies the source of the difficulty: relatively small pivots. A way out of this difficulty is to interchange rows. In our example, if $P$ is the permutation

$$P = \begin{bmatrix} 0 & 1 \\ 1 & 0 \end{bmatrix}$$

then

$$PA = \begin{bmatrix} 1 & 1 \\ .0001 & 1 \end{bmatrix} = \begin{bmatrix} 1 & 0 \\ .0001 & 1 \end{bmatrix} \begin{bmatrix} 1 & 1 \\ 0 & .9999 \end{bmatrix} = LU.$$

Now the triangular factors are comprised of acceptably small elements.

In this section we show how to determine a permuted version of $A$ that has a reasonably stable LU factorization. There are several ways to do this and they each correspond to a different pivoting strategy. We focus on partial pivoting and complete pivoting. The efficient implementation of these strategies and their properties are discussed. We begin with a discussion of permutation matrix manipulation.

### 3.4.1 Permutation Matrices

The stabilizations of Gaussian elimination that are developed in this section involve data movements such as the interchange of two matrix rows. In keeping with our desire to describe all computations in "matrix terms," it is necessary to acquire a familiarity with *permutation matrices*. A permutation matrix is just the identity with its rows re-ordered, e.g.,

$$P = \begin{bmatrix} 0 & 0 & 0 & 1 \\ 1 & 0 & 0 & 0 \\ 0 & 0 & 1 & 0 \\ 0 & 1 & 0 & 0 \end{bmatrix}.$$

An $n$-by-$n$ permutation matrix should never be explicitly stored. It is much more efficient to represent a general permutation matrix $P$ with an integer $n$-vector $p$. One way to do this is to let $p(k)$ be the column index of the sole "1" in $P$'s $k$th row. Thus, $p = (4\,1\,3\,2)$ is the appropriate encoding of the above $P$. It is also possible to encode $P$ on the basis of where the "1" occurs in each column, e.g., $p = (2\,4\,3\,1)$.

If $P$ is a permutation and $A$ is a matrix, then $PA$ is a row permuted version of $A$ and $AP$ is a column permuted version of $A$.

Permutation matrices are orthogonal and so if $P$ is a permutation then $P^{-1} = P^T$. A product of permutation matrices is a permutation matrix.

In this section we are particularly interested in *interchange permutations*. These are permutations obtained by merely swapping two rows in the identity, e.g.,

$$E = \begin{bmatrix} 0 & 0 & 0 & 1 \\ 0 & 1 & 0 & 0 \\ 0 & 0 & 1 & 0 \\ 1 & 0 & 0 & 0 \end{bmatrix}.$$

Interchange permutations can be used to describe row and column swapping. With the above 4-by-4 example, $EA$ is $A$ with rows 1 and 4 interchanged. Likewise, $AE$ is $A$ with columns 1 and 4 swapped.

If $P = E_n \cdots E_1$ and each $E_k$ is the identity with rows $k$ and $p(k)$ interchanged, then $p(1{:}n)$ is a useful vector encoding of $P$. Indeed, $x \in \mathbb{R}^n$ can be overwritten by $Px$ as follows:

**for** $k = 1{:}n$
  $x(k) \leftrightarrow x(p(k))$
**end**

Here, the "$\leftrightarrow$" notation means "swap contents":

$$x(k) \leftrightarrow x(p(k)) \quad \Leftrightarrow \quad \tau = x(k),\ x(k) = x(p(k)),\ x(p(k)) = \tau\,.$$

Since each $E_k$ is symmetric and $P^T = E_1 \cdots E_n$, the representation can also be used to overwrite $x$ with $P^T x$:

**for** $k = n{:} -1{:}1$
    $x(k) \leftrightarrow x(p(k))$
**end**

It should be noted that no floating point arithmetic is involved in a permutation operation. However, permutation matrix operations often involve the irregular movement of data and so can represent a significant computational overhead.

### 3.4.2 Partial Pivoting: The Basic Idea

We show how interchange permutations can be used in LU computations to guarantee that no multiplier is greater than one in absolute value. Suppose

$$A = \begin{bmatrix} 3 & 17 & 10 \\ 2 & 4 & -2 \\ 6 & 18 & -12 \end{bmatrix}.$$

To get the smallest possible multipliers in the first Gauss transform using row interchanges we need $a_{11}$ to be the largest entry in the first column. Thus, if $P_1$ is the interchange permutation

$$P_1 = \begin{bmatrix} 0 & 0 & 1 \\ 0 & 1 & 0 \\ 1 & 0 & 0 \end{bmatrix}$$

then

$$P_1 A = \begin{bmatrix} 6 & 18 & -12 \\ 2 & 4 & -2 \\ 3 & 17 & 10 \end{bmatrix}$$

and

$$M_1 = \begin{bmatrix} 1 & 0 & 0 \\ -1/3 & 1 & 0 \\ -1/2 & 0 & 1 \end{bmatrix} \quad \Longrightarrow \quad M_1 P_1 A = \begin{bmatrix} 6 & 18 & -12 \\ 0 & -2 & 2 \\ 0 & 8 & 16 \end{bmatrix}.$$

Now to get the smallest possible multiplier in $M_2$ we need to swap rows 2 and 3. Thus, if

$$P_2 = \begin{bmatrix} 1 & 0 & 0 \\ 0 & 0 & 1 \\ 0 & 1 & 0 \end{bmatrix} \quad \text{and} \quad M_2 = \begin{bmatrix} 1 & 0 & 0 \\ 0 & 1 & 0 \\ 0 & 1/4 & 1 \end{bmatrix}$$

then

$$M_2P_2M_1P_1A = \begin{bmatrix} 6 & 18 & -12 \\ 0 & 8 & 16 \\ 0 & 0 & 6 \end{bmatrix}.$$

The example illustrates the basic idea behind the row interchanges. In general we have:

**for** $k = 1{:}n-1$
  Determine an interchange matrix $E_k$ such that if $z$ is the
    $k$th column of $E_kA$ , then $|z(k)| = \| z(k{:}n) \|_\infty$.
  $A = E_kA$
  Determine the Gauss transform $M_k$ such that if $v$ is the
    $k$th column of $M_kA$, then $v(k+1{:}n) = 0$ .
  $A = M_kA$
**end**

This particular row interchange strategy is called *partial pivoting.* Upon completion we emerge with $M_{n-1}E_{n-1}\cdots M_1E_1A = U$, an upper triangular matrix.

As a consequence of the partial pivoting, no multiplier is larger than one in absolute value. This is because

$$|(E_kM_{k-1}\cdots M_1E_1A)_{kk}| = \max_{k\le i\le n} |(E_kM_{k-1}\cdots M_1E_1A)_{ik}|$$

for $k = 1{:}n-1$. Thus, partial pivoting effectively guards against arbitrarily large multipliers.

### 3.4.3 Partial Pivoting Details

We are now set to detail the overall Gaussian Elimination (G.E.) with partial pivoting algorithm.

**Algorithm 3.4.1 (G.E. with Partial Pivoting: Outer Product Version)** If $A \in \mathbb{R}^{n\times n}$ then the following algorithm computes Gauss transforms $M_1, \cdots M_{n-1}$ and interchange permutations $E_1, \cdots E_{n-1}$ such that $M_{n-1}E_{n-1}\cdots M_1E_1A = U$ is upper triangular. No multiplier is bigger than 1 in absolute value. $A(1{:}k, k)$ is overwritten by $U(1{:}k, k)$, $k = 1{:}n$. $A(k+1{:}n, k)$ is overwritten by $-M_k(k+1{:}n, k)$, $k = 1{:}n-1$. The integer vector $piv(1{:}n-1)$ defines the interchange permutations. In particular, $E_k$ interchanges rows $k$ and $piv(k)$, $k = 1{:}n-1$.

```
for k = 1:n − 1
    Determine μ with k ≤ μ ≤ n so |A(μ, k)| = || A(k:n, k) ||∞
    A(k, k:n) ↔ A(μ, k:n); piv(k) = μ
    if A(k, k) ≠ 0
        t = gauss(A(k:n, k)); A(k + 1:n, k) = t
        A(k:n, k + 1:n) = gauss.app(A(k:n, k + 1:n), t)
    end
end
```

Note that if $\| A(k{:}n, k \|_\infty = 0$ in step $k$, then in exact arithmetic the first $k$ columns of $A$ are linearly dependent. In contrast to Algorithm 3.2.3, this poses no difficulty. We merely skip over the zero pivot.

The overhead associated with partial pivoting is minimal from the standpoint of floating point arithmetic as there are only $O(n^2)$ comparisons associated with the search for the pivots. The overall algorithm involves $2n^3/3$ flops.

To solve the linear system $Ax = b$ after invoking Algorithm 3.4.1 we

- Compute $y = M_{n-1}E_{n-1} \cdots M_1 E_1 b$.
- Solve the upper triangular system $Ux = y$.

All the information necessary to do this is contained in the array $A$ and the integer vector $piv$.

**Example 3.4.1** If Algorithm 3.4.1 is applied to the matrix $A$ in §3.4.2, then upon exit

$$A = \begin{bmatrix} 6 & 18 & -12 \\ 1/3 & 8 & 16 \\ 1/2 & 1/4 & 6 \end{bmatrix}$$

and $piv = (3\ \ 3)$.

### 3.4.4 Where is L?

Gaussian elimination with partial pivoting computes the LU factorization of a row permuted version of $A$. The proof is a messy subscripting argument.

**Theorem 3.4.1** *If Gaussian elimination with partial pivoting is used to compute the upper triangularization*

$$M_{n-1}E_{n-1} \cdots M_1 E_1 A \;=\; U \qquad (3.4.2)$$

*via Algorithm 3.4.1, then*

$$PA = LU$$

*where $P = E_{n-1} \cdots E_1$ and $L$ is a unit lower triangular matrix with $|\ell_{ij}| \le 1$. The $k$th column of $L$ below the diagonal is a permuted version of the* *kth Gauss vector. In particular, if* $M_k = I - t^{(k)}e_k^T$ *then* $L(k+1{:}n, k) = g(k+1{:}n)$ *where* $g = E_{n-1}\cdots E_{k+1}t^{(k)}$.

**Proof.** A manipulation of (3.4.2) reveals that $\tilde{M}_{n-1}\cdots\tilde{M}_1 PA = U$ where $\tilde{M}_{n-1} = M_{n-1}$ and

$$\tilde{M}_k = E_{n-1}\cdots E_{k+1}M_k E_{k+1}\cdots E_{n-1} \qquad k \le n-2\,.$$

Since each $E_j$ is an interchange permutation involving row $j$ and a row $\mu$ with $\mu \ge j$ we have $E_j(1{:}j-1, 1{:}j-1) = I_{j-1}$ . It follows that each $\tilde{M}_k$ is a Gauss transform with Gauss vector $\tilde{t}^{(k)} = E_{n-1}\cdots E_{k+1}t^{(k)}$. □

As a consequence of the theorem, it is easy to see how to change Algorithm 3.4.1 so that upon completion, $A(i,j)$ houses $L(i,j)$ for all $i > j$. We merely apply each $E_k$ to all the previously computed Gauss vectors. This is accomplished by changing the line "$A(k, k{:}n) \leftrightarrow A(\mu, k{:}n)$" in Algorithm 3.4.1 to "$A(k, 1{:}n) \leftrightarrow A(\mu, 1{:}n)$."

### 3.4.5 The Gaxpy Version

In §3.2 we developed outer product and gaxpy schemes for computing the LU factorization. Having just incorporated pivoting in the outer product version, it is natural to do the same with the gaxpy approach. Recall from (3.2.5) the general structure of the gaxpy LU process:

```
for j = 1:n
    Solve L(1:j-1, 1:j-1)U(1:j-1, j) = A(1:j-1, j)
    v(j:n) = A(j:n, j) - L(j:n, 1:j-1)U(1:j-1, j)
    U(j, j) = v(j)
    L(j+1:n, j) = v(j+1:n)/U(j, j)
end
```

With partial pivoting this now becomes

```
for j = 1:n
    A(:, j) = E_{j-1} ··· E_1 A(:, j)
    Solve L(1:j-1, 1:j-1)U(1:j-1, j) = A(1:j-1, j)
    v(j:n) = A(j:n) - L(j:n, 1:j-1)U(1:j-1, j)
    Determine μ, j ≤ μ ≤ n , so |v(μ)| = || v(j:n) ||_∞.
    Let E_j exchange rows j and μ and swap: v(j) ↔ v(μ).
    if v(j) ≠ 0
        v(j+1:n) = v(j+1:n)/v(j)
    end
    U(j, j) = v(j)
    L(j+1:n, j) = v(j+1:n)
    L(:, 1:j-1) = E_j L(:, 1:j-1)
end
```

If we fill in the details and overwrite $A$ we obtain

**Algorithm 3.4.3 (G.E. with Partial Pivoting: Gaxpy Version)** This algorithm computes unit lower triangular $L$, upper triangular $U$ and a permutation $P$ so $PA = LU$. $A(i,j)$ is overwritten by $L(i,j)$ if $i > j$ and by $U(i,j)$ otherwise. $P = E_{n-1} \cdots E_1$ is encoded in the integer vector $piv(1{:}n-1)$. In particular, $E_j$ is an interchange permutation involving rows $j$ and $piv(j)$.

```
for j = 1:n
    { A(:,j) = E_{j-1} ··· E_1 A(:,j) }
    for k = 1:j − 1
        A(k,j) ↔ A(piv(k),j)
    end
    {Solve L(1:j − 1,1:j − 1)U(1:j − 1,j) = A(1:j − 1,j)}
    for k = 1:j − 1
        for i = k + 1:j − 1
            A(i,j) = A(i,j) − A(i,k)A(k,j)
        end
    end
    { v(j:n) = A(j:n,j) − L(j:n,1:j − 1)U(1:j − 1,j)}
    for k = 1:j − 1
        for i = j:n
            A(i,j) = A(i,j) − A(i,k)A(k,j)
        end
    end
    { Determine E_j and apply to v and L(:,1:j − 1). }
    Find μ with j ≤ μ ≤ n so |A(μ,j)| = || A(j:n,j) ||_∞
    piv(j) = μ
    for k = 1:j
        A(j,k) ↔ A(μ,k)
    end
    { L(j + 1:n,j) = v(j + 1:n)/v(j) }
    if A(j,j) ≠ 0
        for i = j + 1:n
            A(i,j) = A(i,j)/A(j,j)
        end
    end
end
```

As with the outer product version, this procedure requires $2n^3/3$ flops and $O(n^2)$ comparisons. To solve $Ax = b$ after invoking Algorithm 3.4.3 we (a) set $c = Pb$ (a permutation of $b$'s components), (b) solve $Ly = c$ (a lower triangular system), and (c) solve $Ux = y$ (an upper triangular system).

**Example 3.4.2** If Algorithm 3.4.2 is applied to the matrix $A$ in Example 3.4.1, then upon completion

$$A = \begin{bmatrix} 6 & 18 & -12 \\ 1/3 & 8 & 16 \\ 1/2 & 1/4 & 6 \end{bmatrix}$$

and $piv = (3,\ 3)$.

### 3.4.6 Error Analysis

We now examine the stability that is obtained with partial pivoting. This requires an accounting of the rounding errors that are sustained during elimination and during the triangular system solving. Bearing in mind that there are no rounding errors associated with permutation, it is not hard to show using Theorem 3.3.2 that the computed solution $\hat{x}$ satisfies $(A+E)\hat{x} = b$ where

$$|E| \le n\mathbf{u}\left(3|A| + 5\hat{P}^T|\hat{L}||\hat{U}|\right) + O(\mathbf{u}^2). \tag{3.4.3}$$

Here we are assuming that $\hat{P}$, $\hat{L}$, and $\hat{U}$ are the computed analogs of $P$, $L$, and $U$ as produced by the above algorithms. Pivoting implies that the elements of $\hat{L}$ are bounded by one. Thus $\| \hat{L} \|_\infty \le n$ and we obtain the bound

$$\| E \|_\infty \le n\mathbf{u}\left(3\| A \|_\infty + 5n\| \hat{U} \|_\infty\right) + O(\mathbf{u}^2). \tag{3.4.4}$$

The problem now is to bound $\| \hat{U} \|_\infty$. Define the *growth factor* $\rho$ by

$$\rho = \max_{i,j,k} \frac{|\hat{a}_{ij}^{(k)}|}{\| A \|_\infty} \tag{3.4.5}$$

where $\hat{A}^{(k)}$ is the computed version of the matrix $A^{(k)} = M_k E_k \cdots M_1 E_1 A$. It follows that

$$\| E \|_\infty \le 8n^3\rho\| A \|_\infty \mathbf{u} + O(\mathbf{u}^2). \tag{3.4.6}$$

Whether or not this compares favorably with the ideal bound (3.3.1) hinges upon the size of the growth factor of $\rho$. (The factor $n^3$ is not an operating factor in practice and may be ignored in this discussion.) The growth factor measures how large the numbers become during the process of elimination. In practice, $\rho$ is usually of order 10 but it can also be as large as $2^{n-1}$. Despite this, most numerical analysts regard the occurrence of serious element growth in Gaussian elimination with partial pivoting as highly unlikely in practice. The method can be used with confidence.

**Example 3.4.3** If Gaussian elimination with partial pivoting is applied to the problem

$$\begin{bmatrix} .001 & 1.00 \\ 1.00 & 2.00 \end{bmatrix} \begin{bmatrix} x_1 \\ x_2 \end{bmatrix} = \begin{bmatrix} 1.00 \\ 3.00 \end{bmatrix}$$

with $= 10, t = 3$, chopped arithmetic, then

$$P = \begin{bmatrix} 0 & 1 \\ 1 & 0 \end{bmatrix}, \quad \hat{L} = \begin{bmatrix} 1.00 & 0 \\ .001 & 1.00 \end{bmatrix}, \quad \hat{U} = \begin{bmatrix} 1.00 & 2.00 \\ 0 & 1.00 \end{bmatrix}$$

and $\hat{x} = (1.00,\ .996)^T$, which is correct to three digits.

**Example 3.4.4** If $A \in \mathbb{R}^{n \times n}$ is defined by

$$a_{ij} = \begin{cases} 1 & \text{if } i = j \text{ or } j = n \\ -1 & \text{if } i > j \\ 0 & \text{otherwise} \end{cases}$$

then $A$ has an LU factorization with $|\ell_{ij}| \le 1$ and $u_{nn} = 2^{n-1}$.

### 3.4.7 Block Gaussian Elimination

Gaussian Elimination with partial pivoting can be organized so that it is rich in level-3 operations. We detail a block outer product procedure but block gaxpy and block dot product formulations are also possible. See Dayde and Duff (1988).

Assume $A \in \mathbb{R}^{n \times n}$ and for clarity that $n = rN$. Partition $A$ as follows:

$$A = \begin{bmatrix} A_{11} & A_{12} \\ A_{21} & A_{22} \end{bmatrix} \begin{matrix} r \\ n-r \end{matrix}$$
$$\qquad\qquad r \quad n-r$$

The first step in the block reduction is typical and proceeds as follows:

- Use scalar Gaussian elimination with partial pivoting (e.g. rectangular versions of Algorithm 3.4.1 or 3.4.2) to compute permutation $P_1 \in \mathbb{R}^{n \times n}$, unit lower triangular $L_{11} \in \mathbb{R}^{r \times r}$ and upper triangular $U_{11} \in \mathbb{R}^{r \times r}$ so

$$P_1 \begin{bmatrix} A_{11} \\ A_{21} \end{bmatrix} = \begin{bmatrix} L_{11} \\ L_{21} \end{bmatrix} U_{11}.$$

- Apply the $P_1$ across the rest of $A$:

$$\begin{bmatrix} \tilde{A}_{12} \\ \tilde{A}_{22} \end{bmatrix} = P_1 \begin{bmatrix} A_{12} \\ A_{22} \end{bmatrix}.$$

- Solve the lower triangular multiple right hand side problem

$$L_{11}U_{12} = \tilde{A}_{12}.$$

- Perform the level-3 update

$$\tilde{A} = \tilde{A}_{22} - L_{21}U_{12}.$$

With these computations we obtain the factorization

$$P_1 A = \begin{bmatrix} L_{11} & 0 \\ L_{21} & I_{n-r} \end{bmatrix} \begin{bmatrix} I_r & 0 \\ 0 & \tilde{A} \end{bmatrix} \begin{bmatrix} U_{11} & U_{12} \\ 0 & I_{n-r} \end{bmatrix}.$$

The process is then repeated on the first $r$ columns of $\tilde{A}$.

In general, during step $k$ ($1 \leq k \leq N-1$) of the block algorithm we apply scalar Gaussian elimination to a matrix of size $(n-(k-1)r)$-by-$r$. An $r$-by-$(n-kr)$ multiple right hand side system is solved and a level 3 update of size $(n-kr)$-by-$(n-kr)$ is performed. The level 3 fraction for the overall process is approximately given by $1 - 3/(2N)$. Thus, for large $N$ the procedure is rich in matrix multiplication.

### 3.4.8 Complete Pivoting

Another pivot strategy called *complete pivoting* has the property that the associated growth factor bound is considerably smaller than $2^{n-1}$. Recall that in partial pivoting, the $k$th pivot is determined by scanning the current subcolumn $A(k{:}n, k)$. In complete pivoting, the largest entry in the current submatrix $A(k{:}n, k{:}n)$ is permuted into the $(k, k)$ position. Thus, we compute the upper triangularization $M_{n-1}E_{n-1} \cdots M_1 E_1 A F_1 \cdots F_{n-1} = U$ with the property that in step $k$ we are confronted with the matrix

$$A^{(k-1)} = M_{k-1}E_{k-1} \cdots M_1 E_1 A F_1 \cdots F_{k-1}$$

and determine interchange permutations $E_k$ and $F_k$ such that

$$\left| \left( E_k A^{(k-1)} F_k \right)_{kk} \right| = \max_{k \leq i,j \leq n} \left| \left( E_k A^{(k-1)} F_k \right)_{ij} \right|.$$

We have the analog of Theorem 3.4.1

**Theorem 3.4.2** *If Gaussian elimination with complete pivoting is used to compute the upper triangularization*

$$M_{n-1}E_{n-1} \cdots M_1 E_1 A F_1 \cdots F_{n-1} = U \qquad (3.4.7)$$

*then*

$$PAQ = LU$$

*where* $P = E_{n-1} \cdots E_1$ , $Q = F_1 \cdots F_{n-1}$ *and* $L$ *is a unit lower triangular matrix with* $|\ell_{ij}| \leq 1$. *The* $k$*th column of* $L$ *below the diagonal is a permuted version of the* $k$*th Gauss vector. In particular, if* $M_k = I - t^{(k)} e_k^T$ *then* $L(k+1{:}n, k) = g(k+1{:}n)$ *where* $g = E_{n-1} \cdots E_{k+1} t^{(k)}$ .

**Proof.** The proof is similar to the proof of Theorem 3.4.1. Details are left to the reader. □

Here is Gaussian elimination with complete pivoting in detail:

**Algorithm 3.4.3 (G.E. with Complete Pivoting)** This algorithm computes the complete pivoting factorization $PAQ = LU$ where $L$ is unit lower triangular and $U$ is upper triangular. $P = E_{n-1} \cdots E_1$ and $Q = F_1 \cdots F_{n-1}$ are products of interchange permutations. $A(1{:}k, k)$ is overwritten by $U(1{:}k, k), k = 1{:}n$. $A(k+1{:}n, k)$ is overwritten by $L(k+1{:}n, k), k = 1{:}n-1$. $E_k$ interchanges rows $k$ and $p(k)$. $F_k$ interchanges columns $k$ and $q(k)$.

```
for k = 1:n - 1
    Determine μ and λ so
        |A(μ, λ)| = max{ |A(i, j)| : i = k:n, j = k:n }
    A(k, 1:n) ↔ A(μ, 1:n); A(1:n, k) ↔ A(1:n, λ)
    p(k) = μ; q(k) = λ
    if A(k, k) ≠ 0
        t = gauss(A(k:n, k))
        A(k + 1:n, k) = t
        A(k:n, k + 1:n) = gauss.app(A(k:n, k + 1:n), t)
    end
end
```

This algorithm requires $2n^3/3$ flops and an equal number of comparisons. Unlike partial pivoting, complete pivoting represents a significant overhead because of the two-dimensional search at each stage.

### 3.4.9 Comments on Complete Pivoting

Suppose $\text{rank}(A) = r < n$. It follows that at the beginning of step $r+1$, $A(r+1{:}n, r+1{:}n) = 0$. This implies that $E_k = F_k = M_k = I$ for $k = r+1{:}n$ and so the algorithm can be terminated after step $r$ with the following factorization in hand:

$$PAQ \;=\; LU \;=\; \begin{bmatrix} L_{11} & 0 \\ L_{21} & I_{n-r} \end{bmatrix} \begin{bmatrix} U_{11} & U_{12} \\ 0 & 0 \end{bmatrix}.$$

Here $L_{11}$ and $U_{11}$ are $r$-by-$r$ and $L_{21}$ and $U_{12}^T$ are $(n-r)$-by-$r$. Thus, Gaussian elimination with complete pivoting can in principle be used to determine the rank of a matrix. Yet roundoff errors make the probability of encountering an exactly zero pivot remote. In practice one would have to "declare" $A$ to have rank $k$ if the pivot element in step $k+1$ was sufficiently small. The numerical rank determination problem is discussed in detail in §5.4.

Wilkinson (1961) has shown that in exact arithmetic the elements of the matrix $A^{(k)} = M_k E_k \cdots M_1 E_1 A F_1 \cdots F_k$ satisfy

$$|a_{ij}^{(k)}| \le k^{1/2}(2 \cdot 3^{1/2} \cdots k^{1/k-1})^{1/2} \max|a_{ij}|. \qquad (3.4.8)$$

The upper bound is a rather slow-growing function of $k$. This fact coupled with vast empirical evidence suggesting that $\rho$ is always modestly sized (e.g, $\rho = 10$) permit us to conclude that *Gaussian elimination with complete pivoting is stable.* The method solves a nearby linear system $(A+E)\hat{x} = b$ exactly in the sense of (3.3.1). However, there appears to be no practical justification for choosing complete pivoting over partial pivoting except in cases where rank determination is an issue.

**Example 3.4.5** If Gaussian elimination with complete pivoting is applied to the problem

$$\begin{bmatrix} .001 & 1.00 \\ 1.00 & 2.00 \end{bmatrix} \begin{bmatrix} x_1 \\ x_2 \end{bmatrix} = \begin{bmatrix} 1.00 \\ 3.00 \end{bmatrix}$$

with $\beta = 10, t = 3$, chopped arithmetic, then

$$P = \begin{bmatrix} 0 & 1 \\ 1 & 0 \end{bmatrix}, \quad \Pi = \begin{bmatrix} 0 & 1 \\ 1 & 0 \end{bmatrix}, \quad \hat{L} = \begin{bmatrix} 1.00 & 0.00 \\ .500 & 1.00 \end{bmatrix}, \quad \hat{U} = \begin{bmatrix} 2.00 & 1.00 \\ 0.00 & .499 \end{bmatrix}$$

and $\hat{x} = (1.00,\ 1.00)^T$. This is correct to two signficant digits in each component. Compare with Example 3.4.3.

## 3.4.10 The Avoidance of Pivoting

For certain classes of matrices it is not necessary to pivot. It is important to identify such classes because pivoting usually degrades performance. To illustrate the kind of analysis required to prove that pivoting can be safely avoided, we consider the case of diagonally dominant matrices. We say that $A \in \mathbb{R}^{n \times n}$ is *strictly diagonally dominant* if

$$|a_{ii}| > \sum_{\substack{j=1 \\ j \ne i}}^{n} |a_{ij}| \qquad i = 1{:}n\,.$$

The following theorem shows how this property can ensure a nice, no-pivoting LU factorization.

**Theorem 3.4.3** *If $A^T$ is diagonally dominant then $A$ has an $LU$ factorization and $|l_{ij}| \leq 1$. In other words, if Algorithm 3.4.1 is applied, $P = I$.*

**Proof.** Partition $A$ as follows

$$A = \begin{bmatrix} \alpha & w^T \\ v & C \end{bmatrix}$$

where $\alpha$ is 1-by-1 and note that after one step of the outer product LU process we have the factorization

$$\begin{bmatrix} \alpha & w^T \\ v & C \end{bmatrix} = \begin{bmatrix} 1 & 0 \\ v/\alpha & I \end{bmatrix} \begin{bmatrix} 1 & 0 \\ 0 & C - vw^T/\alpha \end{bmatrix} \begin{bmatrix} \alpha & w^T \\ 0 & I \end{bmatrix}.$$

The theorem follows by induction on $n$ if we can show that the transpose of $B = C - vw^T/\alpha$ is strictly diagonally dominant since $B = L_1U_1$ implies that

$$A = \begin{bmatrix} 1 & 0 \\ v/\alpha & L_1 \end{bmatrix} \begin{bmatrix} \alpha & w^T \\ 0 & U_1 \end{bmatrix} \equiv LU .$$

But the proof that $B^T$ is strictly diagonally dominant is straight forward. From the definitions we have

$$\begin{aligned} \sum_{\substack{i=1 \\ i \neq j}}^{n-1} |b_{ij}| &= \sum_{\substack{i=1 \\ i \neq j}}^{n-1} |c_{ij} - v_i w_j/\alpha| \leq \sum_{\substack{i=1 \\ i \neq j}}^{n-1} |c_{ij}| + \frac{|w_j|}{|\alpha|} \sum_{\substack{i=1 \\ i \neq j}}^{n-1} |v_i| \\ &\leq (|c_{jj}| - |w_j|) + \frac{|w_j|}{|\alpha|}(|\alpha| - |v_j|) \\ &\leq \left| c_{jj} - \frac{w_j v_j}{\alpha} \right| = |b_{jj}| . \square \end{aligned}$$

### 3.4.11 Some Applications

We conclude with some examples that illustrate how to think in terms of matrix factorizations when confronted with various linear equation situations.

Suppose $A$ is nonsingular and $n$-by-$n$ and that $B$ is $n$-by-$p$. Consider the problem of finding $X$ ($n$-by-$p$) so $AX = B$, i.e., the *multiple right hand side problem.* If $X = [\, x_1, \ldots, x_p \,]$ and $B = [\, b_1, \ldots, b_p \,]$ are column partitions then

```
Compute PA = LU.
for k = 1:p
    Solve Ly = Pb_k
    Solve Ux_k = y                                  (3.4.9)
end
```

Note that $A$ is factored just once. If $B = I_n$ then we emerge with a computed $A^{-1}$

As another example of getting the LU factorization "outside the loop," suppose we want to solve the linear system $A^k x = b$ where $A \in \mathbb{R}^{n\times n}$, $b \in \mathbb{R}^n$, and $k$ is a positive integer. One approach is to compute $C = A^k$ and then solve $Cx = b$. This involves $O(kn^3)$ flops. A much more efficient procedure is the following:

Compute $PA = LU$
**for** $j = 1{:}k$
  Overwrite $b$ with the solution to $Ly = Pb$. (3.4.10)
  Overwrite $b$ with the solution to $Ux = b$.
**end**

As a final example we show how to avoid the pitfall of explicit inverse computation. Suppose we are given $A \in \mathbb{R}^{n\times n}$, $d \in \mathbb{R}^n$, and $c \in \mathbb{R}^n$ and that we want to compute $s = c^T A^{-1} d$. One approach is to compute $X = A^{-1}$ as suggested above and then compute $s = c^T X d$. A more economical procedure is to compute $PA = LU$ and then solve the triangular systems $Ly = Pd$ and $Ux = y$. It follows that $s = c^T x$. The point of this example is to stress that when a matrix inverse is encountered in a formula, we must think in terms of solving equations rather than in terms of explicit inverse formation.

### Problems

**P3.4.1** Let $A = LU$ be the LU factorization of $n$-by-$n$ $A$ with $|\ell_{ij}| \leq 1$. Let $a_i^T$ and $u_i^T$ denote the $i$th rows of $A$ and $U$, respectively. Verify the equation

$$u_i^T = a_i^T - \sum_{j=1}^{i-1} \ell_{ij} u_j^T$$

and use it to show that $\| U \|_\infty \leq 2^{n-1} \| A \|_\infty$ . (Hint: Take norms and use induction.)

**P3.4.2** Show that if $PAQ$ = LU is obtained via Gaussian elimination with complete pivoting, then no element of $U(i, i{:}n)$ is larger in absolute value than $|u_{ii}|$.

**P3.4.3** Suppose $A \in \mathbb{R}^{n\times n}$ has an $LU$ factorization and that $L$ and $U$ are known. Give an algorithm which can compute the $(i,j)$ entry of $A^{-1}$ in approximately $(n-j)^2+(n-i)^2$ flops.

**P3.4.4** Show that if $\hat{X}$ is the computed inverse obtained via (3.4.9), then $A\hat{X} = I + F$, where $\| F \|_1 \leq n\mathbf{u}\left(3 + 5n^2\rho\right) \| A \|_\infty \| \hat{X} \|_1$.

**P3.4.5** Prove Theorem 3.4.2.

**P3.4.6** Extend Algorithm 3.4.3 so that it can factor an arbitrary rectangular matrix.

**P3.4.7** Write a detailed version of the block elimination algorithm outlined in §3.4.7.

### Notes and References for Sec. 3.4

Fortran codes for solving general linear systems may be found in Linpack (Chapter 1) and Algol versions in

H.J. Bowdler, R.S. Martin, G. Peters, and J.H. Wilkinson (1966). "Solution of Real and Complex Systems of Linear Equations," *Numer. Math. 8,* 217–34. See also *HACLA*, pp. 93–110.

The conjecture that $|a_{ij}^{(k)}| \leq n \max|a_{ij}|$ when complete pivoting is used has been proven in the real $n = 4$ case in

C.W. Cryer (1968). "Pivot Size in Guassian Elimination," *Numer. Math. 12,* 335–45.

Other papers concerned with element growth and pivoting include

P.A. Businger (1971). "Monitoring the Numerical Stability of Gaussian Elimination," *Numer. Math. 16,* 360–61.
A.M. Cohen (1974). "A Note on Pivot Size in Gaussian Elimination," *Lin. Alg. and Its Applic. 8,* 361–68.
J. Day and B. Peterson (1988). "Growth in Gaussian Elimination," *Amer. Math. Monthly 95,* 489-513.
A.M. Erisman and J.K. Reid (1974). "Monitoring the Stability of the Triangular Factorization of a Sparse Matrix," *Numer. Math. 22,* 183–86.
N.J. Higham and D.J. Higham (1988). "Large Growth Factors in Gaussian Elimination with Pivoting," Report 152, Department of Mathematics, University of Manchester, M13 9PL, England, to appear *SIAM J. Matrix Analysis and Applications.*
J.K. Reid (1971). "A Note on the Stability of Gaussian Elimination," *J.Inst. Math. Applics. 8,* 374–75.
L.N. Trefethen and R.S. Schreiber (1987). "Average-Case Stability of Gaussian Elimination," Numer. Anal. Report 88-3, Department of Mathematics, MIT. (To appear *SIAM J. Matrix Anal. & Applic.*)
J.H. Wilkinson (1961). "Error Analysis of Direct Methods of Matrix Inversion," *J.ACM 8,* 281–330.

The designers of sparse Gaussian elimination codes are interested in the topic of element growth because multipliers greater than unity are sometimes tolerated for the sake of minimizing fill-in. See

I.S. Duff, A.M. Erisman, and J.K. Reid (1986). *Direct Methods for Sparse Matrices,* Oxford University Press.

The connection between small pivots and near singularity is reviewed in

T.F. Chan (1985). "On the Existence and Computation of LU Factorizations with small pivots," *Math. Comp. 42,* 535–548.

A pivot strategy that we did not discuss is *pairwise pivoting.* In this approach, 2-by-2 Gauss transformations are used to zero the lower triangular portion of $A$. The technique is appealing in certain multiprocessor environments because only adjacent rows are combined in each step. See

D. Sorensen (1985). "Analysis of Pairwise Pivoting in Gaussian Elimination," *IEEE Trans. on Computers C-34,* 274–278.

As a sample paper in which a class of matrices is identified that require no pivoting, see

S. Serbin (1980). "On Factoring a Class of Complex Symmetric Matrices Without Pivoting," *Math. Comp. 35,* 1231-1234.

Just as there are six "conventional" versions of scalar Gaussian elimination, there are also six conventional block formulations of Gaussian elimination. For a discussion of these procedures and their implementation see

M.J. Dayde and I.S. Duff (1988). "Use of Level-3 BLAS in LU Factorization on the Cray-2, the ETA-10P, and the IBM 3090-200/VF," Report CSS-229, Computer Science and Systems Division, Harwell Laboratory, Oxon OX11 ORA, England.

Further practical and theoretical details about block LU may be found in

D.A. Calihan (1986). "Block-Oriented, Local-Memory-Based Linear Equation Solution on the Cray-2: Uniprocessor Algorithms," *Proceedings of the 1986 Conference on Parallel Processing*, pp. 375–378.
K. Gallivan, W. Jalby, U. Meier, and A.H. Sameh (1988). "Impact of Hierarchical Memory Systems on Linear Algebra Algorithm Design," *Int'l J. Supercomputer Applic. 2*, 12–48.

# 3.5 Improving and Estimating Accuracy

Suppose Gaussian elimination with partial pivoting is used to solve the $n$-by-$n$ system $Ax = b$. Assume $t$-digit, base $\beta$ floating point arithmetic is used. Equation (3.4.4) essentially says that if the growth factor is modest then the computed solution $\hat{x}$ satisfies

$$(A + E)\hat{x} = b, \qquad \| E \|_\infty \approx \mathbf{u}\| A \|_\infty, \quad \mathbf{u} = \beta^{-t}. \tag{3.5.1}$$

In this section we explore the practical ramifications of this result. We begin by stressing the distinction that should be made between residual size and accuracy. This is followed by a discussion of scaling, iterative improvement, and condition estimation.

We make two notational remarks at the outset. The infinity norm is used throughout since it is very handy in roundoff error analysis and in practical error estimation. Second, whenever we refer to "Gaussian elimination" in this section we really mean Gaussian elimination with some stablizing pivot strategy such as partial pivoting.

## 3.5.1 Residual Size Versus Accuracy

The *residual* of a computed solution $\hat{x}$ to the linear system $Ax = b$ is the vector $b - A\hat{x}$. A small residual means that $A\hat{x}$ effectively "predicts" the right hand side $b$. From (3.5.1) we have $\| b - A\hat{x} \|_\infty \approx \mathbf{u}\| A \|_\infty \| \hat{x} \|_\infty$ and so we obtain

**Heuristic I.** Gaussian elimination produces a solution $\hat{x}$ with a relatively small residual.

Small residuals do not imply high accuracy. Combining (3.5.1) with Theorem 2.7.1 we see that

$$\frac{\| \hat{x} - x \|_\infty}{\| x \|_\infty} \approx \mathbf{u}\kappa_\infty(A)\,. \tag{3.5.2}$$

This justifies a second guiding principle.

**Heuristic II.** If the unit roundoff and condition satisfy $\mathbf{u} \approx 10^{-d}$ and $\kappa_\infty(A) \approx 10^q$, then Gaussian elimination produces a solution $\hat{x}$ that has about $d - q$ correct decimal digits.

As an illustration of the Heuristics I and II, consider the system

$$\begin{bmatrix} .986 & .579 \\ .409 & .237 \end{bmatrix} \begin{bmatrix} x_1 \\ x_2 \end{bmatrix} = \begin{bmatrix} .235 \\ .107 \end{bmatrix}$$

in which $\kappa_\infty(A) \approx 700$ and $x = (2, -3)^T$. Here is what we find for various machine precisions:

| $\beta$ | $t$ | $\hat{x}_1$ | $\hat{x}_2$ | $\frac{\| \hat{x} - x \|_\infty}{\| x \|_\infty}$ | $\frac{\| b - A\hat{x} \|_\infty}{\| A \|_\infty \| \hat{x} \|_\infty}$ |
|---|---|---|---|---|---|
| 10 | 3 | 2.11 | -3.17 | $5 \cdot 10^{-2}$ | $2.0 \cdot 10^{-3}$ |
| 10 | 4 | 1.986 | -2.975 | $8 \cdot 10^{-3}$ | $1.5 \cdot 10^{-4}$ |
| 10 | 5 | 2.0019 | -3.0032 | $1 \cdot 10^{-3}$ | $2.1 \cdot 10^{-6}$ |
| 10 | 6 | 2.00025 | -3.00094 | $3 \cdot 10^{-4}$ | $4.2 \cdot 10^{-7}$ |

Whether or not one is content with the computed solution $\hat{x}$ depends on the requirements of the underlying source problem. In many applications accuracy is not important but small residuals are. In such a situation, the $\hat{x}$ produced by Gaussian elimination is probably adequate. On the other hand, if the number of correct digits in $\hat{x}$ is an issue then the situation is more complicated and the discussion in the remainder of this section is relevant.

### 3.5.2 Scaling

Let $\beta$ be the machine base and define the diagonal matrices $D_1$ and $D_2$ by

$$\begin{aligned} D_1 &= \text{diag}(\beta^{r_1} \ldots \beta^{r_n}) \\ D_2 &= \text{diag}(\beta^{c_1} \ldots \beta^{c_n})\,. \end{aligned}$$

The solution to the $n$-by-$n$ linear system $Ax = b$ can be found by solving the *scaled system* $(D_1^{-1}AD_2)y = D_1^{-1}b$ using Gaussian elimination and then setting $x = D_2 y$. The scalings of $A$, $b$, and $y$ require only $O(n^2)$ flops and may be accomplished without roundoff. Note that $D_1$ scales equations and $D_2$ scales unknowns.

It follows from Heuristic II that if $\hat{x}$ and $\hat{y}$ are the computed versions of $x$ and $y$, then

$$\frac{\| D_2^{-1}(\hat{x} - x) \|_\infty}{\| D_2^{-1} x \|_\infty} = \frac{\| \hat{y} - y \|_\infty}{\| y \|_\infty} \approx \mathbf{u}\kappa_\infty(D_1^{-1} A D_2) . \qquad (3.5.3)$$

Thus, if $\kappa_\infty(D_1^{-1}AD_2)$ can be made considerably smaller than $\kappa_\infty(A)$, then we might expect a correspondingly more accurate $\hat{x}$, provided errors are measured in the "$D_2$" norm defined by $\| z \|_{D_2} = \| D_2^{-1} z \|_\infty$. This is the objective of scaling. Note that it encompasses two issues: the condition of the scaled problem and the appropriateness of appraising error in the $D_2$-norm.

An interesting but very difficult mathematical problem concerns the exact minimization of $\kappa_p(D_1^{-1}AD_2)$ for general diagonal $D_i$ and various $p$. What results there are in this direction are not very practical. This is hardly discouraging, however, when we recall that (3.5.3) is heuristic and it makes little sense to minimize exactly a heuristic bound. What we seek is a fast, approximate method for improving the quality of the computed solution $\hat{x}$.

One technique of this variety is *simple row scaling.* In this scheme $D_2$ is the identity and $D_1$ is chosen so that each row in $D_1^{-1}A$ has approximately the same $\infty$-norm. Row scaling reduces the likelihood of adding a very small number to a very large number during elimination - an event that can greatly diminish accuracy.

Slightly more complicated than simple row scaling is *row-column equilibration.* Here, the object is to choose $D_1$ and $D_2$ so that the $\infty$-norm of each row and column of $D_1^{-1}AD_2$ belongs to the interval $[1/\beta, 1]$ where $\beta$ is the base of the floating point system. For work along these lines see McKeeman (1962).

It cannot be stressed too much that simple row scaling and row-column equilibration do not "solve" the scaling problem. Indeed, either technique can render a worse $\hat{x}$ than if no scaling whatever is used. The ramifications of this point are thoroughly discussed in Forsythe and Moler (SLE, chapter 11). The basic recommendation is that the scaling of equations and unknowns must proceed on a problem-by-problem basis. General scaling strategies are unreliable. It is best to scale (if at all) on the basis of what the source problem proclaims about the significance of each $a_{ij}$. Measurement units and data error may have to be considered.

**Example 3.5.1** (Forsythe and Moler [SLE, pp. 34, 40]) . If

$$\begin{bmatrix} 10 & 100,000 \\ 1 & 1 \end{bmatrix} \begin{bmatrix} x_1 \\ x_2 \end{bmatrix} = \begin{bmatrix} 100,000 \\ 2 \end{bmatrix}$$

and the equivalent row-scaled problem

$$\begin{bmatrix} .0001 & 1 \\ 1 & 1 \end{bmatrix} \begin{bmatrix} x_1 \\ x_2 \end{bmatrix} = \begin{bmatrix} 1 \\ 2 \end{bmatrix}$$

are each solved using $\beta = 10, t = 3$ arithmetic, then solutions $\hat{x} = (0.00,\ 1.00)^T$ and $\hat{x} = (1.00,\ 1.00)^T$ are respectively computed. Note that $x = (1.0001\ldots,\ .9999\ldots)^T$ is the exact solution.

### 3.5.3 Iterative Improvement

Suppose $Ax = b$ has been solved via the partial pivoting factorization $PA = LU$ and that we wish to improve the accuracy of the computed solution $\hat{x}$. If we execute

$$\begin{aligned} &r = b - A\hat{x} \\ &\text{Solve } Ly = Pr. \\ &\text{Solve } Uz = y. \\ &x_{new} = \hat{x} + z \end{aligned} \tag{3.5.4}$$

then in exact arithmetic $Ax_{new} = A\hat{x} + Az = (b - r) + r = b$. Unfortunately, the naive floating point execution of these formulae renders an $x_{new}$ that is no more accurate than $\hat{x}$. This is to be expected since $\hat{r} = fl(b - A\hat{x})$ has few, if any, correct significant digits. (Recall Heuristic I.) Consequently, $\hat{z} = fl(A^{-1}r) \approx A^{-1} \cdot \text{noise} \approx \text{noise}$ is a very poor correction *from the standpoint of improving the accuracy of* $\hat{x}$. However, Skeel (1980) has done an error analysis that indicates when (3.5.4) gives an improved $x_{new}$ *from the standpoint of backwards error.* In particular, if the quantity

$$\tau = (\| \, |A| \, |A^{-1}| \, \|_\infty) \left( \max_i \ (|A||x|)_i \,/ \min_i \ (|A||x|)_i \right)$$

is not too big, then (3.5.4) produces an $x_{new}$ such that $(A + E)x_{new} = b$ for very small $E$. Of course, if Gaussian elimination with partial pivoting is used then the computed $\hat{x}$ already solves a nearby system. However, this may not be the case for some of the pivot strategies that are used to preserve sparsity. In this situation, the *fixed precision iterative improvement* step (3.5.4) can be very worthwhile and cheap. See Arioli, Demmel, and Duff (1988).

For (3.5.4) to produce a more accurate $x$, it is necessary to compute the residual $b - A\hat{x}$ with extended precision floating point arithmetic. Typically, this means that if $t$-digit arithmetic is used to compute $PA = LU$, $x$, $y$, and $z$, then $2t$-digit arithmetic is used to form $b - A\hat{x}$, i.e., double precision. The process can be iterated. In particular, once we have computed $PA = LU$ and initialize $x = 0$, we repeat the following:

$$\begin{aligned} & r = b - Ax \text{ (Double Precision)} \\ & \text{Solve } Ly = Pr \text{ for } y. \\ & \text{Solve } Uz = y \text{ for } z. \\ & x = x + z \end{aligned} \tag{3.5.5}$$

We refer to this process as *mixed precision iterative improvement*. The original $A$ must be used in the double precision computation of $r$. The basic result concerning the performance of (3.5.5) is summarized in the following heuristic:

**Heuristic III.** If the machine precision $\mathbf{u}$ and condition satisfy $\mathbf{u} = 10^{-d}$ and $\kappa_\infty(A) \approx 10^q$, then after $k$ executions of (3.5.5), $x$ has approximately $\min(d, k(d-q))$ correct digits.

Roughly speaking, if $\mathbf{u}\kappa_\infty(A) \leq 1$, then iterative improvement can ultimately produce a solution that is correct to full (single) precision. Note that the process is relatively cheap. Each improvement costs $O(n^2)$, to be compared with the original $O(n^3)$ investment in the factorization $PA = LU$. Of course, no improvement may result if $A$ is badly enough conditioned with respect to the machine precision.

The primary drawback of mixed precision iterative improvement is that its implementation is somewhat machine-dependent. This discourages its use in software that it is intended for wide distribution. The need for retaining an original copy of $A$ is another aggravation associated with the method.

On the other hand, mixed precision iterative improvement is usually very easy to implement on a given machine that has provision for the accumulation of inner products, i.e., provision for the double precision calculation of inner products between the rows of $A$ and $x$. In a short mantissa computing environment the presence of an iterative improvement routine can significantly widen the class of solvable $Ax = b$ problems.

**Example 3.5.2** If (3.5.5) is applied to the system

$$\begin{bmatrix} .986 & .579 \\ .409 & .237 \end{bmatrix} \begin{bmatrix} x_1 \\ x_2 \end{bmatrix} = \begin{bmatrix} .235 \\ .107 \end{bmatrix}$$

and $\beta = 10$ and $t = 3$, then iterative improvement produces the following sequence of computed solutions:

$$\hat{x} = \begin{bmatrix} 2.11 \\ -3.17 \end{bmatrix}, \begin{bmatrix} 1.99 \\ -2.99 \end{bmatrix}, \begin{bmatrix} 2.00 \\ -3.00 \end{bmatrix}, \ldots$$

The exact solution is $x = (2, -3)^T$.

### 3.5.4 Condition Estimation

Suppose that we have solved $Ax = b$ via $PA = LU$ and that we now wish to ascertain the number of correct digits in the computed solution $\hat{x}$. It follows from Heuristic II that in order to do this we need an estimate of the condition $\kappa_\infty(A) = \| A \|_\infty \| A^{-1} \|_\infty$. Computing $\| A \|_\infty$ poses no problem as we merely use the formula

$$\| A \|_\infty = \max_{1 \le i \le n} \sum_{j=1}^{n} |a_{ij}|.$$

The challenge is with respesct to the factor $\| A^{-1} \|_\infty$. Conceivably, we could estimate this quantity by $\| \hat{X} \|_\infty$, where $\hat{X} = [\, \hat{x}_1, \ldots, \hat{x}_n \,]$ and $\hat{x}_i$ is the computed solution to $Ax_i = e_i$. (See §3.4.9.) The trouble with this approach is its expense: $\hat{\kappa}_\infty = \| A \|_\infty \| \hat{X} \|_\infty$ costs about three times as much as $\hat{x}$.

The central problem of *condition estimation* is how to estimate the condition number in $O(n^2)$ flops assuming the availability of $PA = LU$ or some other factorizations that are presented in subsequent chapters. An approach described in Forsythe and Moler (SLE, p. 51) is based on iterative improvement and the heuristic $\mathbf{u}\kappa_\infty(A) \approx \| z \|_\infty / \| x \|_\infty$ where $z$ is the first correction of $x$ in (3.5.5). While the resulting condition estimator is $O(n^2)$, it suffers from the shortcoming of iterative improvement, namely, machine dependency.

Cline, Moler, Stewart, and Wilkinson (1979) have proposed a very successful approach to the condition estimation problem without this flaw. It is based on exploitation of the implication

$$Ay = d \quad \Longrightarrow \quad \| A^{-1} \|_\infty \ \ge \ \| y \|_\infty / \| d \|_\infty.$$

The idea behind their estimator is to choose $d$ so that the solution $y$ is large in norm and then set

$$\hat{\kappa}_\infty \ = \ \| A \|_\infty \| y \|_\infty / \| d \|_\infty.$$

The success of this method hinges on how close the ratio $\| y \|_\infty / \| d \|_\infty$ is to its maximum value $\| A^{-1} \|_\infty$.

Consider the case when $A = T$ is upper triangular. The relation between $d$ and $y$ is completely specified by the following column version of back substitution:

$$\begin{array}{l} p(1{:}n) = 0 \\ \textbf{for } k = n{:} -1{:}1 \\ \quad \text{Choose } d(k). \\ \quad y(k) = (d(k) - p(k))/T(k,k) \\ \quad p(1{:}k-1) = p(1{:}k-1) + y(k)T(1{:}k-1,k) \\ \textbf{end} \end{array} \qquad (3.5.6)$$

Normally, we use this algorithm to solve a *given* triangular system $Ty = d$. Now, however, we are free to pick the right-hand side $d$ subject to the "constraint" that $y$ is large relative to $d$.

One way to encourage growth in $y$ is to choose $d(k)$ from the set $\{-1, +1\}$ so as to maximize $y(k)$. If $p(k) \geq 0$ then set $d(k) = -1$. If $p(k) < 0$ then set $d(k) = +1$ . In other words, (3.5.6) is invoked with $d(k)$ $= \text{sign}(p(k))$. Since $d$ is then a vector of the form $d(1{:}n) = (\pm 1, \ldots, \pm 1)^T$, we obtain the estimator $\hat{\kappa}_\infty = \| T \|_\infty \| y \|_\infty$.

A more reliable estimator results if $d(k) \in \{-1, +1\}$ is chosen so as to encourage growth both in $y(k)$ and the updated running sum given by $p(1{:}k-1, k) + T(1{:}k-1, k)y(k)$. In particular, at step $k$ we compute

$$y(k)^+ = (1 - p(k))/T(k,k)$$

$$s(k)^+ = |y(k)^+| \; + \; \| p(1{:}k-1) + T(1{:}k-1, k)y(k)^+ \|_1$$

$$y(k)^- = (-1 - p(k))/T(k,k)$$

$$s(k)^- = |y(k)^-| \; + \; \| p(1{:}k-1) + T(1{:}k-1, k)y(k)^- \|_1$$

We then set $y(k)$ to $y(k)^+$ if $s(k)^+ \geq s(k)^-$ and to $y(k)^-$ otherwise. This gives

**Algorithm 3.5.1 (Condition Estimator)** Let $T \in \mathbb{R}^{n \times n}$ be a nonsingular upper triangular matrix. This algorithm computes unit $\infty$-norm $y$ and a scalar $\kappa$ so $\| Ty \|_\infty \approx 1/\| T^{-1} \|_\infty$ and $\kappa \approx \kappa_\infty(T)$

$p(1{:}n) = 0$
**for** $k = n{:} -1{:}1$
  $y(k)^+ = (1 - p(k))/T(k,k)$; $y(k)^- = (-1 - p(k))/T(k,k)$
  $p(k)^+ = p(1{:}k-1) + T(1{:}k-1, k)y(k)^+$
  $p(k)^- = p(1{:}k-1) + T(1{:}k-1, k)y(k)^-$
  **if** $|y(k)^+| + \| p(k)^+ \|_1 \geq |y(k)^-| + \| p(k)^- \|_1$
    $y(k) = y(k)^+$; $p(1{:}k-1) = p(k)^+$
  **else**
    $y(k) = y(k)^-$; $p(1{:}k-1) = p(k)^-$
  **end**
**end**
$\kappa = \| y \|_\infty \| T \|_\infty$

The algorithm involves several times the work of ordinary back substitution.

We are now in a position to describe a procedure for estimating the condition of a square nonsingular matrix $A$ whose $PA = LU$ factorization we know:

- Apply the lower triangular version of Algorithm 3.5.1 to $U^T$ and obtain a large norm solution to $U^T y = d$.
- Solve the triangular systems $L^T r = y$, $Lw = Pr$, and $Uz = w$.
- $\hat{\kappa}_\infty = \| A \|_\infty \| z \|_\infty / \| r \|_\infty$.

Note that $\| z \|_\infty \le \| A^{-1} \|_\infty \| r \|_\infty$. The method is based on several heuristics. First, if $A$ is ill-conditioned and $PA = LU$, then it is usually the case that $U$ is correspondingly ill-conditioned. The lower triangle $L$ tends to be fairly well-conditioned. Thus, it is more profitable to apply the condition estimator to $U$ than to $L$. The vector $r$, because it solves $A^T P^T r = d$, tends to be rich in the direction of the left singular vector associated with $\sigma_{min}(A)$. Righthand sides with this property render large solutions to the problem $Az = r$.

In practice, it is found that the condition estimation technique that we have outlined produces good order-of-magnitude estimates of the actual condition number. See Linpack.

**Example 3.5.3** If Algorithm 3.5.1 is applied to the matrix $B_n$ given in (2.7.9), then $y = (2^{n-1}, 2^{n-2}, \ldots, 2, 1)^T$ and $\| y \|_\infty = \| B_n^{-1} \|_\infty$.

### Problems

**P3.5.1** Show by example that there may be more than one way to equilibrate a matrix.

**P3.5.2** Using $\beta = 10, t = 2$ arithmetic, solve

$$\begin{bmatrix} 11 & 15 \\ 5 & 7 \end{bmatrix} \begin{bmatrix} x_1 \\ x_2 \end{bmatrix} = \begin{bmatrix} 7 \\ 3 \end{bmatrix}$$

using Gaussian elimination with partial pivoting. Do one step of iterative improvement using $t = 4$ arithmetic to compute the residual. (Do not forget to round the computed residual to two digits.)

**P3.5.3** Suppose $P(A + E) = \hat{L}\hat{U}$, where $P$ is a permutation, $\hat{L}$ is lower triangular with $|\hat{\ell}_{ij}| \le 1$, and $\hat{U}$ is upper triangular. Show that

$$\hat{\kappa}_\infty(A) \quad \ge \quad \frac{\| A \|_\infty}{\| E \|_\infty + \mu}$$

where $\mu = \min |\hat{u}_{ii}|$. Conclude that if a small pivot is encountered when Gaussian elimination with pivoting is applied to $A$, then $A$ is ill-conditioned. The converse is not true. (Let $A = B_n$).

**P3.5.4** (Kahan 1966) The system $Ax = b$ where

$$A = \begin{bmatrix} 2 & -1 & 1 \\ -1 & 10^{-10} & 10^{-10} \\ 1 & 10^{-10} & 10^{-10} \end{bmatrix} \qquad b = \begin{bmatrix} 2(1 + 10^{-10} \\ -10^{-10} \\ 10^{-10} \end{bmatrix}$$

has solution $x = (10^{-10} \;\; -1 \;\; 1)^T$. (a) Show that if $(A + E)y = b$ and $|E| \le 10^{-8}|A|$, then $|x - y| \le 10^{-7}|x|$. That is, small relative changes in $A$'s entries do not induce large changes in $x$ even though $\kappa_\infty(A) = 10^{10}$. (b) Define $D = \text{diag}(10^{-5}, 10^5, 10^5)$. Show $\kappa_\infty(DAD) \leq 5$. (c) Explain what is going on in terms of Theorem 2.7.3.

**P3.5.5** Consider the matrix:

$$T = \begin{bmatrix} 1 & 0 & M & -M \\ 0 & 1 & -M & M \\ 0 & 0 & 1 & 0 \\ 0 & 0 & 0 & 1 \end{bmatrix} \quad M \in \mathbf{R}.$$

What estimate of $\kappa_\infty(T)$ is produced when (3.5.6) is applied with $d(k) = -\text{sign}(p(k))$? What estimate does Algorithm 3.5.1 produce? What is the true $\kappa_\infty(T)$?

### Notes and References for Sec. 3.5

The following papers are concerned with the scaling of $Ax = b$ problems:

F.L. Bauer (1963). "Optimally Scaled Matrices," *Numer. Math. 5,* 73–87.

P.A. Businger (1968). "Matrices Which Can be Optimally Scaled," *Numer. Math. 12,* 346–48.

T. Fenner and G. Loizou (1974). "Some New Bounds on the Condition Numbers of Optimally Scaled Matrices," *J. ACM 21,* 514–24.

G.H. Golub and J.M. Varah (1974). "On a Characterization of the Best $L_2$-Scaling of a Matrix," *SIAM J. Num. Anal. 11,* 472–79.

C. McCarthy and G. Strang (1973). "Optimal Conditioning of Matrices," *SIAM J. Num. Anal. 10,* 370–88.

R. Skeel (1979). "Scaling for Numerical Stability in Gaussian Elimination," *J. ACM. 26,* 494–526.

R. Skeel (1981). "Effect of Equilibration on Residual Size for Partial Pivoting," *SIAM J. Num. Anal. 18,* 449–55.

A. van der Sluis (1969). "Condition Numbers and Equilibration Matrices," *Numer. Math. 14,* 14–23.

A. van der Sluis (1970). "Condition, Equilibration, and Pivoting in Linear Algebraic Systems," *Numer. Math. 15,* 74–86.

Part of the difficulty in scaling concerns the selection of a norm in which to measure errors. An interesting discussion of this frequently overlooked point appears in

W. Kahan (1966). "Numerical Linear Algebra," *Canadian Math. Bull. 9,* 757–801.

For a rigorous analysis of iterative improvement and related matters, see

M. Jankowski and M. Wozniakowski (1977). "Iterative Refinement Implies Numerical Stability," *BIT 17,* 303–311.

C.B. Moler (1967). "Iterative Refinement in Floating Point," *J. ACM 14,* 316-71.

R.D. Skeel (1980). "Iterative Refinement Implies Numerical Stability for Gaussian Elimination," *Math. Comp. 35,* 817–832.

G.W. Stewart (1981). "On the Implicit Deflation of Nearly Singular Systems of Linear Equations," *SIAM J. Sci. and Stat. Comp. 2,* 136-140.

The condition estimator that we described is given in

A.K. Cline, C.B. Moler, G.W. Stewart, and J.H. Wilkinson (1979). "An Estimate for the Condition Number of a Matrix," *SIAM J. Num. Anal.* 16, 368-75.

and is incorporated in Linpack (pp. 1.10-1.13). Other references concerned with the condition estimation problem include

C.G. Broyden (1973). "Some Condition Number Bounds for the Gaussian Elimiantion Process," *J. Inst. Math. Applic. 12,* 273–86.

A.K. Cline, A.R. Conn, and C. Van Loan (1982). "Generalizing the LINPACK Condition Estimator," in *Numerical Analysis* , ed., J.P. Hennart, Lecture Notes in Mathematics no. 909, Springer-Verlag, New York.

A.K. Cline and R.K. Rew (1983). "A Set of Counter examples to Three Condition Number Estimators," *SIAM J. Sci. and Stat. Comp. 4,* 602–611.

R.G. Grimes and J.G. Lewis (1981). "Condition Number Estimation for Sparse Matrices," *SIAM J. Sci. and Stat. Comp. 2,* 384–88.

W. Hager (1984). "Condition Estimates," *SIAM J. Sci. and Stat. Comp. 5,* 311–316.

N.J. Higham (1987). "A Survey of Condition Number Estimation for Triangular Matrices," *SIAM Review 29,* 575–596.

N.J. Higham (1988). "Fortran Codes for Estimating the One-norm of a Real or Complex Matrix, with Applications to Condition Estimation," *ACM Trans. Math. Soft. 14,* 381–396.

F. Lemeire (1973). "Bounds for Condition Numbers of Triangular Value of a Matrix," *Lin. Alg. and Its Applic. 11,* 1–2.

D.P. O'Leary (1980). "Estimating Matrix Condition Numbers,"*SIAM J. Sci. Stat. Comp. 1,* 205–9.

G.W. Stewart (1980). "The Efficient Generation of Random Orthogonal Matrices with an Application to Condition Estimators," *SIAM J. Num. Anal. 17,* 403–9.

C. Van Loan (1987). "On Estimating the Condition of Eigenvalues and Eigenvectors," *Lin. Alg. and Its Applic.* 88/89, 715–732.

R.S. Varga (1976). "On Diagonal Dominance Arguments for Bounding $\| A^{-1} \|_\infty$," *Lin. Alg. and Its Applic. 14,* 211–17.

# Chapter 4

# Special Linear Systems



It is a basic tenet of numerical analysis that structure should be exploited whenever solving a problem. In numerical linear algebra, this translates into an expectation that algorithms for general matrix problems can be streamlined in the presence of such properties as symmetry, definiteness, and sparsity. This is the central theme of the current chapter, where our principal aim is to devise special algorithms for computing special variants of the LU factorization.

We begin by pointing out the connection between the triangular factors $L$ and $U$ when $A$ is symmetric. This is achieved by examining the $\text{LDM}^\text{T}$ factorization in §4.1. We then turn our attention to the important case when $A$ is both symmetric and positive definite, deriving the stable Cholesky factorization in §4.2. Unsymmetric positive definite systems are also investigated in this section. In §4.3, banded versions of Gaussian elimination and other factorization methods are discussed. We then examine the interesting situation when $A$ is symmetric but indefinite. Our treatment of this problem in §4.4 highlights the numerical analyst's ambivalence towards pivoting. We love pivoting for the stability it induces but despise it for the structure that it can destroy. Fortunately, there is a happy resolution to this conflict in the symmetric indefinite problem.

Any block banded matrix is also banded and so the methods of §4.3 are applicable. Yet, there are occasions when it pays not to adopt this point of view. To illustrate this we consider the important case of block tridiagonal systems in §4.5.

In the final two sections we examine some very interesting $O(n^2)$ algorithms that can be used to solve Vandermonde and Toeplitz systems.

## 4.1 The LDM$^\mathrm{T}$ and LDL$^\mathrm{T}$ Factorizations

We want to develop a structure-exploiting method for solving symmetric $Ax = b$ problems. To do this we establish a variant of the LU factorization in which $A$ is factored into a three-matrix product $LDM^T$ where $D$ is diagonal and $L$ and $M$ are unit lower triangular. Once this factorization is obtained, the solution to $Ax = b$ may be found in $O(n^2)$ flops by solving $Ly = b$ (forward elimination), $Dz = y$, and $M^Tx = z$ (back substitution). The reason for developing the LDM$^\mathrm{T}$ factorization is to set the stage for the symmetric case for if $A = A^T$ then $L = M$ and the work associated with the factorization is half of that required by Gaussian elimination. The issue of pivoting is taken up in subsequent sections.

### 4.1.1 The LDM$^\mathrm{T}$ Factorization

Our first result connects the LDM$^\mathrm{T}$ factorization with the LU factorization.

**Theorem 4.1.1** *If all the leading principal submatrices of $A \in \mathbb{R}^{n\times n}$ are nonsingular, then there exist unique unit lower triangular matrices $L$ and $M$ and a unique diagonal matrix $D = \mathrm{diag}(d_1, \ldots, d_n)$ such that $A = LDM^T$.*

**Proof.** By Theorem 3.2.1 we know that $A$ has an LU factorization $A = LU$. Set $D = \mathrm{diag}(d_1, \ldots, d_n)$ with $d_i = u_{ii}$ for $i = 1{:}n$. Notice that $D$ is nonsingular and that $M^T = D^{-1}U$ is unit upper triangular. Thus, $A = LU = LD(D^{-1}U) = LDM^T$. Uniqueness follows from the uniqueness of the LU factorization as described in Theorem 3.2.1. □

The proof shows that the LDM$^\mathrm{T}$ factorization can be found by using Gaussian elimination to compute $A = LU$ and then determining $D$ and $M$ from the equation $U = DM^T$. However, an interesting alternative algorithm can be derived by computing $L$, $D$, and $M$ directly.

Assume that we know the first $j-1$ columns of $L$, diagonal entries $d_1, \ldots, d_{j-1}$ of $D$, and the first $j-1$ rows of $M$ for some $j$ with $1 \le j \le n$. To develop recipes for $L(j+1{:}n, j)$, $M(j, 1{:}j-1)$, and $d_j$ we equate $j$th columns in the equation $A = LDM^T$. In particular,

$$A(1:n, j) = Lv \tag{4.1.1}$$

where

$$v = DM^T e_j.$$

The "top" half of (4.1.1) defines $v(1:j)$ as the solution of a known lower triangular system:

$$L(1:j, 1:j)v(1:j) = A(1:j, j).$$

Once we know $v$ then we compute

$$\begin{aligned} d(j) &= v(j) \\ m(j,i) &= v(i)/d(i) \qquad i = 1:j-1. \end{aligned}$$

The "bottom" half of (4.1.1) says

$$L(j+1:n, 1:j)v(1:j) = A(j+1:n, j)$$

which can be rearranged to obtain a recipe for the $j$th column of $L$:

$$L(j+1:n, j)v(j) = A(j+1:n, j) - L(j+1:n, 1:j-1)v(1:j-1).$$

Thus, $L(j+1:n, j)$ is a scaled gaxpy operation and overall we obtain

$$\begin{array}{l}
\textbf{for } j = 1:n \\
\quad \text{Solve } L(1:j, 1:j)v(1:j) = A(1:j, j) \text{ for } v(1:j). \\
\quad \textbf{for } i = 1:j-1 \\
\quad\quad M(j,i) = v(i)/d(i) \\
\quad \textbf{end} \\
\quad d(j) = v(j) \\
\quad L(j+1:n, j) = \\
\quad\quad (A(j+1:n, j) - L(j+1:n, 1:j-1)v(1:j-1))\,/v(j) \\
\textbf{end}
\end{array} \tag{4.1.2}$$

As with the LU factorization, it is possible to overwrite $A$ with the $L$, $D$, and $M$ factors. If the column version of forward elimination is used to solve for $v(1:j)$ then we obtain the following procedure:

**Algorithm 4.1.1 (LDM$^T$)** If $A \in \mathbb{R}^{n \times n}$ has an LU factorization then this algorithm computes unit lower triangular matrices $L$ and $M$ and a diagonal matrix $D = \text{diag}(d_1, \ldots, d_n)$ such that $A = LDM^T$. The entry $a_{ij}$ is overwritten with $\ell_{ij}$ if $i > j$, with $d_i$ if $i = j$, and with $m_{ji}$ if $i < j$.

```
for j = 1:n
    { Solve L(1:j,1:j)v(1:j) = A(1:j,j). }
    v(1:j) = A(1:j,j)
    for k = 1:j − 1
        v(k + 1:j) = v(k + 1:j) − v(k)A(k + 1:j,k)
    end
    { Compute M(j,1:j − 1) and store in A(1:j − 1,j). }
    for i = 1:j − 1
        A(i,j) = v(i)/A(i,i)
    end
    { Store d(j) in A(j,j). }
    A(j,j) = v(j)
    { Compute L(j + 1:n,j) and store in A(j + 1:n,j) }
    for k = 1:j − 1
        A(j + 1:n,j) = A(j + 1:n,j) − v(k)A(j + 1:n,k)
    end
    A(j + 1:n,j) = A(j + 1:n,j)/v(j)
end
```

This algorithm involves the same amount of work as the LU factorization, about $2n^3/3$ flops.

The computed solution $\hat{x}$ to $Ax = b$ obtained via Algorithm 4.1.1 and the usual triangular system solvers of §3.1 can be shown to satisfy a perturbed system $(A+E)\hat{x} = b$, where

$$|E| \leq n\mathbf{u}\left(3|A| + 5|\hat{L}||\hat{D}||\hat{M}^T|\right) + O(\mathbf{u}^2) \tag{4.1.3}$$

and $\hat{L}$, $\hat{D}$, and $\hat{M}$ are the computed versions of $L$, $D$, and $M$, respectively.

As in the case of the LU factorization considered in the previous chapter, the upper bound in (4.1.3) is without limit unless some form of pivoting is done. Hence, for Algorithm 4.1.1 to be a practical procedure, it must be modified so as to compute a factorization of the form $PA = LDM^T$, where $P$ is a permutation matrix chosen so that the entries in $L$ satisfy $|\ell_{ij}| \leq 1$. The details of this are not pursued here since they are straightforward and since our main object for introducing the LDM$^T$ factorization is to motivate special methods for symmetric systems.

**Example 4.1.1**

$$A = \begin{bmatrix} 10 & 10 & 20 \\ 20 & 25 & 40 \\ 30 & 50 & 61 \end{bmatrix} = \begin{bmatrix} 1 & 0 & 0 \\ 2 & 1 & 0 \\ 3 & 4 & 1 \end{bmatrix} \begin{bmatrix} 10 & 0 & 0 \\ 0 & 5 & 0 \\ 0 & 0 & 1 \end{bmatrix} \begin{bmatrix} 1 & 1 & 2 \\ 0 & 1 & 0 \\ 0 & 0 & 1 \end{bmatrix}$$

and upon completion, Algorithm 4.1.1 overwrites $A$ as follows:

$$A = \begin{bmatrix} 10 & 1 & 2 \\ 2 & 5 & 0 \\ 3 & 4 & 1 \end{bmatrix}.$$

### 4.1.2 Symmetry and the LDL$^T$ Factorization

There is redundancy in the LDM$^T$ factorization if $A$ is symmetric.

**Theorem 4.1.2** *If $A = LDM^T$ is the $LDM^T$ factorization of a nonsingular symmetric matrix $A$, then $L = M$.*

**Proof.** The matrix $M^{-1}AM^{-T} = M^{-1}LD$ is both symmetric and lower triangular and therefore diagonal. Since $D$ is nonsingular, this implies that $M^{-1}L$ is also diagonal. But $M^{-1}L$ is unit lower triangular and so $M^{-1}L = I.\square$

In view of this result, it is possible to halve the work in Algorithm 4.1.1 when it is applied to a symmetric matrix. In the $j$th step we already know $M(j, 1{:}j-1)$ since $M = L$ and we presume knowledge of $L$'s first $j-1$ columns. Recall that in the $j$th step of (4.1.2) the vector $v$ is defined by $v = DM^Te_j$. Since $M = L$ this says that

$$v = \begin{bmatrix} d(1)L(j,1) \\ \vdots \\ d(j-1)L(j,j-1) \\ d(j) \end{bmatrix}.$$

Hence, the vector $v(1{:}j-1)$ can be obtained by a simple scaling of $L$'s $j$th row. A formula for the $j$th component of $v$ can be derived from the $j$th equation in $L(1{:}j, 1{:}j)v = A(1{:}j, j)$,

$$v(j) = A(j,j) - L(j, 1{:}j-1)v(1{:}j-1)$$

and we therefore obtain:

```
for j = 1:n
    for i = 1:j - 1
        v(i) = L(j,i)d(i)
    end
    v(j) = A(j,j) - L(j,1:j - 1)v(1:j - 1)
    d(j) = v(j)
    L(j + 1:n,j) =
        (A(j + 1:n,j) - L(j + 1:n,1:j - 1)v(1:j - 1))/v(j)
end
```

With overwriting this becomes

**Algorithm 4.1.2 (LDL$^T$)** If $A \in \mathbb{R}^{n\times n}$ is symmetric and has an LU factorization then this algorithm computes a unit lower triangular matrix $L$ and a diagonal matrix $D = \text{diag}(d_1, \ldots, d_n)$ so $A = LDL^T$. The entry $a_{ij}$ is overwritten with $\ell_{ij}$ if $i > j$ and with $d_i$ if $i = j$.

```
for j = 1:n
    { Compute v(1:j). }
    for i = 1:j - 1
        v(i) = A(j,i)A(i,i)
    end
    v(j) = A(j,j) - L(j,1:j - 1)v(1:j - 1)
    { Store d(j) and compute L(j + 1:n, j). }
    A(j,j) = v(j)
    A(j + 1:n, j) =
        (A(j + 1:n, j) - A(j + 1:n, 1:j - 1)v(1:j - 1))/v(j)
end
```

This algorithm requires $n^3/3$ flops, about half the number of flops involved in Gaussian elimination.

In the next section, we show that if $A$ is both symmetric and positive definite, then Algorithm 4.1.2 not only runs to completion, but is extremely stable. If $A$ is symmetric but not positive definite, then pivoting may be necessary and the methods of §4.4 are relevant.

**Example 4.1.2**

$$A = \begin{bmatrix} 10 & 20 & 30 \\ 20 & 45 & 80 \\ 30 & 80 & 171 \end{bmatrix} = \begin{bmatrix} 1 & 0 & 0 \\ 2 & 1 & 0 \\ 3 & 4 & 1 \end{bmatrix} \begin{bmatrix} 10 & 0 & 0 \\ 0 & 5 & 0 \\ 0 & 0 & 1 \end{bmatrix} \begin{bmatrix} 1 & 2 & 3 \\ 0 & 1 & 4 \\ 0 & 0 & 1 \end{bmatrix}$$

and so if Algorithm 4.1.2 is applied, $A$ is overwritten by

$$A = \begin{bmatrix} 10 & 20 & 30 \\ 2 & 5 & 80 \\ 3 & 4 & 1 \end{bmatrix}.$$

### Problems

**P4.1.1** Show that the $LDM^T$ factorization of a nonsingular $A$ is unique if it exists.

**P4.1.2** Modify Algorithm 4.1.1 so that it computes a factorization of the form $PA = LDM^T$, where $L$ and $M$ are both unit lower triangular, $D$ is diagonal, and $P$ is a permutation that is chosen so $|\ell_{ij}| \leq 1$.

**P4.1.3** Suppose the $n$-by-$n$ symmetric matrix $A = (a_{ij})$ is stored in a vector $c$ as follows: $c = (a_{11}, a_{21}, \ldots, a_{n1}, a_{22}, \ldots, a_{n2}, \ldots, a_{nn})$. Rewrite Algorithm 4.1.2 with $A$ stored in this fashion. Get as much indexing outside the inner loops as possible.

**P4.1.4** Rewrite Algorithm 4.1.2 for $A$ stored by diagonal. See §1.2.8.

### Notes and References for Sec. 4.1

Algorithm 4.1.1 is related to the methods of Crout and Doolittle in that outer product updates are avoided. See Chapter 4 of

L. Fox (1964). *An Introduction to Numerical Linear Algebra,* Oxford University Press, Oxford.

as well as Stewart (IMC, pp. 131-149). An Algol procedure may be found in

H.J. Bowdler, R.S. Martin, G. Peters, and J.H. Wilkinson (1966), "Solution of Real and Complex Systems of Linear Equations," *Numer. Math.* *8*, 217–234. See also *HACLA*, pp. 93–110.

See also

G.E. Forsythe (1960). "Crout with Pivoting," *Comm. ACM 3*, 507–08.
W.M. McKeeman (1962). "Crout with Equilibration and Iteration," *Comm. ACM 5*, 553–55.

## 4.2 Positive Definite Systems

A matrix $A \in \mathbb{R}^{n\times n}$ is *positive definite* if $x^TAx > 0$ for all nonzero $x \in \mathbb{R}^n$. Positive definite systems constitute one of the most important classes of special $Ax = b$ problems. Consider the 2-by-2 symmetric case. If

$$A = \begin{bmatrix} a_{11} & a_{12} \\ a_{21} & a_{22} \end{bmatrix}$$

is positive definite then

$$\begin{aligned} x &= (1,\ 0)^T & \Rightarrow\quad x^TAx &= a_{11} > 0 \\ x &= (0,1)^T & \Rightarrow\quad x^TAx &= a_{22} > 0 \\ x &= (1,\ 1)^T & \Rightarrow\quad x^TAx &= a_{11} + 2a_{12} + a_{22} > 0 \\ x &= (1,\ -1)^T & \Rightarrow\quad x^TAx &= a_{11} - 2a_{12} + a_{22} > 0\,. \end{aligned}$$

The last two equations imply $|a_{12}| \le (a_{11} + a_{22})/2$. From these results we see that the largest entry in $A$ is on the diagonal and that it is positive. This turns out to be true in general. A symmetric positive definite matrix has a "weighty" diagonal. The mass on the diagonal is not blatantly obvious as in the case of diagonal dominance (c.f. §3.4.9) but it has the same effect in that it precludes the need for pivoting.

We begin with a few comments about the property of positive definiteness and what it implies in the unsymmetric case with respect to pivoting. We then focus on the efficient organization of the Cholesky procedure which can be used to safely factor a symmetric positive definite $A$. Gaxpy, outer product, and block versions are developed. The section concludes with a few comments about the semidefinite case.

### 4.2.1 Positive Definiteness

Suppose $A \in \mathbb{R}^{n\times n}$ is positive definite. It is obvious that a positive definite matrix is nonsingular for otherwise we could find a nonzero $x$ so $x^TAx = 0$. However, much more is implied by the positivity of the *quadratic form* $x^TAx$ as the following results show.

**Theorem 4.2.1** *If $A \in \mathbb{R}^{n\times n}$ is positive definite and $X \in \mathbb{R}^{n\times k}$ has rank $k$, then $B = X^TAX \in \mathbb{R}^{k\times k}$ is also positive definite.*

**Proof.** If $z \in \mathbb{R}^k$ satisfies $0 \geq z^TBz = (Xz)^TA(Xz)$ then $Xz = 0$. But since $X$ has full column rank, this implies that $z = 0$. □

**Corollary 4.2.2** *If $A$ is positive definite then all its principal submatrices are positive definite. In particular, all the diagonal entries are positive.*

**Proof.** If $v \in \mathbb{R}^k$ is an integer vector with $1 \leq v_1 < \cdots < v_k \leq n$, then $X = I_n(:, v)$ is a rank $k$ matrix made up columns $v_1, \ldots, v_k$ of the identity. It follows from Theorem 4.2.1 that $A(v, v) = X^TAX$ is positive definite. □

**Corollary 4.2.3** *If $A$ is positive definite then the factorization $A = LDM^T$ exists and $D = \text{diag}(d_1, \ldots, d_n)$ has positive diagonal entries.*

**Proof.** From Corollary 4.2.2, it follows that the submatrices $A(1{:}k, 1{:}k)$ are nonsingular for $k = 1{:}n$ and so from Theorem 4.1.1 the factorization $A = LDM^T$ exists. If we apply Theorem 4.2.1 with $X = L^{-T}$ then $B = DM^TL^{-T} = L^{-1}AL^{-T}$ is positive definite. Since $M^TL^{-T}$ is unit upper triangular, $B$ and $D$ have the same diagonal and it must be positive. □

There are several typical situations that give rise to positive definite matrices in practice:

- The quadratic form is an energy function whose positivity is guaranteed from physical principles.
- The matrix $A$ equals a *cross-product* $X^TX$ where $X$ has full column rank. (Positive definiteness follows by setting $A = I_n$ in Theorem 4.2.1.)
- Both $A$ and $A^T$ are diagonally dominant and each $a_{ii}$ is positive.

The proof that the third condition implies positive definiteness is left to the reader.

### 4.2.2 Unsymmetric Positive Definite Systems

The mere existence of an LDM$^{\text{T}}$ factorization does not mean that its computation is advisable because the resulting factors may have unacceptably large elements. For example, if $\epsilon > 0$ then the matrix

$$A = \begin{bmatrix} \epsilon & m \\ -m & \epsilon \end{bmatrix} = \begin{bmatrix} 1 & 0 \\ -m/\epsilon & 1 \end{bmatrix} \begin{bmatrix} \epsilon & 0 \\ 0 & \epsilon + m^2/\epsilon \end{bmatrix} \begin{bmatrix} 1 & m/\epsilon \\ 0 & 1 \end{bmatrix}$$

is positive definite. But if $m/\epsilon \gg 1$, then pivoting is recommended.

The following result suggests when to expect element growth in the LDM$^{\text{T}}$ factorization of a positive definite matrix.

**Theorem 4.2.4** *Let $A \in \mathbb{R}^{n\times n}$ be positive definite and set $T = (A+A^T)/2$ and $S = (A - A^T)/2$. If $A = LDM^T$, then*

$$\| \, |L||D||M^T| \, \|_F \; \leq \; n \left( \| T \|_2 + \| ST^{-1}S \|_2 \right) \tag{4.2.1}$$

**Proof.** See Golub and Van Loan (1979). □

The theorem suggests when it is safe not to pivot. Assume that the computed factors $\hat{L}$, $\hat{D}$, and $\hat{M}$ satisfy:

$$\| \, |\hat{L}||\hat{D}||\hat{M}^T| \, \|_F \; \leq \; c \| \, |L||D||M^T| \, \|_F, \tag{4.2.2}$$

where $c$ is a constant of modest size. It follows from (4.2.1) and the analysis in §3.3 that if these factors are used to compute a solution to $Ax = b$, then the computed solution $\hat{x}$ satisfies $(A + E)\hat{x} = b$ with

$$\| E \|_F \; \leq \; \mathbf{u} \left( 3n \| A \|_F + 5cn^2 \left( \| T \|_2 + \| ST^{-1}S \|_2 \right) \right) + O(\mathbf{u}^2). \tag{4.2.3}$$

It is easy to show that $\| T \|_2 \leq \| A \|_2$, and so it follows that it if

$$\Omega \; = \; \frac{\| ST^{-1}S \|_2}{\| A \|_2} \tag{4.2.4}$$

is not too large then it is safe not to pivot. In other words, the norm of the skew part $S$ has to be modest relative to the condition of the symmetric part $T$. Sometimes it is possible to estimate $\Omega$ in an application. This is trivially the case when $A$ is symmetric for then $\Omega = 0$.

### 4.2.3 Symmetric Positive Definite Systems

When we apply the above results to a symmetric positive definite system we know that the factorization $A = LDL^T$ exists and moreover is stable to compute. However, in this situation another factorization is available.

**Theorem 4.2.5 (Cholesky Factorization )** *If $A \in \mathbb{R}^{n\times n}$ is symmetric positive definite, then there exists a unique lower triangular $G \in \mathbb{R}^{n\times n}$ with positive diagonal entries such that $A = GG^T$.*

**Proof.** From Theorem 4.1.2, there exists a unit lower triangular $L$ and a diagonal $D = \text{diag}(d_1, \ldots, d_n)$ such that $A = LDL^T$. Since the $d_k$ are positive, the matrix $G = L\,\text{diag}(\sqrt{d_1}, \ldots, \sqrt{d_n})$ is real lower triangular with positive diagonal entries. It also satisfies $A = GG^T$. Uniqueness follows from the uniqueness of the $\text{LDL}^\text{T}$ factorization. □

The factorization $A = GG^T$ is known as the *Cholesky factorization* and $G$ is referred to as the *Cholesky triangle*. Note that if we compute the Cholesky factorization and solve the triangular systems $Gy = b$ and $G^T x = y$, then $b = Gy = G(G^T x) = (GG^T)x = Ax$.

Our proof of the Cholesky factorization in Theorem 4.2.5 is constructive. However, more effective methods for computing the Cholesky triangle can be derived by manipulating the equation $A = GG^T$. This can be done in several ways as we show in the next few subsections.

**Example 4.2.1** The matrix

$$\begin{bmatrix} 2 & -2 \\ -2 & 5 \end{bmatrix} = \begin{bmatrix} 1 & 0 \\ -1 & 1 \end{bmatrix} \begin{bmatrix} 2 & 0 \\ 0 & 3 \end{bmatrix} \begin{bmatrix} 1 & -1 \\ 0 & 1 \end{bmatrix} = \begin{bmatrix} \sqrt{2} & 0 \\ -\sqrt{2} & \sqrt{3} \end{bmatrix} \begin{bmatrix} \sqrt{2} & -\sqrt{2} \\ 0 & \sqrt{3} \end{bmatrix}$$

is positive definite.

### 4.2.4 Gaxpy Cholesky

We first derive an implementation of Cholesky that is rich in the gaxpy operation. If we compare $j$th columns in the equation $A = GG^T$ then we obtain

$$A(:,j) = \sum_{k=1}^{j} G(j,k)G(:,k) .$$

This says that

$$G(j,j)G(:,j) \;=\; A(:,j) - \sum_{k=1}^{j-1} G(j,k)G(:,k) \;\equiv\; v . \tag{4.2.5}$$

If we know the first $j-1$ columns of $G$ then $v$ is computable. It follows by equating components in (4.2.5) that $G(j{:}n,j) = v(j{:}n)/\sqrt{v(j)}$. This is a scaled gaxpy operation and so we obtain the following gaxpy-based method for computing the Cholesky factorization:

```
for j = 1:n
    v(j:n) = A(j:n, j)
    for k = 1:j − 1
        v(j:n) = v(j:n) − G(j, k)G(j:n, k)
    end
    G(j:n, j) = v(j:n)/√v(j)
end
```

It is possible to arrange the computations so that $G$ overwrites the lower triangle of $A$.

**Algorithm 4.2.1 (Cholesky: Gaxpy Version)** Given a symmetric positive definite $A \in \mathbb{R}^{n\times n}$, the following algorithm computes a lower triangular $G \in \mathbb{R}^{n\times n}$ such that $A = GG^T$. For all $i \geq j$, $G(i,j)$ overwrites $A(i,j)$.

```
for j = 1:n
    if j > 1
        A(j:n, j) = A(j:n, j) - A(j:n, 1:j-1)A(j, 1:j-1)^T
    end
    A(j:n, j) = A(j:n, j)/sqrt(A(j, j))
end
```

This algorithm requires $n^3/3$ flops.

### 4.2.5 Outer Product Cholesky

An alternative Cholesky procedure based on outer product (rank-1) updates can be derived from the partitioning

$$A = \begin{bmatrix} \alpha & v^T \\ v & B \end{bmatrix} = \begin{bmatrix} \beta & 0 \\ v/\beta & I_{n-1} \end{bmatrix} \begin{bmatrix} 1 & 0 \\ 0 & B - vv^T/\alpha \end{bmatrix} \begin{bmatrix} \beta & v^T/\beta \\ 0 & I_{n-1} \end{bmatrix}. \tag{4.2.6}$$

Here, $\beta = \sqrt{\alpha}$ and we know that $\alpha > 0$ because $A$ is positive definite. Note that $B - vv^T/\alpha$ is positive definite because it is a principal submatrix of $X^T AX$ where

$$X = \begin{bmatrix} 1 & -v^T/\alpha \\ 0 & I_{n-1} \end{bmatrix}.$$

Thus, if we have the Cholesky factorization $G_1 G_1^T = B - vv^T/\alpha$ then from (4.2.6) it follows that $A = GG^T$ with

$$G = \begin{bmatrix} \beta & 0 \\ v/\beta & G_1 \end{bmatrix}.$$

The idea, then, is to perform repeatedly the reduction (4.2.6) on ever smaller submatrices, much as in the $kji$ Gaussian elimination algorithm. If we overwrite $A$'s lower triangular part with $G$ we obtain

**Algorithm 4.2.2 (Cholesky: Outer product Version)** Given a symmetric positive definite $A \in \mathbb{R}^{n\times n}$, the following algorithm computes a lower triangular $G \in \mathbb{R}^{n\times n}$ such that $A = GG^T$. For all $i \geq j$, $G(i,j)$ overwrites $A(i,j)$.

```
for k = 1:n
    A(k, k) = sqrt(A(k, k))
    A(k+1:n, k) = A(k+1:n, k)/A(k, k)
    for j = k+1:n
        A(j:n, j) = A(j:n, j) - A(j:n, k)A(j, k)
    end
end
```

This algorithm involves $n^3/3$ flops. Note that the $j$-loop computes the lower triangular part of the outer product update

$$A(k+1{:}n, k+1{:}n) \;=\; A(k+1{:}n, k+1{:}n) \;-\; A(k+1{:}n, k)A(k+1{:}n, k)^T.$$

Recalling our discussion in §1.4.8 about gaxpy versus outer product updates, it is easy to show that Algorithm 4.2.1 involves fewer vector touches than Algorithm 4.2.2 by a factor of two.

### 4.2.6 Block Dot Product Cholesky

Suppose $A \in \mathbb{R}^{n\times n}$ is symmetric positive definite. Regard $A = (A_{ij})$ and its Cholesky factor $G = (G_{ij})$ as $N$-by-$N$ block matrices with square diagonal blocks. Equating $(i, j)$ blocks in the equation $A = GG^T$ with $i \geq j$ we obtain $A_{ij} = \sum_{k=1}^{j} G_{ik}G_{jk}^T$. Defining

$$S \;=\; A_{ij} - \sum_{k=1}^{j-1} G_{ik}G_{jk}^T$$

we see that $G_{jj}G_{jj}^T = S$ if $i = j$ and that $G_{ij}G_{jj}^T = S$ if $i > j$. Properly sequenced, these equations can be arranged to compute all the $G_{ij}$:

**Algorithm 4.2.3 (Cholesky: Block Dot Product Version)** Given a symmetric positive definite $A \in \mathbb{R}^{n\times n}$, the following algorithm computes a lower triangular $G \in \mathbb{R}^{n\times n}$ such that $A = GG^T$. The lower triangular part of $A$ is overwritten by the lower triangular part of $G$. $A$ is regarded as an $N$-by-$N$ block matrix with square diagonal blocks.

**for** $j = 1{:}N$
    **for** $i = j{:}N$

$$S \;=\; A_{ij} \;-\; \sum_{k=1}^{j-1} G_{ik}G_{jk}^T$$

        **if** $i = j$
            Compute Cholesky factorization $S = G_{jj}G_{jj}^T$.
        **else**
            Solve $G_{ij}G_{jj}^T = S$ for $G_{ij}$
        **end**
        Overwrite $A_{ij}$ with $G_{ij}$.
    **end**
**end**

The overall process involves $n^3/3$ flops like the other Cholesky procedures that we have developed. The procedure is rich in matrix multiplication assuming a suitable blocking of the matrix $A$. For example, if $n = rN$ and each $A_{ij}$ is $r$-by-$r$, then the level-3 fraction is approximately $1 - (1/N^2)$.

Algorithm 4.2.3 is incomplete in the sense that we have not specified how the products $G_{ik}G_{jk}$ are formed or how the $r$-by-$r$ Cholesky factorizations $S = G_{jj}G_{jj}^T$ are computed. These important details would have to be worked out carefully in order to extract high performance.

Another block procedure can be derived from the gaxpy Cholesky algorithm. After $r$ steps of Algorithm 4.2.1 we know the matrices $G_{11} \in \mathbb{R}^{r\times r}$ and $G_{21} \in \mathbb{R}^{(n-r)\times r}$ in

$$\begin{bmatrix} A_{11} & A_{12} \\ A_{21} & A_{22} \end{bmatrix} = \begin{bmatrix} G_{11} & 0 \\ G_{21} & I_{n-r} \end{bmatrix} \begin{bmatrix} I_r & 0 \\ 0 & \tilde{A} \end{bmatrix} \begin{bmatrix} G_{11} & 0 \\ G_{21} & I_{n-r} \end{bmatrix}^T .$$

We then perform $r$ more steps of gaxpy Cholesky not on $A$ but on the reduced matrix $\tilde{A} = A_{22} - G_{21}G_{21}^T$ which we *explicitly* form exploiting symmetry. Continuing in this way we obtain a block Cholesky algorithm whose $k$th step involves $r$ gaxpy Cholesky steps on a matrix of order $n - (k-1)r$ followed a level-3 computation having order $n - kr$. The level-3 fraction is approximately equal to $1 - 3/(2N)$ if $n \approx rN$.

### 4.2.7 Stability of the Cholesky Process

In exact arithmetic, we know that a symmetric positive definite matrix has a Cholesky factorization. Conversely, if the Cholesky process runs to completion with strictly positive square roots, then $A$ is positive definite. Thus, to find out if a matrix $A$ is positive definite, we merely try to compute its Cholesky factorization using any of the methods given above.

The situation in the context of roundoff error is more interesting. The numerical stability of the Cholesky algorithm follows from the inequality

$$g_{ij}^2 \le \sum_{k=1}^{i} g_{ik}^2 = a_{ii} .$$

This shows that the entries in the Cholesky triangle are nicely bounded. The same conclusion can be reached from the equation $\| G \|_2^2 = \| A \|_2$.

The roundoff errors associated with the Cholesky factorization have been extensively studied in a classical paper by Wilkinson (1968). Using the results in this paper, it can be shown that if $\hat{x}$ is the computed solution to $Ax = b$, obtained via any of our Cholesky procedures then $\hat{x}$ solves the perturbed system $(A+E)\hat{x} = b$ where $\| E \|_2 \le c_n \mathbf{u} \| A \|_2$ and $c_n$ is a small constant depending upon $n$. Moreover, Wilkinson shows that if $q_n \mathbf{u} \kappa_2(A) \le 1$ where $q_n$ is another small constant, then the Cholesky process runs to completion, i.e, no square roots of negative numbers arise.

**Example 4.2.2** If Algorithm 4.2.2 is applied to the positive definite matrix

$$A = \begin{bmatrix} 100 & 15 & .01 \\ 15 & 2.3 & .01 \\ .01 & .01 & 1.00 \end{bmatrix}$$

and $\beta = 10$, $t = 2$, rounded arithmetic used, then $\hat{g}_{11} = 10$, $\hat{g}_{21} = 1.5$, $\hat{g}_{31} = .001$ and $\hat{g}_{22} = 0.00$. The algorithm then breaks down trying to compute $g_{32}$.

### 4.2.8 The Semidefinite Case

A matrix is said to be *positive semidefinite* if $x^TAx \geq 0$ for all vectors $x$. Symmetric positive semidefinite (*sps*) matrices are important and we briefly discuss some Cholesky-like manipulations that can be used to solve various *sps* problems. Results about the diagonal entries in an *sps* matrix are needed first.

**Theorem 4.2.6** *If $A \in \mathbb{R}^{n\times n}$ is symmetric positive semidefinite, then*

$$|a_{ij}| \leq (a_{ii} + a_{jj})/2 \qquad (4.2.7)$$

$$|a_{ij}| \leq \sqrt{a_{ii}a_{jj}} \qquad (i \neq j) \qquad (4.2.8)$$

$$\max_{i,j} |a_{ij}| = \max_i a_{ii} \qquad (4.2.9)$$

$$a_{ii} = 0 \Rightarrow A(i,:) = 0,\ A(:,i) = 0 \qquad (4.2.10)$$

**Proof.** If $x = e_i + e_j$ then $0 \leq x^TAx = a_{ii} + a_{jj} + 2a_{ij}$ while $x = e_i - e_j$ implies $0 \leq x^TAx = a_{ii} + a_{jj} - 2a_{ij}$. Inequality (4.2.7) follows from these two results. Equation (4.2.9) is an easy consequence of (4.2.7).

To prove (4.2.8) assume without loss of generality that $i = 1$ and $j = 2$ and consider the inequality

$$0 \leq \begin{bmatrix} x \\ 1 \end{bmatrix}^T \begin{bmatrix} a_{11} & a_{12} \\ a_{21} & a_{22} \end{bmatrix} \begin{bmatrix} x \\ 1 \end{bmatrix} = a_{11}x^2 + 2a_{12}x + a_{22}$$

which holds since $A(1{:}2, 1{:}2)$ is also semidefinite. This is a quadratic equation in $x$ and for the inequality to hold, the discriminant $4a_{12}^2 - 4a_{11}a_{22}$ must be negative. Implication (4.2.10) follows from (4.2.8). □

Consider what happens when outer product Cholesky is applied to an *sps* matrix. If a zero $A(k,k)$ is encountered then from (4.2.10) $A(k{:}n,k)$ is zero and there is "nothing to do" and we obtain

```
for k = 1:n
    if A(k,k) > 0
        A(k,k) = sqrt(A(k,k)); A(k+1:n,k) = A(k+1:n,k)/A(k,k)
        for j = k+1:n
            A(j:n,j) = A(j:n,j) - A(j:n,k)A(j,k)        (4.2.11)
        end
    end
end
```

Thus, a simple change makes Algorithm 4.2.2 applicable to the semidefinite case. However, in practice rounding errors preclude the generation of exact zeros and it may be preferable to incorporate pivoting.

### 4.2.9 Symmetric Pivoting

To preserve symmetry in a symmetric $A$ we only consider data reorderings of the form $PAP^T$ where $P$ is a permutation. Row permutations ($A \leftarrow PA$) or column permutations ($A \leftarrow AP$) alone destroy symmetry. A permutation update of the form $A \leftarrow PAP^T$ is called a *symmetric permutation* of $A$. Note that such an operation *does not* move off-diagonal elements to the diagonal. The diagonal of $PAP^T$ is a reordering of the diagonal of $A$.

Suppose at the beginning of the $k$th step in (4.2.11) we symmetrically permute the largest diagonal entry of $A(k{:}n, k{:}n)$ into the lead position. If that largest diagonal entry is zero then $A(k{:}n, k{:}n) = 0$ by virtue of (4.2.10). In this way we can compute the factorization $PAP^T = GG^T$ where $G \in \mathbb{R}^{n \times (k-1)}$ is lower triangular.

**Algorithm 4.2.4** Suppose $A \in \mathbb{R}^{n \times n}$ is symmetric positive semidefinite and that $\text{rank}(A) = r$. The following algorithm computes a permutation $P$, the index $r$, and an $n$-by-$r$ lower triangular matrix $G$ such that $PAP^T = GG^T$. The lower triangular part of $A(:, 1{:}r)$ is overwritten by the lower triangular part of $G$. $P = P_r \cdots P_1$ where $P_k$ is the identity with rows $k$ and $piv(k)$ interchanged.

$r = 0$
**for** $k = 1{:}n$
    Find $q$ $(k \le q \le n)$ so $A(q,q) = \max\,\{A(k,k), .., A(n,n)\}$
    **if** $A(q,q) > 0$
        $r = r + 1$
        $piv(k) = q$
        $A(k,:) \leftrightarrow A(q,:)$
        $A(:,k) \leftrightarrow A(:,q)$
        $A(k,k) = \sqrt{A(k,k)}$
        $A(k+1{:}n, k) = A(k+1{:}n, k)/A(k,k)$
        **for** $j = k+1{:}n$
            $A(j{:}n, j) = A(j{:}n, j) - A(j{:}n, k)A(j,k)$
        **end**
    **end**
**end**

In practice, a tolerance is used to detect small $A(k,k)$. However, the situation is quite tricky and the reader should consult Higham (1989). In addition, §5.5 has a discussion of tolerances in the rank detection problem.

Finally, we remark that a truly efficient implementation of Algorithm 4.2.4 would only access the lower triangular portion of $A$.

### Problems

**P4.2.1** Suppose that $H = A + iB$ is Hermitian and positive definite with $A, B \in \mathbb{R}^{n\times n}$. This means that $x^H Hx > 0$ whenever $x \neq 0$. (a) Show that

$$C = \begin{bmatrix} A & -B \\ B & A \end{bmatrix}$$

is symmetric and positive definite. (b) Formulate a $8n^3/3$ algorithm for solving $(A + iB)(x + iy) = (b + ic)$, where $b$, $c$, $x$, and $y$ are in $\mathbb{R}^n$. How much storage is required?

**P4.2.2** Suppose $A \in \mathbb{R}^{n\times n}$ is symmetric and positive definite. Give an algorithm for computing an upper triangular matrix $R \in \mathbb{R}^{n\times n}$ such that $A = RR^T$.

**P4.2.3** Let $A \in \mathbb{R}^{n\times n}$ be positive definite and set $T = (A + A^T)/2$ and $S = (A - A^T)/2$. (a) Show that $\| A^{-1} \|_2 \leq \| T^{-1} \|_2$ and $x^T A^{-1} x \leq x^T T^{-1} x$ for all $x \in \mathbb{R}^n$. (b) Show that if $A = LDM^T$ then $d_k \geq 1/\| T^{-1} \|_2$ for $k = 1{:}n$

**P4.2.4** Find a 2-by-2 real matrix $A$ with the property that $x^T Ax > 0$ for all real nonzero 2-vectors but which is not positive definite when regarded as a member of $\mathbb{C}^{2\times 2}$.

**P4.2.5** Suppose $A \in \mathbb{R}^{n\times n}$ has a positive diagonal. Show that if both $A$ and $A^T$ are strictly diagonally dominant then $A$ is positive definite.

**P4.2.6** Show that the function $f(x) = (x^T Ax)/2$ is a vector norm on $\mathbb{R}^n$ if and only if $A$ is positive definite.

**P4.2.7** Modify Algorithm 4.2.1 so that if the square root of a negative number is encountered, the algorithm finds a unit vector $x$ so $x^T Ax < 0$ and terminates.

**P4.2.8** The numerical range $W(A)$ of a complex matrix $A$ is defined to be the set $W(A) = \{x^H Ax : x^H x = 1\}$. Show that if $W(A)$ does not contain the origin then $A$ has an LU factorization.

**P4.2.9** Use the SVD to show that if $A \in \mathbb{R}^{m\times n}$ with $m \geq n$ then there exist $Q \in \mathbb{R}^{m\times n}$ and $P \in \mathbb{R}^{n\times n}$ such that $A = QP$, where $Q^T Q = I_n$ and $P$ is symmetric and nonnegative definite. This decomposition is sometimes referred to as the *polar decomposition* because it is analogous to the complex number factorization $z = e^{i\,arg(z)}|z|$.

**P4.2.10** Suppose $A = I + uu^T$ where $A \in \mathbb{R}^{n\times n}$ and $\| u \|_2 = 1$. Give explicit formulae for the diagonal and subdiagonal of $A$'s Cholesky factor.

### Notes and References for Sec. 4.2

The definiteness of the quadratic form $x^T Ax$ can frequently be established by considering the mathematics of the underlying problem. For example, the discretization of certain partial differential operators gives rise to provably positive definite matrices. The question of whether pivoting is necessary when solving positive definite systems is considered in

G.H. Golub and C. Van Loan (1979). "Unsymmetric Positive Definite Linear Systems," *Lin. Alg. and Its Applic. 28*, 85–98.

The results in this paper were motivated by the earlier work on linear systems of the form $(I + S)x = b$, where $S$ is skew symmetric

A. Buckley (1974). "A Note on Matrices $A = I + H$, $H$ Skew-Symmetric," *Z. Angew. Math. Mech. 54*, 125–26.

A. Buckley (1977). "On the Solution of Certain Skew-Symmetric Linear Systems," *SIAM J. Num. Anal. 14*, 566–70.

Symmetric positive definite systems constitute the most important class of special $Ax = b$ problems. Algol programs for both the $GG^T$ and $LDL^T$ factorizations are given in

R.S. Martin, G. Peters, and J.H. Wilkinson (1965). "Symmetric Decomposition of a Positive Definite Matrix," *Numer. Math. 7*, 362-83. See also *HACLA*, pp. 9–30.

The technique of iterative improvement which we discussed in §3.5.3 in connection with the LU factorization, can also be implemented with the factorizations $GG^T$ and $LDL^T$. See

R.S. Martin, G. Peters, and J.H. Wilkinson (1966). "Iterative Refinement of the Solution of a Positive Definite System of Equations," *Numer. Math. 8*, 203–16. See also *HACLA*, pp. 31–44.

The roundoff errors associated with the Cholesky factorization are discussed in

N.J. Higham (1989) "Analysis of the Cholesky Decomposition of a Semi-definite Matrix," in *Reliable Numerical Computation*, eds. M.G. Cox and S.J. Hammarling, Oxford University Press.
A. Kielbasinski (1987). "A Note on Rounding Error Analysis of Cholesky Factorization," *Lin. Alg. and Its Applic. 88/89*, 487–494.
J. Meinguet (1983). "Refined Error Analyses of Cholesky Factorization," *SIAM J. Numer. Anal. 20*, 1243–1250.
J.H. Wilkinson (1968). "À Priori Error Analysis of Algebraic Processes," *Proc. International Congress Math.* (Moscow: Izdat. Mir, 1968), pp. 629–39.

The question of how the Cholesky triangle $G$ changes when $A = GG^T$ is perturbed is analyzed in

G.W. Stewart (1977b). "Perturbation Bounds for the $QR$ Factorization of a Matrix," *SIAM J. Num. Anal. 14*, 509-18.

An Algol procedure for inverting a symmetric positive definite matrix without any additional storage is given in

F.L. Bauer and C. Reinsch (1970). "Inversion of Positive Definite Matrices by the Gauss-Jordan Method," in *HACLA*, pp. 45–49.

Fortan programs for solving symmetric positive definite systems are in LINPACK, chapters 3 and 8.

## 4.3 Banded Systems

In many applications that involve linear systems, the matrix of coefficients is *banded.* This is the case whenever the equations can be ordered so that each unknown $x_i$ appears in only a few equations in a "neighborhood" of the $i$th equation. Formally, we say that $A = (a_{ij})$ has *upper bandwidth* $q$ if $a_{ij} = 0$ whenever $j > i + q$ and *lower bandwidth* $p$ if $a_{ij} = 0$ whenever $i > j + p$. Substantial economies can be realized when solving banded systems because the triangular factors in $LU$, $GG^T$, $LDM^T$, etc., are also banded.

Before proceeding the reader is advised to review §1.2 where several aspects of band matrix manipulation are discussed.

### 4.3.1 Band LU Factorization

Our first result shows that if $A$ is banded and $A = LU$ then $L(U)$ inherits the lower (upper) bandwidth of $A$.

**Theorem 4.3.1** *Suppose $A \in \mathbb{R}^{n\times n}$ has an LU factorization $A = LU$. If $A$ has upper bandwidth $q$ and lower bandwidth $p$, then $U$ has upper bandwidth $q$ and $L$ has lower bandwidth $p$.*

**Proof.** The proof is by induction on $n$. Writing the factorization

$$A = \begin{bmatrix} \alpha & w^T \\ v & B \end{bmatrix} = \begin{bmatrix} 1 & 0 \\ v/\alpha & I_{n-1} \end{bmatrix} \begin{bmatrix} 1 & 0 \\ 0 & B - vw^T/\alpha \end{bmatrix} \begin{bmatrix} \alpha & w^T \\ 0 & I_{n-1} \end{bmatrix}.$$

It is clear that $B - vw^T/\alpha$ has upper bandwidth $q$ and lower bandwidth $p$ because only the first $q$ components of $w$ and the first $p$ components of $v$ are nonzero. Let $L_1U_1$ be the $LU$ factorization of this matrix. Using the induction hypothesis and the sparsity of $w$ and $v$, it follows that

$$L = \begin{bmatrix} 1 & 0 \\ v/\alpha & L_1 \end{bmatrix} \quad \text{and} \quad U = \begin{bmatrix} \alpha & w^T \\ 0 & U_1 \end{bmatrix}$$

have the desired bandwidth properties and satisfy $A = LU$. □

The specialization of Gaussian elimination to banded matrices having an LU factorization is straightforward.

**Algorithm 4.3.1 (Band Gaussian Elimination: Outer Product Version)** Given $A \in \mathbb{R}^{n\times n}$ with upper bandwidth $q$ and lower bandwidth $p$, the following algorithm computes the factorization $A = LU$, assuming it exists. $A(i,j)$ is overwritten by $L(i,j)$ if $i > j$ and by $U(i,j)$ otherwise.

```
for k = 1:n − 1
    for i = k + 1:min(k + p, n)
        A(i, k) = A(i, k)/A(k, k)
    end
    for j = k + 1:min(k + q, n)
        for i = k + 1:min(k + p, n)
            A(i, j) = A(i, j) − A(i, k)A(k, j)
        end
    end
end
```

If $n \gg p$ and $n \gg q$ then this algorithm involves about $2npq$ flops. Band versions of Algorithm 4.1.1 (LDM$^T$) and all the Cholesky procedures also exist, but we leave their formulation to the exercises.

### 4.3.2 Band Triangular System Solving

Analogous savings can also be made when solving banded triangular systems.

**Algorithm 4.3.2 (Band Forward Substitution: Column Version)** Let $L \in \mathbb{R}^{n \times n}$ be a unit lower triangular matrix having lower bandwidth $p$. Given $b \in \mathbb{R}^n$, the following algorithm overwrites $b$ with the solution to $Lx = b$.

```
for j = 1:n
    for i = j + 1:min(j + p, n)
        b(i) = b(i) - L(i, j)b(j)
    end
end
```

If $n \gg p$ then this algorithm requires about $2np$ flops.

**Algorithm 4.3.3 (Band Back-Substitution: Column Version)** Let $U \in \mathbb{R}^{n \times n}$ be a nonsingular upper triangular matrix having upper bandwidth $q$. Given $b \in \mathbb{R}^n$, the following algorithm overwrites $b$ with the solution to $Ux = b$.

```
for j = n: - 1:1
    b(j) = b(j)/U(j, j)
    for i = max(1, j - q):j - 1
        b(i) = b(i) - L(i, j)b(j)
    end
end
```

If $n \gg q$ then this algorithm requires about $2nq$ flops.

### 4.3.3 Band Matrix Data Structures

The above algorithms are written as if the matrix $A$ is conventionally stored in an $n$-by-$n$ array. In practice, a band linear equation solver would be organized around a data structure that takes advantage of the many zeroes in $A$. Recall from §1.2.6 that if $A$ has lower bandwidth $p$ and upper bandwidth $q$ it can be represented in a $(p + q + 1)$-by-$n$ array $A.band$ where band entry $a_{ij}$ is stored in $A.band(i - j + q + 1, j)$. In this arrangement, the nonzero portion of $A$'s $j$th column is housed in the $j$th column of $A.band$.

Another possible band matrix data structure that we discussed in §1.2.8 involves storing $A$ by diagonal in a 1-dimensional array $A.diag$.

Regardless of the data structure adopted, the design of a matrix computation with a band storage arrangement requires care in order to minimize subscripting overheads. To illustrate, here is an $A.band$ implementation of Algorithm 4.3.1 with references to $A$ replaced with suitable references to $A.band$:

```
for k = 1:n − 1
    for i = k + 1:min(k + p, n)
        A.band(i − k + q + 1, k) =
            A.band(i − k + q + 1, k)/A.band(q + 1, k)
    end
    for j = k + 1:min(k + q, n)
        for i = k + 1:min(k + p, n)
            A.band(i − j + q + 1, j) = A.band(i − j + q + 1, j)
                −A.band(i − k + q + 1, k)A.band(k − j + q + 1, j)
        end
    end
end
```

Although correct, this literal translation of Algorithm 4.3.1 is not optimal because there is an excess of subscripting arithmetic. For example, the first $i$-loop is better written as follows:

```
τ = q + 1
for i = q + 2:q + p + 1
    A.band(i, k) = A.band(i, k)/A(τ, k)
end
```

(This works because of the "corner zeros" that we assume to be part of the $A.band$ format, c.f §1.2.6.) The design of efficient band procedures requires this kind of manipulation. Smart compilers help but the fastest code may require the programmer to completely oversee the integer calculations. This is one reason why band matrix software is unpleasant to read.

### 4.3.4 Band Gaussian Elimination With Pivoting

Gaussian elimination with partial pivoting can also be specialized to exploit band structure in $A$. If, however, $PA = LU$, then the band properties of $L$ and $U$ are not quite so simple. For example, if $A$ is tridiagonal and the first two rows are interchanged at the very first step of the algorithm, then $u_{13}$ is nonzero. Consequently, row interchanges expand bandwidth. Precisely how the band enlarges is the subject of the following theorem.

**Theorem 4.3.2** *Suppose $A \in \mathbb{R}^{n\times n}$ is nonsingular and has upper and lower bandwidths $q$ and $p$, respectively. If Gaussian elimination with partial pivoting is used to compute Gauss transformations*

$$M_j = I - \alpha^{(j)} e_j^T \qquad j = 1{:}n-1$$

*and permutations $P_1, \ldots, P_{n-1}$ such that $M_{n-1}P_{n-1}\cdots M_1P_1A = U$ is upper triangular, then $U$ has upper bandwidth $p+q$ and $\alpha_i^{(j)} = 0$ whenever $i \le j$ or $i > j+p$.*

**Proof.** Let $PA = LU$ be the factorization computed by Gaussian elimination with partial pivoting and recall that $P = P_{n-1}\cdots P_1$. Write $P^T = [\, e_{s_1}, \ldots, e_{s_n} \,]$, where $\{s_1, ..., s_n\}$ is a permutation of $\{1, 2, ..., n\}$. If $s_i > i+p$ then it follows that the leading $i$-by-$i$ principal submatrix of $PA$ is singular, since

$$(PA)_{ij} = a_{s_i,j} \qquad j = 1{:}s_i - p - 1$$

and $s_i - p - 1 \ge i$. This implies that $U$ and $A$ are singular, a contradiction. Thus, $s_i \le i + p$ for $i = 1{:}n$ and therefore, $PA$ has upper bandwidth $p+q$. It follows from Theorem 4.3.1 that $U$ has upper bandwidth $p+q$.

The assertion about the $\alpha^{(j)}$ can be verified by observing that $M_j$ need only zero elements $(j+1, j), \ldots, (j+p, j)$ of the partially reduced matrix $P_jM_{j-1}P_{j-1}\cdots{}_1\, P_1A$. □

Thus, pivoting destroys band structure in the sense that $U$ becomes "fatter" than $A$'s upper triangle, while nothing at all can be said about the bandwidth of $L$. However, since the $j$th column of $L$ is a permutation of the $j$th Gauss vector $\alpha_j$, it follows that $L$ has at most $p+1$ nonzero elements per column.

### 4.3.5 Hessenberg LU

As an example of an unsymmetric band matrix computation, we show how Gaussian elimination with partial pivoting can be applied to factor an upper Hessenberg matrix $H$. (Recall that if $H$ is upper Hessenberg then $h_{ij} = 0$, $i > j+1$). After $k-1$ steps of Gaussian elimination with partial pivoting we are left with an upper Hessenberg matrix of the form:

$$\begin{bmatrix} \times & \times & \times & \times & \times & \times \\ 0 & \times & \times & \times & \times & \times \\ 0 & 0 & \times & \times & \times & \times \\ 0 & 0 & \times & \times & \times & \times \\ 0 & 0 & 0 & \times & \times & \times \\ 0 & 0 & 0 & 0 & \times & \times \end{bmatrix} \qquad k = 3, n = 6$$

By virtue of the special structure of this matrix, we see that the next permutation, $P_3$, is either the identity or the identity with rows 3 and 4 interchanged. Moreover, the next Gauss transformation $M_k$ has a single nonzero multiplier in the $(k+1,k)$ position. This illustrates the $k$th step of the following algorithm.

**Algorithm 4.3.4 (Hessenberg LU)** Given an upper Hessenberg matrix $H \in \mathbb{R}^{n\times n}$, the following algorithm computes the upper triangular matrix $M_{n-1}P_{n-1}\cdots M_1P_1A = U$ where each $P_k$ is a permutation and each $M_k$ is a Gauss transformation whose entries are bounded by unity. $H(i,k)$ is overwritten with $U(i,k)$ if $i \le k$ and by $(M_k)_{k+1,k}$ if $i = k+1$. An integer vector $piv(1{:}n-1)$ encodes the permutations. If $P_k = I$, then $piv(k) = 0$. If $P_k$ interchanges rows $k$ and $k+1$, then $piv(k) = 1$.

```
for k = 1:n - 1
    if |H(k, k)| < |H(k + 1, k)|
        piv(k) = 1; H(k, k:n) ↔ H(k + 1:k:n)
    else
        piv(k) = 0
    end
    if H(k, k) ≠ 0
        t = -H(k + 1, k)/H(k, k)
        for j = k + 1:n
            H(k + 1, j) = H(k + 1, j) + tH(k, j)
        end
        H(k + 1, k) = t
    end
end
```

This algorithm requires $n^2$ flops.

### 4.3.6 Band Cholesky

The rest of this section is devoted to banded $Ax = b$ problems where the matrix $A$ is also symmetric positive definite. The fact that pivoting is unnecessary for such matrices leads to some very compact, elegant algorithms. In particular, it follows from Theorem 4.3.1 that if $A = GG^T$ is the Cholesky factorization of $A$, then $G$ has the same lower bandwidth as $A$. This leads to the following banded version of Algorithm 4.2.1, gaxpy-based Cholesky

**Algorithm 4.3.5 (Band Cholesky: Gaxpy Version)** Given a symmetric positive definite $A \in \mathbb{R}^{n\times n}$ with bandwidth $p$, the following algorithm computes a lower triangular matrix $G$ with lower bandwidth $p$ such that $A = GG^T$. For all $i \ge j$, $G(i,j)$ overwrites $A(i,j)$.

```
for j = 1:n
    for k = max(1, j - p):j - 1
        λ = min(k + p, n)
        A(j:λ, j) = A(j:λ, j) - A(j, k)A(j:λ, k)
    end
    λ = min(j + p, n)
    A(j:λ, j) = A(j:λ, j)/sqrt(A(j, j))
end
```

If $n \gg p$ then this algorithm requires about $n(p^2 + 3p)$ flops and $n$ square roots. Of course, in a serious implementation an appropriate data structure for $A$ should be used. For example, if we just store the nonzero lower triangular part, then a $(p+1)$-by-$n$ array would suffice. (See §1.2.6)

If our band Cholesky procedure is coupled with appropriate band triangular solve routines then approximately $np^2 + 7np + 2n$ flops and $n$ square roots are required to solve $Ax = b$. For small $p$ it follows that the square roots represent a significant portion of the computation and it is preferable to use the LDL$^\text{T}$ approach. Indeed, a careful flop count of the steps $A = LDL^T$, $Ly = b$, $Dz = y$, and $L^T x = z$ reveals that $np^2 + 8np + n$ flops and no square roots are needed.

### 4.3.7 Tridiagonal System Solving

As a sample narrow band $LDL^T$ solution procedure, we look at the case of symmetric positive definite tridiagonal systems. Setting

$$L = \begin{bmatrix} 1 & & & \cdots & 0 \\ e_1 & 1 & & & \vdots \\ & \ddots & \ddots & & \\ \vdots & & \ddots & & \\ 0 & \cdots & & e_{n-1} & 1 \end{bmatrix}$$

and $D = \text{diag}(d_1, \ldots, d_n)$ we deduce from the equation $A = LDL^T$ that:

$$\begin{aligned} a_{11} &= d_1 & \\ a_{k,k-1} &= e_{k-1}d_{k-1} & k = 2{:}n \\ a_{kk} &= d_k + e_{k-1}^2 d_{k-1} = d_k + e_{k-1}a_{k,k-1} & k = 2{:}n \end{aligned}$$

Thus, the $d_i$ and $e_i$ can be resolved as follows:

```
d_1 = a_11
for k = 2:n
    e_{k-1} = a_{k,k-1}/d_{k-1}; d_k = a_kk - e_{k-1} a_{k,k-1}
end
```

To obtain the solution to $Ax = b$ we solve $Ly = b$, $Dz = y$, and $L^T x = z$. With overwriting we obtain

**Algorithm 4.3.6 (Symmetric, Tridiagonal, Positive Definite System Solver)** Given an $n$-by-$n$ symmetric, tridiagonal, positive definite matrix $A$ and $b \in \mathbb{R}^n$, the following algorithm overwrites $b$ with the solution to $Ax = b$. It is assumed that the diagonal of $A$ is stored in $d(1{:}n)$ and the superdiagonal in $e(1{:}n-1)$.

```
for k = 2:n
    t = e(k − 1); e(k − 1) = t/d(k − 1); d(k) = d(k) − te(k − 1)
end
for k = 2:n
    b(k) = b(k) − e(k − 1)b(k − 1)
end
b(n) = b(n)/d(n)
for k = n − 1: − 1:1
    b(k) = b(k)/d(k) − e(k)b(k + 1)
end
```

This algorithm requires $8n$ flops.

### 4.3.8 Vectorization Issues

The tridiagonal example brings up a sore point: narrow band problems and vector/pipeline architectures do not mix well. The narrow band implies short vectors. However, it is sometimes the case that large, independent sets of such problems must be solved at the same time. Let us look at how such a computation should be arranged in light of the issues raised in §1.4.

For simplicity, assume that we must solve the $n$-by-$n$ unit lower bidiagonal systems

$$A^{(k)}x^{(k)} = b^{(k)} \qquad k = 1{:}m$$

and that $m \gg n$. Suppose we have arrays $E(1{:}n-1, 1{:}m)$ and $B(1{:}n, 1{:}m)$ with the property that $E(1{:}n-1, k)$ houses the subdiagonal of $A^{(k)}$ and $B(1{:}n, k)$ houses the $k$th right hand side $b^{(k)}$. We can overwrite $b^{(k)}$ with the solution $x^{(k)}$ as follows:

```
for k = 1:m
    for i = 2:n
        B(i, k) = B(i, k) − E(i − 1, k)B(i − 1, k)
    end
end
```

The problem with this algorithm, which sequentially solves each bidiagonal system in turn, is that the inner loop does not vectorize. This is because of the dependence of $B(i,k)$ on $B(i-1,k)$. If we interchange the $k$ and $i$ loops we get

$$
\begin{array}{l}
\textbf{for } i = 2{:}n \\
\quad \textbf{for } k = 1{:}m \\
\quad\quad B(i,k) = B(i,k) - E(i-1,k)B(i-1,k) \\
\quad \textbf{end} \\
\textbf{end}
\end{array}
\qquad (4.3.1)
$$

Now the inner loop vectorizes well as it involves a vector multiply and a vector add. Unfortunately, (4.3.1) is not a unit stride procedure. However, this problem is easily rectified if we store the subdiagonals and right-hand-sides by row. That is, we use the arrays $E(1{:}m, 1{:}n-1)$ and $B(1{:}m, 1{:}n-1)$ and store the subdiagonal of $A^{(k)}$ in $E(k, 1{:}n-1)$ and $b^{(k)^T}$ in $B(k, 1{:}n)$. The computation (4.3.1) then transforms to

$$
\begin{array}{l}
\textbf{for } i = 2{:}n \\
\quad \textbf{for } k = 1{:}m \\
\quad\quad B(k,i) = (B(k,i) - E(k,i-1)B(k,i-1) \\
\quad \textbf{end} \\
\textbf{end}
\end{array}
$$

illustrating once again the effect of data structure on performance.

## Problems

**P4.3.1** Derive a banded $\text{LDM}^\text{T}$ procedure similar to Algorithm 4.3.1.

**P4.3.2** Show how the output of Algorithm 4.3.4 can be used to solve the upper Hessenberg system $Hx = b$.

**P4.3.3** Give an algorithm for solving an unsymmetric tridiagonal system $Ax = b$ that uses Gaussian elimination with partial pivoting and which requires only four $n$-vectors of floating point storage.

**P4.3.4** For $C \in \mathbb{R}^{n \times n}$ define the *profile indices* $m(C,i) = \min\{j{:}c_{ij} \neq 0\}$, where $i = 1{:}n$. Show that if $A = GG^T$ is the Cholesky factorization of $A$, then $m(A,i) = m(G,i)$ for $i = 1{:}n$. (We say that $G$ has the same *profile* as $A$.)

**P4.3.5** Suppose $A \in \mathbb{R}^{n \times n}$ is symmetric positive definite with profile indices $m_i = m(A,i)$ where $i = 1{:}n$. Assume that $A$ is stored in a one-dimensional array $v$ as follows: $v = (a_{11}, a_{2,m_2}, \ldots, a_{22}, a_{3,m_3}, \ldots, a_{33}, \ldots, a_{n,m_n}, \ldots, a_{nn})$. Write an algorithm that overwrites $v$ with the corresponding entries of the Cholesky factor $G$ and then uses this factorization to solve $Ax = b$. How many flops are required?

**P4.3.6** For $C \in \mathbb{R}^{n \times n}$ define $p(C,i) = \max\{j{:}c_{ij} \neq 0\}$. Suppose that $A \in \mathbb{R}^{n \times n}$ has an LU factorization $A = LU$ and that:

$$
\begin{array}{ccccccc}
m(A,1) & \leq & m(A,2) & \leq & \cdots & \leq & m(A,n) \\
p(A,1) & \leq & p(A,2) & \leq & \cdots & \leq & p(A,n)
\end{array}
$$

Show that $m(A,i) = m(L,i)$ and $p(A,i) = p(U,i)$ for $i = 1{:}n$. Recall the definition of $m(A,i)$ from P4.3.4.

**P4.3.7** Develop a gaxpy version of Algorithm 4.3.1.

**P4.3.8** Develop a unit stride, vectorizable algorithm for solving the symmetric positive definite tridiagonal systems $A^{(k)}x^{(k)} = b^{(k)}$. Assume that the diagonals, superdiagonals, and right hand sides are stored by row in arrays $D$, $E$, and $B$ and that $b^{(k)}$ is overwritten with $x^{(k)}$.

**P4.3.9** Develop a version of Algorithm 4.3.1 in which $A$ is stored by diagonal.

**P4.3.10** Give an example of a 3-by-3 symmetric positive definite matrix whose tridiagonal part is not positive definite.

### Notes and References for Sec. 4.3

We wish to stress again that our flop counts are meant only to guide our appraisals of work. The reader should not assign too much meaning to their precise value, especially in the band matrix area where so much depends on the cleverness of the implementation.

Fortran codes for banded linear systems may be found in Linpack, chapters 2, 4, and 7. The literature concerned with banded systems is immense. Some representative papers include

E.L. Allgower (1973). "Exact Inverses of Certain Band Matrices," *Numer. Math. 21*, 279–84.

Z. Bohte (1975). "Bounds for Rounding Errors in the Gaussian Elimiantion for Band Systems," *J. Inst. Math. Applic. 16*, 133–42.

I.S. Duff (1977). "A Survey of Sparse Matrix Research," *Proc. IEEE 65*, 500–535.

R.S. Martin and J.H. Wilkinson (1965). "Symmetric Decomposition of Positive Definite Band Matrices," *Numer. Math. 7*, 355–61. See also *HACLA*, pp. 50–56.

R. S. Martin and J.H. Wilkinson (1967). "Solution of Symmetric and Unsymmetric Band Equations and the Calculation of Eigenvalues of Band Matrices," *Numer. Math. 9*, 279–301. See also *HACLA*, pp. 70–92.

A topic of considerable interest in the area of banded matrices deals with methods for reducing the width of the band. See

E. Cuthill (1972). "Several Strategies for Reducing the Bandwidth of Matrices," in *Sparse Matrices and Their Applications*, ed. D.J. Rose and R.A. Willoughby, Plenum Press, New York.

N.E. Gibbs, W.G. Poole, Jr., and P.K. Stockmeyer (1976). "An Algorithm for Reducing the Bandwidth and Profile of a Sparse Matrix," *SIAM J. Num. Anal. 13*, 236–50.

N.E. Gibbs, W.G. Poole, Jr., and P.K. Stockmeyer (1976). "A Comparison of Several Bandwidth and Profile Reduction Algorithms," *ACM Trans. Math. Soft. 2*, 322–30.

As we mentioned, tridiagonal systems arise with particular frequency. Thus, it is not surprising that a great deal of attention has been focused on special methods for this class of banded problems

C. Fischer and R.A. Usmani (1969). "Properties of Some Tridiagonal Matrices and Their Application to Boundary Value Problems," *SIAM J. Num. Anal. 6*, 127–42.

N.J. Higham (1986c). "Efficient Algorithms for computing the condition number of a tridiagonal matrix," *SIAM J. Sci. and Stat. Comp. 7*, 150–165.

D. Kershaw(1982). "Solution of Single Tridiagonal Linear Systems and Vectorization of the ICCG Algorithm on the Cray-1," in *Parallel Computation*, ed. G. Roderigue, Academic Press, NY, 1982.

J. Lambiotte and R.G. Voigt (1975). "The Solution of Tridiagonal Linear Systems of the CDC-STAR 100 Computer," *ACM Trans. Math. Soft. 1*, 308–29.

M.A. Malcolm and J. Palmer (1974). "A Fast Method for Solving A Class of Tridiagonal Systems of Linear Equations," *Comm. ACM 17*, 14–17.

D.J. Rose (1969), "An Algorithm for Solving a Special Class of Tridiagonal Systems of Linear Equations," *Comm. ACM 12*, 234–36.
H.S. Stone (1973). "An Efficient Parallel Algorithm for the Solution of a Tridiagonal Linear System of Equations," *J. ACM 20*, 27–38.
H.S. Stone (1975). "Parallel Tridiagonal Equation Solvers," *ACM Trans. Math. Soft.1*, 289–307.

Chapter 4 of

J.A. George and J.W. Liu (1981). *Computer Solution of Large Sparse Positive Definite Systems*, Prentice-Hall, Englewood Cliffs, New Jersey.

contains a nice survey of band methods for positive definite systems.

## 4.4 Symmetric Indefinite Systems

A symmetric matrix whose quadratic form $x^T Ax$ takes on both positive and negative values is called *indefinite*. Although an indefinite $A$ may have an $\text{LDL}^\text{T}$ factorization, the entries in the factors can have arbitrary magnitude:

$$\begin{bmatrix} \epsilon & 1 \\ 1 & 0 \end{bmatrix} = \begin{bmatrix} 1 & 0 \\ 1/\epsilon & 1 \end{bmatrix} \begin{bmatrix} \epsilon & 0 \\ 0 & -1/\epsilon \end{bmatrix} \begin{bmatrix} 1 & 0 \\ 1/\epsilon & 1 \end{bmatrix}^T .$$

Of course, any of the pivot strategies in §3.4 could be invoked. However, they destroy symmetry and with it, the chance for a "Cholesky speed" indefinite system solver. Symmetric pivoting, i.e., data reshufflings of the form $A \leftarrow PAP^T$, must be used as we discussed in §4.2.9. Unfortunately, symmetric pivoting does not always stabilize the $\text{LDL}^\text{T}$ computation. If $\epsilon_1$ *and* $\epsilon_2$ are small then regardless of $P$, the matrix

$$\tilde{A} = P \begin{bmatrix} \epsilon_1 & 1 \\ 1 & \epsilon_2 \end{bmatrix} P^T$$

has small diagonal entries and large numbers surface in the factorization. With symmetric pivoting, the pivots are always selected from the diagonal and trouble results if these numbers are small relative to what must be zeroed off the diagonal. Thus, $\text{LDL}^\text{T}$ with symmetric pivoting cannot be recommended as a reliable approach to symmetric indefinite system solving. It seems that the challenge is to involve the off-diagonal entries in the pivoting process while at the same time maintaining symmetry.

In this section we discuss two ways to do this. The first method is due to Aasen(1971) and it computes the factorization

$$PAP^T = LTL^T \tag{4.4.1}$$

where $L = (\ell_{ij})$ is unit lower triangular and $T$ is tridiagonal. $P$ is a permutation chosen such that $|\ell_{ij}| \leq 1$. In contrast, the *diagonal pivoting method* computes a permutation $P$ such that

$$PAP^T = LDL^T \tag{4.4.2}$$

where $D$ is a direct sum of 1-by-1 and 2-by-2 pivot blocks. Again, $P$ is chosen so that the entries in the unit lower tiangular $L$ satisfy $|\ell_{ij}| \leq 1$. Both factorizations involve $n^3/3$ flops and once computed, can be used to solve $Ax = b$ with $O(n^2)$ work:

$$PAP^T = LTL^T, Lz = Pb, Tw = z, L^Ty = w, x = Py \quad \Rightarrow \quad Ax = b$$

$$PAP^T = LDL^T, Lz = Pb, Dw = z, L^Ty = w, x = Py \quad \Rightarrow \quad Ax = b$$

The only thing "new" to discuss in these solution procedures are the $Tw = z$ and $Dw = z$ systems.

In Aasen's method, the symmetric indefinite tridiagonal system $Tw = z$ is solved in $O(n)$ time using band Gaussian elimination with pivoting. Note that there is no serious price to pay for the disregard of symmetry at this level since the overall process is $O(n^3)$.

In the diagonal pivoting approach, the $Dw = z$ system amounts to a set of 1-by-1 and 2-by-2 symmetric indefinite systems. The 2-by-2 problems can be handled via Gaussian elimination with pivoting. Again, there is no harm in disregarding symmetry during this $O(n)$ phase of the calculation.

Thus, the central issue in this section is the efficient computation of the factorizations (4.4.1) and (4.4.2).

### 4.4.1 The Parlett-Reid Algorithm

Parlett and Reid (1970) show how to compute (4.4.1) using Gauss transforms. Their algorithm is sufficiently illustrated by displaying the $k = 2$ step for the case $n = 5$. At the beginning of this step the matrix $A$ has been transformed to

$$A^{(1)} = M_1P_1AP_1^TM_1^T = \begin{bmatrix} \alpha_1 & \beta_1 & 0 & 0 & 0 \\ \beta_1 & \alpha_2 & v_3 & v_4 & v_5 \\ 0 & v_3 & \times & \times & \times \\ 0 & v_4 & \times & \times & \times \\ 0 & v_5 & \times & \times & \times \end{bmatrix}$$

where $P_1$ is a permutation chosen so that the entries in the Gauss transformation $M_1$ are bounded by unity in modulus. Scanning the vector $(v_3\ v_4\ v_5)^T$ for its largest entry, we now determine a 3-by-3 permutation $\tilde{P}_2$ such that

$$\tilde{P}_2 \begin{bmatrix} v_3 \\ v_4 \\ v_5 \end{bmatrix} = \begin{bmatrix} \tilde{v}_3 \\ \tilde{v}_4 \\ \tilde{v}_5 \end{bmatrix} \quad \Rightarrow \quad |\tilde{v}_3| = \max\{|\tilde{v}_3|, |\tilde{v}_4|, |\tilde{v}_5|\}\,.$$

If this maximal element is zero, we set $M_2 = P_2 = I$ and proceed to the next step. Otherwise, we set $P_2 = \text{diag}(I_2, \tilde{P}_2)$ and $M_2 = I - \alpha^{(2)}e_3^T$ with

$$\alpha^{(2)} = \left(\ 0 \quad 0 \quad 0 \quad \tilde{v}_4/\tilde{v}_3 \quad \tilde{v}_5/\tilde{v}_3\ \right)^T$$

and observe that

$$A^{(2)} = M_2P_2A^{(1)}P_2^TM_2^T = \begin{bmatrix} \alpha_1 & \beta_1 & 0 & 0 & 0 \\ \beta_1 & \alpha_2 & \tilde{v}_3 & 0 & 0 \\ 0 & \tilde{v}_3 & \times & \times & \times \\ 0 & 0 & \times & \times & \times \\ 0 & 0 & \times & \times & \times \end{bmatrix}.$$

In general, the process continues for $n-2$ steps leaving us with a tridiagonal matrix

$$T = A^{(n-2)} = (M_{n-2}P_{n-2}\cdots M_1P_1)A(M_{n-2}P_{n-2}\cdots M_1P_1)^T .$$

It can be shown that (4.4.1) holds with $P = P_{n-2}\cdots P_1$ and

$$L = (M_{n-2}P_{n-2}\cdots M_1P_1P^T)^{-1} .$$

Analysis of $L$ reveals that its first column is $e_1$ and that its subdiagonal entries in column $k$ with $k > 1$ are "made up" of the multipliers in $M_{k-1}$.

The efficient implementation of the Parlett-Reid method requires care when computing the update

$$A^{(k)} = M_k(P_kA^{(k-1)}P_k^T)M_k^T. \tag{4.4.3}$$

To see what is involved with a minimum of notation, suppose $B = B^T$ has order $n-k$ and that we wish to form: $B_+ = (I - we_1^T)B(I - we_1^T)^T$ where $w \in \mathbb{R}^{n-k}$ and $e_1$ is the first column of $I_{n-k}$. Such a calculation is at the heart of (4.4.3). If we set

$$u = Be_1 - \frac{b_{11}}{2}w,$$

then the lower half of the symmetric matrix $B_+ = B - wu^T - uw^T$ can be formed in $2(n-k)^2$ flops. Summing this quantity as $k$ ranges from 1 to $n-2$ indicates that the Parlett-Reid procedure requires $2n^3/3$ flops—twice what we would like.

**Example 4.4.1** If the Parlett-Reid algorithm is applied to

$$A = \begin{bmatrix} 0 & 1 & 2 & 3 \\ 1 & 2 & 2 & 2 \\ 2 & 2 & 3 & 3 \\ 3 & 2 & 3 & 4 \end{bmatrix}$$

then

$$\begin{aligned} P_1 &= [\, e_1\ e_4\ e_3\ e_2 \,] \\ M_1 &= I_4 - (0,\ 0,\ 2/3,\ 1/3,\ )^Te_2^T \\ P_2 &= [\, e_1\ e_2\ e_4\ e_3 \,] \\ M_2 &= I_4 - (0,\ 0,\ 0,\ 1/2)^Te_3^T \end{aligned}$$

and $PAP^T = LTL^T$ , where $P = [\, e_1, \, e_3, \, e_4, \, e_2 \,]$,

$$L = \begin{bmatrix} 1 & 0 & 0 & 0 \\ 0 & 1 & 0 & 0 \\ 0 & 1/3 & 1 & 0 \\ 0 & 2/3 & 1/2 & 1 \end{bmatrix} \quad \text{and} \quad T = \begin{bmatrix} 0 & 3 & 0 & 0 \\ 3 & 4 & 2/3 & 0 \\ 0 & 2/3 & 10/9 & 0 \\ 0 & 0 & 0 & 1/2 \end{bmatrix}.$$

### 4.4.2 The Method of Aasen

An $n^3/3$ approach to computing (4.4.1) due to Aasen (1971) can be derived by reconsidering some of the computations in the Parlett-Reid approach. We need a notation for the tridiagonal $T$:

$$T = \begin{bmatrix} \alpha_1 & \beta_1 & & \cdots & 0 \\ \beta_1 & \alpha_2 & \ddots & & \vdots \\ & \ddots & \ddots & \ddots & \\ \vdots & & \ddots & \ddots & \beta_{n-1} \\ 0 & \cdots & & \beta_{n-1} & \alpha_n \end{bmatrix}.$$

For clarity, we temporarily ignore pivoting and assume that the factorization $A = LTL^T$ exists where $L$ is unit lower triangular with $L(:,1) = e_1$. Aasen's method is organized as follows:

$$\begin{array}{l}
\textbf{for } j = 1{:}n \\
\quad \text{Compute } h(1{:}j) \text{ where } h = TL^T e_j = He_j. \\
\quad \text{Compute } \alpha(j). \\
\quad \textbf{if } j \le n-1 \\
\quad\quad \text{Compute } \beta(j) \\
\quad \textbf{end} \\
\quad \textbf{if } j \le n-2 \\
\quad\quad \text{Compute } L(j+2{:}n, j+1). \\
\quad \textbf{end} \\
\textbf{end}
\end{array} \qquad (4.4.4)$$

Thus, the mission of the $j$th Aasen step is to compute the $j$th column of $T$ and the $(j+1)$-st column of $L$. The algorithm exploits the fact that the matrix $H = TL^T$ is upper Hessenberg. As can be deduced from (4.4.4), the computation of $\alpha(j)$, $\beta(j)$, and $L(j+2{:}n, j+1)$ hinges upon the vector $h(1{:}j) = H(1{:}j, j)$. Let us see why.

Consider the $j$th column of the equation $A = LH$:

$$A(:,j) = L(:,1{:}j+1)h(1{:}j+1)\,. \qquad (4.4.5)$$

This says that $A(:,j)$ is a linear combination of the first $j+1$ columns of $L$. In particular,

$$A(j+1{:}n,j) \;=\; L(j+1{:}n,1{:}j)h(1{:}j) \;+\; L(j+1{:}n,j+1)h(j+1)\,.$$

It follows that if we compute

$$v(j+1{:}n) = A(j+1{:}n,j) - L(j+1{:}n,1{:}j)h(1{:}j)\,,$$

then

$$L(j+1{:}n,j+1)h(j+1) \;=\; v(j+1{:}n)\,. \tag{4.4.6}$$

Thus, $L(j+2{:}n,j+1)$ is a scaling of $v(j+2{:}n)$. Since $L$ is unit lower triangular we have from (4.4.6) that

$$v(j+1) = h(j+1)$$

and so from that same equation we obtain the following recipe for the $(j+1)$-st column of $L$:

$$L(j+2{:}n,j+1) = v(j+2{:}n)/v(j+1)\,.$$

Note that $L(j+2{:}n,j+1)$ is a scaled gaxpy.

We next develop formulae for $\alpha(j)$ and $\beta(j)$. Compare the $(j,j)$ and $(j+1,j)$ entries in the equation $H = TL^T$. With the convention $\beta(0) = 0$ we find that $h(j) = \beta(j-1)L(j,j-1) + \alpha(j)$ and $h(j+1) = v(j+1)$ and so

$$\begin{aligned} \alpha(j) &= h(j) - \beta(j-1)L(j,j-1) \\ \beta(j) &= v(j+1)\,. \end{aligned}$$

With these recipes we can completely describe the Aasen procedure:

**for** $j = 1{:}n$
  Compute $h(1{:}j)$ where $h = TL^Te_j$.
  **if** $j = 1 \vee j = 2$
    $\alpha(j) = h(j)$
  **else**
    $\alpha(j) = h(j) - \beta(j-1)L(j,j-1)$
  **end**
  **if** $j \le n-1$ (4.4.7)
    $v(j+1{:}n) = A(j+1{:}n,j) - L(j+1{:}n,1{:}j)h(1{:}j)$
    $\beta(j) = v(j+1)$
  **end**
  **if** $j \le n-2$
    $L(j+2{:}n,j+1) = v(j+2{:}n)/v(j+1)$
  **end**
**end**

To complete the description we must detail the computation of $h(1{:}j)$. From (4.4.5) it follows that

$$A(1{:}j, j) = L(1{:}j, 1{:}j)h(1{:}j)\,. \tag{4.4.8}$$

This lower triangular system can be solved for $h(1{:}j)$ since we know the first $j$ columns of $L$. However, a much more efficient way to compute $H(1{:}j, j)$ is obtained by exploiting the $j$th column of the equation $H = TL^T$. In particular, with the convention that $\beta(0)L(j, 0) = 0$ we have

$$h(k) = \beta(k-1)L(j, k-1) + \alpha(k)L(j, k) + \beta(k)L(j, k+1)\,.$$

for $k = 1{:}j$. These are working formulae except in the case $k = j$ because we have not yet computed $\alpha(j)$ and $\beta(j)$. However, once $h(1{:}j-1)$ is known we can obtain $h(j)$ from the last row of the triangular system (4.4.8), i.e.,

$$h(j) \;=\; A(j, j) \;-\; \sum_{k=1}^{j-1} L(j, k)h(k)\,.$$

Collecting results and using a work array $\ell(1{:}n)$ for $L(j, 1{:}j)$ we see that the computation of $h(1{:}j)$ in (4.4.7) can be organized as follows:

$$
\begin{array}{l}
\textbf{if } j = 1 \\
\quad h(1) = A(1,1) \\
\textbf{elseif } j = 2 \\
\quad h(1) = \beta(1);\; h(2) = A(2,2) \\
\textbf{else} \\
\quad \ell(0) = 0;\; \ell(1) = 0;\; \ell(2{:}j-1) = L(j, 2{:}j-1);\; \ell(j) = 1 \\
\quad h(j) = A(j,j) \\
\quad \textbf{for } k = 1{:}j-1 \\
\qquad h(k) = \beta(k-1)\ell(k-1) + \alpha(k)\ell(k) + \beta(k)\ell(k+1) \\
\qquad h(j) = h(j) - \ell(k)h(k) \\
\quad \textbf{end} \\
\textbf{end}
\end{array}
\tag{4.4.9}
$$

Note that with this $O(j)$ method for computing $h(1{:}j)$, the gaxpy calculation of $v(j+1{:}n)$ is the dominant operation in (4.4.7). During the $j$th step this gaxpy involves about $2j(n-j)$ flops. Summing this for $j = 1{:}n$ shows that Aasen's method requires $n^3/3$ flops. Thus, the Aasen and Cholesky algorithms entail the same amount of arithmetic.

### 4.4.3 Pivoting in Aasen's Method

As it now stands, the columns of $L$ are scalings of the $v$-vectors in (4.4.7). If any of these scalings are large, i.e., if any of the $v(j+1)$'s are small, then we are in trouble. To circumvent this problem we need only permute the largest component of $v(j+1{:}n)$ to the top position. Of course, this permutation must be suitably applied to the unreduced portion of $A$ and the previously computed portion of $L$.

**Algorithm 4.4.1 (Aasen's Method)** If $A \in \mathbb{R}^{n\times n}$ is symmetric then the following algorithm computes a permutation $P$, a unit lower triangular $L$, and a tridiagonal $T$ such that $PAP^T = LTL^T$ with $|L(i,j)| \leq 1$. The permutation $P$ is encoded in an integer vector $piv$. In particular, $P = P_1 \cdots P_{n-2}$ where $P_j$ is the identity with rows $piv(j)$ and $j+1$ interchanged. The diagonal and subdiagonal of $T$ are stored in $\alpha(1{:}n)$ and $\beta(1{:}n-1)$, respectively. Only the subdiagonal portion of $L(2{:}n, 2{:}n)$ is computed.

```
for j = 1:n
    Compute h(1:j) via (4.4.9).
    if j = 1 ∨ j = 2
        α(j) = h(j)
    else
        α(j) = h(j) − β(j − 1)L(j, j − 1)
    end
    if j ≤ n − 1
        v(j + 1:n) = A(j + 1:n, j) − L(j + 1:n, 1:j)h(1:j)
        Find q so |v(q)| = || v(j + 1:n) ||∞ with j + 1 ≤ q ≤ n.
        piv(j) = q; v(j + 1) ↔ v(q); L(j + 1, 2:j) ↔ L(q, 2:j)
        A(j + 1, j + 1:n) ↔ A(q, j + 1:n)
        A(j + 1:n, j + 1) ↔ A(j + 1:n, q)
        β(j) = v(j + 1)
    end
    if j ≤ n − 2
        L(j + 2:n, j + 1) = v(j + 2:n)
        if v(j + 1) ≠ 0
            L(j + 2:n, j + 1) = L(j + 2:n, j + 1)/v(j + 1)
        end
    end
end
```

Aasen's method is stable in the same sense that Gaussian elimination with partial pivoting is stable. That is, the exact factorization of a matrix near $A$ is obtained provided $\| \hat{T} \|_2 / \| A \|_2 \approx 1$, where $\hat{T}$ is the computed version of the tridiagonal matrix $T$. In general, this is almost always the case.

In a practical implementation of the Aasen algorithm, the lower triangular portion of $A$ would be overwritten with $L$ and $T$. Here is $n = 5$

case:

$$A \leftarrow \begin{bmatrix} \alpha_1 & & & & \\ \beta_1 & \alpha_2 & & & \\ \ell_{32} & \beta_2 & \alpha_3 & & \\ \ell_{42} & \ell_{43} & \beta_3 & \alpha_4 & \\ \ell_{52} & \ell_{53} & \ell_{54} & \beta_4 & \alpha_5 \end{bmatrix}$$

Notice that the columns of $L$ are shifted left in this arrangement.

### 4.4.4 Diagonal Pivoting Methods

We next describe the computation of the block $LDL^T$ factorization (4.4.2). We follow the discussion in Bunch and Parlett (1971). Suppose

$$P_1 A P_1^T = \begin{bmatrix} E & C^T \\ C & B \end{bmatrix} \begin{matrix} s \\ n-s \end{matrix}$$
$$\qquad\qquad s \quad n-s$$

where $P_1$ is a permutation matrix and $s = 1$ or 2. If $A$ is nonzero, then it is always possible to choose these quantities so that $E$ is nonsingular thereby enabling us to write

$$P_1 A P_1^T = \begin{bmatrix} I_s & 0 \\ CE^{-1} & I_{n-s} \end{bmatrix} \begin{bmatrix} E & 0 \\ 0 & B - CE^{-1}C^T \end{bmatrix} \begin{bmatrix} I_s & E^{-1}C^T \\ 0 & I_{n-s} \end{bmatrix}$$

For the sake of stability, the $s$-by-$s$ "pivot" $E$ should be chosen so that the entries in

$$\tilde{A} = (\tilde{a}_{ij}) \equiv B - CE^{-1}C^T \qquad (4.4.10)$$

are suitably bounded. To this end, let $\alpha \in (0,1)$ be given and define the size measures

$$\mu_0 = \max_{i,j} |a_{ij}| \qquad\qquad \mu_1 = \max_{i} |a_{ii}|.$$

The Bunch-Parlett pivot strategy is as follows:

**if** $\mu_1 \geq \alpha\mu_0$
  $s = 1$
  Choose $P_1$ so $|e_{11}| = \mu_1$.
**else**
  $s = 2$
  Choose $P_1$ so $|e_{21}| = \mu_0$.
**end**

It is easy to verify from (4.4.10) that if $s = 1$ then

$$|\tilde{a}_{ij}| \leq (1 + \alpha^{-1})\mu_0 \tag{4.4.11}$$

while $s = 2$ implies

$$|\tilde{a}_{ij}| \leq \frac{3 - \alpha}{1 - \alpha}\mu_0 . \tag{4.4.12}$$

By equating $(1 + \alpha^{-1})^2$, the growth factor associated with two $s = 1$ steps, and $(3 - \alpha)/(1 - \alpha)$, the corresponding $s = 2$ factor, Bunch and Parlett conclude that $\alpha = (1 + \sqrt{17})/8$ is optimum from the standpoint of minimizing the bound on element growth.

The reductions outlined above are then repeated on the $n - s$ order symmetric matrix $\tilde{A}$. A simple induction argument establishes that the factorization (4.4.2) exists and that $n^3/3$ flops are required if the work associated with pivot determination is ignored.

### 4.4.5 Stability and Efficiency

Diagonal pivoting with the above strategy is shown by Bunch (1971a) to be as stable as Gaussian elimination with complete pivoting. Unfortunately, the overall process requires between $n^3/12$ and $n^3/6$ comparisons, since $\mu_0$ involves a two-dimensional search at each stage of the reduction. The actual number of comparisons depends on the total number of 2-by-2 pivots but in general the Bunch-Parlett method for computing (4.4.2) is considerably slower than the technique of Aasen. See Barwell and George(1976).

This is not the case with the diagonal pivoting method of Bunch and Kaufman (1977). In their scheme, it is only necessary to scan two columns at each stage of the reduction. The strategy is fully illustrated by considering the very first step in the reduction:

$\alpha = (1 + \sqrt{17})/8$; $\lambda = |a_{r1}| = \max\{|a_{21}|, \ldots, |a_{n1}|\}$
**if** $\lambda > 0$
  **if** $|a_{11}| \geq \alpha\lambda$
    $s = 1; P_1 = I$
  **else**
    $\sigma = |a_{pr}| = \max\{|a_{1r}, \ldots, |a_{r-1,r}|, |a_{r+1,r}|, \ldots, |a_{nr}|\}$
    **if** $\sigma|a_{11}| \geq \alpha\lambda^2$
      $s = 1, P_1 = I$
    **elseif** $|a_{rr}| \geq \alpha\sigma$
      $s = 1$ and choose $P_1$ so $(P_1^T A P_1)_{11} = a_{rr}$.
    **else**
      $s = 2$ and choose $P_1$ so $(P_1^T A P_1)_{21} = a_{rp}$.
    **end**
  **end**
**end**

Overall, the Bunch-Kaufman algorithm requires $n^3/3$ flops, $O(n^2)$ comparisons, and, like all the methods of this section, $n^2/2$ storage.

**Example 4.4.2** If the Bunch-Kaufman algorithm is applied to

$$A = \begin{bmatrix} 1 & 10 & 20 \\ 10 & 1 & 30 \\ 20 & 30 & 1 \end{bmatrix}$$

then in the first step $\lambda = 20$, $r = 3$, $\sigma = 1$, and $p = 3$. The permutation $P = [\ e_1\ e_3\ e_2\ ]$ is applied giving

$$PAP^T = \begin{bmatrix} 1 & 20 & 10 \\ 20 & 1 & 30 \\ 10 & 30 & 1 \end{bmatrix}$$

A 2-by-2 pivot is then used to produce the reduction

$$PAP^T = \begin{bmatrix} 1 & 0 & 0 \\ 0 & 1 & 0 \\ \frac{590}{399} & \frac{170}{399} & 1 \end{bmatrix} \begin{bmatrix} 1 & 20 & 0 \\ 20 & 1 & 0 \\ 0 & 0 & \frac{-10601}{399} \end{bmatrix} \begin{bmatrix} 1 & 0 & 0 \\ 0 & 1 & 0 \\ \frac{590}{399} & \frac{170}{399} & 1 \end{bmatrix}^T$$

## 4.4.6 Aasen's Method Versus Diagonal Pivoting

Because "future" columns must be scanned in the pivoting process, it is awkward (but possible) to obtain a gaxpy-rich diagonal pivoting algorithm. On the other hand, Aasen's method is naturally rich in gaxpy's making it perhaps a better choice in vector-pipeline environments. Finally we mention that block versions of both procedures are possible.

### Problems

**P4.4.1** Show that if all the 1-by-1 and 2-by-2 principal submatrices of an $n$-by-$n$ symmetric matrix $A$ are singular, then $A$ is zero.

**P4.4.2** Show that no 2-by-2 pivots can arise in the Bunch-Kaufman algorithm if $A$ is positive definite.

**P4.4.3** Arrange Algorithm 4.4.1 so that only the lower triangular portion of $A$ is referenced and so that $\alpha(j)$ overwrites $A(j,j)$ for $j = 1{:}n$, $\beta(j)$ overwrites $A(j+1,j)$ for $j = 1{:}n-1$, and $L(i,j)$ overwrites $A(i,j-1)$ for $j = 2{:}n-1$ and $i = j+1{:}n$.

**P4.4.4** Suppose $A \in \mathbb{R}^{n\times n}$ is nonsingular, symmetric, and diagonally dominant. Give an algorithm that computes the factorization

$$\Pi A \Pi^T = \begin{bmatrix} R & 0 \\ S & -M \end{bmatrix} \begin{bmatrix} R^T & S^T \\ 0 & M^T \end{bmatrix}$$

where $R \in \mathbb{R}^{k\times k}$ and $M \in \mathbb{R}^{(n-k)\times(n-k)}$ are lower triangular and nonsingular and $\Pi$ is a permutation.

### Notes and References for Sec. 4.4

The basic references for computing (4.4.1) are

J.O. Aasen (1971). "On the Reduction of a Symmetric Matrix to Tridiagonal Form," *BIT 11*, 233–42.
B.N. Parlett and J.K. Reid (1970). "On the Solution of a System of Linear Equations Whose Matrix is Symmetric but not Definite," *BIT 10*, 386–97.

The diagonal pivoting literature includes

J.R. Bunch (1971a). "Analysis of the Diagonal Pivoting Method," *SIAM J. Num. Anal. 8*, 656–80.
J.R. Bunch (1974). "Partial Pivoting Strategies for Symmetric Matrices," *SIAM J. Num. Anal. 11*, 521–28.
J.R. Bunch and L. Kaufman (1977). "Some Stable Methods for Calculating Inertia and Solving Symmetric Linear Systems," *Math. Comp. 31*, 162–79.
J.R. Bunch, L. Kaufman, and B.N. Parlett (1976). "Decomposition of a Symmetric Matrix," *Numer. Math. 27*, 95–109.
J.R. Bunch and B.N. Parlett (1971). "Direct Methods for Solving Symmetric Indefinite Systems of Linear Equations," *SIAM J. Num. Anal. 8*, 639–55.

The second to last reference contains an ALGOL version of the diagonal pivoting method contained in Linpack, chapter 5.

The question of whether Aasen's method is to be preferred to the Bunch-Kaufman algorithm is studied in

V. Barwell and J.A. George (1976). "A Comparison of Algorithms for Solving Symmetric Indefinite Systems of Linear Equations," *ACM Trans. Math. Soft. 2*, 242–51.

They suggest that the two algorithms behave similarly on conventional processors with perhaps a slight edge to Aasen when $n$ is larger than 200. The performance data in this paper is the basis of an interesting statistical analysis in

D. Hoaglin (1977). "Mathematical Software and Exploratory Data Analysis," in *Mathematical Software III*, ed. John Rice, Academic Press, New York, pp. 139–59.

The small advantage of Aasen's method is perhaps due to its simpler pivot strategy.

Another idea for a cheap pivoting strategy utilizes error bounds based on more liberal interchange criteria, an idea borrowed from some work done in the area of sparse elimination methods. See

R. Fletcher (1976). "Factorizing Symmetric Indefinite Matrices," *Lin. Alg. and Its Applic. 14*, 257–72.

We also mention the paper

A. Dax and S. Kaniel (1977). "Pivoting Techniques for Symmetric Gaussian Elimination," *Numer. Math. 28*, 221–42.

in which diagonal pivot entries are "built up" if necessary by using upper triangular multiplier matrices. Unfortunately, the technique appears to require $O(n^3)$comparisons.

Before using any symmetric $Ax = b$ solver, it may be advisable to equilibrate $A$. An $O(n^2)$ algorithm for accomplishing this task is given in

J.R. Bunch (1971b). "Equilibration of Symmetric Matrices in the Max-Norm," *J. ACM 18*, 566–72.

Finally we mention that analogues of the methods in this section exist for skew-symmetric systems, see

J.R. Bunch (1982). "A Note on the Stable Decomposition of Skew Symmetric Matrices," *Math. Comp. 158*, 475–480.

## 4.5 Block Tridiagonal Systems

In many application areas the matrices that arise have exploitable block structure. For example, in constrained optimization linear systems of the form

$$\begin{bmatrix} A & B \\ B^T & 0 \end{bmatrix} \begin{bmatrix} y \\ z \end{bmatrix} = \begin{bmatrix} c \\ d \end{bmatrix} \tag{4.5.1}$$

must frequently be solved where $A$ is symmetric positive definite and $B$ has full column rank. In this situation, it pays to exploit this structure rather than to treat (4.5.1) as "just another" symmetric indefinite system. See Heath (1978) for a detailed discussion.

As a case study in how to exploit block structure, we have chosen to look at the solution of block tridiagonal systems of the form

$$\begin{bmatrix} D_1 & F_1 & & \cdots & 0 \\ E_1 & D_2 & \ddots & & \vdots \\ & \ddots & \ddots & \ddots & \\ \vdots & & \ddots & \ddots & F_{n-1} \\ 0 & \cdots & & E_{n-1} & D_n \end{bmatrix} \begin{bmatrix} x_1 \\ x_2 \\ \vdots \\ \vdots \\ x_n \end{bmatrix} = \begin{bmatrix} b_1 \\ b_2 \\ \vdots \\ \vdots \\ b_n \end{bmatrix} \tag{4.5.2}$$

Here we assume that all blocks are $q$-by-$q$ and that the $x_i$ and $b_i$ are in $\mathbb{R}^q$. In this section we discuss both a block LU approach to this problem as well as a divide and conquer scheme known as *cyclic reduction*. Other aspects of block tridiagonal systems are discussed in §10.3.3

### 4.5.1 Block LU Factorization

We begin by considering a block LU factorization for the matrix in (4.5.2). Define the block tridiagonal matrices $A_k$ by

$$A_k = \begin{bmatrix} D_1 & F_1 & & \cdots & 0 \\ E_1 & D_2 & \ddots & & \vdots \\ & \ddots & \ddots & \ddots & \\ \vdots & & \ddots & \ddots & F_{k-1} \\ 0 & \cdots & & E_{k-1} & D_k \end{bmatrix} \qquad k = 1{:}n\,. \tag{4.5.3}$$

Comparing blocks in

$$A_n = \begin{bmatrix} I & & \cdots & 0 \\ L_1 & I & & \vdots \\ & \ddots & \ddots & \\ \vdots & & \ddots & \\ 0 & \cdots & L_{n-1} & I \end{bmatrix} \begin{bmatrix} U_1 & F_1 & & \cdots & 0 \\ 0 & U_2 & \ddots & & \vdots \\ & \ddots & \ddots & \ddots & \\ \vdots & & \ddots & \ddots & F_{n-1} \\ 0 & \cdots & & 0 & U_n \end{bmatrix} \qquad (4.5.4)$$

we formally obtain the following algorithm for the $L_i$ and $U_i$:

$$\begin{array}{l} U_1 = D_1 \\ \textbf{for } i = 2{:}n \\ \quad \text{Solve } L_{i-1}U_{i-1} = E_{i-1} \text{ for } L_{i-1}. \\ \quad U_i = D_i - L_{i-1}F_{i-1} \\ \textbf{end} \end{array} \qquad (4.5.5)$$

The procedure is defined so long as the $U_i$ are nonsingular. This is assured, for example, if the matrices $A_1, \ldots, A_n$ are nonsingular.

Having computed the factorization (4.5.4), the vector $x$ in (4.5.2) can be obtained via block forward and back substitution:

$$\begin{array}{l} y_1 = b_1 \\ \textbf{for } i = 2{:}n \\ \quad y_i = b_i - L_{i-1}y_{i-1} \\ \textbf{end} \\ \text{Solve } Ux_n = y_n \text{ for } x_n. \\ \textbf{for } i = n-1{:}-1{:}1 \\ \quad \text{Solve } U_i x_i = y_i - F_i x_{i+1} \text{ for } x_i. \\ \textbf{end} \end{array} \qquad (4.5.6)$$

To carry out both (4.5.5) and (4.5.6), each $U_i$ must be factored since linear systems involving these submatrices are solved. This could be done using Gaussian elimination with pivoting. However, this does not guarantee the stability of the overall process. To see this just consider the case when the block size $q$ is unity.

## 4.5.2 Block Diagonal Dominance

In order to obtain satisfactory bounds on the $L_i$ and $U_i$ it is necessary to make additional assumptions about the underlying block matrix. For example, if for $i = 1{:}n$ we have the block diagonal dominance relations

$$\| D_i^{-1} \|_1 (\| F_{i-1} \|_1 + \| E_i \|_1) < 1 \qquad E_n \equiv F_0 \equiv 0 \qquad (4.5.7)$$

then the factorization (4.5.3) exists and it is possible to show that the $L_i$ and $U_i$ satisfy the inequalities

$$\begin{aligned} \| L_i \|_1 &\leq 1 \qquad (4.5.8) \\ \| U_i \|_1 &\leq \| A \|_1 \qquad (4.5.9) \end{aligned}$$

### 4.5.3 Block Versus Band Solving

At this point it is reasonable to ask why we do not simply regard the matrix $A$ in (4.5.2) as a $qn$-by-$qn$ matrix having scalar entries and bandwidth $2q - 1$. Band Gaussian elimination as described in §4.3 could be applied. The effectiveness of this course of action depends on such things as the dimensions of the blocks and the sparsity patterns within each block.

To illustrate this in a very simple setting, suppose that we wish to solve

$$\begin{bmatrix} D_1 & F_1 \\ E_1 & D_2 \end{bmatrix} \begin{bmatrix} x_1 \\ x_2 \end{bmatrix} = \begin{bmatrix} b_1 \\ b_2 \end{bmatrix} \qquad (4.5.10)$$

where $D_1$ and $D_2$ are diagonal and $F_1$ and $E_1$ are tridiagonal. Assume that each of these blocks is $n$-by-$n$ and that it is "safe" to solve (4.5.10) via (4.5.4) and (4.5.6). Note that

$$\begin{aligned} U_1 &= D_1 && \text{(diagonal)} \\ L_1 &= E_1 U_1^{-1} && \text{(tridiagonal)} \\ U_2 &= D_2 - L_1 F_1 && \text{(pentadiagonal)} \\ y_1 &= b_1 \\ y_2 &= b_2 - E_1(D_1^{-1} y_1) \\ U_2 x_2 &= y_2 \\ D_1 x_1 &= y_1 - F_1 x_2. \end{aligned}$$

Consequently, some very simple $n$-by-$n$ calculations with the original banded blocks renders the solution.

On the other hand, the naive application of band Gaussian elimination to the system (4.5.10) would entail a great deal of unnecessary work and storage as the system has bandwidth $n + 1$. However, we mention that by permuting the rows and columns of the system via the permutation

$$P = [e_1, e_{n+1}, e_2, \ldots, e_n, e_{2n}] \qquad (4.5.11)$$

we find (in the $n = 5$ case) that

$$PAP^T = \begin{bmatrix} \times & \times & 0 & \times & 0 & 0 & 0 & 0 & 0 & 0 \\ \times & \times & \times & 0 & 0 & 0 & 0 & 0 & 0 & 0 \\ 0 & \times & \times & \times & 0 & \times & 0 & 0 & 0 & 0 \\ \times & 0 & \times & \times & \times & 0 & 0 & 0 & 0 & 0 \\ 0 & 0 & 0 & \times & \times & \times & 0 & \times & 0 & 0 \\ 0 & 0 & \times & 0 & \times & \times & \times & 0 & 0 & 0 \\ 0 & 0 & 0 & 0 & 0 & \times & \times & \times & 0 & \times \\ 0 & 0 & 0 & 0 & \times & 0 & \times & \times & \times & 0 \\ 0 & 0 & 0 & 0 & 0 & 0 & 0 & \times & \times & \times \\ 0 & 0 & 0 & 0 & 0 & 0 & \times & 0 & \times & \times \end{bmatrix}$$

This matrix has bandwidth three and so a very reasonable solution procedure results by applying band Gaussian elimination to this permuted version of $A$.

The subject of bandwidth-reducing permutations is important. See George and Liu (1981, Chapter 4). We also refer to the reader to Varah (1972) and George (1974) for further details concerning the solution of block tridiagonal systems.

### 4.5.4 Block Cyclic Reduction

We next describe the method of *block cyclic reduction* that can be used to solve some important special instances of the block tridiagonal system (4.5.2). For simplicity, we assume that $A$ has the form

$$A = \begin{bmatrix} D & F & & \cdots & 0 \\ F & D & \ddots & & \vdots \\ & \ddots & \ddots & \ddots & \\ \vdots & & \ddots & \ddots & F \\ 0 & \cdots & & F & D \end{bmatrix} \in \mathbb{R}^{nq \times nq} \qquad (4.5.12)$$

where $F$ and $D$ are $q$-by-$q$ matrices that satisfy $DF = FD$. We also assume that $n = 2^k - 1$. These conditions hold in certain important applications such as the discretization of Poisson's equation on a rectangle. In that situation,

$$D = \begin{bmatrix} 4 & -1 & & \cdots & 0 \\ -1 & 4 & \ddots & & \vdots \\ & \ddots & \ddots & \ddots & \\ \vdots & & \ddots & \ddots & -1 \\ 0 & \cdots & & -1 & 4 \end{bmatrix} \qquad (4.5.13)$$

and $F = -I_q$. The integer $n$ is determined by the size of the mesh and can often be chosen to be of the form $n = 2^k - 1$. (Sweet (1977) shows how to proceed when the dimension is not of this form.)

The basic idea behind cyclic reduction is to halve the dimension of the problem on hand repeatedly until we are left with a single $q$-by-$q$ system for the unknown subvector $x_{2^{k-1}}$. This system is then solved by standard means. The previously eliminated $x_i$ are found by a back-substitution process.

The general procedure is adequately motivated by considering the case $n = 7$:

$$\begin{array}{lcl} b_1 &=& Dx_1 + Fx_2 \\ b_2 &=& Fx_1 + Dx_2 + Fx_3 \\ b_3 &=& Fx_2 + Dx_3 + Fx_4 \\ b_4 &=& Fx_3 + Dx_4 + Fx_5 \\ b_5 &=& Fx_4 + Dx_5 + Fx_6 \\ b_6 &=& Fx_5 + Dx_6 + Fx_7 \\ b_7 &=& Fx_6 + Dx_7 \end{array} \tag{4.5.14}$$

For $i = 2$, 4, and 6 we multiply equations $i-1$, $i$, and $i+1$ by $F$, $-D$, and $F$, respectively, and add the resulting equations to obtain

$$\begin{array}{lcl} (2F^2 - D^2)x_2 + F^2x_4 &=& F(b_1 + b_3) - Db_2 \\ F^2x_2 + (2F^2 - D^2)x_4 + F^2x_6 &=& F(b_3 + b_5) - Db_4 \\ F^2x_4 + (2F^2 - D^2)x_6 &=& F(b_5 + b_7) - Db_6 \end{array}$$

Thus, with this tactic we have removed the odd-indexed $x_i$ and are left with a reduced block tridiagonal system of the form

$$\begin{array}{lcl} D^{(1)}x_2 + F^{(1)}x_4 &=& b_2^{(1)} \\ F^{(1)}x_2 + D^{(1)}x_4 + F^{(1)}x_6 &=& b_4^{(1)} \\ F^{(1)}x_4 + D^{(1)}x_6 &=& b_6^{(1)} \end{array}$$

where $D^{(1)} = 2F^2 - D^2$ and $F^{(1)} = F^2$ commute. Applying the same elimination strategy as above, we multiply these three equations respectively by $F^{(1)}$, $-D^{(1)}$, and $F^{(1)}$. When these transformed equations are added together, we obtain the single equation

$$\left(2[F^{(1)}]^2 - D^{(1)^2}\right)x_4 = F^{(1)}\left(b_2^{(1)} + b_6^{(1)}\right) - D^{(1)}b_4^{(1)}$$

which we write as

$$D^{(2)}x_4 = b^{(2)}.$$

This completes the cyclic reduction. We now solve this (small) $q$-by-$q$ system for $x_4$. The vectors $x_2$ and $x_6$ are then found by solving the systems

$$\begin{array}{lcl} D^{(1)}x_2 &=& b_2^{(1)} - F^{(1)}x_4 \\ D^{(1)}x_6 &=& b_6^{(1)} - F^{(1)}x_4 \end{array}$$

Finally, we use the first, third, fifth, and seventh equations in (4.5.14) to compute $x_1$, $x_3$, $x_5$, and $x_7$, respectively.

For general $n$ of the form $n = 2^k - 1$ we set $D^{(0)} = D$, $F^{(0)} = F$, $b^{(0)} = b$ and compute:

$$
\begin{array}{l}
\textbf{for } p = 1{:}k-1 \\
\quad D^{(p)} = 2[F^{(p-1)}]^2 - [D^{(p-1)}]^2 \\
\quad F^{(p)} = [F^{(p-1)}]^2 \\
\quad r = 2^p \\
\quad \textbf{for } j = 1{:}2^{k-p} - 1 \\
\qquad b_{jr}^{(p)} = F^{(p-1)}\left(b_{jr-r/2}^{(p-1)} + b_{jr+r/2}^{(p-1)}\right) - D^{(p-1)} b_{jr}^{(p-1)} \\
\quad \textbf{end} \\
\textbf{end}
\end{array}
\qquad (4.5.15)
$$

The $x_i$ are then computed as follows:

$$
\begin{array}{l}
\text{Solve } D^{(k-1)} x_{2^{k-1}} = b_1^{(k-1)} \text{ for } x_{2^{k-1}}. \\
\textbf{for } p = k-2{:}-1{:}0 \\
\quad r = 2^p \\
\quad \textbf{for } j = 1{:}2^{k-p-1} \\
\qquad \textbf{if } j = 1 \\
\qquad\quad c = b_{(2j-1)r}^{(p)} - F^{(p)} x_{2jr} \\
\qquad \textbf{elseif } j = 2^{k-p+1} \\
\qquad\quad c = b_{(2j-1)r}^{(p)} - F^{(p)} x_{(2j-2)r} \\
\qquad \textbf{else} \\
\qquad\quad c = b_{(2j-1)r}^{(p)} - F^{(p)}\left(x_{2jr} + x_{(2j-2)r}\right) \\
\qquad \textbf{end} \\
\qquad \text{Solve } D^{(p)} x_{(2j-1)r} = c \text{ for } x_{(2j-1)r} \\
\quad \textbf{end} \\
\textbf{end}
\end{array}
\qquad (4.5.16)
$$

The amount of work required to perform these recursions depends greatly upon the sparsity of the $D^{(p)}$ and $F^{(p)}$. In the worse case when these matrices are full, the overall flop count has order $\log(n)q^3$. Care must be exercised in order to ensure stability during the reduction. For further details, see Buneman (1969).

**Example 4.5.1** Suppose $q = 1$, $D = (4)$, and $F = (-1)$ in (4.5.14) and that we wish to solve:

$$
\begin{bmatrix}
4 & -1 & 0 & 0 & 0 & 0 & 0 \\
-1 & 4 & -1 & 0 & 0 & 0 & 0 \\
0 & -1 & 4 & -1 & 0 & 0 & 0 \\
0 & 0 & -1 & 4 & -1 & 0 & 0 \\
0 & 0 & 0 & -1 & 4 & -1 & 0 \\
0 & 0 & 0 & 0 & -1 & 4 & -1 \\
0 & 0 & 0 & 0 & 0 & -1 & 4
\end{bmatrix}
\begin{bmatrix} x_1 \\ x_2 \\ x_3 \\ x_4 \\ x_5 \\ x_6 \\ x_7 \end{bmatrix}
=
\begin{bmatrix} 2 \\ 4 \\ 6 \\ 8 \\ 10 \\ 12 \\ 22 \end{bmatrix}
$$

By executing (4.5.15) we obtain the reduced systems:

$$\begin{bmatrix} -14 & 1 & 0 \\ 1 & -14 & 1 \\ 0 & 1 & -14 \end{bmatrix} \begin{bmatrix} x_2 \\ x_4 \\ x_6 \end{bmatrix} = \begin{bmatrix} -24 \\ -48 \\ -80 \end{bmatrix} \qquad p = 1$$

and

$$\begin{bmatrix} -194 \end{bmatrix} = \begin{bmatrix} x_4 \end{bmatrix} \begin{bmatrix} -776 \end{bmatrix} \qquad p = 2$$

The $x_i$ are then determined via (4.5.16):

$$\begin{array}{lllll} p = 2: & x_4 = 4 & & & \\ p = 1: & x_2 = 2 & x_6 = 6 & & \\ p = 0: & x_1 = 1 & x_3 = 3 & x_5 = 5 & x_7 = 7 \end{array}$$

Cyclic reduction is an example of a divide and conquer algorithm. Other divide and conquer procedures are discussed in §1.3.8 and §8.6.

### Problems

**P4.5.1** Show that a block diagonally dominant matrix is nonsingular.

**P4.5.2** Verify that (4.5.7) implies (4.5.8) and (4.5.9).

**P4.5.3** Suppose block cyclic reduction is applied with $D$ given by (4.5.13) and $F = -I_q$. What can you say about the band structure of the matrices $F^{(p)}$ and $D^{(p)}$ that arise?

### Notes and References for Sec. 4.5

A discussion of the symmetric indefinite system mentioned at the beginning of this section may be found in

M.T. Heath (1978). "Numerical Algorithms for Nonlinearly Constrained Optimization," Report STAN-CS 78-656, Department of Computer Science, Stanford University, (Ph.D. thesis), Stanford, California.

The following papers provide insight into the various nuances of block matrix computations:

R. Fourer (1984). "Staircase Matrices and Systems," *SIAM Review 26*, 1-71.
J.A. George (1974). "On Block Elimination for Sparse Linear Systems," *SIAM J. Num. Anal. 11*, 585–603.
J.A. George and J.W. Liu (1981). *Computer Solution of Large Sparse Positive Definite Systems*, Prentice-Hall, Englewood Cliffs, New Jersey.
M.L. Merriam (1985). "On the Factorization of Block Tridiagonals With Storage Constraints," *SIAM J. Sci. and Stat. Comp. 6*, 182-192.
J.M. Varah (1972). "On the Solution of Block-Tridiagonal Systems Arising from Certain Finite-Difference Equations," *Math. Comp. 26*, 859–68.

The property of block diagonal dominance and its various implications is the central theme in

D.G. Feingold and R.S. Varga (1962). "Block Diagonally Dominant Matrices and Generalizations of the Gershgorin Circle Theorem," *Pacific J. Math. 12*, 1241–50.

Early methods that involve the idea of cyclic reduction are described in

B.L. Buzbee, G.H. Golub, and C.W. Nielson (1970). "On Direct Methods for Solving Poisson's Equations," *SIAM J. Num. Anal. 7,* 627-56.
R.W. Hockney (1965). "A Fast Direct Solution of Poisson's Equation Using Fourier Analysis, "*J. ACM 12,* 95-113.

The accumulation of the right-hand side must be done with great care, for otherwise there would be a significant loss of accuracy. A stable way of doing this is described in

O. Buneman (1969). "A Compact Non-Iterative Poisson Solver," Report 294, Stanford University Institute for Plasma Research, Stanford, California.

Other literature concerned with cyclic reduction includes

B.L. Buzbee, F.W. Dorr, J.A. George, and G.H. Golub (1971). "The Direct Solution of the Discrete Poisson Equation on Irregular Regions," *SIAM J. Num. Anal. 8,* 722–36.
B.L. Buzbee and F.W. Dorr (1974). "The Direct Solution of the Biharmonic Equation on Rectangular Regions and the Poisson Equation on Irregular Regions," *SIAM J. Num. Anal. 11,* 753–63.
P. Concus and G.H. Golub (1973). "Use of Fast Direct Methods for the Efficient Numerical Solution of Nonseparable Elliptic Equations," *SIAM J. Num. Anal. 10,* 1103–20.
F.W. Dorr (1970). "The Direct Solution of the Discrete Poisson Equation on a Rectangle," *SIAM Review 12,* 248–63.
F.W. Dorr (1973). "The Direct Solution of the Discrete Poisson Equation in $O(n^2)$ Operations," *SIAM Review 15,* 412–415.
D. Heller (1976). "Some Aspects of the Cyclic Reduction Algorithm for Block Tridiagonal Linear Systems," *SIAM J. Num. Anal. 13,* 484–96.

Various generalizations and extensions to cyclic reduction have been proposed to handle the problem of irregular boundaries,

M.A. Diamond and D.L.V. Ferreira (1976). "On a Cyclic Reduction Method for the Solution of Poisson's Equation," *SIAM J. Num. Anal. 13,* 54–70.

the problem of arbitrary dimension,

R.A. Sweet (1974). "A Generalized Cyclic Reduction Algorithm," *SIAM J. Num. Anal. 11,* 506–20.
R.A. Sweet (1977). "A Cyclic Reduction Algorithm for Solving Block Tridiagonal Systems of Arbitrary Dimension," *SIAM J. Num. Anal. 14,* 706–20.

and the problem of periodic end conditions,

P.N. Swarztrauber and R.A. Sweet (1973). "The Direct Solution of the Discrete Poisson Equation on a Disk," *SIAM J. Num. Anal. 10,* 900–907.

For certain matrices that arise in conjunction with elliptic partial differential equations, block elimination corresponds to rather natural operations on the underlying mesh. A classical example of this is the method of nested dissection described in

A. George (1973). "Nested Dissection of a Regular Finite Element Mesh," *SIAM J. Num. Anal. 10,* 345–63.

Finally, we mention the general survey in

J.R. Bunch (1976). "Block Methods for Solving Sparse Linear Systems," in *Sparse Matrix Computations* , ed. J. R. Bunch and D.J. Rose, Academic Press, New York.

## 4.6 Vandermonde Systems

Suppose $x(0:n) \in \mathbb{R}^{n+1}$. A matrix $V \in \mathbb{R}^{(n+1)\times(n+1)}$ of the form

$$V = V(x_0, \ldots, x_n) = \begin{bmatrix} 1 & 1 & \cdots & 1 \\ x_0 & x_1 & \cdots & x_n \\ \vdots & \vdots & & \vdots \\ x_0^n & x_1^n & \cdots & x_n^n \end{bmatrix}$$

is said to be a *Vandermonde matrix*. In this section, we show how the systems $V^T a = f = f(0:n)$ and $Vz = b = b(0:n)$ can be solved in $O(n^2)$ flops. For convenience, vectors and matrices are subscripted from 0.

### 4.6.1 Polynomial Interpolation: $\mathbf{V^T a = f}$

Vandermonde systems arise in many approximation and interpolation problems. Indeed, the key to obtaining a fast Vandermonde solver is to recognize that solving $V^T a = f$ is equivalent to polynomial interpolation. This follows because if $V^T a = f$ and

$$p(x) = \sum_{j=0}^{n} a_j x^j \tag{4.6.1}$$

then $p(x_i) = f_i$ for $i = 0:n$.

Recall that if the $x_i$ are distinct then there is a unique polynomial of degree $n$ that interpolates $(x_0, f_0), \ldots, (x_n, f_n)$. Consequently, $V$ is nonsingular so long as the $x_i$ are distinct. We assume this throughout the section.

The first step in computing the $a_j$ of (4.6.1) is to calculate the Newton represention of the interpolating polynomial $p$:

$$p(x) = \sum_{k=0}^{n} c_k \left( \prod_{i=0}^{k-1} (x - x_i) \right) . \tag{4.6.2}$$

The constants $c_k$ are divided differences and may be determined as follows:

$$\begin{array}{l} c(0:n) = f(0:n) \\ \textbf{for } k = 0:n-1 \\ \quad \textbf{for } i = n:-1:k+1 \\ \quad\quad c_i = (c_i - c_{i-1})/(x_i - x_{i-k-1}) \\ \quad \textbf{end} \\ \textbf{end} \end{array} \tag{4.6.3}$$

See Conte and de Boor (1980, chapter 2).

The next task is to generate $a(0{:}n)$ from $c(0{:}n)$. Define the polynomials $p_n(x), \ldots, p_0(x)$ by the iteration

$$
\begin{array}{l}
p_n(x) = c_n \\
\textbf{for } k = n-1{:}-1{:}0 \\
\quad p_k(x) = c_k + (x - x_k)p_{k+1}(x) \\
\textbf{end}
\end{array}
$$

and observe that $p_0(x) = p(x)$. Writing

$$p_k(x) = a_k^{(k)} + a_{k+1}^{(k)}x + \cdots + a_n^{(k)}x^{n-k}$$

and equating like powers of $x$ in the equation $p_k = c_k + (x - x_k)p_{k+1}$ gives the following recursion for the coefficients $a_i^{(k)}$:

$$
\begin{array}{l}
a_n^{(n)} = c_n \\
\textbf{for } k = n-1{:}-1{:}0 \\
\quad a_k^{(k)} = c_k - x_k a_{k+1}^{(k+1)} \\
\quad \textbf{for } i = k+1{:}n-1 \\
\quad\quad a_i^{(k)} = a_i^{(k+1)} - x_k a_{i+1}^{(k+1)} \\
\quad \textbf{end} \\
\quad a_n^{(k)} = a_n^{(k+1)} \\
\textbf{end}
\end{array}
$$

Consequently, the coefficients $a_i = a_i^{(0)}$ can be calculated as follows:

$$
\begin{array}{l}
a(0{:}n) = c(0{:}n) \\
\textbf{for } k = n-1{:}-1{:}0 \\
\quad \textbf{for } i = k{:}n-1 \\
\quad\quad a_i = a_i - x_k a_{i+1} \\
\quad \textbf{end} \\
\textbf{end}
\end{array}
\qquad (4.6.4)
$$

Combining this iteration with (4.6.3) renders the following algorithm:

**Algorithm 4.6.1** Given $x(0{:}n) \in \mathbb{R}^{n+1}$ with distinct entries and $f = f(0{:}n) \in \mathbb{R}^{n+1}$, the following algorithm overwrites $f$ with the solution $a = a(0{:}n)$ to the Vandermonde system $V(x_0, \ldots, x_n)^T a = f$.

$$
\begin{array}{l}
\textbf{for } k = 0{:}n-1 \\
\quad \textbf{for } i = n{:}-1{:}k+1 \\
\quad\quad f(i) = (f(i) - f(i-1))/(x(i) - x(i-k-1)) \\
\quad \textbf{end} \\
\textbf{end}
\end{array}
$$

**for** $k = n-1:-1:0$
    **for** $i = k:n-1$
        $f(i) = f(i) - f(i+1)x(k)$
    **end**
**end**

This algorithm requires $5n^2/2$ flops.

**Example 4.6.1** Suppose Algorithm 4.6.1 is used to solve

$$\begin{bmatrix} 1 & 1 & 1 & 1 \\ 1 & 2 & 4 & 8 \\ 1 & 3 & 9 & 27 \\ 1 & 4 & 16 & 64 \end{bmatrix}^T \begin{bmatrix} a_0 \\ a_1 \\ a_2 \\ a_3 \end{bmatrix} = \begin{bmatrix} 10 \\ 26 \\ 58 \\ 112 \end{bmatrix}.$$

The first $k$-loop computes the Newton representation of $p(x)$:

$$p(x) = 10 + 16(x-1) + 8(x-1)(x-2) + (x-1)(x-2)(x-3).$$

The second $k$-loop computes $a = (4\ 3\ 2\ 1)^T$ from $(10\ 16\ 8\ 1)^T$.

### 4.6.2 The System Vz = b

Now consider the system $Vz = b$. To derive an efficient algorithm for this problem, we describe what Algorithm 4.6.1 does in matrix-vector language. Define the lower bidiagonal matrix $L_k(\alpha) \in \mathbb{R}^{(n+1)\times(n+1)}$ by

$$L_k(\alpha) = \left[\begin{array}{c|ccccc} I_k & & & 0 & & \\ \hline & 1 & & \cdots & & 0 \\ & -\alpha & 1 & & & \\ & & \ddots & \ddots & & \\ 0 & \vdots & & \ddots & \ddots & \vdots \\ & & & & \ddots & 1 \\ & 0 & & \cdots & -\alpha & 1 \end{array}\right]$$

and the diagonal matrix $D_k$ by

$$D_k = \text{diag}(\underbrace{1,\ldots,1}_{k+1}, x_{k+1} - x_0, \ldots, x_n - x_{n-k-1}).$$

With these definitions it is easy to verify from (4.6.3) that if $f = f(0:n)$ and $c = c(0:n)$ is the vector of divided differences then $c = U^T f$ where $U$ is the upper triangular matrix defined by

$$U^T = D_{n-1}^{-1} L_{n-1}(1) \cdots D_0^{-1} L_0(1).$$

Similarly, from (4.6.4) we have

$$a = L^T c,$$

where $L$ is the unit lower triangular matrix defined by:

$$L^T \;=\; L_0(x_0)^T \cdots L_{n-1}(x_{n-1})^T.$$

Thus, $a = L^T U^T f$ where $V^{-T} = L^T U^T$. In other words, Algorithm 4.6.1 solves $V^T a = f$ by tacitly computing the "UL" factorization of $V^{-1}$.

Consequently, the solution to the system $Vz = b$ is given by

$$\begin{aligned} z &= V^{-1}b \;=\; U(Lb) \\ &= \left(L_0(1)^T D_0^{-1} \cdots L_{n-1}(1)^T D_{n-1}^{-1}\right)\left(L_{n-1}(x_{n-1}) \cdots L_0(x_0)b\right) \end{aligned}$$

This observation gives rise to the following algorithm:

**Algorithm 4.6.2** Given $x(0{:}n) \in \mathbb{R}^{n+1}$ with distinct entries and $b = b(0{:}n) \in \mathbb{R}^{n+1}$, the following algorithm overwrites $b$ with the solution $z = z(0{:}n)$ to the Vandermonde system $V(x_0, \ldots, x_n)z = b$.

```
for k = 0:n - 1
    for i = n: - 1:k + 1
        b(i) = b(i) - x(k)b(i - 1)
    end
end
for k = n - 1: - 1:0
    for i = k + 1:n
        b(i) = b(i)/(x(i) - x(i - k - 1))
    end
    for i = k:n - 1
        b(i) = b(i) - b(i + 1)
    end
end
```

This algorithm requires $5n^2/2$ flops.

**Example 4.6.2** Suppose Algoritm 4.6.2 is used to solve

$$\begin{bmatrix} 1 & 1 & 1 & 1 \\ 1 & 2 & 3 & 4 \\ 1 & 4 & 9 & 16 \\ 1 & 8 & 27 & 64 \end{bmatrix} \begin{bmatrix} z_0 \\ z_1 \\ z_2 \\ z_3 \end{bmatrix} = \begin{bmatrix} 0 \\ -1 \\ 3 \\ 35 \end{bmatrix}.$$

The first $k$-loop computes the vector

$$L_3(3)L_2(2)L_1(1) \begin{bmatrix} 0 \\ -1 \\ 3 \\ 35 \end{bmatrix} = \begin{bmatrix} 0 \\ -1 \\ 6 \\ 6 \end{bmatrix}.$$

The second $k$-loop then calculates

$$L_0(1)^T D_0^{-1} L_1(1)^T D_1^{-1} L_2(1)^T D_2^{-1} \begin{bmatrix} 0 \\ -1 \\ 3 \\ 35 \end{bmatrix} = \begin{bmatrix} 3 \\ -4 \\ 0 \\ 1 \end{bmatrix}.$$

### 4.6.3 Stability

Algorithms 4.6.1 and 4.6.2 are discussed and analyzed in Björck and Pereyra (1970). Their experience is that these algorithms frequently produce surprisingly accurate solutions, even when $V$ is ill-conditioned. They also show how to update the solution when a new coordinate pair $(x_{n+1}, f_{n+1})$ is added to the set of points to be interpolated, and how to solve *confluent Vandermonde systems*, i.e., systems involving matrices like

$$V = V(x_0, x_1, x_1, x_3) = \begin{bmatrix} 1 & 1 & 0 & 1 \\ x_0 & x_1 & 1 & x_3 \\ x_0^2 & x_1^2 & 2x_1^2 & x_3^2 \\ x_0^3 & x_1^3 & 3x_1^2 & x_3^3 \end{bmatrix}$$

**Problems**

**P4.6.1** Show that if $V = V(x_0, \ldots, x_n)$, then

$$\det(V) = \prod_{n \geq i > j \geq 0} (x_i - x_j).$$

**P4.6.2** (Gautschi 1975a) Verify the following inequality for the $n = 1$ case:

$$\| V^{-1} \|_\infty \leq \max_{0 \leq k \leq n} \prod_{\substack{i=0 \\ i \neq k}}^{n} \frac{1 + |x_i|}{|x_k - x_i|}.$$

Equality results if the $x_i$ are all on the same ray in the complex plane.

**Notes and References for Sec. 4.6**

Our discussion of Vandermonde linear systems is drawn from the papers

A. Björck and V. Pereyra (1970). "Solution of Vandermonde Systems of Equations," *Math. Comp. 24*, 893–903.

A. Björck and T. Elfving (1973). "Algorithms for Confluent Vandermonde Systems," *Numer. Math. 21*, 130–37.

The latter reference includes an Algol procedure. The most perceptive analysis of Vandermonde system solving may be found in

N.J. Higham (1987b). "Error Analysis of the Björck-Pereyra Algorithms for Solving Vandermonde Systems," *Numer. Math. 50*, 613–632.
N.J. Higham (1988a). "Fast Solution of Vandermonde-like Systems Involving Orthogonal Polynomials," *IMA J. Numer. Anal. 8*, 473-486

See also

G. Galimberti and V. Pereyra (1970). "Numerical Differentiation and the Solution of Multidimensional Vandermonde Systems," *Math. Comp. 24*, 357–64.
G. Galimberti and V. Pereyra (1971). "Solving Confluent Vandermonde Systems of Hermitian Type," *Numer. Math. 18*, 44–60.
H. Van de Vel (1977). "Numerical Treatment of a Generalized Vandermonde systems of Equations," *Lin. Alg. and Its Applic. 17*, 149–74.

Interesting theoretical results concerning the condition of Vandermonde systems may be found in

W. Gautschi (1975a). "Norm Estimates for Inverses of Vandermonde Matrices," *Numer. Math. 23*, 337–47.
W. Gautschi (1975b). "Optimally Conditioned Vandermonde Matrices," *Numer. Math. 24*, 1–12.

In

G.H. Golub and W.P Tang (1981). "The Block Decomposition of a Vandermonde Matrix and Its Applications," *BIT 21*, 505–17.

a block Vandermonde algorithm is given that enables one to circumvent complex arithmetic in certain interpolation problems.

The divided difference computations we discussed are detailed in chapter 2 of

S.D. Conte and C. de Boor (1980). *Elementary Numerical Analysis: An Algorithmic Approach*, 3rd ed., McGraw-Hill, New York.

# 4.7 Toeplitz Systems

Matrices whose entries are constant along each diagonal arise in many applications and are called *Toeplitz matrices.* Formally, $T \in \mathbb{R}^{n\times n}$ is Toeplitz if there exist scalars $r_{-n+1}, \ldots, r_0, \ldots, r_{n-1}$ such that $a_{ij} = r_{j-i}$ for all $i$ and $j$. Thus,

$$T = \begin{bmatrix} r_0 & r_1 & r_2 & r_3 \\ r_{-1} & r_0 & r_1 & r_2 \\ r_{-2} & r_{-1} & r_0 & r_1 \\ r_{-3} & r_{-2} & r_{-1} & r_0 \end{bmatrix}$$

is Toeplitz.

Toeplitz matrices belong to the larger class of *persymmetric matrices.* We say that $B \in \mathbb{R}^{n\times n}$ is persymmetric if it symmetric about its northeast-southwest diagonal, i.e., $b_{ij} = b_{n-j+1,n-i+1}$ for all $i$ and $j$. This is equivalent to requiring $B = EB^TE$ where

$$E = [\, e_n, \ldots, e_1 \,] \;=\; I_n(:, n\!: -1\!:\!1)$$

is the $n$-by-$n$ *exchange matrix*. It is easy to verify that (a) Toeplitz matrices are persymmetric and (b) the inverse of a nonsingular Toeplitz matrix is persymmetric. In this section we show how the careful exploitation of (b) can enable one to solve Toeplitz systems in $O(n^2)$ time. The discussion is restricted to the important case when $T$ is also symmetric and positive definite.

### 4.7.1 Three Problems

Assume that we have scalars $r_1, \ldots, r_n$ such that for $k = 1{:}n$ the matrices

$$T_k = \begin{bmatrix} 1 & r_1 & \cdots & r_{k-2} & r_{k-1} \\ r_1 & 1 & \ddots & & r_{k-2} \\ \vdots & \ddots & \ddots & \ddots & \vdots \\ r_{k-2} & & \ddots & \ddots & r_1 \\ r_{k-1} & r_{k-2} & \cdots & r_1 & 1 \end{bmatrix}$$

are positive definite. (There is no loss of generality in normalizing the diagonal. ) Three algorithms are described in this section:

- Durbin's algorithm for the *Yule-Walker problem* $T_n y = -(r_1, \ldots, r_n)^T$.
- Levinson's algorithm for the general righthand side problem $T_n x = b$.
- Trench's algorithm for computing $B = T_n^{-1}$.

In deriving these methods, we denote the $k$-by-$k$ exchange matrix by $E_k$, i.e., $E_k = I_k(:, k{:} -1{:}1)$.

### 4.7.2 Solving the Yule-Walker Equations

We begin by presenting Durbin's algorithm for the Yule-Walker equations which arise in conjunction with certain linear prediction problems. Suppose for some $k$ that satisfies $1 \le k \le n-1$ we have solved the $k$-th order Yule-Walker system $T_k y = -r = -(r_1, \ldots, r_k)^T$. We now show how the $(k+1)$-st order Yule-Walker system

$$\begin{bmatrix} T_k & E_k r \\ r^T E_k & 1 \end{bmatrix} \begin{bmatrix} z \\ \alpha \end{bmatrix} = -\begin{bmatrix} r \\ r_{k+1} \end{bmatrix}$$

can be solved in $O(k)$ flops. First observe that

$$z = T_k^{-1}(-r - \alpha E_k r) = y - \alpha T_k^{-1} E_k r$$

and

$$\alpha = -r_{k+1} - r^T E_k z.$$

Since $T_k^{-1}$ is persymmetric, $T_k^{-1}E_k = E_kT_k^{-1}$ and thus,

$$z = y - \alpha E_k T_k^{-1} r = y + \alpha E_k y.$$

By substituting this into the above expression for $\alpha$ we find

$$\alpha = -r_{k+1} - r^T E_k(y + \alpha E_k y) = -(r_{k+1} + r^T E_k y)/(1 + r^T y).$$

The denominator is positive because $T_{k+1}$ is positive definite and because

$$\begin{bmatrix} I & E_k y \\ 0 & 1 \end{bmatrix}^T \begin{bmatrix} T_k & E_k r \\ r^T E_k & 1 \end{bmatrix} \begin{bmatrix} I & E_k y \\ 0 & 1 \end{bmatrix} = \begin{bmatrix} T_k & 0 \\ 0 & 1 + r^T y \end{bmatrix}.$$

We have illustrated the $k$th step of an algorithm proposed by Durbin (1960). It proceeds by solving the Yule-Walker systems

$$T_k y^{(k)} = -r^{(k)} = -(r_1, \ldots, r_k)^T$$

for $k = 1{:}n$ as follows:

$$\begin{array}{l} y^{(1)} = -r_1 \\ \textbf{for } k = 1{:}n-1 \\ \quad \beta_k = 1 + [r^{(k)}]^T y^{(k)} \\ \quad \alpha_k = -(r_{k+1} + r^{(k)^T} E_k y^{(k)})/\beta_k \\ \quad z^{(k)} = y^{(k)} + \alpha_k E_k y^{(k)} \\ \quad y^{(k+1)} = \begin{bmatrix} z^{(k)} \\ \alpha_k \end{bmatrix} \\ \textbf{end} \end{array} \tag{4.7.1}$$

As it stands, this algorithm would require $3n^2$ flops to generate $y = y^{(n)}$. It is possible, however, to reduce the amount of work even further by exploiting some of the above expressions:

$$\begin{aligned} \beta_k &= 1 + [r^{(k)}]^T y^{(k)} \\ &= 1 + \begin{bmatrix} r^{(k-1)^T} & r_k \end{bmatrix} \begin{bmatrix} y^{(k-1)} + \alpha_{k-1} E_{k-1} y^{(k-1)} \\ \alpha_{k-1} \end{bmatrix} \\ &= (1 + [r^{(k-1)}]^T y^{(k-1)}) + \alpha_{k-1} \left( [r^{(k-1)}]^T E_{k-1} y^{(k-1)} + r_k \right) \\ &= \beta_{k-1} + \alpha_{k-1}(-\beta_{k-1}\alpha_{k-1}) \\ &= (1 - \alpha_{k-1}^2)\beta_{k-1}. \end{aligned}$$

Using this recursion we obtain the following algorithm:

**Algorithm 4.7.1. (Durbin)** Given real numbers $1 = r_0, r_1, \ldots, r_n$ such that $T = (r_{|i-j|}) \in \mathbb{R}^{n \times n}$ is positive definite, the following algorithm computes $y \in \mathbb{R}^n$ such that $Ty = -(r_1, \ldots, r_n)^T$.

$y(1) = -r(1);\ \beta = 1;\ \alpha = -r(1)$
**for** $k = 1{:}n-1$
  $\beta = (1-\alpha^2)\beta$
  $\alpha = -\left(r(k+1) + r(k{:}-1{:}1)^T y(1{:}k)\right)/\beta$
  **for** $i = 1{:}k$
    $z(i) = y(i) + \alpha y(k+1-i)$
  **end**
  $y(1{:}k) = z(1{:}k);\ y(k+1) = \alpha$
**end**

This algorithm requires $2n^2$ flops. We have included an auxiliary vector $z$ for clarity, but it can be avoided.

**Example 4.7.1** Suppose we wish to solve the Yule-Walker system

$$\begin{bmatrix} 1 & .5 & .2 \\ .5 & 1 & .5 \\ .2 & .5 & 1 \end{bmatrix} \begin{bmatrix} y_1 \\ y_2 \\ y_3 \end{bmatrix} = - \begin{bmatrix} .5 \\ .2 \\ .1 \end{bmatrix}$$

using Algorithm 4.7.1. After one pass through the loop we obtain

$$\alpha = 1/15, \qquad \beta = 3/4, \qquad y = \begin{bmatrix} -8/15 \\ 1/15 \end{bmatrix}$$

We then compute

$$\begin{aligned} \beta &= (1-\alpha^2)\beta = 56/75 \\ \alpha &= -(r_3 + r_2 y_1 + r_1 y_2)/\beta = -1/28 \\ z_1 &= y_1 + \alpha y_2 = -225/420 \\ z_2 &= y_2 + \alpha y_1 = -36/420, \end{aligned}$$

giving the final solution $y = (-75,\ 12,\ -5)^T/140$.

### 4.7.3 The General Right Hand Side Problem

With a little extra work, it is possible to solve a symmetric positive definite Toeplitz system that has an arbitrary right-hand side. Suppose that we have solved the system

$$T_k x = b = (b_1, \ldots, b_k)^T \tag{4.7.2}$$

for some $k$ satisfying $1 \le k < n$ and that we now wish to solve

$$\begin{bmatrix} T_k & E_k r \\ r^T E_k & 1 \end{bmatrix} \begin{bmatrix} v \\ \mu \end{bmatrix} = \begin{bmatrix} b \\ b_{k+1} \end{bmatrix}. \tag{4.7.3}$$

Here, $r = (r_1, \ldots, r_k)^T$ as above. Assume also that the solution to the $k$th order Yule-Walker system $T_k y = -r$ is available. Since

$$v = T_k^{-1}(b - \mu E_k r) = x + \mu E_k y$$

it follows that

$$\begin{aligned}\mu &= b_{k+1} - r^T E_k v = b_{k+1} - r^T E_k x - \mu r^T y \\ &= \left(b_{k+1} - r^T E_k x\right) / \left(1 + r^T y\right).\end{aligned}$$

Consequently, we can effect the transition from (4.7.2) to (4.7.3) in $O(k)$ flops.

Overall, we can efficiently solve the system $T_n x = b$ by solving the systems $T_k x^{(k)} = b^{(k)} = (b_1, \ldots, b_k)^T$ and $T_k y^{(k)} = -r^{(k)} = (r_1, \ldots, r_k)^T$ "in parallel" for $k = 1{:}n$. This is the gist of the following algorithm:

**Algorithm 4.7.2 (Levinson)** Given $b \in \mathbb{R}^n$ and real numbers $1 = r_0, r_1, \ldots, r_n$ such that $T = (r_{|i-j|}) \in \mathbb{R}^{n \times n}$ is positive definite, the following algorithm computes $x \in \mathbb{R}^n$ such that $Tx = b$.

```
y(1) = −r(1); x(1) = b(1); β = 1; α = −r(1)
for k = 1:n − 1
    β = (1 − α²)β; μ = (b(k + 1) − r(1:k)^T x(k: − 1:1)) /β
    v(1:k) = x(1:k) + μy(k: − 1:1)
    x(1:k) = v(1:k); x(k + 1) = μ
    if k < n − 1
        α = − (r(k + 1) + r(1:k)^T y(k: − 1:1)) /β
        z(1:k) = y(1:k) + αy(k: − 1:1)
        y(1:k) = z(1:k); y(k + 1) = α
    end
end
```

This algorithm requires $4n^2$ flops. The vectors $z$ and $v$ are for clarity and may be dispensed.

**Example 4.7.2** Suppose we wish to solve the symmetric positive definite Toeplitz system

$$\begin{bmatrix} 1 & .5 & .2 \\ .5 & 1 & .5 \\ .2 & .5 & 1 \end{bmatrix} \begin{bmatrix} x_1 \\ x_2 \\ x_3 \end{bmatrix} = - \begin{bmatrix} 4 \\ -1 \\ 3 \end{bmatrix}$$

using the above algorithm. After one pass through the loop we obtain

$$\alpha = 1/15, \qquad \beta = 3/4, \qquad y = \begin{bmatrix} -8/15 \\ 1/15 \end{bmatrix} \qquad x = \begin{bmatrix} 6 \\ -4 \end{bmatrix}.$$

We then compute

$$\begin{aligned} \beta &= (1 - \alpha^2)\beta = 56/75 & \mu &= (b_3 - r_1 x_2 - r_2 x_1)/\beta = 285/56 \\ v_1 &= x_1 + \mu y_2 = 355/56 & v_2 &= x_2 + \mu y_1 = -376/56 \end{aligned}$$

giving the final solution $x = (355,\ -376,\ 285)^T/56$.

### 4.7.4 Computing the Inverse

One of the most surprising properties of a symmetric positive definite Toeplitz matrix $T_n$ is that its complete inverse can be calculated in $O(n^2)$ flops. To derive the algorithm for doing this, partition $T_n^{-1}$ as follows

$$T_n^{-1} = \begin{bmatrix} A & Er \\ r^T E & 1 \end{bmatrix}^{-1} = \begin{bmatrix} B & v \\ v^T & \gamma \end{bmatrix} \qquad (4.7.4)$$

where $A = T_{n-1}$, $E = E_{n-1}$, and $r = (r_1, \ldots, r_{n-1})^T$. From the equation

$$\begin{bmatrix} A & Er \\ r^T E & 1 \end{bmatrix} \begin{bmatrix} v \\ \gamma \end{bmatrix} = \begin{bmatrix} 0 \\ 1 \end{bmatrix}$$

it follows that $Av = -\gamma Er = -\gamma E(r_1, \ldots, r_{n-1})^T$ and $\gamma = 1 - r^T Ev$. If $y$ solves the $(n-1)$-st order Yule-Walker system $Ay = -r$, then these expressions imply that

$$\begin{aligned} \gamma &= 1/(1 + r^T y) \\ v &= \gamma E y \,. \end{aligned}$$

Thus, the last row and column of $T_n^{-1}$ are readily obtained.

It remains for us to develop working formulae for the entries of the submatrix $B$ in (4.7.4). Since $AB + Erv^T = I_{n-1}$, it follows that

$$B = A^{-1} - (A^{-1}Er)v^T = A^{-1} + \frac{vv^T}{\gamma} \,.$$

Now since $A = T_{n-1}$ is nonsingular and Toeplitz, its inverse is persymmetric. Thus,

$$\begin{aligned} b_{ij} &= (A^{-1})_{ij} + \frac{v_i v_j}{\gamma} \\ &= (A^{-1})_{n-j,n-i} + \frac{v_i v_j}{\gamma} \\ &= b_{n-j,n-i} - \frac{v_{n-j} v_{n-i}}{\gamma} + \frac{v_i v_j}{\gamma} \\ &= b_{n-j,n-i} + \frac{1}{\gamma}(v_i v_j - v_{n-j} v_{n-i}) \,. \end{aligned} \qquad (4.7.5)$$

This indicates that although $B$ is not persymmetric, we can readily compute an element $b_{ij}$ from its reflection across the northeast-southwest axis. Coupling this with the fact that $A^{-1}$ is persymmetric enables us to determine $B$ from its "edges" to its "interior."

Because the order of operations is rather cumbersome to describe, we preview the formal specification of the algorithm pictorially. To this end, assume that we know the last column and row of $A^{-1}$:

$$A^{-1} = \begin{bmatrix} u & u & u & u & u & k \\ u & u & u & u & u & k \\ u & u & u & u & u & k \\ u & u & u & u & u & k \\ u & u & u & u & u & k \\ k & k & k & k & k & k \end{bmatrix}$$

Here $u$ and $k$ denote the unknown and the known entries respectively, and $n = 6$. Alternately exploiting the persymmetry of $A^{-1}$ and the recursion (4.7.5), we can compute $B$ as follows:

$$\overset{persym.}{\longrightarrow} \begin{bmatrix} k & k & k & k & k & k \\ k & u & u & u & u & k \\ k & u & u & u & u & k \\ k & u & u & u & u & k \\ k & u & u & u & u & k \\ k & k & k & k & k & k \end{bmatrix} \overset{(4.7.5)}{\longrightarrow} \begin{bmatrix} k & k & k & k & k & k \\ k & u & u & u & k & k \\ k & u & u & u & k & k \\ k & u & u & u & k & k \\ k & k & k & k & k & k \\ k & k & k & k & k & k \end{bmatrix}$$

$$\overset{persym.}{\longrightarrow} \begin{bmatrix} k & k & k & k & k & k \\ k & k & k & k & k & k \\ k & k & u & u & k & k \\ k & k & u & u & k & k \\ k & k & k & k & k & k \\ k & k & k & k & k & k \end{bmatrix} \overset{(4.7.5)}{\longrightarrow} \begin{bmatrix} k & k & k & k & k & k \\ k & k & k & k & k & k \\ k & k & u & k & k & k \\ k & k & k & k & k & k \\ k & k & k & k & k & k \\ k & k & k & k & k & k \end{bmatrix}$$

$$\overset{persym.}{\longrightarrow} \begin{bmatrix} k & k & k & k & k & k \\ k & k & k & k & k & k \\ k & k & k & k & k & k \\ k & k & k & k & k & k \\ k & k & k & k & k & k \\ k & k & k & k & k & k \end{bmatrix}$$

Of course, when computing a matrix that is both symmetric and persymmetric, such as $A^{-1}$, it is only necessary to compute the "upper wedge" of the matrix—e.g.,

$$\begin{array}{cccccc} \times & \times & \times & \times & \times & \times \\ & \times & \times & \times & \times & \\ & & \times & \times & & \end{array} \qquad (n = 6)$$

With this last observation, we are ready to present the overall algorithm.

**Algorithm 4.7.3 (Trench)** Given real numbers $1 = r_0, r_1, \ldots, r_n$ such that $T = (r_{|i-j|}) \in \mathbb{R}^{n \times n}$ is positive definite, the following algorithm computes $B = T_n^{-1}$. Only those $b_{ij}$ for which $i \le j$ and $i + j \le n + 1$ are computed.

```
Use Algorithm 4.7.1 to solve T_{n-1}y = -(r_1, ..., r_{n-1})^T.
γ = 1/(1 + r(1:n-1)^T y(1:n-1))
v(1:n-1) = γ y(n-1:-1:1)
B(1,1) = γ
B(1,2:n) = v(n-1:-1:1)^T
for i = 2:floor((n-1)/2) + 1
    for j = i:n-i+1
        B(i,j) = B(i-1,j-1)+
            (v(n+1-j)v(n+1-i) - v(i-1)v(j-1)) / γ
    end
end
```

This algorithm requires $13n^2/4$ flops.

**Example 4.7.3** If the above algorithm is applied to compute the inverse $B$ of the positive definite Toeplitz matrix

$$\begin{bmatrix} 1 & .5 & .2 \\ .5 & 1 & .5 \\ .2 & .5 & 1 \end{bmatrix}$$

then we obtain $\gamma = 75/56$, $b_{11} = 75/56$, $b_{12} = -5/7$, $b_{13} = 5/56$, and $b_{22} = 12/7$.

### 4.7.5 Stability Issues

Error analyses for the above algorithms have been performed by Cybenko (1978), and we briefly report on some of his findings.

The key quantities turn out to be the $\alpha_k$ in (4.7.1). In exact arithmetic these scalars satisfy

$$|\alpha_k| < 1$$

and can be used to bound $\| T^{-1} \|_1$:

$$\max\left\{ \frac{1}{\prod_{j=1}^{n-1}(1-\alpha_j^2)}, \frac{1}{\prod_{j=1}^{n-1}(1-\alpha_j)} \right\} \le \| T_n^{-1} \| \le \prod_{j=1}^{n-1} \frac{1+|\alpha_j|}{1-|\alpha_j|} \quad (4.7.7)$$

Moreover, the solution to the Yule-Walker system $T_n y = -r(1:n)$ satisfies

$$\| y \|_1 = \left( \prod_{k=1}^{n-1}(1+\alpha_k) \right) - 1 \quad (4.7.8)$$

provided all the $\alpha_k$ are non-negative.

Now if $\hat{x}$ is the computed Durbin solution to the Yule-Walker equations then $r_D = T_n\hat{x} + r$ can be bounded as follows

$$\| r_D \| \approx \mathbf{u} \prod_{k=1}^{n}(1 + |\hat{\alpha}_k|)$$

where $\hat{\alpha}_k$ is the computed version of $\alpha_k$. By way of comparison, since each $|r_i|$ is bounded by unity, it follows that $\| r_C \| \approx \mathbf{u} \| y \|_1$ where $r_C$ is the residual associated with the computed solution obtained via Cholesky. Note that the two residuals are of comparable magnitude provided (4.7.7) holds. Experimental evidence suggests that this is the case even if some of the $\alpha_k$ are negative. Similar comments apply to the numerical behavior of the Levinson algorithm.

For the Trench method, the computed inverse $\hat{B}$ of $T_n^{-1}$ can be shown to satisfy

$$\frac{\| T_n^{-1} - \hat{B} \|_1}{\| T_n^{-1} \|_1} \approx \mathbf{u} \prod_{k=1}^{n} \frac{1 + |\hat{\alpha}_k|}{1 - |\hat{\alpha}_k|} .$$

In light of (4.5.7) we see that the right-hand side is an approximate upper bound for $\mathbf{u} \| T_n^{-1} \|$ which is approximately the size of the relative error when $T_n^{-1}$ is calculated using the Cholesky factorization.

### Problems

**P4.7.1** For any $v \in \mathbb{R}^n$ define the vectors $v_+ = (v + E_n v)/2$ and $v_- = (v - E_n v)/2$. Suppose $A \in \mathbb{R}^{n \times n}$ is symmetric and persymmetric. Show that if $Ax = b$ then $Ax_+ = b_+$ and $Ax_- = b_-$.

**P4.7.2** Let $U \in \mathbb{R}^{n \times n}$ be the unit upper triangular matrix with the property that: $U(1{:}k-1, k) = E_{k-1} y^{(k-1)}$ where $y^{(k)}$ is defined by (4.7.1). Show that

$$U^T T_n U = \text{diag}(1, \beta_1, \ldots, \beta_{n-1}).$$

**P4.7.3** Suppose $z \in \mathbb{R}^n$ and that $S \in \mathbb{R}^{n \times n}$ is orthogonal. Show that if

$$X = \left[ z, \ Sz, \ \ldots, S^{n-1} z \right]$$

then $X^T X$ is Toeplitz.

**P4.7.4** Consider the $\text{LDL}^\text{T}$ factorization of an $n$-by-$n$ symmetric, tridiagonal, positive definite Toeplitz matrix. Show that $d_n$ and $\ell_{n,n-1}$ converge as $n \to \infty$.

**P4.7.5** Show that the product of two lower triangular Toeplitz matrices is Toeplitz.

**P4.7.6** Give an algorithm for determining $\mu \in \mathbb{R}$ such that

$$T_n + \mu \left( e_n e_1^T + e_1 e_n^T \right)$$

is singular. Assume $T_n = (r_{|i-j|})$ is positive definite, with $r_0 = 1$.

**P4.7.7** Rewrite Algorithm 4.7.2 so that it does not require the vectors $z$ and $v$.

**P4.7.8** Give an algorithm for computing $\kappa_\infty(T_k)$ for $k = 1{:}n$.

### Notes and References for Sec. 4.7

Anyone who ventures into the vast Toeplitz method literature should first read

J.R. Bunch (1985). "Stability of Methods for Solving Toeplitz Systems of Equations," *SIAM J. Sci. Stat. Comp. 6*, 349–364.

for a clarification of stability issues. As is true with the "fast algorithms" area in general, unstable Toeplitz techniques abound and caution must be exercised.

The original references for the three algorithms described in this section are:

J. Durbin (1960). "The Fitting of Time Series Models," *Rev. Inst. Int. Stat. 28* 233–43.
N. Levinson (1947). "The Weiner RMS Error Criterion in Filter Design and Prediction," *J. Math. Phys. 25,* 261–78.
W.F. Trench (1964). "An Algorithm for the Inversion of Finite Toeplitz Matrices," *J. SIAM 12,* 515–22.

A more detailed description of the nonsymmetric Trench algorithm is given in

S. Zohar (1969). "Toeplitz Matrix Inversion: The Algorithm of W.F. Trench," *J. ACM 16,* 592–601.

Other references pertaining to Toeplitz matrix inversion include

W.F. Trench (1974). "Inversion of Toeplitz Band Matrices," *Math. Comp. 28,* 1089–95.
G.A. Watson (1973). "An Algorithm for the Inversion of Block Matrices of Toeplitz Form," *J. ACM 20,* 409–15.

The error bounds that we cited were taken from the roundoff analysis in

G. Cybenko (1978). "Error Analysis of Some Signal Processing Algorithms," Ph.D. thesis, Princeton University.
G. Cybenko (1980). "The Numerical Stability of the Levinson-Durbin Algorithm for Toeplitz Systems of Equations," *SIAM J. Sci. and Stat. Comp. 1,* 303-19.

$O(n^3)$ triangular factorization methods for Toeplitz systems also exist

J.L. Phillips (1971). "The Triangular Decomposition of Hankel Matrices," *Math. Comp. 25,* 599-602.
J. Rissanen (1973). "Algorithms for Triangular Decomposition of Block Hankel and Toeplitz Matrices with Application to Factoring Positive Matrix Polynomials," *Math. Comp. 27,* 147-54.

We mention the following references which described some important Toeplitz matrix applications:

J. Makhoul (1975). "Linear Prediction: A Tutorial Review," *Proc. IEEE* 63(4), 561-80.
J. Markel and A. Gray (1976). *Linear Prediction of Speech,* Springer-Verlag, Berlin and New York.
A.V. Oppenheim (1978). *Applications of Digital Signal Processing* , Prentice-Hall, Englewood Cliffs.

# Chapter 5

# Orthogonalization and Least Squares



This chapter is primarily concerned with the least squares solution of overdetermined systems of equations, i.e., the minimization of $\| Ax - b \|_2$ where $A \in \mathbb{R}^{m \times n}$ with $m \geq n$ and $b \in \mathbb{R}^m$. The most reliable solution procedures for this problem involve the reduction of $A$ to various canonical forms via orthogonal transformations. Householder reflections and Givens rotations are central to this process and we begin the chapter with a discussion of these important transformations. In §5.2 we discuss the computation of the factorization $A = QR$ where $Q$ is orthogonal and $R$ is upper triangular. This amounts to finding an orthonormal basis for range($A$). The QR factorization can be used to solve the full rank least squares problem as we show in §5.3. The technique is compared with the method of normal equations after an illuminating perturbation theory is developed.

In §5.4 and §5.5 we consider methods for handling the difficult situation when $A$ is rank deficient (or nearly so). QR with column pivoting and the SVD are featured.

In §5.6 we discuss several steps that can be taken to improve the quality of a computed least squares solution. Some remarks about underdetermined systems are offered in §5.7.

## 5.1 Householder and Givens Matrices

Recall that an $n$-by-$n$ matrix $Q$ is *orthogonal* if $Q^TQ = I_n$. Orthogonal matrices have a important role to play in least squares and eigenvalue computations. In this section we introduce the keys to effective orthogonal matrix computations: Householder reflections and Givens rotations.

### 5.1.1 A 2-by-2 Preview

It is instructive to examine the geometry associated with rotations and reflections at the $n = 2$ level. A 2-by-2 orthogonal matrix $Q$ is a *rotation* if it has the form

$$Q = \begin{bmatrix} \cos(\theta) & \sin(\theta) \\ -\sin(\theta) & \cos(\theta) \end{bmatrix}.$$

If $y = Q^Tx$ then $y$ is obtained by rotating $x$ counter clockwise through an angle $\theta$.

A 2-by-2 orthogonal matrix $Q$ is a *reflection* if it has the form

$$Q = \begin{bmatrix} \cos(\theta) & \sin(\theta) \\ \sin(\theta) & -\cos(\theta) \end{bmatrix}.$$

If $y = Q^Tx = Qx$ then $y$ is obtained by reflecting the vector $x$ across the line defined by

$$S = \text{span}\left\{\begin{bmatrix} \cos(\theta/2) \\ \sin(\theta/2) \end{bmatrix}\right\}.$$

Reflections and rotations are computationally attractive because they are easily constructed and because they can be used to introduce zeros in a vector by properly choosing the rotation angle or the reflection plane.

**Example 5.1.1** Suppose $x = (1, \sqrt{3})^T$. If we set

$$Q = \begin{bmatrix} \cos(-60^o) & \sin(-60^o) \\ -\sin(-60^o) & \cos(-60^o) \end{bmatrix} = \begin{bmatrix} 1/2 & -\sqrt{3}/2 \\ \sqrt{3}/2 & 1/2 \end{bmatrix}$$

then $Q^Tx = (2, 0)^T$. Thus, a rotation of $-60^o$ zeros the second component of $x$. If

$$Q = \begin{bmatrix} \cos(30^o) & \sin(30^o) \\ \sin(30^o) & -\cos(30^o) \end{bmatrix} = \begin{bmatrix} \sqrt{3}/2 & 1/2 \\ 1/2 & -\sqrt{3}/2 \end{bmatrix}$$

then $Q^Tx = (2, 0)^T$. Thus, by reflecting $x$ across the $30^o$ line we can zero its second component.

### 5.1.2 Householder Reflections

Let $v \in \mathbb{R}^n$ be nonzero. An $n$-by-$n$ matrix $P$ of the form

$$P = I - 2vv^T/v^Tv \tag{5.1.1}$$

is called a *Householder reflection.* (Synonyms: Householder matrix, Householder transformation.) The vector $v$ is called a *Householder vector.* When a vector $x$ is multiplied by $P$, it is reflected in the hyperplane span$\{v\}^\perp$. It is easy to verify that Householder matrices are symmetric and orthogonal.

Householder reflections are similar in two ways to Gauss transformations, which we introduced in §3.2.1. They are rank-1 modifications of the identity and they can be used to zero selected components of a vector. In particular, suppose we are given $0 \neq x \in \mathbb{R}^n$ and want $Px$ to be a multiple of $e_1$, the first column of $I_n$. For any $x \in \mathbb{R}^n$ we have

$$Px \;=\; \left(I \;-\; \frac{2vv^T}{v^Tv}\right)x \;=\; x \;-\; \frac{2v^Tx}{v^Tv}v$$

and thus, $Px \in \text{span}\{e_1\}$ implies $v \in \text{span}\{x, e_1\}$. Setting $v = x + \alpha e_1$ gives

$$v^Tx = x^Tx + \alpha x_1$$

and

$$v^Tv = x^Tx + 2\alpha x_1 + \alpha^2,$$

and therefore

$$Px \;=\; \left(1 - 2\frac{x^Tx + \alpha x_1}{x^Tx + 2\alpha x_1 + \alpha^2}\right)x \;-\; 2\alpha\frac{v^Tx}{v^Tv}e_1\,.$$

In order for the coefficient of $x$ to be zero, we need only set $\alpha = \pm\| x \|_2$. Thus,

$$v = x \pm \| x \|_2 e_1 \;\Rightarrow\; Px = \left(I - 2\frac{vv^T}{v^Tv}\right)x = \mp\| x \|_2 e_1. \tag{5.1.2}$$

It is this simple determination of $v$ that makes the Householder reflections so useful.

**Example 5.1.2** If $x = (3,\ 1,\ 5,\ 1)^T$ and $v = (9,\ 1,\ 5,\ 1)^T$, then

$$P \;=\; I - 2\frac{vv^T}{v^Tv} \;=\; \frac{1}{54}\begin{bmatrix} -27 & -9 & -45 & -9 \\ -9 & 53 & -5 & -1 \\ -45 & -5 & 29 & -5 \\ -9 & -1 & -5 & 53 \end{bmatrix}$$

has the property that $Px = (\,-6,\ 0,\ 0,\ 0,\,)^T$.

### 5.1.3 Computing the Householder Vector

There are a number of important practical details associated with the determination of a Householder matrix, i.e., the determination of a Householder vector. One concerns the choice of sign in the definition of $v$ in (5.1.2). If $x$ is close to a multiple of $e_1$, then $v = x - \text{sign}(x_1)\| x \|_2 e_1$ has small norm. Consequently, large relative error can be expected in the factor $\beta = 2/v^T v$. This kind of difficulty can be avoided merely by choosing the sign of $\alpha$ to be the same as the sign of $x$'s first component:

$$v = x + \text{sign}(x_1)\| x \|_2 e_1.$$

This ensures that $\| v \|_2 \geq \| x \|_2$ and guarantees near-perfect orthogonality in the computed $P$ (see below). Also note that $|v_1| = \| v \|_\infty$.

A less critical but useful guideline followed throughout the text is to normalize $v$ so that $v(1) = 1$. This is a nonstandard normalization but it simplifies the presentation of several important algorithms that require Householder vector storage. We then refer to $v(2{:}n)$ as the *essential part* of $v$. Note that the essential part of $v$ can be stored in the zeroed portion of $x$.

Here is an encapsulation of the Householder vector computation:

**Algorithm 5.1.1 (Householder Vector)** Given an $n$-vector $x$, this function computes an $n$-vector $v$ with $v(1) = 1$ such that $(I - 2vv^T/v^Tv)x$ is zero in all but the first component.

```
function: v = house(x)
    n = length(x); μ = || x ||_2; v = x
    if μ ≠ 0
        β = x(1) + sign(x(1))μ
        v(2:n) = v(2:n)/β
    end
    v(1) = 1
end house
```

This algorithm involves about $3n$ flops.

### 5.1.4 Applying Householder Matrices

It is critical to exploit structure when applying a Householder reflection to a matrix. If $A$ is a matrix and $P = I - 2vv^T/v^Tv$ then

$$PA = \left(I - 2\frac{vv^T}{v^Tv}\right)A = A + vw^T$$

where $w = \beta A^T v$ with $\beta = -2/v^T v$. Thus, a Householder update of a matrix involves a matrix-vector multiplication followed by an outer product update. Failure to recognize this and to treat $P$ as a general matrix increases work by an order of magnitude. *Householder updates never entail the explicit formation of the Householder matrix.* The following two functions make this clear.

**Algorithm 5.1.2 (Householder Pre-Multiplication)** Given an $m$-by-$n$ matrix $A$ and a nonzero $m$-vector $v$ with $v(1) = 1$, the following algorithm overwrites $A$ with $PA$ where $P = I - 2vv^T/v^T v$.

**function:** $A =$ **row.house**$(A, v)$
$\beta = -2/v^T v$
$w = \beta A^T v$
$A = A + vw^T$
**end row.house**

**Algorithm 5.1.3 (Householder Post-Multiplication)** Given an $m$-by-$n$ matrix $A$ and an $n$-vector $v$ with $v(1) = 1$, the following algorithm overwrites $A$ with $AP$ where $P = I - 2vv^T/v^T v$.

**function:** $A =$ **col.house**$(A, v)$
$\beta = -2/v^T v$
$w = \beta A v$
$A = A + wv^T$
**end col.house**

Each of the above Householder update procedures requires about $4mn$ flops. Note that both **row.house** and **col.house** can be designed to exploit the $v(1) = 1$ normalization. This feature can be important in **row.house** when $m$ is small and important in **col.house** when $n$ is small as it saves a multiplication.

Typically, **house**, **row.house**, and **col.house** are used to zero a designated subcolumn or subrow of a matrix. As an example, suppose we want to overwrite $A \in \mathbb{R}^{m \times n}$ $(m \geq n)$ with $B = Q^T A$ where $Q$ is an orthogonal matrix chosen so that $B(j+1{:}m, j) = 0$ for some $j$ that satisfies $1 \leq j \leq n$. In addition, suppose $A(j{:}m, 1{:}j-1) = 0$ and that we want to store the nontrivial portion of the Householder vector in $A(j+1{:}m, j)$. The following instructions accomplish this task:

$v(j{:}m) =$ **house**$(A(j{:}m, j))$
$A(j{:}m, j{:}n) =$ **row.house**$(A(j{:}m, j{:}n), v(j{:}m))$
$A(j+1{:}m, j) = v(j+1{:}m)$

From the computational point of view, we have applied an order $m-j+1$ Householder matrix $\tilde{P} = I - 2\tilde{v}\tilde{v}^T/\tilde{v}^T\tilde{v}$ to the bottom $m-j+1$ rows of $A$. However, mathematically we have also applied the $m$-by-$m$ Householder matrix

$$P = \begin{bmatrix} I_{j-1} & 0 \\ 0 & \tilde{P} \end{bmatrix} = I - 2\frac{vv^T}{v^Tv} \qquad v = \begin{bmatrix} 0 \\ \tilde{v} \end{bmatrix}$$

to $A$ in its entirety. Regardless, the "essential" part of the Householder vector has been recorded in the zeroed portion of $A$.

### 5.1.5 Roundoff Properties

The roundoff properties associated with Householder matrices are very favorable. Wilkinson (AEP, pp. 152-62) shows that **house** produces a Householder vector $\hat{v}$ very near the exact $v$. If $\hat{P} = I - 2\hat{v}\hat{v}^T/\hat{v}^T\hat{v}$ then

$$\| \hat{P} - P \|_2 = O(\mathbf{u})$$

Moreover, the computed updates with $\hat{P}$ are close to the exact updates with $P$ :

$$\begin{aligned} fl(\hat{P}A) &= P(A+E) \qquad \| E \|_2 = O(\mathbf{u}\| A \|_2) \\ fl(A\hat{P}) &= (A+E)P \qquad \| E \|_2 = O(\mathbf{u}\| A \|_2) \end{aligned}$$

### 5.1.6 Factored Form Representation

Many Householder based factorization algorithms that are presented in the following sections compute products of Householder matrices

$$Q = Q_1Q_2\cdots Q_r \qquad Q_j = I - 2\frac{v^{(j)}v^{(j)T}}{v^{(j)^T}v^{(j)}} \tag{5.1.3}$$

where $r \leq n$ and each $v^{(j)}$ has the form

$$v^{(j)} = (\underbrace{0,\, 0, \cdots 0,}_{j-1}\, 1\, v_{j+1}^{(j)}, \;\cdots\; , v_n^{(j)})^T .$$

It is usually not necessary to compute $Q$ explicitly even if it is involved in subsequent calculations. For example, if $C \in \mathbb{R}^{n\times q}$ and we wish to compute $Q^TC$ , then we merely execute the loop

$\quad$ **for** $j = 1{:}r$
$\quad\quad C = Q_jC$
$\quad$ **end**

The storage of the Householder vectors $v^{(1)} \cdots v^{(r)}$ amounts to a *factored form* representation of $Q$. To illustrate the economies of the factored form representation, suppose that we have an array $A$ and that $A(j+1{:}n, j)$ houses $v^{(j)}(j+1{:}n)$, the essential part of the $j$th Householder vector. The overwriting of $C \in \mathbb{R}^{n \times q}$ with $Q^T C$ can then be implemented as follows:

$$\begin{array}{l} \textbf{for } j = 1{:}r \\ \quad v(j) = 1;\ v(j+1{:}n) = A(j+1{:}n, j) \\ \quad C(j{:}n, :) = \textbf{row.house}(C(j{:}n, :), v(j{:}n)) \\ \textbf{end} \end{array} \tag{5.1.4}$$

This involves about $2qr(2n - r)$ flops. If $Q$ is explicitly represented as an $n$-by-$n$ matrix, $Q^T C$ would involve $2n^2 q$ flops.

Of course, in some applications, it is necessary to explicitly form $Q$ (or parts of it). Two possible algorithms for computing the Householder product matrix $Q$ in (5.1.3) are *forward accumulation*,

$$\begin{array}{l} Q = I_n \\ \textbf{for } j = 1{:}r \\ \quad Q = QP_j \\ \textbf{end} \end{array}$$

and *backward accumulation*,

$$\begin{array}{l} Q = I_n \\ \textbf{for } j = r{:} -1{:}1 \\ \quad Q = P_j Q \\ \textbf{end} \end{array}$$

Recall that the leading $(j-1)$-by-$(j-1)$ portion of $Q_j$ is the identity. Thus, at the beginning of backward accumulation, $Q$ is "mostly the identity" and it gradually becomes full as the iteration progresses. This pattern can be exploited to reduce the number of required flops. In contrast, $Q$ is full in forward accumulation after the first step.

Because of this, it is less economical to accumulate the $Q_i$ in forward order. Thus, backward accumulation is the strategy of choice and the details are as follows:

$$\begin{array}{l} Q = I_n \\ \textbf{for } j = r{:} -1{:}1 \\ \quad v(j) = 1;\ v(j+1{:}n) = A(j+1{:}n, j) \\ \quad Q(j{:}n, j{:}n) = \textbf{row.house}(Q(j{:}n, j{:}n), v(j{:}n)) \\ \textbf{end} \end{array} \tag{5.1.5}$$

This involves about $4(n^2 r - nr^2 + r^3/3)$ flops.

### 5.1.7 A Block Representation

Suppose $Q = Q_1 \cdots Q_r$ is a product of $n$-by-$n$ Householder matrices as in (5.1.4). Since each $Q_j$ is a rank-one modification of the identity, it follows from the structure of the Householder vectors that $Q$ is a rank-$r$ modification of the identity and can be written in the form

$$Q = I + WY^T \qquad (5.1.6)$$

where $W$ and $Y$ are $n$-by-$r$ matrices. The key to computing the *block representation* (5.1.6) is the following lemma.

**Lemma 5.1.1** *Suppose $Q = I + WY^T$ is an $n$-by-$n$ orthogonal matrix with $W, Y \in \mathbb{R}^{n \times j}$. If $P = I - 2vv^T/v^Tv$ with $v \in \mathbb{R}^n$ and $z = -2Qv/v^Tv$ then*

$$Q_+ \;=\; QP \;=\; I + W_+Y_+^T$$

*where $W_+ = [\, W \; z \,]$ and $Y_+ = [\, Y \; v \,]$ are each $n$-by-$(j+1)$.*

**Proof.**

$$\begin{aligned} QP &= (I + WY^T)\left(I - 2\frac{vv^T}{v^Tv}\right) = I + WY^T - 2\frac{Qvv^T}{v^Tv} \\ &= I + WY^T + zv^T = [\, W \; z \,][\, Y \; v \,]^T \quad \square \end{aligned}$$

By repeatedly applying the lemma, we can generate the block representation of $Q$ in (5.1.3) from the factored form representation as follows:

**Algorithm 5.1.4** Suppose $Q = Q_1 \cdots Q_r$ is a product of $n$-by-$n$ Householder matrices as described in (5.1.3). This algorithm computes matrices $W, Y \in \mathbb{R}^{n \times r}$ such that $Q = I + WY^T$.

$Y = v^{(1)}$
$W = -2v^{(1)}/[v^{(1)}]^T v^{(1)}$
**for** $j = 2{:}r$
  $z = -2(I + WY^T)v^{(j)}/[v^{(j)}]^T v^{(j)}$
  $W = [W \; z]$
  $Y = [\, Y \; v^{(j)} \,]$
**end**

This algorithm involves about $2r^2n - 2r^3/3$ flops if the zeros in the $v^{(j)}$ are exploited. Note that $Y$ is merely the matrix of Householder vectors and is therefore unit lower triangular. Clearly, the central task in the generation of the WY representation (5.1.6) is the computation of the $W$ matrix.

The block representation for products of Householder matrices is attractive in situations where $Q$ must be applied to a matrix. Suppose $C \in \mathbb{R}^{n \times q}$. It follows that the operation

$$C \leftarrow Q^T C = (I + WY^T)^T C = C + Y(W^T C)$$

is rich in level-3 operations. On the other hand, if $Q$ is in factored form, $Q^T C$ is just rich in the level-2 operations of matrix-vector multiplication and outer product updates. Of course, in this context the distinction between level-2 and level-3 diminishes as $C$ gets narrower.

We mention that the "WY" representation is not a generalized Householder transformation from the geometric point of view. True block reflectors have the form $Q = I - 2VV^T$ where $V \in \mathbb{R}^{n \times r}$ satisfies $V^T V = I_r$. See Schreiber and Parlett (1987) and also Schreiber and Van Loan (1989).

**Example 5.1.3** If $n = 4$, $r = 2$, and $(\,1,\ .6,\ 0,\ .8\,)^T$ and $(\,0,\ 1,\ .8,\ .6\,)^T$ are the Householder vectors associated with $Q_1$ and $Q_2$ respectively, then

$$Q_1 Q_2 = I + \begin{bmatrix} -1 & 1.080 \\ -.6 & -.352 \\ 0 & -.800 \\ -.8 & .264 \end{bmatrix} \begin{bmatrix} 1 & .6 & 0 & .8 \\ 0 & 1 & .8 & .6 \end{bmatrix}.$$

### 5.1.8 Givens Rotations

Householder reflections are exceedingly useful for introducing zeros on a grand scale, e.g., the annihilation of all but the first component of a vector. However, in calculations where it is necessary to zero elements more selectively, *Givens rotations* are the transformation of choice. These are rank-two corrections to the identity of the form

$$G(i,k,\theta) = \begin{bmatrix} 1 & \cdots & 0 & \cdots & 0 & \cdots & 0 \\ \vdots & \ddots & \vdots & & \vdots & & \vdots \\ 0 & \cdots & c & \cdots & s & \cdots & 0 \\ \vdots & & \vdots & \ddots & \vdots & & \vdots \\ 0 & \cdots & -s & \cdots & c & \cdots & 0 \\ \vdots & & \vdots & & \vdots & \ddots & \vdots \\ 0 & \cdots & 0 & \cdots & 0 & \cdots & 1 \end{bmatrix} \begin{matrix} \\ \\ i \\ \\ k \\ \\ \\ \end{matrix} \qquad (5.1.7)$$

$$\qquad\qquad\qquad\qquad i \qquad\quad k$$

where $c = \cos(\theta)$ and $s = \sin(\theta)$ for some $\theta$. Givens rotations are clearly orthogonal.

Premultiplication by $G(i,k,\theta)^T$ amounts to a counterclockwise rotation of $\theta$ radians in the $(i,k)$ coordinate plane. Indeed, if $x \in \mathbb{R}^n$ and $y = G(i,k,\theta)^T x$, then

$$y_j = \begin{cases} cx_i - sx_k & j = i \\ sx_i + cx_k & j = k \\ x_j & j \neq i,k \end{cases}$$

From these formulae it is clear that we can force $y_k$ to be zero by setting

$$c = \frac{x_i}{\sqrt{x_i^2 + x_k^2}} \qquad s = \frac{-x_k}{\sqrt{x_i^2 + x_k^2}} \tag{5.1.8}$$

Thus, it is a simple matter to zero a specified entry in a vector by using a Givens rotation.

In practice, there are better ways to compute $c$ and $s$ than (5.1.8). The following algorithm, for example, guards against overflow.

**Algorithm 5.1.5** Given scalars $a$ and $b$, this function computes $c = \cos(\theta)$ and $s = \sin(\theta)$ so

$$\begin{bmatrix} c & s \\ -s & c \end{bmatrix}^T \begin{bmatrix} a \\ b \end{bmatrix} = \begin{bmatrix} r \\ 0 \end{bmatrix}.$$

```
function: [c, s] = givens(a, b)
    if b = 0
        c = 1; s = 0
    else
        if |b| > |a|
            τ = -a/b; s = 1/√(1 + τ²); c = sτ
        else
            τ = -b/a; c = 1/√(1 + τ²); s = cτ
        end
    end
end givens
```

This algorithm requires 5 flops and a single square root. Note that it does not compute $\theta$ and so it does not involve inverse trigonometric functions.

**Example 5.1.4** If $x = (1,\ 2,\ 3,\ 4)^T$, $\cos(\theta) = 1/\sqrt{5}$, and $\sin(\theta) = 2/\sqrt{5}$, then $G(2,4,\theta)^T x = (1,\ \sqrt{20},\ 3,\ 0)^T$.

### 5.1.9 Applying Givens Rotations

It is critical that the simple structure of $G(i,k,\theta)$ be exploited when computing matrix products of the form $G(i,k,\theta)^T A$ and $AG(i,k,\theta)$. Note that only rows $i$ and $k$ are affected by the premultiplication and columns $i$ and $k$ by postmultiplication. We can effect these updates with the following functions.

**Algorithm 5.1.6** Given $A \in \mathbb{R}^{2\times q}$, $c = \cos(\theta)$, and $s = \sin(\theta)$, the following algorithm overwrites $A$ with the matrix $\begin{bmatrix} c & s \\ -s & c \end{bmatrix}^T A$.

**function:** $A = \textbf{row.rot}(A, c, s)$
  $q = \textbf{cols}(A)$
  **for** $j = 1{:}q$
    $\tau_1 = A(1, j);\ \tau_2 = A(2, j)$
    $A(1, j) = c\tau_1 - s\tau_2;\ A(2, j) = s\tau_1 + c\tau_2$
  **end**
**end row.rot**

This algorithm requires $6q$ flops.

**Algorithm 5.1.7** Given $A \in \mathbb{R}^{q\times 2}$, $c = \cos(\theta)$, and $s = \sin(\theta)$, the following algorithm overwrites $A$ with the matrix $A\begin{bmatrix} c & s \\ -s & c \end{bmatrix}$.

**function:** $A = \textbf{col.rot}(A, c, s)$
  $q = \textbf{rows}(A)$
  **for** $i = 1{:}q$
    $\tau_1 = A(i, 1);\ \tau_2 = A(i, 2)$
    $A(i, 1) = c\tau_1 - s\tau_2;\ A(i, 2) = s\tau_1 + c\tau_2$
  **end**
**end col.rot**

This algorithm requires $6q$ flops.

Recalling that $A([\,i\ k\,], :)$ specifies rows $i$ and $k$ of $A$, we see that

$$A([\,i\ k\,], :) \;=\; \textbf{row.rot}(A([\,i\ k\,], :), c, s)$$

performs the update $A \leftarrow G(i, k, \theta)^T A$. Likewise,

$$A(:, [i\ k]) \;=\; \textbf{col.rot}(A(:, [i\ k]), c, s)$$

performs the update $A \leftarrow AG(i, k, \theta)$.

To illustrate further a typical use of Givens procedures, suppose $A$ is an $m$-by-$n$ matrix with $A(i-1{:}i, 1{:}j-1) = 0$ and that we want to zero $A(i, j)$ by rotating rows $i - 1$ and $i$. Here is how this can be accomplished:

$[\,c,\ s\,] = \textbf{givens}(A(i-1, j), A(i, j))$
$A(i-1{:}i, j{:}n) = \textbf{row.rot}(A(i-1{:}i, j{:}n), c, s)$

### 5.1.10 Roundoff Properties

The numerical properties of Givens rotations are as favorable as those for Householder reflections. In particular, it can be shown that the computed $\hat{c}$ and $\hat{s}$ in **givens** satisfy

$$\begin{array}{rclcrcl} \hat{c} & = & c(1+\epsilon_c) & \quad & \epsilon_c & = & O(\mathbf{u}) \\ \hat{s} & = & s(1+\epsilon_s) & \quad & \epsilon_s & = & O(\mathbf{u}). \end{array}$$

If $\hat{c}$ and $\hat{s}$ are subsequently used in a Givens update, then the computed update is the exact update of a nearby matrix:

$$\begin{array}{rclc} fl[\hat{G}(i,k,\theta)^T A] & = & G(i,k,\theta)^T(A+E) & \quad \| E \|_2 \approx \mathbf{u} \| A \|_2 \\ fl[A\hat{G}(i,k,\theta)] & = & (A+E)G(i,k,\theta) & \quad \| E \|_2 \approx \mathbf{u} \| A \|_2. \end{array}$$

A detailed error analysis of Givens rotations may be found in Wilkinson (AEP, pp. 131-39).

### 5.1.11 Representing Products of Givens Rotations

Suppose $Q = G_1 \cdots G_t$ is a product of Givens rotation. As we have seen in connection with Householder reflections, it is more economical to keep the orthogonal matrix $Q$ in factored form than to compute explicitly the product of the rotations. Using a technique demonstrated by Stewart (1976), it is possible to do this in a very compact way. The idea is to associate a single floating point number $\rho$ with each rotation. Specifically, if

$$Z = \begin{bmatrix} c & s \\ -s & c \end{bmatrix} \qquad c^2 + s^2 = 1$$

then we define the scalar $\rho$ by

$$\begin{array}{l} \textbf{if } c = 0 \\ \qquad \rho = 1 \\ \textbf{elseif } |s| < |c| \\ \qquad \rho = \text{sign}(c)s/2 \\ \textbf{else} \\ \qquad \rho = 2\text{sign}(s)/c \\ \textbf{end} \end{array} \qquad (5.1.9)$$

Essentially, this amounts to storing $s/2$ if the sine is smaller and $2/c$ if the cosine is smaller. With this encoding, it is possible to reconstruct $\pm Z$ as follows:

$$
\begin{array}{l}
\textbf{if } \rho = 1 \\
\qquad c = 0;\ s = 1 \\
\textbf{elseif } |\rho| < 1 \\
\qquad s = 2\rho;\ c = \sqrt{1-s^2} \\
\textbf{else} \\
\qquad c = 2/\rho;\ s = \sqrt{1-c^2} \\
\textbf{end}
\end{array}
\tag{5.1.10}
$$

That $-Z$ may be generated is usually of no consequence for if $Z$ zeros a particular matrix entry, so does $-Z$. The reason for essentially storing the smaller of $c$ and $s$ is that the formula $\sqrt{1-x^2}$ renders poor results if $x$ is near unity. More details may be found in Stewart (1976). Of course, to "reconstruct" $G(i,k,\theta)$ we need $i$ and $k$ in addition to the associated $\rho$. This usually poses no difficulty as we discuss in §5.2.3.

## 5.1.12 Error Propagation

We offer some remarks about the propagation of roundoff error in algorithms that involve sequences of Householder/Givens updates. To be precise, suppose $A = A_0 \in \mathbb{R}^{m\times n}$ is given and that matrices $A_1, \ldots, A_p = B$ are generated via the formula

$$A_k = fl(\hat{Q}_k A_{k-1} \hat{Z}_k) \qquad k = 1{:}p\,.$$

Assume that the above Householder and Givens algorithms are used for both the generation and application of the $\hat{Q}_k$ and $\hat{Z}_k$. Let $Q_k$ and $Z_k$ be the orthogonal matrices that would be produced in the absence of roundoff. It can be shown that

$$B \;=\; (Q_p \cdots Q_1)(A+E)(Z_1 \cdots Z_p), \tag{5.1.11}$$

where $\| E \|_2 \le c\mathbf{u} \| A \|_2$ and $c$ is a constant that depends mildly on $n$, $m$, and $p$. In plain English, $B$ is an exact orthogonal update of a matrix near to $A$.

## 5.1.13 Fast Givens Transformations

The ability to introduce zeros in a selective fashion makes Givens rotations an important zeroing tool in certain structured problems. This has led to the development of "fast Givens" procedures. The fast Givens idea amounts to a clever representation of $Q$ when $Q$ is the product of Givens rotations. In particular, $Q$ is represented by a matrix pair $(M, D)$ where $M^T M = D = \text{diag}(d_i)$ and each $d_i$ is positive. The matrices $Q$, $M$, and $D$ are connected through the formula

$$Q \;=\; MD^{-1/2} \;=\; M\text{diag}(1/\sqrt{d_i}).$$

Note that $(MD^{-1/2})^T(MD^{-1/2}) = D^{-1/2}DD^{-1/2} = I$ and so the matrix $MD^{-1/2}$ is orthogonal.

The details are best explained at the 2-by-2 level. Let $x = (x_1\ x_2)^T$ and $D = \text{diag}(d_1, d_2)$ be given and assume that $d_1$ and $d_2$ are positive. Define

$$M_1 = \begin{bmatrix} \beta_1 & 1 \\ 1 & \alpha_1 \end{bmatrix} \tag{5.1.14}$$

and observe that

$$M_1^T x = \begin{bmatrix} \beta_1 x_1 + x_2 \\ x_1 + \alpha_1 x_2 \end{bmatrix}$$

and

$$M_1^T D M_1 = \begin{bmatrix} d_2 + \beta_1^2 d_1 & d_1\beta_1 + d_2\alpha_1 \\ d_1\beta_1 + d_2\alpha_1 & d_1 + \alpha_1^2 d_2 \end{bmatrix} \equiv D_1 .$$

If $x_2 \neq 0$ and $\alpha_1 = -x_1/x_2$ and $\beta_1 = -\alpha_1 d_2/d_1$, then

$$M_1^T x = \begin{bmatrix} x_2(1+\gamma_1) \\ 0 \end{bmatrix}$$

$$M_1^T D M_1 = \begin{bmatrix} d_2(1+\gamma_1) & 0 \\ 0 & d_1(1+\gamma_1) \end{bmatrix}$$

where $\gamma_1 = -\alpha_1\beta_1 = (d_2/d_1)(x_1/x_2)^2$.

Analogously, if we assume $x_1 \neq 0$ and define $M_2$ by

$$M_2 = \begin{bmatrix} 1 & \alpha_2 \\ \beta_2 & 1 \end{bmatrix} \tag{5.1.15}$$

where $\alpha_2 = -x_2/x_1$ and $\beta_2 = -(d_1/d_2)\alpha_2$, then

$$M_2^T x = \begin{bmatrix} x_1(1+\gamma_2) \\ 0 \end{bmatrix}$$

and

$$M_2^T D M_2 = \begin{bmatrix} d_1(1+\gamma_2) & 0 \\ 0 & d_2(1+\gamma_2) \end{bmatrix} \equiv D_2,$$

where $\gamma_2 = -\alpha_2\beta_2 = (d_1/d_2)(x_2/x_1)^2$.

It is easy to show that for either $i = 1$ or 2, the matrix $J = D^{1/2}M_iD_i^{-1/2}$ is orthogonal and that it is designed so that the second component of $J^T(D^{-1/2}x)$ is zero. ($J$ may actually be a reflection and thus it is half-correct to use the popular term "fast Givens.")

Notice that the $\gamma_i$ satisfy $\gamma_1\gamma_2 = 1$ Thus, we can always select $M_i$ in the above so that the "growth factor" $(1+\gamma_i)$ is bounded by 2. Matrices of the form

$$M_1 = \begin{bmatrix} \beta_1 & 1 \\ 1 & \alpha_1 \end{bmatrix} \qquad M_2 = \begin{bmatrix} 1 & \alpha_2 \\ \beta_2 & 1 \end{bmatrix}$$

that satisfy $-1 \le \alpha_i\beta_i \le 0$ are 2-by-2 *fast Givens transformations.* Notice that premultiplication by a fast Givens transformation involves half the number of multiplies as premultiplication by an "ordinary" Givens transformation. Also, the zeroing is carried out without an explicit square root which is sometimes important in VLSI settings.

In the $n$-by-$n$ case, everything "scales up" as with ordinary Givens rotations. The "type 1" transformations have the form

$$F(i,k,\alpha,\beta) = \begin{bmatrix} 1 & \cdots & 0 & \cdots & 0 & \cdots & 0 \\ \vdots & \ddots & \vdots & & \vdots & & \vdots \\ 0 & \cdots & \beta & \cdots & 1 & \cdots & 0 \\ \vdots & & \vdots & \ddots & \vdots & & \vdots \\ 0 & \cdots & 1 & \cdots & \alpha & \cdots & 0 \\ \vdots & & \vdots & & \vdots & \ddots & \vdots \\ 0 & \cdots & 0 & \cdots & 0 & \cdots & 1 \end{bmatrix} \begin{matrix} \\ \\ i \\ \\ k \\ \\ \\ \end{matrix} \qquad (5.1.16)$$
$$\qquad\qquad\qquad i \qquad\quad k$$

while the "type 2" transformations are structured as follows:

$$F(i,k,\alpha,\beta) = \begin{bmatrix} 1 & \cdots & 0 & \cdots & 0 & \cdots & 0 \\ \vdots & \ddots & \vdots & & \vdots & & \vdots \\ 0 & \cdots & 1 & \cdots & \alpha & \cdots & 0 \\ \vdots & & \vdots & \ddots & \vdots & & \vdots \\ 0 & \cdots & \beta & \cdots & 1 & \cdots & 0 \\ \vdots & & \vdots & & \vdots & \ddots & \vdots \\ 0 & \cdots & 0 & \cdots & 0 & \cdots & 1 \end{bmatrix} \begin{matrix} \\ \\ i \\ \\ k \\ \\ \\ \end{matrix} \qquad (5.1.17)$$
$$\qquad\qquad\qquad i \qquad\quad k$$

Thus if $M \in \mathbb{R}^{n\times n}$ and $D = \text{diag}(d_i)$ satisfy $M^TM = D$ and if $F$ is an $n$-by-$n$ fast Givens transformation, with $F^TDF = D_{new}$ diagonal, then $M_{new}^TM_{new} = D_{new}$ where $M_{new} = MF$. Thus, it is possible to update the fast Givens representation $(M, D)$ to obtain $(M_{new}, D_{new})$.

Corresponding to the function **givens** we have

**Algorithm 5.1.8** Given $x \in \mathbb{R}^2$ and positive $d \in \mathbb{R}^2$, the following algorithm computes a 2-by-2 fast Givens transformation $M$ such that the second component of $M^Tx$ is zero and $M^TDM = D_1$ is diagonal where $D = \text{diag}(d_1, d_2)$. If *type* = 1 then $M$ has the form (5.1.14) while if *type* = 2 then $M$ has the form (5.1.15). The diagonal elements of $D_1$ overwrite $d$.

```
function: [α, β, type] = fast.givens(x, d)
    if x(2) ≠ 0
        α = −x(1)/x(2); β = −αd(2)/d(1); γ = −αβ
        if γ ≤ 1
            type = 1
            τ = d(1); d(1) = (1 + γ)d(2); d(2) = (1 + γ)τ
        else
            type = 2
            α = 1/α; β = 1/β; γ = 1/γ
            d(1) = (1 + γ)d(1); d(2) = (1 + γ)d(2)
        end
    else
        type = 2
        α = 0; β = 0
    end
end fast.givens
```

The algorithm for applying a fast Givens transformation is similar to **row.rot** and **col.rot**. For row updates we have

**Algorithm 5.1.9** Given $A \in \mathbb{R}^{2\times q}$, and a fast givens transformation $M$ with type = 1 or 2, the following algorithm overwrites $A$ with $M^T A$.

```
function: A = row.fast.rot(A, α, β, type)
    if type = 1
        for j = 1:q
            τ1 = A(1, j); τ2 = A(2, j)
            A(1, j) = βτ1 + τ2; A(2, j) = τ1 + ατ2
        end
    else
        for j = 1:q
            τ1 = A(1, j); τ2 = A(2, j)
            A(1, j) = τ1 + βτ2; A(2, j) = ατ1 + τ2
        end
    end
end row.fast.rot
```

This algorithm requires $4q$ flops. Note that

$$A([\,i\ k\,], :) = \textbf{row.fast.rot}(A([\,i\ k\,], :), \alpha, \beta, type)$$

performs the update $A \leftarrow F(i, k, \alpha, \beta)^T A$ of the appropriate type.

Although **fast.givens** chooses the appropriate type of transformation, the growth factor $1 + \gamma$ may still be as large as two. Thus, $2^s$ growth can occur in the entries of $D$ and $M$ after $s$ updates. This means that the diagonal $D$ must be monitored during a fast givens procedure to avoid overflow. This is a nontrivial overhead that can degrade performance. Nevertheless, element growth in $M$ and $D$ is controlled because at all times we have $MD^{-1/2}$ orthogonal. The roundoff properties of a fast givens procedure are what we would expect of a Givens matrix technique. For example, if we computed $\hat{Q} = fl(\hat{M}\hat{D}^{-1/2})$ where $\hat{M}$ and $\hat{D}$ are the computed $M$ and $D$, then $\hat{Q}$ is orthogonal to working precision: $\| \hat{Q}^T\hat{Q} - I \|_2 \approx \mathbf{u}$.

### Problems

**P5.1.1** Execute **house** with $x = (1, 7, 2, 3, -1)^T$.

**P5.1.2** Let $x$ and $y$ be nonzero vectors in $\mathbf{R}^n$. Give an algorithm for determining a Householder matrix $P$ such that $Px$ is a multiple of $y$.

**P5.1.3** Suppose $x \in \mathbb{C}^n$ and that $x_1 = |x_1|e^{i\theta}$ with $\theta \in \mathbf{R}$. Assume $x \neq 0$ and define $u = x + e^{i\theta}\| x \|_2 e_1$ Show that $P = I - 2uu^H/u^Hu$ is unitary and that $Px = -e^{i\theta}\| x \|_2 e_1$.

**P5.1.4** Use Householder matrices to show that $\det(I + xy^T) = 1 + x^Ty$ where $x$ and $y$ are given $n$-vectors.

**P5.1.5** Suppose $x \in \mathbb{C}^2$. Give an algorithm for determining a unitary matrix of the form

$$Q = \begin{bmatrix} c & \bar{s} \\ -s & c \end{bmatrix} \qquad c \in \mathbf{R},\ c^2 + |s|^2 = 1$$

such that the second component of $Q^Hx$ is zero.

**P5.1.6** Suppose $x$ and $y$ are unit vectors in $\mathbf{R}^n$. Give an algorithm using Givens transformations which computes an orthogonal $Q$ such that $Q^Tx = y$.

**P5.1.7** Determine $c = \cos(\theta)$ and $s = \sin(\theta)$ such that

$$\begin{bmatrix} c & s \\ -s & c \end{bmatrix}^T \begin{bmatrix} 5 \\ 12 \end{bmatrix} = \begin{bmatrix} 13 \\ 0 \end{bmatrix}.$$

**P5.1.8** Suppose that $Q = I + YTY^T$ is orthogonal where $Y \in \mathbf{R}^{n\times j}$ and $T \in \mathbf{R}^{j\times j}$ is upper triangular. Show that if $Q_+ = QP$ where $P = I - 2vv^T/v^Tv$ is a Householder matrix, then $Q_+$ can be expressed in the form $Q_+ = I + Y_+T_+Y_+^T$ where $Y_+ \in \mathbf{R}^{n\times(j+1)}$ and $T_+ \in \mathbf{R}^{(j+1)\times(j+1)}$ is upper triangular.

**P5.1.9** Give a detailed implementation of Algorithm 5.1.4 with the assumption that $v^{(j)}(j+1{:}n)$, the essential part of the the $j$th Householder vector, is stored in $A(j+1{:}n, j)$. Since $Y$ is effectively represented in $A$, your procedure need only set up the $W$ matrix.

**P5.1.10** Show that if $S$ is skew-symmetric ($S^T = -S$), then $Q = (I + S)(I - S)^{-1}$ is orthogonal. ($Q$ is called the *Cayley transform* of $S$.) Construct a rank-2 $S$ so that if $x$ is a vector then $Q^Tx$ is zero except in the first component.

### Notes and References for Sec. 5.1

The transformations that we have discussed in this chapter are extensively used as building blocks in more complicated algorithms. Thus, there is a premium on doing the computations right at this elementary level. See

C.L. Lawson, R.J. Hanson, F.T. Krough, and D.R. Kincaid (1979). "Basic Linear Algebra Subprograms for FORTRAN Usage," *ACM Trans. Math. Soft.* 5, 308–23.

The basic linear algebra subroutines, or BLAS, are described in detail in Linpack.

Householder matrices are named after A.S. Householder, who popularized their use in numerical analysis. However, the properties of these matrices have been known for quite some time. See

H.W. Turnbull and A.C. Aitken (1961). *An Introduction to the Theory of Canonical Matrices*, Dover Publications, New York, pp. 102–5.

Other references concerned with Householder transformations include

J.J.M. Cuppen (1984). "On Updating Triangular Products of Householder Matrices," *Numer. Math. 45,* 403-410.

A.R. Gourlay (1970). "Generalization of Elementary Hermitian Matrices," *Comp. J. 13,* 411–12.

L. Kaufman (1987). "The Generalized Householder Transformation and Sparse Matrices," *Lin. Alg. and Its Applic. 90,* 221–234.

B.N. Parlett (1971). "Analysis of Algorithms for Reflections in Bisectors," *SIAM Review 13,* 197–208.

N.K. Tsao (1975). "A Note on Implementing the Householder Transformations." *SIAM J. Num. Anal. 12,* 53–58.

A detailed error analysis of Householder transformations is given in Lawson and Hanson (SLS, pp. 83–89).

The basic references for block Householder representations and the associated computations include

C.H. Bischof and C. Van Loan (1987). "The WY Representation for Products of Householder Matrices," *SIAM J. Sci. and Stat. Comp. 8,* s2–s13.

N.J. Higham and R.S. Schreiber (1988). "Fast Polar Decomposition of an Arbitrary Matrix," Report 88-942, Dept. of Computer Science, Cornell University, Ithaca, NY 14853.

R. Schreiber and B.N. Parlett (1987). "Block Reflectors: Theory and Computation," *SIAM J. Numer. Anal. 25,* 189-205.

R. Schreiber and C. Van Loan (1989). "A Storage Efficient WY Representation for Products of Householder Transformations," *SIAM J. Sci. and Stat. Comp.*, to appear.

Givens rotations, named after W. Givens, are also referred to as Jacobi rotations. Jacobi devised a symmetric eigenvalue algorithm based on these transformations in 1846. See §8.6. The Givens rotation storage scheme discussed in the text is detailed in

G.W. Stewart (1976). "The Economical Storage of Plane Rotations," *Numer. Math. 25,* 137–38.

Fast Givens transformations are also referred to as "square-root-free" Givens transformations. (Recall that a square root must ordinarily be computed during the formation of Givens transformation.) There are several ways fast Givens calculations can be arranged. See

M. Gentleman (1973). "Least Squares Computations by Givens Transformations without Square Roots," *J. Inst. Math. Appl. 12,* 329–36.

S. Hammarling (1974). "A Note on Modifications to the Givens Plane Rotation," *J. Inst. Math. Appl. 13,* 215–18.

C.F. Van Loan (1973). "Generalized Singular Values With Algorithms and Applications," Ph.D. thesis, University of Michigan, Ann Arbor.

J.H. Wilkinson (1977). "Some Recent Advances in Numerical Linear Algebra," in *The State of the Art in Numerical Analysis,* ed. D.A.H. Jacobs, Academic Press, New York, pp. 1–53.

We wish to repeat that a nontrivial amount of monitoring is necessary in order to successfully implement fast Givens transformations.

# 5.2 The QR Factorization

We now show how Householder and Givens transformations can be used to compute various factorizations, beginning with the QR factorization. The QR factorization of an $m$-by-$n$ matrix $A$ is given by

$$A = QR$$

where $Q \in \mathbb{R}^{m \times m}$ is orthogonal and $R \in \mathbb{R}^{m \times n}$ is upper triangular. In this section we assume $m \geq n$. If $A$ has full column rank then the first $n$ columns of $Q$ form an orthonormal basis for range($A$). Thus, calculation of the QR factorization is one way to compute an orthonormal basis for a set of vectors. This computation can be arranged in several ways. We give methods based on Householder, block Householder, Givens, and fast Givens transformations. The Gram-Schmidt orthogonalization process and a numerically more stable variant called modified Gram-Schmidt are also discussed.

## 5.2.1 Householder QR

We begin with a QR factorization method that utilizes Householder transformations. The essence of the algorithm can be conveyed by a small example. Suppose $m = 6$, $n = 5$, and assume that Householder matrices $H_1$ and $H_2$ have been computed so that

$$H_2H_1A = \begin{bmatrix} \times & \times & \times & \times & \times \\ 0 & \times & \times & \times & \times \\ 0 & 0 & \boldsymbol{\times} & \times & \times \\ 0 & 0 & \boldsymbol{\times} & \times & \times \\ 0 & 0 & \boldsymbol{\times} & \times & \times \\ 0 & 0 & \boldsymbol{\times} & \times & \times \end{bmatrix}.$$

Concentrating on the boldfaced entries, we next determine a Householder matrix $\tilde{H}_3 \in \mathbb{R}^{4 \times 4}$ such that

$$\tilde{H}_3 \begin{bmatrix} \boldsymbol{\times} \\ \boldsymbol{\times} \\ \boldsymbol{\times} \\ \boldsymbol{\times} \end{bmatrix} = \begin{bmatrix} \times \\ 0 \\ 0 \\ 0 \end{bmatrix}.$$

If $H_3 = \text{diag}(I_2, \tilde{H}_3)$, then

$$H_3H_2H_1A = \begin{bmatrix} \times & \times & \times & \times & \times \\ 0 & \times & \times & \times & \times \\ 0 & 0 & \times & \times & \times \\ 0 & 0 & 0 & \times & \times \\ 0 & 0 & 0 & \times & \times \\ 0 & 0 & 0 & \times & \times \end{bmatrix}.$$

After $n$ such steps we obtain an upper triangular $H_nH_{n-1}\cdots H_1A = R$ and so by setting $Q = H_1 \cdots H_n$ we obtain $A = QR$. Formally, we have

**Algorithm 5.2.1 (Householder QR)** Given $A \in \mathbb{R}^{m\times n}$ with $m \geq n$, the following algorithm finds Householder matrices $H_1, \ldots, H_n$ such that if $Q = H_1 \cdots H_n$, then $Q^TA = R$ is upper triangular. The upper triangular part of $A$ is overwritten by the upper triangular part of $R$ and components $j+1{:}m$ of the $j$th Householder vector are stored in $A(j+1{:}m, j), j < m$.

```
for j = 1:n
    v(j:m) = house(A(j:m, j))
    A(j:m, j:n) = row.house(A(j:m, j:n), v(j:m))
    if j < m
        A(j + 1:m, j) = v(j + 1:m)
    end
end
```

This algorithm requires $2n^2(m - n/3)$ flops.

To clarify how $A$ is overwritten, if

$$v^{(j)} = [\underbrace{0, \ldots, 0}_{j-1}, 1, v_{j+1}^{(j)}, \ldots, v_m^{(j)}]^T$$

is the $j$th Householder vector then upon completion

$$A = \begin{bmatrix} r_{11} & r_{12} & r_{13} & r_{14} & r_{15} \\ v_2^{(1)} & r_{22} & r_{23} & r_{24} & r_{25} \\ v_3^{(1)} & v_3^{(2)} & r_{33} & r_{34} & r_{35} \\ v_4^{(1)} & v_4^{(2)} & v_4^{(3)} & r_{44} & r_{45} \\ v_5^{(1)} & v_5^{(2)} & v_5^{(3)} & v_5^{(4)} & r_{55} \\ v_6^{(1)} & v_6^{(2)} & v_6^{(3)} & v_6^{(4)} & v_6^{(5)} \end{bmatrix}$$

If the matrix $Q = H_1 \cdots H_n$ is required, then it can be accumulated using (5.1.5). This accumulation requires $4(m^2n - mn^2 + n^3/3)$ flops.

The computed upper triangular matrix $\hat{R}$ is the exact $R$ for a nearby $A$ in the sense that $Z^T(A+E) = \hat{R}$ where $Z$ is some exact orthogonal matrix and $\| E \|_2 \approx \mathbf{u}\| A \|_2$.

### 5.2.2 Block Householder QR Factorization

Algorithm 5.2.1 is rich in the level-2 operations of matrix-vector multiplication and outer product updates. By reorganizing the computation and using the block Householder representation discussed in §5.1.7 we can obtain a level-3 procedure. The idea is to apply clusters of Householder transformations that are represented in the WY form of §5.1.7.

A small example illustrates the main idea. Suppose $n = 12$ and that the "blocking parameter" $r$ has the value $r = 3$. The first step is to generate Householders $H_1$, $H_2$, and $H_3$ as in Algorithm 5.2.1. However, unlike Algorithm 5.2.1 where the $H_i$ are applied to all of $A$, we only apply $H_1$, $H_2$, and $H_3$ to $A(:,1{:}3)$. After this is accomplished we generate the block representation $H_1H_2H_3 = I + W_1Y_1^T$ and then perform the level-3 update

$$A(:,4{:}12) \;=\; (I + WY^T)A(:,4{:}12)\,.$$

Next, we generate $H_4$, $H_5$, and $H_6$ as in Algorithm 5.2.1. However, these transformations are not applied to $A(:,7{:}12)$ until their block representation $H_4H_5H_6 = I + W_2Y_2^T$ is found. This illustrates the general pattern.

$\lambda = 1;\ k = 0$
**while** $\lambda \leq n$
  $\tau = \min(\lambda + r - 1, n);\ k = k + 1$
  Using Algorithm 5.2.1, upper triangularize $A(\lambda{:}m, \lambda{:}n)$
    generating Householder matrices $H_\lambda, \ldots, H_\tau$.     (5.2.1)
  Use Algorithm 5.1.4 to get the block representation
    $I + W_kY_k = H_\lambda, \ldots, H_\tau$.
  $A(\lambda{:}m, \tau + 1{:}n) \;=\; (I + W_kY_k^T)^TA(\lambda{:}m, \tau + 1{:}n)$
  $\lambda = \tau + 1$
**end**

The zero-nonzero structure of the Householder vectors that define the matrices $H_\lambda, \ldots, H_\tau$ implies that the first $\lambda - 1$ rows of $W_k$ and $Y_k$ are zero. This fact would be exploited in a practical implementation.

The proper way to regard (5.2.1) through the block column partitioning

$$A \;=\; [\,A_1, \ldots, A_N\,] \qquad N \;=\; \text{ceil}(n/r)$$

where block column $A_k$ is processed during the $k$th step. In the $k$th step of (5.2.1), a block Householder is formed that zeros the subdiagonal portion of $A_k$. The remaining block columns are then updated.

The roundoff properties of (5.2.1) are essentially the same as those for Algorithm 5.2.1. There is a slight increase in the number of flops required because of the $W$-matrix computations. However, as a result of the blocking, all but a small fraction of the flops occur in the context of matrix multiplication. In particular, the level-3 fraction of (5.2.1) is approximately $1 - 2/N$. See Bischof and Van Loan (1987) for further details.

### 5.2.3 Givens QR Methods

Givens rotations can also be used to compute the QR factorization. The 4-by-3 case illustrates the general idea:

$$\begin{bmatrix} \times & \times & \times \\ \times & \times & \times \\ \times & \times & \times \\ \times & \times & \times \end{bmatrix} \xrightarrow{(3,4)} \begin{bmatrix} \times & \times & \times \\ \times & \times & \times \\ \times & \times & \times \\ 0 & \times & \times \end{bmatrix} \xrightarrow{(2,3)} \begin{bmatrix} \times & \times & \times \\ \times & \times & \times \\ 0 & \times & \times \\ 0 & \times & \times \end{bmatrix} \xrightarrow{(1,2)}$$

$$\begin{bmatrix} \times & \times & \times \\ 0 & \times & \times \\ 0 & \times & \times \\ 0 & \times & \times \end{bmatrix} \xrightarrow{(3,4)} \begin{bmatrix} \times & \times & \times \\ 0 & \times & \times \\ 0 & \times & \times \\ 0 & 0 & \times \end{bmatrix} \xrightarrow{(2,3)} \begin{bmatrix} \times & \times & \times \\ 0 & \times & \times \\ 0 & 0 & \times \\ 0 & 0 & \times \end{bmatrix} \xrightarrow{(3,4)} R$$

Here we have highlighted the 2-vectors that define the underlying Givens rotations. Clearly, if $G_j$ denotes the $j$th Givens rotation in the reduction, then $Q^T A = R$ is upper triangular where $Q = G_1 \cdots G_t$ and $t$ is the total number of rotations. For general $m$ and $n$ we have:

**Algorithm 5.2.2 (Givens QR)** Given $A \in \mathbb{R}^{m \times n}$ with $m \geq n$, the following algorithm overwrites $A$ with $Q^T A = R$, where $R$ is upper triangular and $Q$ is orthogonal.

```
for j = 1:n
    for i = m: - 1:j + 1
        [c, s] = givens(A(i - 1, j), A(i, j))
        A(i - 1:i, j:n) = row.rot(A(i - 1:i, j:n), c, s)
    end
end
```

This algorithm requires $3n^2(m - n/3)$ flops. Note that we could use (5.1.9) to encode $(c, s)$ in a single number $\rho$ which could then be stored in the zeroed entry $A(i, j)$. An operation such as $x \leftarrow Q^T x$ could then be implemented by using (5.1.10), taking care to reconstruct the rotations in the proper order.

Other sequences of rotations can be used to upper triangularize $A$. For example, if we replace the **for** statements in Algorithm 5.2.2 with

```
for i = m: - 1:2
    for j = 1:min{i - 1, n}
```

then the zeros in $A$ are introduced row-by-row.

Another parameter in a Givens QR procedure concerns the planes of rotation that are involved in the zeroing of each $a_{ij}$. For example, instead of rotating rows $i-1$ and $i$ to zero $a_{ij}$ as in Algorithm 5.2.2, we could use rows $j$ and $i$, i.e.,

```
for j = 1:n
    for i = m: - 1:j + 1
        [c, s] = givens(A(j, j), A(i, j))
        A([j i], j:n) = row.rot(A([j i], j:n), c, s)
    end
end
```

### 5.2.4 Hessenberg QR via Givens

As an example of how Givens rotations can be used in structured problems, we show how they can be effectively used to compute the QR factorization of an upper Hessenberg matrix. A small example illustrates the general idea. Suppose $n = 6$ and that after two steps we have computed

$$G(2,3,\theta_2)^T G(1,2,\theta_1)^T A = \begin{bmatrix} \times & \times & \times & \times & \times & \times \\ 0 & \times & \times & \times & \times & \times \\ 0 & 0 & \times & \times & \times & \times \\ 0 & 0 & \times & \times & \times & \times \\ 0 & 0 & 0 & \times & \times & \times \\ 0 & 0 & 0 & 0 & \times & \times \end{bmatrix}$$

We then compute $G(3,4,\theta_3)$ to zero the current (4,3) entry thereby obtaining

$$G(3,4,\theta_3)^T G(2,3,\theta_2)^T G(1,2,\theta_1)^T A = \begin{bmatrix} \times & \times & \times & \times & \times & \times \\ 0 & \times & \times & \times & \times & \times \\ 0 & 0 & \times & \times & \times & \times \\ 0 & 0 & 0 & \times & \times & \times \\ 0 & 0 & 0 & \times & \times & \times \\ 0 & 0 & 0 & 0 & \times & \times \end{bmatrix}$$

Overall we have

**Algorithm 5.2.3 (Hessenberg QR)** If $A \in \mathbb{R}^{n \times n}$ is upper Hessenberg, then the following algorithm overwrites $A$ with $Q^T A = R$ where $Q$ is orthogonal and $R$ is upper triangular. $Q = G_1 \cdots G_{n-1}$ is a product of Givens rotations where $G_j$ has the form $G_j = G(j, j+1, \theta_j)$.

```
for j = 1:n-1
    [c s] = givens(A(j,j), A(j+1,j))
    A(j:j+1, j:n) = row.rot(A(j:j+1, j:n), c, s)
end
```

This algorithm requires about $3n^2$ flops.

### 5.2.5 Fast Givens QR

We can use the fast Givens transformations described in §5.1.13 to compute an $(M, D)$ representation of $Q$. In particular, if $M$ is nonsingular and $D$ is diagonal such that $M^T A = T$ is upper triangular and $M^T M = D$ is diagonal, then $Q = MD^{-1/2}$ is orthogonal and $Q^T A = D^{-1/2}T \equiv R$ is upper triangular. Recalling the functions **fast.givens** and **fast.row.rot** that we developed in §5.1.13, we have the following analog of the Givens QR procedure:

**Algorithm 5.2.4 (Fast Givens QR)** Given $A \in \mathbb{R}^{m\times n}$ with $m \geq n$, the following algorithm computes nonsingular $M \in \mathbb{R}^{m\times m}$ and positive $d(1{:}m)$ such that $M^T A = T$ is upper triangular, and $M^T M = \text{diag}(d_1, \ldots, d_m)$. $A$ is overwritten by $T$.

```
for i = 1:n
    d(i) = 1
end
for j = 1:n
    for i = m:-1:j+1
        [α, β, type] = fast.givens(A(i-1:i, j), d(i-1:i))
        A(i-1:i, j:n) = fast.row.rot(A(i-1:i, j:n), α, β, type)
    end
end
```

This algorithm requires $2n^2(m - n/3)$ flops. As we mentioned in the previous section, it is necessary to guard against overflow in fast Givens algorithms such as the above. This means that $M$, $D$, and $A$ must be periodically scaled if their entries become large.

When the QR factorization of a narrow band matrix is required, the fast Givens approach is attractive because it involves no square roots. (We found $\text{LDL}^\text{T}$ preferable to Cholesky in the narrow band case for the same reason; see §4.3.7.) In particular, if $A \in \mathbb{R}^{m\times n}$ has upper bandwidth $q$ and lower bandwidth $p$, then $Q^T A = R$ has upper bandwidth $p + q$. In this case Givens $QR$ requires about $O(np(p+q))$ flops and $O(np)$ square roots. Thus, the square roots are a significant portion of the overall computation if $p, q \ll n$.

### 5.2.6 Properties of the QR Factorization

The above algorithms "prove" that the QR factorization exists. Now we relate the columns of $Q$ to range($A$) and range$(A)^\perp$ and examine the uniqueness question.

**Theorem 5.2.1** *If $A = QR$ is a QR factorization of a full column rank $A \in \mathbb{R}^{m\times n}$ and $A = [\, a_1, \ldots, a_n \,]$ and $Q = [\, q_1, \ldots, q_m \,]$ are column partitionings, then*

$$\text{span}\{a_1, \ldots, a_k\} \;=\; \text{span}\{q_1, \ldots, q_k\} \qquad k = 1{:}n\,.$$

*In particular, if $Q_1 = Q(1{:}m, 1{:}n)$ and $Q_2 = Q(1{:}m, n+1{:}m)$ then*

$$\begin{aligned} \text{range}(A) &= \text{range}(Q_1) \\ \text{range}(A)^\perp &= \text{range}(Q_2) \end{aligned}$$

*and $A = Q_1R_1$ with $R_1 = R(1{:}n, 1{:}n)$.*

**Proof.** Comparing $k$th columns in $A = QR$ we conclude that

$$a_k \;=\; \sum_{i=1}^{k} r_{ik}q_i \;\in\; \text{span}\{q_1, \ldots, q_k\}\;. \tag{5.2.2}$$

Thus, $\text{span}\{a_1, \ldots, a_k\} \subseteq \text{span}\{q_1, \ldots, q_k\}$. However, since $\text{rank}(A) = n$ it follows that $\text{span}\{a_1, \ldots, a_k\}$ has dimension $k$ and so must equal $\text{span}\{q_1, \ldots, q_k\}$ The rest of the theorem follows trivially. □

The matrices $Q_1 = Q(1{:}m, 1{:}n)$ and $Q_2 = Q(1{:}m, n+1{:}m)$ can be easily computed from a factored form representation of $Q$.

**Theorem 5.2.2** *Suppose $A \in \mathbb{R}^{m\times n}$ has full column rank. The "skinny" QR factorization*

$$A = Q_1R_1$$

*is unique where $Q_1 \in \mathbb{R}^{m\times n}$ has orthonormal columns and $R_1$ is upper triangular with positive diagonal entries. Moreover, $R_1 = G^T$ where $G$ is the lower triangular Cholesky factor of $A^TA$.*

**Proof.** Since $A^TA = (Q_1R_1)^T(Q_1R_1) = R_1^TR_1$ we see that $G = R_1^T$ is the Cholesky factor of $A^TA$. This factor is unique by Theorem 4.2.5. Since $Q_1 = AR_1^{-1}$ it follows that $Q_1$ is also unique. □

Whenever we wish to refer to the "skinny" QR factorization of a matrix $A$, as in Theorem 5.2.2, we write $A = Q_1R_1$.

### 5.2.7 Classical Gram-Schmidt

We now discuss two alternative methods that can be used to compute the $A = Q_1R_1$ factorization directly. If rank$(A) = n$ then equation (5.2.2) can be solved for $q_k$:

$$q_k = \left(a_k - \sum_{i=1}^{k-1} r_{ik}q_i\right) \Big/ r_{kk} .$$

Thus, we can think of $q_k$ as a unit 2-norm vector in the direction of

$$z_k = a_k - \sum_{i=1}^{k-1} r_{ik}q_i$$

where to ensure $z_k \in \text{span}\{q_1, \ldots, q_{k-1}\}^{\perp}$ we choose

$$r_{ik} = q_i^T a_k \qquad i = 1{:}k-1 .$$

This leads to the *classical Gram-Schmidt* (CGS) algorithm for computing $A = Q_1R_1$.

$$\begin{array}{l} \textbf{for } k = 1{:}n \\ \quad R(1{:}k-1, k) = Q(1{:}m, 1{:}k-1)^T A(1{:}m, k) \\ \quad z = A(1{:}m, k) - Q(1{:}m, 1{:}k-1)R(1{:}k-1, k) \\ \quad R(k,k) = \| z \|_2 \\ \quad Q(1{:}m, k) = z/R(k,k) \\ \textbf{end} \end{array} \qquad (5.2.3)$$

In the $k$th step of CGS, the $k$th columns of both $Q$ and $R$ are generated.

### 5.2.8 Modified Gram-Schmidt

Unfortunately, the CGS method has very poor numerical properties in that there is typically a severe loss of orthogonality among the computed $q_i$. Interestingly, a rearrangement of the calculation, known as *modified Gram-Schmidt* (MGS), yields a much sounder computational procedure. In the $k$th step of MGS, the $k$th column of $Q$ (denoted by $q_k$) and the $k$th row of $R$ (denoted by $r_k^T$) are determined. To derive the MGS method, define the matrix $A^{(k)} \in \mathbb{R}^{m\times(n-k+1)}$ by

$$A - \sum_{i=1}^{k-1} q_i r_i^T = \sum_{i=k}^{n} q_i r_i^T = [\, 0 \; A^{(k)} \,] . \qquad (5.2.4)$$

It follows that if

$$A^{(k)} = [\underset{1}{z} \quad \underset{n-k}{B}\,]$$

then $r_{kk} = \| z \|_2$, $q_k = z/r_{kk}$ and $(r_{k,k+1} \cdots r_{kn}) = q_k^T B$. We then compute the outer product $A^{(k+1)} = B - q_k (r_{k,k+1} \cdots r_{kn})$ and proceed to the next step. This completely describes the $k$th step of MGS.

**Algorithm 5.2.5 (Modified Gram-Schmidt)** Given $A \in \mathbb{R}^{m \times n}$ with $\text{rank}(A) = n$, the following algorithm computes the factorization $A = Q_1 R_1$ where $Q_1 \in \mathbb{R}^{m \times n}$ has orthonormal columns and $R_1 \in \mathbb{R}^{n \times n}$ is upper triangular.

```
for k = 1:n
    R(k,k) = || A(1:m,k) ||_2
    Q(1:m,k) = A(1:m,k)/R(k,k)
    for j = k+1:n
        R(k,j) = Q(1:m,k)^T A(1:m,j)
        A(1:m,j) = A(1:m,j) - Q(1:m,k)R(k,j)
    end
end
```

This algorithm requires $2mn^2$ flops. It is not possible to overwrite $A$ with both $Q_1$ and $R_1$. Typically, the MGS computation is arranged so that $A$ is overwritten by $Q_1$ and the matrix $R_1$ is stored in a separate array.

### 5.2.9 Work and Accuracy

If one is interested in computing an orthonormal basis for range($A$), then the Householder approach requires $2mn^2 - 2n^3/3$ flops to get $Q$ in factored form and another $2mn^2 - 2n^3/3$ flops to get the first $n$ columns of $Q$. Therefore, for the problem of finding an orthonormal basis for range($A$), MGS is about twice as efficient as Householder orthogonalization. However, Björck (1967b) has shown that MGS produces a computed $\hat{Q}_1 = [\,\hat{q}_1, \ldots, \hat{q}_n\,]$ that satisfies

$$\hat{Q}_1^T \hat{Q}_1 = I + E_{MGS} \qquad \| E_{MGS} \|_2 \approx \mathbf{u}\kappa_2(A)$$

whereas the corresponding result for the Householder approach is of the form

$$\hat{Q}_1^T \hat{Q}_1 = I + E_H \qquad \| E_H \|_2 \approx \mathbf{u}\,.$$

Thus, if orthonormality is critical, then MGS should be used to compute orthonormal bases only when the vectors to be orthogonalized are fairly independent.

**Example 5.2.1** If modified Gram-Schmidt is applied to

$$A = \begin{bmatrix} 1 & 1 \\ 10^{-3} & 0 \\ 0 & 10^{-3} \end{bmatrix} \qquad \kappa_2(A) \approx 1.4 \cdot 10^3$$

with 6-digit decimal arithmetic, then

$$[\hat{q}_1\ \hat{q}_2] = \begin{bmatrix} 1.00000 & 0 \\ .001 & -707107 \\ 0 & .707100 \end{bmatrix}.$$

## Problems

**P5.2.1** Adapt the Householder QR algorithm so that it can efficiently handle the case when $A \in \mathbb{R}^{m \times n}$ has lower bandwidth $p$ and upper bandwidth $q$.

**P5.2.2** Adapt the Householder QR algorithm so that it computes the factorization $A = QL$ where $L$ is lower triangular and $Q$ is orthogonal. Assume that $A$ is square. This involves rewriting the Householder vector function $v = \mathbf{house}(x)$ so that $(I - 2vv^T/v^Tv)x$ is zero everywhere but its bottom component.

**P5.2.3** Adapt the Givens QR factorization algorithm so that the zeros are introduced by diagonal. That is, the entries are zeroed in the order $(m, 1)$, $(m-1, 1)$, $(m, 2)$, $(m-2, 1)$, $(m-1, 2)$, $(m, 3)$ , etc.

**P5.2.4** Adapt the fast Givens QR factorization algorithm so that it efficiently handles the case when $A$ is $n$-by-$n$ and tridiagonal. Assume that the subdiagonal, diagonal, and superdiagonal of $A$ are stored in $e(1{:}n-1)$, $a(1{:}n)$, $f(1{:}n-1)$ respectively. Design your algorithm so that these vectors are overwritten by the nonzero portion of $T$.

**P5.2.5** Suppose $L \in \mathbb{R}^{m \times n}$ with $m \geq n$ is lower triangular. Show how Householder matrices $H_1 \ldots H_n$ can be used to determine a lower triangular $L_1 \in \mathbb{R}^{n \times n}$ so that

$$H_n \cdots H_1 L = \begin{bmatrix} L_1 \\ 0 \end{bmatrix}$$

Hint: The second step in the 6-by-3 case involves finding $H_2$ so that

$$P_2 \begin{bmatrix} \times & 0 & 0 \\ \times & \times & 0 \\ \times & \times & \times \\ \times & \times & 0 \\ \times & \times & 0 \\ \times & \times & 0 \end{bmatrix} = \begin{bmatrix} \times & 0 & 0 \\ \times & \times & 0 \\ \times & \times & \times \\ \times & 0 & 0 \\ \times & 0 & 0 \\ \times & 0 & 0 \end{bmatrix}$$

with the property that rows 1 and 3 are left alone.

**P5.2.6** Show that if

$$A = \begin{bmatrix} R & w \\ 0 & v \end{bmatrix} \begin{matrix} k \\ m-k \end{matrix} \qquad b = \begin{bmatrix} c \\ d \end{bmatrix} \begin{matrix} k \\ m-k \end{matrix}$$
$$\qquad\ k \quad n-k$$

and $A$ has full rank, then $\min \| Ax - b \|_2^2 = \| d \|_2^2 - \left(v^T d / \| v \|_2\right)^2$.

**P5.2.7** Suppose $A \in \mathbb{R}^{n \times n}$ and $D = \text{diag}(d_1, \ldots, d_n) \in \mathbb{R}^{n \times n}$. Show how to construct an orthogonal $Q$ such that $Q^T A - DQ^T = R$ is upper triangular. Do not worry about efficiency—this is just an exercise in QR manipulation.

### Notes and References for Sec. 5.2

The idea of using Householder transformations to solve the LS problem was proposed in

A.S. Householder (1958). "Unitary Triangularization of a Nonsymmetric Matrix," *J. ACM.* 5, 339–42.

The practical details were worked out in

P. Businger and G.H. Golub (1965). "Linear Least Squares Solutions by Householder Transformations," *Numer. Math. 7*, 269–76. See also *HACLA*, pp. 111–18.
G.H. Golub (1965). "Numerical Methods for Solving Linear Least Squares Problems," *Numer. Math. 7*, 206–16.

The basic references on QR via Givens rotations include

M. Gentleman (1973). "Error Analysis of QR Decompositions by Givens Transformations," *Lin. Alg. and Its Appl. 10*, 189–97.
W. Givens (1958). "Computation of Plane Unitary Rotations Transforming a General Matrix to Triangular Form," *SIAM J. App. Math. 6*, 26–50.

For a discussion of how the QR factorization can be used to solve numerous problems in statistical computation, see

G.H. Golub (1969). "Matrix Decompositions and Statistical Computation," in *Statistical Computation* , ed. R.C. Milton and J.A. Nelder, Academic Press, New York, pp. 365–97.

If a matrix $A$ is perturbed, how does its $QR$ factorization change? This question is analyzed in

G.W. Stewart (1977). "Perturbation Bonds for the QR Factorization of a Matrix," *SIAM J. Num. Anal. 14*, 509–18.

where the main result is that the changes in $Q$ and $R$ are bounded by the condition of $A$ times the relative change in $A$.

Various aspects of the Gram-Schmidt process are discussed in

N.N. Abdelmalek (1971). "Roundoff Error Analysis for Gram-Schmidt Method and Solution of Linear Least Squares Problems," *BIT 11*, 345–68.
A. Björck (1967b). "Solving Linear Least Squares Problems by Gram-Schmidt Orthogonalization," *BIT 7*, 1–21.
J.R. Rice (1966). "Experiments on Gram-Schmidt Orthogonalization," *Math. Comp. 20*, 325–28.

The QR factorization of a structured matrix is usually structured itself. See

A.W. Bojanczyk, R.P. Brent, and F.R. de Hoog (1986). "QR Factorization of Toeplitz Matrices," *Numer. Math. 49*, 81-94.

# 5.3 The Full Rank Least Squares Problem

Consider the problem of finding a vector $x \in \mathbb{R}^n$ such that $Ax = b$ where the *data matrix* $A \in \mathbb{R}^{m \times n}$ and the *observation vector* $b \in \mathbb{R}^m$ are given and $m \geq n$. When there are more equations than unknowns, we say that the system $Ax = b$ is *overdetermined.* Usually an overdetermined system has no exact solution since $b$ must be an element of range($A$), a proper subspace of $\mathbb{R}^m$.

This suggests that we strive to minimize $\| Ax - b \|_p$ for some suitable choice of $p$. Different norms render different optimum solutions. For example, if $A = [\, 1\ 1\ 1\,]^T$ and $b = (b_1\ b_2\ b_3)^T$ with $b_1 \geq b_2 \geq b_3 \geq 0$, then it can be verified that

$$\begin{array}{rcrcrcl} p & = & 1 & \Rightarrow & x_{opt} & = & b_2 \\ p & = & 2 & \Rightarrow & x_{opt} & = & (b_1 + b_2 + b_3)/3 \\ p & = & \infty & \Rightarrow & x_{opt} & = & (b_1 + b_3)/2. \end{array}$$

Minimization in the 1-norm and $\infty$ -norm is complicated by the fact that the function $f(x) = \| Ax - b \|_p$ is not differentiable for these values of $p$. However, much progress has been made in this area, and there are several good techniques available for 1-norm and $\infty$ -norm minimization. See Bartels, Conn, and Sinclair (1978) and Bartels, Conn, and Charalambous (1978).

In contrast to general $p$-norm minimization, the *least squares* (LS) problem

$$\min_{x \in \mathbb{R}^n} \| Ax - b \|_2 \tag{5.3.1}$$

is more tractable for two reasons:

- $\phi(x) = \frac{1}{2}\| Ax - b \|_2^2$ is a differentiable function of $x$ and so the minimizers of $\phi$ satisfy the gradient equation $\nabla\phi(x) = 0$. This turns out to be an easily constructed symmetric linear system which is positive definite if $A$ has full column rank.

- The 2-norm is preserved under orthogonal transformation. This means that we can seek an orthogonal $Q$ such that the equivalent problem of minimizing $\| (Q^T A)x - (Q^T b) \|_2$ is "easy" to solve.

In this section we pursue these two solution approaches for the case when $A$ has full column rank. Methods based on normal equations and the QR factorization are detailed and compared.

## 5.3.1 Implications of Full Rank

Suppose $x \in \mathbb{R}^n$, $z \in \mathbb{R}^n$ , and $\alpha \in \mathbb{R}$ and consider the equality

$$\| A(x + \alpha z) - b \|_2^2 \;=\; \| Ax - b \|_2^2 \;+ 2\alpha z^T A^T (Ax - b) + \alpha^2 \| Az \|_2^2$$

where $A \in \mathbb{R}^{m \times n}$ and $b \in \mathbb{R}^m$. If $x$ solves the LS problem (5.3.1) then we must have $A^T(Ax - b) = 0$. Otherwise, if $z = -A^T(Ax - b)$ and we make $\alpha$ small enough, then we obtain the contradictory inequality $\| A(x + \alpha z) - b \|_2 < \| Ax - b \|_2$. We may also conclude that if $x$ *and* $x + \alpha z$ are LS minimizers then $z \in \text{null}(A)$.

Thus, if $A$ has full column rank then there is a unique LS solution $x_{LS}$ and it solves the symmetric positive definite linear system

$$A^T A x_{LS} = A^T b\,.$$

These are called the *normal equations*. Since $\nabla\phi(x) = A^T(Ax - b)$ where $\phi(x) = \frac{1}{2}\| Ax - b \|_2^2$ , we see that solving the normal equations is tantamount to solving the gradient equation $\nabla\phi = 0$.

We call

$$r_{LS} = b - Ax_{LS}$$

the *minimum residual* and we use the notation

$$\rho_{LS} = \| Ax_{LS} - b \|_2$$

to denote its size. Note that if $\rho_{LS}$ is small then we can "predict" $b$ with the columns of $A$.

So far we have been assuming that $A \in \mathbb{R}^{m\times n}$ has full column rank. This assumption is dropped in §5.5. However, even if $\text{rank}(A) = n$ then we can expect trouble in the above procedures if $A$ is nearly rank deficient.

When assessing the quality of a computed LS solution $\hat{x}_{LS}$, there are two important issues to bear in mind:

- How close is $\hat{x}_{LS}$ to $x_{LS}$?
- How small is $\hat{r}_{LS} = b - A\hat{x}_{LS}$ when compared to $r_{LS} = b - Ax_{LS}$?

The relative importance of these two criteria varies from application to application. In any case it is important to understand how $x_{LS}$ and $r_{LS}$ are affected by perturbations in $A$ and $b$. Our intuition tells us that if the columns of $A$ are nearly dependent, then these quantities may be quite sensitive. To clarify LS sensitivity issues we need to extend the notion of condition to rectangular matrices.

### 5.3.2 Condition of a Rectangular Matrix

In analyzing the least squares problem it is natural to work with the 2-norm. Recall from §2.7.3 that the 2-norm condition of a square nonsingular matrix is the ratio of the largest and smallest singular values. For rectangular matrices with full column rank we continue with this definition:

$$A \in \mathbb{R}^{m\times n}, \text{rank}(A) = n \implies \kappa_2(A) = \frac{\sigma_{max}(A)}{\sigma_{min}(A)} .$$

If the columns of $A$ are nearly dependent then $\kappa_2(A)$ is large. The following example suggests that ill-conditioned LS problems have sensitive solutions and sensitive minimum residuals.

**Example 5.3.1** Suppose

$$A = \begin{bmatrix} 1 & 0 \\ 0 & 10^{-6} \\ 0 & 0 \end{bmatrix}, \quad \delta A = \begin{bmatrix} 0 & 0 \\ 0 & 0 \\ 0 & 10^{-8} \end{bmatrix}, \quad b = \begin{bmatrix} 1 \\ 0 \\ 1 \end{bmatrix}, \quad \delta b = \begin{bmatrix} 0 \\ 0 \\ 0 \end{bmatrix},$$

and that $x_{LS}$ and $\hat{x}_{LS}$ minimize $\| Ax - b \|_2$ and $\| (A + \delta A)x - (b + \delta b) \|_2$ respectively. Let $r_{LS}$ and $\hat{r}_{LS}$ be the corresponding minimum residuals. Then

$$x_{LS} = \begin{bmatrix} 1 \\ 0 \end{bmatrix}, \ \hat{x}_{LS} = \begin{bmatrix} 1 \\ .9999 \cdot 10^4 \end{bmatrix}, \ r_{LS} = \begin{bmatrix} 0 \\ 0 \\ 1 \end{bmatrix}, \ \hat{r}_{LS} = \begin{bmatrix} 0 \\ -.9999 \cdot 10^{-2} \\ .9999 \cdot 10^0 \end{bmatrix}$$

Since $\kappa_2(A) = 10^6$ we have

$$\frac{\| \hat{x}_{LS} - x_{LS} \|_2}{\| x_{LS} \|_2} \approx .9999 \cdot 10^4 \le \kappa_2(A)^2 \frac{\| \delta A \|_2}{\| A \|_2} = 10^{12} \cdot 10^{-8}$$

and

$$\frac{\| \hat{r}_{LS} - r_{LS} \|_2}{\| b \|_2} \approx .7070 \cdot 10^{-2} \le \kappa_2(A) \frac{\| \delta A \|_2}{\| A \|_2} = 10^6 \cdot 10^{-8} .$$

Note from Example 5.3.1 that the sensitivity of $x_{LS}$ seems to depend upon $\kappa_2(A)^2$. At the end of this section we develop a perturbation theory for the LS problem and we use it to assess the normal equation and QR solution procedures that we now describe.

### 5.3.3 The Method of Normal Equations

The most widely used method for solving the full rank LS problem is the method of normal equations.

**Algorithm 5.3.1 (Normal Equations)** Given $A \in \mathbb{R}^{m \times n}$ with the property that $\text{rank}(A) = n$ and $b \in \mathbb{R}^m$, this algorithm computes the solution $x_{LS}$ to the LS problem $\min \| Ax - b \|_2$ where $b \in \mathbb{R}^m$.

Compute the lower triangular portion of $C = A^T A$.
$d = A^T b$
Compute the Cholesky factorization $C = GG^T$.
Solve $Gy = d$ and $G^T x_{LS} = y$.

This algorithm requires $(m + n/3)n^2$ flops. The normal equation approach is convenient because it relies on standard algorithms: Cholesky factorization, matrix-matrix multiplication, and matrix-vector multiplication. The compression of the $m$-by-$n$ data matrix $A$ into the much smaller $n$-by-$n$ cross-product matrix $C$ when $m \gg n$ is attractive.

Let us consider the accuracy of the computed normal equations solution $\hat{x}_{LS}$. For clarity, assume that no roundoff errors occur during the formation of $C = A^T A$ and $d = A^T b$. (Typically, inner products are accumulated during this portion of the computation and so this is not a terribly unfair assumption.) It follows from what we know about the roundoff properties of the Cholesky factorization (cf. §4.2.7) that

$$(A^T A + E)\hat{x}_{LS} = A^T b,$$

where $\| E \|_2 \approx \mathbf{u}\| A^T \|_2 \| A \|_2 \approx \mathbf{u}\| A^T A \|_2$ and thus we can expect

$$\frac{\| \hat{x}_{LS} - x_{LS} \|_2}{\| x_{LS} \|_2} \approx \mathbf{u}\kappa_2(A^T A) = \mathbf{u}\kappa_2(A)^2 . \qquad (5.3.2)$$

In other words, the accuracy of the computed normal equations solution depends on the square of the condition. This seems to be consistent with Example 5.3.1 but more refined comments follow in §5.3.9.

**Example 5.3.2** It should be noted that the formation of $A^T A$ can result in a severe loss of information.

$$A = \begin{bmatrix} 1 & 1 \\ 10^{-3} & 0 \\ 0 & 10^{-3} \end{bmatrix} \text{ and } b = \begin{bmatrix} 2 \\ 10^{-3} \\ 10^{-3} \end{bmatrix}$$

then $\kappa_2(A) \approx 1.4 \cdot 10^3$, $x_{LS} = (1\ 1)^T$, and $\rho_{LS} = 0$. If the normal equations method is executed with base 10, $t = 6$ chopped arithmetic, then a divide exception occurs during the solution process, since

$$fl(A^T A) = \begin{bmatrix} 1 & 1 \\ 1 & 1 \end{bmatrix}$$

is exactly singular. On the other hand, if 7-digit, chopped arithmetic is used, then $\hat{x}_{LS} = (2.000001,\ 0)^T$ and $\| \hat{x}_{LS} - x_{LS} \|_2 / \| x_{LS} \|_2 \approx \mathbf{u}\kappa_2(A)^2$.

### 5.3.4 LS Solution Via QR Factorization

Let $A \in \mathbb{R}^{m \times n}$ with $m \geq n$ and $b \in \mathbb{R}^m$ be given and suppose that an orthogonal matrix $Q \in \mathbb{R}^{m \times m}$ has been computed such that

$$Q^T A = R = \begin{bmatrix} R_1 \\ 0 \end{bmatrix} \begin{matrix} n \\ m-n \end{matrix} \qquad (5.3.3)$$

is upper triangular. If

$$Q^T b = \begin{bmatrix} c \\ d \end{bmatrix} \begin{matrix} n \\ m-n \end{matrix}$$

then

$$\| Ax - b \|_2^2 = \| Q^T Ax - Q^T b \|_2^2 = \| R_1 x - c \|_2^2 + \| d \|_2^2$$

for any $x \in \mathbb{R}^n$. Clearly, if $\text{rank}(A) = \text{rank}(R_1) = n$, then $x_{LS}$ is defined by the upper triangular system $R_1 x_{LS} = c$. Note that

$$\rho_{LS} = \| d \|_2 .$$

We conclude that the full rank LS problem can be readily solved once we have computed the QR factorization of $A$. Details depend on the exact QR procedure. When Householder matrices are used and $Q^T$ is applied in factored form to $b$ then we obtain

**Algorithm 5.3.2 (Householder LS Solution)** If $A \in \mathbb{R}^{m\times n}$ has full column rank and $b \in \mathbb{R}^m$, then the following algorithm computes a vector $x_{LS} \in \mathbb{R}^n$ such that $\| Ax_{LS} - b \|_2$ is minimum.

Use Algorithm 5.2.1 to overwrite $A$ with its QR factorization.
**for** $j = 1{:}n$
  { Apply the $j$-th Householder transformation. }
  $v(j) = 1;\ v(j+1{:}m) = A(j+1{:}m, j)$
  $b(j{:}m) = \textbf{row.house}(b(j{:}m, v(j{:}m))$
**end**
Solve $R(1{:}n, 1{:}n)x_{LS} = b(1{:}n)$ using back substitution.

This method for solving the full rank LS problem requires $2n^2(m - n/3)$ flops. The $O(mn)$ flops associated with the updating of $b$ and the $O(n^2)$ flops associated with the back substitution are not significant compared to the work required to factor $A$.

It can be shown that the computed $\hat{x}_{LS}$ solves

$$\min \| (A + \delta A)x - (b + \delta b) \|_2 \tag{5.3.4}$$

where

$$\| \delta A \|_F \le (6m - 3n + 41)n\mathbf{u} \| A \|_F + O(\mathbf{u}^2) \tag{5.3.5}$$

and

$$\| \delta b \|_2 \le (6m - 3n + 40)nu \| b \|_2 + O(\mathbf{u}^2). \tag{5.3.6}$$

These inequalities are established in Lawson and Hanson (SLS, pp. 90 ff.) and show that $\hat{x}_{LS}$ satisfies a "nearby" LS problem. (We cannot address the relative error in $\hat{x}_{LS}$ without an LS perturbation theory, to be discussed shortly.) We mention that similar results hold if Givens QR is used.

### 5.3.5 Breakdown in Near-Rank Deficient Case

Like the method of normal equations, the Householder method for solving the LS problem breaks down in the back substitution phase if $\text{rank}(A) < n$. Numerically, trouble can be expected whenever $\kappa_2(A) = \kappa_2(R) \approx 1/\mathbf{u}$. This is in contrast to the normal equations approach, where completion of the Cholesky factorization becomes problematical once $\kappa_2(A)$ is in the neighborhood of $1/\sqrt{\mathbf{u}}$. (See Example 5.3.2.) Hence the claim in Lawson and Hanson (SLS, pp. 126–27) that for a fixed machine precision, a wider class of LS problems can be solved using Householder orthogonalization.

### 5.3.6 A Note on the MGS Approach

Note that MGS does not produce an orthonormal basis for range$(A)^\perp$ as does Householder orthogonalization. However, it does render enough information to solve the LS problem. In particular, MGS computes the factorization $A = Q_1 R_1$ where $Q_1 \in \mathbb{R}^{m\times n}$ has orthonormal columns and $R_1 \in \mathbb{R}^{n\times n}$ is upper triangular. Thus, the normal equations $(A^T A)x = A^T b$ transform to the upper triangular system $R_1 x = Q_1^T b$.

MGS can be readily adapted to solve the LS problem and is as stable as the Householder approach. See Björck (1967b). However, the MGS method is slightly more expensive because in it always manipulates $m$-vectors whereas the Householder QR procedure deals with ever shorter vectors.

### 5.3.7 Fast Givens LS Solver

The LS problem can also be solved using fast Givens transformations. Suppose $M^T M = D$ is diagonal and

$$M^T A = \begin{bmatrix} S_1 \\ 0 \end{bmatrix} \begin{matrix} n \\ m-n \end{matrix}$$

is upper triangular. If

$$M^T b = \begin{bmatrix} c \\ d \end{bmatrix} \begin{matrix} n \\ m-n \end{matrix}$$

then

$$\begin{aligned} \| Ax - b \|_2^2 &= \| D^{-1/2} M^T A x - D^{-1/2} M^T b \|_2^2 \\ &= \left\| D^{-1/2} \left( \begin{bmatrix} S_1 \\ 0 \end{bmatrix} x - \begin{bmatrix} c \\ d \end{bmatrix} \right) \right\|_2^2 \end{aligned}$$

for any $x \in \mathbb{R}^n$. Clearly, $x_{LS}$ is obtained by solving the nonsingular upper triangular system $S_1 x = c$.

The computed solution $\hat{x}_{LS}$ obtained in this fashion can be shown to solve a nearby LS problem in the sense of (5.3.4)-(5.3.6). This may seem surprising since large numbers can arise during the calculation. An entry in the scaling matrix $D$ can double in magnitude after a single fast Givens update. However, largeness in $D$ must be exactly compensated for by largeness in $M$, since $D^{-1/2}M$ is orthogonal at all stages of the computation. It is this phenomenon that enables one to push through a favorable error analysis.

However, as we discuss in §5.1.13, the possibility of element growth requires continual monitoring of $D$ to avoid overflow. Because of the non-trivial overhead involved, the resulting algorithm can be slower than the Householder approach and is certainly more complicated to implement.

### 5.3.8 The Sensitivity of the LS Problem

We now develop a perturbation theory that assists in the comparison of the normal equations and QR approaches to the LS problem. The theorem below examines how the LS solution and its residual are affected by changes in $A$ and $b$. In so doing, the condition of the LS problem is identified.

Two easily established facts are required in the analysis:

$$\begin{aligned} \| A \|_2 \, \| (A^T A)^{-1} A^T \|_2 &= \kappa_2(A) \\ \| A \|_2^2 \; \| (A^T A)^{-1} \|_2 &= \kappa_2(A)^2 \end{aligned} \tag{5.3.7}$$

These equations can be verified using the SVD.

**Theorem 5.3.1** *Suppose $x$, $r$, $\hat{x}$, and $\hat{r}$ satisfy*

$$\| Ax - b \|_2 = \min \qquad r = b - Ax$$

$$\| (A + \delta A)\hat{x} - (b + \delta b) \|_2 = \min \qquad \hat{r} = (b + \delta b) - (A + \delta A)\hat{x}$$

*where $A$ and $\delta A$ are in $\mathbb{R}^{m \times n}$ with $m \geq n$ and $0 \neq b$ and $\delta b$ are in $\mathbb{R}^m$. If*

$$\epsilon = \max\left\{ \frac{\| \delta A \|_2}{\| A \|_2}, \frac{\| \delta b \|_2}{\| b \|_2} \right\} < \frac{\sigma_n(A)}{\sigma_1(A)}$$

*and*

$$\sin(\theta) = \frac{\rho_{LS}}{\| b \|_2} \neq 1$$

*where $\rho_{LS} = \| Ax_{LS} - b \|_2$, then*

$$\frac{\| \hat{x} - x \|_2}{\| x \|_2} \leq \epsilon \left\{ \frac{2\kappa_2(A)}{\cos(\theta)} + \tan(\theta)\kappa_2(A)^2 \right\} + O(\epsilon^2) \tag{5.3.8}$$

$$\frac{\| \hat{r} - r \|_2}{\| b \|_2} \leq \epsilon \, (1 + 2\kappa_2(A)) \min(1, m - n) + O(\epsilon^2) \,. \tag{5.3.9}$$

**Proof.** Let $E$ and $f$ be defined by $E = \delta A/\epsilon$ and $f = \delta b/\epsilon$. By hypothesis $\| \delta A \|_2 < \sigma_n(A)$ and so by Theorem 2.5.2 we have $\text{rank}(A + tE) = n$ for all $t \in [0, \epsilon]$. It follows that the solution $x(t)$ to

$$(A + tE)^T (A + tE) x(t) = (A + tE)^T (b + tf) \tag{5.3.10}$$

is continuously differentiable for all $t \in [0, \epsilon]$. Since $x = x(0)$ and $\hat{x} = x(\epsilon)$, we have

$$\hat{x} = x + \epsilon\dot{x}(0) + O(\epsilon^2).$$

The assumptions $b \neq 0$ and $\sin(\theta) \neq 1$ ensure that $x$ is nonzero and so

$$\frac{\| \hat{x} - x \|_2}{\| x \|_2} = \epsilon \frac{\| \dot{x}(0) \|_2}{\| x \|_2} + O(\epsilon^2). \tag{5.3.11}$$

In order to bound $\| \dot{x}(0) \|_2$, we differentiate (5.3.10) and set $t = 0$ in the result. This gives

$$E^T Ax + A^T Ex + A^T A\dot{x}(0) = A^T f + E^T b$$

i.e.,

$$\dot{x}(0) = (A^T A)^{-1} A^T (f - Ex) + (A^T A)^{-1} E^T r. \tag{5.3.12}$$

By substituting this result into (5.3.11), taking norms, and using the easily verified inequalities $\| f \|_2 \leq \| b \|_2$ and $\| E \|_2 \leq \| A \|_2$ we obtain

$$\begin{aligned}\frac{\| \hat{x} - x \|_2}{\| x \|_2} \leq\ & \epsilon \left\{ \| A \|_2 \| (A^T A)^{-1} A^T \|_2 \left( \frac{\| b \|_2}{\| A \|_2 \| x \|_2} + 1 \right) \right. \\ & \left. + \frac{\rho_{LS}}{\| A \|_2 \| x \|_2} \| A \|_2^2 \, \| (A^T A)^{-1} \|_2 \right\} + O(\epsilon^2).\end{aligned}$$

Since $A^T(Ax - b) = 0$, $Ax$ is orthogonal to $Ax - b$ and so

$$\| b - Ax \|_2^2 + \| Ax \|_2^2 = \| b \|_2^2 .$$

Thus,

$$\| A \|_2^2 \, \| x \|_2^2 \geq \| b \|_2^2 - \rho_{LS}^2$$

and so by using (5.3.7)

$$\frac{\| \hat{x} - x \|_2}{\| x \|_2} \leq \epsilon \left\{ \kappa_2(A) \left( \frac{1}{\cos(\theta)} + 1 \right) + \kappa_2(A)^2 \frac{\sin(\theta)}{\cos(\theta)} \right\} + O(\epsilon^2)$$

thereby establishing (5.3.8).

To prove (5.3.9), we define the differentiable vector function $r(t)$ by

$$r(t) = (b + tf) - (A + tE)x(t)$$

and observe that $r = r(0)$ and $\hat{r} = r(\epsilon)$. Using (5.3.12) it can be shown that

$$\dot{r}(0) = \left(I - A(A^T A)^{-1} A^T\right)(f - Ex) - A(A^T A)^{-1} E^T r .$$

Since $\| \hat{r} - r \|_2 = \epsilon \| \dot{r}(0) \|_2 + O(\epsilon^2)$ we have

$$\begin{aligned}\frac{\| \hat{r} - r \|_2}{\| b \|_2} &= \epsilon \frac{\| \dot{r}(0) \|_2}{\| b \|_2} + O(\epsilon^2) \\ &\leq \epsilon \left\{ \| I - A(A^TA)^{-1}A^T \|_2 \left( 1 + \frac{\| A \|_2 \| x \|_2}{\| b \|_2} \right) \right. \\ &\quad \left. + \| A(A^TA)^{-1} \|_2 \| A \|_2 \frac{\rho_{LS}}{\| b \|_2} \right\} + O(\epsilon^2) .\end{aligned}$$

Inequality (5.3.9) now follows because

$$\| A \|_2 \| x \|_2 = \| A \|_2 \| A^+ b \|_2 \leq \kappa_2(A) \| b \|_2 ,$$

$$\rho_{LS} = \| (I - A(A^TA)^{-1}A^T)b \|_2 \leq \| I - A(A^TA)^{-1}A^T \|_2 \| b \|_2 ,$$

and

$$\| (I - (A^TA)^{-1}A^T \|_2 = \min(m - n, 1). \; \square$$

An interesting feature of the upper bound in (5.3.8) is the factor

$$\tan(\theta)\kappa_2(A)^2 = \frac{\rho_{LS}}{\sqrt{\| b \|_2^2 - \rho_{LS}^2}} \kappa_2(A)^2 .$$

Thus, in nonzero residual problems it is the square of the condition that measures the sensitivity of $x_{LS}$. In contrast, residual sensitivity depends just linearly on $\kappa_2(A)$. These dependencies are confirmed by Example 5.3.1.

### 5.3.9 Normal Equations Versus QR

It is instructive to compare the normal equation and Householder QR approaches to the LS problem. Recall the following main points from our discussion:

- The sensitivity of the LS solution is roughly proportional to the quantity $\kappa_2(A) + \rho_{LS}\kappa_2(A)^2$.
- The method of normal equations produces an $\hat{x}_{LS}$ whose relative error depends on the square of the condition.
- The Householder QR approach solves a nearby LS problem and therefore produces a solution that has a relative error approximately given by $\mathbf{u}(\kappa_2(A) + \rho_{LS}\kappa_2(A)^2)$.

Thus, we may conclude that if $\rho_{LS}$ is small and $\kappa_2(A)$ is large, then the method of normal equations does not solve a nearby problem and will usually render an LS solution that is less accurate than the Householder approach. Conversely, the two methods produce comparably inaccurate results when applied to large residual, ill-conditioned problems.

Finally, we mention two other factors that figure in the debate about QR versus normal equations:

- The normal equations approach involves about half of the arithmetic when $m \gg n$ and does not require as much storage.

- The Householder QR approach is applicable to a wider class of matrices because the Cholesky process applied to $A^TA$ breaks down "before" the back substitution process on $Q^TA = R$.

At the very minimum, this discussion should convince you how difficult it can be to choose the "right" algorithm!

### Problems

**P5.3.1** Assume $A^TAx = A^Tb$, $(A^TA + F)\hat{x} = A^Tb$, and $2\| F \|_2 \leq \sigma_n(A)^2$. Show that if $r = b - Ax$ and $\hat{r} = b - A\hat{x}$, then $\hat{r} - r = \left(I - A(A^TA + F)^{-1}A^T\right)Ax$ and as a consequence,

$$\| \hat{r} - r \|_2 \leq 2\kappa_2(A)\frac{\| F \|_2}{\| A \|_2}\| x \|_2$$

**P5.3.2** Let $A \in \mathbb{R}^{m\times n}$ with $m > n$ and $y \in \mathbb{R}^m$ and define the matrix $\bar{A} = [A\ y] \in \mathbb{R}^{m\times(n+1)}$. Show $\sigma_1(\bar{A}) \geq \sigma_1(A)$ and $\sigma_{n+1}(\bar{A}) \leq \sigma_n(A)$. Thus, the condition grows if a column is added to a matrix.

**P5.3.3** Let $A \in \mathbb{R}^{m\times n}$, $w \in \mathbb{R}^n$, and define

$$B = \left[ \begin{array}{c} A \\ w^T \end{array} \right].$$

Show $\sigma_n(B) \geq \sigma_n(A)$ and $\sigma_1(B) \leq \sqrt{\| A \|_2^2 + \| w \|_2^2}$. Thus, the condition of a matrix may increase or decrease if a row is added.

**P5.3.4** Give an algorithm for solving the LS problem which uses MGS. No additional two-dimensional arrays are required if $a_{ij}$ is overwritten with $r_{ji}$ with $i \geq j$ and if the quantities $q_i^Tb$ are found as the orthonormal $q_i$ are generated.

**P5.3.5** (Cline 1973) Suppose that $A \in \mathbb{R}^{m\times n}$ has rank $n$ and that Gaussian elimination with partial pivoting is used to compute the factorization $PA = LU$, where $L \in \mathbb{R}^{m\times n}$ is unit lower triangular, $U \in \mathbb{R}^{n\times n}$ is upper triangular, and $P \in \mathbb{R}^{m\times m}$ is a permutation. Explain how the decomposition in P5.2.5 can be used to find a vector $z \in \mathbb{R}^n$ such that $\| Lz - Pb \|_2$ is minimized. Show that if $Ux = z$, then $\| Ax - b \|_2$ is minimum. Show that this method of solving the LS problem is more efficient than Householder QR from the flop point of view whenever $m \leq 5/3n$.

**P5.3.6** The matrix $C = (A^TA)^{-1}$, where $\text{rank}(A) = n$, arises in many statistical applications and is known as the *variance-covariance matrix*. Assume that the factorization $A = QR$ is available. (a) Show $C = (R^TR)^{-1}$. (b) Give an algorithm for computing the diagonal of $C$ that requires $n^3/3$ flops. (c) Show that

$$R = \left[ \begin{array}{cc} \alpha & v^T \\ 0 & S \end{array} \right] \quad \Rightarrow \quad C = (R^TR)^{-1} = \left[ \begin{array}{cc} (1 + v^TC_1v)/\alpha & -v^TC_1/\alpha \\ -C_1v/\alpha & C_1 \end{array} \right].$$

where $C_1 = (S^TS)^{-1}$. (d) Using (c), give an algorithm that overwrites the upper triangular portion of $R$ with the upper triangular portion of $C$. Your algorithm should require $2n^3/3$ flops.

### Notes and References for Sec. 5.3

Our restriction to least squares approximation is not a vote against minimization in other norms. There are occasions when it is advisable to minimize $\| Ax - b \|_p$ for $p = 1$ and $\infty$. Some algorithms for doing this are described in

I. Barrodale and F.D.K. Roberts (1973). "An Improved Algorithm for Discrete $L_1$ Linear Approximation," *SIAM J. Num. Anal. 10,* 839–48.

I. Barrodale and C. Phillips (1975). "Algorithm 495: Solution of An Overdetermined System of Linear Equations in the Chebychev Norm," *ACM Trans. Math. Soft. 1,* 264–70.

R.H. Bartels, A.R. Conn, and J.W. Sinclair (1978). "Minimization Techniques for Piecewise Differentiable Functions: The $L_1$ Solution to An Overdetermined Linear System," *SIAM J. Num. Anal. 15,* 224–41.

R.H. Bartels, A.R. Conn, and C. Charalambous (1978). "On Cline's Direct Method for Solving Overdetermined Linear Systems in the $L_\infty$ Sense," *SIAM J. Num. Anal. 15,* 255–70.

A.K. Cline (1976a). "A Descent Method for the Uniform Solution to Overdetermined Systems of Equations," *SIAM J. Num. Anal. 13,* 293–309.

Two excellent books on the linear least squares problem include Lawson and Hanson (SLE) and

Å. Björck (1988). *Least Squares Methods: Handbook of Numerical Analysis Vol. 1 Solution of Equations in* $\mathbf{R}^N$, Elsevier North Holland.

The use of Gauss transformations to solve the LS problem has attracted some attention because they are cheaper to use than Householder or Givens matrices. See

A.K. Cline (1973). "An Elimination Method for the Solution of Linear Least Squares Problems," *SIAM J. Num. Anal. 10,* 283–89.

G. Peters and J.H. Wilkinson (1970). "The Least Squares Problem and Pseudo-Inverses," *Comp. J. 13,* 309–16.

R.J. Plemmons (1974). "Linear Least Squares by Elimination and MGS," *J. Assoc. Comp. Mach. 21,* 581–85.

Conditioning aspects of the LS problem are discussed in

G.H. Golub and J.H. Wilkinson (1966). "Note on the Iterative Refinement of Least Squares Solution," *Numer. Math. 9,* 139–48.

Y. Saad (1986). "On the Condition Number of Some Gram Matrices Arising from Least Squares Approximation in the Complex Plane," *Numer. Math. 48,* 337–348.

A. van der Sluis (1975). "Stability of the Solutions of Linear Least Squares Problem," *Numer. Math. 23,* 241–54.

See also

Å. Björck (1987). "Stability Analysis of the Method of Seminormal Equations," *Lin. Alg. and Its Applic. 88/89,* 31–48.

The "seminormal" equations are given by $R^T Rx = A^T b$ where $A = QR$. In the above paper it is shown that by solving the seminormal equations an acceptable LS solution is obtained if one step of fixed precision iterative improvement is performed.

Fortran programs for solving the LS problem via Householder transformations may be found in Linpack, chapter 9. An Algol implementation of the MGS method for solving the LS problem appears in

F.L. Bauer (1965). "Elimination with Weighted Row Combinations for Solving Linear Equations and Least Squares Problems," *Numer. Math. 7,* 338–52. See also *HACLA,* 119–33.

Least squares problems often have special structure which, of course, should be exploited. Two of the many papers having this theme include

M.G. Cox (1981). "The Least Squares Solution of Overdetermined Linear Equations having Band or Augmented Band Structure," *IMA J. Numer. Anal. 1*, 3–22.

G. Cybenko (1984). "The Numerical Stability of the Lattice Algorithm for Least Squares Linear Prediction Problems," *BIT 24*, 441–455.

The use of Householder matrices to solve sparse LS problems requires careful attention to avoid excessive fill-in. References include

I.S. Duff and J.K. Reid (1976). "A Comparison of Some Methods for the Solution of Sparse Over-Determined Systems of Linear Equations," *J. Inst. Math. Applic. 17*, 267–80.

P.E. Gill and W. Murray (1976). "The Orthogonal Factorization of a Large Sparse Matrix," in *Sparse Matrix Compautations*, ed. J.R. Bunch and D.J. Rose, Academic Press, New York, pp. 177-200.

L. Kaufman (1979). "Application of Dense Householder Transformations to a Sparse Matrix," *ACM Trans. Math. Soft. 5*, 442–51.

J.K. Reid (1967). "A Note on the Least Squares Solution of a Band System of Linear Equations by Householder Reductions," *Comp. J. 10*, 188–89.

Although the computation of the QR factorization is more efficient with Householder reflections, there are some settings where the Givens approach is advantageous. For example, if $A$ is sparse, then the careful application of Givens rotations can minimize fill-in. See

I.S. Duff (1974). "Pivot Selection and Row Ordering in Givens Reduction on Sparse Matrices," *Computing 13*, 239–48.

J.A. George and M.T. Heath (1980). "Solution of Sparse Linear Least Squares Problems Using Givens Rotations," *Lin. Alg. and Its Applic. 34*, 69–83.

# 5.4 Other Orthogonal Factorizations

If $A$ is rank deficient then the QR factorization need not give a basis for range($A$). This problem can be corrected by computing the QR factorization of a column-permuted version of $A$, i.e., $A\Pi = QR$ where $\Pi$ is a permutation.

The "data" in $A$ can be compressed further if we permit right multiplication by a general orthogonal matrix $Z$: $Q^T AZ = T$. There are interesting choices for $Q$ and $Z$ and these, together with the column pivoted QR factorization, are discussed in this section.

## 5.4.1 Rank Deficiency: QR with Column Pivoting

If $A \in \mathbb{R}^{m \times n}$ and rank($A$) $< n$, then the QR factorization does not necessarily produce an orthonormal basis for range($A$). For example, if $A$ has three columns and

$$A = \begin{bmatrix} a_1 & a_2 & a_3 \end{bmatrix} = \begin{bmatrix} q_1 & q_2 & q_3 \end{bmatrix} \begin{bmatrix} 1 & 1 & 1 \\ 0 & 0 & 1 \\ 0 & 0 & 1 \end{bmatrix}$$

is its QR factorization, then $\text{rank}(A) = 2$ but $\text{range}(A)$ does not equal $\text{span}\{q_1, q_2\}$, $\text{span}\{q_1, q_3\}$, or $\text{span}\{q_2, q_3\}$.

Fortunately, the Householder QR factorization procedure (Algorithm 5.2.1) can be modified in a simple way to produce an orthonormal basis for $\text{range}(A)$. The modified algorithm computes the factorization

$$Q^T A\Pi = \begin{bmatrix} R_{11} & R_{12} \\ 0 & 0 \end{bmatrix} \begin{matrix} r \\ m-r \end{matrix} \qquad (5.4.1)$$
$$\qquad\qquad \begin{matrix} r & n-r \end{matrix}$$

where $r = \text{rank}(A)$, $Q$ is orthogonal, $R_{11}$ is upper triangular and nonsingular, and $\Pi$ is a permutation. If we have the column partitionings $A\Pi = [\, a_{c_1}, \ldots, a_{c_n} \,]$ and $Q = [\, q_1, \ldots, q_m \,]$ then for $k = 1{:}n$ we have

$$a_{c_k} = \sum_{i=1}^{\min\{r,k\}} r_{ik} q_i \in \text{span}\{q_1, \ldots, q_r\}$$

implying $\text{range}(A) = \text{span}\{q_1, \ldots, q_r\}$.

The matrices $Q$ and $\Pi$ are products of Householder matrices and interchange matrices respectively. Assume for some $k$ that we have computed Householder matrices $H_1, \ldots, H_{k-1}$ and permutations $\Pi_1, \ldots, \Pi_{k-1}$ such that

$$(H_{k-1} \cdots H_1) A (\Pi_1 \cdots \Pi_{k-1}) = \qquad (5.4.2)$$

$$R^{(k-1)} = \begin{bmatrix} R_{11}^{(k-1)} & R_{12}^{(k-1)} \\ 0 & R_{22}^{(k-1)} \end{bmatrix} \begin{matrix} k-1 \\ m-k+1 \end{matrix}$$
$$\qquad\qquad \begin{matrix} k-1 & n-k+1 \end{matrix}$$

where $R_{11}^{(k-1)}$ is a nonsingular and upper triangular matrix. Now suppose that $R_{22}^{(k-1)} = \left[ z_k^{(k-1)}, \ldots, z_n^{(k-1)} \right]$ is a column partitioning and let $p$ be the smallest index satisfying $(k \le p \le n)$ such that

$$\| z_p^{(k-1)} \|_2 = \max\left\{ \| z_k^{(k-1)} \|_2, \cdots, \| z_n^{(k-1)} \|_2 \right\}. \qquad (5.4.3)$$

Note that if $k-1 = \text{rank}(A)$, then this maximum is zero and we are finished. Otherwise, let $\Pi_k$ be the $n$-by-$n$ identity with columns $p$ and $k$ interchanged and determine a Householder matrix $H_k$ such that if $R^{(k)} = H_k R^{(k-1)} \Pi_k$

then $R^{(k)}(k+1{:}m,k) = 0$. In other words, $\Pi_k$ moves the largest column in $R_{22}^{(k-1)}$ to the lead position and $\tilde{H}_k$ zeroes all of its subdiagonal components.

The column norms do not have to be recomputed at each stage if we exploit the property

$$Q^T z = \begin{bmatrix} \alpha \\ w \end{bmatrix} \begin{matrix} 1 \\ s-1 \end{matrix} \qquad \Longrightarrow \qquad \| w \|_2^2 = \| z \|_2^2 - \alpha^2,$$

which holds for any orthogonal matrix $Q \in \mathbb{R}^{s \times s}$. This reduces the overhead associated with column pivoting from $O(mn^2)$ flops to $O(mn)$ flops because we can get the new column norms by updating the old column norms, e.g., $\| z^{(j)} \|_2^2 = \| z^{(j-1)} \|_2^2 - r_{kj}^2$.

Combining all of the above we obtain the following algorithm established by Businger and Golub (1965):

**Algorithm 5.4.1 (Householder QR With Column Pivoting)** Given $A \in \mathbb{R}^{m \times n}$ with $m \geq n$, the following algorithm computes $r = \text{rank}(A)$ and the factorization (5.4.1) with $Q = H_1 \cdots H_r$ and $\Pi = \Pi_1 \cdots \Pi_r$. The upper triangular part of $A$ is overwritten by the upper triangular part of $R$ and components $j+1{:}m$ of the $j$th Householder vector are stored in $A(j+1{:}m, j)$. The permutation $\Pi$ is encoded in an integer vector $piv$. In particular, $\Pi_j$ is the identity with rows $j$ and $piv(j)$ interchanged.

```
for j = 1:n
    c(j) = A(1:m,j)^T A(1:m,j)
end
r = 0; τ = max{c(j),... c(n)}
Find smallest k with 1 ≤ k ≤ n so c(k) = τ
while τ > 0
    r = r + 1
    piv(r) = k; A(1:m,r) ↔ A(1:m,k); c(r) ↔ c(k)
    v(r:m) = house(A(r:m,r))
    A(r:m,r:n) = row.house(A(r:m,r:n), v(r:m))
    A(r+1:m,r) = v(r+1:m)
    for i = r+1:n
        c(i) = c(i) − A(r,i)^2
    end
    if r < n
        τ = max{c(r+1),... c(n)}
        Find smallest k with r+1 ≤ k ≤ n so c(k) = τ.
    else
        τ = 0
    end
end
```

This algorithm requires $4mnr - 2r^2(m+n) + 4r^3/3$ flops where $r = \text{rank}(A)$. As with the nonpivoting procedure, Algorithm 5.2.1, the orthogonal matrix $Q$ is stored in factored form in the subdiagonal portion of $A$.

**Example 5.4.1** If Algorithm 5.4.1 is applied to

$$A = \begin{bmatrix} 1 & 2 & 3 \\ 1 & 5 & 6 \\ 1 & 8 & 9 \\ 1 & 11 & 12 \end{bmatrix}$$

then $\Pi = [e_3\ e_2\ e_1]$ and to three significant digits we obtain

$$A\Pi = QR = \begin{bmatrix} -.182 & -.816 & .514 & .191 \\ -.365 & .408 & -.827 & .129 \\ .548 & .000 & .113 & -.829 \\ -.730 & .408 & .200 & .510 \end{bmatrix} \begin{bmatrix} -16.4 & -14.600 & -1.820 \\ 0.0 & .816 & -.816 \\ 0.0 & .000 & 0.000 \end{bmatrix}.$$

## 5.4.2 Complete Orthogonal Decompositions

The matrix $R$ produced by Algorithm 5.4.1 can be further reduced if it is post-multiplied by an appropriate sequence of Householder matrices. In particular, we can use Algorithm 5.2.1 to compute

$$Z_r \cdots Z_1 \begin{bmatrix} R_{11}^T \\ R_{12}^T \end{bmatrix} = \begin{bmatrix} T_{11}^T \\ 0 \end{bmatrix} \begin{matrix} r \\ n-r \end{matrix} \tag{5.4.4}$$

where the $Z_i$ are Householder transformations and $T_{11}^T$ is upper triangular. It then follows that

$$Q^T A Z = T = \begin{bmatrix} T_{11} & 0 \\ 0 & 0 \end{bmatrix} \begin{matrix} r \\ m-r \end{matrix} \quad . \tag{5.4.5}$$
$$\qquad\qquad r \quad n-r$$

where $Z = \Pi Z_1 \cdots Z_r$. We refer to any decomposition of this form as a *complete orthogonal decomposition*. Note that $\text{null}(A) = \text{range}(Z(1{:}n, r+1{:}n))$. See P5.2.5 for details about the exploitation of structure in (5.4.4).

## 5.4.3 Bidiagonalization

Suppose $A \in \mathbb{R}^{m \times n}$ and $m \geq n$. We next show how to compute orthogonal $U$ ($m$-by-$m$) and $V$ ($n$-by-$n$) such that

$$U^T A V = \begin{bmatrix} d_1 & f_1 & 0 & \cdots & 0 \\ 0 & d_2 & f_2 & & 0 \\ \vdots & \ddots & \ddots & \ddots & \vdots \\ 0 & \cdots & & d_{n-1} & f_{n-1} \\ 0 & \cdots & & 0 & d_n \end{bmatrix}. \tag{5.4.6}$$

$U = U_1 \cdots U_n$ and $V = V_1 \cdots V_{n-2}$ can each be determined as a product of Householder matrices:

$$\begin{bmatrix} \times & \times & \times & \times \\ \times & \times & \times & \times \\ \times & \times & \times & \times \\ \times & \times & \times & \times \\ \times & \times & \times & \times \end{bmatrix} \xrightarrow{U_1} \begin{bmatrix} \times & \times & \times & \times \\ 0 & \times & \times & \times \\ 0 & \times & \times & \times \\ 0 & \times & \times & \times \\ 0 & \times & \times & \times \end{bmatrix} \xrightarrow{V_1}$$

$$\begin{bmatrix} \times & \times & 0 & 0 \\ 0 & \times & \times & \times \\ 0 & \times & \times & \times \\ 0 & \times & \times & \times \\ 0 & \times & \times & \times \end{bmatrix} \xrightarrow{U_2} \begin{bmatrix} \times & \times & 0 & 0 \\ 0 & \times & \times & \times \\ 0 & 0 & \times & \times \\ 0 & 0 & \times & \times \\ 0 & 0 & \times & \times \end{bmatrix} \xrightarrow{V_2}$$

$$\begin{bmatrix} \times & \times & 0 & 0 \\ 0 & \times & \times & 0 \\ 0 & 0 & \times & \times \\ 0 & 0 & \times & \times \\ 0 & 0 & \times & \times \end{bmatrix} \xrightarrow{U_3} \begin{bmatrix} \times & \times & 0 & 0 \\ 0 & \times & \times & 0 \\ 0 & 0 & \times & \times \\ 0 & 0 & 0 & \times \\ 0 & 0 & 0 & \times \end{bmatrix}$$

$$\xrightarrow{U_4} \begin{bmatrix} \times & \times & 0 & 0 \\ 0 & \times & \times & 0 \\ 0 & 0 & \times & \times \\ 0 & 0 & 0 & \times \\ 0 & 0 & 0 & 0 \end{bmatrix}.$$

In general, $U_k$ introduces zeros into the $k$th column, while $V_k$ zeros the appropriate entries in row $k$. Overall we have:

**Algorithm 5.4.2 (Householder Bidiagonalization)** Given $A \in \mathbb{R}^{m \times n}$ with $m \geq n$, the following algorithm overwrites the upper bidiagonal part of $A$ with the upper bidiagonal part of $U^T AV = B$ where $B$ is upper bidiagonal and $U = U_1 \cdots U_n$ and $V = V_1 \cdots V_{n-2}$. The essential part of $U_j$'s Householder vector is stored in $A(j+1{:}m, j)$ while the essential part of $V_j$'s Householder vector is stored in $A(j, j+2{:}n)$.

```
for j = 1:n
    v(j:m) = house(A(j:m, j))
    A(j:m, j:n) = row.house(A(j:m, j:n), v(j:m))
    A(j + 1:m, j) = v(j + 1:m)
    if j ≤ n − 2
        v(j + 1:n) = house(A(j, j + 1:n)^T)
        A(j:m, j + 1:n) = col.house(A(j:m, j + 1:n), v(j + 1:n))
        A(j, j + 2:n) = v(j + 2:n)^T
    end
end
```

This algorithm requires $4mn^2 - 4n^3/3$ flops. Such a technique is used in Golub and Kahan (1965), where bidiagonalization is first described. If the matrices $U_B$ and $V_B$ are explicitly desired, then they can be accumulated in $4m^2n - 4n^3/3$ and $4n^3/3$ flops, respectively (cf. §5.1.6.) We mention that the bidiagonalization of $A$ is related to the tridiagonalization of $A^TA$. See §8.2.1.

**Example 5.4.2** If Algorithm 5.4.2 is applied to

$$A = \begin{bmatrix} 1 & 2 & 3 \\ 4 & 5 & 6 \\ 7 & 8 & 9 \\ 10 & 11 & 12 \end{bmatrix}$$

then to three significant digits we obtain

$$\hat{B} = \begin{bmatrix} 12.8 & 21.8 & 0 \\ 0 & 2.24 & -.613 \\ 0 & 0 & 0 \\ 0 & 0 & 0 \end{bmatrix} \qquad \hat{V}_B = \begin{bmatrix} 1.00 & 0.00 & 0.00 \\ 0.00 & -.667 & -.745 \\ 0.00 & -.745 & .667 \end{bmatrix}$$

$$\hat{U}_B = \begin{bmatrix} -.0776 & -.833 & .392 & -.383 \\ -.3110 & -.451 & -.238 & .802 \\ -.5430 & -.069 & .701 & -.457 \\ -.7760 & .312 & .547 & .037 \end{bmatrix}.$$

### 5.4.4 R-Bidiagonalization

A faster method of bidiagonalizing when $m \gg n$ results if we upper triangularize $A$ first before applying Algorithm 5.4.2. In particular, suppose we compute an orthogonal $Q \in \mathbb{R}^{m \times m}$ such that

$$Q^T A = \begin{bmatrix} R_1 \\ 0 \end{bmatrix}$$

is upper triangular. We then bidiagonalize the square matrix $R_1$,

$$U_R^T R_1 V = B_1 .$$

Here $U_R$ and $V$ are $n$-by-$n$ orthogonal and $B_1$ is $n$-by-$n$ upper bidiagonal. If $U = Q\text{diag}(U_R, I_{m-n})$ then

$$U^T A V = \begin{bmatrix} B_1 \\ 0 \end{bmatrix} \equiv B$$

is a bidiagonalization of $A$.

The idea of computing the bidiagonalization in this manner is mentioned in Lawson and Hanson (SLS, p.119) and more fully analyzed in Chan (1982a). We refer to this method as $R$-bidiagonalization. By comparing its flop count $(2mn^2 + 2n^3)$ with that for Algorithm 5.4.2 $(4mn^2 - 4n^3/3)$ we see that it involves fewer computations (approximately) whenever $m \geq 5n/3$.

### 5.4.5 The SVD and its Computation

Once the bidiagonalization of $A$ has been achieved, the next step in the Golub-Reinsch SVD algorithm is to zero the superdiagonal elements in $B$. This is an iterative process and is accomplished by an algorithm due to Golub and Kahan (1965). Unfortunately, we must defer our discussion of this iteration until §8.3 as it requires an understanding of the symmetric eigenvalue problem. Suffice it to say here that it computes orthogonal matrices $U_\Sigma$ and $V_\Sigma$ such that

$$U_\Sigma^T B V_\Sigma \;=\; \Sigma \;=\; \text{diag}(\sigma_1, \ldots, \sigma_n) \;\in\; \mathbb{R}^{m\times n}\;.$$

By defining $U = U_B U_\Sigma$ and $V = V_B V_\Sigma$ we see that $U^T AV = \Sigma$ is the SVD of $A$. The flop counts associated with this portion of the algorithm depend upon "how much" of the SVD is required. For example, when solving the LS problem, $U^T$ need never be explicity formed but merely applied to $b$ as it is developed. In other applications, only the matrix $U_1 = U(:, 1{:}n)$ is required. Altogether there are six possibilities and the total amount of work required by the SVD algorithm in each case is summarized in the table below. Because of the two possible bidiagonalization schemes, there are two columns of flop counts. If the bidiagonalization is achieved via Algorithm 5.4.2, the Golub-Reinsch (1970) SVD algorithm results, while if $R$-bidiagonalization is invoked we obtain the $R$-SVD algorithm detailed in Chan (1982a). By comparing the entries in this table (which are meant only as approximate estimates of work), we conclude that the $R$-SVD approach is more efficient unless $m \approx n$.

| Required | Golub-Reinsch SVD | $R$-SVD |
|---|---|---|
| $\Sigma$ | $4mn^2 - 4n^3/3$ | $2mn^2 + 2n^3$ |
| $\Sigma, V$ | $4mn^2 + 8n^3$ | $2mn^2 + 11n^3$ |
| $\Sigma, U$ | $4m^2n - 8mn^2$ | $4m^2n + 13n^3$ |
| $\Sigma, U_1$ | $14mn^2 - 2n^3$ | $6mn^2 + 11n^3$ |
| $\Sigma, U, V$ | $4m^2n + 8mn^2 + 9n^3$ | $4m^2n + 22n^3$ |
| $\Sigma, U_1, V$ | $14mn^2 + 8n^3$ | $6mn^2 + 20n^3$ |

**Problems**

**P5.4.1** Suppose $A \in \mathbb{R}^{m\times n}$ with $m < n$. Give an algorithm for computing the factorization

$$U^T AV = [\,B\;O\,]$$

where $B$ is an $m$-by-$m$ upper bidiagonal matrix. (Hint: Obtain the form

$$\begin{bmatrix} \times & \times & 0 & 0 & 0 & 0 \\ 0 & \times & \times & 0 & 0 & 0 \\ 0 & 0 & \times & \times & 0 & 0 \\ 0 & 0 & 0 & \times & \times & 0 \end{bmatrix}.$$

using Householder matrices and then "chase" the $(m, m+1)$ entry up the $(m+1)$st column by applying Givens rotations from the right.)

**P5.4.2** Show how to efficiently bidiagonalize an $n$-by-$n$ upper triangular matrix using Givens rotations.

**P5.4.3** Show how to upper bidiagonalize a tridiagonal matrix $T \in \mathbb{R}^{n \times n}$ using Givens rotations.

**P5.4.4** Let $A \in \mathbb{R}^{m \times n}$ and assume that $0 \neq v$ satisfies $\| Av \|_2 = \sigma_n(A) \| v \|_2$ Let $\Pi$ be a permutation such that if $\Pi^T v = w$, then $|w_n| = \| w \|_\infty$. Show that if $A\Pi = QR$ is the QR factorization of $A\Pi$, then $|r_{nn}| \leq \sqrt{n}\sigma_n(A)$. Thus, there always exists a permutation $\Pi$ such that the QR factorization of $A\Pi$ "displays" near rank deficiency.

**P5.4.5** Let $x, y \in \mathbb{R}^m$ and $Q \in \mathbb{R}^{m \times m}$ be given with $Q$ orthogonal. Show that if

$$Q^T x = \begin{bmatrix} \alpha \\ u \end{bmatrix} \begin{matrix} 1 \\ m-1 \end{matrix} \qquad Q^T y = \begin{bmatrix} \beta \\ v \end{bmatrix} \begin{matrix} 1 \\ m-1 \end{matrix}$$

then $u^T v = x^T y - \alpha\beta$.

**P5.4.6** Let $A = [\, a_1, \ldots, a_n \,] \in \mathbb{R}^{m \times n}$ and $b \in \mathbb{R}^m$ be given. For any subset of $A$'s columns $\{a_{c_1}, \ldots, a_{c_k}\}$ define

$$\text{res}\,[\, a_{c_1}, \ldots, a_{c_k} \,] = \min_{x \in \mathbb{R}^k} \| [\, a_{c_1}, \ldots, a_{c_k} \,] x - b \|_2$$

Describe an alternative pivot selection procedure for Algorithm 5.4.1 such that if $QR = A\Pi = \left[\, a_{c_q}, \ldots, a_{c_n} \,\right]$ in the final factorization, then for $k = 1{:}n$:

$$\text{res}\,[\, a_{c_1}, \ldots, a_{c_k} \,] = \min_{i \geq k} \text{res}[a_{c_1}, \ldots, a_{c_{k-1}}, a_{c_i}]$$

### Notes and References for Sec. 5.4

Aspects of the complete orthogonal decomposition are discussed in

G.H. Golub and V. Pereyra (1976). "Differentiation of Pseduo-Inverses, Separable Nonlinear Least Squares Problems and Other Tales," in *Generalized Inverses and Applications* , ed. M.Z. Nashed, Academic Press, New York, pp. 303–24.

R.J. Hanson and C.L. Lawson (1969). "Extensions and Applications of the Householder Algorithm for Solving Linear Least Square Problems," *Math. Comp. 23,* 787–812.

P.A. Wedin (1973). "On the Almost Rank-Deficient Case of the Least Squares Problem," *BIT 13,* 344–54.

The computation of this decomposition is detailed in

P.A. Businger and G.H. Golub (1969). "Algorithm 358: Singular Value Decomposition of the Complex Matrix," *Comm. ACM 12,* 564–65.

T.F. Chan (1982). "An Improved Algorithm for Computing the Singular Value Decomposition," *ACM Trans. Math. Soft. 8,* 72–83.

G.H. Golub and W. Kahan (1965). "Calculating the Singular Values and Pseudo-Inverse of a Matrix," *SIAM J. Num. Anal. 2,* 205–24.

G.H. Golub and C. Reinsch (1970). "Singular Value Decomposition and Least Squares Solutions," *Numer. Math. 14*, 403–20. See also *HACLA*, pp. 1334–51.

An interesting approach to rank estimation that uses QR with column pivoting and condition estimation is discussed in

T.F. Chan (1987). "Rank Revealing QR Factorizations," *Lin. Alg. and Its Applic. 88/89*, 67–82.

# 5.5 The Rank Deficient LS Problem

If $A$ is rank deficient then there are an infinite number of solutions to the LS problem and we must resort to special techniques. These techniques must address the difficult problem of numerical rank determination.

After some SVD preliminaries, we show how QR with column pivoting can be used to determine a minimizer $x_B$ with the property that $Ax_B$ is a linear combination of $r = \text{rank}(A)$ columns. We then discuss the minimum 2-norm solution that can be obtained from the SVD.

## 5.5.1 The Minimum Norm Solution

Suppose $A \in \mathbb{R}^{m\times n}$ and $\text{rank}(A) = r < n$. The rank deficient LS problem has an infinite number of solutions, for if $x$ is a minimizer and $z \in \text{null}(A)$ then $x + z$ is also a minimizer. The set of all minimizers

$$\mathcal{X} = \{x \in \mathbb{R}^n : \| Ax - b \|_2 = \min \}.$$

is convex for if $x_1, x_2 \in \mathcal{X}$ and $\lambda \in [0, 1]$ then

$$\begin{aligned} \| A(\lambda x_1 + (1-\lambda)x_2) - b \|_2 &\le \lambda\| Ax_1 - b \|_2 + (1-\lambda)\| Ax_2 - b \|_2 \\ &= \min \| Ax - b \|_2. \end{aligned}$$

Thus, $\lambda x_1 + (1-\lambda)x_2 \in \mathcal{X}$. It follows that $\mathcal{X}$ has a unique element having minimum 2-norm and we denote this solution by $x_{LS}$. (Note that in the full rank case, there is only one LS solution and so it must have minimal 2-norm. Thus, we are consistent with the notation in §5.3.)

## 5.5.2 Complete Orthogonal Factorization and $x_{LS}$

Any complete orthogonal factorization can be used to compute $x_{LS}$. In particular, if $Q$ and $Z$ are orthogonal matrices such that

$$Q^TAZ = T = \begin{bmatrix} T_{11} & 0 \\ 0 & 0 \end{bmatrix} \begin{matrix} r \\ m-r \end{matrix}$$
$$\qquad\qquad\quad r \quad n-r$$

then

$$\| Ax - b \|_2^2 = \| (Q^T AZ)Z^T x - Q^T b \|_2^2 = \| T_{11} w - c \|_2^2 + \| d \|_2^2$$

where

$$Z^T x = \begin{bmatrix} w \\ y \end{bmatrix} \begin{matrix} r \\ n-r \end{matrix} \qquad Q^T b = \begin{bmatrix} c \\ d \end{bmatrix} \begin{matrix} r \\ m-r \end{matrix} .$$

Clearly, if $x$ is to minimize the sum of squares, then we must have $w = T^{-1}c$. For $x$ to have minimal 2-norm, $y$ must be zero, and thus,

$$x_{LS} = Z \begin{bmatrix} T_{11}^{-1} c \\ 0 \end{bmatrix}$$

### 5.5.3 The SVD and the LS Problem

Of course, the SVD is a particularly revealing complete orthogonal decomposition. It provides a neat expression for $x_{LS}$ and the norm of the minimum residual $\rho_{LS} = \| Ax_{LS} - b \|_2$.

**Theorem 5.5.1** *Suppose $U^T AV = \Sigma$ is the SVD of $A \in \mathbb{R}^{m \times n}$ with $r =$ rank$(A)$. If $U = [\, u_1, \ldots, u_m \,]$ and $V = [\, v_1, \ldots, v_n \,]$ are column partitionings and $b \in \mathbb{R}^m$, then*

$$x_{LS} = \sum_{i=1}^{r} \frac{u_i^T b}{\sigma_i} v_i \tag{5.5.1}$$

*minimizes $\| Ax - b \|_2$ and has the smallest 2-norm of all minimizers. Moreover*

$$\rho_{LS}^2 = \| Ax_{LS} - b \|_2^2 = \sum_{i=r+1}^{m} (u_i^T b)^2. \tag{5.5.2}$$

**Proof.** For any $x \in \mathbb{R}^n$ we have:

$$\begin{aligned} \| Ax - b \|_2^2 &= \| (U^T AV)(V^T x) - U^T b \|_2^2 = \| \Sigma\alpha - U^T b \|_2^2 \\ &= \sum_{i=1}^{r} (\sigma_i \alpha_i - u_i^T b)^2 + \sum_{i=r+1}^{m} (u_i^T b)^2 \end{aligned}$$

where $\alpha = V^T x$. Clearly, if $x$ solves the LS problem, then $\alpha_i = (u_i^T b/\sigma_i)$ for $i = 1{:}r$. If we set $\alpha(r + 1{:}n) = 0$, then then the resulting $x$ clearly has minimal 2-norm. □

### 5.5.4 The Pseudo-Inverse

Note that if we define the matrix $A^+ \in \mathbb{R}^{n\times m}$ by $A^+ = V\Sigma^+U^T$ where

$$\Sigma^+ = \text{diag}\left(\frac{1}{\sigma_1},\ldots,\frac{1}{\sigma_r},0,\ldots,0\right) \in \mathbb{R}^{n\times m}$$

then $x_{LS} = A^+b$ and $\rho_{LS} = \| (I - AA^+)b \|_2$. $A^+$ is referred to as the *pseudo-inverse* of $A$. It is the unique minimal $F$-norm solution to the problem

$$\min_{X \in \mathbb{R}^{n\times m}} \| AX - I_m \|_F . \tag{5.5.3}$$

If rank$(A) = n$, then $A^+ = (A^TA)^{-1}A^T$, while if $m = n = \text{rank}(A)$, then $A^+ = A^{-1}$. Typically, $A^+$ is defined to be the unique matrix $X \in \mathbb{R}^{n\times m}$ that satisfies the four *Moore-Penrose conditions:*

$$\begin{array}{llll} \text{(i)} & AXA = A & \text{(iii)} & (AX)^T = AX \\ \text{(ii)} & XAX = X & \text{(iv)} & (XA)^T = XA \end{array}$$

These conditions amount to the requirement that $AA^+$ and $A^+A$ be orthogonal projections onto range$(A)$ and range$(A^T)$, respectively. Indeed, $AA^+ = U_1U_1^T$ where $U_1 = U(1{:}m, 1{:}r)$ and $A^+A = V_1V_1^T$ where $V_1 = V(1{:}n, 1{:}r)$.

### 5.5.5 Some Sensitivity Issues

In §5.3 we examined the sensitivity of the full rank LS problem. The behavior of $x_{LS}$ in this situation is summarized in Theorem 5.3.1. If we drop the full rank assumptions then $x_{LS}$ is not even a continuous function of the data and small changes in $A$ and $b$ can induce arbitrarily large changes in $x_{LS} = A^+b$. The easiest way to see this is to consider the behavior of the pseudo inverse. If $A$ and $\delta A$ are in $\mathbb{R}^{m\times n}$, then Wedin (1973) and Stewart (1975a) show that

$$\| (A + \delta A)^+ - A^+ \|_F \le 2\| \delta A \|_F \max\left\{\| A^+ \|_2^2 \;,\; \| (A + \delta A)^+ \|_2^2 \right\}.$$

This inequality is a generalization of Theorem 2.3.4 in which perturbations in the matrix inverse are bounded. However, unlike the square nonsingular case, the upper bound does not necessarily tend to zero as $\delta A$ tends to zero. If

$$A = \begin{bmatrix} 1 & 0 \\ 0 & 0 \\ 0 & 0 \end{bmatrix} \qquad \text{and} \qquad \delta A = \begin{bmatrix} 0 & 0 \\ 0 & \epsilon \\ 0 & 0 \end{bmatrix}$$

then

$$A^+ = \begin{bmatrix} 1 & 0 & 0 \\ 0 & 0 & 0 \end{bmatrix} \qquad \text{and} \qquad (A + \delta A)^+ = \begin{bmatrix} 1 & 0 & 0 \\ 1 & 1/\epsilon & 0 \end{bmatrix}$$

and $\| A^+ - (A + \delta A)^+ \|_2 = 1/\epsilon$. The numerical determination of an LS minimizer in the presence of such discontinuities is a major challenge.

### 5.5.6 QR with Column Pivoting and Basic Solutions

Suppose $A \in \mathbb{R}^{m \times n}$ has rank $r$. QR with column pivoting (Algorithm 5.4.1) produces the factorization $A\Pi = QR$ where

$$R = \begin{bmatrix} R_{11} & R_{12} \\ 0 & 0 \end{bmatrix} \begin{matrix} r \\ m-r \end{matrix} . \qquad \begin{matrix} r & n-r \end{matrix}$$

Given this reduction, the LS problem can be readily solved. Indeed, for any $x \in \mathbb{R}^n$ we have

$$\begin{aligned} \| Ax - b \|_2^2 &= \| (Q^T A \Pi)(\Pi^T x) - (Q^T b) \|_2^2 \\ &= \| R_{11} y - (c - R_{12} z) \|_2^2 + \| d \|_2^2 , \end{aligned}$$

where

$$\Pi^T x = \begin{bmatrix} y \\ z \end{bmatrix} \begin{matrix} r \\ n-r \end{matrix} \quad \text{and} \quad Q^T b = \begin{bmatrix} c \\ d \end{bmatrix} \begin{matrix} r \\ m-r \end{matrix} .$$

Thus, if $x$ is an LS minimizer then we must have

$$x = \Pi \begin{bmatrix} R_{11}^{-1}(c - R_{12} z) \\ z \end{bmatrix} .$$

If $z$ is set to zero in this expression, then we obtain the *basic solution*

$$x_B = \Pi \begin{bmatrix} R_{11}^{-1} c \\ 0 \end{bmatrix} .$$

Notice that $x_B$ has at most $r$ nonzero components and so $Ax_B$ involves a subset of $A$'s columns.

The basic solution is not the mimimal 2-norm solution unless the submatrix $R_{12}$ is zero since

$$\| x_{LS} \|_2 = \min_{z \in \mathbb{R}^{n-r}} \left\| x_B - \Pi \begin{bmatrix} R_{11}^{-1} R_{12} \\ -I_{n-r} \end{bmatrix} z \right\|_2 . \qquad (5.5.4)$$

Indeed, this characterization of $\| x_{LS} \|_2$ can be used to show

$$1 \le \frac{\| x_B \|_2}{\| x_{LS} \|_2} \le \sqrt{1 + \| R_{11}^{-1} R_{12} \|_2^2} . \qquad (5.5.5)$$

See Golub and Pereyra (1976) for details.

### 5.5.7 Numerical Rank Determination with $A\Pi = QR$

If Algorithm 5.4.1 is used to compute $x_B$, then care must be exercised in the determination of rank($A$). In order to appreciate the difficulty of this, suppose

$$fl(H_k \cdots H_1 A \Pi_1 \cdots \Pi_k) = \hat{R}^{(k)} = \begin{bmatrix} \hat{R}_{11}^{(k)} & \hat{R}_{12}^{(k)} \\ 0 & \hat{R}_{22}^{(k)} \end{bmatrix} \begin{matrix} k \\ m-k \end{matrix}$$
$$\qquad\qquad\qquad\qquad\qquad k \qquad n-k$$

is the matrix computed after $k$ steps of the algorithm have been executed in floating point. Suppose rank$(A) = k$. Because of roundoff error, $\hat{R}_{22}^{(k)}$ will not be exactly zero. However, if $\hat{R}_{22}^{(k)}$ is suitably small in norm then it is reasonable to terminate the reduction and declare $A$ to have rank $k$. A typical termination criteria might be

$$\| \hat{R}_{22}^{(k)} \|_2 \le \epsilon_1 \| A \|_2 \qquad (5.5.6)$$

for some small machine-dependent parameter $\epsilon_1$. In view of the roundoff properties associated with Householder matrix computation (cf. §5.1.12), we know that $\hat{R}^{(k)}$ is the exact $R$ factor of a matrix $A + E_k$, where

$$\| E_k \|_2 \le \epsilon_2 \| A \|_2 \qquad \epsilon_2 = O(\mathbf{u}) .$$

Using Corollary 2.3.3 we have

$$\sigma_{k+1}(A + E_k) = \sigma_{k+1}(R^{(k)}) \le \| \hat{R}_{22}^{(k)} \|_2 .$$

Since $\sigma_{k+1}(A) \le \sigma_{k+1}(A + E_k) + \| E_k \|_2$, it follows that

$$\sigma_{k+1}(A) \le (\epsilon_1 + \epsilon_2) \| A \|_2 .$$

In other words, a relative perturbation of $O(\epsilon_1 + \epsilon_2)$ in $A$ can yield a rank-$k$ matrix. With this termination criterion, we conclude that QR with column pivoting "discovers" rank degeneracy if in the course of the reduction $\hat{R}_{22}^{(k)}$ is small for some $k < n$.

Unfortunately, this is not always the case. A matrix can be nearly rank deficient without a single $\hat{R}_{22}^{(k)}$ being particularly small. Thus, QR with column pivoting *by itself* is not entirely reliable as a method for detecting near rank deficiency. However, if a good condition estimator is applied to $R$ it is practically impossible for near rank deficiency to go unnoticed.

**Example 5.5.1** Let $T_n(c)$ be the matrix

$$T_n(c) = \text{diag}(1, s, \ldots, s^{n-1}) \begin{bmatrix} 1 & -c & -c & \cdots & -c \\ 0 & 1 & -c & \cdots & -c \\ & & \ddots & & \vdots & \vdots \\ \vdots & & & & 1 & -c \\ 0 & & \cdots & & & 1 \end{bmatrix}$$

with $c^2 + s^2 = 1$ with $c, s > 0$ (See Lawson and Hanson [SLS, p. 31].) These matrices are unaltered by Algorithm 5.4.1 and thus $\| R_{22}^{(k)} \|_2 \geq s^{n-1}$ for $k = 1{:}n-1$ . This inequality implies (for example) that the matrix $T_{100}(.2)$ has no particularly small trailing principal submatrix since $s^{99} \approx .13$. However, it can be shown that $\sigma_n = O(10^{-8})$.

### 5.5.8 Numerical Rank and the SVD

We now focus our attention on the ability of the SVD to handle rank-deficiency in the presence of roundoff. Our claims apply to either the Golub-Reinsch or the Chan version. Recall that if $A = U\Sigma V^T$ is the SVD of $A$, then

$$x_{LS} = \sum_{i=1}^{r} \frac{u_i^T b}{\sigma_i} v_i . \tag{5.5.7}$$

where $r = \text{rank}(A)$. Denote the computed versions of $U$, $V$, and $\Sigma = \text{diag}(\sigma_i)$ by $\hat{U}$, $\hat{V}$, and $\hat{\Sigma} = \text{diag}(\hat{\sigma}_i)$. Assume that both sequences of singular values range from largest to smallest. It can be shown that

$$\hat{U} = W + \Delta U \qquad W^T W = I_m \qquad \| \Delta U \|_2 \leq \epsilon \tag{5.5.8}$$

$$\hat{V} = Z + \Delta V \qquad Z^T Z = I_n \qquad \| \Delta V \|_2 \leq \epsilon \tag{5.5.9}$$

$$\hat{\Sigma} = W^T(A + \Delta A)Z \qquad \| \Delta A \|_2 \leq \epsilon \| A \|_2 \tag{5.5.10}$$

where $\epsilon$ is a small multiple of $\mathbf{u}$, the machine precision. In plain English, the SVD algorithm computes the singular values of the "nearby" matrix $A + \Delta A$.

Note that $\hat{U}$ and $\hat{V}$ are not necessarily close to their exact counterparts. However, we can show that $\hat{\sigma}_k$ is close to $\sigma_k$. Using (5.5.10) and Corollary 2.3.3 we have

$$\begin{aligned} \sigma_k &= \min_{\text{rank}(B)=k-1} \| A - B \|_2 \\ &= \min_{\text{rank}(B)=k-1} \| (\hat{\Sigma} - B) - W^T(\Delta A)Z \|_2 . \end{aligned}$$

Since $\| W^T(\Delta A)Z \|_2 \leq \epsilon \| A \|_2 = \epsilon\sigma_1$ and

$$\min_{\text{rank}(B)=k-1} \| \hat{\Sigma}_k - B \|_2 = \hat{\sigma}_k$$

it follows that $|\sigma_k - \hat{\sigma}_k| \leq \epsilon\sigma_1$ for $k = 1{:}n$. Thus, if $A$ has rank $r$ then we can expect $n - r$ of the computed singular values to be small. Near rank deficiency in $A$ cannot escape detection when the SVD of $A$ is computed.

**Example 5.5.2** If the Linpack SVD algorithm is applied with machine precision $\mathbf{u} \approx$

$10^{-17}$ to the matrix $T_{100}(.2)$ in Example 5.5.1, then

$$\hat{\sigma}_n = .367805646308792467 \cdot 10^{-8}$$

One approach to estimating $r = \text{rank}(A)$ from the computed singular values is to have a tolerance $\delta > 0$ and a convention that $A$ has "numerical rank" $\hat{r}$ if the $\hat{\sigma}_i$ satisfy

$$\hat{\sigma}_1 \geq \cdots \geq \hat{\sigma}_{\hat{r}} > \delta \geq \hat{\sigma}_{\hat{r}+1} \geq \cdots \geq \hat{\sigma}_n .$$

The tolerance $\delta$ should be consistent with the machine precison, e.g. $\delta = \mathbf{u}\| A \|_\infty$. However, if the general level of relative error in the data is larger than $\mathbf{u}$, then $\delta$ should be correspondingly bigger, e.g., $\delta = 10^{-2}\| A \|_\infty$ if the entries in $A$ are correct to two digits.

If $\hat{r}$ is accepted as the numerical rank then we can regard

$$x_{\hat{r}} = \sum_{i=1}^{\hat{r}} \frac{\hat{u}_i^T b}{\hat{\sigma}_i} \hat{v}_i$$

as an approximation to $x_{LS}$. Since $\| x_{\hat{r}} \|_2 \approx 1/\sigma_{\hat{r}} \leq 1/\delta$ then $\delta$ may also be chosen with the intention of producing an approximate LS solution with suitabily small norm. In §12.1, we discuss more sophisticated methods for doing this.

If $\hat{\sigma}_{\hat{r}} \gg \delta$, then we have reason to be comfortable with $x_{\hat{r}}$ because $A$ can then be unambiguously regarded as a rank$(A_{\hat{r}})$ matrix (modulo $\delta$).

On the other hand, $\{\hat{\sigma}_1, \ldots, \hat{\sigma}_n\}$ might not clearly split into subsets of small and large singular values, making the determination of $\hat{r}$ by this means somewhat arbitrary. This leads to more complicated methods for estimating rank which we now discuss in the context of the LS problem.

For example, suppose $r = n$, and assume for the moment that $\Delta A = 0$ in (5.5.10). Thus $\sigma_i = \hat{\sigma}_i$ for $i = 1{:}n$. Denote the $i$th columns of the matrices $\hat{U}$, $\hat{W}$, $\hat{V}$, and $Z$ by $u_i$, $w_i$, $v_i$, and $z_i$, respectively. Subtracting $x_{\hat{r}}$ from $x_{LS}$ and taking norms we obtain

$$\| x_{\hat{r}} - x_{LS} \|_2 \leq \sum_{i=1}^{\hat{r}} \frac{\| (w_i^T b) z_i - (u_i^T b) v_i \|_2}{\sigma_i} + \sqrt{\sum_{i=\hat{r}+1}^{n} \left( \frac{w_i^T b}{\sigma_i} \right)^2} .$$

From (5.5.8) and (5.5.9) it is easy to verify that

$$\| (w_i^T b) z_i - (u_i^T b) v_i \|_2 \leq 2(1+\epsilon)\epsilon \| b \|_2 \tag{5.5.11}$$

and therefore

$$\| x_{\hat{r}} - x_{LS} \|_2 \leq \frac{\hat{r}}{\sigma_{\hat{r}}} 2(1+\epsilon)\epsilon \| b \|_2 + \sqrt{\sum_{i=\hat{r}+1}^{n} \left( \frac{w_i^T b}{\sigma_i} \right)^2} .$$

The parameter $\hat{r}$ can be determined as that integer which minimizes the upper bound. Notice that the first term in the bound increases with $\hat{r}$, while the second decreases.

On occasions when minimizing the residual is more important than accuracy in the solution, we can determine $\hat{r}$ on the basis of how close we surmise $\| b - Ax_{\hat{r}} \|_2$ is to the true minimum. Paralleling the above analysis, it can be shown that

$$\| b - Ax_{\hat{r}} \|_2 - \| b - Ax_{LS} \|_2 \leq (n - \hat{r})\| b \|_2 + \epsilon \| b \|_2 \left( \hat{r} + \frac{\hat{\sigma}_1}{\hat{\sigma}_{\hat{r}}}(1 + \epsilon) \right).$$

Again $\hat{r}$ could be chosen to minimize the upper bound.

These techniques can be incorporated in a computer program by approximating $\sigma_i$ and $w_i$ by $\hat{\sigma}_i$ and $\hat{u}_i$, respectively. The analysis can also be extended to the case when $\Delta A \neq 0$. See Varah (1973) for details.

### 5.5.9 Some Comparisons

As we mentioned, when solving the LS problem via the SVD, only $\Sigma$ and $V$ have to be computed. The following table compares the efficiency of this approach with the other algorithms that we have presented.

| LS Algorithm | Flop Count |
|---|---|
| Normal Equations | $mn^2 + n^3/3$ |
| Householder Orthogonalization | $2mn^2 - 2n^3/3$ |
| Modified Gram Schmidt | $2mn^2$ |
| Givens Orthogonalization | $3mn^2 - n^3$ |
| Householder Bidiagonalization | $4mn^2 - 4n^3/2$ |
| R-Bidiagonalization | $2mn^2 + 2n^3$ |
| Golub-Reinsch SVD | $4mn^2 + 8n^3$ |
| Chan SVD | $2mn^2 + 11n^3$ |

**Problems**

**P5.5.1** Show that if

$$A = \begin{bmatrix} T & S \\ 0 & 0 \end{bmatrix} \begin{matrix} r \\ m-r \end{matrix} \quad \text{(column blocks } r,\ n-r\text{)}$$

where $r = \text{rank}(A)$ and $T$ is nonsingular, then

$$X = \begin{bmatrix} T^{-1} & 0 \\ 0 & 0 \end{bmatrix} \begin{matrix} r \\ n-r \end{matrix} \quad \text{(column blocks } r,\ m-r\text{)}$$

satisfies $AXA = A$ and $(AX)^T = (AX)$. In this case, we say that $X$ is a (1,3) *pseudo-inverse* of $A$. Show that for general $A$, $x_B = Xb$ where $X$ is a (1,3) pseudo-inverse of $A$.

**P5.5.2** Define $B(\lambda) \in \mathbb{R}^{n \times m}$ by $B(\lambda) = (A^TA + \lambda I)^{-1}A^T$, where $\lambda > 0$. Show

$$\| B(\lambda) - A^+ \|_2 \;=\; \frac{\lambda}{\sigma_r(A)[\sigma_r(A)^2 + \lambda]} \qquad r = \text{rank}(A)$$

and therefore that $B(\lambda) \to A^+$ as $\lambda \to 0$.

**P5.5.3** Consider the rank deficient LS problem

$$\min_{\substack{y \in \mathbb{R}^r \\ z \in \mathbb{R}^{n-r}}} \left\| \begin{bmatrix} R & S \\ 0 & 0 \end{bmatrix} \begin{bmatrix} y \\ z \end{bmatrix} - \begin{bmatrix} c \\ d \end{bmatrix} \right\|_2$$

where $R \in \mathbb{R}^{r \times r}$, $S \in \mathbb{R}^{r \times n-r}$, $y \in \mathbb{R}^r$, and $z \in \mathbb{R}^{n-r}$. Assume that $R$ is upper triangular and nonsingular. Show how to obtain the minimum norm solution to this problem by computing an appropriate QR factorization without pivoting and then solving for the appropriate $y$ and $z$.

**P5.5.4** Show that if $A_k \to A$ and $A_k^+ \to A^+$, then there exists an integer $k_0$ such that rank$(A_k)$ is constant for all $k \geq k_0$.

**P5.5.5** Show that if $A \in \mathbb{R}^{m \times n}$ has rank $n$, then so does $A + E$ if we have the inequality $\| E \|_2 \| A^+ \|_2 < 1$.

### Notes and References for Sec. 5.5

A Fortran code for solving the LS problem via the SVD or QR with column pivoting may be found in Linpack.

The pseudo-inverse literature is vast, as evidenced by the 1,775 references in

M.Z. Nashed (1976). *Generalized Inverses and Applications*, Academic Press, New York.

The differentiation of the pseudo-inverse is further discussed in

C.L. Lawson and R.J. Hanson (1969). "Extensions and Applications of the Householder Algorithm for Solving Linear Least Squares Problems," *Math. Comp. 23*, 787–812.

G.H. Golub and V. Pereyra (1973). "The Differentiation of Pseudo-Inverses and Nonlinear Least Squares Problems Whose Variables Separate," *SIAM J. Num. Anal. 10*, 413–32.

Survey treatments of LS perturbation theory may be found in Lawson and Hanson (SLS, chapters 7–9) as well as in

P.A. Wedin (1973). "Perturbation Theory for Pseudo-Inverses," *BIT 13*, 217–32.

G.W. Stewart (1977a). "On the Perturbation of Pseudo-Inverses, Projections, and Linear Least Squares," *SIAM Review 19*, 634–62.

QR with column pivoting was first discussed in

P.A. Businger and G.H. Golub (1965). "Linear Least Squares Solutions by Householder Transformations," *Numer. Math. 7*, 269–76. See also *HACLA*, pp. 11–18.

Even for full rank problems, column pivoting seems to produce more accurate solutions. The error analysis in the following paper attempts to explain why.

L.S. Jennings and M.R. Osborne (1974). "A Direct Error Analysis for Least Squares," *Numer. Math. 22*, 322–32.

Knowing when to stop in Algorithm 5.4.1 is difficult but important. In questions of rank deficiency, it is helpful to obtain information about the smallest singular value of the upper triangular matrix $R$. This can be done using the techniques of §3.5.4 or those that are discussed in

N. Anderson and I. Karasalo (1975). "On Computing Bounds for the Least Singular Value of a Triangular Matrix," *BIT 15*, 1–4.
I. Karasalo (1974). "A Criterion for Truncation of the QR Decomposition Algorithm for the Singular Linear Least Squares Problem," *BIT 14*, 156–66.

See also Lawson and Hanson (SLS, chapter 6). Various other aspects rank deficiency are discussed in

L.V. Foster (1986). "Rank and Null Space Calculations Using Matrix Decomposition without Column Interchanges," *Lin. Alg. and Its Applic. 74*, 47–71.
P.C. Hansen (1987). "The Truncated SVD as a Method for Regularization," *BIT 27*, 534–553.
G.W. Stewart (1984). "Rank Degeneracy," *SIAM J. Sci. and Stat. Comp. 5*, 403–413.
G.W. Stewart (1987). "Collinearity and Least Squares Regression," *Statistical Science 2*, 68–100.
J.M. Varah (1973). "On the Numerical Solution of Ill-Conditioned Linear Systems with Applications to Ill-Posed Problems," *SIAM J. Num. Anal. 10*, 257–67.

We have more to say on the subject in §12.1 and §12.2.

## 5.6 Weighting and Iterative Improvement

In §3.5, the concepts of scaling and iterative improvement were introduced in the context of square linear systems. Generalizations of these ideas that are applicable to the least squares problem are now offered.

### 5.6.1 Column Weighting

Suppose $G \in \mathbb{R}^{n\times n}$ is nonsingular. A solution to the LS problem

$$\min \| Ax - b \|_2 \qquad A \in \mathbb{R}^{m\times n},\ b \in \mathbb{R}^m \tag{5.6.1}$$

can be obtained by finding the minimum 2-norm solution $y_{LS}$ to

$$\min \| (AG)y - b \|_2 \tag{5.6.2}$$

and then setting $x_G = Gy_{LS}$. If rank$(A) = n$, then $x_G = x_{LS}$. Otherwise, $x_G$ is the minimum $G$-norm solution to (5.6.1), where the $G$-norm is defined by $\| z \|_G = \| G^{-1}z \|_2$.

The choice of $G$ is important. Sometimes its selection can be based on à priori knowledge of the uncertainties in $A$. See Lawson and Hanson

(SLS, pp.185-88). On other occasions, it may be desirable to normalize the columns of $A$ by setting

$$G = G_0 \equiv \text{diag}(1/\| A(:,1) \|_2, \ldots, 1/\| A(:,n) \|_2) .$$

Van der Sluis (1969) has shown that

$$\mu = \min_{G \text{ diagonal}} \kappa_2(AG) \Rightarrow \mu \le \kappa_2(AG_0) \le \sqrt{n}\mu .$$

Since the computed accuracy of $y_{LS}$ depends on $\kappa_2(AG)$, a case can be made for setting $G = G_0$.

We remark that column weighting affects singular values. Consequently, a scheme for determining numerical rank may not return the same estimates when applied to $A$ and $AG$. See Stewart (1984b).

### 5.6.2 Row Weighting

Let $D = \text{diag}(d_1, \ldots, d_m)$ be nonsingular and consider the *weighted least squares problem*

$$\text{minimize } \| D(Ax - b) \|_2 \qquad A \in \mathbb{R}^{m \times n},\ b \in \mathbb{R}^m . \tag{5.6.3}$$

Assume rank$(A) = n$ and that $x_D$ solves (5.6.3). It follows that the solution $x_{LS}$ to (5.6.1) satisfies

$$x_D - x_{LS} = (A^T D^2 A)^{-1} A^T (D^2 - I)(b - Ax_{LS}) . \tag{5.6.4}$$

This shows that row weighting in the LS problem affects the solution. (An important exception occurs when $b \in \text{range}(A)$ for then $x_D = x_{LS}$.)

One way of determining $D$ is to let $d_k$ be some measure of the uncertainty in $b_k$, e.g., the reciprocal of the standard deviation in $b_k$. The tendency is for $r_k = e_k^T(b - Ax_D)$ to be small whenever $d_k$ is large. The precise effect of $d_k$ on $r_k$ can be clarified as follows. Define

$$D(\delta) = \text{diag}(d_1, \ldots, d_{k-1},\ d_k\sqrt{1+\delta}\,, d_{k+1}, \ldots, d_m)$$

where $\delta > -1$. If $x(\delta)$ minimizes $\| D(\delta)(Ax - b) \|_2$ and $r_k(\delta)$ is the $k$-th component of $b - Ax(\delta)$, then it can be shown that

$$r_k(\delta) = \frac{r_k}{1 + \delta d_k^2 e_k^T A (A^T D^2 A)^{-1} A^T e_k} . \tag{5.6.5}$$

This explicit expression shows that $r_k(\delta)$ is a monotone decreasing function of $\delta$. Of course, how $r_k$ changes when all the weights are varied is much more complicated.

Further discussion of row weighting may be found in Lawson and Hanson (SLS, pp. 183–85).

**Example 5.6.1** Suppose

$$A = \begin{bmatrix} 1 & 2 \\ 3 & 4 \\ 5 & 6 \\ 7 & 8 \end{bmatrix} \qquad b = \begin{bmatrix} 1 \\ 0 \\ 0 \\ 0 \end{bmatrix}$$

If $D = I_4$ then $x_D = (\ -1,\ .85\ )^T$ and $r = b - Ax_D = (\ .3,\ -.4,\ -.1,\ .2\ )^T$. On the other hand, if $D = \text{diag}(\ 1000,\ 1,\ 1,\ 1\ )$ then we have $x_D \approx (\ -1.43,\ 1.21\ )^T$ and $r = b - Ax_D = .000428\ -.571428\ -.142853\ .285714]^T$.

## 5.6.3 Generalized Least Squares

In many estimation problems, the vector of observations $b$ is related to $x$ through the equation

$$b = Ax + w \tag{5.6.6}$$

where the *noise vector* $w$ has zero mean and a symmetric positive definite *variance-covariance* matrix $\sigma^2 W$. Assume that $W$ is known and that $W = BB^T$ for some $B \in \mathbb{R}^{m \times m}$. The matrix $B$ might be given or it might be $W$'s Cholesky triangle. In order that all the equations in (5.6.6) contribute equally to the determination of $x$, statisticians frequently solve the LS problem

$$\min \|\ B^{-1}(Ax - b)\ \|_2\ . \tag{5.6.7}$$

An obvious computational approach to this problem is to form $\tilde{A} = B^{-1}A$ and $\tilde{b} = B^{-1}b$ and then apply any of our previous techniques to minimize $\|\ \tilde{A}x - \tilde{b}\ \|_2$. Unfortunately, $x$ will be poorly determined by such a procedure if $B$ is ill-conditioned.

A much more stable way of solving (5.6.7) using orthogonal transformations has been suggested by Paige (1979a, 1979b). It is based on the idea that (5.6.7) is equivalent to the *generalized least squares* problem,

$$\min_{b = Ax + Bv} \ v^T v\ . \tag{5.6.8}$$

Notice that this problem is defined even if $A$ and $B$ are rank deficient. Although Paige's technique can be applied when this is the case, we shall describe it under the assumption that both these matrices have full rank.

The first step is to compute the QR factorization of $A$:

$$Q^T A = \begin{bmatrix} R_1 \\ 0 \end{bmatrix} \qquad Q = [\ \underset{n}{Q_1} \quad \underset{m-n}{Q_2}\ ]$$

An orthogonal matrix $Z \in \mathbb{R}^{m \times m}$ is then determined so that

$$Q_2^T B Z = \underset{\quad n \quad m-n}{[\ 0 \quad S\ ]} \qquad Z = \underset{\quad n \quad m-n}{[\ Z_1 \quad Z_2\ ]}$$

where $S$ is upper triangular. With the use of these orthogonal matrices the constraint in (5.6.8) transforms to

$$\begin{bmatrix} Q_1^T b \\ Q_2^T b \end{bmatrix} = \begin{bmatrix} R_1 \\ 0 \end{bmatrix} x + \begin{bmatrix} Q_1^T B Z_1 & Q_1^T B Z_2 \\ 0 & S \end{bmatrix} \begin{bmatrix} Z_1^T v \\ Z_2^T v \end{bmatrix}.$$

Notice that the "bottom half" of this equation determines $v$,

$$Su = Q_2^T b \qquad v = Z_2 u, \tag{5.6.9}$$

while the "top half" prescribes $x$:

$$R_1 x = Q_1^T b - (Q_1^T B Z_1 Z_1^T + Q_1^T B Z_2 Z_2^T) v = Q_1^T b - Q_1^T B Z_2 u. \tag{5.6.10}$$

The attractiveness of this method is that all potential ill-conditioning is concentrated in triangular systems (5.6.9) and (5.6.10). Moreover, Paige (1979b) has shown that the above procedure is numerically stable, something that is not true of any method that explicitly forms $B^{-1}A$.

### 5.6.4 Iterative Improvement

A technique for refining an appoximate LS solution has been analyzed by Björck (1967, 1968). It is based on the idea that if

$$\begin{bmatrix} I_m & A \\ A^T & 0 \end{bmatrix} \begin{bmatrix} r \\ x \end{bmatrix} = \begin{bmatrix} b \\ 0 \end{bmatrix} \qquad A \in \mathbb{R}^{m \times n},\ b \in \mathbb{R}^m \tag{5.6.11}$$

then $\| b - Ax \|_2 = \min$. This follows because $r + Ax = b$ and $A^T r = 0$ imply $A^T A x = A^T b$. The above augmented system is nonsingular if $\text{rank}(A) = n$, which we hereafter assume.

By casting the LS problem in the form of a square linear system, the iterative improvement scheme (3.5.5) can be applied:

$r^{(0)} = 0$; $x^{(0)} = 0$
**for** $k = 0, 1,$

$$\begin{bmatrix} f^{(k)} \\ g^{(k)} \end{bmatrix} = \begin{bmatrix} b \\ 0 \end{bmatrix} - \begin{bmatrix} I & A \\ A^T & 0 \end{bmatrix} \begin{bmatrix} r^{(k)} \\ x^{(k)} \end{bmatrix}$$

$$\begin{bmatrix} I & A \\ A^T & 0 \end{bmatrix} \begin{bmatrix} p^{(k)} \\ z^{(k)} \end{bmatrix} = \begin{bmatrix} f^{(k)} \\ g^{(k)} \end{bmatrix}$$

$$\begin{bmatrix} r^{(k+1)} \\ x^{(k+1)} \end{bmatrix} = \begin{bmatrix} r^{(k)} \\ x^{(k)} \end{bmatrix} + \begin{bmatrix} p^{(k)} \\ z^{(k)} \end{bmatrix}$$

**end**

The residuals $f^{(k)}$ and $g^{(k)}$ must be computed in higher precision and an original copy of $A$ must be around for this purpose.

If the QR factorization of $A$ is available, then the solution of the augmented system is readily obtained. In particular, if $A = QR$ and $R_1 = R(1{:}n, 1{:}n)$, then a system of the form

$$\begin{bmatrix} I & A \\ A^T & 0 \end{bmatrix} \begin{bmatrix} p \\ z \end{bmatrix} = \begin{bmatrix} f \\ g \end{bmatrix}$$

transforms to

$$\begin{bmatrix} I_n & 0 & R_1 \\ 0 & I_{m-n} & 0 \\ R_1^T & 0 & 0 \end{bmatrix} \begin{bmatrix} h \\ f_2 \\ z \end{bmatrix} = \begin{bmatrix} f_1 \\ f_2 \\ g \end{bmatrix}$$

where

$$Q^T f = \begin{bmatrix} f_1 \\ f_2 \end{bmatrix} \begin{matrix} n \\ m-n \end{matrix} \qquad Q^T p = \begin{bmatrix} h \\ f_2 \end{bmatrix} \begin{matrix} n \\ m-n \end{matrix} \; .$$

Thus, $p$ and $z$ can be determined by solving the triangular systems $R_1^T h = g$ and $R_1 z = f_1 - h$ and setting $p = Q \begin{bmatrix} h \\ f_2 \end{bmatrix}$. Assuming that $Q$ is stored in factored form, each iteration requires $8mn - 2n^2$ flops.

The key to the iteration's success is that both the LS residual and solution are updated—not just the solution. Björck (1968) shows that if $\kappa_2(A) \approx \beta^q$ and $t$-digit , $\beta$-base arithmetic used, then $x^{(k)}$ has approximately $k(t-q)$ correct base $\beta$ digits, provided the residuals are computed in double precision. Notice that it is $\kappa_2(A)$, not $\kappa_2(A)^2$, that appears in this heuristic.

### Problems

**P5.6.1** Verify (5.6.4).

**P5.6.2** Let $A \in \mathbb{R}^{m \times n}$ have full rank and define the diagonal matrix

$$\Delta = \text{diag}(\underbrace{1,\ldots,1}_{k-1}, (1+\delta), \underbrace{1,\ldots,1}_{m-k})$$

for $\delta > -1$. Denote the LS solution to min $\| \Delta(Ax - b) \|_2$ by $x(\delta)$ and its residual by $r(\delta) = b - Ax(\delta)$. (a) Show

$$r(\delta) \;=\; \left(I \;-\; \delta\frac{A(A^TA)^{-1}A^Te_ke_k^T}{1+\delta e_k^TA(A^TA)^{-1}A^Te_k}\right) r(0)\,.$$

(b) Letting $r_k(\delta)$ stand for the $k$th component of $r(\delta)$, show

$$r_k(\delta) \;=\; \frac{r_k(0)}{1+\delta e_k^TA(A^TA)^{-1}A^Te_k}$$

(c) Use (b) to verify (5.6.5).

**P5.6.3** Show how the SVD can be used to solve the generalized LS problem when the matrices $A$ and $B$ in (5.6.8) are rank deficient.

**P5.6.4** Let $A \in \mathbb{R}^{m \times n}$ have rank $n$ and for $\alpha \geq 0$ define

$$M(\alpha) \;=\; \begin{bmatrix} \alpha I_m & A \\ A^T & 0 \end{bmatrix}.$$

Show that

$$\sigma_{m+n}(M(\alpha)) \;=\; \min\left\{\alpha\,,\; -\frac{\alpha}{2} + \sqrt{\sigma_n(A)^2 + \left(\frac{\alpha}{2}\right)^2}\right\}$$

and determine the value of $\alpha$ that minimizes $\kappa_2(M(\alpha))$.

**P5.6.5** Another iterative improvement method for LS problems is the following:

$x^{(0)} = 0$
**for** $k = 0, 1, \ldots$
  $r^{(k)} = b - Ax^{(k)}$  (double precision)
  $\| Az^{(k)} - r^{(k)} \|_2 = \min$
  $x^{(k+1)} = x^{(k)} + z^{(k)}$
**end**

(a) Assuming that the $QR$ factorization of $A$ is available, how many flops per iteration are required? (b) Show that the above iteration results by setting $g^{(k)} = 0$ in the iterative improvement scheme given in §5.6.4.

### Notes and References for Sec. 5.6

Row and column weighting in the LS problem is discussed in Lawson and Hanson (SLS, pp. 180-88). The various effects of scaling are discussed in

A. van der Sluis (1969). "Condition Numbers and Equilibration of Matrices," *Numer. Math. 14*, 14–23.

G.W. Stewart (1984b). "On the Asymptotic Behavior of Scaled Singular Value and QR Decompositions," *Math. Comp. 43*, 483–490.

The theoretical and computational aspects of the generalized least squares problem appear in

S. Kourouklis and C.C. Paige (1981). "A Constrained Least Squares Approach to the General Gauss-Markov Linear Model," *J. Amer. Stat. Assoc. 76,* 620–25.
C.C. Paige (1979a). "Computer Solution and Perturbation Analysis of Generalized Least Squares Problems," *Math. Comp. 33,* 171–84.
C.C. Paige (1979b). "Fast Numerically Stable Computations for Generalized Linear Least Squares Problems," *SIAM J. Num. Anal. 16,* 165–71.
C.C. Paige (1985). "The General Limit Model and the Generalized Singular Value Decomposition," *Lin. Alg. and Its Applic. 70,* 269–284.

Iterative improvement in the LS problem is discussed in

Å. Björck (1967a). "Iterative Refinement of Linear Least Squares Solutions I," *BIT 7,* 257–78.
Å. Björck (1968). "Iterative Refinement of Linear Least Squares Solutions II," *BIT 8,* 8–30.
Å. Björck (1987). "Stability Analysis of the Method of Seminormal Equations for Linear Least Squares Problems," *Linear Alg. and Its Applic. 88/89,* 31–48.
Å. Björck and G.H. Golub (1967). "Iterative Refinement of Linear Least Squares Solutions by Householder Transformation," *BIT 7,* 322–37.
G.H. Golub and J.H. Wilkinson (1966). "Note on Iterative Refinement of Least Squares Solutions," *Numer. Math. 9,* 139–48.

## 5.7 Square and Underdetermined Systems

The orthogonalization methods developed in this chapter can be applied to square systems and also to systems in which there are fewer equations than unknowns. In this brief section we discuss some of the various possibilities.

### 5.7.1 Using QR and SVD to Solve Square Systems

The least squares solvers based on the QR factorization and the SVD can be used to solve square linear systems: just set $m = n$. However, from the flop point of view, Gaussian elimination is the cheapest way to solve a square linear system as evidenced by the following table:

| Method | Flops |
|---|---|
| Gaussian Elimination | $2n^3/3$ |
| Householder Orthogonalization | $4n^3/3$ |
| Modified Gram-Schmidt | $2n^3$ |
| Bidiagonalization | $8n^3/3$ |
| Singular Value Decomposition | $12n^3$ |

Nevertheless, there are three reasons why orthogonalization methods might be considered:

- The flop counts tend to exaggerate the Gaussian elimination advantage. When memory traffic and vectorization overheads are considered, the QR approach is comparable in efficiency.

- The orthogonalization methods have guaranteed stability; there is no "growth factor" to worry about as in Gaussian elimination.

- In cases of ill-conditioning, the orthogonal methods give an added measure of reliability. $QR$ with condition estimation is very dependable and, of course, SVD is unsurpassed when it comes to producing a meaningful solution to a nearly singular system.

We are not expressing a strong preference for orthogonalization methods but merely suggesting viable alternatives to Gaussian elimination.

We also mention that the SVD entry in Table 5.7.1 assumes the availability of $b$ at the time of decomposition. Otherwise, $20n^3$ flops are required because it then becomes necessary to accumulate the $U$ matrix.

### 5.7.2 Underdetermined Systems

We say that a linear system

$$Ax = b \qquad A \in \mathbb{R}^{m \times n},\ b \in \mathbb{R}^m \tag{5.7.1}$$

is *underdetermined* whenever $m < n$. Notice that such a system either has no solution or has an infinity of solutions. In the second case, it is important to distinguish between algorithms that find the minimum 2-norm solution and those that do not. The first algorithm we present is in the latter category.

**Algorithm 5.7.1** Given $A \in \mathbb{R}^{m \times n}$ with $\text{rank}(A) = m$ and $b \in \mathbb{R}^m$, the following algorithm finds an $x \in \mathbb{R}^n$ such that $Ax = b$.

$Q^T A\Pi = R$ (QR with column pivoting.)
Solve $R(1{:}m, 1{:}m)z = Q^T b$.
Set $x = \Pi \begin{bmatrix} z \\ 0 \end{bmatrix}$.

This algorithm requires $2m^2n - m^3/3$ flops.

If we work with $A^T$ then we can obtain the minimum norm solution.

**Algorithm 5.7.2** Given $A \in \mathbb{R}^{m \times n}$ with $\text{rank}(A) = m$ and $b \in \mathbb{R}^m$, the following algorithm finds the minimal 2-norm solution to $Ax = b$.

$A^T = QR$ (QR factorization)
Solve $R(1{:}m, 1{:}m)^T z = b$.
$x = Q(:, 1{:}m)z$

This algorithm requires at most $2m^2n - 2m^3/3$

The SVD can also be used to compute the minimal norm solution of an underdetermined $Ax = b$ problem. If

$$A = \sum_{i=1}^{r} \sigma_i u_i v_i^T \qquad r = \text{rank}(A)$$

is $A$'s singular value expansion, then

$$x = \sum_{i=1}^{r} \frac{u_i^T b}{\sigma_i} v_i \ .$$

As in the least squares problem, the SVD approach is desirable whenever $A$ is nearly rank deficient.

### 5.7.3 Perturbed Underdetermined Systems

We conclude with a perturbation result for full-rank underdetermined systems.

**Theorem 5.7.1** *Suppose* $\text{rank}(A) = m \le n$ *and that* $A \in \mathbb{R}^{m\times n}$ , $\delta A \in \mathbb{R}^{m\times n}$, $0 \ne b \in \mathbb{R}^m$, *and* $\delta b \in \mathbb{R}^m$ *satisfy*

$$\epsilon = \max\{\epsilon_A, \epsilon_b\} < \sigma_m(A),$$

*where* $\epsilon_A = \| \delta A \|_2 / \| A \|_2$ *and* $\epsilon_b = \| \delta b \|_2 / \| b \|_2$. *If* $x$ *and* $\hat{x}$ *are minimum norm solutions that satisfy*

$$Ax = b \qquad\qquad (A + \delta A)\hat{x} = b + \delta b$$

*then*

$$\frac{\| \hat{x} - x \|_2}{\| x \|_2} \le \kappa_2(A) \left(\epsilon_A \min\{2, n - m + 1\} + \epsilon_b\right) + O(\epsilon^2).$$

**Proof.** Let $E$ and $f$ be defined by $\delta A/\epsilon$ and $\delta b/\epsilon$. Note that $\text{rank}(A + tE) = m$ for all $0 < t < \epsilon$ and that

$$x(t) = (A + tE)^T \left((A + tE)(A + tE)^T\right)^{-1} (b + tf)$$

satisfies $(A + tE)x(t) = b + tf$. By differentiating this expression with respect to $t$ and setting $t = 0$ in the result we obtain

$$\dot{x}(0) = \left(I - A^T(AA^T)^{-1}A\right) E^T(AA^T)^{-1}b + A^T(AA^T)^{-1}(f - Ex).$$

Since

$$\| x \|_2 = \| A^T(AA^T)^{-1}b \|_2 \geq \sigma_m(A)\| (AA^T)^{-1}b \|_2,$$

$$\| I - A^T(AA^T)^{-1}A \|_2 = \min(1, n - m),$$

and

$$\frac{\| f \|_2}{\| x \|_2} \leq \frac{\| f \|_2 \| A \|_2}{\| b \|_2},$$

we have

$$\frac{\| \hat{x} - x \|_2}{\| x \|_2} = \frac{x(\epsilon) - x(0)}{\| x(0) \|_2} = \epsilon \frac{\| \dot{x}(0) \|_2}{\| x \|_2} + O(\epsilon^2)$$

$$\leq \epsilon \min(1, n - m) \left\{ \frac{\| E \|_2}{\| A \|_2} + \frac{\| f \|_2}{\| b \|_2} + \frac{\| E \|_2}{\| A \|_2} \right\} \kappa_2(A) + O(\epsilon^2)$$

from which the theorem follows. □

Note that there is no $\kappa_2(A)^2$ factor as in the case of overdetermined systems.

### Problems

**P5.7.1** Derive the above expression for $\dot{x}(0)$.

**P5.7.2** Find the minimal norm solution to the system $Ax = b$ where $A = [\,1\ 2\ 3\,]$ and $b = 1$.

### Notes and References for Sec. 5.7

Interesting aspects concerning singular systems are discussed in

T.F. Chan (1984). "Deflated Decomposition Solutions of Nearly Singular Systems," *SIAM J. Numer. Anal. 21*, 738–754.

G.H. Golub and C.D. Meyer (1986). "Using the QR Factorization and Group Inversion to Compute, Differentiate, and estimate the Sensitivity of Stationary Probabilities for Markov Chains," *SIAM J. Alg. and Dis. Methods*, 7, 273–281.

Papers on underdetermined systems include

M. Arioli and A. Laratta (1985). "Error Analysis of an Algorithm for Solving an Underdetermined System," *Numer. Math. 46*, 255–268.

R.E. Cline and R.J. Plemmons (1976). "$L_2$-Solutions to Underdetermined Linear Systems," *SIAM Review 18*, 92-106.

# Chapter 6

# Parallel Matrix Computations



The parallel matrix computation area is an area of intense research. Since the field is so new, the literature is redundant and it is dominated by the "case study." We circumvent the difficulties of machine dependence by focussing on a sufficiently high level of algorithmic detail. The distributed and shared memory paradigms are considered. Hardware details are kept to a minimum and we gloss over several important language and system problems in order to get at the central issues. As a consequence, the chapter is thin on specific advice–you will not find firm recommendations on how to solve matrix problem X on parallel machine Y.

The gaxpy operation is used to introduce key ideas in §6.1 and §6.2. In §6.3 parallel matrix multiplication for both distributed and shared memory systems is discussed. The parallel computation of various matrix factorizations is then considered in §6.4, §6.5 and §6.6. The emphasis is on the Cholesky factorization but pointers on how to proceed with LU and QR are given throughout.

# 6.1 Distributed Memory Gaxpy

In this section a notation is developed for the expression of distributed memory matrix algorithms. The gaxpy operation,

$$z = y + Ax, \qquad A \in \mathbb{R}^{n \times n}, x, y, z \in \mathbb{R}^n \tag{6.1.1}$$

is used to illustrate the notation and to identify the key issues. More complicated computations are considered in §6.3-§6.5.

## 6.1.1 Distributed Memory Systems

In a *distributed memory* multiprocessor we have a network of processors each with its own local memory and program. The processors communicate with one another by sending and receiving messages. Important factors in the design of an effective distributed memory algorithm include (a) the number of processors and the capacity of the local memories, (b) how the processors are interconnected, and (c) the speed of computation relative to the speed of interprocessor communication. Our goal is to show, quite informally, how these factors interact and affect algorithm design.

Within the distributed memory framework there are two models of computation to consider: the fully synchronous "systolic" model and the asynchronous message-passing model. In the systolic[1] setting, the processors pace their computations and communications according to the tick of a global clock. The fully synchronous, lockstep nature of systolic computation makes it relatively easy to visualize the flow of computation. For this reason, it is an excellent setting for introducing the reader to parallel matrix computations.

After developing a systolic solution to (6.1.1) we reconsider the same problem in the message-passing context. In this style of distributed memory computation, the processors coordinate their activity on the basis of the received messages. There is no global clock.

## 6.1.2 Processor Networks

In a distributed memory multiprocessor each processor has a *local memory* and executes its own *node program* which leads to changes in the values of the *local variables.* It is our custom throughout this chapter to let $p$ designate the number of processors and refer to the pattern of their interconnection as the *network topology.* Popular topologies include the ring, the mesh, and the torus, depicted in Figures 6.1.1-6.1.3. Each of the distibuted memory algorithms that we present in this chapter is tailored to one

[1] We use the word "systolic" when referring to a distributed memory computation that is highly regimented and synchronous. We capture, some, but not all of the aspects of systolic array computing as formulated by Kung (1982).

of these three topologies. However, there are several other very important interconnection schemes including the hypercube (for its generality, optimality, and commercial availability) and the tree (for its handling of divide and conquer procedures). See Ortega and Voigt (1985) for a discussion of the possibilities.

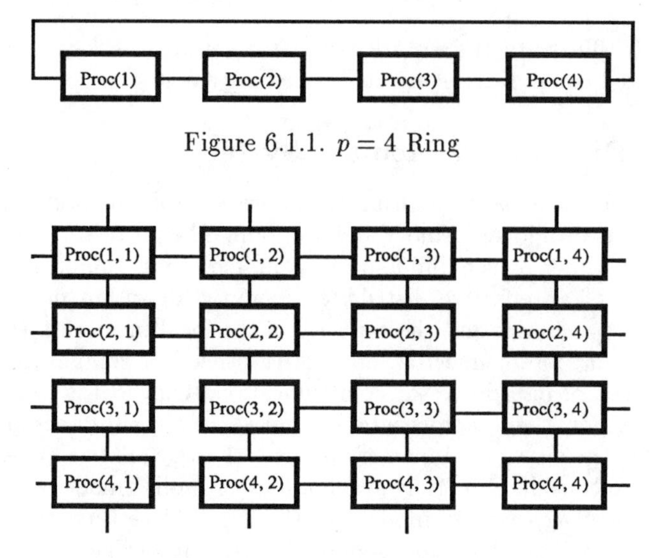


Figure 6.1.1. $p = 4$ Ring

Figure 6.1.2. 4-by-4 Mesh

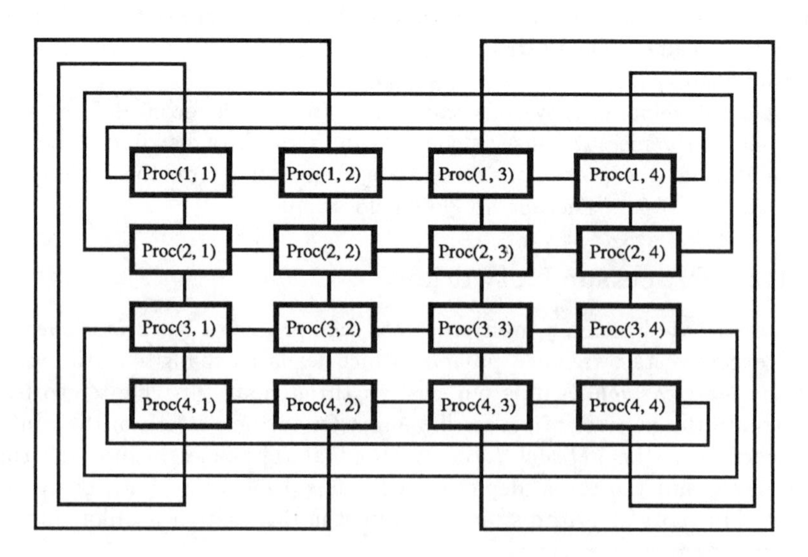


Figure 6.1.3. 4-by-4 Torus

Note that in a torus, each column and row of processors make up a ring.

### 6.1.3 Naming Neighbors

In a multiprocessor each processor has a numerical *identification* or *id*. For rings we use the notation Proc($\mu$) to designate the $\mu$th processor. In two dimensional networks a double subscript is used: Proc($i,j$).

We say that Proc($\lambda$) is a *neighbor* of Proc($\mu$) if there is a direct physical connection between them. Thus, in a 4-processor ring, Proc(2) and Proc(4) are neighbors of Proc(3). In a 4-by-4 torus, Proc(1,3) has neighbors Proc(1,2), Proc(1,4), Proc(2,3) and Proc(4,3).

A processor's neighbors usually figure quite heavily in its node program and so we adopt a handy, non-numeric notation for their specification. In a ring *left* and *right* refer to the indices of the left and right neighbors. Thus, in a 4-processor ring, Proc(4) has local variables *left* and *right* with the values 3 and 1 respectively.

For two-dimensional networks, each individual processor Proc($i,j$) has local variables *north*, *east*, *south*, and *west* that contain the *id*'s of its four neighbors, i.e., the values $(i-1,j)$, $(i,j+1)$, $(i+1,j)$, and $(i,j-1)$ respectively. For a mesh, these may refer to nonexistent processors if Proc($i,j$) is on the edge, but we gloss over that detail. For a torus, "wrap-around" arithmetic must be used to ascertain the neighbors.

### 6.1.4 Initialization and Termination

Like programs in the single processor setting, node programs begin with initializations. For example, a matrix, its dimensions, and a set of block parameters may have to be set up before the start of serious computation. However, in a distributed memory multiprocessor information about the network must also be made available. Indeed, the interesting part of a node program may refer to

- The total number of processors $p$ in the network.
- The processor *id*, i.e., the index or indices that name the executing processor.
- The appropriate values for *left* and *right* if the processor is part of a ring and the appropriate values for *north*, *east*, *south*, and *west* if it is part of a torus or mesh.

For expository purposes we lump all of these initializations into a single **loc.init** (local initialization) command. To illustrate, if the $\mu$th processor in a $p$-processor ring is to house the submatrix $A(1{:}m, , \mu{:}p{:}n)$ in a local array $A_{loc}$, then the node program would begin with a statement of the form

$$\textbf{loc.init}[\,p,\ \mu = my.id,\ left,\ right,\ m,\ n,\ A.loc = A(\mu{:}p{:}n)\,]\,.$$

All node programs that we develop begin in this fashion. The "$\mu = my.id$" expression is merely a way to indicate that the local variable $\mu$ houses the numerical identification of the processor. As we have seen in the case of the mesh and the torus, this *id* might be a 2-tuple: $(\alpha, \beta) = my.id$.

The **loc.init** statement is a handy way to group together all the distracting aspects of the node program, thereby allowing us to focus on higher level algorithmic issues.

All node programs terminate with a **quit** statement. This notation stresses the fact that a processor becomes inactive after the execution of its own node program is completed. Other processors may still be executing.

### 6.1.5 Communication

In a distributed memory environment, the processors must communicate among one another. We need a language to describe this activity and to this end we adopt a simple send/receive notation:

**send**( {*matrix*} , {destination} )

**recv**( {*matrix*} , {origin} )

Thus, if Proc$(i, j)$ executes the instruction **send**$(A, south)$ then a copy of the locally stored matrix $A$ is sent to Proc$(i+1, j)$ . Of course, a matrix can be 1-by-1 or $n$-by-1 in which case only a scalar or a vector is sent. We assume that whenever a matrix is named in a **send** or **recv** statement that its elements are contiguous in local memory. This may not be necessary in practice but it forces us to think in a healthy way about data structure. Thus, **send**$(A(i,:), left)$ is illegal if $A$ is a local matrix stored in column major order.

The node programs that collectively define the distributed computation look like ordinary, sequential programs except that they include communication primitives such as **send** and **recv** whose semantics we now discuss. If Proc$(\mu)$ executes the instruction **send**$(V, \lambda)$ then a copy of the local matrix $V$ is sent to Proc$(\lambda)$. The execution of the node program in Proc$(\mu)$ resumes immediately. We assume (as mentioned above) that the elements of the the matrix $V$ are contiguous in the local memory of Proc$(\mu)$. We also assume that it is legal for a processor to send a message to itself.

If Proc$(\mu)$ executes the instruction **recv**$(V, \lambda)$ then the execution of the node program is suspended until a message is received from Proc$(\lambda)$. Once received, the message is placed in a local matrix $V$ and Proc$(\mu)$ resumes execution of its node program. It is assumed that the received $V$ has the same data structure as it does in Proc$(\lambda)$.

In the design of a distributed memory program, care are must be taken to ensure that requested data is actually sent. Otherwise, a node program may be "hung up" waiting for a nonexistent message.

Although our send/receive notation is perfectly adequate for exposition, it does suppress a number of important details that are relevant to performance:

- The length of a message in bytes or some other unit. Longer messages take longer to process. Our notation does not capture this important aspect of distributed memory computing.
- The path of the message through the network. When a message is sent between two processors it may follow a circuitous path if the routing software deems it appropriate. This makes it difficult to predict communication overheads.
- The buffering of received messages. Because nodes have finite memory, there may not be enough local space to house an onslaught of received messages. We ignore this possibility.

### 6.1.6 Some Distributed Data Structures

Suppose $x \in \mathbb{R}^n$ and that we are to distribute $x$ among the local memories of a $p$-processor network. Two "canonical" approaches to this problem are store-by-row and store-by-column. Assume for the time being that $n = rp$. In store-by-column we regard the vector $x$ as an $r$-by-$p$ matrix,

$$x_{r\times p} = \left[\; x(1{:}r) \quad x(r+1{:}2r) \quad \cdots \quad x((p-1)r+1{:}n) \;\right],$$

and store each column in a processor, i.e., $x((\mu-1)r+1{:}\mu r) \in \text{Proc}(\mu)$. (In this context "$\in$" means "is stored in.") Note that each processor houses a contiguous portion of $x$.

In the store-by-row scheme we regard $x$ as a $p$-by-$r$ matrix

$$x_{p\times r} = \left[\; x(1{:}p) \quad x(p+1{:}2p) \quad \cdots \quad x((r-1)p+1{:}n) \;\right],$$

and store each *row* in a processor, i.e., $x(\mu{:}p{:}n) \in \text{Proc}(\mu)$. Store-by-rows is sometimes referred to as the *wrap* method of distributing a vector because the components of $x$ can be thought of as cards in a deck that are "dealt" to the processors in wrap-around fashion.

If $n$ is not an exact multiple of $p$ then these ideas go through with minor modification. Consider store-by-column with $n = 14$ and $p = 4$:

$$x^T = [\underbrace{x_1\, x_2\, x_3\, x_4}_{\text{Proc}(1)} \mid \underbrace{x_5\, x_6\, x_7\, x_8}_{\text{Proc}(2)} \mid \underbrace{x_9\, x_{10}\, x_{11}}_{\text{Proc}(3)} \mid \underbrace{x_{12}\, x_{13}\, x_{14}}_{\text{Proc}(4)}].$$

In general, if $n = pr + q$ with $0 \le q < p$ then Proc(1),...,Proc($q$) can each house $r+1$ components and Proc($q+1$),..., Proc($p$) can house $r$ components. In store-by-row we simply let Proc($\mu$) house $x(\mu{:}p{:}n)$.

Next consider the distribution of an $n$-by-$n$ matrix $A$ over a ring network. Again, assume that $n = rp$ for clarity. There are four obvious possibilities:

| Orientation | Style | What is in Proc($\mu$) |
|---|---|---|
| Column | Contiguous | $A(:, (\mu - 1)r + 1:\mu r)$ |
| Column | Wrap | $A(:, \mu:p:n)$ |
| Row | Contiguous | $A((\mu - 1)r + 1:\mu r, :)$ |
| Row | Wrap | $A(\mu:p:n, :)$ |

The distribution of a matrix over a mesh or torus can be accomplished in a number of ways. Given a $p_1$-by-$p_2$ network the simplest distributed data structure is to regard $A = (A_{ij})$ as a $p_1$-by-$p_2$ block matrix and store $A_{ij}$ in Proc$(i, j)$. Two dimensional wrap mappings are also possible.

### 6.1.7 The Systolic Model

We now consider systolic computation. In a systolic network the processors operate in a completely synchronous fashion. During a "tick" of a global clock, each processor communicates *with its neighbors* and then performs some local computation.

In this simple, highly structured model of distributed memory computing, we assume that each processor executes a node program of the following form:

**for** $t = 1:t_{final}$
    Send data to neighbors (if necessary).
    Receive data from neighbors (if necessary). (6.1.2)
    Perform calculations on local data (if necessary).
**end**
**quit**

Think of $t$ as a global counter that paces the computation across the network. First, all the processors do their $t = 1$ duties. Then all the processors perform their $t = 2$ duties and so on. We have arranged things so that across the network, first all the messages are sent, then they are all received, and then the local computations are performed. This kind of regimentation makes it easier to rigorously establish the correctness of a distributed computation, but it may actually make the development of that algorithm more difficult. This is dramatized in §6.5.

To illustrate our notation, let us consider the gaxpy computation (6.1.1) on a $p$-processor systolic ring. Assume that $n = rp$ for clarity and partition the problem as follows:

$$\begin{bmatrix} z_1 \\ \vdots \\ z_p \end{bmatrix} = \begin{bmatrix} y_1 \\ \vdots \\ y_p \end{bmatrix} + \begin{bmatrix} A_{11} & \cdots & A_{1p} \\ \vdots & & \vdots \\ A_{p1} & \cdots & A_{pp} \end{bmatrix} \begin{bmatrix} x_1 \\ \vdots \\ x_p \end{bmatrix}. \qquad (6.1.3)$$

Here, $A_{ij} \in \mathbb{R}^{r \times r}$ and $x_i, y_i, z_i \in \mathbb{R}^r$. Let us assume that Proc($\mu$) houses the $\mu$th block row of $A$, $x_\mu$, and $y_\mu$ at the start of the computation and that its mission is to overwrite $y_\mu$ with $z_\mu$. Notice that the computation of

$$z_\mu = y_\mu + \sum_{\tau=1}^{p} A_{\mu\tau} x_\tau$$

involves local data ($A_{\mu\tau}, y_\mu, x_\mu$) and nonlocal data (the other $x_\tau$). A way to make $x$ available throughout the ring is to circulate the $x_\tau$ in merry-go-round fashion. The $p = 3$ case is illustrative. Here is "where" the $x_\tau$ are located during each step of the procedure:

| step | Proc(1) | Proc(2) | Proc(3) |
|---|---|---|---|
| 1 | $x_3$ | $x_1$ | $x_2$ |
| 2 | $x_2$ | $x_3$ | $x_1$ |
| 3 | $x_1$ | $x_2$ | $x_3$ |

This method of circulating the $x_\tau$ is particularly attractive in a ring since all communications are between neighbors. Here is a precise formulation of the procedure.

**Algorithm 6.1.1** Suppose $A \in \mathbb{R}^{n \times n}$, $x \in \mathbb{R}^n$ and $y \in \mathbb{R}^n$ are given and that $z = y + Ax$. If each processor in a $p$-processor systolic ring executes the following node program and $n = rp$, then upon completion Proc($\mu$) houses $z(1 + (\mu - 1)r{:}\mu r)$ in $y_{loc}$.

**loc.init**[ $p$, $\mu = my.id$, $left$, $right$, $n$, $r = n/p$, $row = 1 + (\mu - 1)r{:}\mu r$,
  $A_{loc} = A(row, :)$, $x_{loc} = x(row)$, $y_{loc} = y(row)$ ]
**for** $t = 1{:}p$
  **send**($x_{loc}, right$); **recv**($x_{loc}, left$)
  $\tau = \mu - t$
  **if** $\tau \leq 0$
    $\tau = \tau + p$
  **end**
  { $x_{loc} = x(1 + (\tau - 1)r{:}\tau r)$ }
  $y_{loc} = y_{loc} + A_{loc}(:, 1 + (\tau - 1)r{:}\tau r)x_{loc}$
**end**
**quit**

The integer arithmetic in this procedure is typical of a distributed memory matrix computation. The index $\tau$ names the currently available part of $x$. Once $\tau$ is known then the update of the locally housed portion of $y$ can be performed. Note that $A_{loc}(i,j)$ does not house $A(i,j)$.

### 6.1.8 Attributes of a Parallel Algorithm

How can we anticipate the performance of a parallel procedure such as Algorithm 6.1.1? We faced a comparable situation when we discussed vector computing in §1.4. In that section we solved the expository problem with a rather informal discussion about "what to think about" when developing a computation for a vector computer.

Our philosophy here is similar. We are not going to build a detailed model of parallel computation that can be used to predict execution time. Instead, we mention a few key issues that one should bear in mind when designing a parallel matrix computation. Our task is not unlike that of a manager who oversees a university admissions office. Has the workload been evenly distributed among the staff? Are there bottlenecks that lead to idleness? Is more time spent shuffling the applications from desk to desk instead of reading them? How efficient is the office? What would be the effect of adding more staff?

### 6.1.9 Load Balancing

We start with the attribute of *load balancing*. We say that a parallel computation is load balanced if each processor has roughly the same amount of *useful* computation and communication to perform each time step. Algorithm 6.1.1 is perfectly load balanced in that at each time step each processor has an identically sized computation and communication.

To illustrate load *imbalance* in Algorithm 6.1.1, suppose that the matrix $A$ is block lower triangular with $p$-by-$p$ blocks and that $r = p$. It is then appropriate to update $y_{loc}$ only if $A_{loc}(:, 1 + (\tau - 1)p{:}\tau p)$ is nonzero:

**if** $\tau \le \mu$
  $y_{loc} = y_{loc} + A_{loc}(:, 1 + (\tau - 1)p{:}\tau p)x_{loc}$
**end**

With this exploitation of structure the resulting procedure is not load balanced because (for example) when $t = p-1$, only Proc($p$) has an update to perform. Another way to see the load imbalance is to observe that $z_\mu$ requires $2\mu p^2$ flops and thus, Proc($p$) has more flops to perform than Proc(1) by a factor of $p$.

However, to achieve a load balanced procedure we merely resort to a wrap mapping of the data:

**Algorithm 6.1.2** Suppose $A \in \mathbb{R}^{n\times n}$, $x \in \mathbb{R}^n$ and $y \in \mathbb{R}^n$ are given and that $z = y + Ax$. Assume that $n = rp$ and that $A$ is block lower triangular with $r$-by-$r$ blocks. If each processor in a $p$-processor systolic ring executes the following node program, then upon completion Proc($\mu$) houses $z(\mu{:}p{:}n)$ in $y_{loc}$.

```
loc.init[p, μ = my.id, left, right, n, r = n/p,
        A_loc = A(μ:p:n, :), x_loc = x(μ:p:n), y_loc = y(μ:p:n) ]
for t = 1:p
        send(x_loc, right); recv(x_loc, left)
        τ = μ − t
        if τ ≤ 0
                τ = τ + p
        end
        { x_loc = x(τ:p:n) }
        for j = 1:r
                y_loc(j:r) = y_loc(j:r) + A_loc(j:r, τ + (j − 1)p)x_loc(j)
        end
end
quit
```

Each processor has a set of identically sized saxpy's to perform each step and so this is a load balanced procedure.

The hardest thing to understand about Algorithm 6.1.2 is the indexing. We think in terms of $A(i,j)$, $x(j)$ and $y(i)$ but these do not correspond to $A_{loc}(i,j)$, $x_{loc}(j)$, and $y_{loc}(i)$. Thus, a considerable amount of "index thinking" is associated with the derivation of this algorithm and that, unfortunately, is typical in distributed memory matrix computations unless a smart compiler is available.

### 6.1.10 The Cost of Communication

Communication overheads can be quantified if we model the cost of sending a message. To this end we assume that the sending of $m$ floating point numbers between two processors requires

$$\text{comm}(m) = \alpha_d + \beta_d m \qquad (6.1.4)$$

seconds. Here $\alpha_d$ is the time required to initiate a message and $\beta_d$ is the rate that a message can be transferred. During each timestep in Algorithm 6.1.1 an $r$-vector is sent and received and $2r^2$ flops are performed. If the computation proceeds at $R$ flops per second and there is no idle waiting associated with the **recv**, then each time step requires $(2r^2/R)+2(\alpha_d+\beta_d r)$ seconds.

A more instructive statistic is the *computation/communication ratio.* For Algorithm 6.1.1 this is prescribed by

$$\frac{\text{Time spent computing}}{\text{Time spent communicating}} \approx \frac{2r^2/R}{2(\alpha_d+\beta_d r)}.$$

This fraction quantifies the overhead of communication relative to the volume of computation. Clearly, as $r = n/p$ grows the fraction of time spent computing increases.

### 6.1.11 Efficiency and Speed-Up

The *efficiency* of a $p$-processor parallel algorithm is given by

$$E = \frac{T(1)}{pT(p)}$$

where $T(k)$ is the time required to execute the program on $k$ processors. If computation proceeds at $R$ flops/sec and communication is modeled by (6.1.4), then a reasonable estimate of $T(k)$ for Algorithm 6.1.1 is given by

$$T(k) = \sum_{t=1}^{k} 2(n/k)^2/R + 2(\alpha_d + \beta_d(n/k)) = \frac{2n^2}{Rk} + 2\alpha_d k + 2\beta_d n\,.$$

This assumes no idle waiting. With these assumptions we find that

$$E = \frac{1+\dfrac{\alpha_d R}{n^2}+\dfrac{\beta_d R}{n}}{1+\dfrac{\alpha_d R p^2}{n^2}+\dfrac{\beta_d R p}{n}}.$$

Note that efficiency improves with increasing $n$ and fixed $p$ and degradates with increasing $p$ and fixed $n$. That kind of useful observation is just about all that can be gleaned from our crude model of computation and communication. In practice, benchmarking is the only dependable way to acquire an intuition about a particular distributed memory matrix computation.

A concept related to efficiency is *speed-up.* We say that a parallel algorithm for a particular problem achieves speed-up $S$ if

$$S = T_{seq}/T_{par}$$

where $T_{par}$ is the time required for execution of the parallel program and $T_{seq}$ is the time required by one processor when the best uniprocessor procedure is used. For some problems, the fastest sequential algorithm does not parallelize and so two distinct algorithms are involved in the speed-up assessment.

## 6.1.12 Granularity

Another important aspect of a parallel algorithm is its *granularity*. This qualitative term refers to the amount of computation that takes place in between synchronization points. In a systolic algorithm the processors synchronize each time step so in this model, the granularity merely refers to the amount of computation each step. For example, each timestep in Algorithm 6.1.1 involves an $r$-by-$r$ gaxpy operation. A finer granularity results if messages are restricted to a single floating point number. Here is the resulting procedure.

$$
\begin{array}{l}
\textbf{loc.init}[\,p,\ \mu = my.id,\ left,\ right,\ n,\ r = n/p,\ row = 1 + (\mu-1)r{:}\mu r,\\
\qquad A_{loc} = A(row,:),\ x_{loc} = x(row),\ y_{loc} = y(row)\,]\\
\textbf{for } t = 1{:}n\\
\qquad \textbf{send}(x_{loc}(\text{ceil}(t/p)), right);\ \textbf{recv}(x_{loc}(\text{ceil}(t/p)), left)\\
\qquad \rho = \text{ceil}(t/p);\ t_1 = t - (\rho - 1)p\\
\qquad \tau = \mu - t_1\\
\qquad \textbf{if } \tau \le 0\\
\qquad\qquad \tau = \tau + p\\
\qquad \textbf{end}\\
\qquad y_{loc} = y_{loc} + A_{loc}(:, (\tau-1)r + \rho)x_{loc}(\rho)\\
\textbf{end}\\
\textbf{quit}
\end{array}
\qquad (6.1.5)
$$

Instead of the merry-go-round revolving once as in Algorithm 6.1.1, it revolves $r$ times. During revolution $\rho$, $x(\rho{:}p{:}n)$ circulates around the ring. At the time of the $y_{loc}$ update, $x_{loc}(\rho)$ houses $x(\rho + (\tau - 1)r)$.

Notice that (6.1.5) has more timesteps than Algorithm 6.1.1 by a factor of $r = n/p$. However, there is less work per step as a saxpy is performed instead of a gaxpy. Thus, (6.1.5) has a finer granularity than Algorithm 6.1.1.

It is important to see that communication overheads change with granularity. In Algorithm 6.1.1 a processor spends $2(p\alpha_d + \beta_d n)$ seconds communicating while the corresponding figure for (6.1.5) is $2(n\alpha_d + \beta_d n)$. Thus, a system that supports fine-grained parallelism should feature a small message start-up factor $\alpha_d$.

Fine grained parallelism often makes load balancing easier. However, if granularity is too fine in a system with high-performance nodes, then it may be impossible for the node programs to perform at level-2 or level-3 speeds simply because there just is not enough local linear algebra.

The choice of a distributed data structure can also effect granularity. Here is a reorganization of Algorithm 6.1.1 with the assumption that Proc($\mu$) houses the $\mu$th block column of the $A$ matrix in (6.1.1) instead of the $\mu$th block row.

**Algorithm 6.1.3** Suppose $A \in \mathbb{R}^{n\times n}$, $x \in \mathbb{R}^n$ and $y \in \mathbb{R}^n$ are given and that $z = y + Ax$. If each processor in a $p$-processor systolic ring executes the following node program and $n = rp$, then upon completion Proc($\mu$) houses $z(1 + (\mu - 1)r{:}\mu r)$ in $y_{loc}$.

```
loc.init[p, μ = my.id, left, right, n, r = n/p, col = 1 + (μ − 1)r:μr,
        A_loc = A(:, col), x_loc = x(col), y_loc = y(col) ]
for t = 0:p
    if t = 0
        w = A_loc x_loc
    else
        send(w, right); recv(w, left)
        τ = μ − t
        if τ ≤ 0
            τ = τ + p
        end
        { w = A_τ x_τ }
        y_loc = y_loc + w(col)
    end
end
quit
```

Note that in this procedure the $w$ vectors circulate, not the $x_{loc}$ vectors as in Algorithm 6.1.1. Because the $w$ vectors have length $n$ and the $x_{loc}$ vectors have length $r = n/p$, we see that there is a larger communication overhead associated with Algorithm 6.1.3.

With respect to granularity, instead of $2r^2$ flops in between communication points as in Algorithm 6.1.1, we now have just $r$ flops. The change in granularity is due to the mass of computation that takes place during the $t = 0$ step.

### 6.1.13 The Message Passing Model

We now discuss distributed memory computation without the presence of a global clock. Here, the coordination among the processors is achieved by the sending and receiving of messages. Unlike the systolic model, a processor can send a message at any time and does not have to wait until the next tick of the global clock.

Our message passing language "does not care" how the processors are interconnected. Any processor can send a message to any other processor. However, in practice it is good to arrange a distributed computation so that messages are sent to "nearby" processors in the network. This will be the case in most of our examples.

We begin with a message-passing version of Algorithm 6.1.3.

**Algorithm 6.1.4** Suppose $A \in \mathbb{R}^{n\times n}$, $x \in \mathbb{R}^n$ and $y \in \mathbb{R}^n$ are given and that $z = y + Ax$. If each processor in a $p$-processor ring executes the following node program and $n = rp$, then upon completion Proc($\mu$) houses $z(1 + (\mu - 1)r{:}\mu r)$ in $y_{loc}$.

```
loc.init[p, μ = my.id, left, right, n, r = n/p, col = 1 + (μ − 1)r:μr,
        A_loc = A(:, col), x_loc = x(col), y_loc = y(col) ]
w = A_loc x_loc
k = 1
while k ≤ p
        send(w, right); recv(w, left)
        τ = τ − k
        if τ ≤ 0
                τ = τ + p
        end
        { w = A_τ x_τ }
        y_loc = y_loc + w(col)
        k = k + 1
end
quit
```

Although this looks very much like its systolic analog Algorithm 6.1.3, the underlying logic is entirely different. In particular, Proc($\mu$) knows what to do on the basis of the index $k$ which keeps track of the number of received messages.

### 6.1.14 Further Message Passing Examples

It is instructive to consider some variations of Algorithm 6.1.4 to illustrate further the asynchronous message passing paradigm.

As a first example, let us assume that $x$ and $y$ are completely housed in Proc(1) and that Proc(1) is to overwrite $y$ with $z = y + Ax$. (We continue to assume that $A$ is distributed by block column.) One approach to this problem is to have Proc(1) send a copy of $x_\tau$ to Proc($\tau$). Proc($\tau$) then computes $A_\tau x_\tau$ and returns the result to Proc(1) where it is used in the update of $y$. Here is one way to accomplish this:

```
{ Distribute x throughout the network. }
if μ = 1
        for τ = 1:p
                send(x(1 + (τ − 1)r:τr), τ)
        end
end
{ Receive x_μ, compute w = A_μ x_μ, send w to Proc(1). }
recv(x_loc, 1); w = A_loc x_loc; send(w, 1)
```

```
{ Proc(1) collects the w vectors and updates y. }
if μ = 1
    for τ = 1:p
        recv(w, τ)
        y = y + w
    end
quit
```

Note that Proc(1) sends a message to itself, a perfectly legal thing to do in our model. We also mention that the communication overhead is more difficult to predict than in Algorithm 6.1.1 because non-neighbors communicate. The system software that routes the messages would figure heavily in an assessment of performance.

Now let us suppose that Proc($\mu$) houses $x(col)$ and $A(:,col)$ where $col = 1+(\mu-1)r{:}\mu r$ and that we want to overwrite $y$ with $z = y+Ax$. We assume that $y$ is housed in Proc(1) at the beginning and end of the computation. It seems reasonable to proceed as follows:

```
if μ = 1
    y = y + A_loc x_loc
    send(y, right); recv(y, left)                              (6.1.6)
else
    recv(y, left); y = y + A_loc x_loc; send(y, right)
end
quit
```

The logic behind this approach is that $y$ circulates around the ring incorporating $A_{loc}x_{loc}$ at each stop. The problem is that there is no concurrent computation. At any instant, only one processor is busy doing arithmetic.

To correct this we merely make sure Proc($\mu$) does its computation before it waits for the arrival of the circulating $y$:

```
if μ = 1
    y = y + A_loc x_loc
    send(y, right); recv(y, left)
else
    w = A_loc x_loc
    recv(y, left); y = y + w; send(y, right)
end
quit
```

The level of idle waiting is now much reduced compared to (6.1.6).

**Problems**

**P6.1.1** Modify Algorithm 6.1.1 so that it does not require $n$ to be an integral multiple of $p$.

**P6.1.2** Modify algorithm 6.1.1 so that it efficiently handles the case when $A$ is tridiagonal.

**P6.1.3** Modify Algorithm 6.1.2 so that it can handle the case when $A$ is (a) lower triangular and (b) block upper triangular.

**P6.1.4** Modify algorithms 6.1.3 and 6.1.4 so that they overwrite $y$ with $z = y + A^m x$ for a given positive integer $m$ that is available to each processor.

**P6.1.5** Modify Algorithms 6.1.3 and 6.1.4 so that $y$ is overwritten by $z = y + A^T Ax$.

**P6.1.6** Modify Algorithm 6.1.3 so that upon completion, the local array $A_{loc}$ in Proc($\mu$) houses the $\mu$th block column of $A + xy^T$.

### Notes and References for Sec. 6.1

General references on parallel matrix computations include

J.J. Dongarra and D.C. Sorensen (1986). "Linear Algebra on High Performance Computers," *Appl. Math. and Comp. 20,* 57–88.

V. Fadeeva and D. Fadeev (1977). "Parallel Computations in Linear Algebra," *Kibernetica 6,* 28–40.

D. Heller (1978). "A Survey of Parallel Algorithms in Numerical Linear Algebra," *SIAM Review 20,* 740-777.

R.W. Hockney and C.R. Jesshope (1988). *Parallel Computers 2*, Adam Hilger, Bristol and Philadelphia.

J.J. Modi (1988). *Parallel Algorithms and Matrix Computation,* Oxford University Press, Oxford.

J.M. Ortega and R.G. Voigt (1985). "Solution of Partial Differential Equations on Vector and Parallel Computers," *SIAM Review 27,* 149-240.

The modeling of performance in a distributed memory environment is an important and difficult problem. See

L. Adams and T. Crockett (1984). "Modeling Algorithm Execution Time on Processor Arrays," *Computer 17,* 38–43.

D. Gannon and J. Van Rosendale (1984). "On the Impact of Communication Complexity on the Design of Parallel Numerical Algorithms," *IEEE Trans. Comp. C-33,* 1180–1194.

S.L. Johnsson (1987). "Communication Efficient Basic Linear Algebra Computations on Hypercube Multiprocessors," *J. Parallel and Distributed Computing*, No. 4, 133–172.

Y. Saad and M.H. Schultz (1985). "Data Communication in Hypercubes," Report YALEU/DCS/RR-428, Dept. of Computer Science, Yale University, New Haven, CT.

Y. Saad and M.H. Schultz (1985). "Topological Properties of Hypercubes," Report YALEU/DCS/RR-389, Dept. of Computer Science, Yale University, New Haven, CT.

Aspects of implementation in parallel environments are discussed in

K. Connolly, J.J. Dongarra, D. Sorensen, and J. Patterson (1988). "Programming Methodology and Performance Issues for Advanced Computer Architectures," *Parallel Computing 5,* 41-58.

J. Dongarra and D.C. Sorensen (1987). "A Portable Environment for Developing Parallel Programs," *Parallel Computing 5,* 175–186.

D.P. O'Leary and G.W. Stewart (1986). "Assignment and Scheduling in Parallel Matrix Factorization," *Lin. Alg. and Its Applic. 77,* 275–300.

Matrix computations on 2-dimensional arrays are discussed in

H.T. Kung (1982). "Why Systolic Architectures?," *Computer 15*, 37-46.
D.P. O'Leary and G.W. Stewart (1985). "Data Flow Algorithms for Parallel Matrix Computations," *Comm. ACM 28*, 841-853.

For a discussion of basic linear algebra computation on a distributed memory system see

S.L. Johnsson and C.T. Ho (1987). "Matrix Transposition on Boolean n-cube Configured Ensemble Architectures," Report YALEU/DCS/RR-574, Dept. of Computer Science, Yale University, New Haven, CT., to appear in *SIAM J. Alg. and Discrete Methods.*
O. McBryan and E.F. van de Velde (1987). "Hypercube Algorithms and Implementations," *SIAM J. Sci. and Stat. Comp. 8*, s227–s287.
C.B. Moler (1986). "Matrix Computations on Distributed Memory Multiprocessors," in *Proceedings of First SIAM Conference on Hypercube Multiprocessors*, ed. M.T. Heath, SIAM Press, Philadelphia, Pa., 1986.

The proceedings of the hypercube conference are a good source of material on distributed memory matrix computations. See

M.T. Heath (ed) (1986). *Proceedings of First SIAM Conference on Hypercube Multiprocessors*, SIAM Publications, Philadelphia, Pa.
M.T. Heath (ed) (1987). *Hypercube Multiprocessors*, SIAM Publications, Philadelphia, Pa.
G. Fox (ed) (1988). *The Third Conference on Hypercube Concurrent Computers and Applications, Vol. II - Applications*, ACM Press, New York.

## 6.2 Shared Memory Gaxpy

We now discuss the gaxpy problem for shared memory multiprocessors. In this environment, each processor has access to a common, shared memory as depicted in the following figure:

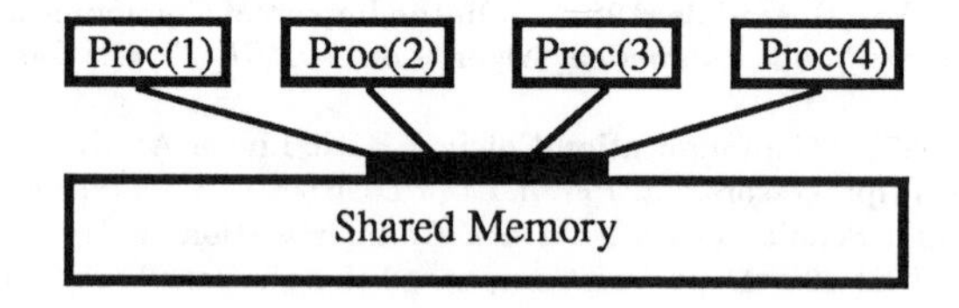


Figure 6.2.1 A 4-Processor Shared Memory System

Communication between processors is achieved by reading and writing to *shared variables* that reside in the shared memory. Each processor executes its own *local program* and has its own *local memory*. Data flows to and from the shared memory during execution. For example, in the gaxpy computation $z = y + Ax$, $x$,$y$, and $A$ might reside in shared memory. A participating processor might (a) copy $x$, the $i$th block row $A_i$, and the $i$th subvector $y_i$ from shared memory into its local memory, (b) compute $z_i = y_i + A_i x$, and (c) copy $z_i$ from its local memory into shared memory.

All the concerns that attend distributed memory computation are with us in modified form. The overall procedure must be *load balanced* as we discussed in the previous section. The computations should be arranged so that the individual processors have to wait as little as possible for something useful to compute. The traffic between the shared and local memories must be managed carefully as we assume that such data transfers are a significant overhead. (This corresponds to interprocessor communication in the distributed memory setting.) The nature of the physical connection between the processors and the shared memory is very important and can effect algorithmic development. For example, if only one processor can access shared memory at a given instant, then care must be taken to minimize the chance of access collisions. For the most part we suppress this kind of detail preferring to regard this aspect of the system as a black box as depicted in Figure 6.2.1.

All local computations are subject to traditional efficiency concerns. If the individual processors are vector/pipeline processors then as discussed in §1.4, issues such as stride, vector length, local memory touches, and data reuse can be important.

Once again, parallel gaxpy computation is a useful setting in which to introduce shared memory matrix computations. We first develop a gaxpy algorithm in which each processor's computational task is worked out in advance. For less regular computations it is necessary to assign tasks as the computation progresses. Special synchronization tools that permit the dynamic assignment of tasks are necessary and to this end we introduce the concept of a monitor. With monitors we can elegantly solve a wide range of scheduling problems.

### 6.2.1 A Statically Scheduled Gaxpy

We assume throughout that our model shared memory computer has $p$ processors and that Proc($\mu$) designates the $\mu$th processor. As in the previous section we consider the following partitioning of the $n$-by-$n$ gaxpy problem $z = y + Ax$:

$$\begin{bmatrix} z_1 \\ \vdots \\ z_p \end{bmatrix} = \begin{bmatrix} y_1 \\ \vdots \\ y_p \end{bmatrix} + \begin{bmatrix} A_1 \\ \vdots \\ A_p \end{bmatrix} x \,. \qquad (6.2.1)$$

Here we assume that $n = rp$ and that $A_\mu \in \mathbb{R}^{r \times n}$, $y_\mu \in \mathbb{R}^r$, and $z_\mu \in \mathbb{R}^r$. We introduce our shared memory notations with the following algorithm in which Proc($\mu$) overwrites $y_\mu$ with $z_\mu = y_\mu + A_\mu x$:

**Algorithm 6.2.1** Suppose $A \in \mathbb{R}^{n \times n}$, $x \in \mathbb{R}^n$, and $y \in \mathbb{R}^n$ reside in a shared memory accessible to $p$ processors. If $n = rp$ and each processor executes the following algorithm, then upon completion, $y$ is overwritten by $z = y + Ax$.

**shared variables**$[A(1{:}n, 1{:}n),\ x(1{:}n),\ y(1{:}n)]$
**glob.init**$[\,A,\ x,\ y\,]$
**loc.init**$[\,p,\ \mu = my.id,\ n\,]$
$r = n/p;\ row = 1 + (\mu - 1)r{:}\mu r$
**get**$(x, x_{loc})$; **get**$(y(row), y_{loc})$
**for** $j = 1{:}n$
  **get**$(A(row, j), a_{loc})$; $y_{loc} = y_{loc} + a_{loc}x(j)$
**end**
**put**$(y_{loc}, y(row))$
**end**
**quit**

We assume that a copy of this program resides in each processor. All variables are local except those that are identified by the **shared variables** statement. We often stress that a variable is local with a *loc* subscript. The **glob.init** statement indicates shared variables that must be initialized. The **local.init** statement indicates local variables that must be initialized including the processor identification $\mu$. We do not care how these initializations are accomplished, or how the local programs are down-loaded, or how they are all prodded into parallel execution. These are important implementation details but they do not contribute our high-level overview. The **quit** statement reminds us that the termination of a processor program does not mean that the overall computation is over.

The actual local program to be executed follows the shared variable identification and the local and global initializations. Data is transferred to and from shared memory by means of **get** and **put** statements. For example, **get**$(A(row, j), a_{loc})$ copies the subcolumn $A(row, j)$ from shared memory into the local workspace $a_{loc}$. Likewise, after Proc($\mu$) completes the computation of $y_\mu + A_\mu x$ it returns the value to shared memory via the statement **put**$(y_{loc}, y(row))$

Note that in Algorithm 6.2.1, only one processor writes to a given shared memory location in $y$. Thus, there is no need for synchronization. Each processor has a completely independent part of the overall gaxpy operation and does not have to monitor the progress of the other processors. We say that the algorithm is *statically scheduled* because the partitioning of work is determined before execution.

### 6.2.2 Shared Memory Traffic

It is important to recognize the overheads associated with **get** and **put** statements. These correspond to the **send**and **recv**overheads in distributed memory computation. In general, we assume that **put** and **get** have the form

**get**( {shared memory matrix } , {local memory matrix })

**put**( {local memory matrix } , {shared memory matrix } )

We assume that the shared and local matrices have the same size and that they are *contiguous* in their respective memories. In Algorithm 6.2.1, we assume that $A$ is stored by column in shared memory. Thus, $A_\mu$ is not contiguous and so has to be copied into local memory column-by-column.

If the **get** or **put** involves $m$ floating point numbers then we model the transfer time by

$$\text{trans}(m) \;=\; \alpha_s + \beta_s m\,. \tag{6.2.2}$$

Here, $\alpha_s$ represents a start-up overhead and $\beta_s$ is the reciprocal transfer rate. Note the similarity of this overhead with our model execution time for a vector operation in §1.4.3 and our model of communication in the distributed memory environment in (6.1.4).

Accounting for all the **gets** and **puts** in Algorithm 6.2.1 we see that the time that each processor spends communicating with shared memory is estimated by

$$T_{6.2.1} \;=\; (n+3)\alpha_s + (n + 2r + nr)\beta_s\,. \tag{6.2.3}$$

However, there may be hidden delays associated with each **get** and **put** because processors may have to compete for access to the shared memory. Thus, it is important to bear in mind that (6.2.3) is an *estimate* of the shared memory traffic overhead in Algorithm 6.2.1.

### 6.2.3 Load Balancing

Algorithm 6.2.1 is nicely load balanced. However, if $A$ is lower triangular then a redistribution of the workload would be necessary to ensure that each processor had roughly the same amount of useful work. One way to accomplish this is to assign Proc($\mu$) the computation of $z(\mu{:}p{:}n) = y(\mu{:}p{:}n) + A(\mu{:}p{:}n,:)x$. Details are left to the reader as the desired procedure corresponds very closely to Algorithm 6.1.2.

### 6.2.4 Barrier Synchronization

Suppose that we want to overwrite $y$ with a unit vector in the direction of $z = y + Ax$, which we assume to be nonzero. One approach to this problem is to compute $z$ in parallel using Algorithm 6.2.1 and then have one processor, say Proc(1), perform the scaling $z/\| z \|_2$. At first glance, the following seems to be a reasonable approach to this scaled gaxpy problem:

```
shared variables[A(1:n,1:n), x(1:n), y(1:n), s]
glob.init[ A, x, y ]
loc.init[ p, μ = my.id, n ]
r = n/p; row = 1 + (μ − 1)r:μr
get(x, x_loc); get(y(row), y_loc)
for j = 1:n
    get(A(row, j), a_loc); y_loc = y_loc + a_loc x(j)
end
put(y_loc, y(row))
if μ = 1
    get(y, u_loc); s_loc = || u_loc ||_2; put(s_loc, s)
end
get(s, s_loc); y_loc = y_loc/s; put(y_loc, y(row))
quit
```

However, there are two serious flaws in this approach. We have no guarantee that $y$ contains $z$ when Proc(1) begins the 2-norm computation and we have no guarantee that $s = \| z \|_2$ when Proc(2),..., Proc($p$) begin the scaling.

To rectify these problems we need a synchronization construct that can delay the 2-norm computation until all the processors have computed and stored their portion of $z$ and which can delay the scalings until $s$ is available. To this end we introduce the **barrier** construct:

**Algorithm 6.2.2** Suppose $A \in \mathbb{R}^{n \times n}$, $x \in \mathbb{R}^n$, and $y \in \mathbb{R}^n$ reside in a shared memory accessible to $p$ processors. If $n = rp$ and each processor executes the following algorithm, then upon completion, $y$ is overwritten by $z/\| z \|_2$ where $z = y + Ax$ is assumed to be nonzero.

```
shared variables[A(1:n,1:n), x(1:n), y(1:n), s]
glob.init[ A, x, y ]
loc.init[ p, μ = my.id, n ]
r = n/p; row = 1 + (μ − 1)r:μr
get(x, x_loc); get(y(row), y_loc)
for j = 1:n
    get(A(row, j), a_loc); y_loc = y_loc + a_loc x(j)
end
put(y_loc, y(row))
barrier
if μ = 1
    get(y, u_loc); s_loc = || u_loc ||_2; put(s_loc, s)
end
barrier
get(s, s_loc); y_loc = y_loc/s; put(y_loc, y(row))
quit
```

To understand the **barrier**, it is convenient to regard a processor as either blocked or free. A processor is blocked and suspends execution when it executes the **barrier**. After the $p$th processor is blocked, all the processors return to the "free state" and resume execution. More informally, a **barrier** is a hurdle. All $p$ processors congregate at the hurdle and then they jump in unison to the other side.

In Algorithm 6.2.2, Proc($\mu$) computes $z_\mu$ and is immediately blocked. We cannot predict the order in which the processors are blocked. We only know that once the last processor reaches the barrier, execution resumes in each processor. Proc(1) then goes on to compute $s = \| z \|_2$ but the remaining processors are immediately blocked again. Finally, when $s$ is available each processor is released and can proceed with its portion of the scaling.

### 6.2.5 The Pool-of-Tasks Paradigm

For load balanced, coarse-grained computations the **barrier** is a useful construct. However, it should be used sparingly as it enforces a global idleness. A more flexible approach to synchronization is needed if we are to handle situations where the equitable distribution of work is difficult to arrange in advance.

To motivate the key ideas we develop a parallel algorithm for the $n$-by-$n$ gaxpy $z = y + Ax$ that is based upon a general block row partitioning, i.e.,

$$\begin{bmatrix} z_1 \\ \vdots \\ z_N \end{bmatrix} = \begin{bmatrix} y_1 \\ \vdots \\ y_N \end{bmatrix} + \begin{bmatrix} A_1 \\ \vdots \\ A_N \end{bmatrix} x \qquad (6.2.4)$$

where $y_i, z_i \in \mathbb{R}^{n_i}$ and $A_i \in \mathbb{R}^{n_i \times n}$ for $i = 1{:}N$. Assume that the entries of $A_i$ are contiguous in shared memory. Define the integer vectors $first(1{:}N)$ and $last(1{:}N)$ by

$$\begin{aligned} first(i) &= 1 + n_1 + \cdots + n_{i-1} \\ last(i) &= n_1 + \cdots + n_i \,. \end{aligned}$$

Here is a sequential algorithm based on the blocking (6.2.4) that overwrites $y$ with $z = y + Ax$:

$$\begin{aligned} &\textbf{for } i = 1{:}N \\ &\qquad row = first(i){:}last(i) \\ &\qquad y(row) = y(row) + A(row,:)x \\ &\textbf{end} \end{aligned} \qquad (6.2.5)$$

Note that the loop specifies $N$ independent computations and so there is no problem developing a parallel algorithm. We could assign the computation of $z_i, i = \mu{:}p{:}N$ to Proc($\mu$). However, the resulting procedure may not be load balanced if the $A_i$ varied in row dimension.

This prompts us to develop a dynamic method for distributing the work. Until now, we have only considered "nice" blockings of the gaxpy computation that permit us to work out in advance a satisfactory work distribution. This is called *static scheduling.* Our approach to *dynamic scheduling* is based upon the *pool-of-tasks* idea. To illustrate this approach we regard the computation associated with the $i$th pass through the loop in (6.2.5) as task($i$). In the pool-of-tasks approach a list of remaining tasks is maintained. When a processor completes a task it checks the list. If it is empty, then there is no work to be assigned. If the list is nonempty, then the processor selects a task from the "pool" of remaining tasks. The chosen task is then removed from the pool.

A reasonable way to apply this idea to (6.2.5) is to have a global variable *count* that maintains the value of the last task selected. This can be accomplished (or so it would seem) by initializing *count* to zero and then incrementing its value each time a processor takes on a new task:

$$
\begin{array}{l}
\textbf{shared variables}[A(1{:}n,1{:}n),\ x(1{:}n),\ y(1{:}n),\ count\,] \\
\textbf{glob.init}[\,A,\ x,\ y,\ count = 0\,] \\
\textbf{loc.init}[\,p,\ \mu = my.id,\ n\ ,N\ ,first(1{:}N),\ last(1{:}N)\,] \\
\textbf{get}(x, x_{loc}) \\
\textbf{get}(count, i);\ i = i+1;\ \textbf{put}(i, count) \\
\textbf{while}\ i \le N \\
\qquad row = first(i){:}last(i) \\
\qquad \textbf{get}(A(row,:), A_{loc});\ \textbf{get}(y(row), y_{loc}) \\
\qquad y_{loc} = y_{loc} + A_{loc}x_{loc} \\
\qquad \textbf{put}(y_{loc}, y(row)) \\
\qquad \textbf{get}(count, i);\ i = i+1;\ \textbf{put}(i, count) \\
\textbf{end}
\end{array}
$$

The trouble with this approach to dynamic scheduling is that the updating of *count* is uncontrolled. Two different processors could "grab" the same value of *count* if their respective executions of

$$\textbf{get}(count, i);\ i = i+1;\ \textbf{put}(i, count) \qquad (6.2.6)$$

overlap. Thus, we need a way to ensure that at most one processor is engaged in the update (6.2.6) at any one instant. We say that (6.2.6) is a *critical section* of the algorithm.

## 6.2.6 Monitors

*Monitors* can be usd to solve the critical section problem. We illustrate the concept by considering the parallel execution of (6.2.5) using the pool-of-tasks idea. The key to the solution is the design of a special procedure

**nexti.index**$(p, N, more, i)$

that has as input $p$ (the number of processors assigned to the parallel execution of the loop) and $N$ (the number of tasks). The output consists of a logical value in *more* and an index in $i$ with the property that

$$\begin{array}{lcl} more = true & \Rightarrow & \text{task}(i) \text{ is assigned} \\ more = false & \Rightarrow & \text{no tasks remaining} \end{array}$$

If we can ensure that only one processor executes **nexti.index** at any instant, then

```
call nextq.index(p, N, more, i)
while more = true
    Overwrite y_i with z_i = y_i + A_i x                    (6.2.7)
    call nextq.index(p, N, more, i)
end
```

correctly performs the computation. To specify this kind of execution we place **nexti.index** inside a *monitor*:

```
monitor: nexti
    protected variables: flag, count, idle.proc
    condition variables: waiti
    initializations: flag = 0
    procedure nexti.index(p, N, more, i)
        if flag = 0
            count = 0; flag = 1; idle.proc = 0
        end
        count = count + 1
        if count ≤ N
            more = true; i = count
        else
            more = false; idle.proc = idle.proc + 1
            if idle.proc ≤ p − 1
                delay(waiti)
            else
                flag = 0
            end
            continue(waiti)
        end
    end nexti.index
end nexti
```

From the example we deduce that a monitor has a name and consists of *protected variables*, *condition variables*, *initializations*, and one or more *monitor procedures.* This particular monitor has one procedure, **nexti.index**. Its primary mission is to carefully maintain a counter *count* that steps from 1 to $N$, the number of tasks that are to be processed by the $p$ processors. When no more tasks remain then the monitor initiates a "winding down" process. The variable *idle.proc* keeps track of the number of processors that have failed to acquire a new task. When this counter reaches $p$ the monitor's job is over and it is restored to its initial state. The mechanics of this process are best described after we see the monitor in action.

**Algorithm 6.2.3** Suppose $A \in \mathbb{R}^{n\times n}$, $x \in \mathbb{R}^n$, and $y \in \mathbb{R}^n$ reside in a shared memory accessible to $p$ processors. Assume that $A$ is stored by block row and that the $i$th block row is prescribed by $A(first(i){:}last(i), :)$. If each processor executes the following algorithm, then upon completion, $y$ is overwritten by $z = y + Ax$.

```
shared variables[A(1:n, 1:n), x(1:n), y(1:n)]
glob.init[ A, x, y ]
loc.init[p, μ = my.id, N, first(1:N), last(1:N) ]
get(x, x_loc)
call nexti.index(p, N, more, i)
while more = true
    row = first(i):last(i)
    get(A(row, :), A_loc); get(y(row), y_loc)
    y_loc = y_loc + A_loc x_loc
    put(y_loc, y(row))
    call nexti.index(p, N, more, i)
end
quit
```

We use this example to explain the main features of monitor use.

- A processor tries to "enter" a monitor by calling one of its procedures. Thus, "call **nexti.index**($p, N, more, i$)" represents an attempt by the executing processor to enter **nexti**.

- When a processor is in a monitor it is either *active* or *delayed.* At any instant, there can be a most one active processor in a monitor. Thus, if Proc(1) is in **nexti** and Proc(2) calls **nexti.index**, then Proc(2) is delayed at the point of the call in its local program.

- A processor is delayed within a monitor when it encounters a statement of the form **delay**({ *condition variable* }). Conceptually, the condition variable names a queue and when a processor is delayed, its *id* is placed on the queue. When a processor is delayed the monitor becomes free and if any processors are waiting to enter the monitor, exactly one will gain admission. In **nexti**, if Proc($\mu$) enters the monitor and *count* has the value of $N$ indicating that there are no more available tasks, then Proc($\mu$) is delayed. Note that before it is delayed Proc($\mu$) sets $more = false$ and increments *idle.proc*. Thus, when control is passed back to the local program that invoked **nexti.index**, it will 'know" that no more tasks remain. This is what makes the **while** loop work in Algorithm 6.2.3.

- When the currently active processor executes a statement of the form **continue**({ *condition variable* }), it immediately leaves the monitor with several possible ramifications. If the queue named in the **continue** is nonempty, then precisely one of the delayed processors becomes the next active processor. Should the named queue be empty then the monitor is available for entry. If this happens and processors are waiting to enter the monitor, exactly one will be allowed to enter. In **nexti.index** when the protected variable *idle.proc* has value $p$ then all computational tasks have been completed. The first processor to discover that there are no more tasks resets $flag$ to zero and triggers the awakening of the $p-1$ delayed processors by executing the statement **continue**(*waiti*). When a processor is awakened it resumes execution. In **nexti.index** an awakend processor immediately executes a **continue** thereby awakening yet another delayed processor. This activation process continues until all $p$ processors exit the monitor.

- The value of a protected variable in a monitor can only be changed by a procedure in the monitor. Note that the calling programs do not alter $flag$, *count*, or *idle.proc*.

- Protected variables can be initialized once, before any reference is made to the monitor. In **nexti**, $flag$ is initialized to zero. While **nexti** oversees the distribution of tasks, $flag$ has a value of 1. It is reset to zero by the first processor to discover that there are no remaining tasks.

It is worth spending a little time stepping through this somewhat involved logic so as to gain an appreciation for the reasoning that underlies a dynamically scheduled matrix computation.

### 6.2.7 The Column Gaxpy Approach

As a second illustration of monitor use and the pool-of-task idea, we develop a parallel gaxpy computation that is based upon the following block column partitioning of $z = y + Ax$:

$$z = y + [A_1, \ldots, A_N] \begin{bmatrix} x_1 \\ \vdots \\ x_N \end{bmatrix} = y + A_1x_1 + \cdots + A_Nx_N . \qquad (6.2.8)$$

We make no assumptions about the block column widths. A reasonable parallel algorithm based upon this blocking is to assign Proc($\mu$) the computation of $w_k = A_kx_k$ where $k = \mu{:}p{:}N$. These partial vector sums would be summed in order to obtain the final gaxpy. At first glance, the following procedure appears to represent the correct solution:

**shared variables**[$A(1{:}n, 1{:}n)$, $x(1{:}n)$, $y(1{:}n)$]
**glob.init**[$A$, $x$, $y$]
**loc.init**[$p$, $\mu = my.id$, $N$]
**for** $k = \mu{:}p{:}N$
  **get**($x_k, x_{loc}$); **get**($A_k, A_{loc}$)
  $w_{loc} = A_{loc}x_{loc}$
  **get**($y, y_{loc}$); $y_{loc} = y_{loc} + w_{loc}$; **put**($y_{loc}, y$)
**end**
**quit**

However, problems arise if execution of the update

$$\textbf{get}(y, y_{loc});\ y_{loc} = y_{loc} + w_{loc};\ \textbf{put}(y_{loc}, y) \qquad (6.2.9)$$

is not properly controlled. For example, the contribution of Proc(1) could be lost if Proc(2) reads $y$ while Proc(1) is midway through the execution of (6.2.9). Things would work out if the processors did not intermingle their access to $y$. The situation calls for a monitor that controls the computation of a vector sum that is built up by several processors:

**monitor: gax**
  **procedure gax.add**($u, v$)
    **get**($v, v_{loc}$); $v_{loc} = v_{loc} + u$; **put**($v_{loc}, v$)
  **end gax.add**
**end gax**

This simple monitor has no protected or condition variables. The procedure **gax.add** has two arguments. The first is a local vector that the calling processor wishes to incorporate in the global running sum named by the second argument. With this monitor we can practice safe updating during the gaxpy computation.

**Algorithm 6.2.4** Suppose $A \in \mathbb{R}^{n\times n}$, $x \in \mathbb{R}^n$, and $y \in \mathbb{R}^n$ reside in a shared memory accessible to $p$ processors and that we have the blocking (6.2.8). If each processor executes the following algorithm, then upon completion, $y$ is overwritten by $z = y + Ax$.

**shared variables**$[A(1{:}n, 1{:}n),\ x(1{:}n),\ y(1{:}n)]$
**glob.init**$(\ A,\ x, y\ )$
**loc.init**$[\ p,\ \mu = my.id,\ N\ ]$
call **nexti.index**$(p, N, more, k)$
**while** $more = true$
  **get**$(A_k, A_{loc})$; **get**$(x_k, x_{loc})$; $w_{loc} = A_{loc}x_{loc}$
  call **gax.add**$(w_{loc}, y)$
  call **nexti.index**$(p, n, more, k)$
**end**
quit

Thus, when, Proc($\mu$) calls **gax.add** another vector is added into $y$. The properties of a monitor ensure that only one processor is updating $y$ at any given time. Note that the pool-of-tasks in Algorithm 6.2.4 is controlled by the monitor procedure **nexti.index** that we developed in §6.2.6.

There are some interconnections between granularity, load balancing, and level of performance that Algorithm 6.2.4 highlights. As $N$ increases the volume of computation associated with a task decreases. This implies greater load balancing as there is less idle waiting at the end of the computation when processors are delayed upon the discover that no tasks remain. On the other hand, as $N$ increases the local computation $w_{loc} = A_{loc}x_{loc}$ looks less and less like a level-2 operation as the column width of $A_{loc}$ diminishes. Thus, in this problem there is a tension between load balancing and high level linear algebra. The mediation of this conflict might call for a clever variable width blocking in (6.2.8), e.g., skinnier block columns towards the right edge of the matrix.

Problems

**P6.2.1** Modify Algorithms 6.2.2 and 6.2.3 so that $A$ is overwritten by the outer product update $A + xy^T$.

**P6.2.2** Modify Algorithm 6.2.2 so that $x$ is overwritten with $A^2x$.

**P6.2.3** Modify Algorithm 6.2.2 so that $y$ is overwritten by a unit 2-norm vector in the direction of $y + A^kx$.

**P6.2.4** Modify Algorithm 6.2.1 so that it efficiently handles the case when $A$ is lower triangular. See §6.2.3.

**P6.2.5** This problem concerns changing the definition of "next task" in Algorithm 6.2.4. Consider the $n$-by-$n$ gaxpy problem $y \leftarrow y + Ax$ and let $m \le n$ be given. Suppose that the computation $A(:, 1{:}last)x(1{:}last)$ has been assigned where $1 \le last \le n$. Define the "next task" to be the incorporation of $A(:, last + 1{:}last + k)x(last + 1{:}last + k)$ into the running vector sum $y$ where $k = \min(m, \max(1, \text{ceil}((n - last)/p)))$. Modify the monitor **nexti** and Algorithm 6.2.4 so that this new definition of next task is implemented. Hint: You'll need a protected variable *last* and a monitor procedure of the form **nexti.indices**$(p, n, more, i, j)$ that returns indices $i$ and $j$ identifying $y \leftarrow y + A(:, i{:}j)x(i{:}j)$ as the next assigned task.

**Notes and References for Sec. 6.2**

A good discussion of monitors and associated notations can be found in either of the following references:

G. Andrews and F.B. Schneider (1983). "Concepts and Notations for Concurrent Programming," *Computing Surveys 15*, 1–43.
J. Boyle, R. Butler, T. Disz, B. Glickfield, E. Lusk, R. Overbeek, J. Patterson, and R. Stevens (1987). *Portable Programs for Parallel Processors*, Holt, Rinehart and Winston.

## 6.3 Parallel Matrix Multiplication

In this section we consider the matrix-matrix multiplication problem $D = C + AB$ for both distributed memory and shared memory multiprocessors. Block gaxpy and block dot product procedures are considered.

### 6.3.1 Block Gaxpy Procedures

Suppose $A$, $B$, and $C$ are $n$-by-$n$ matrices and that we wish to compute

$$D = C + AB.$$

For clarity, assume that $n = rp$ where $p$ is the number of processors and that we have the partitioning

$$[\, D_1, \ldots, D_p \,] \;=\; [\, C_1, \ldots, C_p \,] \;+\; [\, A_1, \ldots, A_p \,][\, B_1, \ldots, B_p \,] \qquad (6.3.1)$$

where each block column has width $r$. If

$$B_j \;=\; \begin{bmatrix} B_{1j} \\ \vdots \\ B_{pj} \end{bmatrix}, \qquad B_{ij} \in \mathbb{R}^{r \times r},$$

then one way to distribute the workload is to assign Proc($\mu$) the computation of

$$D_\mu \;=\; C_\mu + AB_\mu \;=\; C_\mu \;+\; \sum_{\tau=1}^{p} A_\tau B_{\tau\mu}\,.$$

We first develop a distributed memory implementation under the assumption that Proc($\mu$) houses $A_\mu$, $B_\mu$, and $C_\mu$ and that upon completion, $D_\mu$ overwrites $C_\mu$. Note that the computation of $D_\mu$ involves local data ($A_\mu$,$B_\mu$ and $C_\mu$ ) and nonlocal data (the other $A_\tau$). Thus, during the course of execution, each processor must have access to every other $A$-column.

A simple way to handle this is to circulate the $A_\tau$ around the ring in "merry-go-round" fashion just as we circulated the $x$ subvectors in our distributed gaxpy procedure, Algorithm 6.1.1. Here is what we obtain:

**Algorithm 6.3.1** Suppose $A$, $B$, and $C$ are $n$-by-$n$ matrices and that $D = C + AB$. If each processor in a $p$-processor ring executes the following program and $n = rp$, then upon completion Proc($\mu$) houses $D_\mu$ in the local array $C_{loc}$.

> **loc.init**[ $p$, $\mu = my.id$, $left$, $right$, $n$, $r = n/p$, $col = 1 + (\mu - 1)r{:}\mu r$
> $\quad A_{loc} = A(:, col)$, $B_{loc} = B(:, col)$, $C_{loc} = C(:, col)$ ]
> **for** $k = 1{:}p$
> $\quad$ **send**($A_{loc}, right$); **recv**($A_{loc}, left$)
> $\quad \tau = \mu - k$
> $\quad$ **if** $\tau \leq 0$
> $\quad\quad \tau = \tau + p$
> $\quad$ **end**
> $\quad \{ A_{loc} = A_\tau = A(:, 1 + (\tau - 1)r{:}\tau r \}$
> $\quad C_{loc} = C_{loc} + A_{loc}B_{loc}(1 + (\tau - 1)r{:}\tau r, :)$
> **end**
> **quit**

This procedure is load balanced. If we assume no idle waiting and no overlap of computation and communication, then the time each processor spends communicating is given by

$$T_{6.3.1} = 2p(\alpha_d + \beta_d n^2/p) = 2p\alpha_d + 2\beta_d n^2 \qquad (6.3.2)$$

as there are $p$ **send**'s, $p$ **recv**'s, and each involves a message of length $nr = n^2/p$.

Now let us develop a shared memory procedure based upon the same blocking (6.3.1). Following Algorithm 6.3.1, we arrange matters so that Proc($\mu$) computes $D_\mu$ from the summation

$$D_\mu = C_\mu + A_{\mu-1}B_{\mu-1,\mu} + \cdots + A_1 B_{1\mu} + A_p B_{p\mu} + \cdots + A_\mu B_{\mu\mu} \quad (6.3.3)$$

The "staggered" access to $A$ is done to reduce the chance of two processors accessing the same $A_j$ at the same time.

Here are the details:

**Algorithm 6.3.2** Suppose $A$, $B$, and $C$ are $n$-by-$n$ matrices residing in a shared memory accessible to $p$ processors. Assume that $n = pr$. If each processor executes the following algorithm, then upon completion $C$ is overwritten by $D = C + AB$.

**shared variables** $[A(1{:}n,1{:}n),\ B(1{:}n,1{:}n),\ C(1{:}n,1{:}n)]$
**glob.init**$[\,A,\ B,\ C\,]$
**loc.init**$[\,p,\ \mu = my.id,\ n,\ r = n/p\,]$
**get**$(B(:,1+(\mu-1)r{:}\mu r), B_{loc})$
**get**$(C(:,1+(\mu-1)r{:}\mu r), C_{loc})$
**for** $k = 1{:}p$
    $\tau = \mu - k$
    **if** $\tau \le 0$
        $\tau = \tau + p$
    **end**
    **get**$(A(:,1+(\tau-1)r{:}\tau r), A_{loc})$
    $\{\ A_{loc} = A_\tau = A(:,1+(\tau-1)r{:}\tau r)\ \}$
    $C_{loc} = C_{loc} + A_{loc}B_{loc}(1+(\tau-1)r{:}\tau r,:)$
**end**
**put**$(C_{loc}, C(:,1+(\mu-1)r{:}\mu r))$
**quit**

Note that in this load balanced procedure, each $C_\mu$ involves $(p+2)$ **get** statements and one **put**. It follows that the time each processor spends communicating with the shared memory is about

$$T_{6.3.2} = (p+3)(\alpha_s + \beta_s n^2/p) \tag{6.3.4}$$

The width $r$ of the block columns in (6.3.1) is completely determined by the problem size and the number of processors: $r = n/p$. It is important to understand that this rigid connection is neither necessary nor desirable. For example, $nr$ may exceed the length of a permissable message in a distributed memory system. In a shared memory system it may be the case that the three $n$-by-$r$ arrays required in Algorithm 6.3.2 do not fit in local memory. To circumvent these possible constraints we outline the tailoring of our two algorithms to the general blocking

$$[\,D_1,\ldots,D_N\,] \;=\; [\,C_1,\ldots,C_N\,] + [\,A_1,\ldots,A_N\,][\,B_1,\ldots,B_N\,] \tag{6.3.5}$$

where each block column has width $r = n/N$. The parameter $N$ can be set large enough so that messages or local buffers are small enough.

To modify Algorithm 6.3.1 with the simplifying assumption that $N = qp$, we assume that Proc($\mu$) houses the block columns $A_j$, $B_j$, and $C_j$ for $j = \mu{:}p{:}N$. The $A_i$ are again circulated via the merry-go-round principle only $q$ revolutions are required. During the $k$th revolution, block columns $A_{i+(k-1)p}$, $i = 1{:}p$ are circulated. The communication overhead associated with this procedure is given by

$$\tilde{T}_{6.3.1} = 2qp(\alpha_d + \beta_d nr) \tag{6.3.6}$$

To modify Algorithm 6.3.2 so that only three $n$-by-$r$ local buffers are required, we assign Proc($\mu$) the task of computing $D_j$ for $j = \mu{:}p{:}N$. For each $D_j$ computation, a processor begins by loading $B_j$ and $C_j$. All the $A_j$ must then be accessed during the modificationof $C_j$. After $C_j$ is completely updated it is restored to shared memory. The shared memory traffic overhead is given by

$$\tilde{T}_{6.3.2} = q(3+N)(\alpha_s + \beta_s nr)\,. \tag{6.3.7}$$

### 6.3.2 Load Balancing Concerns

Suppose the matrix $B$ in (6.3.1) is lower triangular. Since $D_1$ involves much more work that $D_p$ it is clear that a redistribution of work is needed in order to load balance Algorithms 6.3.1 and 6.3.2.

The solution to this problem lies in a wrap mapping of the data. In particular, we assign Proc($\mu$) the computation of

$$D(:,\mu{:}p{:}n) = C(:,\mu{:}p{:}n) + AB(:,\mu{:}p{:}n) = \sum_{\tau=1}^{p} A(:,\tau{:}p{:}n)B(\tau{:}p{:}n,\mu{:}p{:}n) \tag{6.3.8}$$

### 6.3.3 Some Granularity Concerns

As is obvious in Algorithms 6.3.1 and 6.3.2, granularity can be determined with an appropriate blocking. To illustate some additional aspects of the granularity issue, consider the revision of Algorithm 6.3.1 so that it efficiently handles a $C + AB$ computation of the form

$$[\,C_1\; C_2\; C_3\; C_4\,] + [\,A_1\; A_2\; A_3\; A_4\,]\begin{bmatrix} B_{11} & 0 & 0 & B_{14} \\ B_{21} & B_{22} & 0 & 0 \\ 0 & B_{32} & B_{33} & 0 \\ 0 & 0 & B_{43} & B_{44} \end{bmatrix}.$$

Note that the only nonlocal $A_\tau$ required by Proc($\mu$) is housed in its right neighbor. Thus, this particular $C + AB$ computation can carried out as follows:

$$\begin{array}{l} \textbf{send}(A_{loc}, left) \\ \textbf{recv}(W, right) \\ C_{loc} = C_{loc} + A_{loc}B_{\mu\mu} + WB_{\mu+1,\mu} \end{array} \tag{6.3.9}$$

where we adopt the convention that $B_{p+1,p} = B_{1p}$. Observe that a 2-dimensional workspace for $W$ is required.

A way of reducing this workspace requirement is to compute $W = A_{\mu+1}B_{\mu+1,\mu}$ as a sum of outer products. That way, Proc($\mu$) need only request $A_{\mu+1}$ from its right neighbor one column at a time:

$C_{loc} = C_{loc} + A_{loc}B_{loc}(1 + (\mu - 1)r{:}\mu r)$
**for** $t = 1{:}r$
    **send**$(A_{loc}(:,t), left)$
    **recv**$(w, right)$
    **if** $\mu < p$
        $i = t + \mu r$
    **else**
        $i = t$
    **end**
    $\{\ w = A(:,i)\ \}$
    $C_{loc} = C_{loc} + wB_{loc}(i,:)$
**end**

This is a more fine grained procedure than (6.3.9) thereby illustrating that granularity and buffer requirements can be related.

### 6.3.4 3-by-3 Systolic Multiplication

We now show how to implement the $N$-by-$N$ matrix multiplication problem $D = C + AB$ on an $N$-by-$N$ systolic torus. (See Figure 6.1.3.) Other network topologies could be used but we have chosen the torus for pedagogical reasons.

We begin with the $N = 3$ case. Assume that we have a 3-by-3 systolic torus which we depict in cellular form as follows:

| Proc(1,1) | Proc(1,2) | Proc(1,3) |
|---|---|---|
| Proc(2,1) | Proc(2,2) | Proc(2,3) |
| Proc(3,1) | Proc(3,2) | Proc(3,3) |

We show how to organize a 3-by-3 matrix multiplication $D = C + AB$ on this systolic network. The mission of Proc$(i,j)$ is to overwrite $c_{ij}$ with $d_{ij}$. At each timestep Proc$(i,j)$ incorporates one of the products $a_{ik}b_{kj}$ into a local running sum $c$ which houses $d_{ij}$ after three timesteps.

We first focus attention on Proc(1,1) and the calculation of

$$d_{11} = c_{11} + a_{11}b_{11} + a_{12}b_{21} + a_{13}b_{31} \ .$$

Suppose the six inputs that define this dot product are positioned within the torus as follows:

| $a_{11}$ $b_{11}$ | $a_{12}$ · | $a_{13}$ · |
|---|---|---|
| · $b_{21}$ | · · | · · |
| · $b_{31}$ | · · | · · |

(Pay no attention to the "dots." They are later be replaced by various $a_{ij}$ and $b_{ij}$).

Now in our systolic model each processor executes a program of the form (6.1.1). Thus, we can "ratchet" the first row of $A$ and the first column of $B$ through Proc(1,1) so that at each time step a new $\{a_{1k}, b_{k1}\}$ pair is incorporated in a running sum $c$ used for the accumulation of $c_{11}$. The data flow for this particular portion of the overall computation is illustrated as follows:

| $a_{12}$ $b_{21}$ | $a_{13}$ · | $a_{11}$ · |
|---|---|---|
| · $b_{31}$ | · · | · · |
| · $b_{11}$ | · · | · · |

$t = 1$
$c = c + a_{12}b_{21}$

| $a_{13}$ $b_{31}$ | $a_{11}$ · | $a_{12}$ · |
|---|---|---|
| · $b_{11}$ | · · | · · |
| · $b_{21}$ | · · | · · |

$t = 2$
$c = c + a_{13}b_{31}$

| $a_{11}$ $b_{11}$ | $a_{12}$ · | $a_{13}$ · |
|---|---|---|
| · $b_{21}$ | · · | · · |
| · $b_{31}$ | · · | · · |

$t = 3$
$c = c + a_{11}b_{11}$

Thus, after three steps, the local variable $c$ in Proc(1,1) houses $d_{11}$.

We have organized the flow of data so that the $a_{1j}$ migrate westwards and the $b_{i1}$ migrate northwards through the torus. It is thus apparent that Proc(1,1) must execute a node program of the form:

**for** $t = 1{:}3$
    **send**($a$, *west*); **send**($b$, *north*)
    **recv**($a$, *east*); **recv**($b$, *south*)
    $c = c + ab$
**end**

We next consider the activity in Proc(1,2), Proc(1,3), Proc(2,1), and Proc(3,1). At this point in the development these processors are merely help circulate the elements of $A(1,1{:}3)$ and $B(1{:}3,1)$. They can do more. Consider Proc(1,2). During $t = 1{:}3$ its houses $a_{13}$, $a_{11}$ and $a_{12}$ respectively. If $b_{32}$, $b_{12}$, and $b_{22}$ flowed through Proc(1,2) during these timesteps, then $d_{12} = c_{12} + a_{13}b_{32} + a_{11}b_{12} + a_{12}b_{22}$ could be formed. Likewise, Proc(1,3) could compute $d_{13} = c_{13} + a_{11}b_{13} + a_{12}b_{23} + a_{13}b_{33}$ if $b_{13}$, $b_{23}$, and $b_{33}$ are available during $t = 1{:}3$. To this end we initialize the torus as follows

| | | |
|---|---|---|
| $a_{11}$ $b_{11}$ | $a_{12}$ $b_{22}$ | $a_{13}$ $b_{33}$ |
| · $b_{21}$ | · $b_{32}$ | · $b_{13}$ |
| · $b_{31}$ | · $b_{12}$ | · $b_{23}$ |

With northward flow of the $b_{ij}$ we get

| | | | |
|---|---|---|---|
| $a_{12}$ $b_{21}$ | $a_{13}$ $b_{32}$ | $a_{11}$ $b_{13}$ | |
| · $b_{31}$ | · $b_{12}$ | · $b_{23}$ | $t = 1$ |
| · $b_{11}$ | · $b_{22}$ | · $b_{33}$ | |

| | | | |
|---|---|---|---|
| $a_{13}$ $b_{31}$ | $a_{11}$ $b_{12}$ | $a_{12}$ $b_{23}$ | |
| · $b_{11}$ | · $b_{22}$ | · $b_{33}$ | $t = 2$ |
| · $b_{21}$ | · $b_{32}$ | · $b_{13}$ | |

| | | | |
|---|---|---|---|
| $a_{11}$ $b_{11}$ | $a_{12}$ $b_{22}$ | $a_{13}$ $b_{33}$ | |
| · $b_{21}$ | · $b_{32}$ | · $b_{13}$ | $t = 3$ |
| · $b_{31}$ | · $b_{12}$ | · $b_{23}$ | |

Thus, if $B$ is mapped onto the torus in a "staggered start" fashion, we can arrange for the first row of processors to compute the first row of $C$.

If we stagger the second and third rows of $A$ in a similar fashion, we can arrange for all nine processors to perform a multiply-add at each step. In particular, if we set

| | | | | | |
|---|---|---|---|---|---|
| $a_{11}$ | $b_{11}$ | $a_{12}$ | $b_{22}$ | $a_{13}$ | $b_{33}$ |
| $a_{22}$ | $b_{21}$ | $a_{23}$ | $b_{32}$ | $a_{21}$ | $b_{13}$ |
| $a_{33}$ | $b_{31}$ | $a_{31}$ | $b_{12}$ | $a_{32}$ | $b_{23}$ |

then with westward flow of the $a_{ij}$ and northward flow of the $b_{ij}$ we obtain

$t = 1$

| | | | | | |
|---|---|---|---|---|---|
| $a_{12}$ | $b_{21}$ | $a_{13}$ | $b_{32}$ | $a_{11}$ | $b_{13}$ |
| $a_{23}$ | $b_{31}$ | $a_{21}$ | $b_{12}$ | $a_{22}$ | $b_{23}$ |
| $a_{31}$ | $b_{11}$ | $a_{32}$ | $b_{22}$ | $a_{33}$ | $b_{33}$ |

$t = 2$

| | | | | | |
|---|---|---|---|---|---|
| $a_{13}$ | $b_{31}$ | $a_{11}$ | $b_{12}$ | $a_{12}$ | $b_{23}$ |
| $a_{21}$ | $b_{11}$ | $a_{22}$ | $b_{22}$ | $a_{23}$ | $b_{33}$ |
| $a_{32}$ | $b_{21}$ | $a_{33}$ | $b_{32}$ | $a_{31}$ | $b_{13}$ |

$t = 3$

| | | | | | |
|---|---|---|---|---|---|
| $a_{11}$ | $b_{11}$ | $a_{12}$ | $b_{22}$ | $a_{13}$ | $b_{33}$ |
| $a_{22}$ | $b_{21}$ | $a_{23}$ | $b_{32}$ | $a_{21}$ | $b_{13}$ |
| $a_{33}$ | $b_{31}$ | $a_{31}$ | $b_{12}$ | $a_{32}$ | $b_{23}$ |

From this example we are ready to specify the general $N$ algorithm.

### 6.3.5 The General Case

To describe the distribution of the $A$ and $B$ matrices for the general $N$-by-$N$ case it is handy to define an integer function $< \cdot >_N$ as follows

$$\left.\begin{array}{l} a = \alpha N + \beta \\ 0 \le \beta < N \end{array}\right\} \Longrightarrow \; < a >_N = \left\{\begin{array}{ll} \beta & \beta \ne 0 \\ N & \beta = 0 \end{array}\right.$$

**Algorithm 6.3.3** Suppose $A \in \mathbb{R}^{N\times N}$, $B \in \mathbb{R}^{N\times N}$, and $C \in \mathbb{R}^{N\times N}$ are given and that $D = C + AB$. If each processor in an $N$-by-$N$ systolic torus executes the following algorithm, then upon completion Proc$(i,j)$ houses $d_{ij}$ in local variable $c$.

**loc.init**[ $N$, $(i,j) = my.id$, $a = a_{i,<i+j-1>_N}$, $b = b_{<i+j-1>_N,j}$ ]
**for** $t = 1{:}N$
  **send**$(a, west)$; **send**$(b, north)$
  **recv**$(a, east)$; **recv**$(b, south)$
  $c = c + ab$
**end**
**quit**

It is not hard to prove that this algorithm works. First observe that at time $t$, Proc$(i,j)$ houses $a_{i,<i+j-1+t>_N}$ and $b_{<i+j-1+t>_N,j}$. Upon termination

$$c \;=\; c_{ij} + \sum_{t=1}^{N} a_{i,<i+j-1+t>_N} b_{<i+j-1+t>_N,j} \;=\; c_{ij} + \sum_{k=1}^{N} a_{ik}b_{kj} \;=\; d_{ij}\ .$$

One of the appeals of the systolic model is that the regimentation makes node program verification relatively straight forward.

### 6.3.6 A Block Analog

Suppose $A$, $B$, and $C$ are $N$-by-$N$ block matrices with $r$-by-$r$ blocks. It is easy to modify Algorithm 6.3.3 so that it computes

$$D = \begin{bmatrix} C_{11} & \cdots & C_{1N} \\ \vdots & & \vdots \\ C_{N1} & \cdots & C_{NN} \end{bmatrix} + \begin{bmatrix} A_{11} & \cdots & A_{1N} \\ \vdots & & \vdots \\ A_{N1} & \cdots & A_{NN} \end{bmatrix} \begin{bmatrix} B_{11} & \cdots & B_{1N} \\ \vdots & & \vdots \\ B_{N1} & \cdots & B_{NN} \end{bmatrix}$$

All we have to do is replace the references to $a_{ij}$, $b_{ij}$, and $c_{ij}$ with references to blocks $A_{ij}$, $B_{ij}$, and $C_{ij}$.

### 6.3.7 An Asynchronous Toroidal Procedure

We next develop an asynchronous analog of Algorithm 6.3.3 for computing $D = C + AB$. Assume $A \in \mathbb{R}^{N\times N}$ and $B \in \mathbb{R}^{N\times N}$ and that as in Algorithm 6.3.4, the mission of Proc$(i,j)$ is to compute $d_{ij}$. To make things interesting we assume that Proc$(i,j)$ initially houses $a_{ij}$, $b_{ij}$, and $c_{ij}$. In order to obtain the staggering of the data that proved critical to Algorithm 6.3.3, the rows of $A$ will circulate westward across the rows of the torus while the columns of $B$ circulate northward up the columns of the torus.

**Algorithm 6.3.4** Suppose $A \in \mathbb{R}^{N\times N}$, $B \in \mathbb{R}^{N\times N}$, and $C \in \mathbb{R}^{N\times N}$ are given and that $D = C + AB$. If each processor in an $N$-by-$N$ torus executes the following algorithm, then upon completion Proc$(i,j)$ houses $d_{ij}$ in local variable $c$.

```
loc.init[N, (μ, λ) = my.id, a = a_{μ,λ}, b = b_{μ,λ}, c = c_{μ,λ}]
for k = 1:μ − 1
    send(a, west); recv(a, east)
end
for k = 1:λ − 1
    send(b, north); recv(b, south)
end
for k = 1:n
    c = c + ab
    send(a, west); recv(a, east)
    send(b, north); recv(b, south)
end
for k = 1:μ − 1
    send(a, east); recv(a, west)
end
for k = 1:λ − 1
    send(b, south); recv(b, north)
end
quit
```

### 6.3.8 Block Dot Approach Via Pool-of-Tasks

We now examine the shared memory computation of the $n$-by-$n$ matrix $D = C + AB$ that is based upon block dot products. Assume that $n = Nr$ and that $A = (A_{ij})$, $B = (B_{ij})$, $C = (C_{ij})$, and $D = (D_{ij})$ are $N$-by-$N$ block matrices with $r$-by-$r$ blocks. Thus,

$$D_{ij} = C_{ij} + \sum_{k=1}^{N} A_{ik} B_{kj} \ .$$

We regard the computation of $D_{ij}$ as task$(i, j)$ and develop a parallel algorithm based on the pool-of-tasks approach. To make things interesting, let us assume that $B$ is to be overwritten by $D$.

We organize the computation so that $D$ is computed block-by-block in column major order. After the $j$th block column of $D$ is computed, then it is safe to overwite the $j$th block column of $B$. Thus, we seek a procedure of the following form

```
Repeat until no more tasks:
    Get the next task, call it task(i, j).
    Compute D_ij
    If it is safe, overwrite B_ij with D_ij.
end
```

As in our pool-of-task gaxpy procedures, we need a monitor whose mission is to assign tasks. The situation here is two-dimensional in that tasks are most conveniently labeled with a pair of indices that identify the $D_{ij}$ to be computed:

$$(1,1), \ldots, (N,1), (1,2), \ldots, (N,2), \ldots, (1,N), \ldots, (N,N).$$

So analogous to **nexti** we have

```
monitor: nextij
    protected variables: flag, row, col, idle.proc
    condition variables: waiti
    initializations: flag = 0
    procedure nextij.index(N, p, more, i, j)
        if flag = 0
            row = 0; col = 1; idle.proc = 0; flag = 1
        end
        if row ≤ N − 1
            row = row + 1
        else
            row = 1; col = col + 1
        end
        if col ≤ N
            i = row; j = col; more = true
        else
            more = false; idle.proc = idle.proc + 1
            if idle.proc ≤ p − 1
                delay(waiti)
            else
                flag = 0
            end
            continue(waiti)
        end
    end nextij
```

Safe overwriting can be managed by another monitor with a procedure **prod.canwrite**. In particular, if **prod.canwrite**$(W, B, i, j, r, N)$ safely overwrites block $B_{ij}$ with the $r$-by-$r$ matrix $W$, then we obtain

**Algorithm 6.3.5** Suppose $n = Nr$ and that $A = (A_{ij})$, $B = (B_{ij})$, and $C = (C_{ij})$, are $N$-by-$N$ block matrices with $r$-by-$r$ blocks. If these matrices are stored by block in shared memory and each processor executes the following algorithm, then upon completion $B_{ij}$ is overwritten with $C_{ij} + \sum_{k=1}^{N} A_{ik}B_{kj}$ for $i = 1{:}N$ and $j = 1{:}N$. It is assumed that $p \geq N$.

```
call nextij.index(N, p, more, i, j)
while more = true
    get(C_ij, C_loc)
    for k = 1:N
        get(A_ik, A_loc); get(B_kj, B_loc); C_loc = C_loc + A_loc B_loc
    end
    call prod.canwrite(C_loc, i, j, N)
    call nextij.index(N, p, more, i, j)
end
```

The monitor **prod** that ensures safe overwriting is structured as follows:

```
monitor: prod
    protected variables: flag, col.done(·)
    condition variables: waitq(·)
    initializations: flag = 0
    procedure prod.canwrite (W, i, j, N)
        if flag = 0
            flag = 1; col.done(1:N) = 0
        end
        col.done(j) = col.done(j) + 1
        if col.done(j) < N
            delay(waitq(j))
        end
        put(W, B_ij)
        continue(waitq(j))
    end canwrite
end prod
```

The protected vector *col.done* is used to count the number of blocks that have been found in each block column of $D$. When a processor calls **prod.canwrite** with $D_{ij}$ stored in $C_{loc}$, *col.done*$(j)$ is incremented. If the resulting value is less than $N$ then it is not safe to overwrite $B_{ij}$ as another task may yet require its value. When this is the case the calling processor is placed on a queue controlled by the condition variable *waitq*$(j)$. After the $N$th block in block column $j$ is available, it is stored in the appropriate $B_{ij}$ and the processors waiting on *waitq*$(j)$ are released to proceed with the storage of their $D_{ij}$.

The assumption that $A$, $B$, and $C$ are stored by block (c.f. §1.3), implies that the $A_{ij}$ and $B_{ij}$ can be retrieved by a single **get** or stored by a single **put**.

Without the assumption $p \geq N$ in Algorithm 6.3.5 the processors would all get delayed on *waitq*$(1)$ as the $N$th block in the first block column would never be computed.

If we assume no idle waiting and that each processor handles approximately $N^2/p$ tasks, then each processor spends

$$T_{6.3.5} = \frac{N^2}{p}(2N+2)(\alpha_s + \beta_s r^2) \tag{6.3.9}$$

seconds communicating with shared memory.

This approach to the $D = C + AB$ problem can involve less shared memory traffic than a block saxpy algorithm based on (6.3.5). The argument is very similar to our data-reuse discussion in §1.4.9. That discussion applies here with local memory and shared memory corresponding to cache and main memory. Of course, the local memories in a shared memory system may have their own hierarchy.

### Problems

**P6.3.1** Extend algorithms 6.3.1 and 6.3.2 so that they overwrite $C$ with $C + ABA^T$.

**P6.3.2** Develop detailed versions of Algorithms 6.3.1 and 6.3.2 based upon (6.3.5).

**P6.3.3** Give a detailed, structure-exploiting algorithm for the computation $C_{loc} = C_{loc} + A_{loc}B_{loc}(\tau{:}p{:}n,:)$ in Algorithm 6.3.3.

**P6.3.4** Develop a pool-of-tasks algorithm in which $A$ is overwritten by $D = C + AB$ where $B$ is lower triangular. Let task($j$) be the computation of $D(:,j)$.

**P6.3.5** Modify Algorithm 6.3.4 so that $C$ is overwritten by $D = C + A^kB$.

**P6.3.6** Develop a load balanced version of Algorithm 6.3.1 to handle the case when $B$ is lower triangular.

**P6.3.7** Specialize Algorithm 6.3.1 to handle the case when $B$ is tridiagonal. Let $e(1{:}n)$, $d(1{:}n)$, and $f(1{:}n)$ denote the subdiagonal, diagonal, and superdiagonal part with $e(n) = f(1) = 0$. Assume that Proc($\mu$) houses $e(col)$, $d(col)$, and $f(col)$ where $col = 1 + (\mu - 1)r{:}\mu r$.

**P6.3.8** An upper triangular matrix can be overwritten with its square without any additional workspace. Write a dynamically scheduled procedure for doing this.

### Notes and References for Sec. 6.3

Various aspects of parallel matrix multiplication are discussed in

K.H. Cheng and S. Sahni (1987). "VLSI Systems for Band Matrix Multiplication," *Parallel Computing 4*, 239–258.

A. Elster and A.P. Reeves (1988). "Block Matrix Operations Using Orthogonal Trees," in G. Fox (ed), *The Third Conference on Hypercube Concurrent Computers and Applications, Vol. II - Applications*, ACM Press, New York, pp. 1554–1561.

G. Fox, M. Johnson, G. Lyzenga, S. Otto, J. Salmon, and D. Walker (1988). *On Concurrent Processors Vol I: General Techiques and Regular Problems*, Prentice-Hall, Englewood Cliffs, NJ.

G. Fox, S.W. Otto, and A.J. Hey (1987). "Matrix Algorithms on a Hypercube I: Matrix Multiplication," *Parallel Computing 4*, 17–31.

S.L. Johnsson and C.T. Ho (1987). "Algorithms for Multiplying Matrices of Arbitrary Shapes Using Shared Memory Primatives on a Boolean Cube," Report YALEU/DCS/RR-569, Dept. of Computer Science, Yale University, New Haven, CT.

# 6.4 Ring Factorization Procedures

In this section we discuss how various matrix factorizations can be organized on a ring. After presenting a parallel implementation for gaxpy-based Cholesky, we show how the same scheduling logic works for the factorizations $PA = LU$ and $A = QR$. We conclude with a discussion about the parallel solution of triangular systems.

## 6.4.1 A Ring Cholesky Algorithm: The n=p Case

Let us see how the Cholesky factorization procedure can be distributed on a ring of $p$ processors Proc(1), ..., Proc($p$). The starting point is the equation

$$G(j,j)G(j{:}n,j) \;=\; A(j{:}n,j) - \sum_{k=1}^{j-1} G(j,k)G(j{:}n,k) \;\equiv\; v(j{:}n)\,.$$

This equation is obtained by equating the $j$th columns in the $n$-by-$n$ equation $A = GG^T$. Once the vector $v(j{:}n)$ is found then $G(j{:}n,j)$ is a simple scaling: $G(j{:}n,j) = v(j{:}n)/\sqrt{v(j)}$. For clarity, we first assume that $n = p$ and that Proc($\mu$) initially houses $A(\mu{:}n,\mu)$. Upon completion, each processor overwrites its $A$-column with the corresponding $G$-column. For Proc($\mu$) this process involves $\mu - 1$ saxpy updates of the form $A(\mu{:}n,\mu) \longleftarrow A(\mu{:}n,\mu) - G(\mu,j)G(\mu{:}n,j)$ followed by a square root and a scaling. The general structure of Proc($\mu$)'s node program is therefore as follows:

> **for** $j = 1{:}\mu - 1$
>     Receive a $G$-column from the left neighbor.
>     Send a copy of the received $G$-column to the right neighbor.
>     Update $A(\mu{:}n,\mu)$ .
> **end**
> Generate $G(\mu{:}n,\mu)$ and send it to the right neighbor.

Thus Proc(1) immediately computes $G(1{:}n,1) = A(1{:}n,1)/\sqrt{A(1,1)}$ and sends it to Proc(2). As soon as Proc(2) receives this column it can generate $G(2{:}n,2)$ and pass it to Proc(3) etc.. With this pipelining arrangement we can assert that once a processor computes its $G$-column, it can quit. It also follows that each processor receives $G$-columns in ascending order, i.e., $G(:,1)$, $G(:,2)$, etc. Based on these observations we have

**Algorithm 6.4.1** Suppose $A \in \mathbb{R}^{n\times n}$ is symmetric and positive definite and that $A = GG^T$ is its Cholesky factorization. If each node in an $n$-processor ring executes the following program, then upon completion Proc($\mu$) houses $G(\mu{:}n,\mu)$ in the local array $A_{loc}(\mu{:}n)$.

```
loc.init[n, μ = my.id, left, right, A_loc(μ:n) = A(μ:n, μ)]
j = 1
while j < μ
    recv(g(j:n), left)
    if μ < n
        send(g(j:n), right)
    end
    A_loc(μ:n) = A_loc(μ:n) − g(μ)g(μ:n)
    j = j + 1
end
A_loc(μ:n) = A_loc(μ:n)/sqrt(A_loc(μ))
if μ < n
    send(A_loc(μ:n), right)
end
quit
```

Note that the number of received $G$-columns is given by $j-1$. If $j = \mu$, then it is time for Proc($\mu$) to generate and send $G(\mu{:}n, \mu)$.

### 6.4.2 Ring Cholesky: The General Case

We now extend Algorithm 6.4.1 to the general $n$ case. There are two obvious ways to distribute the computation. We could require each processor to compute a contiguous set of $G$-columns. For example, if $n = 11$, $p = 3$, and $A = [\, a_1, \ldots, a_{11} \,]$ then we could distribute $A$ as follows

$$\Big[\, \underbrace{a_1\ a_2\ a_3\ a_4}_{\text{Proc(1)}} \mid \underbrace{a_5\ a_6\ a_7\ a_8}_{\text{Proc(2)}} \mid \underbrace{a_9\ a_{10}\ a_{11}}_{\text{Proc(3)}} \,\Big] .$$

Each processor could then proceed to find the corresponding $G$ columns. The trouble with this approach is that (for example) Proc(1) is idle after the fourth column of $G$ is found even though much work remains.

Greater load balancing results if we distribute the computational tasks using the wrap mapping, i.e.,

$$\Big[\, \underbrace{a_1\ a_4\ a_7\ a_{10}}_{\text{Proc(1)}} \mid \underbrace{a_2\ a_5\ a_8\ a_{11}}_{\text{Proc(2)}} \mid \underbrace{a_3\ a_6\ a_9}_{\text{Proc(3)}} \,\Big] .$$

In this scheme Proc($\mu$) carries out the construction of $G(:, \mu{:}p{:}n)$. When a given processor finishes computing its $G$-columns, each of the other processors has at most one more $G$ column to find. Thus if $n/p \gg 1$ then all of the processors are busy most of the time.

Let us examine the details of a wrap-distributed Cholesky procedure. Each processor maintains a pair of counters. The counter $j$ is the index of the next $G$-column to be received by Proc($\mu$). A processor also needs to know the index of the next $G$-column that it is to produce. Note that if $col = \mu{:}p{:}n$ then Proc($\mu$) is responsible for $G(:, col)$ and that $L = \textbf{length}(col)$ is the number of the $G$-columns that it must compute. We use $q$ to indicate the status of $G$-column production. At any instant, $col(q)$ is the index of the next $G$-column to be produced.

**Algorithm 6.4.2** Suppose $A \in \mathbb{R}^{n \times n}$ is symmetric and positive definite and that $A = GG^T$ is its Cholesky factorization. If each node in a $p$-processor ring executes the following program, then upon completion Proc($\mu$) houses $G(k{:}n, k)$ for $k = \mu{:}p{:}n$ in a local array $A_{loc}(1{:}n, L)$ where $L = \textbf{length}(col)$ and $col = \mu{:}p{:}n$. In particular, $G(col(q){:}n, col(q))$ is housed in $A_{loc}(col(q){:}n, q)$ for $q = 1{:}L$.

```
loc.init[p, μ = pid(·), left, right, n, A_loc = A(1:n, μ:p:n)]
j = 1; q = 1; col = μ:p:n; L = length(col)
while q ≤ L
     if j = col(q)
          { Form G(j:n, j) }
          A_loc(j:n, q) = A_loc(j:n, q)/√A_loc(j, q)
          if j < n
               send(A_loc(j:n, q), right)
          end
          j = j + 1
          { Update local columns. }
          for k = q + 1:L
               r = col(k)
               A_loc(r:n, k) = A_loc(r:n, k) − A_loc(r, q)A_loc(r:n, q)
          end
          q = q + 1
     else
          recv(g(j:n), left)
          Compute α where Proc(α) generated the received G-column.
          Compute β = index of Proc(right)'s final column.
          if right ≠ α ∧ j < β
               send(g(j:n), right)
          end
          { Update local columns. }
          for k = q:L
               r = col(k); A.loc(r:n, k) = A_loc(r:n, k) − g(r)g(r:n)
          end
          j = j + 1
     end
end
quit
```

To illustrate the logic of the pointer system we consider a sample 3-processor situation with $n = 11$. Assume that the three local values of $q$ are 3,2, and 2 and that the corresponding values of $col(q)$ are 7, 5, and 6 :

$$\Big[\; \underbrace{a_1\ a_4\ \overset{\downarrow}{a_7}\ a_{10}}_{\text{Proc}(1)} \;\Big|\; \underbrace{a_2\ \overset{\downarrow}{a_5}\ a_8\ a_{11}}_{\text{Proc}(2)} \;\Big|\; \underbrace{a_3\ \overset{\downarrow}{a_6}\ a_9}_{\text{Proc}(3)} \;\Big]$$

Proc(2) now generates the fifth $G$-column and increment its $q$ to 3.

The logic associated with the decision to pass a received $G$-column to the right neighbor needs to be explained. Two conditions must be fulfilled:

- The right neighbor must not be the processor which generated the $G$ column. This way the circulation of the received $G$-column is properly terminated.
- The right neighbor must still have more $G$-columns to generate. Otherwise, a $G$-column will be sent to an inactive processor.

This kind of reasoning is quite typical in distributed memory matrix computations.

Let us examine the behavior of Algorithm 6.4.2 under the assumption that $n \gg p$. It is not hard to show that Proc($\mu$) performs

$$f(\mu) = \sum_{k=1}^{L} 2(n - (\mu + (k-1)p))(\mu + (k-1)p) \approx n^3/p$$

flops. Note that $f(1)/f(p) = 1 + O(p^2/n^2)$ and so we have approximate load balancing from the flop point of view.

Each processor receives and sends just about every $G$-column. Using our communication overhead model (6.1.4) we see that the time each processor spends communicating is given by

$$\sum_{j=1}^{n} 2(\alpha_d + \beta_d(n-j)) \approx 2\alpha_d n + \beta_d n^2 .$$

If we assume that computation proceeds at $R$ flops per second then the computation/communication ratio for Algorithm 6.4.2 is approximately given by $(n/p)(2/3R\beta_d)$. Thus, communication overheads diminish in importance as $n/p$ grows.

## 6.4.3 Ring Procedures for Other Factorizations

It is very easy to modify Algorithm 6.4.2 so that it computes other factorizations such as $PA = LU$ and $A = QR$. This is because in the $n$-by-$n$ case the algorithms for these factorizations can be structured as follows:

```
for j = 1:n − 1
    Compute a transformation T_j based on the current A(:, j).
    Update A(j:n, j:n).
end
```

Sample $T_j$ include Gauss and Householder transformations and sample updates include the saxpy operation.

To stress the commonality between the different ring factorization procedures we present a procedure "template" based on Algorithm 6.4.2:

```
loc.init[p, μ = my.id, left, right, n, A_loc = A(1:n, μ:p:n)]
j = 1; q = 1; col = μ:p:n; L = length(col)
while q ≤ L
    if j = col(q)
        Compute the transformation T_j.
        if j < n                                          (6.4.1)
            Send T_j to right neighbor.
        end
        Apply T_j to local columns.
        Save T_j.
        j = j + 1
        q = q + 1
    else
        Receive T_j from left neighbor.
        Send T_j to right neighbor if necessary.
        Apply T_j to local columns.
        j = j + 1
    end
end
quit
```

It is important to stress that in practice an appropriate "representation" of $T_j$ is manipulated at all times. For example, in the Cholesky factorization, $T_j$ is encoded in the vector $G(j{:}n, j)$ . In LU it is the $j$th Gauss vector. In QR it is the $j$th Householder vector, etc. These "vector" representations are passed around the ring, not the explicit transformations.

### 6.4.4 Parallel Triangular System Solving

Usually after a matrix is factored the factors are used to solve linear systems. Triangular system solving with a single processor has some interesting aspects which we covered in §3.1, but it has never been a "hot topic" simply because factorizations cost $O(n^3)$ while back substitution and forward elimination are $O(n^2)$. However, with a distributed memory multiprocessor it is not obvious that a parallel $Ax = b$ solver is dominated by the cost of the parallel factorization. Compared to (say) the Cholesky factorization, there is much less "room" for parallelizing forward and back substitution. The point of this subsection is to detail one particular approach to triangular system solving on a ring that is highly efficient. Used in conjunction with the above parallel factorization schemes, we can continue to maintain the view that triangular system solving is a cheap "postscript" to the factorization computation.

For simplicity, we focus on the $n$-by-$n$ unit lower triangular problem $Lx = b$. Assume that we have a $p$-processor ring and that $n$ is an integral multiple of $p$, $n = pr$. We also assume that $L$ and $b$ are wrap mapped, i.e., that Proc($\mu$) houses $L(:, \mu{:}p{:}n)$ and $b(\mu{:}p{:}n)$. How to proceed without these assumptions is pursued in the exercises and in Li and Coleman (1988) upon which our discussion is based.

The starting point is to write saxpy-based forward substitution as follows:

```
j = 0; z(1:n) = 0
while j < n
      j = j + 1; b(j) = b(j) - z(j)                              (6.4.2)
      z(j + 1:n) = z(j + 1:n) + L(j + 1:n, j)b(j)
end
```

In the ring procedure to be developed, the mission of Proc($\mu$) is to overwrite $b(\mu{:}p{:}n)$ with $x(\mu{:}p{:}n)$. To accomplish this we circulate the $z$ vector in (6.4.2) around the ring $r$ times beginning at Proc(1). We number these $z$-revolutions from 0 to $r - 1$. When $z$ "visits" Proc($\mu$) during the $k$th revolution, it is used to compute $x(\mu + kp)$. Formally we have

```
loc.init[p, μ = my.id, left, right, L_loc = L(1:n, μ:p:n),
      b_loc = b(μ:p:n) ]
r = n/p; col = μ:p:n; k = 0; z(1:n) = 0
while k < r - 1
      { Proc(1) receives nothing at the start of revolution 0. }
      if k ≠ 0 ∨ μ > 1
            recv(z, left)
      end                                                         (6.4.3)
      k = k + 1; j = col(k)
      { Compute x_j and, if necessary, update z and send it right. }
      b_loc(k) = b_loc(k) - z(j)
      if k ≠ r - 1 ∨ μ < p
            z(j + 1:n) = z(j + 1:n) + L_loc(j + 1:n, k)b_loc(k)
            send(z, right)
      end
end
quit
```

Although this algorithm is distributed, it has the property that only one processor is computing at any given instant. To introduce a measure of parallelism we notice that the $z$ vector need not be completely updated before its right neighbor can begin solving for the next unknown.. For example, Proc(2) can begin work on $x(2)$ as soon as it receives $x(1)$ from Proc(1). Thus, Proc(1) should forward $x(1)$ *and then* update $z(2{:}n)$. However, Proc(3), which will compute $x(3) = b(3) - L(3,1)x(1) - L(3,2)x(2)$ needs Proc(1) to also update $z(3)$. In general, after Proc($\mu$) computes $x(j)$, the quantity $L(j+1{:}j+p-1)x(j)$ must be available to processors $\mu+1,\ldots,p,1,\ldots,\mu-1$.

What this suggests is that something less than the complete $z$ vector should circulate around the ring. Let us call that something "$w$" and try to discover what it must be. Assume that $x(\mu+kp)$ is determined when $w$ visits Proc($\mu$) during the $k$th revolution. We also assume that Proc($\lambda$) houses the vector $z_\lambda^{(k)} = L(:,\lambda{:}p{:}\lambda+kp)x(\lambda{:}p{:}\lambda+kp)$ after the $k$th revolution. Now let us look more carefully at what Proc($\mu$) needs to compute $x(\mu+kp)$ in terms of the $z_\lambda^{(k)}$. For $j = \mu + kp$ we have

$$x(j) = b(j) - \sum_{i=1}^{j-1} L(j,i)x(i) = b(j) - \sum_{i=1}^{\min(p,j-1)} L(j,i{:}p{:}j-1)x(i{:}p{:}j-1)\,.$$

At the time when this calculation is to be performed at Proc($\mu$),

$$L(j,i{:}p{:}j-1)x(i{:}p{:}j-1) = \begin{cases} z_i^{(k)}(j) & i = 1{:}\mu-1 \\ z_i^{(k-1)}(j) & i = \mu{:}p \end{cases}$$

and so

$$x(j) = b(j) - \sum_{i=1}^{\mu-1} z_i^{(k)}(j) - z_\mu^{(k-1)}(j) - \sum_{i=\mu+1}^{p} z_i^{(k-1)}(j)\,.$$

This suggests that the circulating $w$ be a $p$-vector with the property that

$$w(\mu) = \sum_{i=1}^{\mu-1} z_i^{(k)}(j) + \sum_{i=\mu+1}^{p} z_i^{(k-1)}(j)$$

upon arrival at Proc($\mu$) during the $k$th revolution. Note that the values that make up the sum can be "picked up" during the $p-1$ stops that the circulating $w$ makes prior to arrival at Proc($\mu$). Here is how these ideas can be implemented:

**Algorithm 6.4.3** Suppose $L \in \mathbb{R}^{n \times n}$ is unit lower triangular and that $x \in \mathbb{R}^n$ solves $Lx = b$ where $b \in \mathbb{R}^n$. If $n = pr$ and each processor in a $p$-processor ring executes the following program, then upon completion Proc($\mu$) houses $x(\mu{:}p{:}n)$ in $b_{loc}(1{:}r)$.

```
loc.init[p, μ = my.id, left, right, L_loc = L(1:n, μ:p:n),
        b_loc = b(μ:p:n) ]
r = n/p; col = μ:p:n; k = 0; z(1:n) = 0; w(1:p) = 0
while k < r − 1
    if k ≠ 0 ∨ μ > 1
        recv(w, left)
    end
    k = k + 1; j = col(k)
    b_loc(k) = b_loc(k) − z(j) − w(μ)
    if k ≠ r − 1 ∨ μ < p
        { Update the urgently needed part of z. }
        q = min(n, j + p − 1)
        z(j + 1:q) = z(j + 1:q) + L_loc(j + 1:q, k)b_loc(k)
        { Update the circulating vector w. }
        if k < r − 1
            w(1:μ − 1) = w(1:μ − 1) + z(j + p − μ + 1:j + p − 1)
            w(μ) = 0;
        end
        w(μ + 1:p) = w(μ + 1:p) + z(j + 1:j + p − μ)
        send(w, right)
        { Update the rest of z. }
        if q < n
            z(q + 1:n) = z(q + 1:n) + L_loc(q + 1:n, k)b_loc(k)
        end
    end
end
quit
```

We refer the reader to Li and Coleman (1988) for evidence that close to linear speed-up can be attained with this algorithm.

### Problems

**P6.4.1** Complete the template (6.4.1) so that it computes the QR factorization. Upon completion Proc($\mu$) should house both the Householder vectors and the $R$-columns that it generates.

**P6.4.2** Complete the template (6.4.1) so that it computes the $PA = LU$ factorization. Upon completion, $Proc(\mu)$ should house a suitable encoding of $P$ and the portions of $L$ and $U$ that it generated.

**P6.4.3** Explain why the messages (the $G$-vectors) can be made shorter in Algorithm 6.4.2.

**P6.4.4** Modify Algorithm 6.4.3 so that it can handle general lower triangular systems of arbitrary dimension.

**P6.4.5** Modify Algorithm 6.4.3 so that it solves the unit upper triangular system $Ux = b$ of dimension $n = pr$.

**P6.4.6** Modify Algorithm 6.4.3 so that it solves $L^T x = b$.

### Notes and References for Sec. 6.4

General comments on distributed memory factorization procedures may be found in

G.A. Geist and M.T. Heath (1986). "Matrix Factorization on a Hypercube," in M.T. Heath (ed) (1986). *Proceedings of First SIAM Conference on Hypercube Multiprocessors*, SIAM Publications, Philadelphia, Pa.
in *Hypercube Multiprocessors* , ed. M.T. Heath, SIAM Press, 161–180.

I.C.F. Ipsen, Y. Saad, and M. Schultz (1986). "Dense Linear Systems on a Ring of Processors," *Lin. Alg. and Its Applic. 77,* 205–239.

D.P. O'Leary and G.W. Stewart (1986). "Assignment and Scheduling in Parallel Matrix Factorization," *Lin. Alg. and Its Applic. 77,* 275–300.

Papers specifically concerned with LU, Cholesky and QR include

C.H. Bischof (1988). "QR Factorization Algorithms for Coarse Grain Distributed Systems," PhD Thesis, Dept. of Computer Science, Cornell University, Ithaca, NY.

G.J. Davis (1986). "Column LU Pivoting on a Hypercube Multiprocessor," *SIAM J. Alg. and Disc. Methods* 7, 538–550.

J.M. Delosme and I.C.F. Ipsen (1986). "Parallel Solution of Symmetric Positive Definite Systems with Hyperbolic Rotations," *Lin. Alg. and Its Applic. 77,* 75–112.

G.A. Geist and M.T. Heath (1985). "Parallel Cholesky Factorization on a Hypercube Multiprocessor," Report ORNL 6190, Oak Ridge Laboratory, Oak Ridge, TN.

G.A. Geist and C.H. Romine (1988). "LU Factorization Algorithms on Distributed Memory Multiprocessor Architectures," *SIAM J. Sci. and Stat. Comp. 9,* 639–649.

R.N. Kapur and J.C. Browne (1984). "Techniques for Solving Block Tridiagonal Systems on Reconfigurable Array Computers," *SIAM J. Sci. and Stat. Comp. 5,* 701-719.

J.M. Ortega and C.H. Romine (1988). "The *ijk* Forms of Factorization Methods II: Parallel Systems," *Parallel Computing 7,* 149–162.

A. Pothen, S. Jha, and U. Vemapulati (1987). "Orthogonal Factorization on a Distributed Memory Multiprocessor," in *Hypercube Multiprocessors*, ed. M.T. Heath, SIAM Press, 1987.

D.W. Walker, T. Aldcroft, A. Cisneros, G. Fox, and W. Furmanski (1988). "LU Decomposition of Banded Matrices and the Solution of Linear Systems on Hypercubes," in G. Fox (ed) (1988). *The Third Conference on Hypercube Concurrent Computers and Applications, Vol. II – Applications,* ACM Press, New York. pp. 1635–1655.

Parallel triangular system solving is discussed in

C.H. Romine and J.M. Ortega (1988). "Parallel Solution of Triangular Systems of Equations," *Parallel Computing 6,* 109–114.

M.T. Heath and C.H. Romine (1988). "Parallel Solution of Triangular Systems on Distributed Memory Multiprocessors," *SIAM J. Sci. and Stat. Comp. 9,* 558–588.

G. Li and T. Coleman (1988). "A Parallel Triangular Solver for a Distributed-Memory Multiprocessor," *SIAM J. Sci. and Stat. Comp. 9,* 485–502.

S.C Eisenstat, M.T. Heath, C.S. Henkel, and C.H. Romine (1988). "Modified Cyclic Algorithms for Solving Triangular Systems on Distributed Memory Multiprocessors," *SIAM J. Sci. and Stat. Comp. 9,* 589–600.

D.J. Evans and R. Dunbar (1983). "The Parallel Solution of Triangular Systems of Equations," *IEEE Trans. Comp. C-32,* 201–204.
R. Montoye and D. Laurie (1982). "A Practical Algorithm for the Solution of Triangular Systems on a Parallel Processing System," *IEEE Trans. Comp. C-31,* 1076–1082.
S.L. Johnsson (1984). "Odd-Even Cyclic Reduction on Ensemble Architectures and the Solution of Tridiagonal Systems of Equations," Report YALEU/CSD/RR-339, Dept. of Comp. Sci., Yale University, New Haven, CT.

# 6.5 Mesh Factorization Procedures

In this section we discuss various mesh algorithms for the Cholesky and QR factorizations. Systolic and asynchronous procedures are given for the Cholesky factorization enabling us to compare the kind of reasoning associated with these two styles of distributed memory computing. The section ends with a brief discussion of a systolic mesh QR factorization procedure.

## 6.5.1 A Mesh Systolic Cholesky Procedure

Recall that the outer product version of the Cholesky factorization involves the repeated application of following reduction:

$$A = \begin{bmatrix} \alpha & v^T \\ v & B \end{bmatrix} = \begin{bmatrix} \beta & 0 \\ w & I_{n-1} \end{bmatrix} \begin{bmatrix} 1 & 0 \\ 0 & C \end{bmatrix} \begin{bmatrix} \beta & w^T \\ 0 & I_{n-1} \end{bmatrix}.$$

Here $\beta = \sqrt{\alpha}$, $w = v/\beta$ , and $C = B - ww^T$. Let us write the outer product $kji$ Cholesky algorithm with superscripts in order to keep track of the stages that $A$ goes through:

$$\begin{array}{l}
A^{(0)} = A \\
\textbf{for } k = 1{:}n \\
\quad a_{ij}^{(k)} = a_{ij}^{(k-1)} \text{ for all } (i,j) \text{ with } i < k \text{ or } j < k \\
\quad a_{kk}^{(k)} = \sqrt{a_{kk}^{(k-1)}} \\
\quad \textbf{for } i = k+1{:}n \\
\qquad a_{ik}^{(k)} = a_{ik}^{(k-1)}/a_{kk}^{(k)};\ a_{ki}^{(k)} = a_{ki}^{(k-1)}/a_{kk}^{(k)} \\
\quad \textbf{end} \\
\quad \textbf{for } j = k+1{:}n \\
\qquad \textbf{for } i = k+1{:}n \\
\qquad\quad a_{ij}^{(k)} = a_{ij}^{(k-1)} - a_{ik}^{(k)} a_{kj}^{(k)} \\
\qquad \textbf{end} \\
\quad \textbf{end} \\
\textbf{end}
\end{array}$$

To derive a systolic version of this procedure we assume for clarity that we have an $n$-by-$n$ systolic mesh and that $A \in \mathbb{R}^{n \times n}$. Proc$(i,j)$ is to be responsible for the generation of $a_{ij}^{(k)}$ for $k = 1{:}n$. Symmetry will be ignored to facillitate exposition.

Assume that Proc$(i, j)$ uses a local variable $a$ for storing the $a_{ij}^{(k)}$. Notice that there are four types of computations associated with the transition from $A^{(k-1)}$ to $A^{(k)}$:

$$\begin{aligned} \text{nothing:} &\quad a \leftarrow a \\ \text{square root :} &\quad a \leftarrow \sqrt{a} \\ \text{divide :} &\quad a \leftarrow a/d \\ \text{update:} &\quad a \leftarrow a - bc \end{aligned}$$

Recalling our model of systolic computation (see §6.1.5) we seek to develop a node program of the form

**for** $t = 1{:}t_{final}$
  Send local data to neighbors.
  Receive data from neighbors. (6.5.1)
  Perform a square root, a divide, an update, or do nothing.
**end**

What takes place in Proc$(i, j)$ at time step $t$ is a function of $t$, $i$, and $j$. Notice that at $t = 1$, only Proc(1,1) can do anything significant and that is compute $\sqrt{A(1,1)}$. This quantity is then needed by the first row and column of processors which compute $A(2{:}n, 1) \leftarrow A(2{:}n, 1)/A(1, 1)$ and $A(1, 2{:}n) \leftarrow A(1, 2{:}n)/A(1, 1)$ respectively. Now in our systolic model we assume that data can only move to a neighbor processor at each time step. Thus, these divide steps can take place at the following times:

$$\begin{bmatrix} 1 & 2 & 3 & 4 & 5 \\ 2 & \times & \times & \times & \times \\ 3 & \times & \times & \times & \times \\ 4 & \times & \times & \times & \times \\ 5 & \times & \times & \times & \times \end{bmatrix}.$$

As the multipliers in row and column one begin to form, we can begin the execution of the update $A(2{:}n, 2{:}n) \leftarrow A(2{:}n, 2{:}n) - A(2{:}n, 1)A(1, 2{:}n)$. Note that Proc$(i, j)$ needs the multipliers housed in Proc$(i, 1)$ and Proc$(1, j)$. These two quantities can be arranged to meet at Proc$(i, j)$ at the same time if Proc$(i, 1)$ sends its multiplier across row $i$ and Proc$(1, j)$ sends its multiplier down column $j$ as soon as it is available. If we define the array

$$T^{(1)} = \begin{bmatrix} 1 & 2 & 3 & 4 & 5 \\ 2 & 3 & 4 & 5 & 6 \\ 3 & 4 & 5 & 6 & 7 \\ 4 & 5 & 6 & 7 & 8 \\ 5 & 6 & 7 & 8 & 9 \end{bmatrix}$$

then $T^{(1)}(i,j)$ is the time step in which $a_{ij}^{(1)}$ is found. Note that at time step $t = 4$ we can compute $a_{22}^{(2)}$, and so by the above reasoning,

$$T^{(2)} = \begin{bmatrix} \cdot & \cdot & \cdot & \cdot & \cdot \\ \cdot & 4 & 5 & 6 & 7 \\ \cdot & 5 & 6 & 7 & 8 \\ \cdot & 6 & 7 & 8 & 9 \\ \cdot & 7 & 8 & 9 & 10 \end{bmatrix}.$$

Here, $T^{(2)}(i,j)$ with $i \geq 2$ and $j \geq 2$ is the time step in which $a_{ij}^{(2)}$ is computed. The "·" entries involve no work or communication and do not figure in the scheduling discussion.

In similar fashion, the time of computation for the "interesting" elements in $A^{(3)}$, $A^{(4)}$, and $A^{(5)}$ are prescribed by

$$T^{(3)} = \begin{bmatrix} \cdot & \cdot & \cdot & \cdot & \cdot \\ \cdot & \cdot & \cdot & \cdot & \cdot \\ \cdot & \cdot & 7 & 8 & 9 \\ \cdot & \cdot & 8 & 9 & 10 \\ \cdot & \cdot & 9 & 10 & 11 \end{bmatrix}, \quad T^{(4)} = \begin{bmatrix} \cdot & \cdot & \cdot & \cdot & \cdot \\ \cdot & \cdot & \cdot & \cdot & \cdot \\ \cdot & \cdot & \cdot & \cdot & \cdot \\ \cdot & \cdot & \cdot & 10 & 11 \\ \cdot & \cdot & \cdot & 11 & 12 \end{bmatrix},$$

and

$$T^{(5)} = \begin{bmatrix} \cdot & \cdot & \cdot & \cdot & \cdot \\ \cdot & \cdot & \cdot & \cdot & \cdot \\ \cdot & \cdot & \cdot & \cdot & \cdot \\ \cdot & \cdot & \cdot & \cdot & \cdot \\ \cdot & \cdot & \cdot & \cdot & 13 \end{bmatrix}.$$

We draw three conclusions from this $n = 5$ example:

- Proc$(i,j)$ has nothing to compute until $t = i + j - 1$.
- If $a_{kk}^{(k)}$ is generated at $t = \tau$ , then $a_{ij}^{(k)}$ can be computed at $t = \tau + (i-k) + (j-k)$ for all $i \geq k$ and $j \geq k$ .
- If $a_{kk}^{(k)}$ is generated time $t = \tau$, then $a_{k+1,k+1}^{(k+1)}$ can be generated at $t = \tau + 3$.

These observations enable us to deduce a number of important facts about scheduling. For example, the diagonal processor Proc$(i,i)$ has nothing to do for $t = 1{:}2i-2$. For $t = 2i-1{:}3i-3$ it has updates to perform followed by a square root at $t = 3i-2$. For $t \geq 3i-1$ it again has nothing to do. Similar conclusions may be reached for the off-diagonal processors and overall we obtain the following agenda for Proc$(i,j)$:

```
for t = 1:3n - 2
    if i = j
        if t ≥ 2i - 1 ∧ t < 3i - 2
            update
        elseif t = 3i - 2
            square root
        end
    elseif i > j
        if t ≥ i + j - 1 ∧ t < i + 2j - 2
            update
        elseif t = i + 2j - 2
            divide
        end
    elseif i < j
        if t ≥ i + j - 1 ∧ t < j + 2i - 2
            update
        elseif t = j + 2i - 2
            divide
        end
    end
end
```

A total of $3n - 2$ steps are required because Proc$(n, n)$ computes its square root, i.e., $g_{nn}$, at $t = 3n - 2$.

Our remaining task is to work out the communications that are associated with these computational activities. There are two observations to make:

- The time step *after* Proc$(i, j)$ computes $g_{ij}$, i.e., $t = i + j + \min\{i, j\} - 1$, all it must do is send $g_{ij}$ east if $i > j$ and $j < n$, south if $j > i$ and $i < n$, and east and south if $i = j < n$.
- Proc$(i, j)$ is updating for $t = i + j - 1{:}i + j + \min\{i, j\} - 3$ and during these timesteps a computation of the form $a \leftarrow a - a_{north}a_{west}$ is performed where $a_{north}$ is received from the north and $a_{west}$ is received from the west. So that these quantities continue their proper migration, Proc$(i, j)$ must send the current $a_{north}$ south and its current $a_{west}$ east.

Overall we obtain the following procedure.

**Algorithm 6.5.1** Suppose $A \in \mathbb{R}^{n \times n}$ is symmetric and positive definite and that $A = GG^T$ is its Cholesky factorization. If each node in an $n$-by-$n$ mesh-connected systolic array executes the following program, then upon completion the local variable $a$ in Proc$(i, j)$ houses $g_{ij}$ for $i \geq j$ and $g_{ji}$ for $j \geq i$.

**loc.init**[ $(i,j) = my.id$, $north$, $east$, $south$, $west$, $n$, $a = A(i,j)$ ]
**for** $t = 1{:}3n-2$
    **if** $i = j$
        { Diagonal Processor Computations }
        **if** $t = 2i-1 \wedge i \neq 1$
            **recv**$(a_{north}, north)$
            **recv**$(a_{west}, west)$
            $a = a - a_{north}a_{west}$
        **elseif** $2i-1 < t < 3i-2$
            **if** $i \neq n$
                **send**$(a_{north}, south)$
                **send**$(a_{west}, east)$
            **end**
            **recv**$(a_{north}, north)$
            **recv**$(a_{west}, west)$
            $a = a - a_{north}a_{west}$
        **elseif** $t = 3i-2$
            $a = \sqrt{a}$
        **elseif** $t = 3i-1 \wedge i \neq n$
            **send**$(a, south)$
            **send**$(a, east)$
        **end**

    **elseif** $i > j$

        { Subdiagonal Processor Computations }
        **if** $t = i+j-1 \wedge j \neq 1$
            **recv**$(a_{north}, north)$; **recv**$(a_{west}, west)$
            $a = a - a_{north}a_{west}$
        **elseif** $i+j-1 < t < i+2j-2$
            **if** $i \neq n$
                **send**$(a_{north}, south)$
            **end**
            **send**$(a_{west}, east)$
            **recv**$(a_{north}, north)$; **recv**$(a_{west,west})$
            $a = a - a_{north}a_{west}$
        **elseif** $t = i+2j-2$
            **recv**$(a_{north}, north)$; $a = a/a_{north}$
        **elseif** $t = i+2j-1$
            **if** $i \neq n$
                **send**$(a_{north}, south)$
            **end**
            **send**$(a, east)$
        **end**

```
        elseif i < j
            { Superdiagonal Processor Computations }
            if t = i + j - 1 ∧ i ≠ 1
                recv(a_north, north); recv(a_west, west)
                a = a - a_north a_west
            elseif i + j - 1 < t < 2i + j - 2
                if j ≠ n
                    send(a_west, east)
                end
                send(a_north, south)
                recv(a_north, north); recv(a_west,west)
                a = a - a_north a_west
            elseif t = 2i + j - 2
                recv(a_west, west); a = a/a_west
            elseif t = 2i + j - 1
                if j ≠ n
                    send(a_west, east)
                end
                send(a, east)
            end
        end
    end
    quit
```

This algorithm (provided by C. Paige) could be compressed but its current form more closely follows its derivation.

Note that Algorithm 6.5.1 is a *linear time* procedure. Of course, not every processor is busy every time step and so there is some inefficiency. In particular, the average number of active processors during a given time step is given by $(n^3/3)/((3n-2)(n^2) \approx 1/9$.

A block version of Algorithm 6.5.1 is easily formulated. This is of interest when the matrix dimension is bigger then the mesh dimension. If the mesh is $N$-by-$N$ then we regard $A$ as an $N$-by-$N$ block matrix. Other wrap-type mappings of an oversized matrix may be approporiate if we have toroidal connections.

It is also possible to formulate a symmetry exploiting version of the algorithm that executes on a lower triangular systolic network.

### 6.5.2 Asynchronous Mesh Cholesky

To derive an asynchronous, message passing Cholesky procedure for a mesh we do not ask "when" $a_{ij}^{(k)}$ is computed as in the systolic case. Instead, we have Proc$(i,j)$ "trigger" the computation based on a local counter whose value is the number of completed update steps $a \leftarrow a - a_{west} \cdot a_{north}$. The following array indicates the number of updates that Proc$(i, j)$ participates in before it does a divide (if $i \neq j$) or a square root (if $i = j$):

$$\begin{bmatrix} 0 & 0 & 0 & 0 & 0 \\ 0 & 1 & 1 & 1 & 1 \\ 0 & 1 & 2 & 2 & 2 \\ 0 & 1 & 2 & 3 & 3 \\ 0 & 1 & 2 & 3 & 4 \end{bmatrix}.$$

Extrapolating from this example we conclude that the $(i, j)$ entry in this matrix is given by $\min(i, j) - 1$ and we obtain:

**Algorithm 6.5.2** Suppose $A \in \mathbb{R}^{n \times n}$ is symmetric and positive definite and that $A = GG^T$ is its Cholesky factorization. If each node in an $n$-by-$n$ mesh-connected array executes the following program, then upon completion the local variable $a$ in Proc$(i, j)$ houses $g_{ij}$ for $i \geq j$ and $g_{ji}$ for $j \geq i$.

```
loc.init[ (i,j) = my.id, north, east, south, west, n, a = A(i,j) ]
{ Update Phase }
k = 1
while k < min(i,j)
      recv(a_north, north); recv(a_west, west)
      a = a - a_west a_north
      if i < n
            send(a_north, south)
      end
      if j < n
            send(a_west, east)
      end
      k = k + 1
end

{ Wrap-Up Phase }
if i = j
      a = sqrt(a)
      if i < n
            send(a, east); send(a, south)
      end
elseif i > j
      recv(a_north, north)
      if i < n
            send(a_north, south)
      end
      a = a/a_north
      send(a, east)
```

```
    elseif i < j
        recv(a_west, west)
        if j < n
            send(a_west:east)
        end
        a = a/a_west
        send(a, a_south)
    end
    quit
```

Note that the derivation of Algorithm 6.5.2 does not involve a detailed synchronization argument as did its systolic counterpart, Algorithm 6.5.1. Still, to establish the correctness of the procedure a few scheduling facts must be proved. For example, if $k < j < i$ then it can be shown $k$th message to arrive from the west at Proc$(i, j)$ is $g_{ik}$. Likewise, $g_{kj}$ is the $k$th message to arrive from the north. Results like these can then be used to rigorously establish that Proc$(i, j)$ houses $g_{ij}$ upon termination.

### 6.5.3 A Mesh Systolic QR Factorization Procedure

Partial pivoting does not map very well onto two-dimensional processor arrays because a row interchange may involve swapping data between distant processors. For this reason, the QR factorization has become quite an attractive method for solving linear equations, especially in the systolic array area. Recall from §5.2.3 that Givens rotations can be used to compute the QR factorization of a matrix $A \in \mathbb{R}^{m \times n}$ and that there are numerous ways that this can be accomplished. We focus on Algorithm 5.2.2 in which adjacent Givens rotations are used:

```
for j = 1:n
    for i = m: - 1:j + 1
        [c, s] = givens(A(i - 1:i, j))                          (6.5.1)
        A(i - 1:i, j:n) = row.rot(A(i - 1:i, j:n), c, s)
    end
end
```

We disregard the accumulation of $Q$ as it just clutters the discussion of parallelization.

To anticipate a parallel version of (6.5.1), we take a look at some intermediate stages in the reduction. Suppose $(m, n) = (6,3)$ and that $a_{61}$ and

$a_{51}$ have been zeroed:

$$\begin{bmatrix} \times & \times & \times \\ \times & \times & \times \\ \times & \times & \times \\ \times & \times & \times \\ 0 & \times & \times \\ 0 & \times & \times \end{bmatrix}.$$

The algorithm then proceeds to zero $a_{41}$. But note, we are also poised to zero $a_{62}$ so let us do both concurrently:

$$\begin{bmatrix} \times & \times & \times \\ \times & \times & \times \\ \times & \times & \times \\ \times & \times & \times \\ 0 & \times & \times \\ 0 & \times & \times \end{bmatrix} \overset{(3,4),(5,6)}{\longrightarrow} \begin{bmatrix} \times & \times & \times \\ \times & \times & \times \\ \times & \times & \times \\ 0 & \times & \times \\ 0 & \times & \times \\ 0 & 0 & \times \end{bmatrix}.$$

At the next "time step" another pair of zeros can be introduced:

$$\begin{bmatrix} \times & \times & \times \\ \times & \times & \times \\ \times & \times & \times \\ 0 & \times & \times \\ 0 & \times & \times \\ 0 & 0 & \times \end{bmatrix} \overset{(2,3),(4,5)}{\longrightarrow} \begin{bmatrix} \times & \times & \times \\ \times & \times & \times \\ 0 & \times & \times \\ 0 & \times & \times \\ 0 & 0 & \times \\ 0 & 0 & \times \end{bmatrix}.$$

At this stage the zeroing of $a_{21}$, $a_{42}$, and $a_{61}$ may proceed concurrently:

$$\begin{bmatrix} \times & \times & \times \\ \times & \times & \times \\ 0 & \times & \times \\ 0 & \times & \times \\ 0 & 0 & \times \\ 0 & 0 & \times \end{bmatrix} \overset{(1,2),(3,4),(5,6)}{\longrightarrow} \begin{bmatrix} \times & \times & \times \\ 0 & \times & \times \\ 0 & \times & \times \\ 0 & 0 & \times \\ 0 & 0 & \times \\ 0 & 0 & 0 \end{bmatrix}.$$

In the language of time-steps, we see that zeros in the 6-by-3 case may be introduced at the following times:

$$\begin{bmatrix} \times & \times & \times \\ 5 & \times & \times \\ 4 & 6 & \times \\ 3 & 5 & 7 \\ 2 & 4 & 6 \\ 1 & 3 & 5 \end{bmatrix}.$$

In general, subdiagonal matrix entry $a_{ij}$ can be zeroed at time

$$t_{ij} = (2j-1) + m - i = m + 2j - i - 1 . \tag{6.5.2}$$

Thus, component $i = m + 2j - t$ of column $j$ can be zeroed at time $t$ provided $m \le i \le j+1$. We thus obtain the following rearrangement of (6.5.1):

```
for t = 1:m + n − 2
    for j = 1:n
        i = m − t + 2j − 1
        if i ≤ m ∧ i ≥ j + 1
            { Generate and apply G_ij. }                    (6.5.3)
            [c, s] = givens(A(i − 1:i, j))
            A(i − 1:i, j:n) = row.rot(A(i − 1:i, j:n), c, s)
        end
    end
end
```

The total number of time steps required is $m + n - 2$ because $a_{n+1,n}$ is the last entry to be zeroed and this happens at $t = m + n - 2$.

From the parallel computation point of view, the key observation to make about (6.5.3) is that all the rotations associated with a given "time step" $t$ are "nonconflicting" and may be performed concurrently. During the steady state period ($2n-1 \le t \le m-1$) one entry per column is zeroed.

A linear time implementation is possible if we have a systolic mesh having dimension equal to $A$:

- Proc$(i, j)$ houses the current $a_{ij}$.
- Proc$(i, j)$ generates Givens rotation $G_{ij}$.
- After it is generated, $G_{ij}$ migrates eastward across processor rows $i - 1$ and $i$.

In such a scheme Proc$(i, j)$ emerges with $r_{ij}$. A trace of Proc$(i, j)$'s activities reveals that it is idle for an initial period. It then is busy receiving rotations associated with the zeroings in rows $i + 1$ and $i$. During this period it updates the housed $a_{ij}$ and passes the received rotations eastward. Eventually it generates $G_{ij}$ and thereafter is idle. We saw this pattern of activity in the Algorithm 6.5.1. As in that setting a significant fraction of the processors are active at any time step and since they are are $n^2$ in number, a linear time procedure results. For details, see Heller and Ipsen(1983) and Gentleman and Kung(1982).

### Problems

**P6.5.1** Modify Algorithms 6.5.1 and 6.5.2 so that they compute (a) the $LDL^T$ factorization and (b) the $A = LU$ factorization. Be sure to indicate the distribution of the output factors.

**Notes and References for Sec. 6.5**

Papers on the parallel computation of the LU and Cholesky factorization include

R.P. Brent and F.T. Luk (1982) "Computing the Cholesky Factorization Using a Systolic Architecture," *Proc. 6th Australian Computer Science Conf.* 295–302.

M. Costnard, M. Marrakchi, and Y. Robert (1988). "Parallel Gaussian Elimination on an MIMD Computer," *Parallel Computing 6,* 275–296.

J.M. Delosme and I.C.F. Ipsen (1986). "Parallel Solution of Symmetric Positive Definite Systems with Hyperbolic Rotations," *Lin. Alg. and Its Applic. 77,* 75–112.

R.E. Funderlic and A. Geist (1986). "Torus Data Flow for Parallel Computation of Missized Matrix Problems," *Lin. Alg. and Its Applic. 77,* 149–164

D.P. O'Leary and G.W. Stewart (1985). "Data Flow Algorithms for Parallel Matrix Computations," *Comm. of the ACM 28,* 841–853.

Parallel methods for banded systems include

S.L. Johnsson (1984). "Odd-Even Cyclic Reduction on Ensemble Architectures and the Solution of Tridiagonal Systems of Equations," Report YALEU/CSD/RR-339, Dept. of Comp. Sci., Yale University, New Haven, CT.

S.L. Johnsson (1985). "Solving Narrow Banded Systems on Ensemble Architectures," *ACM Trans. Math. Soft. 11,* 271–288.

S.L. Johnsson (1986). "Band Matrix System Solvers on Ensemble Architectures," in *Supercomputers: Algorithms, Architectures, and Scientific Computation,* eds. F.A. Matsen and T. Tajima, University of Texas Press, Austin TX., 196–216.

S.L. Johnsson (1987). "Solving Tridiagonal Systems on Ensemble Architectures," *SIAM J. Sci. and Stat. Comp. 8,* 354–392.

S.L. Johnsson and C.T. Ho (1987). "Multiple Tridiagonal Systems, the Alternatine Direction Methods, and Boolean Cube Configured Multiprocessors," Report YALEU/DCS/RR-532, Dept. of Computer Science, Yale University, New Haven, CT.

Parallel QR factorization procedures are of interest in real-time signal processing. Details may be found in

M. Costnard, J.M. Muller, and Y. Robert (1986). "Parallel QR Decomposition of a Rectangular Matrix," *Numer. Math. 48,* 239–250.

L. Eldin (1988) " A Parallel QR Decomposition Algorithm," Report LiTh Mat R 1988-02, Dept. of Math., Linkoping University, Sweden.

L. Eldin and R. Schreiber (1986). "An Application of Systolic Arrays to Linear Discrete Ill-Posed Problems," *SIAM J. Sci. and Stat. Comp. 7,* 892–903.

W.M. Gentleman and H.T. Kung (1982). "Matrix Triangularization by Systolic Arrays," SPIE Proceedings, Vol. 298, 19–26.

D.E. Heller and I.C.F. Ipsen (1983). "Systolic Networks for Orthogonal Decompositions," *SIAM J. Sci. and Stat. Comp. 4,* 261–269.

F.T. Luk (1986). "A Rotation Method for Computing the QR Factorization," *SIAM J. Sci. and Stat. Comp. 7,* 452–459.

J.J. Modi and M.R.B. Clarke (1986). "An Alternative Givens Ordering," *Numer. Math. 43,* 83–90.

## 6.6 Shared Memory Factorization Methods

In this section we look at several different shared memory implementations of the Cholesky factorization. Each implementation has its own synchronization pattern and collectively they illustrate how a factorization such as Cholesky can be organized in a shared memory environment. We have chosen the Cholesky factorization for expository reasons but much of what we say carries over to the other factorizations in our repertoire.

### 6.6.1 A Statically Scheduled Outer Product Cholesky

We begin by parallelizing the outer product Cholesky algorithm:

**for** $k = 1{:}n$
  $A(k{:}n) = A(k{:}n)/\sqrt{A(k,k)}$
  **for** $j = k+1{:}n$
    $A(j{:}n,j) = A(j{:}n,j) - A(j{:}n,k)A(j,k)$
  **end**
**end**

The $j$-loop oversees an outer product update. The $n-k$ saxpy operations that make up its body are independent and easily distributed.

**Algorithm 6.6.1** Suppose $A \in \mathbb{R}^{n\times n}$ is a symmetric positive definite matrix stored in a shared memory accessible to $p$ processors. If each processor executes the following algorithm then upon completion the lower triangular part of $A$ is overwritten with its Cholesky factor.

**shared variables** $[\,A(1{:}n,1{:}n)\,]$
**glob.init**$[\,A\,]$
**loc.init**$[p,\ \mu = my.id,\ n\,]$
**for** $k = 1{:}n$
  **if** $\mu = 1$
    **get**$(A(k{:}n), v(k{:}n))$; $v(k{:}n) = v(k{:}n)/\sqrt{v(k)}$
    **put**$(v(k{:}n), A(k{:}n,k))$
  **end**
  **barrier**
  **get**$(A(k+1{:}n,k), v(k+1{:}n))$
  **for** $j = (k+\mu){:}p{:}n$
    **get**$(A(j{:}n,j), w(j{:}n))$; $w(j{:}n) = w(j{:}n) - v(j)v(j{:}n)$
    **put**$(w(j{:}n), A(j{:}n,j))$
  **end**
  **barrier**
**end**
**quit**

The scaling before the $j$-loop represents very little work compared to the outer product update and so it is reasonable to assign that portion of the computation to a single processor. Notice that two **barrier** statements are required. The first ensures that a processor does not begin working on the $k$th outer product update until the $k$th column of $G$ is made available by Proc(1). The second **barrier** prevents the processing of the $k$+1st step to begin until the $k$th step is completely finished.

It is possible to formulate a block version of Algorithm 6.6.1. Suppose $n = rN$. For $k = 1{:}N$ we (a) have Proc(1) generate $G(:, 1 + (k-1)r{:}kr)$ and (b) have all processors participate in the rank $r$ update of the trailing submatrix $A(kr + 1{:}n, kr + 1{:}n)$. See §4.2.6. The coarser granularity may improve performance if the individual processors like level-3 operations.

## 6.6.2 Other Statically Scheduled Factorizations

The logic behind Algorithm 6.6.1 extends to other factorization schemes such as LU and QR. Just as we did in §6.4.3 for ring factorizations, we can produce a "template" that captures the pattern of synchronization for shared memory outer product factorizations:

```
shared variables [ A , ... ]
glob.init[ A, ... ]
loc.init[p, μ = my.id, ... ]
for k = 1:n
    if μ = 1
        get(A(k:n), v(k:n))
        { Generate the kth transformation T_k. }
        { Record T_k in shared memory. }
    end                                                    (6.6.1)
    barrier
    { Copy T_k into local memory.}
    for j = k + μ:p:n
        { Apply T_k to A(k:n, j) }
    end
    barrier
end
```

Of course, references to $T_k$ are references to some suitable representation, e.g., the $k$th Householder vector.

## 6.6.3 Two Parallel Gaxpy Cholesky Factorizations

We now develop a pair of parallel Cholesky procedures that are based on the gaxpy version of the algorithm, i.e., Algorithm 4.2.1:

```
for j = 1:n
    for k = 1:j − 1
        A(j:n, j) = A(j:n, j) − A(j, k)A(j:n, k)
    end
    A(j:n, j) = A(j:n, j)/√A(j, j)
end
```

We discuss two ways parallelize this procedure. The first approach is to involve each processor in each gaxpy. If we use the monitor procedure **gax.sum** from §6.2.7 to accumulate each processor's contribution then we obtain

**Algorithm 6.6.2** Suppose $A$ is an $n$-by-$n$ symmetric positive definite matrix stored in a shared memory accessible to $p$ processors. If each processor executes the following algorithm then upon completion the lower triangular part of $A$ is overwritten with its Cholesky factor.

```
shared variables [A(1:n, 1:n)]
glob.init[A]
loc.init[p, μ = my.id, n]
for j = 1:n
    u(j:n) = 0
    for k = μ:p:j − 1
        get(A(j:n, k), a_loc(j:n))
        u(j:n) = u(j:n) − a_loc(j)a_loc(j:n)
    end
    call gax.sum(u(j:n), A(j:n, j))
    barrier
    if μ = 1
        get(A(j:n, j), a_loc(j:n))
        a_loc(j:n) = a_loc(j:n)/√a_loc(j)
        put(a_loc(j:n), A(j:n, j))
    end
    barrier
end
```

The first **barrier** ensures that the gaxpy is completed before it is scaled to render $G(j{:}n, j)$. The second **barrier** makes sure that no processor proceeds to the $j+1$st step until $j$th step is complete.

The second approach to parallelizing the gaxpy-based Cholesky procedure is to apply the pool-of-task idea with task($j$) being the computation of $G(j{:}n, j)$. Since this subcolumn of the Cholesky factor is a scaling of

$$v(j{:}n) = A(j{:}n, j) - \sum_{k=1}^{j-1} G(j,k)G(j{:}n,k) \tag{6.6.2}$$

we see that task($j$) cannot be completed until tasks 1 through $j-1$ are completed. Thus we aim for a procedure of the following form:

```
Repeat until no more tasks:
    Get the next task, call it task(j).
    Build up the vector v(j:n) in (6.6.2) as quickly as possible.
    Compute G(j:n, j) and store in shared memory.
    Inform all processors waiting for G(j:n, j) that it is available.
end
```

We can use the monitor **nexti** developed in §6.2.6 for dealing out the tasks. Another monitor **chol** with two procedures **canread** and **write** is required to carry out the acquisition of $G$-columns and the overwriting of $A$:

```
monitor: chol
    protected variables: cols.done
    condition variables: waitq(·)
    initializations: cols.done = 0
    procedure chol.canread(z, k, j, n)
        ifcols.done < k
            delay(waitq(k))
        end
        get(A(j:n, k), z(j:n))
        continue(waitq(k))
    end chol.canread
    procedure chol.write(v, j, n)
        put(v(j:n), A(j:n, j))
        cols.done = cols.done + 1
        if cols.done = n
            cols.done = 0
        else
            continue(waitq(j))
        end
    end chol.write
end chol
```

This monitor has a number of new features when compared to **nexti** and **gax**. First, it has two procedures. A reference to either **chol.canread** or **chol.write** represents an attempt by the calling processor to enter the monitor. However, it is still the case that there can be at most one active processor in the monitor at any instant. The mission of the two procedures is as follows:

- **chol.canread**$(z, k, j, n)$ returns $G(j{:}n, k)$ in $z(j{:}n)$ if it is available. The calling processor is delayed if $G(j{:}n, k)$ is not avilable.
- **chol.write**$(v, j, n)$ copies $v(j{:}n) = G(j{:}n, j)$ into $A(j{:}n, j)$ and informs all processors waiting for this particular $G$-column that it is now available in shared memory.

Note that a vector of condition variables is used. A processor that requests part of $G(:, k)$ is delayed on $waitq(k)$ if $G(:, k)$ is unavailable.

The protected variable *col.done* keeps track of the number of completed columns. Thus, it is a simple matter to check if a particular $G$-column is available. Note that *col.done* is reset to zero after the last $G$-column is computed.

Here is how the the monitor **chol** is used in a parallel, gaxpy-based Cholesky procedure:

**Algorithm 6.6.3** Suppose $A$ is an $n$-by-$n$ symmetric positive definite matrix stored in shared memory accessible to $p$ processors. If each processor executes the following algorithm then upon completion the lower triangular part of $A$ is overwritten with the lower triangular portion of its Cholesky factor.

```
shared variables [A(1:n, 1:n)]
glob.init[A]
loc.init[p, μ = my.id, n]
call nexti.index(p, n, more, j)
while more = true
      get(A(:j:n, j), v(j:n))
      for k = 1:j − 1
            call chol.canread(z, k, j)
            v(j:n) = v(j:n) − z(j)z(j:n)
      end
      v(j:n) = v(j:n)/√v(j)
      call chol.write(v, j, n)
      call nexti.index(p, n, more, j)
end
quit
```

Recall that **nexti.index** is designed so that no processor can exit the **while** loop until all tasks are completed.

Algorithm 6.6.3 is similar to our ring Cholesky procedures in §6.4. In particular, each processor experiences a period of idleness at both the begining and end of the overall computation. However, this does not seriously degrade performance if $n \gg p$.

It is possible to formulate a block column version of Algorithm 6.6.3 that is based upon the partitioning

$$G = [\, G_1, \ldots, G_N \,] \,. \tag{6.6.3}$$

The $i$th task is now the computation of the block column $G_i$. Because of the coarser granularity the processors may be able to exploit the higher level linear algebra that results because of the blocking.

### 6.6.4 Comments on Shared Memory Traffic

It is interesting to estimate the time that each processor spends communicating with shared memory in Algorithms 6.6.1-6.6.3. These methods are all load balanced for large $n/p$ and in this situation it is reasonable to assume that each processor spends $T/p$ seconds communicating with shared memory where $T$ is the sum of all the **get** and **put** overheads. Using our communication model (6.2.2) we find that for Algorithm 6.6.1, each processor spends

$$T_{6.6.1} \approx \frac{n^2}{p}\alpha_s + \frac{2n^3}{3p}\beta_s$$

seconds communicating with shared memory. Algorithms 6.6.2 and 6.6.3, because they are gaxpy-based, involve less communication:

$$T_{6.6.2} \approx T_{6.6.3} \approx \frac{n^2}{2p}\alpha_s + \frac{n^3}{6p}\beta_s \,.$$

Of course, these are just crude estimates. Nevertheless, they point once again to the fact that it is better to rely upon gaxpy's than outer product updates.

### 6.6.5 Block Cholesky Via Pool-of-Tasks

Suppose $A$ is an $n$-by-$n$ symmetric positive definite matrix and that $n = Nr$. Recalling our discussion of block dot product Cholesky in §4.2.6, we have the governing equation

$$G_{ij}G_{jj}^T = A_{ij} - \sum_{k=1}^{j-1} G_{ik}G_{jk}^T \tag{6.6.4}$$

obtained by equating the $(i,j)$ block in $A = GG^T$. Here, each $G_{ij}$ and $A_{ij}$ is $r$-by-$r$. Note that if $i > j$ then the computation of $G_{ij}$ cannot be completed until $G_{i,1}, \ldots, G_{i,j-1}$ and $G_{j,1}, \ldots, G_{jj}$ are available. If $j = i$, then the computation of $G_{jj}$ cannot be completed until $G_{j1}, \ldots, G_{j,j-1}$ are available. In view of these remarks a reasonable order in which to compute the $G_{ij}$ is column-by-column. In the 4-by-4 case this means finding the $G$-blocks in this order:

| | | | |
|---|---|---|---|
| 1 | | | |
| 2 | 5 | | |
| 3 | 6 | 8 | |
| 4 | 7 | 9 | 10 |

Let task$(i, j)$ be the computation of $G_{ij}$. Following the reasoning behind our pool of task gaxpy Cholesky, we need a monitor procedure for dealing out the next task. A simple modification of the monitor procedure **nextij.index** (c.f. §6.3.8) suffices:

```
monitor: cholij
    protected variables row, col, idle.proc
    condition variables waiti
    initializations: flag = 0
    procedure cholij.index(N, p, more, i, j)
        if flag = 0
            row = 0; col = 1
            idle.proc = 0; flag = 1
        end
        if row < N
            row = row + 1
        else
            col = col + 1; row = col
        end
        if col ≤ N
            i = row; j = col
            more = true
        else
            more = false
            idle.proc = idle.proc + 1
            if idle.proc ≤ p − 1
                delay(waiti)
            else
                flag = 0
            end
            continue(waiti)
        end
    end cholij.index
end cholij
```

We can use **cholij.index** to organize our parallel block Cholesky procedure as follows:

> **call cholij.index**$(N, p, more, i, j)$
> **while** $more = true$
>     **get**$(A_{ij}, S)$
>     **for** $k = 1{:}j-1$
>         { If available, get $G_{ik}$ and $G_{jk}$ and compute
>             the update $S = S - G_{ik}G_{jk}^T$. }
>     **end**
>     **if** $i = j$ (6.6.4)
>         { Solve $G_{jj}G_{jj}^T = S$ by applying Algorithm 4.2.1.}
>     **else**
>         { Solve $G_{ij}G_{jj}^T = S$ for $G_{ij}$. }
>     **end**
>     { Write $G_{ij}$ to shared memory and inform all processors
>         that may be waiting for $G_{ij}$ that it is available.}
>     **call cholij.index**$(N, p, more, i, j)$
> **end**

Examination of (6.6.4) reveals that we need another monitor to oversee access to $A$ as it is overwritten by $G$. In particular we need a block version of the monitor procedures **chol.canread**$(z, k, j, n)$ and **chol.write**$(v, j, n)$. In other words we need a monitor with procedures **block.chol.canread** and **block.chol.write** where

- **block.chol.canread**$(W, i, k)$ returns $G_{ik}$ in $W$ if it is available. The calling processor is delayed if $G_{ik}$ is not available.
- **block.chol.write**$(W, i, j)$ copies $W = G_{ij}$ into $A_{ij}$ and informs all processors that it is now available in shared memory.

Here are the details:

> **monitor: block.chol**
>     **protected variables:** $row.progress(\cdot)$
>     **condition variables:** $waitq(\cdot,\cdot)$
>     **initializations:** $row.progress(\cdot) = 0$
>     **procedure block.chol.canread**$(W, i, k)$
>         **if** $row.progress(i) < k$
>             **delay**$(waitq(i, k))$
>         **end**
>         **get**$(A_{ik}, W)$
>         **continue**$(waitq(i, k))$
>     **end block.chol.canread**

```
        procedure block.chol.write(S, i, j, N)
            put(S, A_ij)
            row.progress(i) = row.progress(i) + 1
            if row.progress(N) < N
                continue(waitq(i, j))
            else
                row.progress(·) = 0
            end
        end block.chol.write
    end block.chol
```

This monitor amounts to a two-dimensional version of **chol**. We have a vector counter $row.progress(1{:}N)$ that keeps track of the number of computed $G_{ij}$ in each block row. This vector is reset to zero at the end of the computation. If a processor requests block $G_{ik}$ and finds that it is not yet available, then it is delayed on $waitq(i, k)$. Whenever a processor computes block $G_{ij}$, it begins the process of awakenening all processors that have been delayed on $waitq(i, j)$. Putting it all together we obtain

**Algorithm 6.6.4** Suppose $A$ is an $N$-by-$N$ block matrix with $r$-by-$r$ blocks. Assume that it is symmetric positive definite and stored by block in a shared memory accessible to $p$ processors. If the following algorithm is executed on each processor, then the lower triangular part of $A$ is overwritten by the lower triangular part of $G$, the Cholesky triangle. (References of the form $A_{ij}$ are really a references to $A(1 + (i - 1)r{:}ir, 1 + (j - 1)r{:}jr)$.)

```
    call cholij.index(N, p, more, i, j)
    while more = true
        get(A_ij, S)
        for k = 1:j - 1
            call block.chol.canread(W, i, k)
            call block.chol.canread(Y, j, k)
            S = S - WY^T
        end
        if i = j
            Solve G_jj G_jj^T = S by applying Algorithm 4.2.1 to S.
        else
            Solve G_ij G_jj^T = S for G_ij.
        end
        call block.chol.write(S, i, j, N)
        call cholij.index(N, p, more, i, j)
    end
    quit
```

This algorithm can be extended to handle an arbitrary blocking.

It is not hard to show that each processor involved in the execution of Algorithm 6.6.4 spends about

$$T_{6.6.4} \approx \frac{N^3(\alpha_s + \beta_s r^2)}{3p}$$

seconds communicating with shared memory.

### Problems

**P6.6.1** Using the template (6.6.1), develop a parallel QR factorization procedure assuming that $A$ is $m$-by-$n$ and stored in shared memory.

**P6.6.2** Develop an LU version of Algorithm 6.6.2.

**P6.6.3** Develop an LU with partial pivoting version of Algorithm 6.6.3.

**P6.6.4** Compare the shared memory traffic associated with Algorithms 6.6.3 and 6.6.4.

**P6.6.5** Tailor Algorithm 6.6.4 and the associated monitors to efficiently handle the case when $A$ is block tridiagonal.

### Notes and References for Sec. 6.6

Parallel factorization methods for shared memory machines are discussed in

S. Chen, J. Dongarra, and C. Hsuing (1984). "Multiprocessing Linear Algebra Algorithms on the Cray X-MP-2: Experiences with Small Granularity," *J. Parallel and Distributed Computing 1*, 22–31.

S. Chen, D. Kuck, and A. Sameh (1978). "Practical Parallel Band Triangular Systems Solvers," *ACM Trans. Math. Soft. 4*, 270–277.

J.J. Dongarra and T. Hewitt (1986). "Implementing Dense Linear Algebra Algorithms Using Multitasking on the Cray X-MP-4 (or Approaching the Gigaflop)," *SIAM J. Sci. and Stat. Comp. 7*, 347–350.

J.J. Dongarra and R.E. Hiromoto (1984). "A Collection of Parallel Linear Equation Routines for the Denelcor HEP,' *Parallel Computing 1*, 133–142.

J.J. Dongarra and A.H. Sameh (1984). "On Some Parallel Banded System Solvers," *Parallel Computing 1*, 223–235.

J.J. Dongarra, A. Sameh, and D. Sorensen (1986). "Implementation of Some Concurrent Algorithms for Matrix Factorization," *Parallel Computing 3*, 25–34.

J.J. Dongarra and D.C. Sorensen (1987). "Linear Algebra on High Performance Computers," *Appl. Math. and Comp. 20*, 57–88.

A. George, M.T. Heath, and J. Liu (1986). "Parallel Cholesky Factorization on a Shared Memory Multiprocessor," *Lin. Alg. and Its Applic. 77*, 165–187.

A. Sameh and D. Kuck (1978). "On Stable Parallel Linear System Solvers," *J. Assoc. Comp. Mach. 25*, 81-91.

P. Swarztrauber (1979). "A Parallel Algorithm for Solving General Tridiagonal Equations," *Math. Comp. 33*, 185-199.

# Chapter 7

# The Unsymmetric Eigenvalue Problem



Having discussed linear equations and least squares, we now direct our attention to the third major problem area in matrix computations, the algebraic eigenvalue problem. The unsymmetric problem is considered in this chapter and the more agreeable symmetric case in the next.

Our first task is to present the decompositions of Schur and Jordan along with the basic properties of eigenvalues and invariant subspaces. The contrasting behavior of these two decompositions sets the stage for §7.2 in which we investigate how the eigenvalues and invariant subspaces of a matrix are affected by perturbation. Condition numbers are developed that permit estimation of the errors that can be expected to arise because of roundoff.

The key algorithm of the chapter is the justly famous QR algorithm. This procedure is the most complex algorithm presented in this book and its development is spread over three sections. We derive the basic QR iteration in §7.3 as a natural generalization of the simple power method. The next two sections are devoted to making this basic iteration computationally feasible. This involves the introduction of the Hessenberg decomposition in §7.4 and the notion of origin shifts in §7.5.

The QR algorithm computes the real Schur form of a matrix, a canonical form that displays eigenvalues but not eigenvectors. Consequently, additional computations usually must be performed if information regarding invariant subspaces is desired. In §7.6, which could be subtitled, "What to Do after the Real Schur Form is Calculated," we discuss various invariant subspace calculations that can follow the QR algorithm.

Finally, in the last section we consider the generalized eigenvalue problem $Ax = \lambda Bx$ and a variant of the QR algorithm that has been devised to solve it. This algorithm, called the QZ algorithm, underscores the importance of orthogonal matrices in the eigenproblem, a central theme of the chapter.

It is appropriate at this time to make a remark about complex versus real arithmetic. In this book, we focus on the development of real arithmetic algorithms for real matrix problems. This chapter is no exception even though a real unsymmetric matrix can have complex eigenvalues. However, in the derivation of the practical, real arithmetic QR algorithm and in the mathematical analysis of the eigenproblem itself, it is convenient to work in the complex field. Thus, the reader will find that we have switched to complex notation in §7.1, §7.2, and §7.3. In these sections, we use the complex QR factorization ($A = QR$, $Q$ unitary, $R$ complex upper triangular) and the complex SVD ($A = U\Sigma V^H$, $U$ and $V$ unitary, $\Sigma$ real diagonal). The derivation of these complex factorizations is not sufficiently different from the real case to warrant a separate exposition.

# 7.1 Properties and Decompositions

In this section we survey the mathematical background necessary to develop and analyze the eigenvalue algorithms that follow.

## 7.1.1 Eigenvalues and Invariant Subspaces

The *eigenvalues* of a matrix $A \in \mathbb{C}^{n\times n}$ are the $n$ roots of its *characteristic polynomial* $p(z) = \det(zI - A)$. The set of these roots is called the *spectrum* and is denoted by $\lambda(A)$. If $\lambda(A) = \{\lambda_1, \ldots, \lambda_n\}$, then it follows that

$$\det(A) = \lambda_1\lambda_2\cdots\lambda_n .$$

Moreover, if we define the *trace* of $A$ by

$$\text{trace}(A) = \sum_{i=1}^{n} a_{ii}$$

then trace$(A) = \lambda_1 + \cdots + \lambda_n$. This follows by looking at the coefficient of $z$ in the characteristic polynomial.

If $\lambda \in \lambda(A)$ then the nonzero vectors $x \in \mathbb{C}^n$ that satisfy

$$Ax = \lambda x$$

are referred to as *eigenvectors.* More precisely, $x$ is a *right eigenvector* for $\lambda$ if $Ax = \lambda x$ and a *left eigenvector* if $x^H A = \lambda x^H$. Unless otherwise stated, "eigenvector" means "right eigenvector."

An eigenvector defines a one-dimensional subspace that is invariant with respect to premultiplication by $A$. More generally, a subspace $S \subseteq \mathbb{C}^n$ with the property that

$$x \in S \Longrightarrow Ax \in S$$

is said to be *invariant* (for $A$). Note that if

$$AX = XB, \qquad B \in \mathbb{C}^{k\times k},\ X \in \mathbb{C}^{n\times k}$$

then range$(X)$ is invariant and

$$By = \lambda y \Rightarrow A(Xy) = \lambda(Xy).$$

Thus, if $X$ has full column rank, then $AX = XB$ implies that $\lambda(B) \subseteq \lambda(A)$. If $X$ is square and nonsingular, then $\lambda(A) = \lambda(B)$ and we say that $A$ and $B = X^{-1}AX$ are *similar.* In this context, $X$ is called a *similarity transformation.*

Many eigenvalue computations involve breaking the given problem down into a collection of smaller eigenproblems. The following result is the basis for these reductions.

**Lemma 7.1.1** *If* $T \in \mathbb{C}^{n\times n}$ *is partitioned as follows,*

$$T = \begin{bmatrix} T_{11} & T_{12} \\ 0 & T_{22} \end{bmatrix} \begin{matrix} p \\ q \end{matrix}$$
$$\qquad\quad p \quad\ q$$

*then* $\lambda(T) = \lambda(T_{11}) \cup \lambda(T_{22})$.

**Proof.** Suppose

$$Tx = \begin{bmatrix} T_{11} & T_{12} \\ 0 & T_{22} \end{bmatrix} \begin{bmatrix} x_1 \\ x_2 \end{bmatrix} = \lambda \begin{bmatrix} x_1 \\ x_2 \end{bmatrix}$$

where $x_1 \in \mathbb{C}^p$ and $x_2 \in \mathbb{C}^q$. If $x_2 \neq 0$ then $T_{22}x_2 = \lambda x_2$ and so $\lambda \in \lambda(T_{22})$. If $x_2 = 0$, then $T_{11}x_1 = \lambda x_1$ and so $\lambda \in \lambda(T_{11})$. It follows that $\lambda(T) \in \lambda(T_{11}) \cup \lambda(T_{22})$. But since both $\lambda(T)$ and $\lambda(T_{11}) \cup \lambda(T_{22})$ have the same cardinality, the two sets are equal. $\square$

## 7.1.2 The Basic Unitary Decompositions

By using similarity transformations, it is possible to reduce a given matrix to any one of several canonical forms. The canonical forms differ in how they display the eigenvalues and in the kind of invariant subspace information that they provide. Because of their numerical stability we begin by discussing the reductions that can be achieved with unitary similarity.

**Lemma 7.1.2** *If $A \in \mathbb{C}^{n\times n}$, $B \in \mathbb{C}^{p\times p}$, and $X \in \mathbb{C}^{n\times p}$ satisfy*

$$AX = XB, \qquad \text{rank}(X) = p \tag{7.1.1}$$

*then there exists a unitary $Q \in \mathbb{C}^{n\times n}$ such that*

$$Q^H AQ = T = \begin{bmatrix} T_{11} & T_{12} \\ 0 & T_{22} \end{bmatrix} \begin{matrix} p \\ n-p \end{matrix} \tag{7.1.2}$$
$$\qquad\qquad p \quad n-p$$

*where $\lambda(T_{11}) = \lambda(A) \cap \lambda(B)$.*

**Proof.** Let

$$X = Q\begin{bmatrix} R \\ 0 \end{bmatrix} \qquad Q \in \mathbb{C}^{n\times n},\ R \in \mathbb{C}^{p\times p}$$

be a QR factorization of $X$. By substituting this into (7.1.1) and rearranging we have

$$\begin{bmatrix} T_{11} & T_{12} \\ T_{21} & T_{22} \end{bmatrix}\begin{bmatrix} R \\ 0 \end{bmatrix} = \begin{bmatrix} R \\ 0 \end{bmatrix} B$$

where

$$Q^H AQ = \begin{bmatrix} T_{11} & T_{12} \\ T_{21} & T_{22} \end{bmatrix} \begin{matrix} p \\ n-p \end{matrix} .$$
$$\qquad\qquad p \quad n-p$$

By using the nonsingularity of $R$ and the equations $T_{21}R = 0$ and $T_{11}R = RB$, we can conclude that $T_{21} = 0$ and $\lambda(T_{11}) = \lambda(B)$. The conclusion now follows because from Lemma 7.1.1 $\lambda(A) = \lambda(T) = \lambda(T_{11}) \cup \lambda(T_{22})$. □

**Example 7.1.1** If

$$A = \begin{bmatrix} 67.00 & 177.60 & -63.20 \\ -20.40 & 95.88 & -87.16 \\ 22.80 & 67.84 & 12.12 \end{bmatrix}$$

$X = (20,\ -9,\ -12)^T$ and $B = (25)$, then $AX = XB$. Moreover, if the orthogonal matrix $Q$ is defined by

$$Q = \begin{bmatrix} -.800 & .360 & .480 \\ .360 & .928 & -.096 \\ .480 & -.096 & .872 \end{bmatrix}$$

then $Q^TX = (-25,\ 0,\ 0)^T$ and

$$Q^TAQ = T = \begin{bmatrix} 25 & -90 & 5 \\ 0 & 147 & -104 \\ 0 & 146 & 3 \end{bmatrix}.$$

Note that $\lambda(A) = \{25, 75 + 100i, 75 - 100i\}$.

Lemma 7.1.2 says that a matrix can be reduced to block triangular form using unitary similarity transformations if we know one of its invariant subspaces. By induction we can readily establish the decompositon of Schur (1909).

**Theorem 7.1.3 (Schur Decomposition)** *If $A \in \mathbb{C}^{n\times n}$ then there exists a unitary $Q \in \mathbb{C}^{n\times n}$ such that*

$$Q^HAQ = T = D + N \qquad (7.1.3)$$

*where $D = \text{diag}(\lambda_1, \ldots, \lambda_n)$ and $N \in \mathbb{C}^{n\times n}$ is strictly upper triangular. Furthermore, $Q$ can be chosen so that the eigenvalues $\lambda_i$ appear in any order along the diagonal.*

**Proof.** The theorem obviously holds when $n = 1$. Suppose it holds for all matrices of order $n - 1$ or less. If $Ax = \lambda x$, where $x \neq 0$, then by Lemma 7.1.2 (with $B = (\lambda)$) there exists a unitary $U$ such that:

$$U^HAU = \begin{bmatrix} \lambda & w^H \\ 0 & C \end{bmatrix} \begin{matrix} 1 \\ n-1 \end{matrix} .$$
$$\qquad\qquad 1 \quad n-1$$

By induction there is a unitary $\tilde{U}$ such that $\tilde{U}^HC\tilde{U}$ is upper triangular. Thus, if $Q = U\text{diag}(1, \tilde{U})$, then $Q^HAQ$ is upper triangular. □

**Example 7.1.2** If

$$A = \begin{bmatrix} 78 & 200 \\ -50 & 75 \end{bmatrix} \qquad \text{and} \qquad Q = \begin{bmatrix} .8944i & .4472 \\ -.4472 & -.8944i \end{bmatrix}$$

then $Q$ is unitary and

$$Q^HAQ = \begin{bmatrix} 75 + 100i & -150 \\ 0 & 75 - 100i \end{bmatrix}.$$

If $Q = [\, q_1, \ldots, q_n \,]$ is a column partitioning of the unitary matrix $Q$ in (7.1.3), then the $q_i$ are referred to as *Schur vectors.* By equating columns in the equations $AQ = QT$ we see that the Schur vectors satisfy

$$Aq_k = \lambda_k q_k + \sum_{i=1}^{k-1} n_{ik} q_i \qquad k = 1{:}n . \qquad (7.1.4)$$

From this we conclude that the subspaces

$$S_k = \text{span}\{q_1, \ldots, q_k\} \qquad k = 1{:}n$$

are invariant. Moreover, it is not hard to show that if $Q_k = [\, q_1, \ldots, q_k \,]$, then $\lambda(Q_k^H A Q_k) = \{\lambda_1, \ldots, \lambda_k\}$. Since the eigenvalues in (7.1.3) can be arbitrarily ordered, it follows that there is at least one $k$-dimensional invariant subspace associated with each subset of $k$ eigenvalues.

Another conclusion to be drawn from (7.1.4) is that the Schur vector $q_k$ is an eigenvector if and only if the $k$-th column of $N$ is zero. This turns out to be the case for $k = 1{:}n$ whenever $A^H A = AA^H$. Matrices that satisfy this property are called *normal.*

**Corollary 7.1.4** *$A \in \mathbb{C}^{n\times n}$ is normal if and only if there exists a unitary $Q \in \mathbb{C}^{n\times n}$ such that $Q^H AQ = \text{diag}(\lambda_1, \ldots, \lambda_n)$.*

**Proof.** It is easy to show that if $A$ is unitarily similar to a diagonal matrix, then $A$ is normal. On the other hand, if $A$ is normal and $Q^H AQ = T$ is its Schur decomposition, then $T$ is also normal. The corollary follows by showing that a normal, upper triangular matrix is diagonal. □

Note that if $Q^H AQ = T = \text{diag}(\lambda_i) + N$ is a Schur decomposition of a general $n$-by-$n$ matrix $A$, then $\| N \|_F$ is independent of the choice of $Q$:

$$\| N \|_F^2 = \| A \|_F^2 - \sum_{i=1}^{n} |\lambda_i|^2 \equiv \Delta^2(A).$$

This quantity is referred to as $A$'s *departure from normality*. Thus, to make $T$ "more diagonal," it is necessary to rely on nonunitary similarity transformations.

### 7.1.3 Nonunitary Reductions

To see what is involved in nonunitary similarity reduction, we examine the block diagonalization of a 2-by-2 block triangular matrix.

**Lemma 7.1.5** *Let $T \in \mathbb{C}^{n\times n}$ be partitioned as follows:*

$$T = \begin{bmatrix} T_{11} & T_{12} \\ 0 & T_{22} \end{bmatrix} \begin{matrix} p \\ q \end{matrix} \quad . \\ \qquad\quad p \quad\;\; q$$

*Define the linear transformation $\phi:\mathbb{C}^{p\times q} \rightarrow \mathbb{C}^{p\times q}$ by*

$$\phi(X) = T_{11}X - XT_{22}$$

*where $X \in \mathbb{C}^{p\times q}$. Then $\phi$ is nonsingular if and only if $\lambda(T_{11}) \cap \lambda(T_{22}) = \emptyset$. If $\phi$ is nonsingular and $Y$ is defined by*

$$Y = \begin{bmatrix} I_p & Z \\ 0 & I_q \end{bmatrix} \qquad \phi(Z) = -T_{12}$$

*then* $Y^{-1}TY = \text{diag}(T_{11}, T_{22})$.

**Proof.** Suppose $\phi(X) = 0$ for $X \neq 0$ and that

$$U^H XV = \begin{bmatrix} \Sigma_r & 0 \\ 0 & 0 \end{bmatrix} \begin{matrix} r \\ p-r \end{matrix}$$
$$\qquad\qquad \begin{matrix} r & q-r \end{matrix}$$

is the SVD of $X$ with $\Sigma_r = \text{diag}(\sigma_i)$, $r = \text{rank}(X)$. Substituting this into the equation $T_{11}X = XT_{22}$ gives

$$\begin{bmatrix} A_{11} & A_{12} \\ A_{21} & A_{22} \end{bmatrix} \begin{bmatrix} \Sigma_r & 0 \\ 0 & 0 \end{bmatrix} = \begin{bmatrix} \Sigma_r & 0 \\ 0 & 0 \end{bmatrix} \begin{bmatrix} B_{11} & B_{12} \\ B_{21} & B_{22} \end{bmatrix}$$

where $U^H T_{11} U = (A_{ij})$ and $V^H T_{22} V = (B_{ij})$. By comparing blocks we see that $A_{21} = 0$, $B_{12} = 0$, and $\lambda(A_{11}) = \lambda(B_{11})$. Consequently,

$$\emptyset \neq \lambda(A_{11}) = \lambda(B_{11}) \subseteq \lambda(T_{11}) \cap \lambda(T_{22}).$$

On the other hand, if $\lambda \in \lambda(T_{11}) \cap \lambda(T_{22})$ then we have nonzero vectors $x$ and $y$ so $T_{11}x = \lambda x$ and $y^H T_{22} = \lambda y^H$. A calculation shows that $\phi(xy^H) = 0$. Finally, if $\phi$ is nonsingular then the matrix $Z$ above exists and

$$\begin{aligned} Y^{-1}TY &= \begin{bmatrix} I & -Z \\ 0 & I \end{bmatrix} \begin{bmatrix} T_{11} & T_{12} \\ 0 & T_{22} \end{bmatrix} \begin{bmatrix} I & Z \\ 0 & I \end{bmatrix} \\ &= \begin{bmatrix} T_{11} & T_{11}Z - ZT_{22} + T_{12} \\ 0 & T_{22} \end{bmatrix} = \begin{bmatrix} T_{11} & 0 \\ 0 & T_{22} \end{bmatrix}. \ \square \end{aligned}$$

**Example 7.1.3** If

$$T = \begin{bmatrix} 1 & 2 & 3 \\ 0 & 3 & 8 \\ 0 & -2 & 3 \end{bmatrix} \quad \text{and} \quad Y = \begin{bmatrix} 1.0 & 0.5 & -0.5 \\ 0.0 & 1.0 & 0.0 \\ 0.0 & 0.0 & 1.0 \end{bmatrix}$$

then

$$Y^{-1}TY = \begin{bmatrix} 1 & 0 & 0 \\ 0 & 3 & 8 \\ 0 & -2 & 3 \end{bmatrix}$$

By repeatedly applying Lemma 7.1.5, we can establish the following more general result:

**Theorem 7.1.6 (Block Diagonal Decomposition)** *Suppose*

$$Q^H AQ = T = \begin{bmatrix} T_{11} & T_{12} & \cdots & T_{1q} \\ 0 & T_{22} & \cdots & T_{2q} \\ \vdots & \vdots & \ddots & \vdots \\ 0 & 0 & \cdots & T_{qq} \end{bmatrix} \qquad (7.1.5)$$

*is a Schur decomposition of* $A \in \mathbb{C}^{n\times n}$ *and assume that the* $T_{ii}$ *are square. If* $\lambda(T_{ii}) \cap \lambda(T_{jj}) = \emptyset$ *whenever* $i \neq j$, *then there exists a nonsingular matrix* $Y \in \mathbb{C}^{n\times n}$ *such that*

$$(QY)^{-1}A(QY) = \text{diag}(T_{11}, \ldots, T_{qq}). \qquad (7.1.6)$$

**Proof.** A proof can be obtained by using Lemma 7.1.5 and induction. □

If each diagonal block $T_{ii}$ is associated with a distinct eigenvalue, then we obtain

**Corollary 7.1.7** *If* $A \in \mathbb{C}^{n\times n}$ *then there exists a nonsingular* $X$ *such that*

$$X^{-1}AX = \text{diag}(\lambda_1 I + N_1, \ldots, \lambda_q I + N_q) \qquad N_i \in \mathbb{C}^{n_i \times n_i} \qquad (7.1.7)$$

*where* $\lambda_1, \ldots, \lambda_q$ *are distinct, the integers* $n_1, \ldots, n_q$ *satisfy* $n_1 + \cdots + n_q = n$, *and each* $N_i$ *is strictly upper triangular.*

A number of important terms are connected with decomposition (7.1.7). The integer $n_i$ is referred to as the *algebraic multiplicity* of $\lambda_i$. If $n_i = 1$, then $\lambda_i$ is said to be *simple* . The *geometric multiplicity* of $\lambda_i$ equals the dimensions of null($N_i$), i.e., the number of linearly independent eigenvectors associated with $\lambda_i$. If the algebraic multiplicity of $\lambda_i$ exceeds its geometric multiplicity, then $\lambda_i$ is said to be a *defective eigenvalue.* A matrix with a defective eigenvalue is referred to as a *defective matrix.* Nondefective matrices are also said to be *diagonalizable* in light of the following result:

**Corollary 7.1.8 (Diagonal Form)** $A \in \mathbb{C}^{n\times n}$ *is nondefective if and only if there exists a nonsingular* $X \in \mathbb{C}^{n\times n}$ *such that*

$$X^{-1}AX = \text{diag}(\lambda_1, \ldots, \lambda_n). \qquad (7.1.8)$$

**Proof.** $A$ is nondefective if and only if there exist independent vectors $x_1 \ldots x_n \in \mathbb{C}^n$ and scalars $\lambda_1, \ldots, \lambda_n$ such that $Ax_i = \lambda_i x_i$ for $i = 1{:}n$. This is equivalent to the existence of a nonsingular $X = [\, x_1, \ldots, x_n \,] \in \mathbb{C}^{n\times n}$ such that $AX = XD$ where $D = \text{diag}(\lambda_1, \ldots, \lambda_n)$. □

Note that if $y_i^H$ is the $i$th row of $X^{-1}$, then $y_i^H A = \lambda_i y_i^H$. Thus, the columns of $X^{-T}$ are left eigenvectors and the columns of $X$ are right eigenvectors.

**Example 7.1.4** If

$$A = \begin{bmatrix} 5 & -1 \\ -2 & 6 \end{bmatrix} \quad \text{and} \quad \begin{bmatrix} 1 & 1 \\ 1 & -2 \end{bmatrix}$$

then $X^{-1}AX = \text{diag}(4,7)$.

If we partition the matrix $X$ in (7.1.7),

$$X = [\ \underset{n_1}{X_1}\ ,\ldots,\ \underset{n_q}{X_q}\ ]$$

then $\mathbb{C}^n = \text{range}(X_1)\oplus\ldots\oplus\text{range}(X_q)$, a direct sum of invariant subspaces. If the bases for these subspaces are chosen in a special way, then it is possible to introduce even more zeroes into the upper triangular portion of $X^{-1}AX$.

**Theorem 7.1.9 (Jordan Decomposition)** *If $A \in \mathbb{C}^{n\times n}$, then there exists a nonsingular $X \in \mathbb{C}^{n\times n}$ such that $X^{-1}AX = \text{diag}(J_1,\ldots,J_t)$ where*

$$J_i = \begin{bmatrix} \lambda_i & 1 & & \cdots & 0 \\ 0 & \lambda_i & \ddots & & \vdots \\ & \ddots & \ddots & \ddots & \\ \vdots & & \ddots & \ddots & 1 \\ 0 & \cdots & & 0 & \lambda_i \end{bmatrix}$$

*is $m_i$-by-$m_i$ and $m_1 + \cdots + m_t = n$.*

**Proof.** See Halmos (1958, pp. 112 ff.) □

The $J_i$ are referred to as *Jordan blocks* . The number and dimensions of the Jordan blocks associated with each distinct eigenvalue is unique, although their ordering along the diagonal is not.

### 7.1.4 Some Comments on Nonunitary Similarity

The Jordan block structure of a defective matrix is difficult to determine numerically. The set of $n$-by-$n$ diagonalizable matrices is dense in $\mathbb{C}^{n\times n}$, and thus, small changes in a defective matrix can radically alter its Jordan form. We have more to say about this in §7.6 .

A related difficulty that arises in the eigenvalue problem is that a nearly defective matrix can have a poorly conditioned matrix of eigenvectors. For example, any matrix $X$ that diagonalizes

$$A = \begin{bmatrix} 1+\epsilon & 1 \\ 0 & 1-\epsilon \end{bmatrix} \qquad 0 < \epsilon \ll 1 \tag{7.1.9}$$

has a 2-norm condition of order $1/\epsilon$.

These observations serve to highlight the difficulties associated with ill-conditioned similarity transformations. Since

$$fl(X^{-1}AX) = X^{-1}AX + E, \tag{7.1.10}$$

where

$$\| E \|_2 \approx \mathbf{u}\kappa_2(X)\| A \|_2 \tag{7.1.11}$$

is it clear that large errors can be introduced into an eigenvalue calculation when we depart from unitary similarity.

**Problems**

**P7.1.1** Show that if $T \in \mathbb{C}^{n \times n}$ is upper triangular and normal, then $T$ is diagonal.

**P7.1.2** Verify that if $X$ diagonalizes the 2-by-2 matrix in (7.1.9) and $\epsilon \leq 1/2$ then $\kappa_1(X) \geq 1/\epsilon$.

**P7.1.3** Suppose $A \in \mathbb{C}^{n \times n}$ has distinct eigenvalues. Show that if $Q^H AQ = T$ is its Schur decomposition and $AB = BA$, then $Q^H BQ$ is upper triangular.

**P7.1.4** Show that if $A$ and $B^H$ are in $\mathbb{C}^{m \times n}$ with $m \geq n$, then:

$$\lambda(AB) = \lambda(BA) \cup \{\underbrace{0,\ldots,0}_{m-n}\}.$$

**P7.1.5** Given $A \in \mathbb{C}^{n \times n}$, use the Schur decomposition to show that for every $\epsilon > 0$, there exists a diagonalizable matrix $B$ such that $\| A - B \|_2 \leq \epsilon$. This shows that the set of diagonalizable matrices is dense in $\mathbb{C}^{n \times n}$ and that the Jordan canonical form is not a continuous matrix decomposition.

**P7.1.6** Suppose $A_k \rightarrow A$ and that $Q_k^H A_k Q_k = T_k$ is a Schur decomposition of $A_k$. Show that $\{Q_k\}$ has a converging subsequence $\{Q_{k_i}\}$ with the property that

$$\lim_{i \rightarrow \infty} Q_{k_i} = Q$$

where $Q^H AQ = T$ is upper triangular. This shows that the eigenvalues of a matrix are continuous functions of its entries.

**P7.1.7** Justify (7.1.10) and (7.1.11).

**Notes and References for Sec. 7.1**

The mathematical properties of the algebraic eigenvalue problem are elegantly covered in Wilkinson (AEP, chapter 1) and Stewart (IMC, chapter 6). For those who need further review we also recommend

R. Bellman (1970). *Introduction to Matrix Analysis,* 2nd ed., McGraw-Hill, New York.

I.C. Gohberg, P. Lancaster, and L. Rodman (1986). *Invariant Subspaces of Matrices With Applications,* John Wiley and Sons, New York.

M. Marcus and H. Minc (1964). *A Survey of Matrix Theory and Matrix Inequalities,* Allyn and Bacon, Boston.

L. Mirsky (1963). *An Introduction to Linear Algebra ,* Oxford University Press, Oxford.

The Schur decomposition originally appeared in

I. Schur (1909). "On the Characteristic Roots of a Linear Substitution with an Application to the Theory of Integral Equations." *Math. Ann. 66*, 488-510 (German).

A proof very similar to ours is given on page 105 of

H.W. Turnbill and A.C. Aitken (1961). *An Introduction to the Theory of Canonical Forms*, Dover, New York.

Sometimes it is possible to establish specially structured Schur decompositions for specially structured matrices. See

C.C. Paige and C. Van Loan (1981). "A Schur Decomposition for Hamiltonian Matrices," *Lin. Alg. and Its Applic. 41*, 11–32.

# 7.2 Perturbation Theory

None of the decompositions in the preceding section can be calculated exactly because of roundoff error and because eigenvalue algorithms are iterative and must be terminated after a finite number of steps. Therefore, it is critical that we develop a useful perturbation theory to guide our thinking in subsequent sections.

## 7.2.1 Eigenvalue Sensitivity

Several eigenvalue routines produce a sequence of similarity transformations $X_k$ with the property that the matrices $X_k^{-1}AX_k$ are progessively "more diagonal." The question naturally arises, how well do the diagonal elements of a matrix approximate its eigenvalues?

**Theorem 7.2.1 (Gershgorin Circle Theorem)** *If* $X^{-1}AX = D + F$ *where* $D = \text{diag}(d_1, \ldots, d_n)$ *and* $F$ *has zero diagonal entries, then*

$$\lambda(A) \subseteq \bigcup_{i=1}^{n} D_i$$

*where*

$$D_i = \left\{ z \in \mathbb{C} : |z - d_i| \leq \sum_{j=1}^{n} |f_{ij}| \right\}.$$

**Proof.** Suppose $\lambda \in \lambda(A)$ and assume without loss of generality that $\lambda \neq d_i$ for $i = 1{:}n$. Since $(D - \lambda I) + F$ is singular, it follows from Lemma 2.3.3 that

$$1 \leq \| (D - \lambda I)^{-1} F \|_\infty = \sum_{j=1}^{n} \frac{|f_{kj}|}{|d_k - \lambda|}$$

for some $k$, $1 \leq k \leq n$. But this implies that $\lambda \in D_k$. $\square$

It can also be shown that if the Gershgorin disk $D_i$ is isolated from the other disks, then it contains precisely one of $A$'s eigenvalues. See Wilkinson (AEP, pp. 71 ff.).

**Example 7.2.1** If

$$A = \begin{bmatrix} 10 & 2 & 3 \\ -1 & 0 & 2 \\ 1 & -2 & 1 \end{bmatrix}$$

then $\lambda(A) \approx \{10.226,\ .3870 + 2.2216i,\ .3870 - 2.2216i\}$ and the Gershgorin disks are $D_1 = \{\ |z| : |z - 10| \leq 5\}$, $D_2 = \{\ |z| : |z| \leq 3\}$, and $D_3 = \{\ |z| : |z - 1| \leq 3\}$.

For some very important eigenvalue routines it is possible to show that the computed eigenvalues are the exact eigenvalues of a matrix $A + E$ where $E$ is small in norm. Consequently, we must undersetand how the eigenvalues of a matrix can be affected by small perturbations. A sample result that sheds light on this issue is the following theorem.

**Theorem 7.2.2 (Bauer-Fike)** *If $\mu$ is an eigenvalue of $A + E \in \mathbb{C}^{n \times n}$ and $X^{-1}AX = D = \text{diag}(\lambda_1, \ldots, \lambda_n)$, then*

$$\min_{\lambda \in \lambda(A)} |\lambda - \mu| \leq \kappa_p(X) \| E \|_p$$

*where $\| \cdot \|_p$ denotes any of the p-norms.*

**Proof.** We need only consider the case when $\mu$ is not in $\lambda(A)$. If the matrix $X^{-1}(A + E - \mu I)X$ is singular, then so is $I + (D - \mu I)^{-1}(X^{-1}EX)$. Thus, from Lemma 2.3.3 we obtain

$$1 \leq \| (D - \mu I)^{-1}(X^{-1}EX) \|_p \leq \| (D - \mu I)^{-1} \|_p \| X \|_p \| E \|_p \| X^{-1} \|_p .$$

Since $(D - \mu I)^{-1}$ is diagonal and the $p$-norm of a diagonal matrix is the absolute value of the largest diagonal entry, it follows that

$$\| (D - \mu I)^{-1} \|_p = \min_{\lambda \in \lambda(A)} \frac{1}{|\lambda - \mu|}$$

from which the theorem follows. □

An analogous result can be obtained via the Schur decomposition:

**Theorem 7.2.3** *Let $Q^H AQ = D+N$ be a Schur decomposition of $A \in \mathbb{C}^{n \times n}$ as in (7.1.3). If $\mu \in \lambda(A + E)$ and p is the smallest positive integer such that $|N|^p = 0$, then*

$$\min_{\lambda \in \lambda(A)} |\lambda - \mu| \leq \max(\theta,\ \theta^{1/p})$$

*where*

$$\theta = \| E \|_2 \sum_{k=0}^{p-1} \| N \|_2^k .$$

**Proof.** Define

$$\delta = \min_{\lambda \in \lambda(A)} |\lambda - \mu| = \frac{1}{\| (\mu I - D)^{-1} \|_2} .$$

The theorem is clearly true if $\delta = 0$. If $\delta > 0$ then $I - (\mu I - A)^{-1}E$ is singular and by Lemma 2.3.3 we have

$$\begin{aligned} 1 &\le \| (\mu I - A)^{-1} E \|_2 \le \| (\mu I - A)^{-1} \|_2 \| E \|_2 \qquad (7.2.1) \\ &= \| ((\mu I - D) - N)^{-1} \|_2 \| E \|_2 . \end{aligned}$$

Since $(\mu I - D)^{-1}$ is diagonal and $|N|^p = 0$ it is not hard to show that $((\mu I - D)^{-1} N)^p = 0$. Thus,

$$((\mu I - D) - N)^{-1} = \sum_{k=0}^{p-1} \left((\mu I - D)^{-1} N\right)^k (\mu I - D)^{-1}$$

and so

$$\| ((\mu I - D) - N)^{-1} \|_2 \le \frac{1}{\delta} \sum_{k=0}^{p-1} \left( \frac{\| N \|_2}{\delta} \right)^k .$$

If $\delta > 1$ then

$$\| (\mu I - D) - N)^{-1} \|_2 \le \frac{1}{\delta} \sum_{k=0}^{p-1} \| N \|_2^k$$

and so from (7.2.1), $\delta \le \theta$. If $\delta \le 1$ then

$$\| (\mu I - D) - N)^{-1} \|_2 \le \frac{1}{\delta^p} \sum_{k=0}^{p-1} \| N \|_2^k$$

and so from (7.2.1), $\delta^p \le \theta$. Thus, $\delta \le \max(\theta, \theta^{1/p})$. □

**Example 7.2.2** If

$$A = \begin{bmatrix} 1 & 2 & 3 \\ 0 & 4 & 5 \\ 0 & 0 & 4.001 \end{bmatrix} \quad \text{and} \quad E = \begin{bmatrix} 0 & 0 & 0 \\ 0 & 0 & 0 \\ .001 & 0 & 0 \end{bmatrix}$$

then $\lambda(A + E) \approx \{1.0001, 4.0582, 3.9427\}$ and $A$'s matrix of eigenvectors satisfies $\kappa_2(X) \approx 10^7$. The Bauer-Fike bound in Theorem 7.2.2 has order $10^4$, while the Schur bound in Theorem 7.2.3 has order $10^0$.

Theorems 7.2.2 and 7.2.3 each indicate potential eigenvalue sensitivity if $A$ is non-normal. Specifically, if $\kappa_2(X)$ or $\| N \|_2^{p-1}$ is large, then small changes in $A$ can induce large changes in the eigenvalues.

**Example 7.2.3** If

$$A = \begin{bmatrix} 0 & I_9 \\ 0 & 0 \end{bmatrix} \qquad \text{and} \qquad E = \begin{bmatrix} 0 & 0 \\ 10^{-10} & 0 \end{bmatrix}$$

then for all $\lambda \in \lambda(A)$ and $\mu \in \lambda(A+E)$, $|\lambda - \mu| = 10^{-1}$. In this example a change of order $10^{-10}$ in $A$ results in a change of order $10^{-1}$ in its eigenvalues.

## 7.2.2 The Condition of a Simple Eigenvalue

Extreme eigenvalue sensitivity for a matrix $A$ cannot occur if $A$ is normal. On the other hand, non-normality does not necessarily imply eigenvalue sensitivity. Indeed, a non-normal matrix can have a mixture of well-conditioned and ill-conditioned eigenvalues. For this reason, it is beneficial to refine our perturbation theory so that it is applicable to individual eigenvalues and not the spectrum as a whole.

To this end, suppose that $\lambda$ is a simple eigenvalue of $A \in \mathbb{C}^{n\times n}$ and that $x$ and $y$ satisfy $Ax = \lambda x$ and $y^H A = \lambda y^H$ with $\| x \|_2 = \| y \|_2 = 1$. Using classical results from function theory, it can be shown that in a neighborhood of the origin there exist differentiable $x(\epsilon)$ and $\lambda(\epsilon)$ such that

$$(A + \epsilon F)x(\epsilon) = \lambda(\epsilon)x(\epsilon) \qquad \| F \|_2 = 1 \,, \; \| x(\epsilon) \|_2 \equiv 1$$

where $\lambda(0) = \lambda$ and $x(0) = x$. By differentiating this equation with respect to $\epsilon$ and setting $\epsilon = 0$ in the result, we obtain

$$A\dot{x}(0) + Fx \;=\; \dot{\lambda}(0)x + \lambda\dot{x}(0)\,.$$

Applying $y^H$ to both sides of this equation, dividing by $y^H x$, and taking absolute values gives

$$|\dot{\lambda}(0)| \;=\; \left| \frac{y^H F x}{y^H x} \right| \;\le\; \frac{1}{|y^H x|}.$$

The upper bound is attained if $F = yx^H$. For this reason we refer to the reciprocal of

$$s(\lambda) \;=\; |y^H x|$$

as the *condition of the eigenvalue* $\lambda$.

Roughly speaking, the above analysis shows that if order $\epsilon$ perturbations are made in $A$, then an eigenvalue $\lambda$ may be perturbed by an amount

$\epsilon/s(\lambda)$. Thus, if $s(\lambda)$ is small, then $\lambda$ is appropriately regarded as ill-conditioned. Note that $s(\lambda)$ is the cosine of the angle between the left and right eigenvectors associated with $\lambda$ and is unique only if $\lambda$ is simple.

A small $s(\lambda)$ implies that $A$ is near a matrix having a multiple eigenvalue. In particular, if $\lambda$ is distinct and $s(\lambda) < 1$, then there exists an $E$ such that $\lambda$ is a repeated eigenvalue of $A + E$ and

$$\frac{\| E \|_2}{\| A \|_2} \le \frac{s(\lambda)}{\sqrt{1 - s(\lambda)^2}} .$$

This result is proved in Wilkinson (1972).

**Example 7.2.4** If

$$A = \begin{bmatrix} 1 & 2 & 3 \\ 0 & 4 & 5 \\ 0 & 0 & 4.001 \end{bmatrix} \quad \text{and} \quad E = \begin{bmatrix} 0 & 0 & 0 \\ 0 & 0 & 0 \\ .001 & 0 & 0 \end{bmatrix}$$

then $\lambda(A + E) \approx \{1.0001,\ 4.0582,\ 3.9427\}$ and $s(1) \approx .79 \times 10^0$, $s(4) \approx .16 \times 10^{-3}$, and $s(4.001) \approx .16 \times 10^{-3}$. Observe that $\| E \|_2 / s(\lambda)$ is a good estimate of the perturbation that each eigenvalue undergoes.

## 7.2.3 Sensitivity of Repeated Eigenvalues

If $\lambda$ is a repeated eigenvalue, then the eigenvalue sensitivity question is more complicated. For example, if

$$A = \begin{bmatrix} 1 & a \\ 0 & 1 \end{bmatrix} \quad \text{and} \quad F = \begin{bmatrix} 0 & 0 \\ 1 & 0 \end{bmatrix}$$

then $\lambda(A + \epsilon F) = \{1 \pm \sqrt{\epsilon a}\}$. Note that if $a \neq 0$, then it follows that the eigenvalues of $A + \epsilon F$ are not differentiable at zero; their rate of change at the origin is infinite. In general, if $\lambda$ is a defective eigenvalue of $A$, then $O(\epsilon)$ perturbations in $A$ can result in $O(\epsilon^{1/p})$ perturbations in $\lambda$ if $\lambda$ is associated with a $p$-dimensional Jordan block. See Wilkinson (AEP, pp. 77ff.) for a more detailed discussion.

## 7.2.4 Eigenvector Sensitivity

We now consider how small changes in a matrix affect its invariant subspaces. The case of eigenvectors is dealt with first. Assume $A \in \mathbb{C}^{n \times n}$ has distinct eigenvalues $\lambda_1, \ldots, \lambda_n$ and that $F \in \mathbb{C}^{n \times n}$ satisfies $\| F \|_2 = 1$. A continuity argument ensures that for all $\epsilon$ in some neighborhood of the origin we have

$$\begin{aligned} (A + \epsilon F)x_k(\epsilon) &= \lambda_k(\epsilon)x_k(\epsilon) & \| x_k(\epsilon) \|_2 \equiv 1 \\ y_k(\epsilon)^H(A + \epsilon F) &= \lambda_k(\epsilon)y_k(\epsilon)^H & \| y_k(\epsilon) \|_2 \equiv 1 \end{aligned} \qquad k = 1{:}n$$

where each $\lambda_k(\epsilon)$, $x_k(\epsilon)$, and $y_k(\epsilon)$ is differentiable. If we differentiate the first of these equations with respect to $\epsilon$ and set $\epsilon = 0$ in the result, we obtain

$$A\dot{x}_k(0) + Fx_k = \dot{\lambda}_k(0)x_k + \lambda_k\dot{x}_k(0),$$

where $\lambda_k = \lambda_k(0)$ and $x_k = x_k(0)$. Since $A$'s eigenvectors form a basis we can write

$$\dot{x}_k(0) = \sum_{i=1}^{n} a_i x_i$$

and so we have

$$\sum_{\substack{i=1\\ i\neq k}}^{n} a_i(\lambda_i - \lambda_k)x_i + Fx_k = \dot{\lambda}_k(0)x_k .$$

To obtain expressions for the $a_i$, note that $y_i(0)^H x_k \equiv y_i^H x_k = 0$ whenever $i \neq k$ and thus

$$a_i = \frac{y_i^H F x_k}{(\lambda_k - \lambda_i)y_i^H x_i} \qquad i \neq k .$$

Therefore, the Taylor expansion for $x_k(\epsilon)$ has the following form

$$x_k(\epsilon) \approx x_k + \sum_{\substack{i=1\\ i\neq k}}^{n} \frac{y_i^H F x_k}{(\lambda_k - \lambda_i)y_i^H x_i} x_i + O(\epsilon^2) .$$

Thus, the sensitivity of $x_k$ depends upon eigenvalue sensitivity *and* the separation of $\lambda_k$ from the other eigenvalues.

That the separation of the eigenvalues should have a bearing upon eigenvector sensitivity should come as no surprise. Indeed, if $\lambda$ is a nondefective, repeated eigenvalue, then there are an infinite number of possible eigenvector bases for the associated invariant subspace. The preceding analysis merely indicates that this indeterminancy begins to be felt as the eigenvalues coalesce. In other words, the eigenvectors associated with nearby eigenvalues are "wobbly."

**Example 7.2.5** If

$$A = \begin{bmatrix} 1.01 & 0.01 \\ 0.00 & 0.99 \end{bmatrix}$$

then the eigenvalue $\lambda = .99$ has condition $1/s(.99) \approx 1.118$ and associated eigenvector $x = (.4472,\ -.8944)^T$. On the other hand, the eigenvalue $\hat{\lambda} = 1.00$ of the "nearby" matrix

$$A + E = \begin{bmatrix} 1.01 & 0.01 \\ 0.00 & 1.00 \end{bmatrix}$$

has an eigenvector $\hat{x} = (.7071,\ -.7071)^T$.

### 7.2.5 Invariant Subspace Sensitivity

A collection of sensitive eigenvectors can define an insensitive invariant subspace provided the corresponding cluster of eigenvalues is isolated. To be precise, suppose

$$Q^H AQ = \begin{bmatrix} T_{11} & T_{12} \\ 0 & T_{22} \end{bmatrix} \begin{matrix} p \\ n-p \end{matrix} \qquad (7.2.2)$$
$$\qquad\qquad p \quad n-p$$

is a Schur decomposition of $A$ with

$$Q = [\, Q_1 \quad Q_2 \,] \quad . \qquad (7.2.3)$$
$$\qquad p \quad n-p$$

It is clear from our discussion of eigenvector perturbation that the sensitivity of the invariant subspace range($Q_1$) depends on the distance between $\lambda(T_{11})$ and $\lambda(T_{22})$. The proper measure of this distance turns out to be the smallest singular value of the linear transformation $X \to T_{11}X - XT_{22}$. (Recall that this transformation figures in Lemma 7.1.5.) In particular, if we define the *separation* between the matrices $T_{11}$ and $T_{22}$ by

$$\operatorname{sep}(T_{11}, T_{22}) = \min_{X \neq 0} \frac{\| T_{11}X - XT_{22} \|_F}{\| X \|_F}$$

then we have the following general result:

**Theorem 7.2.4** *Suppose that (7.2.2) and (7.2.3) hold and that for any matrix $E \in \mathbb{C}^{n\times n}$ we partition $Q^H EQ$ as follows:*

$$Q^H EQ = \begin{bmatrix} E_{11} & E_{12} \\ E_{21} & E_{22} \end{bmatrix} \begin{matrix} p \\ n-p \end{matrix} \quad .$$
$$\qquad\qquad p \quad n-p$$

*If* $\delta = \operatorname{sep}(T_{11}, T_{22}) - \| E_{11} \|_2 - \| E_{22} \|_2 > 0$ *and*

$$\| E_{21} \|_2 (\| T_{12} \|_2 + \| E_{21} \|_2) \leq \delta^2/4,$$

*then there exists a* $P \in \mathbb{C}^{(n-k)\times k}$ *such that*

$$\| P \|_2 \leq 2\| E_{21} \|_2/\delta$$

*and such that the columns of* $\hat{Q} = (Q_1 + Q_2P)(I + P^H P)^{-1/2}$ *form an orthonormal basis for a subspace that is invariant for* $A + E$.

**Proof.** See Stewart (1973b, Theorem 4.11). □

Since $Q_1^H \hat{Q}_1 = (I + P^H P)^{-1/2}$, we see that the singular values of $P$ are the tangents of the principal angles between range($\hat{Q}_1$) and range($Q_1$). (See §12.4.3). The theorem shows that these tangents are essentially bounded by $\| E \|_2 / \text{sep}(T_{11}, T_{22})$. Thus the reciprocal of sep($T_{11}, T_{22}$) can be thought of as a condition number that measures the sensitivity of range($Q_1$) as an invariant subspace.

**Example 7.2.6** Suppose

$$T_{11} = \begin{bmatrix} 3 & 10 \\ 0 & 1 \end{bmatrix}, \quad T_{22} = \begin{bmatrix} 0 & -20 \\ 0 & 3.01 \end{bmatrix}, \quad \text{and} \quad T_{12} = \begin{bmatrix} 1 & -1 \\ -1 & 1 \end{bmatrix}$$

and that

$$A = T = \begin{bmatrix} T_{11} & T_{12} \\ 0 & T_{22} \end{bmatrix}.$$

Observe that $AQ_1 = Q_1 T_{11}$ where $Q_1 = [e_1, e_2] \in \mathbb{R}^{4 \times 2}$. A calculation shows that sep($T_{11}, T_{22}$) $\approx .0003$. If

$$E_{21} = 10^{-6} \begin{bmatrix} 1 & 1 \\ 1 & 1 \end{bmatrix}$$

and we examine the Schur decomposition of

$$A + E = \begin{bmatrix} T_{11} & T_{12} \\ E_{21} & T_{22} \end{bmatrix}$$

we find that $Q_1$ gets perturbed to

$$\hat{Q}_1 = \begin{bmatrix} -.9999 & -.0003 \\ .0003 & -.9999 \\ -.0005 & -.0026 \\ .0000 & .0003 \end{bmatrix}.$$

Thus, we have dist($\hat{Q}_1, Q_1$) $\approx .0027 \approx 10^{-6}/\text{sep}(T_{11}, T_{22})$.

## Problems

**P7.2.1** Suppose $Q^H A Q = \text{diag}(\lambda_1) + N$ is a Schur decomposition of $A \in \mathbb{C}^{n \times n}$ and define $\nu(A) = \| A^H A - AA^H \|_F$. The upper and lower bounds in

$$\frac{\nu(A)^2}{6\| A \|_F^2} \le \| N \|_F^2 \le \sqrt{\frac{n^3 - n}{12}} \nu(A)$$

are established by Henrici (1962) and Eberlein (1965), respectively. Verify these results for the case $n = 2$.

**P7.2.2** Suppose $A \in \mathbb{C}^{n \times n}$ has distinct eigenvalues and that

$$X^{-1} A X = \text{diag}(\lambda_1, \ldots, \lambda_n).$$

Show that if the columns of $X$ have unit 2-norm, then

$$\kappa_F(X) = \sqrt{n \sum_{i=1}^{n} \frac{1}{s(\lambda_i)^2}}.$$

**P7.2.3** Suppose $Q^H AQ = \text{diag}(\lambda_i) + N$ is a Schur decomposition of $A$ and that $X^{-1}AX = \text{diag}(\lambda_i)$. Show

$$\kappa_2(X) \geq \sqrt{1 + \frac{\| N \|_F^2}{\| A \|_F^2}} .$$

See Loizou (1969).

**P7.2.4** If $X^{-1}AX = \text{diag}(\lambda_i)$ and $|\lambda_1| \geq \cdots \geq |\lambda_n|$, then

$$\frac{\sigma_i(A)}{\kappa_2(X)} \leq |\lambda_i| \leq \kappa_2(X)\sigma_i(A) .$$

Prove this result for the $n = 2$ case. See Ruhe (1975).

**P7.2.5** Show that if $A = \begin{bmatrix} a & c \\ 0 & b \end{bmatrix}$ then $s(a) = s(b) = (1 + |c/(a-b)|^2)^{-1}$.

**P7.2.6** Show that the condition of an eigenvalue is preserved under unitary similarity transformations.

**Notes and References for Sec. 7.2**

Many of the results presented in this section may be found in Wilkinson (AEP, chapter 2) as well as in

A.S. Householder (1964). *The Theory of Matrices in Numerical Analysis.* Blaisdell, New York.

F.L. Bauer and C.T. Fike (1960). "Norms and Exclusion Theorems," *Numer. Math. 2,* 123–44.

The following papers are concerned with the effect of perturbations on the eigenvalues of a general matrix:

W. Kahan, B.N. Parlett, and E. Jiang (1982). "Residual Bounds on Approximate Eigensystems of Nonnormal Matrices," *SIAM J. Numer. Anal. 19,* 470–484.

A. Ruhe (1970). "Perturbation Bounds for Means of Eigenvalues and Invariant Subspaces," *BIT 10,* 343–54.

A. Ruhe (1970). "Properties of a Matrix with a Very Ill-Conditioned Eigenproblem," *Numer. Math. 15,* 57–60.

J.H. Wilkinson (1972). "Note on Matrices with a Very Ill-Conditioned Eigenproblem," *Numer. Math. 19,* 176–78.

J.H. Wilkinson (1984). "On Neighboring Matrices with Quadratic Elementary Divisors," *Numer. Math. 44,* 1-21.

Wilkinson's work on nearest defective matrices is typical of a growing body of literature that is concerned with "nearness" problems. See

R. Byers (1988). "A Bisection Method for Measuring the Distance of a Stable Matrix to the Unstable Matrices," *SIAM J. Sci. and Stat. Comp. 9,* 875–881.

J.W. Demmel (1987). "On the Distance to the Nearest Ill-Posed Problem," *Numer. Math. 51,* 251–289.

J.W. Demmel (1987). "A Counterexample for two Conjectures About Stability," *IEEE Trans. Auto. Cont. AC-32,* 340–342.

J.W. Demmel (1988). "The Probability that a Numerical Analysis Problem is Difficult," *Math. Comp. 50,* 449–480.

N.J. Higham (1985). "Nearness Problems in Numerical Linear Algebra," PhD Thesis, University of Manchester, England.

N.J. Higham (1988). "Matrix Nearness Problems and Applications," Numerical Analysis Report 161, University of Manchester, England. (To appear in *The Proceedings of the IMA Conference on Applications of Matrix Theory*, S. Barnett and M.J.C. Gover (eds), Oxford University Press.

A. Ruhe (1987). "Closest Normal Matrix Found!," *BIT 27,* 585-598.

C. Van Loan (1985). "How Near is a Stable Matrix to an Unstable Matrix?," Contemporary Mathematics, Vol. 47.,465–477.

The relationship between the eigenvalue condition number, the departure from normality, and the condition of the eigenvector matrix is discussed in

P. Eberlein (1965). "On Measures of Non-Normality for Matrices," *Amer. Math. Soc. Monthly 72,* 995–96.

P. Henrici (1962). "Bounds for Iterates, Inverses, Spectral Variation and Fields of Values of Non-normal Matrices," *Numer. Math. 4,* 24–40.

G. Loizou (1969). "Nonnormality and Jordan Condition Numbers of Matrices," *J. ACM 16,* 580–40.

R.A. Smith (1967). "The Condition Numbers of the Matrix Eigenvalue Problem," *Numer. Math. 10* 232–40.

A. van der Sluis (1975). "Perturbations of Eigenvalues of Non-normal Matrices," *Comm. ACM 18,* 30–36.

The paper by Henrici also contains a result similar to Theorem 7.2.3. Penetrating treatments of invariant subspace perturbation include

C. Davis and W.M. Kahan (1970). "The Rotation of Eigenvectors by a Perturbation, III," *SIAM J. Num. Anal. 7,* 1–46.

G.W. Stewart (1971). "Error Bounds for Approximate Invariant Subspaces of Closed Linear Operators," *SIAM. J. Num. Anal. 8,* 796–808.

G.W. Stewart (1973b). "Error and Perturbation Bounds for Subspaces Associated with Certain Eigenvalue Problems," *SIAM Review 15,* 727–64.

A detailed analyses of the function sep(.,.) and the transformation $X \rightarrow AX + XA^T$ are given in

R. Byers and S.G. Nash (1987). "On the Singular Vectors of the Lyapunov Operator," *SIAM J. Alg. and Disc. Methods 8,* 59–66.

J. Varah (1979). "On the Separation of Two Matrices," *SIAM J. Num. Anal. 16,* 216–22.

Gershgorin's Theorem can be used to derive a comprehensive perturbation theory. See Wilkinson (AEP, chapter 2). The theorem itself can be generalized and extended in various ways; see

R.J. Johnston (1971). "Gershgorin Theorems for Partitioned Matrices," *Lin. Alg. and Its Applic. 4,* 205–20.

R.S. Varga (1970). "Minimal Gershgorin Sets for Partitioned Matrices," *SIAM J. Num. Anal. 7,* 493–507.

Finally, we mention the classical reference

T. Kato (1966). *Perturbation Theory for Linear Operators,* Springer-Verlag, New York.

Chapter 2 of this work is a comprehensive treatment of the finite dimensional case.

## 7.3 Power Iterations

Suppose that we are given $A \in \mathbb{C}^{n\times n}$ and a unitary $U_0 \in \mathbb{C}^{n\times n}$. Assume that Householder orthogonalization (Algorithm 5.2.1) can be extended to complex matrices (it can) and consider the following iteration:

$$
\begin{array}{l}
T_0 = U_0^H A U_0 \\
\textbf{for } k = 1,2,\ldots \\
\quad T_{k-1} = U_k R_k \quad \text{(QR factorization)} \\
\quad T_k = R_k U_k \\
\textbf{end}
\end{array}
\tag{7.3.1}
$$

Since $T_k = R_k U_k = U_k^T(U_k R_k)U_k = U_k^T T_{k-1} U_k$ it follows by induction that

$$T_k = (U_0U_1\cdots U_k)^H A(U_0U_1\cdots U_k). \tag{7.3.2}$$

Thus, each $T_k$ is unitarily similar to $A$. Not so obvious, and what is the central theme of this section, is that the $T_k$ almost always converge to upper triangular form. That is, (7.3.2) almost always "converges" to a Schur decomposition of $A$.

Iteration (7.3.1) is called the *QR iteration,* and it forms the backbone of the most effective algorithm for computing the Schur decomposition. In order to motivate the method and to derive its convergence properties, two other eigenvalue iterations that are important in their own right are presented first: the power method and the method of orthogonal iteration.

### 7.3.1 The Power Method

Suppose $A \in \mathbb{C}^{n\times n}$ is diagonalizable, that $X^{-1}AX = \text{diag}(\lambda_1,\ldots,\lambda_n)$ with $X = [\, x_1,\ldots,x_n\,]$, and $|\lambda_1| > |\lambda_2| \geq \cdots \geq |\lambda_n|$. Given a unit 2-norm $q^{(0)} \in \mathbb{C}^n$, the *power method* produces a sequence of vectors $q^{(k)}$ as follows:

$$
\begin{array}{l}
\textbf{for } k = 1,2,\ldots \\
\quad z^{(k)} = Aq^{(k-1)} \\
\quad q^{(k)} = z^{(k)}/\| z^{(k)} \|_2 \\
\quad \lambda^{(k)} = [q^{(k)}]^H A q^{(k)} \\
\textbf{end}
\end{array}
\tag{7.3.3}
$$

There is nothing special about doing a 2-norm normalization except that it imparts a greater unity on the overall discussion in this section.

Let us examine the convergence properties of the power iteration. If

$$q^{(0)} = a_1x_1 + a_2x_2 + \cdots + a_nx_n$$

and $a_1 \neq 0$, then it follows that

$$A^k q^{(0)} = a_1\lambda_1^k \left( x_1 + \sum_{j=2}^{n} \frac{a_j}{a_1}\left(\frac{\lambda_j}{\lambda_1}\right)^k x_j \right).$$

Since $q^{(k)} \in \text{span}\{A^k q^{(0)}\}$ we conclude that

$$\text{dist}\left(\text{span}\{q^{(k)}\}, \text{span}\{x_1\}\right) = O\left(\left|\frac{\lambda_2}{\lambda_1}\right|^k\right)$$

and moreover,

$$|\lambda_1 - \lambda^{(k)}| = O\left(\left|\frac{\lambda_2}{\lambda_1}\right|^k\right).$$

If $|\lambda_1| > |\lambda_2| \geq \cdots \geq |\lambda_n|$ then we say that $\lambda_1$ is a *dominant eigenvalue*. Thus, the power method converges if $\lambda_1$ is dominant and if $q^{(0)}$ has a component in the direction of the corresponding *dominant eigenvector* $x_1$. The behavior of the iteration without these assumptions is discussed in Wilkinson (AEP, pp. 570 ff.) and Parlett and Poole (1973).

**Example 7.3.1** If

$$A = \begin{bmatrix} -261 & 209 & -49 \\ -530 & 422 & -98 \\ -800 & 631 & -144 \end{bmatrix}$$

then $\lambda(A) = \{10, 4, 3\}$. Applying (7.3.3) with $q^{(0)} = (1, 0, 0)^T$ we find

| $k$ | $\lambda^{(k)}$ |
|---|---|
| 1 | 994.4900 |
| 2 | 13.0606 |
| 3 | 10.7191 |
| 4 | 10.2073 |
| 5 | 10.0633 |
| 6 | 10.0198 |
| 7 | 10.0063 |
| 8 | 10.0020 |
| 9 | 10.0007 |
| 10 | 10.0002 |

In practice, the usefulness of the power method depends upon the ratio $|\lambda_2|/|\lambda_1|$, since it dictates the rate of convergence. The danger that $q^{(0)}$ is deficient in $x_1$ is a less worrisome matter because rounding errors sustained during the iteration typically ensure that the subsequent $q^{(k)}$ have a component in this direction. Moreover, it is typically the case in applications where the dominant eigenvalue and eigenvector are desired that an à priori estimate of $x_1$ is known. Normally, by setting $q^{(0)}$ to be this estimate, the dangers of a small $a_1$ are minimized.

Note that the only thing required to implement the power method is a subroutine capable of computing matrix-vector products of the form $Aq$. It is not necessary to store $A$ in an $n$-by-$n$ array. For this reason, the algorithm can be of interest when $A$ is large and sparse and when there is a sufficient gap between $|\lambda_1|$ and $|\lambda_2|$.

Estimates for the error $|\lambda^{(k)} - \lambda_1|$ can be obtained by applying the perturbation theory developed in the previous section. Define the vector $r^{(k)} = Aq^{(k)} - \lambda^{(k)}q^{(k)}$ and observe that $(A + E^{(k)})q^{(k)} = \lambda^{(k)}q^{(k)}$ where $E^{(k)} = -r^{(k)}[q^{(k)}]^H$. Thus $\lambda^{(k)}$ is an eigenvalue of $A + E^{(k)}$ and

$$| \lambda^{(k)} - \lambda | \approx \frac{\| E^{(k)} \|_2}{s(\lambda_1)} = \frac{\| r^{(k)} \|_2}{s(\lambda_1)} .$$

If we use the power method to generate approximate right *and* left dominant eigenvectors, then it is possible to obtain an estimate of $s(\lambda_1)$. In particular, if $w^{(k)}$ is a unit 2-norm vector in the direction of $(A^H)^k w^{(0)}$, then we can use the approximation $s(\lambda_1) \approx | {w^{(k)}}^H q^{(k)} |$.

## 7.3.2 Orthogonal Iteration

A straightforward generalization of the power method can be used to compute higher-dimensional invariant subspaces. Let $p$ be a chosen integer satisfying $1 \le p \le n$. Given an $n$-by-$p$ matrix $Q_0$ with orthonormal columns, the method of *orthogonal iteration* generates a sequence of matrices $\{Q_k\} \subseteq \mathbb{C}^{n \times p}$ as follows:

**for** $k = 1, 2, \ldots$
  $Z_k = AQ_{k-1}$ (7.3.4)
  $Q_k R_k = Z_k$ (QR factorization)
**end**

Note that if $p = 1$ then this is just the power method. Moreover, the sequence $\{Q_k e_1\}$ is precisely the sequence of vectors produced by the power iteration with starting vector $q^{(0)} = Q_0 e_1$.

In order to analyze the behavior of this iteration, suppose that

$$Q^H A Q = T = \text{diag}(\lambda_i) + N \qquad |\lambda_1| \ge |\lambda_2| \ge \cdots \ge |\lambda_n| \tag{7.3.5}$$

is a Schur decomposition of $A \in \mathbb{C}^{n \times n}$. Assume that $1 \le p < n$ and partition $Q$, $T$, and $N$ as follows:

$$Q = \left[ \begin{array}{cc} Q_\alpha & Q_\beta \end{array} \right] \begin{array}{c} \\ \end{array} \qquad T = \begin{bmatrix} T_{11} & T_{12} \\ 0 & T_{22} \end{bmatrix} \begin{array}{c} p \\ n-p \end{array}$$
$$\qquad\; p \quad n-p \qquad\qquad\qquad\qquad p \quad n-p \tag{7.3.6}$$

$$N = \begin{bmatrix} N_{11} & N_{12} \\ 0 & N_{22} \end{bmatrix} \begin{array}{c} p \\ n-p \end{array} .$$
$$p \quad n-p$$

If $|\lambda_p| > |\lambda_{p+1}|$, then the subspace $D_p(A) = \text{range}(Q_\alpha)$ is said to be a *dominant* invariant subspace. It is the unique invariant subspace associated with the eigenvalues $\lambda_1, \ldots, \lambda_p$. The following theorem shows that with reasonable assumptions, the subspaces range($Q_k$) generated by (7.3.4) converge to $D_p(A)$ at a rate proportional to $|\lambda_{p+1}/\lambda_p|^k$.

**Theorem 7.3.1** *Let the Schur decomposition of $A \in \mathbb{C}^{n\times n}$ be given by (7.3.5) and (7.3.6) with $n \geq 2$. Assume that $|\lambda_p| > |\lambda_{p+1}|$ and that $\theta \geq 0$ satisfies*

$$(1+\theta)|\lambda_p| \ > \ \| N \|_F$$

*If $Q_0 \in \mathbb{C}^{n\times p}$ has orthonormal columns and*

$$d \ = \ \text{dist}(D_p(A^H), \text{range}(Q_0)) \ < \ 1,$$

*then the matrices $Q_k$ generated by (7.3.4) satisfy*

$$\text{dist}(D_p(A), \text{range}(Q_k)) \ \leq$$
$$\frac{(1+\theta)^{n-2}}{\sqrt{1-d^2}}\left(1 \ + \ \frac{\| T_{12} \|_F}{\text{sep}(T_{11}, T_{22})}\right)\left(\frac{|\lambda_{p+1}| + \| N \|_F/(1+\theta)}{|\lambda_p| - \| N \|_F/(1+\theta)}\right)^k .$$

**Proof.** The proof is long and appears at the end of the section. □

The condition $d < 1$ in Theorem 7.3.1 ensures that the initial $Q$ matrix is not deficient in certain eigendirections:

$$d < 1 \ \leftrightarrow \ D_p(A^H)^\perp \cap \text{range}(Q_0) = \{0\}.$$

The theorem essentially says that if this condition holds and if $\theta$ is chosen large enough, then

$$\text{dist}(D_p(A), \text{range}(Q_k)) \ \leq \ c \left|\frac{\lambda_{p+1}}{\lambda_p}\right|^k$$

where $c$ depends on sep($T_{11}, T_{22}$) and $A$'s departure from normality. Needless to say, convergence can be very slow if the gap between $|\lambda_p|$ and $|\lambda_{p+1}|$ is not sufficiently wide.

**Example 7.3.2** If (7.3.4) is applied to the matrix $A$ in Example 7.3.1, with $Q_0 = [e_1, e_2]$, we find:

| $k$ | dist($D_2(A)$, range($Q_k$)) |
|---|---|
| 1 | .0052 |
| 2 | .0047 |
| 3 | .0039 |
| 4 | .0030 |
| 5 | .0023 |
| 6 | .0017 |
| 7 | .0013 |

The error is tending to zero with rate $(\lambda_3/\lambda_2)^k = (3/4)^k$.

It is possible to accelerate the convergence in orthogonal iteration using a technique described in Stewart (1976d). In the accelerated scheme, the approximate eigenvalue $\lambda_i^{(k)}$ satisfies

$$|\lambda_i^{(k)} - \lambda_i| \approx \left|\frac{\lambda_{p+1}}{\lambda_i}\right|^k \qquad i = 1{:}p$$

(Without the acceleration, the right-hand side is $|\lambda_{i+1}/\lambda_i|^k$.) Stewart's algorithm involves the computing the Schur decomposition of the matrices $Q_k^T A Q_k$ every so often. The method can be very useful in situations where $A$ is large and sparse and a few of its largest eigenvalues are required.

### 7.3.3 The QR Iteration

We now "derive" the QR iteration (7.3.1) and examine its convergence. Suppose $p = n$ in (7.3.4) and the eigenvalues of $A$ satisfy

$$|\lambda_1| > |\lambda_2| > \cdots > |\lambda_n|.$$

Partition the matrix $Q$ in (7.3.5) and $Q_k$ in (7.3.4) as follows:

$$Q = [\, q_1, \ldots, q_n \,] \qquad Q_k = \left[\, q_1^{(k)}, \ldots, q_n^{(k)} \,\right]$$

If

$$\text{dist}(D_i(A^H), \text{span}\{q_1^{(0)}, \ldots, q_i^{(0)}) \; < \; 1 \qquad i = 1{:}n \tag{7.3.7}$$

then it follows from Theorem 7.3.1 that

$$\text{dist}(\text{span}\{q_1^{(k)}, \ldots, q_i^{(k)}\}, \text{span}\{q_1, \ldots, q_i\}) \;\rightarrow\; 0$$

for $i = 1{:}n$. This implies that the matrices $T_k$ defined by

$$T_k \;=\; Q_k^H A Q_k$$

are converging to upper triangular form. Thus, it can be said that the method of orthogonal iteration computes a Schur decompositon provided the original iterate $Q_0 \in \mathbb{C}^{n \times n}$ is not deficient in the sense of (7.3.7).

The QR iteration arises naturally by considering how to compute the matrix $T_k$ directly from its predecessor $T_{k-1}$. On the one hand, we have from (7.3.4) and the definition of $T_{k-1}$ that

$$T_{k-1} = Q_{k-1}^H A Q_{k-1} = Q_{k-1}^H (A Q_{k-1}) = (Q_{k-1}^H Q_k) R_k.$$

On the other hand

$$T_k = Q_k^H A Q_k = (Q_k^H A Q_{k-1})(Q_{k-1}^H Q_k) = R_k (Q_{k-1}^H Q_k).$$

Thus, $T_k$ is determined by computing the QR factorization of $T_{k-1}$ and then multiplying the factors together in reverse order. This is precisely what is done in (7.3.1).

**Example 7.3.3** If the iteration:

$$\begin{array}{l} \textbf{for } k = 1,2,\ldots \\ \quad A = QR \\ \quad A = RQ \\ \textbf{end} \end{array}$$

is applied to the matrix of Example 7.3.1, then the strictly lower triangular elements diminish as follows:

| $k$ | $O(\lvert a_{21}\rvert)$ | $O(\lvert a_{31}\rvert)$ | $O(\lvert a_{32}\rvert)$ |
|---|---|---|---|
| 1 | $10^{-1}$ | $10^{-1}$ | $10^{-2}$ |
| 2 | $10^{-2}$ | $10^{-2}$ | $10^{-3}$ |
| 3 | $10^{-2}$ | $10^{-3}$ | $10^{-3}$ |
| 4 | $10^{-3}$ | $10^{-3}$ | $10^{-3}$ |
| 5 | $10^{-3}$ | $10^{-4}$ | $10^{-3}$ |
| 6 | $10^{-4}$ | $10^{-5}$ | $10^{-3}$ |
| 7 | $10^{-4}$ | $10^{-5}$ | $10^{-3}$ |
| 8 | $10^{-5}$ | $10^{-6}$ | $10^{-4}$ |
| 9 | $10^{-5}$ | $10^{-7}$ | $10^{-4}$ |
| 10 | $10^{-6}$ | $10^{-8}$ | $10^{-4}$ |

Note that a single QR iteration is an $O(n^3)$ calculation. Moreover, since convergence is only linear (when it exists), it is clear that the method is a prohibitively expensive way to compute Schur decompositions. Fortunately these practical difficulties can be overcome, as we show in §7.4 and §7.5.

### 7.3.4 LR Iterations

We conclude with some remarks about power iterations that rely on the LU factorization rather than the QR factorizaton. Let $G_0 \in \mathbb{C}^{n\times p}$ have rank $p$. Corresponding to (7.3.4) we have the following iteration:

$$\begin{array}{ll} \textbf{for } k = 1,2,\ldots & \\ \quad Z_k = AG_{k-1} & \\ \quad Z_k = G_k R_k & \text{(LU factorization)} \\ \textbf{end} & \end{array} \qquad (7.3.8)$$

Suppose $p = n$ and that we define the matrices $T_k$ by

$$T_k \;=\; G_k^{-1} A G_k. \qquad (7.3.9)$$

It can be shown that if we set $L_0 = G_0$, then the $T_k$ can be generated as follows:

$$\begin{array}{ll} T_0 = L_0^{-1} A L_0 & \\ \textbf{for } k = 1, 2, \ldots & \\ \quad T_{k-1} = L_k R_k \quad \text{(LU factorization)} & \\ \quad T_k = R_k L_k & \\ \textbf{end} & \end{array} \tag{7.3.10}$$

Iterations (7.3.8) and (7.3.10) are known as *treppeniteration* and the *LR iteration*, respectively. To successfully implement either method, it is necessary to pivot. See Wilkinson (AEP, p.602ff).

## Appendix

In order to establish Theorem 7.3.1 we need the following lemma which is concerned with bounding the powers of a matrix and its inverse.

**Lemma 7.3.2** *Let $Q^H AQ = T = D + N$ be a Schur decomposition of $A \in \mathbb{C}^{n \times n}$ where $D$ is diagonal and $N$ strictly upper triangular. Let $\lambda$ and $\mu$ denote the largest and smallest eigenvalues of $A$ in absolute value. If $\theta \geq 0$ then for all $k \geq 0$ we have*

$$\| A^k \|_2 \leq (1+\theta)^{n-1} \left( |\lambda| + \frac{\| N \|_F}{1+\theta} \right)^k . \tag{7.3.11}$$

*If $A$ is nonsingular and $\theta \geq 0$ satisfies $(1+\theta)|\mu| > \| N \|_F$, then for all $k \geq 0$ we also have*

$$\| A^{-k} \|_2 \leq (1+\theta)^{n-1} \left( \frac{1}{|\mu| - \| N \|_F / (1+\theta)} \right)^k . \tag{7.3.12}$$

**Proof.** For $\theta \geq 0$, define the diagonal matrix $\Delta$ by

$$\Delta = \text{diag}\,(1, (1+\theta), (1+\theta)^2, \ldots, (1+\theta)^{n-1})$$

and note that $\kappa_2(\Delta) = (1+\theta)^{n-1}$. Since $N$ is strictly upper triangular, it is easy to verify that $\| \Delta N \Delta^{-1} \|_F \leq \| N \|_F / (1+\theta)$. Thus,

$$\begin{aligned} \| A^k \|_2 &= \| T^k \|_2 = \| \Delta^{-1} (D + \Delta N \Delta^{-1})^k \Delta \|_2 \\ &\leq \kappa_2(\Delta) \left( \| D \|_2 + \| \Delta N \Delta^{-1} \|_2 \right)^k \\ &\leq (1+\theta)^{n-1} \left( |\lambda| + \frac{\| N \|_F}{1+\theta} \right)^k . \end{aligned}$$

On the other hand, if $A$ is nonsingular and $(1+\theta)|\mu| > \| N \|_F$, then $\| \Delta D^{-1} N \Delta^{-1} \|_2 < 1$ and using Lemma 2.3.3 we obtain

$$\| A^{-k} \|_2 = \| T^{-k} \|_2 = \left\| \Delta^{-1} [I + \Delta D^{-1} N \Delta^{-1})^{-1} D^{-1}]^k \Delta \right\|_2$$

$$\begin{aligned}
&\leq \kappa_2(\Delta)\left(\frac{\| D^{-1} \|_2}{1-\| \Delta D^{-1} N \Delta^{-1} \|_2}\right)^k \\
&\leq (1+\theta)^{n-1}\left(\frac{1}{|\mu|-\| N \|_F/(1+\theta)}\right)^k . \square
\end{aligned}$$

**Proof of Theorem 7.3.1**

It is easy to show by induction that $A^k Q_0 = Q_k(R_k \cdots R_1)$. By substituting (7.3.5) and 7.3.6) into this equality we obtain

$$T^k \begin{bmatrix} V_0 \\ W_0 \end{bmatrix} = \begin{bmatrix} V_k \\ W_k \end{bmatrix} (R_k \cdots R_1)$$

where $V_k = Q_\alpha^H Q_k$ and $W_k = Q_\beta^H Q_k$. Using Lemma 7.1.4 we know that a matrix $X \in \mathbb{C}^{p\times(n-p)}$ exists such that

$$\begin{bmatrix} I_p & X \\ 0 & I_{n-p} \end{bmatrix}^{-1} \begin{bmatrix} T_{11} & T_{12} \\ 0 & T_{22} \end{bmatrix} \begin{bmatrix} I_p & X \\ 0 & I_{n-p} \end{bmatrix} = \begin{bmatrix} T_{11} & 0 \\ 0 & T_{22} \end{bmatrix} .$$

Moreover, since $\text{sep}(T_{11}, T_{22})$ is the smallest singular value of the linear transformation $\phi(X)= T_{11}X - XT_{22}$ it readily follows from the equation $\phi(X) = -T_{12}$ that

$$\| X \|_F \leq \frac{\| T_{12} \|_F}{\text{sep}(T_{11}, T_{22})} . \tag{7.3.13}$$

Thus we have

$$\begin{bmatrix} T_{11}^k & 0 \\ 0 & T_{22}^k \end{bmatrix} \begin{bmatrix} V_0 - XW_0 \\ W_0 \end{bmatrix} = \begin{bmatrix} V_k - XW_k \\ W_k \end{bmatrix} (R_k \cdots R_1) .$$

If we assume that $V_0 - XW_0$ is nonsingular, then this equation can be solved for $W_k$:

$$W_k = T_{22}^k W_0 (V_0 - XW_0)^{-1} T_{11}^{-k} (V_k - XW_k).$$

By taking norms in this expression and using the fact that

$$\text{dist}(D_p(A), \text{range}(Q_k)) = \| Q_\beta^H Q_k \|_2 = \| W_k \|_2,$$

we get

$$\begin{aligned}
&\text{dist}(D_p(A), \text{range}(Q_k)) \leq \\
&\quad \| T_{22}^k \|_2 \, \| (V_0 - XW_0)^{-1} \|_2 \, \| T_{11}^{-k} \|_2 \, (1 + \| X \|_F) .
\end{aligned} \tag{7.3.14}$$

In order for the inequality to be valid, we must show that $V_0 - XW_0$ is nonsingular.

From the equation $A^H Q = QT^H$ it follows that

$$A^H(Q_\alpha - Q_\beta X^H) = (Q_\alpha - Q_\beta X^H) T_{11}^H,$$

which implies that the orthonormal columns of

$$Z = (Q_\alpha - Q_\beta X^H)(I + XX^H)^{1/2}$$

are a basis for $D_p(A^H)$. It can be shown that

$$V_0 - XW_0 = (I + XX^H)^{1/2} Z^H Q_0$$

and therefore

$$\sigma_p(V_0 - XW_0) \geq \sigma_p(Z^H Q_0) =$$
$$\sigma_p(V_0 - XW_0) \geq \sigma_p(Z^H Q_0) = \sqrt{1-d^2} > 0.$$

This shows that $V_0 - XW_0$ is indeed invertible and that

$$\| (V_0 - XW_0)^{-1} \|_2 \leq 1/\sqrt{1-d^2}. \tag{7.3.15}$$

Finally, by using Lemma 7.3.2, it can be shown that

$$\| T_{22}^k \|_2 \leq (1+\theta)^{n-p-1} \left( |\lambda_{p+1}| + \frac{\| N \|_F}{1+\theta} \right)^k$$

and

$$\| T_{11}^{-k} \|_2 \leq (1+\theta)^{p-1} / \left( |\lambda_p| - \frac{\| N \|_F}{1+\theta} \right)^k .$$

The theorem follows by substituting these inequalities along with (7.3.13) and (7.3.15) into (7.3.14). □

**Problems**

**P7.3.1** (a) Show that if $X \in \mathbb{C}^{n\times n}$ is nonsingular, then $\| A \|_X = \| X^{-1}AX \|_2$ defines a matrix norm with the property that $\| AB \|_X \leq \| A \|_X \| B \|_X$. (b) Let $A \in \mathbb{C}^{n\times n}$ and set $\rho = \max |\lambda_i|$. Show that for any $\epsilon > 0$ there exists a nonsingular $X \in \mathbb{C}^{n\times n}$ such that $\| A \|_X = \| X^{-1}AX \|_2 \leq \rho + \epsilon$. Conclude that there is a constant $M$ such that $\| A^k \|_2 \leq M(\rho+\epsilon)^k$ for all non-negative integers $k$. (Hint: Set $X = Q\ \mathrm{diag}(1, a, \ldots, a^{n-1})$ where $Q^H AQ = D + N$ is $A$'s Schur decomposition.)

**P7.3.2** Verify that (7.3.10) calculates the matrices $T_k$ defined by (7.3.9).

**P7.3.3** Suppose $A \in \mathbb{C}^{n\times n}$ is nonsingular and that $Q_0 \in \mathbb{C}^{n\times p}$ has orthonormal columns. The following iteration is referred to as *inverse orthogonal iteration*.

```
for k = 1, 2, ...
    Solve AZ_k = Q_{k-1} for Z_k ∈ C^{n×p}
    Z_k = Q_k R_k        (QR factorization)
end
```

Explain why this iteration can usually be used to compute the $p$ smallest eigenvalues of $A$ in absolute value. Note that to implement this iteration it is necessary to be able to solve linear systems that involve $A$. When $p = 1$, the method is referred to as the *inverse power method.*

**Notes and References for Sec. 7.3**

A detailed, practical discussion of the power method is given in Wilkinson (AEP, chapter 10). Methods are discussed for accelerating the basic iteration, for calculating nondominant eigenvlaues, and for handling complex conjugate eigenvalue pairs. The connections among the various power iterations are discussed in

B.N. Parlett and W.G. Poole (1973). "A Geometric Theory for the QR, LU, and Power Iterations," *SIAM J. Num. Anal. 10,* 389–412.

The QR iteration was concurrently developed in

J.G.F. Francis (1961). "The QR Transformation: A Unitary Analogue to the LR Transformation," *Comp. J. 4,* 265–71, 332–34.
V.N. Kublanovskaya (1961). "On Some Algorithms for the Solution of the Complete Eigenvalue Problem," *USSR Comp. Math. Phys. 3,* 637–57.

As can be deduced from the title of the first paper, the LR iteration predates the QR iteration. The former very fundamental algorithm was proposed by

H. Rutishauser (1958). "Solution of Eigenvalue Problems with the LR Transformation," *Nat. Bur. Stand. App. Math. Ser. 49,* 47–81.

Numerous papers on the convergence of the QR iteration have appeared. Several of these are

B.N. Parlett (1965). "Convergence of the Q-R Algorithm," *Numer. Math. 7,* 187–93. (Correction in *Numer. Math. 10,* 163–64.)
B.N. Parlett (1966). "Singular and Invariant Matrices Under the QR Algorithm," *Math. Comp. 20,* 611–15.
B.N. Parlett (1968). "Global Convaergence of the Basic QR Algorithm on Hessenberg Matrices," *Math. Comp. 22,* 803–17.
J.H. Wilkinson (1965). "Convergence of the LR, QR, and Related Algorithms," *Comp. J. 8,* 77–84.

Wilkinson (AEP, chapter 9) also discusses the convergence theory for this important algorithm.

Deeper insight into the convergence of the QR algorithm can be attained by reading

T. Nanda (1985). "Differential Equations and the QR Algorithm," *SIAM J. Numer. Anal. 22,* 310–321.
D.S. Watkins (1982). "Understanding the QR Algorithm," *SIAM Review 24,* 427–440.

The following papers are concerned with various practical and theoretical aspects of simultaneous iteration:

M. Clint and A. Jennings (1971). "A Simultaneous Iteration Method for the Unsymmetric Eigenvalue Problem," *J. Inst. Math. Applic. 8,* 111–21.
A. Jennings and D.R.L. Orr (1971). "Application of the Simultaneous Iteration Method to Undamped Vibration Problems," *Inst. J. Numer. Math. Eng. 3,* 13–24.
A. Jennings and W.J. Stewart (1975). "Simultaenous Iteration for the Partial Eigensolution of Real Matrices," *J. Inst. Math. Applic. 15,* 351–62.

H. Rutishauser (1970). "Simultaneous Iteration Method for Symmetric Matrices," *Numer. Math. 16,* 205–23. See also HACLA, pp. 284–302.
G.W. Stewart (1975c). "Methods of Simultaneous Iteration for Calculating Eigenvectors of Matrices," in *Topics in Numerical Analysis II* , ed. John J.H. Miller, Academic Presss, New York, pp. 185–96.
G.W. Stewart (1976d). "Simultaneous Iteration for Computing Invariant Subspaces of Non-Hermitian Matrices," *Numer. Math. 25,* 123–36.

See also chapter 10 of
A. Jennings (1977b). *Matrix Computation for Engineers and Scientists,* John Wiley and Sons, New York.

Simultaneous iteration and the Lanczos algorithm (cf. Chapter 9) are the principal methods for finding a few eigenvalues of a general sparse matrix.

## 7.4 The Hessenberg and Real Schur Forms

In this and the next section we show how to make the QR iteration (7.3.1) a fast, effective method for computing Schur decompositions. Because the majority of eigenvalue/invariant subspace problems involve real data, we concentrate on developing the real analog of (7.3.1) which we write as follows:

$$\begin{array}{l} H_0 = U_0^T A U_0 \\ \textbf{for } k = 1, 2, \ldots \\ \qquad H_{k-1} = U_k R_k \qquad \text{(QR factorization)} \\ \qquad H_k = R_k U_k \\ \textbf{end} \end{array} \tag{7.4.1}$$

Here, $A \in \mathbb{R}^{n \times n}$, each $U_k \in \mathbb{R}^{n \times n}$ is orthogonal, and each $R_k \in \mathbb{R}^{n \times n}$ is upper triangular. A difficulty associated with this real iteration is that the $H_k$ can never converge to strict, "eigenvalue revealing," triangular form in the event that $A$ has complex eigenvalues. For this reason, we must lower our expectations and be content with the calculation of an alternative decomposition known as the *real Schur decomposition*. We mention that in numerical linear algebra, there is usually no loss in generality by focusing attention on real matrix problems because most real arithmetic algorithms have obvious complex arithmetic analogs. The QR iteration is no exception to this.

In order to compute the real Schur decomposition efficiently we must carefully choose the initial orthogonal similarity transformation $U_0$ in (7.4.1). In particular, if we choose $U_0$ so that $H_0$ is upper Hessenberg then the amount of work per iteration is reduced from $O(n^3)$ to $O(n^2)$. The initial reduction to Hessenberg form (the $U_0$ computation) is a very important computation in its own right and can be realized by a sequence of Householder matrix operations.

### 7.4.1 The Real Schur Decomposition

The real Schur decomposition amounts to a block upper triangularization with either 1-by-1 or 2-by-2 diagonal blocks.

**Theorem 7.4.1 (Real Schur Decomposition)** *If $A \in \mathbb{R}^{n\times n}$ then there exists an orthogonal $Q \in \mathbb{R}^{n\times n}$ such that*

$$Q^T A Q =; \begin{bmatrix} R_{11} & R_{12} & \cdots & R_{1m} \\ 0 & R_{22} & \cdots & R_{2m} \\ \vdots & \vdots & \ddots & \vdots \\ 0 & 0 & \cdots & R_{mm} \end{bmatrix} \tag{7.4.2}$$

*where each $R_{ii}$ is either a 1-by-1 matrix or a 2-by-2 matrix having complex conjugate eigenvalues.*

**Proof.** The complex eigenvalues of $A$ must come in conjugate pairs, since the characteristic polynomial $\det(zI - A)$ has real coefficients. Let $k$ be the number of complex conjugate pairs in $\lambda(A)$. We prove the theorem by induction on $k$.

Observe first that Lemma 7.1.1 and Theorem 7.1.2 have obvious real analogs. Thus, the theorem holds if $k = 0$. Now suppose that $k \geq 1$. If $\lambda = \gamma + i\mu \in \lambda(A)$ and $\mu \neq 0$, then there exist vectors $y$ and $z$ in $\mathbb{R}^n (z \neq 0)$ such that

$$A(y + iz) = (\gamma + i\mu)(y + iz),$$

i.e.,

$$A \begin{bmatrix} y & z \end{bmatrix} = \begin{bmatrix} y & z \end{bmatrix} \begin{bmatrix} \gamma & \mu \\ -\mu & \gamma \end{bmatrix}.$$

The assumption that $\mu \neq 0$ implies that $y$ and $z$ span a two-dimensional, real invariant subspace for $A$. It then follows from Lemma 7.1.1 that an orthogonal $U \in \mathbb{R}^{n\times n}$ exists such that

$$U^T A U = \begin{bmatrix} T_{11} & T_{12} \\ 0 & T_{22} \end{bmatrix} \begin{matrix} 2 \\ n-2 \end{matrix}$$
$$\qquad\qquad\quad 2 \quad n-2$$

where $\lambda(T_{11}) = \{\lambda, \bar{\lambda}\}$. By induction, there exists an orthogonal $\tilde{U}$ so $\tilde{U}^T T_{22} \tilde{U}$ has the requisite structure. The theorem follows by setting $Q = U \operatorname{diag}(I_2, \tilde{U})$. □

The theorem shows that any real matrix is orthogonally similar to an upper quasi-triangular matrix. It is clear that the real and imaginary part of the complex eigenvalues can be easily obtained from the 2-by-2 diagonal blocks.

### 7.4.2 A Hessenberg QR Step

We now turn our attention to the speedy calculation of a single QR step in (7.4.1). In this regard, the most glaring shortcoming associated with (7.4.1) is that each step requires a full QR factorization costing $O(n^3)$ flops. Fortunately, the amount of work per iteration can be reduced by an order of magnitude if the orthogonal matrix $U_0$ is judiciously chosen. In particular, if $U_0^T A U_0 = H_0 = (h_{ij})$ is upper Hessenberg ($h_{ij} = 0$, $i > j+1$), then each subsequent $H_k$ requires only $O(n^2)$ flops to calculate. To see this we look at the computations $H = QR$ and $H_+ = RQ$ when $H$ is upper Hessenberg. Using the rotation functions **givens**, **row.rot**, and **col.rot** that we described in §5.1.8, we can upper triangularize $H$ with a sequence of $n-1$ Givens rotations. In particular, we can find rotations $G_1, \ldots, G_{n-1}$ with $G_i = G(i, i+1, \theta_i)$ such that $G_{n-1}^T \cdots G_1^T H = R$ is upper triangular:

$$\begin{bmatrix} \times & \times & \times & \times \\ \times & \times & \times & \times \\ 0 & \times & \times & \times \\ 0 & 0 & \times & \times \end{bmatrix} \xrightarrow{G_1} \begin{bmatrix} \times & \times & \times & \times \\ 0 & \times & \times & \times \\ 0 & \times & \times & \times \\ 0 & 0 & \times & \times \end{bmatrix} \xrightarrow{G_2} \begin{bmatrix} \times & \times & \times & \times \\ 0 & \times & \times & \times \\ 0 & 0 & \times & \times \\ 0 & 0 & \times & \times \end{bmatrix}$$

$$\xrightarrow{G_3} \begin{bmatrix} \times & \times & \times & \times \\ 0 & \times & \times & \times \\ 0 & 0 & \times & \times \\ 0 & 0 & 0 & \times \end{bmatrix}$$

See Algorithm 5.2.3.

The computation $RG_1 \cdots G_{n-1}$ is equally easy using the column rotation function **col.rot**:

$$\begin{bmatrix} \times & \times & \times & \times \\ 0 & \times & \times & \times \\ 0 & 0 & \times & \times \\ 0 & 0 & 0 & \times \end{bmatrix} \xrightarrow{G_1} \begin{bmatrix} \times & \times & \times & \times \\ \times & \times & \times & \times \\ 0 & 0 & \times & \times \\ 0 & 0 & 0 & \times \end{bmatrix} \xrightarrow{G_2} \begin{bmatrix} \times & \times & \times & \times \\ \times & \times & \times & \times \\ 0 & \times & \times & \times \\ 0 & 0 & 0 & \times \end{bmatrix}$$

$$\xrightarrow{G_3} \begin{bmatrix} \times & \times & \times & \times \\ \times & \times & \times & \times \\ 0 & \times & \times & \times \\ 0 & 0 & \times & \times \end{bmatrix}$$

Overall we obtain

**Algorithm 7.4.1** If $H$ is an $n$-by-$n$ upper Hessenberg matrix, then this algorithm overwrites $H$ with $H_+ = RQ$ where $H = QR$ is the QR factorization of $H$. Moreover, $Q = G_1 \cdots G_{n-1}$, a product of Givens rotations where $G_k$ is a rotation in planes $k$ and $k+1$.

```
for k = 1:n − 1
    {Generate G_k and then apply it: H = G(k, k+1, θ_k)^T H.}
    [ c(k) s(k) ] = givens(H(k,k), H(k+1,k))
    H(k:k+1, k:n) = row.rot(H(k:k+1, k:n), c(k), s(k))
end
for k = 1:n − 1
    { Update: H = HG(k, k+1, θ_k).}
    H(1:k+1, k:k+1) = col.rot(H(1:k+1, k:k+1), c(k), s(k))
end
```

Since $G_k = G(k, k+1, \theta_k)$ it is easy to confirm that the matrix $Q = G_1 \cdots G_{n-1}$ is upper Hessenberg. Thus, $RQ = H_+$ is also upper Hessenberg. The algorithm requires about $6n^2$ flops and thus is an order-of-magnitude quicker than a full matrix QR step.

**Example 7.4.1** If Algorithm 7.4.1 is applied to:

$$H = \begin{bmatrix} 3 & 1 & 2 \\ 4 & 2 & 3 \\ 0 & .01 & 1 \end{bmatrix}$$

then

$$G_1 = \begin{bmatrix} .6 & -.8 & 0 \\ .8 & .6 & 0 \\ 0 & 0 & 1 \end{bmatrix}, \qquad G_2 = \begin{bmatrix} 1 & 0 & 0 \\ 0 & .9996 & -.0249 \\ 0 & .0249 & .9996 \end{bmatrix}$$

and

$$H_+ = \begin{bmatrix} 4.7600 & -2.5442 & 5.4653 \\ .3200 & .1856 & -2.1796 \\ .0000 & .0263 & 1.0540 \end{bmatrix}.$$

## 7.4.3 The Hessenberg Reduction

It remains for us to show how the *Hessenberg decomposition*

$$U_0^T A U_0 = H \qquad U_0^T U_0 = I \tag{7.4.3}$$

can be computed. The calculation is not difficult. The transformation $U_0$ is a product of Householder matrices $P_1, \ldots, P_{n-2}$. The role of $P_k$ is to zero the $k$th column below the subdiagonal. In the $n = 6$ case, we have

$$A = \begin{bmatrix} \times & \times & \times & \times & \times & \times \\ \times & \times & \times & \times & \times & \times \\ \times & \times & \times & \times & \times & \times \\ \times & \times & \times & \times & \times & \times \\ \times & \times & \times & \times & \times & \times \\ \times & \times & \times & \times & \times & \times \end{bmatrix} \xrightarrow{P_1} \begin{bmatrix} \times & \times & \times & \times & \times & \times \\ \times & \times & \times & \times & \times & \times \\ 0 & \times & \times & \times & \times & \times \\ 0 & \times & \times & \times & \times & \times \\ 0 & \times & \times & \times & \times & \times \\ 0 & \times & \times & \times & \times & \times \end{bmatrix} \xrightarrow{P_2}$$

$$\begin{bmatrix} \times & \times & \times & \times & \times & \times \\ \times & \times & \times & \times & \times & \times \\ 0 & \times & \times & \times & \times & \times \\ 0 & 0 & \times & \times & \times & \times \\ 0 & 0 & \times & \times & \times & \times \\ 0 & 0 & \times & \times & \times & \times \end{bmatrix} \xrightarrow{P_3} \begin{bmatrix} \times & \times & \times & \times & \times & \times \\ \times & \times & \times & \times & \times & \times \\ 0 & \times & \times & \times & \times & \times \\ 0 & 0 & \times & \times & \times & \times \\ 0 & 0 & 0 & \times & \times & \times \\ 0 & 0 & 0 & \times & \times & \times \end{bmatrix}$$

$$\xrightarrow{P_4} \begin{bmatrix} \times & \times & \times & \times & \times & \times \\ \times & \times & \times & \times & \times & \times \\ 0 & \times & \times & \times & \times & \times \\ 0 & 0 & \times & \times & \times & \times \\ 0 & 0 & 0 & \times & \times & \times \\ 0 & 0 & 0 & 0 & \times & \times \end{bmatrix}$$

In general, after $k-1$ steps we have computed $k-1$ Householder matrices $P_1, \ldots, P_{k-1}$ such that

$$(P_1 \cdots P_{k-1})^T A (P_1 \cdots P_{k-1}) = \begin{bmatrix} B_{11} & B_{12} & B_{13} \\ B_{21} & B_{22} & B_{23} \\ 0 & B_{32} & B_{33} \end{bmatrix} \begin{matrix} k-1 \\ 1 \\ n-k \end{matrix}$$
$$\qquad\qquad\qquad\qquad k-1 \quad\; 1 \quad\; n-k$$

is upper Hessenberg through its first $k-1$ columns. Suppose $\bar{P}_k$ is an order $n-k$ Householder matrix such that $\bar{P}_k B_{32}$ is a multiple of $e_1^{(n-k)}$. If $P_k = \text{diag}(I_k, \bar{P}_k)$, then

$$(P_1 \cdots P_k)^T A (P_1 \cdots P_k) = \begin{bmatrix} B_{11} & B_{12} & B_{13}\bar{P}_k \\ B_{21} & B_{22} & B_{23}\bar{P}_k \\ 0 & \bar{P}_k B_{32} & \bar{P}_k B_{33} \bar{P}_k \end{bmatrix}$$

is upper Hessenberg through its first $k$ columns. Using the Householder procedures **house**, **row.house**, and **col.house** of §5.1.4 we obtain

**Algorithm 7.4.2 (Householder Reduction to Hessenberg Form)** Given $A \in \mathbb{R}^{n \times n}$, the following algorithm overwrites $A$ with $H = U_0^T A U_0$ where $H$ is upper Hessenberg and $U_0 = (P_1 \cdots P_{n-2})$ is product of Householder matrices. The essential part of the $k$th Householder vector is stored in $A(k+2{:}n, k)$.

```
for k = 1:n − 2
    v(k + 1:n) = house(A(k + 1:n, k))
    { A = P_k^T A P_k where P_k = diag(I_k, P̄_k).}
    A(k + 1:n, k:n) = row.house(A(k + 1:n, k:n), v(k + 1:n))
    A(1:n, k + 1:n) = col.house(A(1:n, k + 1:n), v(k + 1:n))
    A(k + 2:n, k) = v(k + 2:n)
end
```

This algorithm requires $10n^3/3$ flops. If $U_0$ is explicitly formed, an additional $4n^3/3$ flops are required. See Martin and Wilkinson (1968d) for a detailed description.

The roundoff properties of this method for reducing $A$ to Hessenberg form are very desirable. Wilkinson (AEP, p. 351) states that the computed Hessenberg matrix $\hat{H}$ satisfies $\hat{H} = Q^T(A+E)Q$, where $Q^TQ = I$ and

$$\| E \|_F \le cn^2\mathbf{u}\| A \|_F$$

with $c$ is a small constant.

**Example 7.4.2** If

$$A = \begin{bmatrix} 1 & 5 & 7 \\ 3 & 0 & 6 \\ 4 & 3 & 1 \end{bmatrix} \qquad \text{and} \qquad U_0 = \begin{bmatrix} 1 & 0 & 0 \\ 0 & .6 & .8 \\ 0 & .8 & -.6 \end{bmatrix}$$

then

$$U_0^T A U_0 = H = \begin{bmatrix} 1.00 & 8.60 & -.020 \\ 5.00 & 4.96 & -.072 \\ 0.00 & 2.28 & -.3.96 \end{bmatrix}.$$

### 7.4.4 Level-3 Aspects

The Hessenberg reduction (Algorithm 7.4.2) is rich in level-2 operations: half gaxpys and half outer product updates. We briefly discuss two methods for introducing level-3 computations into the process.

The first approach involves a block reduction to block Hessenberg form and is quite straight forward. Suppose (for clarity) that that $n = rN$ and write

$$A = \begin{bmatrix} A_{11} & A_{12} \\ A_{21} & A_{22} \end{bmatrix} \begin{matrix} r \\ n-r \end{matrix} \quad . \\ \qquad\quad r \qquad n-r$$

Suppose that we have computed the QR factorization $A_{21} = \bar{Q}_1 R_1$ and that $\bar{Q}_1$ is in WY form. That is, we have $W_1, Y_1 \in \mathbb{R}^{(n-r)\times r}$ such that $\bar{Q}_1 = I - W_1Y_1^T$. (See §5.2.2 for details.) If $Q_1 = \text{diag}(I_r, \bar{Q}_1)$ then

$$Q_1^T A Q_1 = \begin{bmatrix} A_{11} & A_{12}\bar{Q}_1 \\ R_1 & \bar{Q}_1^T A_{22}\bar{Q}_1 \end{bmatrix}.$$

Notice that the updates of the (1,2) and (2,2) blocks are rich in level-3 operations given that $\bar{Q}_1$ is in WY form. This fully illustrates the overall process as $Q_1^T AQ$ is block upper Hessenberg through its first block column. We next repeat the computations on the first $r$ columns of $\bar{Q}_1^T A_{22}\bar{Q}_1$. After

$N - 2$ such steps we obtain

$$H = U_0^T A U_0 = \begin{bmatrix} H_{11} & H_{12} & \cdots & \cdots & H_{1N} \\ H_{21} & H_{22} & \cdots & \cdots & H_{2N} \\ 0 & \ddots & \ddots & \cdots & \vdots \\ \vdots & \vdots & \ddots & \ddots & \vdots \\ 0 & 0 & \cdots & H_{N,N-1} & H_{NN} \end{bmatrix}$$

where each $H_{ij}$ is $r$-by-$r$ and $U_0 = Q_1 \cdots Q_{N-2}$ with with each $Q_i$ in WY form. The overall algorithm has a level-3 fraction of the form 1 - $O(1/N)$.

Note that the subdiagonal blocks in $H$ are upper triangular and so the matrix has lower bandwidth $p$. It is possible to reduce $H$ to actual Hessenberg form by using Givens rotations to zero all but the first subdiagonal. How much this "clean-up" operation diminishes level-3 performance is an open question.

Dongarra, Hammarling and Sorenson (1987) have shown how to proceed directly to Hessenberg form using a mixture of gaxpy's and level-3 updates. Their idea involves minimal updating after each Householder transformation is generated. For example, suppose the first Householder $P_1$ has been computed. To generate $P_2$ we need just the second column of $P_1AP_1$, not the full outer product update. To generate $P_3$ we need just the 3rd column of $P_2P_1AP_1P_2$, etc. In this way, the Householder matrices can be determined using only gaxpy operations. No outer product updates are involved. Once a suitable number of Householder matrices are known they can be aggregated and applied in a level-3 fashion.

### 7.4.5 Important Hessenberg Matrix Properties

The Hessenberg decomposition is not unique. If $Z$ is any $n$-by-$n$ orthogonal matrix and we apply Algorithm 7.4.2 to $Z^TAZ$, then $Q^TAQ = H$ is upper Hessenberg where $Q = ZU_0$. However, $Qe_1 = Z(U_0e_1) = Ze_1$ suggesting that $H$ is unique once the first column of $Q$ is specified. This is essentially the case provided $H$ has no zero subdiagonal entries. Hessenberg matrices with this property are said to be *unreduced.* Here is a very important theorem that clarifies the uniqueness of the Hessenberg reduction.

**Theorem 7.4.2 ( Implicit Q Theorem )** *Suppose* $Q = [\, q_1, \ldots, q_n \,]$ *and* $V = [\, v_1, \ldots, v_n \,]$ *are orthogonal matrices with the property that both* $Q^TAQ = H$ *and* $V^TAV = G$ *are upper Hessesnberg where* $A \in \mathbb{R}^{n \times n}$. *Let* $k$ *denote the smallest positive integer for which* $h_{k+1,k} = 0$, *with the convention that* $k = n$ *if* $H$ *is unreduced. If* $v_1 = q_1$ *then* $v_i = \pm q_i$ *and* $|h_{i,i-1}| = |g_{i,i-1}|$ *for* $i = 2{:}k$. *Moreover, if* $k < n$ *then* $g_{k+1,k} = 0$.

**Proof.** Define the orthogonal matrix $W = [\, w_1, \ldots, w_n \,]$ by $W = V^T Q$ and observe the $GW = WH$. Thus, for $i = 2{:}k$ we have

$$h_{i,i-1} w_i \;=\; G w_{i-1} \;-\; \sum_{j=1}^{i-1} h_{j,i-1} w_j \,.$$

Since $w_1 = e_1$ it follows that $[\, w_1, \ldots, w_k \,]$ is upper triangular and thus, $w_i = \pm e_i$ for $i = 2{:}k$. Since $w_i = V^T q_i$ and $h_{i,i-1} = w_i^T G w_{i-1}$ it follows that $v_i = \pm q_i$ and $|h_{i,i-1}| = |g_{i,i-1}|$ for $i = 2{:}k$. If $h_{k+1,k} = 0$ then ignoring signs we have

$$\begin{aligned} g_{k+1,k} &= e_{k+1}^T G e_k \;=\; e_{k+1}^T G W e_k \;=\; (e_{k+1}^T W)(H e_k) \\ &= e_{k+1}^T \sum_{i=1}^{k} h_{ik} W e_i \;=\; \sum_{i=1}^{k} h_{ik} e_{k+1}^T e_i = 0 \,. \;\square \end{aligned}$$

The gist of the implicit $Q$ theorem is that if $Q^T A Q = H$ and $Z^T A Z = G$ are each unreduced upper Hessenberg matrices and $Q$ and $Z$ have the same first column, then $G$ and $H$ are "essentially equal" in the sense that $G = D^{-1} H D$ where $D = \mathrm{diag}(\pm 1, \ldots, \pm 1)$.

Our next theorem shows that there is a connection between the QR factorization of the Krylov matrix $K(A, Q(:,1), n)$ and the Hessenberg reduction $Q^T A Q = H$.

**Theorem 7.4.3** *Suppose $Q \in \mathbb{R}^{n \times n}$ is an orthogonal matrix and $A \in \mathbb{R}^{n \times n}$. Then $Q^T A Q = H$ is an unreduced upper Hessenberg matrix if and only if $Q^T K(A, Q(:,1), n) = R$ is nonsingular and upper triangular.*

**Proof.** Suppose $Q \in \mathbb{R}^{n \times n}$ is orthogonal and set $H = Q^T A Q$. Consider the identity

$$Q^T K(A, Q(:,1), n) \;=\; [\, e_1,\, H e_1, \ldots, H^{n-1} e_1 \,] \;\equiv\; R \,.$$

If $H$ is an unreduced upper Hessenberg matrix then it is clear that $R$ is upper triangular with $r_{ii} = h_{21} h_{32} \cdots h_{i,i-1}$ for $i = 2{:}n$. Since $r_{11} = 1$ it follows that $R$ is nonsingular.

To prove the converse, suppose $R$ is upper triangular and nonsingular. Since $R(:, k+1) = H R(:, k)$ it follows that $H(:, k) \in \mathrm{span}\{:e_1, \ldots, e_{k+1}\}$. This implies that $H$ is upper Hessenberg. Since $r_{nn} = h_{21} h_{32} \cdots h_{n,n-1} \neq 0$ it follows that $H$ is also unreduced. $\square$

Thus, there is more or less a correspondence between nonsingular Krylov matrices $K(x, A, n-1)$ and orthogonal similarity reductions to unreduced Hessenberg form.

Our last result concerns eigenvalues of an unreduced upper Hessenberg matrix.

**Theorem 7.4.4** *If $\lambda$ is an eigenvalue of an unreduced upper Hessenberg matrix $H \in \mathbb{R}^{n\times n}$, then its geometric multiplicity is one.*

**Proof.** For any $\lambda \in \mathbb{C}$ we have $\text{rank}(A - \lambda I) \geq n - 1$ because the first $n - 1$ columns of $H - \lambda I$ are independent. □

### 7.4.6 Companion Matrix Form

Just as the Schur decomposition has a nonunitary analog in the Jordan decomposition, so does the Hessenberg decomposition have a nonunitary analog in the *companion matrix decomposition.* Let $x \in \mathbb{R}^n$ and suppose that the Krylov matrix $K(x, A, n-1)$ is nonsingular. If $c = c(0{:}n-1)$ solves the linear system $K(x, A, n-1)c = -A^n x$ , then it follows that $AK(x, A, n-1) = K(x, A, n-1)C$ where $C$ has the form:

$$C = \begin{bmatrix} 0 & 0 & \cdots & 0 & -c_0 \\ 1 & 0 & \cdots & 0 & -c_1 \\ 0 & 1 & \cdots & 0 & -c_2 \\ \vdots & \vdots & & \vdots & \vdots \\ 0 & 0 & \cdots & 1 & -c_{n-1} \end{bmatrix} \tag{7.4.4}$$

The matrix $C$ is said to be a *companion matrix.* Since

$$\det(zI - C) = c_0 + c_1 z + \cdots + c_{n-1}z^{n-1} + z^n$$

it follows that if $Y = K(x, A, n-1)$ is nonsingular, then the decomposition $Y^{-1}AY = C$ displays $A$'s characteristic polynomial. This, coupled with the sparseness of $C$, has led to "companion matrix methods" in various application areas. These techniques typically involve:

- Computing the Hessenberg decomposition $U_0^T A U_0 = H$.
- Hoping $H$ is unreduced and setting $Y = \left[e_1,\ He_1,\ \ldots,\ H^{n-1}e_1\right]$.
- Solving $YC = HY$ for $C$.

Unfortunately, this calculation can be highly unstable. $A$ is similar to an unreduced Hessenberg matrix only if each eigenvalue has unit geometric multiplicity. Matrices that have this property are called *nonderogatory.* It follows that the matrix $Y$ above can be very poorly conditioned if $A$ is close to a derogatory matrix.

A full discussion of the dangers associated with companion matrix computation can be found in Wilkinson (AEP, pp. 405 ff.).

### 7.4.7 Hessenberg Reduction Via Gauss Transforms

While we are on the subject of nonorthogonal reduction to Hessenberg form, we should mention that Gauss tranformations can be used in lieu of Householder matrices in Algorithm 7.4.2. In particular, suppose permutations $\Pi_1, \ldots, \Pi_{k-1}$ and Gauss transformations $M_1, \ldots, M_{k-1}$ have been determined such that

$$(M_{k-1}\Pi_{k-1} \cdots M_1\Pi_1)A(M_{k-1}\Pi_{k-1} \cdots M_1\Pi_1)^{-1} = B$$

where

$$B = \begin{bmatrix} B_{11} & B_{12} & B_{13} \\ B_{21} & B_{22} & B_{23} \\ 0 & B_{32} & B_{33} \end{bmatrix} \begin{matrix} k-1 \\ 1 \\ n-k \end{matrix}$$
$$\qquad\quad k-1 \quad\ 1 \quad\ n-k$$

is upper Hessenberg through its first $k-1$ columns. A permutation $\bar{\Pi}_k$ of order $n-k$ is then determined such that the first element of $\bar{\Pi}_k B_{32}$ is maximal in absolute value. This makes it possible to determine a stable Gauss transformation $\bar{M}_k = I - z_k e_1^T$ also of order $n-k$, such that all but the first component of $\bar{M}_k(\bar{\Pi}_k B_{32})$ is zero. Defining $\Pi_k = \text{diag}(I_k, \bar{\Pi}_k)$ and $M_k = \text{diag}(I_k, \bar{M}_k)$, we see that

$$(M_k\Pi_k \cdots M_1\Pi_1)A(M_k\Pi_k \cdots M_1\Pi_1)^{-1} =$$
$$\begin{bmatrix} B_{11} & B_{12} & B_{13} \\ B_{21} & B_{22} & B_{23}\bar{\Pi}_k^T\bar{M}_k^{-1} \\ 0 & \bar{M}_k\bar{\Pi}_k B_{32} & \bar{M}_k\bar{\Pi}_k B_{33}\bar{\Pi}_k^T\bar{M}_k^{-1} \end{bmatrix}$$

is upper Hessenberg through its first $k$ columns. Note that $\bar{M}_k^{-1} = I + z_k e_1^T$ and so some very simple rank-one updates are involved in the reduction.

A careful operation count reveals that the Gauss reduction to Hessenberg form requires only half the number of flops of the Householder method. However, as in the case of Gaussian elimination with partial pivoting, there is a (fairly remote) chance of $2^n$ growth. See Businger (1969). Another difficulty associated with the Gauss approach is that the eigenvalue condition numbers — the $s(\lambda)^{-1}$ — are not preserved with nonorthogonal similarity transformations. This somewhat complicates the error estimation schemes presented in §7.6.

#### Problems

**P7.4.1** Suppose $A \in \mathbb{R}^{n \times n}$ and $z \in \mathbb{R}^n$. Give a detailed algorithm for computing an orthogonal $Q$ such that $Q^TAQ$ is upper Hessenberg and $Q^Tz$ is a multiple of $e_1$. (Hint:

Reduce $z$ first and then apply Algorithm 7.4.2.)

**P7.4.2** Specify a complete Gauss reduction to Hessenberg form using **gauss** and **gauss.app** and verify that it only requires $5n^3/3$ flops.

**P7.4.3** In some situations, it is necessary to solve the linear system $(A + zI)x = b$ for many different values of $z \in \mathbf{R}$ and $b \in \mathbf{R}^n$. Show how this problem can be efficiently and stably solved using the Hessenberg decomposition.

**P7.4.4** Give a detailed algorithm for explicitly computing the matrix $U_0$ in Algorithm 7.4.2. Design your algorithm so that $H$ is overwritten by $U_0$.

**P7.4.5** Suppose $H \in \mathbf{R}^{n\times n}$ is an unreduced upper Hessenberg matrix. Show that there exists a diagonal matrix $D$ such that each subdiagonal element of $D^{-1}HD$ is equal to one. What is $\kappa_2(D)$?

**P7.4.6** Suppose $W, Y \in \mathbf{R}^{n\times n}$ and define the matrices $C$ and $B$ by

$$C = W + iY, \qquad B = \begin{bmatrix} W & -Y \\ Y & W \end{bmatrix}$$

Show that if $\lambda \in \lambda(C)$ is real, then $\lambda \in \lambda(B)$. Relate the corresponding eigenvectors.

**P7.4.7** Suppose $A = \begin{bmatrix} w & x \\ y & z \end{bmatrix}$ is a real matrix having eigenvalues $\lambda \pm i\mu$, where $\mu$ is nonzero. Give an algorithm that stably determines $c = \cos(\theta)$ and $s = \sin(\theta)$ such that

$$\begin{bmatrix} c & s \\ -s & c \end{bmatrix}^T \begin{bmatrix} w & x \\ y & z \end{bmatrix} \begin{bmatrix} c & s \\ -s & c \end{bmatrix} = \begin{bmatrix} \lambda & \beta \\ \alpha & \lambda \end{bmatrix}$$

where $\alpha\beta = -\mu^2$.

**P7.4.8** Suppose $(\lambda, x)$ is a known eigenvalue-eigenvector pair for the upper Hessenberg matrix $H \in \mathbf{R}^{n\times n}$. Give an algorithm for computing an orthogonal matrix $P$ such that

$$P^THP = \begin{bmatrix} \lambda & w^T \\ 0 & H_1 \end{bmatrix}$$

where $H_1 \in \mathbf{R}^{(n-1)\times(n-1)}$ is upper Hessenberg. Compute $P$ as a product of Givens rotations.

### Notes and References for Sec. 7.4

The real Schur decomposition was originally presented in

F.D. Murnaghan and A. Wintner (1931). "A Canonical Form for Real Matrices Under Orthogonal Transformations," *Proc. Nat. Acad. Sci. 17*, 417–20.

A thorough treatment of the reduction to Hessenberg form is given in Wilkinson (AEP, chapter 6), and Algol procedures for both the Householder and Gauss methods appear in

R.S. Martin and J.H. Wilkinson (1968d). "Similarity Reduction of a General Matrix to Hessenberg Form," *Numer. Math. 12*, 349–68. See also *HACLA*, pp. 339–58.

Fortran versions of the Algol procedures in the last reference are in Eispack.

Givens rotations can also be used to compute the Hessenberg decomposition. See

W. Rath (1982). "Fast Givens Rotations for Orthogonal Similarity," *Numer. Math. 40*, 47–56.

The high performance computation of the Hessenberg reduction is discussed in

J.J. Dongarra, S. Hammarling, and D.C. Sorensen (1987). "Block Reduction of Matrices to Condensed form for Eigenvalue Computations," ANL-MCS-TM 99, Argonne National Laboratory, Argonne, Illinois.
J.J. Dongarra, L. Kaufman, and S. Hammarling (1986). "Squeezing the Most Out of Eigenvalue Solvers on High Performance Computers," *Lin. Alg. and Its Applic. 77*, 113–136.

The possibility of exponential growth in the Gauss tranformation approach was first pointed out in

P. Businger (1969). "Reducing a Matrix to Hessenberg Form," *Math. Comp. 23*, 819–21.

However, the algorithm should be regarded in the same light as Gaussian elimination with partial pivoting—stable for all practical purposes. See Eispack, pp. 56–58.

Aspects of the Hessenberg decomposition for sparse matrices are discussed in

I.S. Duff and J.K. Reid (1975). "On the Reduction of Sparse Matrices to Condensed Forms by Similarity Transformations," *J. Inst. Math. Applic. 15*, 217–24.

Block reduction to Hessenberg form is discussed in

J.J. Dongarra, S. Hammarling, and D.C. Sorensen (1987). "Block Reduction of Matrices to Condensed form for Eigenvalue Computations," ANL-MCS- TM 99, Argonne National Laboratory, Argonne, Illinois.

Once an eigenvalue of an unreduced upper Hessenberg matrix is known, it is possible to zero the last subdiagonal entry using Givens similarity transformations. See

P.A. Businger (1971). "Numerically Stable Deflation of Hessenberg and Symmetric Tridiagonal Matrices,*BIT 11*, 262–70.

Some interesting mathematical properties of the Hessenberg form may be found in

Y. Ikebe (1979). "On Inverses of Hessenberg Matrices," *Lin. Alg. and Its Applic. 24*, 93–97.
B.N. Parlett (1967). "Canonical Decompositon of Hessenberg Matrices," *Math. Comp. 21*, 223–27.

Although the Hessenberg decomposition is largely appreciated as a "front end" decomposition for the QR iteration, it is increasingly popular as a cheap alternative to the more expensive Schur decomposition in certain problems. For a sampling of applications where it has proven to be very useful, consult

W. Enright (1979). "On the Efficient and Reliable Numerical Solution of Large Linear Systems of O.D.E.'s," *IEEE Trans. Auto. Cont.* AC-24, 905–8.
G.H. Golub, S. Nash and C. Van Loan (1979). "A Hessenberg-Schur Method for the Problem $AX + XB = C$," *IEEE Trans. Auto. Cont. AC-24*, 909–13.
A. Laub (1981). "Efficient Multivariable Frequency Response Computations," *IEEE Trans. Auto. Cont. AC-26*, 407–8.
G. Miminis and C.C. Paige (1982). "An Algorithm for Pole Assignment of Time Invariant Linear Systems," *International J. of Control 35*, 341–354.
C.C. Paige (1981). "Properties of Numerical Algorithms Related to Computing Controllability," *IEEE Trans. Auto. Cont. AC-26*, 130–38.
C. Van Loan (1982b). "Using the Hessenberg Decomposition in Control Theory," in *Algorithms and Theory in Filtering and Control* , D.C. Sorenson and R.J. Wets (eds), Mathematical Programming Study No. 18, North Holland, Amsterdam, pp. 102–11.

# 7.5 The Practical QR Algorithm

We return to the Hessenberg QR iteration which we write as follows:

$$\begin{array}{lll} H = U_0^T A U_0 & \text{(Hessenberg Reduction)} & \\ \textbf{for } k = 1,2,\ldots & & \\ \quad H = UR & \text{(QR factorization)} & \quad (7.5.1) \\ \quad H = RU & & \\ \textbf{end} & & \end{array}$$

Our aim in this section is to describe how the $H$'s converge to upper quasi-triangular form and to show how the convergence rate can be accelerated by incorporating *shifts*.

## 7.5.1 Deflation

Without loss of generality we may assume that each Hessenberg matrix $H$ in (7.5.1) is unreduced. If not, then at some stage we have

$$H = \begin{bmatrix} H_{11} & H_{12} \\ 0 & H_{22} \end{bmatrix} \begin{array}{l} p \\ n-p \end{array}$$
$$\qquad p \quad n-p$$

where $1 \leq p < n$ and the problem *decouples* into two smaller problems involving $H_{11}$ and $H_{22}$. The term *deflation* is also used in this context, usually when $p = n-1$ or $n-2$.

In practice, decoupling occurs whenever a subdiagonal entry in $H$ is suitably small. For example, in Eispack if

$$|h_{p+1,p}| \leq c\mathbf{u}(|h_{pp}| + |h_{p+1,p+1}|) \qquad (7.5.2)$$

for a small constant $c$, then $h_{p+1,p}$ is "declared" to be zero. This is justified since rounding errors of order $\mathbf{u}\| H \|$ are already present throughout the matrix.

## 7.5.2 The Shifted QR Iteration

Let $\mu \in \mathbb{R}$ and consider the iteration:

$$\begin{array}{lll} H = U_0^T A U_0 & \text{(Hessenberg Reduction)} & \\ \textbf{for } k = 1,2,\ldots & & \\ \quad \text{Determine a scalar } \mu. & & \\ \quad H - \mu I = UR & \text{(QR factorization)} & \quad (7.5.3) \\ \quad H = RU + \mu I & & \\ \textbf{end} & & \end{array}$$

The scalar $\mu$ is referred to as a *shift*. Each matrix $H$ generated in (7.5.3) is similar to $A$, since $RU + \mu I = U^T(UR + \mu I)U = U^T HU$. If we order the eigenvalues $\lambda_i$ of $A$ so that

$$|\lambda_1 - \mu| \geq \cdots \geq |\lambda_n - \mu|,$$

and $\mu$ is fixed from iteration to iteration, then the theory of §7.3 says that the $p$th subdiagonal entry in $H$ converges to zero with rate

$$\left|\frac{\lambda_{p+1} - \mu}{\lambda_p - \mu}\right|^k .$$

Of course, if $\lambda_p = \lambda_{p+1}$, then there is no convergence at all. But if, for example, $\mu$ is much closer to $\lambda_n$ than to the other eigenvalues, then the zeroing of the $(n, n-1)$ entry is rapid. In the extreme case we have the following:

**Theorem 7.5.1** *Let $\mu$ be an eigenvalue of an n-by-n unreduced Hessenberg matrix $H$. If $\bar{H} = RU + \mu I$, where $H - \mu I = UR$ is the QR factorization of $H - \mu I$, then $\bar{h}_{n,n-1} = 0$ and $\bar{h}_{nn} = \mu$.*

**Proof.** Since $H$ is an unreduced Hessenberg matrix the first $n-1$ columns of $H - \mu I$ are independent, regardless of $\mu$. Thus, if $UR = (H - \mu I)$ is the QR factorization then $r_{ii} \neq 0$ for $i = 1{:}n-1$. But if $H - \mu I$ is singular then $r_{11} \cdots r_{nn} = 0$. Thus, $r_{nn} = 0$ and $\bar{H}(n,:) = [\,0, \ldots, 0, \mu\,]$ □

The theorem says that if we shift by an exact eigenvalue, then in exact arithmetic deflation occurs in one step.

**Example 7.5.1** If

$$H = \begin{bmatrix} 9 & -1 & -2 \\ 2 & 6 & -2 \\ 0 & 1 & 5 \end{bmatrix},$$

then $6 \in \lambda(H)$. If $UR = H - 6I$ is the QR factorization, then $\bar{H} = RU + 6I$ is given by

$$\bar{H} = \begin{bmatrix} 8.5384 & -3.7313 & -1.0090 \\ 0.6343 & 5.4615 & 1.3867 \\ 0.0000 & 0.0000 & 6.0000 \end{bmatrix}.$$

### 7.5.3 The Single Shift Strategy

Now let us consider varying $\mu$ from iteration to iteration incorporating new information about $\lambda(A)$ as the subdiagonal entries converge to zero. A good heuristic is to regard $h_{nn}$ as the best approximate eigenvalue along the diagonal. If we shift by this quantity during each iteration, we obtain the *single-shift QR iteration*:

**for** $k = 1, 2, \ldots$
$\quad \mu = H(n,n)$
$\quad H - \mu I = UR \qquad$ (QR Factorization) $\qquad$ (7.5.4)
$\quad H = RU + \mu I$
**end**

If the $(n, n-1)$ entry converges to zero, it is likely to do so at a quadratic rate. To see this, we borrow an example from Stewart (IMC, p. 366). Suppose $H$ is an unreduced upper Hessenberg matrix of the form

$$H = \begin{bmatrix} \times & \times & \times & \times & \times \\ \times & \times & \times & \times & \times \\ 0 & \times & \times & \times & \times \\ 0 & 0 & \times & \times & \times \\ 0 & 0 & 0 & \epsilon & h_{nn} \end{bmatrix}$$

and that we perform one step of the single-shift QR algorithm: $UR = H - h_{nn}I$, $\bar{H} = RU + h_{nn}I$. After $n-2$ steps in the reduction of $H - h_{nn}I$ to upper triangular form we obtain a matrix with the following structure:

$$H = \begin{bmatrix} \times & \times & \times & \times & \times \\ 0 & \times & \times & \times & \times \\ 0 & 0 & \times & \times & \times \\ 0 & 0 & 0 & a & b \\ 0 & 0 & 0 & \epsilon & 0 \end{bmatrix}$$

It is not hard to show that the $(n, n-1)$ entry in $\bar{H} = RU + h_{nn}I$ is given by $-\epsilon^2 b/(\epsilon^2 + a^2)$. If we assume that $\epsilon \ll a$, then it is clear that the new $(n, n-1)$ entry has order $\epsilon^2$, precisely what we would expect of a quadratically converging algorithm.

**Example 7.5.2** If

$$H = \begin{bmatrix} 1 & 2 & 3 \\ 4 & 5 & 6 \\ 0 & .001 & 7 \end{bmatrix}$$

and $UR = H - 7I$ is the QR factorization, then $\bar{H} = RU + 7I$ is given by

$$\bar{H} \approx \begin{bmatrix} -0.5384 & 1.6908 & 0.8351 \\ 0.3076 & 6.5264 & -6.6555 \\ 0.0000 & 2 \cdot 10^{-5} & 7.0119 \end{bmatrix}.$$

Near-perfect shifts as above almost always ensure a small $\bar{h}_{n,n-1}$. However, this is just a heuristic. There are examples in which $\bar{h}_{n,n-1}$ is a relatively large matrix entry even though $\sigma_{min}(H - \mu I) \approx \mathbf{u}$.

### 7.5.4 The Double Shift Strategy

Unfortunately, difficulties with (7.5.4) can be expected if at some stage the eigenvalues $a_1$ and $a_2$ of

$$G = \begin{bmatrix} h_{mm} & h_{mn} \\ h_{nm} & h_{nn} \end{bmatrix} \qquad m = n-1 \tag{7.5.5}$$

are complex for then $h_{nn}$ would tend to be a poor approximate eigenvalue.

A way around this difficulty is to perform two single-shift QR steps in succession using $a_1$ and $a_2$ as shifts:

$$\begin{aligned} H - a_1 I &= U_1 R_1 \\ H_1 &= R_1 U_1 + a_1 I \\ H_1 - a_2 I &= U_2 R_2 \\ H_2 &= R_2 U_2 + a_2 I \end{aligned} \tag{7.5.6}$$

These equations can be manipulated to show that

$$(U_1 U_2)(R_2 R_1) = M \tag{7.5.7}$$

where $M$ is defined by

$$M = (H - a_1 I)(H - a_2 I). \tag{7.5.8}$$

Note that $M$ is a real matrix even if $G$'s eigenvalues are complex since

$$M = H^2 - sH + tI$$

where

$$s = a_1 + a_2 = h_{mm} + h_{nn} = \text{trace}(G) \in \mathbb{R}$$

and

$$t = a_1 a_2 = h_{mm} h_{nn} - h_{mn} h_{nm} = \det(G \in \mathbb{R}\,.$$

Thus, (7.5.7) is the QR factorization of a real matrix and we may choose $U_1$ and $U_2$ so that $Z = U_1 U_2$ is real orthogonal. It then follows that

$$H_2 = U_2^H H_1 U_2 = U_2^H (U_1^H H U_1) U_2 = (U_1 U_2)^H H (U_1 U_2) = Z^T H Z.$$

is real.

Unfortunately, roundoff error almost always prevents an exact return to the real field. A real $H_2$ could be guaranteed if we

- explicitly form the real matrix $M = H^2 - sH + tI$,
- compute the real QR factorization $M = ZR$, and
- set $H_2 = Z^T H Z$.

But since the first of these steps requires $O(n^3)$ flops, this is not a practical course of action.

### 7.5.5 The Double Implicit Shift Strategy

Fortunately, it turns out that we can implement the double shift step with $O(n^2)$ flops by appealing to the Implicit Q Theorem. In particular we can effect the transition from $H$ to $H_2$ in $O(n^2)$ flops if we

- compute $Me_1$, the first column of $M$;
- determine a Householder matrix $P_0$ such that $P_0(Me_1)$ is a multiple of $e_1$;
- compute Householder matrices $P_1, \ldots, P_{n-2}$ such that if $Z_1$ is the product $Z_1 = P_0P_1 \cdots P_{n-2}$ then $Z_1^T HZ_1$ is upper Hessenberg and the first columns of $Z$ and $Z_1$ are the same.

Under these circumstances, the the Implicit $Q$ theorem permits us to conclude that if $Z^THZ$ and $Z_1^THZ_1$ are both unreduced upper Hessenberg matrices, then they are essentially equal. Note that if these Hessenberg matrices are not unreduced, then we can effect a decoupling and proceed with smaller unreduced subproblems.

Let us work out the details. Observe first that $P_0$ can be determined in $O(1)$ flops since $Me_1 = (x,\ y,\ z,\ 0, \ldots, 0)^T$ where

$$\begin{aligned} x &= h_{11}^2 + h_{12}h_{21} - sh_{11} + t \\ y &= h_{21}(h_{11} + h_{22} - s) \\ z &= h_{21}h_{32}. \end{aligned}$$

Since a similarity transformation with $P_0$ only changes rows and columns 1, 2, and 3, we see that

$$P_0HP_0 = \begin{bmatrix} \times & \times & \times & \times & \times & \times \\ \times & \times & \times & \times & \times & \times \\ \times & \times & \times & \times & \times & \times \\ \times & \times & \times & \times & \times & \times \\ 0 & 0 & 0 & \times & \times & \times \\ 0 & 0 & 0 & 0 & \times & \times \end{bmatrix}$$

Now the mission of the Householder matrices $P_1, \ldots, P_{n-2}$ is to restore this matrix to upper Hessenberg form. The calculation proceeds as follows:

$$\begin{bmatrix} \times & \times & \times & \times & \times & \times \\ \times & \times & \times & \times & \times & \times \\ \times & \times & \times & \times & \times & \times \\ \times & \times & \times & \times & \times & \times \\ 0 & 0 & 0 & \times & \times & \times \\ 0 & 0 & 0 & 0 & \times & \times \end{bmatrix} \xrightarrow{P_1} \begin{bmatrix} \times & \times & \times & \times & \times & \times \\ \times & \times & \times & \times & \times & \times \\ 0 & \times & \times & \times & \times & \times \\ 0 & \times & \times & \times & \times & \times \\ 0 & \times & \times & \times & \times & \times \\ 0 & 0 & 0 & 0 & \times & \times \end{bmatrix} \xrightarrow{P_2}$$

$$\begin{bmatrix} \times & \times & \times & \times & \times & \times \\ \times & \times & \times & \times & \times & \times \\ 0 & \times & \times & \times & \times & \times \\ 0 & 0 & \times & \times & \times & \times \\ 0 & 0 & \times & \times & \times & \times \\ 0 & 0 & \times & \times & \times & \times \end{bmatrix} \xrightarrow{P_3} \begin{bmatrix} \times & \times & \times & \times & \times & \times \\ \times & \times & \times & \times & \times & \times \\ 0 & \times & \times & \times & \times & \times \\ 0 & 0 & \times & \times & \times & \times \\ 0 & 0 & 0 & \times & \times & \times \\ 0 & 0 & 0 & \times & \times & \times \end{bmatrix}$$

$$\xrightarrow{P_4} \begin{bmatrix} \times & \times & \times & \times & \times & \times \\ \times & \times & \times & \times & \times & \times \\ 0 & \times & \times & \times & \times & \times \\ 0 & 0 & \times & \times & \times & \times \\ 0 & 0 & 0 & \times & \times & \times \\ 0 & 0 & 0 & 0 & \times & \times \end{bmatrix}$$

Clearly, the general $P_k$ has the form $P_k = \text{diag}(I_k, \bar{P}_k, I_{n-k-3})$ where $\bar{P}_k$ is a 3-by-3 Householder matrix. For example,

$$P_2 = \begin{bmatrix} 1 & 0 & 0 & 0 & 0 & 0 \\ 0 & 1 & 0 & 0 & 0 & 0 \\ 0 & 0 & \times & \times & \times & 0 \\ 0 & 0 & \times & \times & \times & 0 \\ 0 & 0 & \times & \times & \times & 0 \\ 0 & 0 & 0 & 0 & 0 & 1 \end{bmatrix}$$

Note that $P_{n-2}$ is an exception to this since $P_{n-2} = \text{diag}(I_{n-2}, \bar{P}_{n-2})$.

The applicability of Theorem 7.4.3 (the Implicit $Q$ theorem) follows from the observation that $P_k e_1 = e_1$ for $k = 1{:}n-2$ and that $P_0$ and $Z$ have the same first column. Hence, $Z_1 e_1 = Z e_1$, and we can assert that $Z_1$ essentially equals $Z$ provided that the upper Hessenberg matrices $Z^T H Z$ and $Z_1^T H Z_1$ are each unreduced.

The implicit determination of $H_2$ from $H$ outlined above was first described by Francis (1961) and we refer to it as a *Francis QR step.* The complete Francis step is summarized as follows:

**Algorithm 7.5.1 (Francis QR Step)** Given the unreduced upper Hessenberg matrix $H \in \mathbb{R}^{n \times n}$ whose trailing 2-by-2 principal submatrix has eigenvalues $a_1$ and $a_2$, this algorithm overwrites $H$ with $Z^T H Z$, where $Z = P_1 \cdots P_{n-2}$ is a product of Householder matrices and $Z^T(H - a_1 I)(H - a_2 I)$ is upper triangular.

```
m = n - 1
{Compute first column of (H - a_1 I)(H - a_2 I).}
s = H(m,m) + H(n,n)
t = H(m,m)H(n,n) - H(m,n)H(n,m)
x = H(1,1)H(1,1) + H(1,2)H(2,1) - sH(1,1) + t
y = H(2,1)(H(1,1) + H(2,2) - s)
z = H(2,1)H(3,2)
for k = 0:n - 3
    {Overwrite H with P_k H P_k^T where P_k = diag(I_k, P̄_k, I_{n-k-3})}
    v = house([x y z]^T)
    H(k + 1:k + 3, k + 1:n) = row.house(H(k + 1:k + 3, k + 1:n), v)
    r = min{k + 4, n}
    H(1:r, k + 1:k + 3) = col.house(H(1:r, k + 1:k + 3), v)
    x = H(k + 2, k + 1)
    y = H(k + 3, k + 1)
    if k < n - 3
        z = H(k + 4, k + 1)
    end
end
{H = P_{n-2} H P_{n-2}^T where P_{n-2} = diag(I_{n-2}, P̄_{n-2})}
v = house([x y]^T)
H(n - 1:n, n - 2:n) = row.house(H(n - 1:n, n - 2:n), v)
H(1:n, n - 1, n) = col.house(H(1:n, n - 1, n), v)
```

This algorithm requires $10n^2$ flops. If $Z$ is accumulated into a given orthogonal matrix, an additional $10n^2$ flops are necessary.

### 7.5.6 The Overall Process

Reducing $A$ to Hessenberg form using Algorithm 7.4.1 and then iterating with Algorithm 7.5.1 to produce the real Schur form is the standard means by which the dense unsymmetric eigenproblem is solved. During the iteration it is necessary to monitor the subdiagonal elements in $H$ in order to spot any possible decoupling. How this is done is illustrated in the following algorithm:

**Algorithm 7.5.2 (QR Algorithm)** Given $A \in \mathbb{R}^{n\times n}$ and a tolerance *tol* greater than the unit roundoff, this algorithm computes the real Schur canonical form $Q^TAQ = T$. $A$ is overwritten with the Hessenberg decomposition. If $Q$ and $T$ are desired then $T$ is stored in $H$. If only the eigenvalues are desired then diagonal blocks in $T$ are stored in the corresponding positions in $H$.

Use Algorithm 7.4.2 to compute the Hessenberg reduction
  $H = U_0^T A U_0$ where $U_0 = P_0 \cdots P_{n-2}$.
If $Q$ is desired form $Q = P_0 \cdots P_{n-2}$. See§5.1.6.
**until** $q = n$
  Set to zero all subdiagonal elements that satisfy:
    $|h_{i,i-1}| \leq tol(|h_{ii}| + |h_{i-1,i-1}|)$.
  Find the largest non-negative $q$ and the smallest
    non-negative $p$ such that

$$H = \begin{bmatrix} H_{11} & H_{12} & H_{13} \\ 0 & H_{22} & H_{23} \\ 0 & 0 & H_{33} \end{bmatrix} \begin{matrix} p \\ n-p-q \\ q \end{matrix}$$
$$\qquad\quad p \qquad n-p-q \qquad q$$

    where $H_{33}$ is upper quasi-triangular and $H_{22}$ is
    unreduced. (Note: either $p$ or $q$ may be zero.)
  **if** $q < n$
    Perform a Francis QR step on $H_{22}$: $H_{22} = Z^T H_{22} Z$
    **if** $Q$ is desired
      $Q = Q\text{diag}(I_p, Z, I_q)$
      $H_{12} = H_{12} Z$
      $H_{23} = Z^T H_{23}$
    **end**
  **end**
**end**
Upper triangularize all 2-by-2 diagonal blocks in $H$ that have
  real eigenvalues and accumulate the transformations
  if necessary.

This algorithm requires $25n^3$ flops if $Q$ and $T$ are computed. If only the eigenvalues are desired, then $10n^3$ flops are necessary. These flops counts are very approximate and are based on the empirical observation that on average only two Francis iterations are required before the lower 1-by-1 or 2-by-2 decouples. A more detailed look at the time required to execute the QR algorithm is given in Eispack, p. 119.

**Example 7.5.3** If Algorithm 7.5.2 is applied to

$$A = H = \begin{bmatrix} 2 & 3 & 4 & 5 & 6 \\ 4 & 4 & 5 & 6 & 7 \\ 0 & 3 & 6 & 7 & 8 \\ 0 & 0 & 2 & 8 & 9 \\ 0 & 0 & 0 & 1 & 10 \end{bmatrix}$$

then the subdiagonal entries converge as follows

| Iteration | $O(\|h_{21}\|)$ | $O(\|h_{32}\|)$ | $O(\|h_{43}\|)$ | $O(\|h_{54}\|)$ |
|---|---|---|---|---|
| 1 | $10^0$ | $10^0$ | $10^0$ | $10^0$ |
| 2 | $10^0$ | $10^0$ | $10^0$ | $10^0$ |
| 3 | $10^0$ | $10^0$ | $10^{-1}$ | $10^0$ |
| 4 | $10^0$ | $10^0$ | $10^{-3}$ | $10^{-3}$ |
| 5 | $10^0$ | $10^0$ | $10^{-6}$ | $10^{-5}$ |
| 6 | $10^{-1}$ | $10^0$ | $10^{-13}$ | $10^{-13}$ |
| 7 | $10^{-1}$ | $10^0$ | $10^{-28}$ | $10^{-13}$ |
| 8 | $10^{-4}$ | $10^0$ | converg. | converg. |
| 9 | $10^{-8}$ | $10^0$ | | |
| 10 | $10^{-8}$ | $10^0$ | | |
| 11 | $10^{-16}$ | $10^0$ | | |
| 12 | $10^{-32}$ | $10^0$ | | |
| 13 | converg. | converg. | | |

The roundoff properties of the QR algorithm are what one would expect of any orthogonal matrix technique. The computed real Schur form $\hat{T}$ is orthogonally similar to a matrix near to $A$, i.e.,

$$Q^T(A+E)Q = \hat{T}$$

where $Q^TQ = I$ and $\| E \|_2 \approx \mathbf{u}\| A \|_2$. The computed $\hat{Q}$ is almost orthogonal in the sense that $\hat{Q}^T\hat{Q} = I + F$ where $\| F \|_2 \approx \mathbf{u}$.

The order of the eigenvalues along $\hat{T}$ is somewhat arbitrary. But as we discuss in §7.6, any ordering can be achieved by using a simple procedure for swappping two adjacent diagonal entries.

### 7.5.7 Balancing

Finally, we mention that if the elements of $A$ have widely varying magnitudes, then $A$ should be *balanced* before applying the QR algorithm. This is an $O(n^2)$ calculation in which a diagonal matrix $D$ is computed so that if

$$D^{-1}AD = [\, c_1, \ldots, c_n \,] = \begin{bmatrix} r_1^T \\ \vdots \\ r_n^T \end{bmatrix}$$

then $\| r_i \|_\infty \approx \| c_i \|_\infty$ for $i = 1{:}n$. The diagonal matrix $D$ is chosen to have the form $D = \text{diag}(\beta^{i_1}, \ldots, \beta^{i_n})$ where $\beta$ is the floating point base. Note that $D^{-1}AD$ can be calculated without roundoff. When $A$ is balanced, the computed eigenvalues are often more accuarate. See Parlett and Reinsch (1969).

**Problems**

**P7.5.1** Show that if $\bar{H} = Q^THQ$ is obtained by performing a single-shift QR step with $\bar{H} = \begin{bmatrix} w & x \\ y & z \end{bmatrix}$, then $|\bar{h}_{21}| \leq |y^2x|[(w-z)^2 + y^2]$.

**P7.5.2** Give a formula for the 2-by-2 diagonal matrix $D$ that minimizes $\| D^{-1}AD \|_F$ where $A = \begin{bmatrix} w & x \\ y & z \end{bmatrix}$.

**P7.5.3** Explain how the single-shift QR step $H - \mu I = UR$, $\bar{H} = RU + \mu I$ can be carried out implicitly. That is, show how the transition from $\bar{H}$ to $H$ can be carried out without subtracting the shift $\mu$ from the diagonal of $H$.

**P7.5.4** Suppose $H$ is upper Hessenberg and that we compute the factorization $PH = LU$ via Gaussian elimination with partial pivoting. (See Algorithm 4.3.4.) Show that $H = U(P^T L)$ is upper Hessenberg and similar to $H$. (This is the basis of the *modified LR algorithm* .)

**P7.5.5** Show that if $H = H_0$ is given and we generate the matrices $H_k$ via $H_k - \mu_k I = U_k R_k$, $H_{k+1} = R_k U_k + \mu_k I$, then

$$(U_1 \cdots U_j)(R_j \cdots R_1) = (H - \mu_1 I) \cdots (H - \mu_j I).$$

**Notes and References for Sec. 7.5**

The development of the practical $QR$ algorithm began with the important paper

H. Rutishauser (1958). "Solution of Eigenvalue Problems with the LR Transformation," *Nat. Bur. Stand. App. Math. Ser. 49,* 47–81.

The algorithm described here was then "orthogonalized" in

J.G.F. Francis (1961). "The QR Transformation: A Unitary Analogue to the LR Transformation, Parts I and II" *Comp. J. 4,* 265–72, 332–45.

Descriptions of the practical QR algorithm may be found in Wilkinson (AEP, chapter 8) and Stewart (IMC, chapter 7). Algol procedures for LR and QR methods are given in

R.S. Martin, G. Peters, and J.H. Wilkinson (1970). "The QR Algorithm for Real Hessenberg Matrices," *Numer. Math. 14,* 219–31. See also *HACLA,* pp. 359–71.
R.S. Martin and J.H. Wilkinson (1968). "The Modified LR Algorithm for Complex Hessenberg Matrices," *Numer. Math. 12,* 369–76. See also *HACLA,* pp. 396–403.

Their Fortran equivalents are in Eispack.
Aspects of the balancing problem are discussed in

E.E. Osborne (1960). "On Preconditioning of Matrices," *JACM 7,* 338–45.
B.N. Parlett and C. Reinsch (1969). "Balancing a Matrix for Calculation of Eigenvalues and Eigenvectors," *Numer. Math. 13,* 292–304. See also *HACLA,* pp. 315-26.

## 7.6 Invariant Subspace Computations

Several important invariant subspace problems can be solved once the real Schur decomposition $Q^T AQ = T$ has been computed. In this section we discuss how to

- compute the eigenvectors associated with some subset of $\lambda(A)$,
- compute an orthonormal basis for a given invariant subspace,

- block-diagonalize $A$ using well-conditioned similarity transformations,
- compute a basis of eigenvectors regardless of their condition, and
- compute an approximate Jordan canonical form of $A$.

Eigenvector/invariant subspace computation for sparse matrices is discussed elsewhere. See §7.3 as well as portions of Chapters 8 and 9.

### 7.6.1 Selected Eigenvectors via Inverse Iteration

Let $q^{(0)} \in \mathbb{C}^n$ be a given unit 2-norm vector and assume that $A - \mu I \in \mathbb{R}^{n\times n}$ is nonsingular. The following is referred to as *inverse iteration*:

$$\begin{array}{l} \textbf{for } k = 1,2,\ldots \\ \quad \text{Solve } (A-\mu I)z^{(k)} = q^{(k-1)} \\ \quad q^{(k)} = z^{(k)}/\| z^{(k)} \|_2 \\ \quad \lambda^{(k)} = q^{(k)^H} A q^{(k)} \\ \textbf{end} \end{array} \tag{7.6.1}$$

Inverse iteration is just the power method applied to $(A-\mu I)^{-1}$.

To analyze the behavior of (7.6.1), assume that $A$ has a basis of eigenvectors $\{x_1,\ldots,x_n\}$ and that $Ax_i = \lambda_i x_i$ for $i = 1{:}n$. If

$$q^{(0)} = \sum_{i=1}^{n} \beta_i x_i$$

then $q^{(k)}$ is a unit vector in the direction of

$$(A-\mu I)^{-k} q^{(0)} = \sum_{i=1}^{n} \frac{\beta_i}{(\lambda_i-\mu)^k} x_i \,.$$

Clearly, if $\mu$ is much closer to an eigenvalue $\lambda_j$ than to the other eigenvalues, then $q^{(k)}$ is rich in the direction of $x_j$ provided $\beta_j \neq 0$.

A sample stopping criterion for (7.6.1) might be to quit as soon as the residual

$$r^{(k)} = (A-\mu I)q^{(k)}$$

satisfies

$$\| r^{(k)} \|_\infty \le c\mathbf{u}\| A \|_\infty \tag{7.6.2}$$

where $c$ is a constant of order unity. Since

$$(A+E_k)q^{(k)} = \mu q^{(k)}$$

with $E_k = -r^{(k)}q^{(k)^T}$, it follows that (7.6.2) forces $\mu$ and $q^{(k)}$ to be an exact eigenpair for a nearby matrix.

Inverse iteration can be used in conjunction with the QR algorithm as follows:

- Compute the Hessenberg decomposition $U_0^T A U_0 = H$.

- Apply the double implicit shift Francis iteration to $H$ *without* accumulating transformations.

- For each computed eigenvalue $\lambda$ whose corresponding eigenvector $x$ is sought, apply (7.6.1) with $A = H$ and $\mu = \lambda$ to produce a vector $z$ such that $Hz \approx \mu z$.

- Set $x = U_0 z$.

Inverse iteration with $H$ is very economical because (1) we do not have to accumulate transformations during the double Francis iteration; (2) we can factor matrices of the form $H - \lambda I$ in $O(n^2)$ flops, and (3) only one iteration is typically required to produce an adequate approximate eigenvector.

This last point is perhaps the most interesting aspect of inverse iteration and requires some justification since $\lambda$ can be comparatively inaccurate if it is ill-conditioned. Assume for simplicity that $\lambda$ is real and let

$$H - \lambda I \;=\; \sum_{i=1}^{n} \sigma_i u_i v_i^T \;=\; U \Sigma V^T$$

be the SVD of $H - \lambda I$. From what we said about the roundoff properties of the QR algorithm in §7.5.6, there exists a matrix $E \in \mathbb{R}^{n \times n}$ such that $H + E - \lambda I$ is singular and $\| E \|_2 \approx \mathbf{u} \| H \|_2$. It follows that $\sigma_n \approx \mathbf{u}\sigma_1$ and $\| (H - \hat{\lambda} I) v_n \|_2 \approx \mathbf{u}\sigma_1$, i.e., $v_n$ is a good approximate eigenvector. Clearly if the starting vector $q^{(0)}$ has the expansion

$$q^{(0)} \;=\; \sum_{i=1}^{n} \gamma_i u_i$$

then

$$z^{(1)} \;=\; \sum_{i=1}^{n} \frac{\gamma_i}{\sigma_i} v_i$$

is "rich" in the direction $v_n$. Note that if $s(\lambda) \approx |u_n^T v_n|$ is small, then $z^{(1)}$ is rather deficient in the direction $u_n$. This explains (heuristically) why another step of inverse iteration is not likely to produce an improved eigenvector approximate, especially if $\lambda$ is ill-conditioned. For more details, see Peters and Wilkinson (1979).

**Example 7.6.1** The matrix

$$A \;=\; \begin{bmatrix} 1 & 1 \\ 10^{-10} & 1 \end{bmatrix}$$

has eigenvalues $\lambda_1 = .99999$ and $\lambda_2 = 1.00001$ and corresponding eigenvectors $x_1 = (1, -10^{-5})^T$ and $x_2 = (1, 10^{-5})^T$. The condition of both eigenvalues is of order $10^5$. The approximate eigenvalue $\mu = 1$ is an exact eigenvalue of $A + E$ where

$$E = \begin{bmatrix} 0 & 0 \\ -10^{-10} & 0 \end{bmatrix}.$$

Thus, the quality of $\mu$ is typical of the quality of an eigenvalue produced by the $QR$ algorithm when executed in 10-digit floating point.

If (7.6.1) is applied with starting vector $q^{(0)} = (0, 1)^T$, then $q^{(1)} = (1, 0)^T$ and $\| Aq^{(1)} - \mu q^{(1)} \|_2 = 10^{-10}$. However, one more step produces $q^{(2)} = (0, 1)^T$ for which $\| Aq^{(2)} - \mu q^{(2)} \|_2 = 1$. This example is discussed in Peters and Wilkinson (1979).

### 7.6.2 Ordering Eigenvalues in the Real Schur Form

Recall that the real Schur decomposition provides information about invariant subspaces. If

$$Q^T AQ = T = \begin{bmatrix} T_{11} & T_{12} \\ 0 & T_{22} \end{bmatrix} \begin{matrix} p \\ q \end{matrix}$$
$$\qquad\qquad\qquad\quad p \qquad q$$

and $\lambda(T_{11}) \cap \lambda(T_{22}) = \emptyset$, then the first $p$ columns of $Q$ span the unique invariant subspace associated with $\lambda(T_{11})$. (See §7.1.4.) Unfortunately, the Francis iteration supplies us with a real Schur decomposition $Q_F^T AQ_F = T_F$ in which the eigenvalues appear somewhat randomly along the diagonal of $T_F$. This poses a problem if we want an orthonormal basis for an invariant subspace whose associated eigenvalues are not at the top of $T_F$'s diagonal. Clearly, we need a method for computing an orthogonal matrix $Q_D$ such that $Q_D^T T_F Q_D$ is upper quasi-triangular with appropriate eigenvalue ordering.

A look at the 2-by-2 case suggests how this can be accomplished. Suppose

$$Q_F^T AQ_F = T_F = \begin{bmatrix} \lambda_1 & t_{12} \\ 0 & \lambda_2 \end{bmatrix} \qquad \lambda_1 \neq \lambda_2$$

and that we wish to reverse the order of the eigenvalues. Note that $T_F x = \lambda_2 x$ where

$$x = \begin{bmatrix} t_{12} \\ \lambda_2 - \lambda_1 \end{bmatrix}.$$

Let $Q_D$ be a Givens rotation such that the second component of $Q_D^T x$ is zero. If $Q = Q_F Q_D$ then

$$(Q^T AQ)e_1 = Q_D^T T_F(Q_D e_1) = \lambda_2 Q_D^T(Q_D e_1) = \lambda_2 e_1$$

and so $Q^T AQ$ must have the form

$$Q^T AQ = \begin{bmatrix} \lambda_2 & \pm t_{12} \\ 0 & \lambda_1 \end{bmatrix}.$$

By systematically interchanging adjacent pairs of eigenvalues using this technique, we can move any subset of $\lambda(A)$ to the top of $T$'s diagonal assuming that no 2-by-2 bumps are encountered along the way.

**Algorithm 7.6.1** Given an orthogonal matrix $Q \in \mathbb{R}^{n\times n}$, an upper triangular matrix $T = Q^TAQ$, and a subset $\Delta = \{\lambda_1, \ldots, \lambda_p\}$ of $\lambda(A)$, the following algorithm computes an orthogonal matrix $Q_D$ such that $Q_D^TTQ_D = S$ is upper triangular and $\{s_{11}, \ldots, s_{pp}\} = \Delta$. The matrices $Q$ and $T$ are overwritten by $QQ_D$ and $S$ respectively.

```
while {t_11, ..., t_pp} ≠ Δ
    for k = 1:n − 1
        if t_kk ∉ Δ and t_k+1,k+1 ∈ Δ
            { Find Q_D and update: T = Q_D^T T Q_D, Q = Q Q_D }
            [c, s] = givens(T(k, k + 1), T(k + 1, k + 1) − T(k, k))
            T(k:k + 1, k:n) = row.rot(T(k:k + 1, k:n), c, s)
            T(1:k + 1, k:k + 1) = col.rot(T(1:k + 1, k:k + 1), c, s)
            Q(1:n, k:k + 1) = col.rot(Q(1:n, k:k + 1), c, s)
        end
    end
end
```

This algorithm requires $k(12n)$ flops, where $k$ is the total number of required swaps. The integer $k$ is never greater than $(n-p)p$.

The swapping gets a little more complicated when $T$ has 2-by-2 blocks along its diagonal. See Ruhe (1970a) and Stewart (1976a) for details. Of course, these interchanging techniques can be used to sort the eigenvalues, say from maximum to minimum modulus.

Computing invariant subspaces by manipulating the real Schur decomposition is extremely stable. If $\hat{Q} = [\,\hat{q}_1, \ldots, \hat{q}_n\,]$ denotes the computed orthogonal matrix $Q$, then $\|\,\hat{Q}^T\hat{Q} - I\,\|_2 \approx \mathbf{u}$ and there exists a matrix $E$ satisfying $\|\,E\,\|_2 \approx \mathbf{u}\|\,A\,\|_2$ such that $(A+E)\hat{q}_i \in \text{span}\{\hat{q}_1, \ldots, \hat{q}_p\}$ for $i = 1{:}p$.

### 7.6.3 Block Diagonalization

Let

$$T = \begin{bmatrix} T_{11} & T_{12} & \cdots & T_{1q} \\ 0 & T_{22} & \cdots & T_{2q} \\ \vdots & \vdots & \ddots & \vdots \\ 0 & 0 & \cdots & T_{qq} \end{bmatrix} \begin{matrix} n_1 \\ n_2 \\ \\ n_q \end{matrix} \qquad (7.6.3)$$
$$\qquad\qquad n_1 \quad n_2 \qquad\quad n_q$$

be a partitioning of some real Schur canonical form $Q^TAQ = T \in \mathbb{R}^{n\times n}$ such that $\lambda(T_{11}), \ldots, \lambda(T_{qq})$ are disjoint. By Theorem 7.1.6 there exists a matrix $Y$ such that $Y^{-1}TY = \text{diag}(T_{11}, \ldots, T_{qq})$. A practical procedure for determining $Y$ is now given together with an analysis of $Y$'s sensitivity as a function of the above partitioning.

Partition $I_n = [\, E_1, \ldots, E_q \,]$ conformably with $T$ and define the matrix $Y_{ij} \in \mathbb{R}^{n \times n}$ as follows:

$$Y_{ij} = I_n + E_i Z_{ij} E_j^T. \qquad i < j,\ Z_{ij} \in \mathbb{R}^{n_i \times n_j}$$

In other words, $Y_{ij}$ looks just like the identity except that $Z_{ij}$ occupies the $(i,j)$ block position. It follows that if $Y_{ij}^{-1}TY_{ij} = \bar{T} = (\bar{T}_{ij})$ then $T$ and $\bar{T}$ are identical except that

$$\begin{array}{ll} \bar{T}_{ij} = T_{ii}Z_{ij} - Z_{ij}T_{jj} + T_{ij} & \\ \bar{T}_{ik} = T_{ik} - Z_{ij}T_{jk} & (k = j+1{:}q) \\ \bar{T}_{kj} = T_{ki}Z_{ij} + T_{kj} & (k = 1{:}i-1) \end{array}$$

Thus, $T_{ij}$ can be zeroed provided we have an algorithm for solving the *Sylvester equation*

$$FZ - ZG = C \tag{7.6.4}$$

where $F \in \mathbb{R}^{p \times p}$ and $G \in \mathbb{R}^{r \times r}$ are given upper quasi-triangular matrices and $C \in \mathbb{R}^{p \times r}$.

Bartels and Stewart (1972) have devised a method for doing this. Let $C = [\, c_1, \ldots, c_r \,]$ and $Z = [\, z_1, \ldots, z_r \,]$ be column partitionings. If $g_{k+1,k} = 0$, then by comparing columns in (7.6.4) we find

$$Fz_k - \sum_{i=1}^{k} g_{ik} z_i = c_k .$$

Thus, once we know $z_1, \ldots, z_{k-1}$ then we can solve the quasi-triangular system

$$(F - g_{kk}I)\, z_k = c_k + \sum_{i=1}^{k-1} g_{ik} z_i$$

for $z_k$. If $g_{k+1,k} \neq 0$, then $z_k$ and $z_{k+1}$ can be simultaneously found by solving the $2p$-by-$2p$ system

$$\begin{bmatrix} F - g_{kk}I & -g_{mk}I \\ -g_{km} & F - g_{mm}I \end{bmatrix} \begin{bmatrix} z_k \\ z_m \end{bmatrix} = \begin{bmatrix} c_k \\ c_m \end{bmatrix} + \sum_{i=1}^{k-1} \begin{bmatrix} g_{ik} z_i \\ g_{im} z_i \end{bmatrix} \tag{7.6.5}$$

where $m = k+1$. By reordering the equations according to the permutation $(1, p+1, 2, p+2, \ldots, p, 2p)$, a banded system is obtained that can be solved in $O(p^2)$ flops. The details may be found in Bartels and Stewart (1972). We summarize the overall process for the simple case when $F$ and $G$ are both triangular.

**Algorithm 7.6.2 (Bartels-Stewart Algorithm)** Given $C \in \mathbb{R}^{p\times r}$ and upper triangular matrices $F \in \mathbb{R}^{p\times p}$ and $G \in \mathbb{R}^{r\times r}$ that satisfy $\lambda(F) \cap \lambda(G) = \emptyset$, the following algorithm overwrites $C$ with the solution to the equation $FZ - ZG = C$.

```
for k = 1:r
    { Compute b = c_k + Σ_{i=1}^{k-1} g_{ik} z_i }
    C(1:p, k) = C(1:p, k) + C(1:p, 1:k − 1)G(1:k − 1, k)
    Solve (F − g_{kk} I)z = b for z.
    C(1:p, k) = z
end
```

This algorithm requires $pr(p+r)$ flops.

Clearly, by zeroing the super diagonal blocks in $T$ in the appropriate order, we can reduce the entire matrix to block diagonal form.

**Algorithm 7.6.3** Given an orthogonal matrix $Q \in \mathbb{R}^{n\times n}$, an upper quasi-triangular matrix $T = Q^T A Q$, and the partitioning (7.6.3), the following algorithm overwrites $Q$ with $QY$ where $Y^{-1}TY = \text{diag}(T_{11}, \ldots, T_{qq})$.

```
for j = 2:q
    for i = 1:j − 1
        Solve T_{ii}Z − ZT_{jj} = −T_{ij} for Z using Algorithm 7.6.2.
        for k = j + 1:q
            T_{ik} = T_{ik} − ZT_{jk}
        end
        for k = 1:q
            Q_{kj} = Q_{ki}Z + Q_{kj}
        end
    end
end
```

The number of flops required by this algorithm is a complicated function of the block sizes in (7.6.3).

The choice of the real Schur form $T$ and its partitioning in (7.6.3) determines the sensitivity of the Sylvester equations that must be solved in Algorithm 7.6.3. This in turn affects the condition of the matrix $Y$ and the overall usefulness of the block diagonalization. The reason for these dependencies is that the relative error of the computed solution $\hat{Z}$ to

$$T_{ii}Z - ZT_{jj} = -T_{ij} \qquad (7.6.6)$$

satisfies

$$\frac{\| \hat{Z} - Z \|_F}{\| Z \|_F} \approx \mathbf{u} \frac{\| T \|_F}{\text{sep}(T_{ii}, T_{jj})} .$$

For details, see Golub, Nash, and Van Loan (1979). Since

$$\text{sep}(T_{ii}, T_{jj}) = \min_{X \neq 0} \frac{\| T_{ii}X - XT_{jj} \|_F}{\| X \|_F} \leq \min_{\substack{\lambda \in \lambda(T_{ii}) \\ \mu \in \lambda(T_{jj})}} |\lambda - \mu| .$$

there can be a substantial loss of accuracy whenever the subsets $\lambda(T_{ii})$ are insufficiently separated. Moreover, if $Z$ satisfies (7.6.6) then

$$\| Z \|_F \leq \frac{\| T_{ij} \|_F}{\text{sep}(T_{ii}, T_{jj})} .$$

Thus, large-norm solutions can be expected if $\text{sep}(T_{ii}, T_{jj})$ is small. This tends to make the matrix $Y$ in Algorithm 7.6.3 ill-conditioned since it is the product of the matrices

$$Y_{ij} = \begin{bmatrix} I & Z \\ 0 & I \end{bmatrix} .$$

Note: $\kappa_F(Y_{ij}) = 2n + \| Z \|_F^2$.

Confronted with these difficulties, Bavely and Stewart (1979) have developed an algorithm for block diagonalizing that dynamically determines the eigenvalue ordering and partitioning in (7.6.3) so that all the $Z$ matrices in Algorithm 7.6.3 are bounded in norm by some user-supplied tolerance. They find that the condition of $Y$ can be controlled by controlling the condition of the $Y_{ij}$.

### 7.6.4 Eigenvector Bases

If the blocks in the partitioning (7.6.3) are all 1-by-1, then Algorithm 7.6.3 produces a basis of eigenvectors. As with the method of inverse iteration, the computed eigenvalue-eigenvector pairs are exact for some "nearby" matrix. A widely followed rule of thumb for deciding upon a suitable eigenvector method is to use inverse iteration whenever fewer than 25% of the eigenvectors are desired.

We point out, however, that the real Schur form can be used to determine selected eigenvectors. Suppose

$$Q^T A Q = \begin{bmatrix} T_{11} & u & T_{13} \\ 0 & \lambda & v^T \\ 0 & 0 & T_{33} \end{bmatrix} \begin{matrix} k-1 \\ 1 \\ n-k \end{matrix}$$

$$\begin{matrix} \quad k-1 & 1 & n-k \end{matrix}$$

is upper quasi-triangular and that $\lambda \notin \lambda(T_{11}) \cup \lambda(T_{33})$. It follows that if we solve the linear systems $(T_{11} - \lambda I)w = -u$ and $(T_{33} - \lambda I)^T z = -v$ then

$$x = Q\begin{bmatrix} w \\ 1 \\ 0 \end{bmatrix} \quad \text{and} \quad y = Q\begin{bmatrix} 0 \\ 1 \\ z \end{bmatrix}$$

are the associated right and left eigenvectors, respectively. Note that the condition of $\lambda$ is prescribed by $1/s(\lambda) = \sqrt{(1 + w^T w)(1 + z^T z)}$.

### 7.6.5 Ascertaining Jordan Block Structures

Suppose that we have computed the real Schur decomposition $A = QTQ^T$, identified clusters of "equal" eigenvalues, and calculated the corresponding block diagonalization $T = Y\text{diag}(T_{11}, \ldots, T_{qq})Y^{-1}$. As we have seen, this can be a formidable task. However, even greater numerical problems confront us if we attempt to ascertain the Jordan block structure of each $T_{ii}$. A brief examination of these difficulties will serve to highlight the limitations of the Jordan decomposition.

Assume for clarity that $\lambda(T_{ii})$ is real. The reduction of $T_{ii}$ to Jordan form begins by replacing it with a matrix of the form $C = \lambda I + N$, where $N$ is the strictly upper triangular portion of $T_{ii}$ and where $\lambda$, say, is the mean of its eigenvalues.

Recall that the dimension of a Jordan block $J(\lambda)$ is the smallest nonnegative integer $k$ for which $[J(\lambda) - \lambda I]^k = 0$. Thus, if $p_i = \dim[\text{null}(N^i)]$, for $i = 0{:}n$, then $p_i - p_{i-1}$ equals the number of blocks in $C$'s Jordan form that have dimension $i$ or greater. A concrete example helps to make this assertion clear and to illustrate the role of the SVD in Jordan form computations.

Assume that $C$ is 7-by-7. Suppose we compute the SVD $U_1^T N V_1 = \Sigma_1$ and "discover" that $N$ has rank 3. If we order the singular values from small to large then it follows that the matrix $N_1 = V_1^T N V_1$ has the form

$$N_1 = \begin{bmatrix} 0 & K \\ 0 & L \end{bmatrix} \begin{matrix} 4 \\ 3 \end{matrix}$$
$$\qquad\quad \begin{matrix} 4 & 3 \end{matrix}$$

At this point, we know that the geometric multiplicity of $\lambda$ is 4—i.e, $C$'s Jordan form has 4 blocks ($p_1 - p_0 = 4 - 0 = 4$).

Now suppose $\tilde{U}_2^T L \tilde{V}_2 = \Sigma_2$ is the SVD of $L$ and that we find that $L$ has unit rank. If we again order the singular values from small to large, then $L_2 = \tilde{V}_2^T L \tilde{V}_2$ clearly has the following structure:

$$L_2 = \begin{bmatrix} 0 & 0 & a \\ 0 & 0 & b \\ 0 & 0 & c \end{bmatrix}$$

However $\lambda(L_2) = \lambda(L) = \{0, 0, 0\}$ and so $c = 0$. Thus, if

$$V_2 = \text{diag}(I_4, \tilde{V}_2)$$

then $N_2 = V_2^T N_1 V_2$ has the following form:

$$N_2 = \begin{bmatrix} 0 & 0 & 0 & 0 & \times & \times & \times \\ 0 & 0 & 0 & 0 & \times & \times & \times \\ 0 & 0 & 0 & 0 & \times & \times & \times \\ 0 & 0 & 0 & 0 & \times & \times & \times \\ 0 & 0 & 0 & 0 & 0 & 0 & a \\ 0 & 0 & 0 & 0 & 0 & 0 & b \\ 0 & 0 & 0 & 0 & 0 & 0 & 0 \end{bmatrix}$$

Besides allowing us to introduce more zeroes into the upper triangle, the SVD of $L$ also enables us to deduce the dimension of the null space of $N^2$. Since

$$N_1^2 = \begin{bmatrix} 0 & KL \\ 0 & L^2 \end{bmatrix} = \begin{bmatrix} 0 & K \\ 0 & L \end{bmatrix} \begin{bmatrix} 0 & K \\ 0 & L \end{bmatrix}$$

and $\begin{bmatrix} K \\ L \end{bmatrix}$ has full column rank,

$$p_2 = \dim(\text{null}(N^2)) = \dim(\text{null}(N_1^2)) = 4 + \dim(\text{null}(L)) = p_1 + 2.$$

Hence, we can conclude at this stage that the Jordan form of $C$ has at least two blocks of dimension 2 or greater.

Finally, it is easy to see that $N_1^3 = 0$, from which we conclude that there is $p_3 - p_2 = 7 - 6 = 1$ block of dimension 3 or larger. If we define $V = V_1 V_2$ then it follows that the decomposition

$$V^T C V = \begin{bmatrix} \lambda & 0 & 0 & 0 & \times & \times & \times \\ 0 & \lambda & 0 & 0 & \times & \times & \times \\ 0 & 0 & \lambda & 0 & \times & \times & \times \\ 0 & 0 & 0 & \lambda & \times & \times & \times \\ 0 & 0 & 0 & 0 & \lambda & \times & a \\ 0 & 0 & 0 & 0 & 0 & \lambda & 0 \\ 0 & 0 & 0 & 0 & 0 & 0 & \lambda \end{bmatrix} \begin{matrix} \left.\begin{matrix} \\ \\ \\ \\ \end{matrix}\right\} \text{4 blocks of order 1 or larger} \\ \left.\begin{matrix} \\ \\ \end{matrix}\right\} \text{2 blocks of order 2 or larger} \\ \} \text{ 1 block of order 3 or larger} \end{matrix}$$

"displays" $C$'s Jordan block structure: 2 blocks of order 1, 1 block of order 2, and 1 block of order 3.

To compute the Jordan decomposition it is necessary to resort to non-orthogonal transformations. We refer the reader to either Golub and Wilkinson (1976) or Kagstrom and Ruhe (1980a, 1980b) for how to proceed with this phase of the reduction.

The above calculations with the SVD amply illustrate that difficult rank decisions must be made at each stage and that the final computed block structure depends critically on those decisions. Fortunately, the stable Schur decomposition can almost always be used in lieu of the Jordan decomposition in practical applications.

### Problems

**P7.6.1** Give a complete algorithm for solving a real, $n$-by-$n$, upper quasi-triangular system $Tx = b$.

**P7.6.2** Suppose $U^{-1}AU = \text{diag}(\alpha_1, \ldots, \alpha_m)$ and $V^{-1}BV = \text{diag}(\beta_1, \ldots, \beta_n)$. Show that if $\phi(X) = AX + XB$, then $\lambda(\phi) = \{\alpha_i + \beta_j : i = 1{:}m,\ j = 1{:}n\}$. What are the corresponding eigenvectors? How can these decompositions be used to solve $AX + XB = C$?

**P7.6.3** Show that if $Y = \begin{bmatrix} I & Z \\ 0 & I \end{bmatrix}$ then $\kappa_2(Y) = [2 + \sigma^2 + \sqrt{4\sigma^2 + \sigma^4}\,]/2$ where $\sigma = \| Z \|_2$.

**P7.6.4** Derive the system (7.6.5).

**P7.6.5** Assume that $T \in \mathbb{R}^{n \times n}$ is block upper triangular and partitioned as follows:

$$T = \begin{bmatrix} T_{11} & T_{12} & T_{13} \\ 0 & T_{22} & T_{23} \\ 0 & 0 & T_{33} \end{bmatrix} \qquad T \in \mathbb{R}^{n \times n}$$

Suppose that the diagonal block $T_{22}$ is 2-by-2 with complex eigenvalues that are disjoint from $\lambda(T_{11})$ and $\lambda(T_{33})$. Give an algorithm for computing the 2-dimensional real invariant subspace associated with $T_{22}$'s eigenvalues.

### Notes and References for Sec. 7.6

Much of the material discussed in this section may be found in the survey paper

G.H. Golub and J.H. Wilkinson (1976). "Ill-Conditioned Eigensystems and the Computation of the Jordan Canonical Form," *SIAM Review 18*, 578–619.

Papers that specifically analyze the method of inverse iteration for computing eigenvectors include

G. Peters and J.H. Wilkinson (1979). "Inverse Iteration, Ill-Conditioned Equations, and Newton's Method," *SIAM Review 21*, 339–60.

J. Varah (1968a). "The Calculation of the Eigenvectors of a General Complex Matrix by Inverse Iteration," *Math. Comp. 22*, 785–91.

J. Varah (1968b). "Rigorous Machine Bounds for the Eigensystem of a General Complex Matrix," *Math. Comp. 22*, 793–801.

J. Varah (1970). "Computing Invariant Subspaces of a General Matrix When the Eigensystem is Poorly Determined," *Math. Comp. 24*, 137–49.

The Algol version of the Eispack inverse iteration subroutine is given in

G. Peters and J.H. Wilkinson (1971). "The Calculation of Specified Eigenvectors by Inverse Iteration," in *HACLA*, pp. 418–39.

The problem of ordering the eigenvalues in the real Schur form is the subject of

A. Ruhe (1970a). "An Algorithm for Numerical Determination of the Structure of a General Matrix," *BIT 10*, 196–216.

G.W. Stewart (1976a). "Algorithm 406: HQR3 and EXCHNG: Fortran Subroutines for Calculating and Ordering the Eigenvalues of a Real Upper Hessenberg Matrix," *ACM Trans. Math. Soft. 2,* 275–80.

Fortran programs for computing block diagonalizations and Jordan forms are described in

C. Bavely and G.W. Stewart (1979). "An Algorithm for Computing Reducing Subspaces by Block Diagonalization," *SIAM J. Num. Anal. 16,* 359–67.
J.W. Demmel (1983a). "A Numerical Analyst's Jordan Canonical Form," Ph.D. Thesis, Berkeley.
B. Kågström and A. Ruhe (1980a). "An Algorithm for Numerical Computation of the Jordan Normal Form of a Complex Matrix," *ACM Trans. Math. Soft. 6,* 398–419.
B. Kågström and A. Ruhe (1980b). "Algorithm 560 JNF: An Algorithm for Numerical Computation of the Jordan Normal Form of a Complex Matrix," *ACM Trans. Math. Soft. 6,* 437–43.

Papers that are concerned with estimating the roundoff error in a computed eigenvalue and/or eigenvector include

S.P. Chan and B.N. Parlett (1977). "Algorithm 517: A Program for Computing the Condition Numbers of Matrix Eigenvalues Without Computing Eigenvectors," *ACM Trans. Math. Soft. 3,* 186–203.
H.J. Symm and J.H. Wilkinson (1980). "Realistic Error Bounds for a Simple Eigenvalue and Its Associated Eigenvector," *Numer. Math. 35,* 113–26.

As we have seen, the sep(.,.) function is of great importance in the assessment of a computed invariant subspace. Aspects of this quantity are discussed in

J. Varah (1979). "On the Separation of Two Matrices," *SIAM J. Num. Anal. 16,* 212–22.

Numerous algorithms have been proposed for the Sylvester equation $FX - XG = C$, but those described in

R.H. Bartels and G.W. Stewart (1972). "Solution of the Equation $AX + XB = C$," *Comm. ACM 15,* 820–26.
G.H. Golub, S. Nash, and C. Van Loan (1979). "A Hessenberg-Schur Method for the Matrix Problem $AX + XB = C$," *IEEE Trans. Auto. Cont. AC-24,* 909–13.

are among the more reliable in that they rely on orthogonal transformations. The Lyapunov problem $FX + XF^T = -C$ where $C$ is non-negative definite has a very important role to play in control theory. See

S. Barnett and C. Storey (1968). "Some Applications of the Lyapunov Matrix Equation," *J. Inst. Math. Applic. 4,* 33–42.

Several authors have considered generalizations of the Sylvester equation, i.e., $\Sigma F_i X G_i = C$. These include

P. Lancaster (1970). "Explicit Solution of Linear Matrix Equations," *SIAM Review 12,* 544–66.
W.J. Vetter (1975). "Vector Structures and Solutions of Linear Matrix Equations," *Lin. Alg. and Its Applic. 10,* 181–88.
H. Wimmer and A.D. Ziebur (1972). "Solving the Matrix Equations $\Sigma f_p(A) g_p(A) = C$," *SIAM Review 14,* 318–23.

Some ideas about improving a computed eigenvalues, eigenvectors, and invariant subspaces may be found in

J.W. Demmel (1987). "Three Methods for Refining Estimates of Invariant Subspaces," *Computing 38,* 43–57.
J.J. Dongarra, C.B. Moler, and J.H. Wilkinson (1983). "Improving the Accuracy of Computed Eigenvalues and Eigenvectors," *SIAM J. Numer. Anal. 20,* 23–46.

Cheap condition estimation techniques have also been developed for the eigenproblem. See

R. Byers (1984). "A Linpack-Style Condition Estimator for the Equation $AX - XB^T = C$," *IEEE Trans. Auto. Cont. AC-29,* 926–928.
C. Van Loan (1987). "On Estimating the Condition of Eigenvalues and Eigenvectors," *Lin. Alg. and Its Applic.* 88/89, 715–732.

## 7.7 The QZ Method for Ax = λBx

Let $A$ and $B$ be two $n$-by-$n$ matrices. The set of all matrices of the form $A - \lambda B$ with $\lambda \in \mathbb{C}$ is said to be a *pencil.* The eigenvalues of the pencil are elements of the set $\lambda(A, B)$ defined by

$$\lambda(A,B) \;=\; \{z \in \mathbb{C} : \det(A - zB) = 0\,\}.$$

If $\lambda \in \lambda(A, B)$ and

$$Ax \;=\; \lambda Bx \qquad x \neq 0 \tag{7.7.1}$$

then $x$ is referred to as an eigenvector of $A - \lambda B$.

In this section we briefly survey some of the mathematical properties of the generalized eigenproblem (7.7.1) and present a stable method for its solution. The important case when $A$ and $B$ are symmetric with the latter positive definite is discussed in §8.7.2.

### 7.7.1 Background

The first thing to observe about the generalized eigenvalue problem is that there are $n$ eigenvalues if and only if $\text{rank}(B) = n$. If $B$ is rank deficient then $\lambda(A, B)$ may be finite, empty, or infinite:

$$A = \begin{bmatrix} 1 & 2 \\ 0 & 3 \end{bmatrix} \quad B = \begin{bmatrix} 1 & 0 \\ 0 & 0 \end{bmatrix} \quad \Rightarrow \quad \lambda(A,B) = \{1\}$$

$$A = \begin{bmatrix} 1 & 2 \\ 0 & 3 \end{bmatrix} \quad B = \begin{bmatrix} 0 & 0 \\ 1 & 0 \end{bmatrix} \quad \Rightarrow \quad \lambda(A,B) = \emptyset$$

$$A = \begin{bmatrix} 1 & 2 \\ 0 & 0 \end{bmatrix} \quad B = \begin{bmatrix} 1 & 0 \\ 0 & 0 \end{bmatrix} \quad \Rightarrow \quad \lambda(A,B) = \mathbb{C}$$

Note that if $0 \neq \lambda \in \lambda(A, B)$ then $(1/\lambda) \in \lambda(B, A)$. Moreover, if $B$ is nonsingular then $\lambda(A, B) = \lambda(B^{-1}A, I) = \lambda(B^{-1}A)$.

This last observation suggests one method for solving the $A - \lambda B$ problem when $B$ is nonsingular:

- Solve $BC = A$ for $C$ using (say) Gaussian eliminaton with pivoting.
- Use the QR algorithm to compute the eigenvalues of $C$.

Note that $C$ will be affected roundoff errors of order $\mathbf{u}\| A \|_2 \| B^{-1} \|_2$. If $B$ is ill-conditioned, then this can rule out the possibility of computing any generalized eigenvalue accurately—even those eigenvalues that may be regarded as well-conditioned.

**Example 7.7.1** If

$$A = \begin{bmatrix} 1.746 & .940 \\ 1.246 & 1.898 \end{bmatrix} \quad \text{and} \quad B = \begin{bmatrix} .780 & .563 \\ .913 & .659 \end{bmatrix}$$

then $\lambda(A, B) = 2,\ 1.07 \times 10^6$. With 7-digit floating point arithmetic, we find $\lambda(fl(AB^{-1})) = \{1.562539,\ 1.01 \times 10^6\}$. The poor quality of the small eigenvalue is because $\kappa_2(B) \approx 2 \times 10^6$. On the other hand, we find that

$$\lambda(I, fl(A^{-1}B)) \approx \{2.000001,\ 1.06 \times 10^6\}.$$

The accuracy of the small eigenvalue is improved because $\kappa_2(A) \approx 4$.

Example 7.7.1 suggests that we seek an alternative approach to the $A - \lambda B$ problem. One idea is to compute well-conditioned $Q$ and $Z$ such that the matrices

$$A_1 = Q^{-1}AZ \qquad B_1 = Q^{-1}BZ \tag{7.7.2}$$

are each in canonical form. Note that $\lambda(A, B) = \lambda(A_1, B_1)$ since

$$Ax = \lambda Bx \quad \Leftrightarrow \quad A_1 y = \lambda B_1 y \qquad x = Zy$$

We say that the pencils $A - \lambda B$ and $A_1 - \lambda B_1$ are *equivalent* if (7.7.2) holds with nonsingular $Q$ and $Z$.

## 7.7.2 The Generalized Schur Decomposition

As in the standard eigenproblem $A - \lambda I$ there is a choice between canonical forms. Analogous to the Jordan form is a decomposition of Kronecker in which both $A_1$ and $B_1$ are block diagonal. The blocks are similar to Jordan blocks. The Kronecker canonical form poses the same numerical difficulties as the Jordan form. However, this decomposition does provide insight into the mathematical properties of the pencil $A - \lambda B$. See Wilkinson (1978) and Demmel and Kågström (1987) for details.

More attractive from the numerical point of view is the following decomposition described in Moler and Stewart (1973).

**Theorem 7.7.1 (Generalized Schur Decomposition)** *If $A$ and $B$ are in $\mathbb{C}^{n\times n}$, then there exist unitary $Q$ and $Z$ such that $Q^HAZ = T$ and $Q^HBZ = S$ are upper triangular. If for some $k$, $t_{kk}$ and $s_{kk}$ are both zero, then $\lambda(A,B) = \mathbb{C}$. Otherwise*

$$\lambda(A,B) = \{t_{ii}/s_{ii} : s_{ii} \neq 0\}.$$

**Proof.** Let $\{B_k\}$ be a sequence of nonsingular matrices that converge to $B$. For each $k$, let $Q_k^H(AB_k^{-1})Q_k = R_k$ be a Schur decomposition of $AB_k^{-1}$. Let $Z_k$ be unitary such that $Z_k^H(B_k^{-1}Q_k) \equiv S_k^{-1}$ is upper triangular. It follows that both $Q_k^HAZ_k = R_kS_k$ and $Q_k^HB_kZ_k = S_k$ are also upper triangular.

Using the Bolzano-Weierstrass theorem, we know that the bounded sequence $\{(Q_k, Z_k)\}$ has a converging subsequence, $\lim(Q_{k_i}, Z_{k_i}) = (Q, Z)$. It is easy to show that $Q$ and $Z$ are unitary and that $Q^HAZ$ and $Q^HBZ$ are upper triangular. The assertions about $\lambda(A,B)$ follow from the identity

$$\det(A - \lambda B) = \det(QZ^H)\prod_{i=1}^{n}(t_{ii} - \lambda s_{ii}). \ \square$$

If $A$ and $B$ are real then the following decomposition, which corresponds to the real schur decomposition (Theorem 7.4.1), is of interest.

**Theorem 7.7.2 (Generalized Real Schur Decomposition)** *If $A$ and $B$ are in $\mathbb{R}^{n\times n}$ then there exist orthogonal matrices $Q$ and $Z$ such that $Q^TAZ$ is upper quasi-triangular and $Q^TBZ$ is upper triangular.*

**Proof.** See Stewart (1972). □

In the remainder of this section we are concerned with the computation of this decomposition and the mathematical insight that it provides.

### 7.7.3 Sensitivity Issues

The generalized Schur decomposition sheds light on the issue of eigenvalue sensitivity for the $A - \lambda B$ problem. Clearly, small changes in $A$ and $B$ can induce large changes in the eigenvalue $\lambda_i = t_{ii}/s_{ii}$ if $s_{ii}$ is small. However, as Stewart (1978) argues, it may not be appropriate to regard such an eigenvalue as "ill-conditioned." The reason is that the reciprocal $\mu_i = s_{ii}/t_{ii}$ might be a very well behaved eigenvalue for the pencil $\mu A - B$. In the Stewart analysis, $A$ and $B$ are treated symmetrically and the eigenvalues are regarded more as ordered pairs $(t_{ii}, s_{ii})$ than as quotients. With this point of view it becomes appropriate to measure eigenvalue perturbations in the *chordal metric* chord $(a,b)$ defined by

$$\text{chord}(a,b) = \frac{|a-b|}{\sqrt{1+a^2}\sqrt{1+b^2}}.$$

Stewart shows that if $\lambda$ is a distinct eigenvalue of $A - \lambda B$ and $\lambda_\epsilon$ is the corresponding eigenvalue of the perturbed pencil $\tilde{A} - \lambda\tilde{B}$ with $\| A - \tilde{A} \|_2 \approx \| B - \tilde{B} \|_2 \approx \epsilon$, then

$$\text{chord}(\lambda, \lambda_\epsilon) \leq \frac{\epsilon}{(y^H Ax)^2 + (y^H Bx)^2} + O(\epsilon^2)$$

where $x$ and $y$ have unit 2-norm and satisfy $Ax = \lambda Bx$ and $y^H = \lambda y^H B$. Note that the denominator in the upper bound is symmetric in $A$ and $B$. The "truly" ill-conditioned eigenvalues are those for which this denominator is small.

The extreme case when $t_{kk} = s_{kk} = 0$ for some $k$ has been studied by Wilkinson (1979). He makes the interesting observation that when this occurs, the remaining quotients $t_{ii}/s_{ii}$ can assume arbitrary values.

### 7.7.4 Hessenberg-Triangular Form

The first step in computing the generalized Schur decomposition of the pair $(A, B)$ is to reduce $A$ to upper Hessenberg form and $B$ to upper triangular form via orthogonal transformations. We first determine an orthogonal $U$ such that $U^T B$ is upper triangular. Of course, to preserve eigenvalues, we must also update $A$ in exactly the same way. Let's trace what happens in the $n = 5$ case.

$$A = U^T A = \begin{bmatrix} \times & \times & \times & \times & \times \\ \times & \times & \times & \times & \times \\ \times & \times & \times & \times & \times \\ \times & \times & \times & \times & \times \\ \times & \times & \times & \times & \times \end{bmatrix}, B = U^T B = \begin{bmatrix} \times & \times & \times & \times & \times \\ 0 & \times & \times & \times & \times \\ 0 & 0 & \times & \times & \times \\ 0 & 0 & 0 & \times & \times \\ 0 & 0 & 0 & 0 & \times \end{bmatrix}$$

Next, we reduce $A$ to upper Hessenberg form while preserving $B$'s upper triangular form. First, a Givens rotation $Q_{45}$ is determined to zero $a_{51}$:

$$A = Q_{45}^T A = \begin{bmatrix} \times & \times & \times & \times & \times \\ \times & \times & \times & \times & \times \\ \times & \times & \times & \times & \times \\ \times & \times & \times & \times & \times \\ 0 & \times & \times & \times & \times \end{bmatrix}, B = Q_{45}^T B = \begin{bmatrix} \times & \times & \times & \times & \times \\ 0 & \times & \times & \times & \times \\ 0 & 0 & \times & \times & \times \\ 0 & 0 & 0 & \times & \times \\ 0 & 0 & 0 & \times & \times \end{bmatrix}$$

The nonzero entry arising in the (5,4) position in $B$ can be zeroed by postmultiplying with an appropriate Givens rotation $Z_{45}$ :

$$A = AZ_{45} = \begin{bmatrix} \times & \times & \times & \times & \times \\ \times & \times & \times & \times & \times \\ \times & \times & \times & \times & \times \\ \times & \times & \times & \times & \times \\ 0 & \times & \times & \times & \times \end{bmatrix}, B = BZ_{45} = \begin{bmatrix} \times & \times & \times & \times & \times \\ 0 & \times & \times & \times & \times \\ 0 & 0 & \times & \times & \times \\ 0 & 0 & 0 & \times & \times \\ 0 & 0 & 0 & 0 & \times \end{bmatrix}$$

Zeros are similarly introduced into the (4, 1) and (3, 1) positions in $A$:

$$A = Q_{34}^T A = \begin{bmatrix} \times & \times & \times & \times & \times \\ \times & \times & \times & \times & \times \\ \times & \times & \times & \times & \times \\ 0 & \times & \times & \times & \times \\ 0 & \times & \times & \times & \times \end{bmatrix}, B = Q_{34}^T B = \begin{bmatrix} \times & \times & \times & \times & \times \\ 0 & \times & \times & \times & \times \\ 0 & 0 & \times & \times & \times \\ 0 & 0 & \times & \times & \times \\ 0 & 0 & 0 & 0 & \times \end{bmatrix}$$

$$A = AZ_{34} = \begin{bmatrix} \times & \times & \times & \times & \times \\ \times & \times & \times & \times & \times \\ \times & \times & \times & \times & \times \\ 0 & \times & \times & \times & \times \\ 0 & \times & \times & \times & \times \end{bmatrix}, B = BZ_{34} = \begin{bmatrix} \times & \times & \times & \times & \times \\ 0 & \times & \times & \times & \times \\ 0 & 0 & \times & \times & \times \\ 0 & 0 & 0 & \times & \times \\ 0 & 0 & 0 & 0 & \times \end{bmatrix}$$

$$A = Q_{23}^T A = \begin{bmatrix} \times & \times & \times & \times & \times \\ \times & \times & \times & \times & \times \\ 0 & \times & \times & \times & \times \\ 0 & \times & \times & \times & \times \\ 0 & \times & \times & \times & \times \end{bmatrix}, B = Q_{23}^T B = \begin{bmatrix} \times & \times & \times & \times & \times \\ 0 & \times & \times & \times & \times \\ 0 & \times & \times & \times & \times \\ 0 & 0 & 0 & \times & \times \\ 0 & 0 & 0 & \times & \times \end{bmatrix}$$

$$A = AZ_{23} = \begin{bmatrix} \times & \times & \times & \times & \times \\ \times & \times & \times & \times & \times \\ 0 & \times & \times & \times & \times \\ 0 & \times & \times & \times & \times \\ 0 & \times & \times & \times & \times \end{bmatrix}, B = BZ_{23} = \begin{bmatrix} \times & \times & \times & \times & \times \\ 0 & \times & \times & \times & \times \\ 0 & 0 & \times & \times & \times \\ 0 & 0 & 0 & \times & \times \\ 0 & 0 & 0 & 0 & \times \end{bmatrix}$$

$A$ is now upper Hessenberg through its first column. The reduction is completed by zeroing $a_{52}$, $a_{42}$, and $a_{53}$. As is evident above, two orthogonal transformations are required for each $a_{ij}$ that is zeroed—one to do the zeroing and the other to restore $B$'s triangularity. Either Givens rotations or 2-by-2 modified Householder transformations can be used. Overall we have:

**Algorithm 7.7.1 (Hessenberg-Triangular Reduction)** Given $A$ and $B$ in $\mathbb{R}^{n \times n}$, the following algorithm overwrites $A$ with an upper Hessenberg matrix $Q^T AZ$ and $B$ with an upper triangular matrix $Q^T BZ$ where both $Q$ and $Z$ are orthogonal.

```
Using Algorithm 5.2.1, overwrite B with Q^T B = R where
      Q is orthogonal and R is upper triangular.
A = Q^T A
for j = 1:n - 2
      for i = n: - 1:j + 2
            { Zero a_ij.}
            [c, s] = givens(A(i - 1, j), A(i, j))
            A(i - 1:i, j:n) = row.rot(A(i - 1:i, j:n), c, s)
            B(i - 1:i, i - 1:n) = row.rot(B(i - 1:i, i - 1:n), c, s)
            [c, s] = givens(B(i, i - 1), B(i, i))
            B(1:i, i - 1:i) = col.rot(B(1:i, i - 1:i), c, s)
            A(1:n, i - 1:i) = col.rot(A(1:n, i - 1:i), c, s)
      end
end
```

This algorithm requires about $8n^3$ flops. The accumulation of $Q$ and $Z$ requires about $4n^3$ and $3n^3$ flops, respectively.

The reduction of $A - \lambda B$ to Hessenberg-triangular form serves as a "front end" decomposition for a generalized QR iteration known as the QZ iteration which we describe next.

**Example 7.7.3** If

$$A = \begin{bmatrix} 10 & 1 & 2 \\ 1 & 2 & -1 \\ 1 & 1 & 2 \end{bmatrix} \quad \text{and} \quad B = \begin{bmatrix} 1 & 2 & 3 \\ 4 & 5 & 6 \\ 7 & 8 & 9 \end{bmatrix}$$

and orthogonal matrices $Q$ and $Z$ are defined by

$$Q = \begin{bmatrix} -.1231 & -.9917 & .0378 \\ -.4924 & .0279 & -.8699 \\ -.8616 & .1257 & .4917 \end{bmatrix} \quad \text{and} \quad Z = \begin{bmatrix} 1.0000 & 0.0000 & 0.0000 \\ 0.0000 & -.8944 & -.4472 \\ 0.0000 & .4472 & -.8944 \end{bmatrix}$$

then $A_1 = Q^T AZ$ and $B_1 = Q^T BZ$ are given by

$$A_1 = \begin{bmatrix} -2.5849 & 1.5413 & 2.4221 \\ -9.7631 & .0874 & 1.9239 \\ 0.0000 & 2.7233 & -.7612 \end{bmatrix} \quad \text{and} \quad B_1 = \begin{bmatrix} -8.1240 & 3.6332 & 14.2024 \\ 0.0000 & 0.0000 & 1.8739 \\ 0.0000 & 0.0000 & .7612 \end{bmatrix}$$

## 7.7.5 Deflation

In describing the QZ iteration we may assume without loss of generality that $A$ is an unreduced upper Hessenberg matrix and that $B$ is a nonsingular upper triangular matrix. The first of these assertions is obvious, for if $a_{k+1,k} = 0$ then

$$A - \lambda B = \begin{bmatrix} A_{11} - \lambda B_{11} & A_{12} - \lambda B_{12} \\ 0 & A_{22} - \lambda B_{22} \end{bmatrix} \begin{matrix} k \\ n-k \end{matrix}$$
$$\qquad\qquad\qquad k \qquad\qquad n-k$$

and we may proceed to solve the two smaller problems $A_{11} - \lambda B_{11}$ and $A_{22} - \lambda B_{22}$. On the other hand, if $b_{kk} = 0$ for some $k$, then it is possible to introduce a zero in $A$'s $(n, n-1)$ position and thereby deflate. Illustrating by example, suppose $n = 5$ and $k = 3$:

$$A = \begin{bmatrix} \times & \times & \times & \times & \times \\ \times & \times & \times & \times & \times \\ 0 & \times & \times & \times & \times \\ 0 & 0 & \times & \times & \times \\ 0 & 0 & 0 & \times & \times \end{bmatrix}, \qquad B = \begin{bmatrix} \times & \times & \times & \times & \times \\ 0 & \times & \times & \times & \times \\ 0 & 0 & 0 & \times & \times \\ 0 & 0 & 0 & \times & \times \\ 0 & 0 & 0 & 0 & \times \end{bmatrix}$$

The zero on $B$'s diagonal can be "pushed down" to the (5,5) position as follows using Givens rotations:

$$A = Q_{34}^T A = \begin{bmatrix} \times & \times & \times & \times & \times \\ \times & \times & \times & \times & \times \\ 0 & \times & \times & \times & \times \\ 0 & \times & \times & \times & \times \\ 0 & 0 & 0 & \times & \times \end{bmatrix}, B = Q_{34}^T B = \begin{bmatrix} \times & \times & \times & \times & \times \\ 0 & \times & \times & \times & \times \\ 0 & 0 & 0 & \times & \times \\ 0 & 0 & 0 & 0 & \times \\ 0 & 0 & 0 & 0 & \times \end{bmatrix}$$

$$A = AZ_{23} = \begin{bmatrix} \times & \times & \times & \times & \times \\ \times & \times & \times & \times & \times \\ 0 & \times & \times & \times & \times \\ 0 & 0 & \times & \times & \times \\ 0 & 0 & 0 & \times & \times \end{bmatrix}, B = BZ_{23} = \begin{bmatrix} \times & \times & \times & \times & \times \\ 0 & \times & \times & \times & \times \\ 0 & 0 & \times & \times & \times \\ 0 & 0 & 0 & 0 & \times \\ 0 & 0 & 0 & 0 & \times \end{bmatrix}$$

$$A = Q_{45}^T A = \begin{bmatrix} \times & \times & \times & \times & \times \\ \times & \times & \times & \times & \times \\ 0 & \times & \times & \times & \times \\ 0 & 0 & \times & \times & \times \\ 0 & 0 & \times & \times & \times \end{bmatrix}, B = Q_{45}^T B = \begin{bmatrix} \times & \times & \times & \times & \times \\ 0 & \times & \times & \times & \times \\ 0 & 0 & \times & \times & \times \\ 0 & 0 & 0 & 0 & \times \\ 0 & 0 & 0 & 0 & 0 \end{bmatrix}$$

$$A = AZ_{34} = \begin{bmatrix} \times & \times & \times & \times & \times \\ \times & \times & \times & \times & \times \\ 0 & \times & \times & \times & \times \\ 0 & 0 & \times & \times & \times \\ 0 & 0 & 0 & \times & \times \end{bmatrix}, B = BZ_{34} = \begin{bmatrix} \times & \times & \times & \times & \times \\ 0 & \times & \times & \times & \times \\ 0 & 0 & \times & \times & \times \\ 0 & 0 & 0 & 0 & \times \\ 0 & 0 & 0 & 0 & 0 \end{bmatrix}$$

$$A = AZ_{45} = \begin{bmatrix} \times & \times & \times & \times & \times \\ \times & \times & \times & \times & \times \\ 0 & \times & \times & \times & \times \\ 0 & 0 & \times & \times & \times \\ 0 & 0 & 0 & 0 & \times \end{bmatrix}, B = BZ_{45} = \begin{bmatrix} \times & \times & \times & \times & \times \\ 0 & \times & \times & \times & \times \\ 0 & 0 & \times & \times & \times \\ 0 & 0 & 0 & \times & \times \\ 0 & 0 & 0 & 0 & 0 \end{bmatrix}$$

This zero-chasing technique is perfectly general and can be used to zero $a_{n,n-1}$ regardless of where the zero appears along $B$'s diagonal.

### 7.7.6 The QZ Step

We are now in a position to describe a QZ step. The basic idea is to update $A$ and $B$ as follows

$$(\bar{A} - \lambda \bar{B}) \;=\; \bar{Q}^T (A - \lambda B)\bar{Z},$$

where $\bar{A}$ is upper Hessenberg, $\bar{B}$ is upper triangular, $\bar{Q}$ and $\bar{Z}$ are each orthogonal, and $\bar{A}\bar{B}^{-1}$ is essentially the same matrix that would result if a Francis QR step (Algorithm 7.5.2) were explicitly applied to $AB^{-1}$. This can be done with some clever zero-chasing and an appeal to the implicit $Q$ theorem.

Let $M = AB^{-1}$ (upper Hessenberg) and let $v$ be the first column of the matrix $(M - aI)(M - bI)$, where $a$ and $b$ are the eigenvalues of $M$'s lower 2-by-2 submatrix. Note that $v$ can be calculated in $O(1)$ flops. If $P_0$ is a Householder matrix such that $P_0 v$ is a multiple of $e_1$, then

$$A \;=\; P_0 A \;=\; \begin{bmatrix} \times & \times & \times & \times & \times & \times \\ \times & \times & \times & \times & \times & \times \\ \times & \times & \times & \times & \times & \times \\ 0 & \times & \times & \times & \times & \times \\ 0 & 0 & 0 & \times & \times & \times \\ 0 & 0 & 0 & 0 & \times & \times \end{bmatrix}$$

$$B \;=\; P_0 B \;=\; \begin{bmatrix} \times & \times & \times & \times & \times & \times \\ \times & \times & \times & \times & \times & \times \\ \times & \times & \times & \times & \times & \times \\ 0 & 0 & 0 & \times & \times & \times \\ 0 & 0 & 0 & 0 & \times & \times \\ 0 & 0 & 0 & 0 & 0 & \times \end{bmatrix}$$

The idea now is to restore these matrices to Hessenberg-triangular form by chasing the unwanted nonzero elements down the diagonal.

To this end, we first determine a pair of Householder matrices $Z_1$ and $Z_2$ to zero $b_{31}$, $b_{32}$, and $b_{21}$:

$$A \;=\; AZ_1 Z_2 \;=\; \begin{bmatrix} \times & \times & \times & \times & \times & \times \\ \times & \times & \times & \times & \times & \times \\ \times & \times & \times & \times & \times & \times \\ \times & \times & \times & \times & \times & \times \\ 0 & 0 & 0 & \times & \times & \times \\ 0 & 0 & 0 & 0 & \times & \times \end{bmatrix}$$

$$B = BZ_1Z_2 = \begin{bmatrix} \times & \times & \times & \times & \times & \times \\ 0 & \times & \times & \times & \times & \times \\ 0 & 0 & \times & \times & \times & \times \\ 0 & 0 & 0 & \times & \times & \times \\ 0 & 0 & 0 & 0 & \times & \times \\ 0 & 0 & 0 & 0 & 0 & \times \end{bmatrix}$$

Then a Householder matrix $P_1$ is used to zero $a_{31}$ and $a_{41}$:

$$A = P_1A = \begin{bmatrix} \times & \times & \times & \times & \times & \times \\ \times & \times & \times & \times & \times & \times \\ 0 & \times & \times & \times & \times & \times \\ 0 & \times & \times & \times & \times & \times \\ 0 & 0 & 0 & \times & \times & \times \\ 0 & 0 & 0 & 0 & \times & \times \end{bmatrix}$$

$$B = P_1B = \begin{bmatrix} \times & \times & \times & \times & \times & \times \\ 0 & \times & \times & \times & \times & \times \\ 0 & \times & \times & \times & \times & \times \\ 0 & \times & \times & \times & \times & \times \\ 0 & 0 & 0 & 0 & \times & \times \\ 0 & 0 & 0 & 0 & 0 & \times \end{bmatrix}$$

Notice that with this step the unwanted nonzero elements have been shifted down and to the right from their original position. This illustrates a typical step in the QZ iteration. Notice that $Q = Q_0Q_1 \cdots Q_{n-2}$ has the same first column as $Q_0$. By the way the initial Householder matrix was determined, we can apply the implicit $Q$ theorem and assert that $AB^{-1} = Q^T(AB^{-1})Q$ is indeed essentially the same matrix that we would obtain by applying the Francis iteration to $M = AB^{-1}$ directly. Overall we have:

**Algorithm 7.7.2 (The QZ Step)** Given an unreduced upper Hessenberg matrix $A \in \mathbb{R}^{n \times n}$ and a nonsingular upper triangular matrix $B \in \mathbb{R}^{n \times n}$, the following algorithm overwrites $A$ with the upper Hessenberg matrix $Q^TAZ$ and $B$ with the upper triangular matrix $Q^TBZ$ where $Q$ and $Z$ are orthogonal and $Q$ has the same first column as the orthogonal similarity transformation in Algorithm 7.5.1 when it is applied to $AB^{-1}$.

Let $M = AB^{-1}$ and compute $(M - aI)(M - bI)e_1 = (x, y, z, 0, \ldots, 0)^T$
    where $a$ and $b$ are the eigenvalues of $M$'s lower 2-by-2.
**for** $k = 1{:}n - 2$
    Find Householder $Q_k$ so $Q_k[\, x\ y\ z\,]^T = [\, *\ 0\ 0\,]^T$.
    $A = \text{diag}(I_{k-1}, Q_k, I_{n-k-2})A$
    $B = \text{diag}(I_{k-1}, Q_k, I_{n-k-2})B$
    Find Householder $Z_{k1}$ so
        $\begin{bmatrix} b_{k+2,k} & b_{k+2,k+1} & b_{k+2,k+2} \end{bmatrix} Z_{k1} = \begin{bmatrix} 0 & 0 & * \end{bmatrix}$.
    $A = A\text{diag}(I_{k-1}, Z_{k1}, I_{n-k-2})$
    $B = B\text{diag}(I_{k-1}, Z_{k1}, I_{n-k-2})$
    Find Householder $Z_{k2}$ so
        $\begin{bmatrix} b_{k+1,k} & b_{k+1,k+1} \end{bmatrix} Z_{k2} = \begin{bmatrix} 0 & * \end{bmatrix}$.
    $A = A\text{diag}(I_{k-1}, Z_{k2}, I_{n-k-1})$
    $B = B\text{diag}(I_{k-1}, Z_{k2}, I_{n-k-1})$
    $x = a_{k+1,k}$; $y = a_{k+1,k}$
    **if** $k < n - 2$
        $z = a_{k+3,k}$
    **end**
**end**
Find Householder $Q_{n-1}$ so $Q_{n-1}\begin{bmatrix} x \\ y \end{bmatrix} = \begin{bmatrix} * \\ 0 \end{bmatrix}$
$A = \text{diag}(I_{n-2}, Q_{n-1})A$
$B = \text{diag}(I_{n-2}, Q_{n-1})B$
Find Householder $Z_{n-1}$ so
    $\begin{bmatrix} b_{n,n-1} & b_{nn} \end{bmatrix} Z_{n-1} = \begin{bmatrix} 0 & * \end{bmatrix}$
$A = A\text{diag}(I_{n-2}, Z_{n-1})$
$B = B\text{diag}(I_{n-2}, Z_{n-1})$

This algorithm requires $22n^2$ flops. $Q$ and $Z$ can be accumulated for an additional $8n^2$ flops and $13n^2$ flops, respectively.

### 7.7.7 The Overall QZ Process

By applying a sequence of QZ steps to the Hessenberg-triangular pencil $A - \lambda B$, it is possible to reduce $A$ to quasi-triangular form. In doing this it is necessary to monitor $A$'s subdiagonal and $B$'s diagonal in order to bring about decoupling whenever possible. The complete process, due to Moler and Stewart (1973), is as follows:

**Algorithm 7.7.3** Given $A \in \mathbb{R}^{n \times n}$ and $B \in \mathbb{R}^{n \times n}$, the following algorithm computes orthogonal $Q$ and $Z$ such that $Q^T AZ = T$ is upper quasi-triangular and $Q^T BZ = S$ is upper triangular. $A$ is overwritten by $T$ and $B$ by $S$.

Using Algorithm 7.7.1, overwrite $A$ with $Q^TAZ$ (upper Hessenberg)
and $B$ with $Q^TBZ$ (upper triangular).
**until** $q = n$
Set all subdiagonal elements in $A$ to zero that satisfy
$|a_{i,i-1}| \leq \epsilon(|a_{i-1,i-1}| + |a_{ii}|)$
Find the largest nonnegative $q$ and the smallest nonnegative $p$
such that if

$$H = \begin{bmatrix} H_{11} & H_{12} & H_{13} \\ 0 & H_{22} & H_{23} \\ 0 & 0 & H_{33} \end{bmatrix} \begin{matrix} p \\ n-p-q \\ q \end{matrix}$$
$$\begin{matrix} p & \quad n-p-q & \quad q \end{matrix}$$

then $A_{33}$ is upper quasi-triangular and $A_{22}$ is unreduced
upper Hessenberg.
Partition $B$ conformably:

$$B = \begin{bmatrix} B_{11} & B_{12} & B_{13} \\ 0 & B_{22} & B_{23} \\ 0 & 0 & B_{33} \end{bmatrix} \begin{matrix} p \\ n-p-q \\ q \end{matrix}$$
$$\begin{matrix} p & \quad n-p-q & \quad q \end{matrix}$$

**if** $q < n$
**if** $B_{22}$ is singular
Zero $a_{n-q,n-q-1}$
**else**
Apply Algorithm 7.7.2 to $A_{22}$ and $B_{22}$
$A = \text{diag}(I_p, Q, I_q)^T A\text{diag}(I_p, Q, I_q)$
$B = \text{diag}(I_p, Q, I_q)^T B\text{diag}(I_p, Q, I_q)$
**end**
**end**
**end**

This algorithm requires $30n^3$ flops. If $Q$ is desired, an additional $16n^3$ are necessary. If $Z$ is required, an additional $20n^3$ are needed. These estimates of work are based on the experience that about two QZ iterations per eigenvalue are necessary. Thus, the convergence properties of QZ are the same as for QR. The speed of the QZ algorithm is not affected by rank deficiency in $B$.

The computed $S$ and $T$ can be shown to satisfy

$$Q_0^T(A+E)Z_0 = T \qquad Q_0^T(B+F)Z_0 = S$$

where $Q_0$ and $Z_0$ are exactly orthogonal and $\| E \|_2 \approx \mathbf{u}\| A \|_2$ and $\| F \|_2 \approx \mathbf{u}\| B \|_2$.

**Example 7.7.5** If the $QZ$ algorithm is applied to

$$A = \begin{bmatrix} 2 & 3 & 4 & 5 & 6 \\ 4 & 4 & 5 & 6 & 7 \\ 0 & 3 & 6 & 7 & 8 \\ 0 & 0 & 2 & 8 & 9 \\ 0 & 0 & 0 & 1 & 10 \end{bmatrix} \quad \text{and} \quad B = \begin{bmatrix} 1 & -1 & -1 & -1 & -1 \\ 0 & 1 & -1 & -1 & -1 \\ 0 & 0 & 1 & -1 & -1 \\ 0 & 0 & 0 & 1 & -1 \\ 0 & 0 & 0 & 0 & 1 \end{bmatrix}$$

then the subdiagonal elements of $A$ converge as follows

| Iteration | $O(\|h_{21}\|)$ | $O(\|h_{32}\|)$ | $O(\|h_{43}\|)$ | $O(\|h_{54}\|)$ |
|---|---|---|---|---|
| 1 | $10_0$ | $10^1$ | $10^0$ | $10^{-1}$ |
| 2 | $10^0$ | $10^0$ | $10^0$ | $10^{-1}$ |
| 3 | $10^0$ | $10^1$ | $10^{-1}$ | $10^{-3}$ |
| 4 | $10^0$ | $10^0$ | $10^{-1}$ | $10^{-8}$ |
| 5 | $10^0$ | $10^1$ | $10^{-1}$ | $10^{-16}$ |
| 6 | $10^0$ | $10^0$ | $10^{-2}$ | converg. |
| 7 | $10^0$ | $10^{-1}$ | $10^{-4}$ | |
| 8 | $10^1$ | $10^{-1}$ | $10^{-8}$ | |
| 9 | $10^0$ | $10^{-1}$ | $10^{-19}$ | |
| 10 | $10^0$ | $10^{-2}$ | converg. | |
| 11 | $10^{-1}$ | $10^{-4}$ | | |
| 12 | $10^{-2}$ | $10^{-11}$ | | |
| 13 | $10^{-3}$ | $10^{-27}$ | | |
| 14 | converg. | converg. | | |

### 7.7.8 Generalized Invariant Subspace Computations

Many of the invariant subspace computations discussed in §7.6 carry over to the generalized eigenvalue problem. For example, approximate eigenvectors can be found via inverse iteration:

$q^{(0)} \in \mathbb{C}^{n \times n}$ given.
**for** $k = 1, 2, \ldots$
  Solve $(A - \mu B)z^{(k)} = Bq^{(k-1)}$
  Normalize: $q^{(k)} = z^{(k)} / \| z^{(k)} \|_2$
  $\lambda^{(k)} = [q^{(k)}]^H A q^{(k)} \; / \; [q^{(k)}]^H A q^{(k)}$
**end**

When $B$ is nonsingular, this is equivalent to applying (7.6.1) with the matrix $B^{-1}A$. Typically, only a single iteration is required if $\mu$ is an approximate eigenvalue computed by the QZ algorithm. By inverse iterating with the Hessenberg-triangular pencil, costly accumulation of the $Z$-transformations during the QZ iteration can be avoided.

Corresponding to the notion of an invariant subspace for a single matrix, we have the notion of a *deflating* subspace for the pencil $A - \lambda B$. In particular, we say that a $k$-dimensional subspace $S \subseteq \mathbb{R}^n$ is "deflating" for the pencil $A - \lambda B$ if the subspace $\{ Ax + By : x, y \in S \}$ has dimension $k$ or less. Note that the columns of the matrix $Z$ in the generalized Schur decomposition define a family of deflating subspaces, for if $Q = [\, q_1, \ldots, q_n \,]$ and $Z = [\, z_1, \ldots, z_n \,]$ then we have $\text{span}\{Az_1, \ldots, Az_k\} \subseteq \text{span}\{q_1, \ldots, q_k\}$ and $\text{span}\{Bz_1, \ldots, Bz_k\} \subseteq \text{span}\{q_1, \ldots, q_k\}$. Properties of deflating subspaces and their behavior under perturbation are described in Stewart (1972).

### Problems

**P7.7.1** Suppose $A$ and $B$ are in $\mathbb{R}^{n \times n}$ and that

$$U^T B V = \begin{bmatrix} D & 0 \\ 0 & 0 \end{bmatrix} \begin{matrix} r \\ n-r \end{matrix} \qquad U = [\, U_1 \quad U_2 \,] \qquad V = [\, V_1 \quad V_2 \,]$$

(with column blocks of widths $r$ and $n-r$) is the SVD of $B$, where $D$ is $r$-by-$r$ and $r = \text{rank}(B)$. Show that if $\lambda(A, B) = C$ then $U_2^T A V_2$ is singular.

**P7.7.2** Define $\mathrm{F} : \mathbb{R}^n \to \mathbb{R}$ by

$$\mathrm{F}(x) = \frac{1}{2} \left\| Ax - \frac{x^T B^T A x}{x^T B^T B x} Bx \right\|_2^2$$

where $A$ and $B$ are in $\mathbb{R}^{m \times n}$. Show that if $\nabla \mathrm{F}(x) = 0$, then $Ax$ is a multiple of $Bx$.

**P7.7.3** Suppose $A$ and $B$ are in $\mathbb{R}^{n \times n}$. Give an algorithm for computing orthogonal $Q$ and $Z$ such that $Q^T A Z$ is upper Hessenberg and $Z^T B Q$ is upper triangular.

**P7.7.4** Show that if $\mu \notin \lambda(A, B)$ then $A_1 = (A - \mu B)^{-1} A$ and $B_1 = (A - \mu B)^{-1} B$ commute.

**P7.7.5** Suppose $A \in \mathbb{C}^{n \times n}$ and $B \in \mathbb{C}^{n \times n}$ and that $A - \mu B$ is nonsingular for some $\mu \in \mathbb{C}$. Assume that $B$ is nonsingular. Show that $\lambda \in \lambda(A, B)$ if and only if $\lambda \in \lambda(B, B(A - \mu B)^{-1} B)$.

### Notes and References for Sec. 7.7

Mathematical aspects of the generalized eigenvalue problem are covered in

I. Erdelyi (1967). "On the Matrix Equation $Ax = \lambda Bx$," *J. Math. Anal. and Applic. 17*, 119–32.

F. Gantmacher (1959). *The Theory of Matrices* , vol. 2, Chelsea, New York.

H.W. Turnbill and A.C. Aitken (1961). *An Introduction to the Theory of Canonical Matrices,* Dover, New York.

A good general volume that covers many aspects of the $A - \lambda B$ problem is

B. Kågström and A. Ruhe (1983). *Matrix Pencils,* Proc. Pite Havsbad, 1982, Lecture Notes in Mathemnatics 973, Springer-Verlag, New York and Berlin.

Stewart deals with questions of eigenvalue sensitivity in

G.W. Stewart (1972). "On the Sensitivity of the Eigenvalue Problem $Ax = \lambda Bx$," *SIAM J. Num. Anal. 9*, 669–86.

G.W. Stewart (1973b). "Error and Perturbation Bounds for Subspaces Associated with Certain Eigenvalue Problems," *SIAM Review 15*, 727–64.

G.W. Stewart (1975b). "Gershgorin Theory for the Generalized Eigenvalue Problem $Ax = \lambda Bx$," *Math. Comp. 29*, 600–606.
G.W. Stewart (1978). "Perturbation Theory for the Generalized Eigenvalue Problem'", in *Recent Advances in Numerical Analysis* , ed. C. de Boor and G.H. Golub, Academic Press, New York.

Rectangular generalized eigenvalue problems arise in certain applications. See

G.L. Thompson and R.L. Weil (1970). "Reducing the Rank of $A - \lambda B$," *Proc. Amer. Math. Sec. 26*, 548–54.
G.L. Thompson and R.L. Weil (1972). "Roots of Matrix Pencils $Ay = \lambda By$: Existence, Calculations, and Relations to Game Theory," *Lin. Alg. and Its Applic. 5*, 207–26.

Nonorthogonal computational procedures for the generalized eigenvalue problem are proposed in

V.N. Kublanovskaja and V.N. Fadeeva (1964). "Computational Methods for the Solution of a Generalized Eigenvalue Problem," *Amer. Math. Soc. Transl. 2*, 271–90.
G. Peters and J.H. Wilkinson (1970a). "$Ax = \lambda Bx$ and the Generalized Eigenproblem," *SIAM J. Num. Anal. 7*, 479–92.

The $QZ$ algorithm is in Eispack and was originally presented in

C.B. Moler and G.W. Stewart (1973). "An Algorithm for Generalized Matrix Eigenvalue Problems," *SIAM J. Num. Anal. 10*, 241–56.

Improvements in the original algorithm are discussed in

K. Kaufman (1977). "Some Thoughts on the QZ Algorithm for Solving the Generalized Eigenvalue Problem," *ACM Trans. Math. Soft. 3*, 65–75.
R.C. Ward (1975). "The Combination Shift QZ Algorithm," *SIAM J. Num. Anal. 12*, 835–53.
R.C. Ward (1981). "Balancing the Generalized Eigenvalue Problem," *SIAM J. Sci and Stat. Comp. 2*, 141–152.

Just as the Hessenberg decomposition is important in its own right, so is the Hessenberg-triangular decomposition that serves as a QZ front end. See

W. Enright and S. Serbin (1978). "A Note on the Efficient Solution of Matrix Pencil Systems," *BIT 18*, 276–81.

A method similar to QZ but relying on Gauss transformations is presented in

K. Kaufman (1974). "The LZ Algorithm to Solve the Generalized Eigenvalue Problem," *SIAM J. Num. Anal. 11*, 997–1024.

while

C.F. Van Loan (1975). "A General Matrix Eigenvalue Algorithm," *SIAM J. Num. Anal. 12*, 819–34.

is concerned with a generalization of the QZ algorithm that handles the problem $A^TCx = \lambda B^TDx$ without inversion or formation of $A^TC$ and $B^TD$.

The behavior of the QZ algorithm on pencils $A - \lambda B$ that are always nearly singular is discussed in

J.H. Wilkinson (1979). "Kronecker's Canonical Form and the QZ Algorithm," *Lin. Alg. and Its Applic. 28*, 285–303.

Other approaches to the problem include

A. Jennings and M.R. Osborne (1977). "Generalized Eigenvalue Problems for Certain Unsymmetric Band Matrices," *Lin. Alg. and Its Applic. 29,* 139–50.
V.N. Kublanovskaya (1984). "AB Algorithm and Its Modifications for the Spectral Problem of Linear Pencils of Matrices," *Numer. Math. 43,* 329–342.
G. Rodrigue (1973). "A Gradient Method for the Matrix Eigenvalue Problem $Ax = \lambda Bx$," *Numer. Math. 22,* 1–16.
H.R. Schwartz (1974). "The Method of Coordinate Relaxation for $(A - \lambda B)x = 0$," *Num. Math. 23,* 135–52.

The general $Ax = \lambda Bx$ problem is central to some important control theory applications. See

W.F. Arnold and A.J. Laub (1984). "Generalized Eigenproblem Algorithms and Software for Algebraic Riccati Equations," *Proc. IEEE 72,* 1746–1754.
J.W. Demmel and B. Kågström (1988). "Accurate Solutions of Ill-Posed Problems in Control Theory," *SIAM J. Matrix Anal. Appl.* 126–145.
P. Van Dooren (1981). "A Generalized Eigenvalue Approach for Solving Riccati Equations," *SIAM J. Sci. and Stat. Comp. 2,* 121–135.
P. Van Dooren (1981). "The Generalized Eigenstructure Problem in Linear System Theory," *IEEE Trans. Auto. Cont. AC-26,* 111–128.

Indeed, various applications in control prompted a deeper look at the Kronecker structure of matrix pencils typified by the following papers:

J.W. Demmel (1983b). "The Condition Number of Equivalence Transformations that Block Diagonalize Matrix Pencils," *SIAM J. Numer. Anal. 20,* 599–610.
J.W. Demmel and B. Kågström (1987). "Computing Stable Eigendecompositions of Matrix Pencils," *Linear Alg. and Its Applic 88/89,* 139–186.
B. Kågström (1985). "The Generalized Singular Value Decomposition and the General $A - \lambda B$ Problem," *BIT 24,* 568–583.
B. Kågström (1986). "RGSVD: An Algorithm for Computing the Kronecker Structure and Reducing Subspaces of Singular $A - \lambda B$ Pencils," *SIAM J. Sci. and Stat. Comp.* 7,185–211.
B. Kågström and L. Westin (1987). "GSYLV- Fortran Routines for the Generalized Schur Method with $\text{dif}^{-1}$ estimators for Solving the Generalized Sylvester Equation," Report UMINF–132.86, Inst. of Inf. Proc., University of Umeå, S-901 87 Umeå, Sweden.
P. Van Dooren (1979). "The Computation of Kronecker's Canonical Form of a Singular Pencil," *Lin. Alg. and Its Applic. 27,* 103–40.
P. Van Dooren (1983). "Reducing Subspaces: definitions, properties and Algorithms," in *Matrix Pencils*, eds. B. Kågström and A. Ruhe, Springer-Verlag, New York, 1983, 58–73
J.H. Wilkinson (1978). "Linear Differential Equations and Kronecker's Canonical Form," in *Recent Advances in Numerical Analysis* , ed. C. de Boor and G.H. Golub, Academic Press, New York, pp. 231–65.

Work on parallel generalized eigenproblem methods has just begun. See

B. Kågström, L. Nyström, and P. Poromaa (1987). "Parallel Algorithms for Solving the Triangular Sylvester Equation on a Hypercube Multiprocessor," Report UMINF–136.87, Inst. of Inf. Proc., University of Umeå, S-901 87 Umeå, Sweden.
B. Kågström, L. Nyström, and P. Poromaa (1988). "Parallel Shared Memory Algorithms for Solving the Triangular Sylvester Equation," Report UMINF–155.88, Inst. of Inf. Proc., University of Umeå, S-901 87 Umeå, Sweden.

# Chapter 8

# The Symmetric Eigenvalue Problem



The perturbation theory and algorithmic developments in the previous chapter undergo considerable simplification when the matrix $A$ is symmetric. Indeed, the symmetric eigenvalue problem with its rich mathematical structure is one of the most aesthetically pleasing problems in numerical algebra.

We begin our presentation with a brief discussion of the mathematical properties that underlie the symmetric eigenproblem. In §8.2, we specialize the algorithms of §7.4 and §7.5 and obtain the elegant symmetric QR algorithm. A variant of this routine capable of computing the singular value decomposition is detailed in §8.3. Because the eigenvalues of a symmetric matrix can be characterized in several ways, there is a host of alternatives to the QR algorithm. Some of these methods are described in §8.4.

One of the earliest matrix algorithms to appear in the literature is Jacobi's method for the symmetric eigenproblem. There is renewed interest in this procedure because of parallel computation and for this reason we have devoted considerable space to the technique in §8.5. Another highly parallel eigensolver, but of the divide and conquer variety, is discussed in §8.6. It can be applied to tridiagonal problems and is particularly interesting because it appears to compete well with the tridiagonal QR iteration, even in single-processor situations.

In the final section we discuss the $A-\lambda B$ problem for the important case when $A$ is symmetric and $B$ is symmetric positive definite. Although no suitable analog of the QZ algorithm exists for this specially structured generalized eigenvalue problem, there are several successful methods that can be applied. The generalized singular value decomposition is also discussed.

Much of the material in this chapter may be found in Parlett (SEP).

## 8.1 Mathematical Background

The symmetric eigenvalue problem has a theory that is both rich and elegant. The important aspects of this theory are summarized in this section.

### 8.1.1 Eigenvalues of Symmetric Matrices

Symmetry simplifies the real eigenvalue problem $Ax = \lambda x$ in two ways. It implies that all of $A$'s eigenvalues are real and it implies that there is an orthonormal basis of eigenvectors. These properties are a consequence of

**Theorem 8.1.1 (Symmetric Real Schur Decomposition)** *If $A$ is a real $n$-by-$n$ symmetric matrix, then there exists a real orthogonal $Q$ such that*

$$Q^T AQ = \text{diag}(\lambda_1, \ldots, \lambda_n).$$

**Proof.** Let $Q^T AQ = T$ be the real Schur decomposition (Theorem 7.4.1) of $A$. Since $T$ is also symmetric, it follows that it must be a direct sum of 1-by-1 and 2-by-2 matrices. However, it is easy to verify that a 2-by-2 symmetric matrix cannot have complex eigenvalues. Consequently, $T$ can have no 2-by-2 blocks along its diagonal. □

**Example 8.1.1** If

$$A = \begin{bmatrix} 6.8 & 2.4 \\ 2.4 & 8.2 \end{bmatrix} \quad \text{and} \quad Q = \begin{bmatrix} .6 & -.8 \\ .8 & .6 \end{bmatrix}$$

then $Q$ is orthogonal and $Q^T AQ = \text{diag}(10,5)$.

### 8.1.2 The Minimax Theorem and Some Consequences

If $A^T = A \in \mathbb{R}^{n\times n}$, then let $\lambda_i(A)$ denote the $i$th largest eigenvalue of $A$. Thus,

$$\lambda_n(A) \le \lambda_{n-1}(A) \le \cdots \le \lambda_2(A) \le \lambda_1(A).$$

It is easy to show that each eigenvalue of a symmetric matrix is a stationary value of the map $x \to x^TAx/x^Tx$ where $x \ne 0$. Moreover, the eigenvalues satisfy the following "minimax" characterization:

**Theorem 8.1.2 (Courant-Fischer Minimax Theorem)** *If $A \in \mathbb{R}^{n\times n}$ is symmetric, then*

$$\lambda_k(A) = \max_{\dim(S)=k} \min_{0\ne y\in S} \frac{y^TAy}{y^Ty}$$

*for $k = 1{:}n$.*

**Proof.** Wilkinson (AEP, pp. 100–101). □

Using this result it is possible to establish several very useful corollaries.

**Corollary 8.1.3** *If $A$ and $A+E$ are n-by-n symmetric matrices, then*

$$\lambda_k(A) + \lambda_n(E) \le \lambda_k(A+E) \le \lambda_k(A) + \lambda_1(E) \qquad k = 1{:}n$$

**Proof.** Wilkinson (AEP, pp. 101–2). □

**Example 8.1.2** If

$$A = \begin{bmatrix} 6.8 & 2.4 \\ 2.4 & 8.2 \end{bmatrix} \quad \text{and} \quad E = \begin{bmatrix} 2.0 & 0.0 \\ 0.0 & 1.0 \end{bmatrix}$$

then $\lambda(A) = \{5,\ 10\}$, $\lambda(E) = \{1,\ 2\}$, and $\lambda(A+E) = \{6.5917,\ 11.4083\}$.

**Corollary 8.1.4 (Interlacing Property)** *If $A_r$ denotes the leading r-by-r principal submatrix of an n-by-n symmetric matrix $A$, then for $r = 1{:}n-1$ the following* interlacing property *holds:*

$$\lambda_{r+1}(A_{r+1}) \le \lambda_r(A_r) \le \lambda_r(A_{r+1}) \le \cdots \le \lambda_2(A_{r+1}) \le \lambda_1(A_r) \le \lambda_1(A_{r+1}).$$

**Proof.** Wilkinson (AEP, pp, 103–4). □

**Example 8.1.3** If

$$A = \begin{bmatrix} 1 & 1 & 1 & 1 \\ 1 & 2 & 3 & 4 \\ 1 & 3 & 6 & 10 \\ 1 & 4 & 10 & 20 \end{bmatrix}$$

then $\lambda(A_1) = \{1\}$, $\lambda(A_2) = \{.3820,\ 2.6180\}$, $\lambda(A_3) = \{.1270,\ 1.0000,\ 7.873\}$, and $\lambda(A_4) = \{.0380,\ .4538,\ 7.2034, 26.3047\}$.

### 8.1.3 Additional Results

If $A$ is perturbed by a rank-one matrix, then the eigenvalues are shifted in a very regular way.

**Theorem 8.1.5** *Suppose $B = A + \tau cc^T$ where $A \in \mathbb{R}^{n\times n}$ is symmetric, $c \in \mathbb{R}^n$ has unit 2-norm and $\tau \in \mathbb{R}$. If $\tau \geq 0$, then*

$$\lambda_i(B) \in [\lambda_i(A),\ \lambda_{i-1}(A)] \qquad i = 2{:}n$$

*while if $\tau \leq 0$ then*

$$\lambda_i(B) \in [\lambda_{i+1}(A), \lambda_i(A)], \qquad i = 1{:}n-1\,.$$

*In either case, there exist nonnegative $m_1, \ldots, m_n$ such that*

$$\lambda_i(B) = \lambda_i(A) + m_i\tau, \qquad i = 1{:}n$$

*with $m_1 + \cdots + m_n = 1$.*

**Proof.** Wilkinson (AEP, pp. 94–97). See also Theorem 8.6.2. □

The effect of unit rank perturbations on the eigenvalues of a symmetric matrix is of interest in several settings, e.g., quasi-Newton methods for unconstrained optimization. See also §8.6 where Theorem 8.1.5 underpins a divide-and-conquer tridiagonal eigenvalue solver.

Our next perturbation result is of a rather different variety. It bounds the difference between $\lambda(A)$ and $\lambda(A+E)$ in terms of $\| E \|_F$.

**Theorem 8.1.6 (Wielandt-Hoffman)** *If $A$ and $A+E$ are n-by-n symmetric matrices, then*

$$\sum_{i=1}^{n} (\lambda_i(A+E) - \lambda_i(A))^2 \;\leq\; \| A \|_F^2\,.$$

**Proof.** A proof can be found in Wilkinson (AEP, pp. 104-8). □

**Example 8.1.4** If

$$A = \begin{bmatrix} 6.8 & 2.4 \\ 2.4 & 8.2 \end{bmatrix} \qquad \text{and} \qquad E = \begin{bmatrix} 2.0 & 0.0 \\ 0.0 & 1.0 \end{bmatrix}$$

then

$$4.5168 = \sum_{i=1}^{2} (\lambda_i(A+E) - \lambda_i(A))^2 \;\leq\; \| E \|_F^2 = 5.$$

See Example 8.1.2.

### 8.1.4 The Sensitivity of Invariant Subspaces

The invariant subspaces of a symmetric matrix are not necessarily well-conditioned; the separation of the eigenvalues is a critical factor just as in the unsymmetric case. However, the bound given in our general perturbation theorem for invariant subspaces (Theorem 7.2.4) undergoes some simplification in the symmetric case.

**Theorem 8.1.7** *Suppose $A$ and $A+E$ are n-by-n symmetric matrices and that*

$$Q = \begin{bmatrix} Q_1 & Q_2 \end{bmatrix} \begin{matrix} \\ \end{matrix}$$
$$\quad\quad k \quad n-k$$

*is an orthogonal matrix such that* range$(Q_1)$ *is an invariant subspace for $A$. Partition the matrices $Q^TAQ$ and $Q^TEQ$ as follows:*

$$Q^TAQ = \begin{bmatrix} A_{11} & 0 \\ 0 & A_{22} \end{bmatrix} \begin{matrix} k \\ n-k \end{matrix}$$
$$\quad\quad\quad\quad k \quad n-k$$

$$Q^TEQ = \begin{bmatrix} E_{11} & E_{12} \\ E_{21} & E_{22} \end{bmatrix} \begin{matrix} k \\ n-k \end{matrix}$$
$$\quad\quad\quad\quad k \quad n-k$$

*If*

$$\delta = \min_{\substack{\lambda \in \lambda(A_{11}) \\ \mu \in \lambda(A_{22})}} |\lambda - \mu| - \| E_{11} \|_2 - \| E_{22} \|_2 > 0$$

*and $\| E_{12} \|_2 \le \delta/2$, then there exists a matrix $P \in \mathbb{R}^{(n-k)\times k}$ satisfying*

$$\| P \|_2 \le \frac{2}{\delta} \| E_{21} \|_2$$

*such that the columns of $Q_1 = (Q_1 + Q_2P)(I + P^TP)^{-1/2}$ form an orthonormal basis for a subspace that is invariant for $A + E$.*

**Proof.** See Stewart (1973b). □

The theorem essentially says that the sensitivity of an invariant subspace of a symmetric matrix depends solely on the separation of the associated eigenvalues.

**Example 8.1.5** If $A = \text{diag}(.999,\ 1.001,\ 2.)$, and

$$E = \begin{bmatrix} 0.00 & 0.01 & 0.01 \\ 0.01 & 0.00 & 0.01 \\ 0.01 & 0.01 & 0.00 \end{bmatrix},$$

then $\hat{Q}^T(A+E)\hat{Q} = \text{diag}(.9899,\ 1.0098,\ 2.0002)$ where

$$\hat{Q} = \begin{bmatrix} -.7418 & .6706 & .0101 \\ .6708 & .7417 & .0101 \\ .0007 & -.0143 & .9999 \end{bmatrix}$$

is orthogonal. Let $\hat{q}_i = \hat{Q}e_i$, $i = 1,2,3$. Thus, $\hat{q}_i$ is the perturbation of $A$'s eigenvector $q_i = e_i$. A calculation shows that

$$\text{dist}\{\text{span}\{q_1\}, \text{span}\{\hat{q}_1\}\} = \text{dist}\{\text{span}\{q_2\}, \text{span}\{\hat{q}_2\}\} = .67$$

Thus, because they are associated with nearby eigenvalues, the eigenvectors $q_1$ and $q_2$ cannot be computed accurately. On the other hand, since $\lambda_1$ and $\lambda_2$ are well separated from $\lambda_3$, they define a two-dimensional subspace that is not particularly sensitive as $\text{dist}\{\text{span}\{q_1, q_2\}, \text{span}\{\hat{q}_1, \hat{q}_2\}\} = .01$.

We next present a collection of results that are concerned with approximate invariant subspaces. These results are useful in subsequent sections and aid in the interpretation of various algorithms.

**Theorem 8.1.8** *Suppose $A \in \mathbb{R}^{n\times n}$ and $S \in \mathbb{R}^{k\times k}$ are symmetric and that*

$$AQ_1 - Q_1S = E_1$$

*where $Q_1 \in \mathbb{R}^{n\times k}$ satisfies $Q_1^TQ_1 = I_k$. Then there exist $\mu_1,\ldots,\mu_k \in \lambda(A)$ such that*

$$|\mu_i - \lambda_i(S)| \le \sqrt{2}\,\|\, E_1 \,\|_2 \qquad i = 1{:}k$$

**Proof.** Let $Q_2 \in \mathbb{R}^{n\times(n-k)}$ be any matrix such that $Q = [\ Q_1,\ Q_2\ ]$ is orthogonal. It follows that

$$Q^TAQ = \begin{bmatrix} S & 0 \\ 0 & Q_2^TAQ_2 \end{bmatrix} + \begin{bmatrix} Q_1^TE_1 & E_1^TQ_2 \\ Q_2^TE_1 & 0 \end{bmatrix} \equiv B + E$$

and so by using Corollary 8.1.3, we have $|\lambda_i(A) - \lambda_i(B)| \le \|\, E \,\|_2$ for $i = 1{:}n$. Since $\lambda(S) \subseteq \lambda(B)$, there exist $\mu_1,\ldots,\mu_k \in \lambda(A)$ such that

$$|\mu_i - \lambda_i(S)| \le \|\, E \,\|_2$$

for $i = 1{:}k$. The theorem follows by noting that for any $x \in \mathbb{R}^k$ and $y \in \mathbb{R}^{n-k}$ we have

$$\left\| E \begin{bmatrix} x \\ y \end{bmatrix} \right\|_2 \le \|\, E_1x \,\|_2 + \|\, E_1^TQ_2y \,\|_2 \le \|\, E_1 \,\|_2 \|\, x \,\|_2 + \|\, E_1 \,\|_2 \|\, y \,\|_2$$

from which we readily conclude that $\|\, E \,\|_2 \le \sqrt{2}\|\, E_1 \,\|_2$. □

**Example 8.1.6** If

$$A = \begin{bmatrix} 6.8 & 2.4 \\ 2.4 & 8.2 \end{bmatrix}, \qquad Q_1 = \begin{bmatrix} .7994 \\ .6007 \end{bmatrix}, \text{ and } S = (5.1) \in \mathbb{R}$$

then

$$AQ_1 - Q_1S = \begin{bmatrix} -.0828 \\ -.0562 \end{bmatrix} = E_1.$$

The theorem predicts that $A$ has an eigenvalue within $\sqrt{2}\,\|\,E_1\,\|_2 \approx .1415$ of 5.1. This is true since $\lambda(A) = \{5, 10\}$.

The eigenvalue bounds in Theorem 8.1.8 depend on the size of the residual of the approximate invariant subspace, i.e., $\|\,AQ_1 - Q_1S\,\|_2$. Given $A$ and $Q_1$, the following theorem indicates how to choose $S$ so that this quantity is minimized when $\|\cdot\| = \|\cdot\|_F$.

**Theorem 8.1.9** *If $A \in \mathbb{R}^{n\times n}$ is symmetric and $Q_1 \in \mathbb{R}^{n\times k}$ satisfies the equation $Q_1^TQ_1 = I_k$, then*

$$\min_{S\in\mathbb{R}^{k\times k}} \|\,AQ_1 - Q_1S\,\|_F = \|\,(I - Q_1Q_1^T)AQ_1\,\|_F\,.$$

**Proof.** Let $Q_2 \in \mathbb{R}^{n\times(n-k)}$ be such that $Q = [\,Q_1,\ Q_2\,]$ is orthogonal. For any $S \in \mathbb{R}^{k\times k}$ we have

$$\begin{aligned} \|\,AQ_1 - Q_1S\,\|_F^2 &= \|\,Q^TAQ_1 - Q^TQ_1S\,\|_F^2 \\ &= \|\,Q_1^TAQ_1 - S\,\|_F^2 + \|\,Q_2^TAQ_1\,\|_F^2. \end{aligned}$$

Clearly, the minimizing $S$ is given by $S = Q_1^TAQ_1$. □

This result enables us to associate any $k$-dimensional subspace range$(Q_1)$, with a set of $k$ "optimal" eigenvalue-eigenvector approximates.

**Theorem 8.1.10** *Suppose $A \in \mathbb{R}^{n\times n}$ is symmetric and that $Q_1 \in \mathbb{R}^{n\times k}$ satisfies $Q_1^TQ_1 = I_k$. If*

$$Z^T(Q_1^TAQ_1)Z = \text{diag}(\theta_1,\ldots,\theta_k) = D$$

*is the Schur decomposition of $Q_1^TAQ_1$ and $Q_1Z = [\,y_1,\ldots,y_k\,]$, then*

$$\|\,Ay_i - \theta_iy_i\,\|_2 = \|\,(I - Q_1Q_1^T)AQ_1Ze_i\,\|_2 = \|\,(I - Q_1Q_1^T)AQ_1\,\|_2$$

*for $i = 1{:}k$. The $\theta_i$ are called* Ritz values, *the $y_i$ are called* Ritz vectors, *and the $(\theta_i, y_i)$ are called* Ritz pairs.

**Proof.**

$$Ay_i - \theta_iy_i = AQ_1Ze_i = Q_1ZDe_i = (AQ_1 - Q_1(Q_1^TAQ_1))Ze_i.$$

The theorem follows by taking norms. □

The usefulness of Theorem 8.1.8 is enhanced if we weaken the assumption that the columns of $Q_1$ are orthonormal. As can be expected, the bounds deteriorate with loss of orthogonality.

**Theorem 8.1.11** *Suppose $A \in \mathbb{R}^{n\times n}$ is symmetric and that*

$$AX_1 - X_1S = F_1,$$

*where $X_1 \in \mathbb{R}^{n\times k}$ and $S = X_1^TAX_1$. If $\sigma_k(X_1) > 0$ and*

$$\| X_1^TX_1 - I_k \|_2 = \tau$$

*then there exist $\mu_1, \ldots, \mu_k \in \lambda(A)$ such that*

$$|\mu_i - \lambda_i(S)| \leq \sqrt{2}\left(\frac{\| F_1 \|_2}{\sigma_k(X_1)} + \tau \| A \|_2\right)$$

*for $k = 1{:}n$.*

**Proof.** Let $X_1 = U_1\Sigma V^T$ be the SVD of $X_1$ with $U_1 \in \mathbb{R}^{n\times k}$, $V \in \mathbb{R}^{k\times k}$, and $\Sigma = \text{diag}(\sigma_1, \ldots, \sigma_k) \in \mathbb{R}^{k\times k}$. By substituting this into the equation $AX_1 - X_1S = F_1$ we get

$$AU_1\Sigma V^T - U_1\Sigma V^T(U_1\Sigma V^T)^TA(U_1\Sigma V^T) = F$$

and so

$$AU_1 - U_1\Sigma^2(U_1^TAU_1) = FV\Sigma^{-1}.$$

From this it follows that

$$AU_1 - U_1(U_1^TAU_1) = E_1 \tag{8.1.1}$$

where $E_1 = FV\Sigma^{-1} + U_1(\Sigma^2 - I_k)U_1^TAU_1$. Noting that

$$\| \Sigma^2 - I_k \|_2 = \| X_1^TX_1 - I_k \|_2 = \tau$$

we obtain

$$\| E_1 \|_2 \leq \| FV\Sigma^{-1} \|_2 + \| U_1(\Sigma^2 - I_k)U_1^TAU_1 \|_2 \leq \frac{\| F \|_2}{\sigma_k(X_1)} + \tau\| A \|_2 .$$

The theorem follows by applying Theorem 8.1.8 with $Q = U_1$ and $S = U_1^TAU_1$. □

### 8.1.5 The Law of Inertia

We conclude the section by proving the Sylvester law of inertia. The *inertia* of a symmetric matrix $A$ is a triplet of nonnegative integers $(m, z, p)$ where $m$, $z$, and $p$ are the number of negative, zero, and positive elements of $\lambda(A)$.

**Theorem 8.1.12 (Sylvester Law of Inertia)** *If $A \in \mathbb{R}^{n\times n}$ is symmetric and $X \in \mathbb{R}^{n\times n}$ is nonsingular, then $A$ and $X^TAX$ have the same inertia.*

**Proof.** Suppose for some $r$ that $\lambda_r(A) > 0$ and define the subspace $S_0 \subseteq \mathbb{R}^n$ by

$$S_0 = \text{span}\{X^{-1}q_1, \ldots, X^{-1}q_r\}, \qquad q_i \neq 0$$

where $Aq_i = \lambda_i(A)q_i$ and $i = 1{:}r$. From the minimax characterization of $\lambda_r(X^TAX)$ we have

$$\lambda_r(X^TAX) = \max_{\dim(S)=r} \min_{y\in S} \frac{y^T(X^TAX)y}{y^Ty}$$

$$\geq \min_{y\in S_0} \frac{y^T(X^TAX)y}{y^Ty}.$$

Now for any $y \in \mathbb{R}^n$ we have

$$\frac{y^T(X^TX)y}{y^Ty} \geq \sigma_n(X)^2$$

while for $y \in S_0$ it is clear that

$$\frac{y^T(X^TAX)y}{y^Ty} \geq \lambda_r(A).$$

Thus,

$$\lambda_r(X^TAX) \geq \min_{y\in S_0} \left\{ \frac{y^T(X^TAX)y}{y^T(X^TX)y} \frac{y^T(X^TX)y}{y^Ty} \right\} \geq \lambda_r(A)\sigma_n(X)^2.$$

An analogous argument with the roles of $A$ and $X^TAX$ reversed shows that

$$\lambda_r(A) \geq \lambda_r(X^TAX)\sigma_n(X^{-1})^2 = \frac{\lambda_r(X^TAX)}{\sigma_1(X)^2}.$$

Thus, $\lambda_r(A)$ and $\lambda_r(X^TAX)$ have the same sign and so we have shown that $A$ and $X^TAX$ have the same number of positive eigenvalues. If we apply this result to $-A$, we conclude that $A$ and $X^TAX$ have the same number of negative eigenvalues. Obviously, the number of zero eigenvalues possessed by each matrix is also the same. □

**Example 8.1.7** If $A = \text{diag}(3, 2, -1)$ and

$$X = \begin{bmatrix} 1 & 4 & 5 \\ 0 & 1 & 2 \\ 0 & 0 & 1 \end{bmatrix},$$

then

$$X^TAX = \begin{bmatrix} 3 & 12 & 15 \\ 12 & 50 & 64 \\ 15 & 64 & 82 \end{bmatrix}$$

and $\lambda(X^TAX) = \{134.769, .3555, -.1252\}$.

**Problems**

**P8.1.1** Show that the eigenvalues of a 2-by-2 symmetric matrix must be real.

**P8.1.2** Compute the Schur decomposition of $A = \begin{bmatrix} 1 & 2 \\ 2 & 3 \end{bmatrix}$.

**P8.1.3** Show that the eigenvalues of a Hermitian matrix ($A^H = A$) are real. For each theorem and corollary in this section, state and prove the corresponding result for Hermitian matrices. Which results have analogs when $A$ is skew-symmetric? (Hint: If $A^T = -A$, then $iA$ is Hermitian.)

**Notes and References for Sec. 8.1**

The perturbation theory for the symmetric eigenvalue problem is surveyed in Wilkinson (AEP, chapter 2) and Parlett (SEP, chapters 10 and 11). Some representative papers in this well-documented area include

P. Deift, T. Nande, and C. Tome (1983). "Ordinary Differerntial Equations and the Symmetric Eigenvalue Problem," *SIAM J. Numer. Anal. 20*, 1–22.

W. Kahan (1975). "Spectra of Nearly Hermitian Matrices," *Proc. Amer. Math. Soc. 48*, 11–17.

W. Kahan (1967). "Inclusion Theorems for Clusters of Eigenvalues of Hermitian Matrices," Computer Science Report, University of Toronto.

C.C. Paige (1974b). "Eigenvalues of Perturbed Hermitian Matrices," *Lin. Alg. and Its Applic* . 8, 1–10.

A. Ruhe (1975). "On the Closeness of Eigenvalues and Singular Values for Almost Normal Matrices," *Lin. Alg. and Its Applic. 11*, 87–94.

A. Schonhage (1979). "Arbitrary Perturbations of Hermitian Matrices," *Lin. Alg. and Its Applic. 24*, 143–49.

D.S. Scott (1985). "On the Accuracy of the Gershgorin Circle Theorem for Bounding the Spread of a Real Symmetric Matrix," *Lin. Alg. and Its Applic. 65*, 147–155

G.W. Stewart (1973b). "Error and Perturbation Bounds for Subspaces Associated with Certain Eigenvalue Problems," *SIAM Review 15*, 727–64.

## 8.2 The Symmetric QR Algorithm

We now investigate how the practical QR algorithm developed in Chapter 7 can be specialized when $A \in \mathbb{R}^{n\times n}$ is symmetric. There are three observations to make immediately:

- If $U_0^TAU_0 = H$ is upper Hessenberg, then $H = T$ must be tridiagonal.

- Symmetry and tridiagonal band structure are preserved when a single-shift QR step is performed.

- There is no need to consider complex shifts since $\lambda(A) \subseteq \mathbb{R}$.

These simplifications together with the nice mathematical properties of the symmetric eigenproblem combine to make the algorithms of this chapter very attractive.

### 8.2.1 Reduction to Tridiagonal Form

We begin by deriving the Householder reduction to tridiagonal form. Suppose that Householder matrices $P_1, \ldots, P_{k-1}$ have been determined such that submatrix $B_{11}$ in

$$A_{k-1} = (P_1 \cdots P_{k-1})^T A (P_1 \cdots P_{k-1})$$

$$= \begin{bmatrix} B_{11} & B_{12} & 0 \\ B_{21} & B_{22} & B_{23} \\ 0 & B_{32} & B_{33} \end{bmatrix} \begin{matrix} k-1 \\ 1 \\ n-k \end{matrix}$$

$$\begin{matrix} k-1 & \quad 1 \quad & n-k \end{matrix}$$

is tridiagonal. If $\bar{P}_k$ is an order $n-k$ Householder matrix such that $\bar{P}_k B_{32}$ is a multiple of $e_1^{(n-k)}$ and if $P_k = \text{diag}(I_k, \bar{P}_k)$, then the leading $k$-by-$k$ principal submatrix of

$$A_k = P_k A_{k-1} P_k = \begin{bmatrix} B_{11} & B_{12} & 0 \\ B_{21} & B_{22} & B_{23}\bar{P}_k \\ 0 & \bar{P}_k B_{32} & \bar{P}_k B_{33} \bar{P}_k \end{bmatrix} \begin{matrix} k-1 \\ 1 \\ n-k \end{matrix}$$

$$\begin{matrix} k-1 & \quad 1 \quad & n-k \end{matrix}$$

is tridiagonal. Clearly, if $U_0 = P_1 \cdots P_{n-2}$, then $U_0^T A U_0 = T$ is tridiagonal.

In the calculation of $A_k$ it is important to exploit symmetry during the formation of the matrix $\bar{P}_k B_{33} \bar{P}_k$. To be specific, suppose that $\bar{P}_k$ has the form

$$\bar{P}_k = I - \frac{2}{v^T v} v v^T \qquad 0 \neq v \in \mathbb{R}^{n-k}.$$

Note that if

$$p = \frac{2}{v^T v} B_{33} v \qquad \text{and} \qquad w = p - \frac{p^T v}{v^T v} v$$

then

$$\bar{P}_k B_{33} \bar{P}_k = B_{33} - v w^T - w v^T.$$

Since only the upper triangular portion of this matrix needs to be calculated, we see that the transition from $A_{k-1}$ to $A_k$ can be accomplished in only $4(n-k)^2$ flops.

Overall we have the following tridiagonalization procedure:

**Algorithm 8.2.1 (Householder Tridiagonalization)** Given an $n$-by-$n$ symmetric matrix $A$, the following algorithm overwrites $A$ with $T = U_0^T A U_0$, where $T$ is tridiagonal and $U_0 = H_1 \cdots H_{n-2}$ is the product of Householder transformations.

$$
\begin{aligned}
&\textbf{for } k = 1{:}n-2 \\
&\qquad v = \textbf{house}(A(k+1{:}n, k)) \\
&\qquad p = 2A(k+1{:}n, k+1{:}n)v/v^Tv \\
&\qquad w = p - (p^Tv)v/v^Tv \\
&\qquad A(k+1{:}n, k+1{:}n) = A(k+1{:}n, k+1{:}n) - vw^T - wv^T \\
&\textbf{end}
\end{aligned}
$$

This algorithm requires $4n^3/3$ flops when symmetry is exploited in calculating the rank-2 update. The matrix $U_0$ can be stored in factored form in the subdiagonal portion of $A$. If $U_0$ is explicitly required, then it can be formed with an additonal $4n^3/3$ flops.

**Example 8.2.1** If

$$
A = \begin{bmatrix} 1 & 3 & 4 \\ 3 & 2 & 8 \\ 4 & 8 & 3 \end{bmatrix} \qquad \text{and} \qquad U_0 = \begin{bmatrix} 1 & 0 & 0 \\ 0 & .6 & .8 \\ 0 & .8 & -.6 \end{bmatrix}
$$

then $U_0^T U_0 = I$ and

$$
U_0^T A U_0 = \begin{bmatrix} 1 & 5 & 0 \\ 5 & 10.32 & 1.76 \\ 0 & 1.76 & -5.32 \end{bmatrix}.
$$

Let $\hat{T}$ denote the computed version of $T$ obtained by Algorithm 8.2.1. It can be shown that $\hat{T} = Q^T(A+E)Q$ where $Q$ is exactly orthogonal and $E$ is a symmetric matrix satisfying $\| E \|_F \le c\mathbf{u}\| A \|_F$ where $c$ is a small constant. See Wilkinson (AEP, p. 297).

## 8.2.2 Explicit Single Shift QR Iteration

We now consider the single-shift QR iteration for symmetric matrices:

$$
\begin{aligned}
&T = U_0^T A U_0 \qquad \text{(tridiagonal)} \\
&\textbf{for } k = 0, 1, \ldots \\
&\qquad \text{Determine real shift } \mu. \\
&\qquad T - \mu I = UR \qquad \text{(QR factorization)} \\
&\qquad T = RU + \mu I \\
&\textbf{end}
\end{aligned}
\tag{8.2.1}
$$

Since this iteration preserves Hessenberg structure (cf. Algorithm 7.4.1) and symmetry, it preserves tridiagonal form. Moreover, the QR factorization of a tridiagonal matrix requires only $O(n)$ flops since $R$ has only two superdiagonals. Thus, there is an order of magnitude less work than in the unsymmetric case. However, if orthogonal matrices are accumulated in (8.2.1), then $O(n^2)$ flops per iteration are required.

Symmetry can also be exploited in conjunction with the calculation of the shift $\mu$. Denote $T$ by

$$T = \begin{bmatrix} a_1 & b_1 & & \cdots & 0 \\ b_1 & a_2 & \ddots & & \vdots \\ & \ddots & \ddots & \ddots & \\ \vdots & & \ddots & \ddots & b_{n-1} \\ 0 & \cdots & & b_{n-1} & a_n \end{bmatrix}.$$

We can obviously set $\mu = a_n$. However, a more effective choice is to shift by the eigenvalue of

$$T(n-1{:}n, n-1{:}n) = \begin{bmatrix} a_{n-1} & b_{n-1} \\ b_{n-1} & a_n \end{bmatrix}$$

that is closer to $a_n$. This is known as the *Wilkinson shift* and it is given by

$$\mu = a_n + d - \text{sign}(d)\sqrt{d^2 + b_{n-1}^2} \tag{8.2.2}$$

where $d = (a_{n-1} - a_n)/2$. Wilkinson (1968b) has shown that (8.2.1) is cubically convergent with either shift strategy, but gives heuristic reasons why (8.2.2) is preferred.

### 8.2.3 Implicit Shift Version

As in the unsymmetric QR iteration, it is possible to shift implicitly in (8.2.1). That is, we can effect the transition from $T$ to $T_+ = RU + \mu I = U^T T U$ without explicitly forming the matrix $T - \mu I$. Let $c = \cos(\theta)$ and $s = \sin(\theta)$ be computed such that

$$\begin{bmatrix} c & s \\ -s & c \end{bmatrix}^T \begin{bmatrix} a_1 - \mu \\ b_1 \end{bmatrix} = \begin{bmatrix} * \\ 0 \end{bmatrix}.$$

If we set $G_1 = G(1,2,\theta)$ then $G_1 e_1 = U e_1$ and

$$T \leftarrow G_1^T T G_1 = \begin{bmatrix} \times & \times & + & 0 & 0 & 0 \\ \times & \times & \times & 0 & 0 & 0 \\ + & \times & \times & \times & 0 & 0 \\ 0 & 0 & \times & \times & \times & 0 \\ 0 & 0 & 0 & \times & \times & \times \\ 0 & 0 & 0 & 0 & \times & \times \end{bmatrix}.$$

We are thus in a position to apply the Implicit $Q$ theorem provided we can compute rotations $G_2, \ldots, G_{n-1}$ with the property that if $Z = G_1G_2 \cdots G_{n-1}$ then $Ze_1 = G_1e_1 = Ue_1$ *and* $Z^TTZ$ is tridiagonal.

Note that the first column of $Z$ and $U$ are identical provided we take each $G_i$ to be of the form $G_i = G(i, i+1, \theta_i)$ , $i = 2{:}n-1$. But $G_i$ of this form can be used to chase the unwanted nonzero element "+" out of the matrix $G_1^TTG_1$ as follows:

$$\xrightarrow{G_2} \begin{bmatrix} \times & \times & 0 & 0 & 0 & 0 \\ \times & \times & \times & + & 0 & 0 \\ 0 & \times & \times & \times & 0 & 0 \\ 0 & + & \times & \times & \times & 0 \\ 0 & 0 & 0 & \times & \times & \times \\ 0 & 0 & 0 & 0 & \times & \times \end{bmatrix} \quad \xrightarrow{G_3} \begin{bmatrix} \times & \times & 0 & 0 & 0 & 0 \\ \times & \times & \times & 0 & 0 & 0 \\ 0 & \times & \times & \times & + & 0 \\ 0 & 0 & \times & \times & \times & 0 \\ 0 & 0 & + & \times & \times & \times \\ 0 & 0 & 0 & 0 & \times & \times \end{bmatrix}$$

$$\xrightarrow{G_4} \begin{bmatrix} \times & \times & 0 & 0 & 0 & 0 \\ \times & \times & \times & 0 & 0 & 0 \\ 0 & \times & \times & \times & 0 & 0 \\ 0 & 0 & \times & \times & \times & + \\ 0 & 0 & 0 & \times & \times & \times \\ 0 & 0 & 0 & + & \times & \times \end{bmatrix} \quad \xrightarrow{G_5} \begin{bmatrix} \times & \times & 0 & 0 & 0 & 0 \\ \times & \times & \times & 0 & 0 & 0 \\ 0 & \times & \times & \times & 0 & 0 \\ 0 & 0 & \times & \times & \times & 0 \\ 0 & 0 & 0 & \times & \times & \times \\ 0 & 0 & 0 & 0 & \times & \times \end{bmatrix}$$

Thus, it follows from the Implicit $Q$ theorem that the tridiagonal matrix $Z^TTZ$ produced by this zero-chasing technique is essentially the same as the tridiagonal matrix $T$ obtained by the explicit method. (We may assume that all tridiagonal matrices in question are unreduced, since otherwise the problem decouples.)

Note that at any stage of the zero-chasing, there is only one nonzero entry outside the tridiagonal band. How this nonzero entry moves down the matrix during the update $T \leftarrow G_k^TTG_k$ is illustrated in the following:

$$\begin{bmatrix} 1 & 0 & 0 & 0 \\ 0 & c & s & 0 \\ 0 & -s & c & 0 \\ 0 & 0 & 0 & 1 \end{bmatrix}^T \begin{bmatrix} a_k & b_k & z_k & 0 \\ b_k & a_p & b_p & 0 \\ z_k & b_p & a_q & b_q \\ 0 & 0 & b_q & a_r \end{bmatrix} \begin{bmatrix} 1 & 0 & 0 & 0 \\ 0 & c & s & 0 \\ 0 & -s & c & 0 \\ 0 & 0 & 0 & 1 \end{bmatrix}$$

$$= \begin{bmatrix} a_k & b_k & 0 & 0 \\ b_k & a_p & b_p & z_p \\ 0 & b_p & a_q & b_q \\ 0 & z_p & b_q & a_r \end{bmatrix}.$$

Here $(p, q, r) = (k+1, k+2, k+3)$. This update can be performed in about 26 flops once $c$ and $s$ have been determined from the equation $b_ks + z_kc = 0$. Overall we obtain

**Algorithm 8.2.2 (Implicit Symmetric QR Step with Wilkinson Shift)** Given an unreduced symmetric tridiagonal matrix $T \in \mathbb{R}^{n \times n}$, the following algorithm overwrites $T$ with $Z^T T Z$, where $Z = G_1 \cdots G_{n-1}$ is a product of Givens rotations with the property that $Z^T(T - \mu I)$ is upper triangular and $\mu$ is that eigenvalue of $T$'s trailing 2-by-2 principal submatrix closer to $t_{nn}$.

$$
\begin{array}{l}
d = (t_{n-1,n-1} - t_{nn})/2 \\
\mu = t_{nn} - t_{n,n-1}^2 / \left( d + \text{sign}(d)\sqrt{d^2 + t_{n,n-1}^2} \right) \\
x = t_{11} - \mu \\
z = t_{21} \\
\textbf{for } k = 1{:}n-1 \\
\quad [\, c, s \,] = \textbf{givens}(x, z) \\
\quad T = G_k^T T G_k, \quad G_k = G(k, k+1, \theta) \\
\quad \textbf{if } k < n-1 \\
\quad\quad x = t_{k+1,k} \\
\quad\quad z = t_{k+2,k} \\
\quad \textbf{end} \\
\textbf{end}
\end{array}
$$

This algorithm requires about $30n$ flops and $n$ square roots. If a given orthogonal matrix $Q$ is overwritten with $QG_1 \cdots G_{n-1}$, then an additional $6n^2$ flops are needed. Of course, in any practical implementation the tridiagonal matrix $T$ would be stored in a pair of $n$-vectors and not in an $n$-by-$n$ array.

**Example 8.2.2** If Algorithm 8.2.2 is applied to

$$
T = \begin{bmatrix} 1 & 1 & 0 & 0 \\ 1 & 2 & 1 & 0 \\ 0 & 1 & 3 & .01 \\ 0 & 0 & .01 & 4 \end{bmatrix},
$$

then the new tridiagonal matrix $T$ is given by

$$
T = \begin{bmatrix} .5000 & .5916 & 0 & 0 \\ .5916 & 1.785 & .1808 & 0 \\ 0 & .1808 & 3.7140 & .0000044 \\ 0 & 0 & .0000044 & 4.002497 \end{bmatrix}.
$$

Algorithm 8.2.2 forms the basis of the symmetric QR algorithm—the standard means for computing the Schur decomposition of a dense symmetric matrix.

**Algorithm 8.2.3 (Symmetric QR Algorithm)** Given an $n$-by-$n$ symmetric matrix $A$ and $\epsilon$, a small multiple of the unit roundoff, the following algorithm overwrites $A$ with $Q^T A Q = D + E$ where $Q$ is orthogonal, $D$ is diagonal, and $E$ satisfies $\| E \|_2 \approx \mathbf{u} \| A \|_2$.

Use Algorithm 8.2.1, compute the tridiagonalization
    $T = (P_1 \cdots P_{n-2})^T A (P_1 \cdots P_{n-2})$
**until** $q = n$
    For $i = 1{:}n-1$, set $a_{i+1,i}$ and $a_{i,i+1}$ to zero if
        $|a_{i+1,i}| = |a_{i,i+1}| \le \epsilon(|a_{ii}| + |a_{i+1,i+1}|)$
    Find the largest $q$ and the smallest $p$ such that if

$$T = \begin{bmatrix} T_{11} & 0 & 0 \\ 0 & T_{22} & 0 \\ 0 & 0 & T_{33} \end{bmatrix} \begin{matrix} p \\ n-p-q \\ q \end{matrix}$$
$$\qquad\quad p \qquad n-p-q \qquad q$$

    then $T_{33}$ is diagonal and $T_{22}$ is unreduced.
    **if** $q < n$
        Apply Algorithm 8.2.2 to $T_{22}$:
            $T = \text{diag}(I_p, \bar{Z}, I_q)^T T\, \text{diag}(I_p, \bar{Z}, I_q)$
    **end**
**end**

This algorithm requires about $4n^3/3$ flops if $Q$ is not accumulated and about $9n^3$ flops if $Q$ is accumulated.

**Example 8.2.3** Suppose Algorithm 8.2.3 is applied to the tridiagonal matrix

$$A = \begin{bmatrix} 1 & 2 & 0 & 0 \\ 2 & 3 & 4 & 0 \\ 0 & 4 & 5 & 6 \\ 0 & 0 & 6 & 7 \end{bmatrix}$$

The subdiagonal entries change as follows during the execution of Algorithm 8.2.3:

| Iteration | $a_{21}$ | $a_{32}$ | $a_{43}$ |
|---|---|---|---|
| 1 | 1.6817 | 3.2344 | .8649 |
| 2 | 1.6142 | 2.5755 | .0006 |
| 3 | 1.6245 | 1.6965 | $10^{-13}$ |
| 4 | 1.6245 | 1.6965 | converg. |
| 5 | 1.5117 | .0150 | |
| 6 | 1.1195 | $10^{-9}$ | |
| 7 | .7071 | converg. | |
| 8 | converg. | | |

Upon completion we find $\lambda(A) = \{-2.4848,\ .7046,\ 4.9366,\ 12.831\}$.

As in the unsymmetric QR algorithm, the computed eigenvalues $\hat{\lambda}_i$ obtained via Algorithm 8.2.3 are the exact eigenvalues of a matrix that is near to $A$, i.e., $Q_0^T(A+E)Q_0 = \text{diag}(\lambda_i)$ where $Q_0^T Q_0 = I$ and $\| E \|_2 \approx \mathbf{u}\| A \|_2$. However, unlike the unsymmetric case, we can claim (via Corollary 8.1.3) that the absolute error in each $\hat{\lambda}_i$ is small in the sense that $|\hat{\lambda}_i - \lambda_i| \approx \mathbf{u}\| A \|_2$. If $\hat{Q} = [\,\hat{q}_1, \ldots, \hat{q}_n\,]$ is the computed matrix of orthonormal eigenvectors, then the accuracy of $\hat{q}_i$ depends on the separation of $\lambda_i$ from the remainder of the spectrum. See Theorem 8.1.7.

If all of the eigenvalues and a few of the eigenvectors are desired, then it is advisable not to accumulate $Q$ in Algorithm 8.2.3. Instead, the desired eigenvectors can be found via inverse iteration with $T$. If only a few eigenvalues and eigenvectors are required, then some of the special techniques in §8.4 are appropriate.

### Problems

**P8.2.1** Let $A = \begin{bmatrix} w & x \\ x & z \end{bmatrix}$ be real and suppose we perform the following shifted QR step: $A - zI = UR$, $\bar{A} = RU + zI$. Show that if $\bar{A} = \begin{bmatrix} \bar{w} & \bar{x} \\ \bar{x} & \bar{z} \end{bmatrix}$ then

$$\begin{aligned} \bar{w} &= w + x^2(w-z)/[(w-z)^2 + x^2] \\ \bar{z} &= z - x^2(w-z)/[(w-z)^2 + x^2] \\ \bar{x} &= -x^3/[(w-z)^2 + x^2]. \end{aligned}$$

**P8.2.2** Suppose $A \in \mathbb{R}^{n\times n}$ is symmetric and positive definite and consider the following iteration:

$$\begin{aligned} & A_0 = A \\ & \textbf{for } k = 1, 2, \ldots \\ & \qquad A_{k-1} = G_k G_k^T \qquad \text{(Cholesky)} \\ & \qquad A_k = G_k^T G_k \\ & \textbf{end} \end{aligned}$$

(a) Show that this iteration is defined. (b) Show that if $A = \begin{bmatrix} a & b \\ b & c \end{bmatrix}$ with $a \geq c$ has eigenvalues $\lambda_1 \geq \lambda_2 > 0$, then the $A_k$ converge to $\text{diag}(\lambda_1, \lambda_2)$.

**P8.2.3** Suppose $A \in \mathbb{R}^{n\times n}$ is skew-symmetric ($A^T = -A$). Show how to construct Householder matrices $P_1, \ldots, P_{n-2}$ such that $(P_1 \cdots P_{n-2})^T A (P_1 \cdots P_{n-2})$ is tridiagonal. How many flops are required by your algorithm?

**P8.2.4** Suppose $A \in \mathbb{C}^{n\times n}$ is Hermitian. Show how to construct unitary $Q$ such that $Q^H A Q = T$ is real, symmetric, and tridiagonal.

**P8.2.5** Show that if $A = B + iC$ is Hermitian, then $M = \begin{bmatrix} B & -C \\ C & B \end{bmatrix}$ is symmetric. Relate the eigenvalues and eigenvectors of $A$ and $M$.

**P8.2.6** Rewrite Algorithm 8.2.2 for the case when $A$ is stored in two $n$-vectors. Justify the given flop count.

### Notes and References for Sec. 8.2

The tridiagonalization of a symmetric matrix is discussed in

N.E. Gibbs and W.G. Poole, Jr. (1974). "Tridiagonalization by Permutations," *Comm. ACM 17*, 20–24.

R.S. Martin and J.H. Wilkinson (1968a). "Householder's Tridiagonalization of a Symmetric Matrix," *Numer. Math. 11*, 181-95. See also *HACLA*, pp. 212–26.
H.R. Schwartz (1968). "Tridiagonalization of a Symmetric Band Matrix," *Numer. Math. 12*, 231–41. See also *HACLA*, pp. 273–83.

The first two references contain Algol programs. Algol procedures for the explicit and implicit tridiagonal QR algorithm are given in

H. Bowdler, R.S. Martin, C. Reinsch, and J.H. Wilkinson (1968). "The QR and QL Algorithms for Symmetric Matrices," *Numer. Math. 11*, 293–306. See also *HACLA*, pp. 227–40.
A. Dubrulle, R.S. Martin, and J.H. Wilkinson (1968). "The Implicit QL Algorithm," *Numer. Math. 12*, 377-83. see also *HACLA*, pp. 241–48.

The "QL" algorithm is identical to the QR algorithm except that at each step the matrix $T - \lambda I$ is factored into a product of an orthogonal matrix and a lower triangular matrix. Other papers concerned with these methods include

G.W. Stewart (1970). "Incorporating Original Shifts into the QR Algorithm for Symmetric Tridiagonal Matrices," *Comm. ACM 13*, 365–67.
A. Dubrulle (1970). "A Short Note on the Implicit QL Algorithm for Symmetric Tridiagonal Matrices," *Numer. Math. 15*, 450.

Extensions to Hermitian and skew-symmetric matrices are described in

D. Mueller (1966). "Householder's Method for Complex Matrices and Hermitian Matrices," *Numer. Math. 8*, 72–92.
R.C. Ward and L.J. Gray (1978). "Eigensystem Computation for Skew-Symmetric and A Class of Symmetric Matrices," *ACM Trans. Math. Soft. 4*, 278–85.

The convergence properties of Algorithm 8.2.3 are detailed in Lawson and Hanson (SLS, app. B), as well as in

T.J. Dekker and J.F. Traub (1971). "The Shifted QR Algorithm for Hermitian Matrices," *Lin. Alg. and Its Applic. 4*, 137–54.
W. Hoffman and B.N. Parlett (1978). "A New Proof of Global Convergence for the Tridiagonal QL Algorithm," *SIAM J. Num. Anal. 15*, 929–37.
J.H. Wilkinson (1968b). "Global Convergence of Tridiagonal QR Algorithm With Origin Shifts," *Lin. Alg. and Its Applic. I*, 409–20.

For an analysis of the method when it is applied to normal matrices see

C.P. Huang (1981). "On the Convergence of the QR Algorithm with Origin Shifts for Normal Matrices," *IMA J. Num. Anal. 1*, 127–33.

Interesting papers concerned with shifting in the tridiagonal QR algorithm include

F.L. Bauer and C. Reinsch (1968). "Rational QR Transformations with Newton Shift for Symmetric Tridiagonal Matrices," *Numer. Math. 11*, 264–72. See also *HACLA*, pp. 257–65.
G.W. Stewart (1970). "Incorporating Origin Shifts into the QR Algorithm for Symmetric Tridiagonal Matrices," *Comm. Assoc. Comp. Mach. 13*, 365–67.

Symmetry is but one example of structure that can be exploited by the QR algorithm. Other specialized applications of the QR algorithm include

R. Byers (1983). "Hamiltonian and Symplectic Algorithms for the Algebraic Riccati Equation," PhD Thesis, Center for Applied Mathematics, Cornell University.

R. Byers (1986) "A Hamiltonian QR Algorithm," *SIAM J. Sci. and Stat. Comp. 7,* 212–229.
W. B. Gragg (1986). "The QR Algorithm for Unitary Hessenberg Matrices," *J. Comp. Appl. Math. 16,* 1–8.
C. Van Loan (1984). "A Symplectic Method for Approximating All the Eigenvalues of a Hamiltonian Matrix," *Lin. Alg. and Its Applic. 61,* 233–252.

Some parallel computation possibilities for the algorithms in this section are discussed in

S. Lo, B. Philippe, and A. Sameh (1987). "A Multiprocessor Algorithm for the Symmetric Tridiagonal Eigenvalue Problem," *SIAM J. Sci. and Stat. Comp. 8,* s155-s165.
H.Y. Chang and M. Salama (1988). "A Parallel Householder Tridiagonalization Stratgem Using Scattered Square Decomposition," *Parallel Computing 6,* 297-312.

## 8.3 Computing the SVD

There are important relationships between the singular value decomposition of a matrix $A$ and the Schur decompositions of the symmetric matrices $A^TA$, $AA^T$, and $\begin{bmatrix} 0 & A^T \\ A & 0 \end{bmatrix}$. Indeed, if

$$U^TAV = \text{diag}(\sigma_1, \ldots, \sigma_n)$$

is the SVD of $A \in \mathbb{R}^{m \times n}$ $(m \geq n)$ then

$$V^T(A^TA)V = \text{diag}(\sigma_1^2, \ldots, \sigma_n^2) \in \mathbb{R}^{n \times n} \tag{8.3.1}$$

and

$$U^T(AA^T)U = \text{diag}(\sigma_1^2, \ldots, \sigma_n^2, \underbrace{0, \ldots, 0}_{m-n}) \in \mathbb{R}^{m \times m} \tag{8.3.2}$$

Moreover, if

$$U = \begin{matrix} [\ U_1 & U_2\ ] \\ n & m-n \end{matrix}$$

and we define the orthogonal matrix $Q \in \mathbb{R}^{(m+n)\times(m+n)}$ by

$$Q = \frac{1}{\sqrt{2}} \begin{bmatrix} V & V & 0 \\ U_1 & -U_1 & \sqrt{2}\,U_2 \end{bmatrix}$$

then

$$Q^T \begin{bmatrix} 0 & A^T \\ A & 0 \end{bmatrix} Q = \text{diag}(\sigma_1, \ldots, \sigma_n, -\sigma_1, \ldots, -\sigma_n, \underbrace{0, \ldots, 0}_{m-n}).$$

These connections to the symmetric eigenproblem allow us to adapt the mathematical and algorithmic developments of the previous two sections to the singular value problem.

### 8.3.1 Perturbation Theory and Properties

We first establish some perturbation results for the SVD based on the theorems of §8.1. Recall that $\sigma_i(A)$ denotes the $i$th largest singular value of $A$.

**Theorem 8.3.1** *If $A \in \mathbb{R}^{m \times n}$ then for $k = 1{:}\min\{m,n\}$*

$$\sigma_k(A) = \max_{\substack{\dim(S)=k \\ \dim(T)=k}} \min_{\substack{x \in S \\ y \in T}} \frac{y^T A x}{\|x\|_2 \|y\|_2} = \max_{\dim(S)=k} \min_{x \in S} \frac{\|Ax\|_2}{\|x\|_2}$$

*Note that in this expression $S \subseteq \mathbb{R}^n$ and $T \subseteq \mathbb{R}^m$ are subspaces.*

**Proof.** The right-most characterization follows by applying Theorem 8.1.2 to $A^T A$. The remainder of the proof we leave as an exercise. □

**Corollary 8.3.2** *If $A$ and $A+E$ are in $\mathbb{R}^{m \times n}$ with $m \geq n$, then for $k = 1{:}n$*

$$|\sigma_k(A+E) - \sigma_k(A)| \leq \sigma_1(E) = \|E\|_2.$$

**Proof.** Apply Corollary 8.1.3 to

$$\begin{bmatrix} 0 & A^T \\ A & 0 \end{bmatrix} \quad \text{and} \quad \begin{bmatrix} 0 & (A+E)^T \\ A+E & 0 \end{bmatrix}. \square$$

**Example 8.3.1** If

$$A = \begin{bmatrix} 1 & 4 \\ 2 & 5 \\ 3 & 6 \end{bmatrix} \quad \text{and} \quad A+E = \begin{bmatrix} 1 & 4 \\ 2 & 5 \\ 3 & 6.01 \end{bmatrix}$$

then $\sigma(A) = \{9.5080, .7729\}$ and $\sigma(A+E) = \{9.5145, .7706\}$. It is clear that for $i = 1{:}2$ we have $|\sigma_i(A+E) - \sigma_i(A)| \leq \|E\|_2 = .01$.

**Corollary 8.3.3** *Let $A = [\, a_1, \ldots, a_n \,] \in \mathbb{R}^{m \times n}$ be a column partitioning with $m \geq n$. If $A_r = [\, a_1, \ldots, a_r \,]$ then for $r = 1{:}n-1$*

$$\sigma_1(A_{r+1}) \geq \sigma_1(A_r) \geq \sigma_2(A_{r+1}) \geq \cdots \geq \sigma_r(A_{r+1}) \geq \sigma_r(A_r) \geq \sigma_{r+1}(A_{r+1}).$$

**Proof.** Apply Corollary 8.1.4 to $A^T A$. □

This last result says that by adding a column to a matrix, the largest singular value increases and the smallest singular value is diminished.

**Example 8.3.2**

$$A = \begin{bmatrix} 1 & 6 & 11 \\ 2 & 7 & 12 \\ 3 & 8 & 13 \\ 4 & 8 & 14 \\ 5 & 10 & 15 \end{bmatrix} \quad \Rightarrow \quad \left\{ \begin{array}{rcl} \sigma(A_1) & = & \{7.4162\} \\ \sigma(A_2) & = & \{19.5377, 1.8095\} \\ \sigma(A_3) & = & \{35.1272, 2.4654, 0.0000\} \end{array} \right.$$

thereby confirming Corollary 8.3.3.

The next result is a Wielandt-Hoffman theorem for singular values:

**Theorem 8.3.4** *If $A$ and $A+E$ are in $\mathbb{R}^{m\times n}$ with $m \geq n$, then*

$$\sum_{k=1}^{n} (\sigma_k(A+E) - \sigma_k(A))^2 \;\leq\; \| E \|_F^2 .$$

**Proof.** Apply Theorem 8.1.5 to $\begin{bmatrix} 0 & A^T \\ A & 0 \end{bmatrix}$ and $\begin{bmatrix} 0 & (A+E)^T \\ A+E & 0 \end{bmatrix}$. □

**Example 8.3.3** If

$$A \;=\; \begin{bmatrix} 1 & 4 \\ 2 & 5 \\ 3 & 6 \end{bmatrix} \qquad \text{and} \qquad A+E \;=\; \begin{bmatrix} 1 & 4 \\ 2 & 5 \\ 3 & 6.01 \end{bmatrix}$$

then

$$\sum_{k=1}^{2} (\sigma_k(A+E) - \sigma_k(A))^2 \;=\; .472 \times 10^{-4} \;\leq\; 10^{-4} \;=\; \| E \|_F^2 .$$

See Example 8.3.1.

For $A \in \mathbb{R}^{m\times n}$ we say that the $k$-dimensional subspaces $S \subseteq \mathbb{R}^n$ and $T \subseteq \mathbb{R}^m$ form a *singular subspace pair* if $x \in S$ and $y \in T$ imply $Ax \in T$ and $A^T y \in S$. The following result is concerned with the perturbation of singular subspace pairs.

**Theorem 8.3.5** *Let $A, E \in \mathbb{R}^{m\times n}$ with $m \geq n$ be given and suppose that $V \in \mathbb{R}^{n\times n}$ and $U \in \mathbb{R}^{m\times m}$ are orthogonal. Assume that*

$$V = [\; \underset{k}{V_1} \quad \underset{n-k}{V_2} \;] \qquad\qquad U = [\; \underset{k}{U_1} \quad \underset{m-k}{U_2} \;]$$

*and that* range$(V_1)$ *and* range$(U_1)$ *form a singular subspace pair for $A$. Let*

$$U^H A V \;=\; \begin{bmatrix} A_{11} & 0 \\ 0 & A_{22} \end{bmatrix} \begin{matrix} k \\ m-k \end{matrix}$$
$$\qquad\qquad\quad\; k \quad n-k$$

$$U^H E V \;=\; \begin{bmatrix} E_{11} & E_{12} \\ E_{21} & E_{22} \end{bmatrix} \begin{matrix} k \\ m-k \end{matrix}$$
$$\qquad\qquad\quad\; k \quad n-k$$

*and define $\epsilon \;=\; \| \,[E_{21} \; E_{12}^T]\, \|_F$. Assume that*

$$0 < \delta \;=\; \min_{\substack{\sigma\in\sigma(A_{11}) \\ \gamma\in\sigma(A_{22})}} |\sigma \;-\; \gamma| - \| E_{11} \|_2 - \| E_{22} \|_2 .$$

*If $\epsilon/\delta \leq 1/2$, then there exist matrices $P \in \mathbb{R}^{(n-k)\times k}$ and $Q \in \mathbb{R}^{(m-k)\times k}$ satisfying*

$$\left\| \begin{bmatrix} Q \\ P \end{bmatrix} \right\|_F \leq 2\,\frac{\epsilon}{\delta}$$

*such that* range$(V_1 + V_2Q)$ *and* range$(U_1 + U_2P)$ *are a pair of singular subspaces for $A + E$.*

**Proof.** See Stewart (1973b), Theorem 6.4. □

Roughly speaking, the theorem says that $O(\epsilon)$ changes in $A$ can alter a singular subspace by an amount $\epsilon/\delta$, where $\delta$ measures the isolation of the relevant singular values.

**Example 8.3.4** The matrix $A = \text{diag}(2.000, 1.001, .999) \in \mathbb{R}^{4\times 3}$ has singular subspace pairs (span$\{v_i\}$, span$\{u_i\}$) for $i$ = 1, 2, 3 where $v_i = e_i^{(3)}$ and $u_i = e_i^{(4)}$. Suppose

$$A + E = \begin{bmatrix} 2.000 & .010 & .010 \\ .010 & 1.001 & .010 \\ .010 & .010 & .999 \\ .010 & .010 & .010 \end{bmatrix}$$

The corresponding columns of the matrices

$$\hat{U} = [\,\hat{u}_1\ \hat{u}_2\ \hat{u}_3\,] = \begin{bmatrix} .9999 & -.0144 & .0007 \\ .0101 & .7415 & .6708 \\ .0101 & .6707 & -.7616 \\ .0051 & .0138 & -.0007 \end{bmatrix}$$

$$\hat{V} = [\,\hat{v}_1\ \hat{v}_2\ \hat{v}_3\,] = \begin{bmatrix} .9999 & -.0143 & .0007 \\ .0101 & .7416 & .6708 \\ .0101 & .6707 & -.7416 \end{bmatrix}$$

define singular subspace pairs for $A+E$. Note that the pair {span$\{\hat{v}_i\}$, span$\{\hat{u}_i\}$}, is close to {span$\{v_i\}$, span$\{u_i\}$} for $i$ = 1 but not for $i$ = 2 or 3. On the other hand, the singular subspace pair {span$\{\hat{v}_2, \hat{v}_3\}$, span$\{\hat{u}_2, \hat{u}_3\}$} is close to {span $\{v_2, v_3\}$, span$\{u_2, u_3\}$}.

## 8.3.2 The SVD Algorithm

We now show how a variant of the QR algorithm can be used to compute the SVD of an $A \in \mathbb{R}^{m\times n}$ with $m \geq n$. At first glance, this appears straightforward. Equation (8.3.1) suggests that we

- form $C = A^TA$,
- use the symmetric QR algorithm to compute $V_1^TCV_1 = \text{diag}(\sigma_i^2)$,
- apply QR with column pivoting to $AV_1$ obtaining $U^T(AV_1)\Pi = R$.

Since $R$ has orthogonal columns, it follows that $U^T A(V_1\Pi)$ is diagonal. However, as we saw in Example 5.3.1, the formation of $A^T A$ can lead to a loss of information. The situation is not quite so bad here, since the original $A$ is used to compute $U$.

A preferable method for computing the SVD is described in Golub and Kahan (1965). Their technique finds $U$ and $V$ simultaneously by *implicitly* applying the symmetric QR algorithm to $A^T A$. The first step is to reduce $A$ to upper bidiagonal form using Algorithm 5.4.2:

$$U_B^T A V_B = \begin{bmatrix} B \\ 0 \end{bmatrix} \qquad B = \begin{bmatrix} d_1 & f_1 & & \cdots & 0 \\ 0 & d_2 & \ddots & & \vdots \\ & \ddots & \ddots & \ddots & \\ \vdots & & \ddots & \ddots & f_{n-1} \\ 0 & \cdots & & 0 & d_n \end{bmatrix} \in \mathbb{R}^{n\times n}.$$

The remaining problem is thus to compute the SVD of $B$. To this end, consider applying an implicit-shift QR step (Algorithm 8.2.2) to the tridiagonal matrix $T = B^T B$:

- Compute the eigenvalue $\lambda$ of

$$T(m{:}n, m{:}n) = \begin{bmatrix} d_m^2 + f_m^2 & d_m f_n \\ d_m f_n & d_n^2 + f_n^2 \end{bmatrix} \qquad m = n-1$$

  that is closer to $d_n^2 + f_m^2$.

- Compute $c_1 = \cos(\theta_1)$ and $s_1 = \sin(\theta_1)$ such that

$$\begin{bmatrix} c_1 & s_1 \\ -s_1 & c_1 \end{bmatrix}^T \begin{bmatrix} d_1^2 - \lambda \\ d_1 f_1 \end{bmatrix} = \begin{bmatrix} * \\ 0 \end{bmatrix}$$

  and set $G_1 = G(1, 2, \theta_1)$.

- Compute Givens rotations $G_2, \ldots, G_{n-1}$ so that if $Q = G_1 \cdots G_{n-1}$ then $Q^T T Q$ is tridiagonal and $Qe_1 = G_1 e_1$.

Note that these calculations require the explicit formation of $B^T B$, which, as we have seen, is unwise from the numerical standpoint.

Suppose instead that we apply the Givens rotation $G_1$ above to $B$ directly. Illustrating with the $n = 6$ case this gives

$$B \leftarrow BG_1 = \begin{bmatrix} \times & \times & 0 & 0 & 0 & 0 \\ + & \times & \times & 0 & 0 & 0 \\ 0 & 0 & \times & \times & 0 & 0 \\ 0 & 0 & 0 & \times & \times & 0 \\ 0 & 0 & 0 & 0 & \times & \times \\ 0 & 0 & 0 & 0 & 0 & \times \end{bmatrix}.$$

We then can determine Givens rotations $U_1$, $V_2$, $U_2$,..., $V_{n-1}$, and $U_{n-1}$ to chase the unwanted nonzero element down the bidiagonal:

$$B \leftarrow U_1^T B = \begin{bmatrix} \times & \times & + & 0 & 0 & 0 \\ 0 & \times & \times & 0 & 0 & 0 \\ 0 & 0 & \times & \times & 0 & 0 \\ 0 & 0 & 0 & \times & \times & 0 \\ 0 & 0 & 0 & 0 & \times & \times \\ 0 & 0 & 0 & 0 & 0 & \times \end{bmatrix}$$

$$B \leftarrow BV_2 = \begin{bmatrix} \times & \times & 0 & 0 & 0 & 0 \\ 0 & \times & \times & 0 & 0 & 0 \\ 0 & + & \times & \times & 0 & 0 \\ 0 & 0 & 0 & \times & \times & 0 \\ 0 & 0 & 0 & 0 & \times & \times \\ 0 & 0 & 0 & 0 & 0 & \times \end{bmatrix}$$

$$B \leftarrow U_2^T B = \begin{bmatrix} \times & \times & 0 & 0 & 0 & 0 \\ 0 & \times & \times & + & 0 & 0 \\ 0 & 0 & \times & \times & 0 & 0 \\ 0 & 0 & 0 & \times & \times & 0 \\ 0 & 0 & 0 & 0 & \times & \times \\ 0 & 0 & 0 & 0 & 0 & \times \end{bmatrix}$$

and so on. The process terminates with a new bidiagonal $\bar{B}$ that is related to $B$ as follows:

$$\bar{B} = (U_{n-1} \cdots U_1^T) B (G_1 V_2 \cdots V_{n-1}) = \bar{U}^T B \bar{V}.$$

Since each $V_i$ has the form $V_i = G(i, i+1, \theta_i)$ where $i = 2{:}n-1$, it follows that $Ve_1 = Qe_1$. By the implicit $Q$ theorem we can assert that $V$ and $Q$ are essentially the same. Thus, we can implicitly effect the transition from $T$ to $\bar{T} = \bar{B}^T \bar{B}$ by working directly on the bidiagonal matrix $B$.

Of course, for these claims to hold it is necessary that the underlying tridiagonal matrices be unreduced. Since the subdiagonal entries of $B^T B$ are of the form $d_{i-1}, f_i$, it is clear that we must search the bidiagonal band for zeros. If $f_k = 0$ for some $k$, then

$$B = \begin{bmatrix} B_1 & 0 \\ 0 & B_2 \end{bmatrix} \begin{matrix} k \\ n-k \end{matrix}$$
$$\quad\;\; \begin{matrix} k & n-k \end{matrix}$$

and the original SVD problem decouples into two smaller problems involving the matrices $B_1$and $B_2$. If $d_k = 0$ for some $k$, then premultiplication by a sequence of Givens transformations can zero $f_k$. For example, if $n =$

6 and $k = 3$ then by rotating in row planes (3,4), (3,5), and (3,6) we can zero the entire third row:

$$B = \begin{bmatrix} \times & \times & 0 & 0 & 0 & 0 \\ 0 & \times & \times & 0 & 0 & 0 \\ 0 & 0 & 0 & \times & 0 & 0 \\ 0 & 0 & 0 & \times & \times & 0 \\ 0 & 0 & 0 & 0 & \times & \times \\ 0 & 0 & 0 & 0 & 0 & \times \end{bmatrix} \xrightarrow{(3,4)} \begin{bmatrix} \times & \times & 0 & 0 & 0 & 0 \\ 0 & \times & \times & 0 & 0 & 0 \\ 0 & 0 & 0 & 0 & + & 0 \\ 0 & 0 & 0 & \times & \times & 0 \\ 0 & 0 & 0 & 0 & \times & \times \\ 0 & 0 & 0 & 0 & 0 & \times \end{bmatrix}$$

$$\xrightarrow{(3,5)} \begin{bmatrix} \times & \times & 0 & 0 & 0 & 0 \\ 0 & \times & \times & 0 & 0 & 0 \\ 0 & 0 & 0 & 0 & 0 & + \\ 0 & 0 & 0 & \times & \times & 0 \\ 0 & 0 & 0 & 0 & \times & \times \\ 0 & 0 & 0 & 0 & 0 & \times \end{bmatrix} \xrightarrow{(3,6)} \begin{bmatrix} \times & \times & 0 & 0 & 0 & 0 \\ 0 & \times & \times & 0 & 0 & 0 \\ 0 & 0 & 0 & 0 & 0 & 0 \\ 0 & 0 & 0 & \times & \times & 0 \\ 0 & 0 & 0 & 0 & \times & \times \\ 0 & 0 & 0 & 0 & 0 & \times \end{bmatrix}$$

Thus, we can decouple whenever $f_1 \cdots f_{n-1} = 0$ or $d_1 \cdots d_{n-1} = 0$.

**Algorithm 8.3.1 (Golub-Kahan SVD Step)** Given a bidiagonal matrix $B \in \mathbb{R}^{m \times n}$ having no zeros on its diagonal or superdiagonal, the following algorithm overwrites $B$ with the bidiagonal matrix $\bar{B} = \bar{U}^T B \bar{V}$ where $\bar{U}$ and $\bar{V}$ are orthogonal and $\bar{V}$ is essentially the orthogonal matrix that would be obtained by applying Algorithm 8.2.2 to $T = B^T B$.

Let $\mu$ be the eigenvalue of the trailing 2-by-2 submatrix of $T = B^T B$ that is closer to $t_{nn}$.

$y = t_{11} - \mu$
$z = t_{12}$
**for** $k = 1{:}n-1$
  Determine $c = \cos(\theta)$ and $s = \sin(\theta)$ such that

$$\begin{bmatrix} y & z \end{bmatrix} \begin{bmatrix} c & s \\ -s & c \end{bmatrix} = \begin{bmatrix} * & 0 \end{bmatrix}$$

  $B = BG(k, k+1, \theta)$
  $y = b_{kk};\ z = b_{k+1,k}$
  Determine $c = \cos(\theta)$ and $s = \sin(\theta)$ such that

$$\begin{bmatrix} c & s \\ -s & c \end{bmatrix}^T \begin{bmatrix} y \\ z \end{bmatrix} = \begin{bmatrix} * \\ 0 \end{bmatrix}$$

  $B = G(k, k+1, \theta)^T B$
  **if** $k < n-1$
    $y = b_{k,k+1};\ z = b_{k,k+2}$
  **end**
**end**

An efficient implementation of this algorithm would store $B$'s diagonal and superdiagonal in vectors $a(1{:}n)$ and $f(1{:}n-1)$ respectively and would require $30n$ flops and $2n$ square roots. Accumulating $U$ requires $6mn$ flops. Accumulating $V$ requires $6n^2$ flops.

Typically, after a few of the above SVD iterations, the superdiagonal entry $f_{n-1}$ becomes negligible. Criteria for smallness within $B$'s band are usually of the form

$$\begin{aligned} |f_i| &\le \epsilon(\,|d_i| + |d_{i+1}|\,) \\ |d_i| &\le \epsilon \,\|\, B \,\| \end{aligned}$$

where $\epsilon$ is a small multiple of the unit roundoff and $\|\cdot\|$ is some computationally convenient norm.

Combining Algorithm 5.4.2 (bidiagonalization), Algorithm 8.3.1, and the decoupling calculations mentioned earlier gives

**Algorithm 8.3.2 (The SVD Algorithm)** Given $A \in \mathbb{R}^{m\times n}$ $(m \ge n)$ and $\epsilon$, a small multiple of the unit roundoff, the following algorithm overwrites $A$ with $U^TAV = D + E$, where $U \in \mathbb{R}^{m\times n}$ is orthogonal, $V \in \mathbb{R}^{n\times n}$ is orthogonal, $D \in \mathbb{R}^{m\times n}$ is diagonal, and $E$ satisfies $\|\, E \,\|_2 \approx \mathbf{u}\|\, A \,\|_2$.

Use Algorithm 5.4.2 to compute the bidiagonalization

$$B \leftarrow (U_1 \cdots U_n)^T A (V_1 \cdots V_{n-2})$$

**until** $q = n$
    Set $a_{i,i+1}$ to zero if $|a_{i,i+1}| \le \epsilon(|a_{ii}| + |a_{i+1,i+1}|)$
        for any $i = 1{:}n-1$.
    Find the largest $q$ and the smallest $p$ such that if

$$B = \begin{bmatrix} B_{11} & 0 & 0 \\ 0 & B_{22} & 0 \\ 0 & 0 & B_{33} \end{bmatrix} \begin{matrix} p \\ n-p-q \\ q \end{matrix}$$
$$\qquad\quad p \qquad n-p-q \qquad q$$

        then $B_{33}$ is diagonal and $B_{22}$ has nonzero superdiagonal.
    **if** $q < n$
        **if** any diagonal entry in $B_{22}$ is zero, then zero
            the superdiagonal entry in the same row.
        **else**
            Apply Algorithm 8.3.1 to $B_{22}$,
            $B = \text{diag}(I_p, U, I_{q+m-n})^T B\,\text{diag}(I_p, V, I_q)$
        **end**
    **end**
**end**

The amount of work required by this algorithm and its numerical properties are discussed in §5.4.5 and §5.5.8.

**Example 8.3.5** If Algorithm 8.3.2 is applied to

$$A = \begin{bmatrix} 1 & 1 & 0 & 0 \\ 0 & 2 & 1 & 0 \\ 0 & 0 & 3 & 1 \\ 0 & 0 & 0 & 4 \end{bmatrix}$$

then the superdiagonal elements converge to zero as follows:

| Iteration | $O(\|a_{21}\|)$ | $O(\|a_{32}\|)$ | $O(\|a_{43}\|)$ |
|---|---|---|---|
| 1 | $10^0$ | $10^0$ | $10^0$ |
| 2 | $10^0$ | $10^0$ | $10^0$ |
| 3 | $10^0$ | $10^0$ | $10^0$ |
| 4 | $10^0$ | $10^{-1}$ | $10^{-2}$ |
| 5 | $10^0$ | $10^{-1}$ | $10^{-8}$ |
| 6 | $10^0$ | $10^{-1}$ | $10^{-27}$ |
| 7 | $10^0$ | $10^{-1}$ | converg. |
| 8 | $10^0$ | $10^{-4}$ | |
| 9 | $10^{-1}$ | $10^{-14}$ | |
| 10 | $10^{-1}$ | converg. | |
| 11 | $10^{-4}$ | | |
| 12 | $10^{-12}$ | | |
| 13 | converg. | | |

Observe the cubic-like convergence.

## Problems

**P8.3.1** Show that if $B \in \mathbb{R}^{n \times n}$ is an upper bidiagonal matrix having a repeated singular value, then $B$ must have a zero on its diagonal or superdiagonal.

**P8.3.2** Give formulae for the eigenvectors of $\begin{bmatrix} 0 & A^T \\ A & 0 \end{bmatrix}$ in terms of the singular vectors of $A \in \mathbb{R}^{m \times n}$ where $m \geq n$.

**P8.3.3** Give an algorithm for reducing a complex matrix $A$ to *real* bidiagonal form using complex Householder transformations.

**P8.3.4** Relate the singular values and vectors of $A = B + iC$ $(B, C \in \mathbb{R}^{m \times n})$ to those of $\begin{bmatrix} B & -C \\ C & B \end{bmatrix}$.

**P8.3.5** Complete the proof of Theorem 8.3.1.

## Notes and References for Sec. 8.3

The SVD and its mathematical properties are discussed in

A.R. Amir-Moez (1965). *Extremal Properties of Linear Transformations and Geometry of Unitary Spaces,* Texas Tech University Mathematics Series, no. 243, Lubbock, Texas.

G.W. Stewart (1973b). "Error and Perturbation Bounds for Subspaces Associated with Certain Eigenvalue Problems," *SIAM Review 15,* 727–64.

P.A. Wedin (1972). "Perturbation Bounds in Connection with the Singular Value Decomposition," *BIT 12*, 99–111.

The last two references are expressly concerned with perturbation properties of the decomposition. Other papers dealing with singular value sensitivity include

A. Ruhe (1975). "On the Closeness of Eigenvalues and Singular Values for Almost Normal Matrices," *Lin. Alg. and Its Applic. 11*, 87–94.
G.W. Stewart (1979). "A Note on the Perturbation of Singular Values," *Lin. Alg. and Its Applic. 28*, 213–16.
G.W. Stewart (1984). "A Second Order Perturbation Expansion for Small Singular Values," *Lin. Alg. and Its Applic. 56*, 231–236.

The idea of adapting the symmetric QR algorithm to compute the SVD first appeared in

G.H. Golub and W. Kahan (1965). "Calculating the Singular Values and Pseudo-Inverse of a Matrix," *SIAM J. Num. Anal. Ser. B 2*, 205–24.

and the first working program in

P.A. Businger and G.H. Golub (1969). "Algorithm 358: Singular Value Decomposition of a Complex Matrix," *Comm. Assoc. Comp. Mach. 12*, 564–65.

The Algol procedure in the following article is the basis for the Eispack 2 and Linpack SVD subroutines:

G.H. Golub and C. Reinsch (1970). "Singular Value Decomposition and Least Squares Solutions," *Numer. Math. 14*, 403–20. See also *HACLA*, pp. 134–51.

Other Fortran SVD programs are given in Lawson and Hanson (SLS, chapter 9) and in

G.E. Forsythe, M. Malcolm and C.B. Moler (1977). *Computer Methods for Mathematical Computations*, Prentice-Hall, Englewood Cliffs, NJ.

Interesting algorithmic developments associated with the SVD appear in

J.J.M. Cuppen (1983). "The Singular Value Decomposition in Product Form," *SIAM J. Sci. and Stat. Comp. 4*, 216–222.
J.J. Dongarra (1983). "Improving the Accuracy of Computed Singular Values," *SIAM J. Sci. and Stat. Comp. 4*, 712–719.
S. Van Huffel, J. Vandewalle, and A. Haegemans (1987). "An Efficient and Reliable Algorithm for Computing the Singular Subspace of a Matrix Associated with its Smallest Singular Values," *J. Comp. and Appl. Math. 19*, 313–330.

The SVD has many applications in statistical computation. See

S.J. Hammarling (1985). "The Singular Value Decomposition in Multivariate Statistics," *ACM SIGNUM Newsletter 20*, 2-25.

Recall the connection made in P4.2.9 between the polar decomposition $A = QP$ and the SVD $A = U\Sigma V^T$. Of course, Algorithm 8.3.2 could be used to compute $Q = UV^T$. However, it is often preferable to invoke the algorithm described in

N.J. Higham (1986a). "Computing the Polar Decomposition—with Applications," *SIAM J. Sci. and Stat. Comp. 7*, 1160–1174.

Finally, we mention that there has been some research into special SVD procedures for specially structured matrices. See

A. Bunse-Gerstner and W.B. Gragg (1988). "Singular Value Decompositions of Complex Symmetric Matrices," *J. Comp. Applic. Math. 21*, 41–54.

# 8.4 Some Special Methods

By exploiting the rich mathematical structure of the symmetric eigenproblem, it is possible to devise useful alternatives to the symmetric QR algorithm. Many of these techniques are appropriate when only a few eigenvalues and/or eigenvectors are desired. Three such methods are described in this section: bisection, Rayleigh quotient iteration, and orthogonal iteration with Ritz acceleration.

## 8.4.1 Bisection

Let $T_r$ denote the leading $r$-by-$r$ principal submatrix of

$$T = \begin{bmatrix} a_1 & b_1 & & \cdots & 0 \\ b_1 & a_2 & \ddots & & \vdots \\ & \ddots & \ddots & \ddots & \\ \vdots & & \ddots & \ddots & b_{n-1} \\ 0 & \cdots & & b_{n-1} & a_n \end{bmatrix} \qquad (8.4.1)$$

and define the polynomials $p_r(x) = \det(T_r - xI)$, $r = 1{:}n$. A simple determinantal expansion can be used to show that

$$p_r(x) = (a_r - x)p_{r-1}(x) - b_{r-1}^2 p_{r-2}(x) \qquad (8.4.2)$$

for $r = 2{:}n$ if we set $p_0(x) = 1$. Because $p_n(x)$ can be evaluated in $O(n)$ flops, it is feasible to find its roots by using the method of bisection. For example, if $p_n(y)p_n(z) < 0$ and $y < z$, then the iteration

```
while |y − z| > ε(|y| + |z|)
    x = (y + z)/2
    if pn(x)pn(y) < 0
        z = x
    else
        y = x
    end
end
```

is guaranteed to terminate with $(y+z)/2$ an approximate zero of $p_n(x)$, i.e., an approximate eigenvalue of $T$. The iteration converges linearly in that the error is approximately halved at each step.

Sometimes it is necessary to compute the $k$th largest eigenvalue of $T$ for some prescribed value of $k$. This can be done efficiently by using the bisection idea and the following classical result:

**Theorem 8.4.1 (Sturm Sequence Property)** *If the tridiagonal matrix in (8.4.1) is unreduced, then the eigenvalues of $T_{r-1}$ strictly separate the eigenvalues of $T_r$:*

$$\lambda_r(T_r) < \lambda_{r-1}(T_{r-1}) < \lambda_{r-1}(T_r) < \cdots < \lambda_2(T_r) < \lambda_1(T_{r-1}) < \lambda_1(T_r).$$

*Moreover, if $a(\lambda)$ denotes the number of sign changes in the sequence*

$$\{\, p_0(\lambda),\ p_1(\lambda), \ldots,\ p_n(\lambda)\,\}$$

*then $a(\lambda)$ equals the number of $T$'s eigenvalues that are less than $\lambda$. Here, the polynomials $p_r(x)$ are defined by (8.5.2) and we have the convention that $p_r(\lambda)$ has the opposite sign of $p_{r-1}(\lambda)$ if $p_r(\lambda) = 0$.*

**Proof.** It follows from Corollary 8.1.4 that the eigenvalues of $T_{r-1}$ weakly separate those of $T_r$. To prove that the separation must be strict, suppose that $p_r(\mu) = p_{r-1}(\mu) = 0$ for some $r$ and $\mu$. It then follows from (8.4.2) and the assumption that $T$ is unreduced that $p_0(\mu) = p_1(\mu) = \cdots = p_r(\mu) = 0$, a contradiction. Thus, we must have strict separation.

The assertion about $a(\lambda)$ is established in Wilkinson (AEP, pp. 300-301). □

**Example 8.4.1** If

$$T = \begin{bmatrix} 1 & -1 & 0 & 0 \\ -1 & 2 & -1 & 0 \\ 0 & -1 & 3 & -1 \\ 0 & 0 & -1 & 4 \end{bmatrix}$$

then $\lambda(T) \approx \{.254,\ 1.82,\ 3.18,\ 4.74\}$. The sequence

$$\{\, p_0(2),\ p_1(2),\ p_2(2),\ p_3(2),\ p_4(2)\,\} \;=\; \{\, 1,\ -1,\ -1,\ 0,\ 1\,\}$$

confirms that there are two eigenvalues less than $\lambda = 2$.

Suppose we wish to compute $\lambda_k(T)$. From the Gershgorin circle theorem (Theorem 7.2.1) it follows that $\lambda_k(T) \in [y, z]$ where

$$y = \min_{1\le i\le n} a_i - |b_i| - |b_{i-1}| \qquad z = \max_{1\le i\le n} a_i + |b_i| + |b_{i-1}|$$

if we define $b_0 = b_n = 0$. With these starting values, it is clear from the Sturm sequence property that the iteration

```
while |z - y| > u(|y| + |z|)
    x = (y + z)/2
    if a(x) >= k                                    (8.4.3)
        z = x
    else
        y = x
    end
end
```

produces a sequence of subintervals that are repeatedly halved in length but which always contain $\lambda_k(T)$.

**Example 8.4.2** If (8.4.3) is applied to the matrix of Example 8.4.1 with $k = 2$, then the values shown in Table 8.4.1 are generated. We conclude from the output that $\lambda_2(T) \in [\,1.7969,\, 1.9375\,]$. Note: $\lambda_2(T) \approx 1.82$.

| $y$ | $z$ | $x$ | $a(x)$ |
|---|---|---|---|
| 0.0000 | 5.0000 | 2.5000 | 2 |
| 0.0000 | 2.5000 | 1.2500 | 1 |
| 1.2500 | 2.5000 | 1.3750 | 1 |
| 1.3750 | 2.5000 | 1.9375 | 2 |
| 1.3750 | 1.9375 | 1.6563 | 1 |
| 1.6563 | 1.9375 | 1.7969 | 1 |

During the execution of (8.4.3), information about the location of other eigenvalues is obtained. By systematically keeping track of this information it is possible to devise an efficient scheme for computing "contiguous" subsets of $\lambda(T)$, e.g., $\lambda_k(T), \lambda_{k+1}(T), \ldots, \lambda_{k+j}(T)$. See Barth, Martin, and Wilkinson (1967).

If selected eigenvalues of a general symmetric matrix $A$ are desired, then it is necessary first to compute the tridiagonalization $T = U_0^T T U_0$ before the above bisection schemes can be applied. This can be done using Algorithm 8.2.1 or by the Lanczos algorithm discussed in the next chapter. In either case, the corresponding eigenvectors can be readily found via inverse iteration (see §7.6), since tridiagonal systems can be solved in $O(n)$ flops.

In those applications where the original matrix $A$ already has tridiagonal form, bisection computes eigenvalues with small relative error, regardless of their magnitude. This is in contrast to the tridiagonal QR iteration, where the computed eigenvalues $\tilde{\lambda}_i$ can be guaranteed only to have small absolute error: $|\tilde{\lambda}_i - \lambda_i(T)| \approx \mathbf{u}\|\, T\,\|_2$

Finally, it is possible to compute specific eigenvalues of a symmetric matrix by using the $LDL^T$ factorization (see §4.2) and exploiting the Sylvester inertia theorem (Theorem 8.1.12). If

$$A - \mu I \;=\; LDL^T \qquad A = A^T \in \mathbb{R}^{n\times n}$$

is the $LDL^T$ factorization of $A - \mu I$ with $D = \text{diag}(d_1, \ldots, d_n)$, then the number of negative $d_i$ equals the number of $\lambda_i(A)$ that are less than $\mu$. See Parlett (SEP, pp. 46 ff.) for details.

### 8.4.2 Rayleigh Quotient Iteration

Suppose $A \in \mathbb{R}^{n \times n}$ is symmetric and that $x$ is a given nonzero $n$-vector. A simple differentiation reveals that

$$\lambda = r(x) \equiv \frac{x^T A x}{x^T x}$$

minimizes $\| (A - \lambda I)x \|_2$. The scalar $r(x)$ is called the *Rayleigh quotient* of $x$. Clearly, if $x$ is an approximate eigenvector, then r($x$) is a reasonable choice for the corresponding eigenvalue. On the other hand, if $\lambda$ is an approximate eigenvalue, then inverse iteration theory tells us that the solution to $(A - \lambda I)x = b$ will almost always be a good approximate eigenvector.

Combining these two ideas in the natural way gives rise to the *Rayleigh quotient iteration:*

$$\begin{array}{l} x_0 \text{ given, } \| x_0 \|_2 = 1 \\ \textbf{for } k = 0, 1, \ldots \\ \quad \mu_k = r(x_k) \\ \quad \text{Solve } (A - \mu_k I)z_{k+1} = x_k \text{ for } z_{k+1} \\ \quad x_{k+1} = z_{k+1}/\| z_{k+1} \|_2 \\ \textbf{end} \end{array} \qquad (8.4.4)$$

Note that for any $k$ we have $(A+E_k)z_{k+1} = \mu_k z_{k+1}$ where the perturbation matrix $E_k$ is given by $E_k = -x_k z_{k+1}^T/\| z_{k+1} \|_2^2$ . It follows from Corollary 8.1.3 that $|\mu_k - \lambda| \leq 1/\| z_{k+1} \|_2$ for some $\lambda \in \lambda(A)$.

**Example 8.4.3** If (8.4.4) is applied to

$$A = \begin{bmatrix} 1 & 1 & 1 & 1 & 1 & 1 \\ 1 & 2 & 3 & 4 & 5 & 6 \\ 1 & 3 & 6 & 10 & 15 & 21 \\ 1 & 4 & 10 & 20 & 35 & 56 \\ 1 & 5 & 15 & 35 & 70 & 126 \\ 1 & 6 & 21 & 56 & 126 & 252 \end{bmatrix}$$

with $x_0 = (1, 1, 1, 1, 1, 1)^T/6$, then

| $k$ | $\mu_k$ |
|---|---|
| 0 | 153.8333 |
| 1 | 120.0571 |
| 2 | 49.5011 |
| 3 | 13.8687 |
| 4 | 15.4959 |
| 5 | 15.5534 |

The iteration is converging to the eigenvalue $\lambda = 15.5534732737$.

Parlett (1974b) has shown that (8.4.4) converges globally and that the convergence is ultimately cubic. We demonstrate this for the case $n = 2$. Without loss of generality, we may assume that $A = \text{diag}(\lambda_1, \lambda_2)$, with $\lambda_1 > \lambda_2$. Denoting $x_k$ by

$$x_k = \begin{bmatrix} c_k \\ s_k \end{bmatrix} \qquad c_k^2 + s_k^2 = 1$$

it follows that $\mu_k = \lambda_1 c_k^2 + \lambda_1 s_k^2$ and

$$z_{k+1} = \frac{1}{\lambda_1 - \lambda_2} \begin{bmatrix} c_k/s_k^2 \\ -s_k/c_k^2 \end{bmatrix}.$$

A calculation shows that

$$c_{k+1} = \frac{c_k^3}{\sqrt{c_k^6 + s_k^6}} \qquad s_{k+1} = \frac{-s_k^3}{\sqrt{c_k^6 + s_k^6}}.$$

From these equations it is clear that the $x_k$ converge cubically to either $\text{span}\{e_1\}$ or $\text{span}\{e_2\}$ provided $|c_k| \neq |s_k|$.

Details associated with the practical implementation of the Rayleigh quotient iteration may be found in Parlett (1974b).

It is interesting to note the connection between Rayleigh quotient iteration and the symmetric QR algorithm. Suppose we apply the latter to the tridiagonal matrix $T \in \mathbb{R}^{n\times n}$ with shift $\sigma = e_n^T T e_n = t_{nn}$. If $T - \sigma I = QR$, then we obtain $T = RQ + \sigma I$. From the equation $(T - \sigma I)Q = R^T$ it follows that

$$(T - \sigma I)q_n = r_{nn}e_n,$$

where $q_n$ is the last column of the orthogonal matrix $Q$. Thus, if we apply (8.4.4) with $x_0 = e_n$, then $x_1 = q_n$.

### 8.4.3 Orthogonal Iteration with Ritz Acceleration

Recall the method of orthogonal iteration from §7.3:

$Q_0 \in \mathbb{R}^{n\times p}$ given with $Q_0^T Q_0 = I_p$
**for** $k = 1, 2, \ldots$ (8.4.5)
  $Z_k = AQ_{k-1}$
  $Z_k = Q_k R_k$  (QR factorization)
**end**

Suppose $A$ is symmetric with Schur decomposition $Q^TAQ = \text{diag}(\lambda_i)$ where $Q = [\,q_1,\ldots,q_n\,]$ and $|\lambda_1| > |\lambda_2| > \cdots > |\lambda_n|$. It follows from Theorem 7.3.2 that if $d = \text{dist}\{D_p(A), \text{range}(Q_0)\} < 1$, then

$$\text{dist}\{D_p(A), \text{range}(Q_k)\} \;\le\; \frac{1}{\sqrt{1-d^2}}\left|\frac{\lambda_{p+1}}{\lambda_p}\right|^k$$

where $D_p(A) = \text{span}\,\{q_1,\ldots,q_p\}$. Moreover, from the analysis in §7.3 we know that if $R_k = (r_{ij}^{(k)})$, then

$$|r_{ii}^{(k)} - \lambda_i| \;=\; O\left(\left|\frac{\lambda_{i+1}}{\lambda_i}\right|^k\right) \qquad i = 1{:}p\,.$$

This can be an unacceptably slow rate of convergence if $\lambda_i$ and $\lambda_{i+1}$ are of nearly equal modulus.

This difficulty can be partially surmounted by replacing $Q_k$ with its Ritz vectors at each step:

$Q_0 \in \mathbb{R}^{n\times p}$ given with $Q_0^TQ_0 = I_p$
**for** $k = 1, 2, \ldots$
  $Z_k = AQ_{k-1}$
  $\tilde{Q}_kR_k = Z_k$ (QR factorization)
  $S_k = \tilde{Q}_k^TA\tilde{Q}_k$ (8.4.6)
  $U_k^TS_kU_k = D_k$ (Schur decomposition)
  $Q_k = \tilde{Q}_kU_k$
**end**

It can be shown that if

$$D_k \;=\; \text{diag}(\theta_1^{(k)},\ldots,\theta_p^{(k)}) \qquad |\theta_1^{(k)}| \ge \cdots \ge |\theta_p^{(k)}|$$

then

$$|\theta_i^{(k)} - \lambda_i(A)| \;=\; O\left(\left|\frac{\lambda_{p+1}}{\lambda_i}\right|^k\right) \qquad i = 1{:}p$$

Thus, the Ritz values $\theta_i^{(k)}$ converge at a much more favorable rate than the $r_{ii}^{(k)}$ in (8.4.5). For details, see Stewart (1969).

**Example 8.4.4** If we apply (8.4.6) with

$$A \;=\; \begin{bmatrix} 100 & 1 & 1 & 1 \\ 1 & 99 & 1 & 1 \\ 1 & 1 & 2 & 1 \\ 1 & 1 & 1 & 1 \end{bmatrix} \quad \text{and} \quad Q_0 \;=\; \begin{bmatrix} 1 & 0 \\ 0 & 1 \\ 0 & 0 \\ 0 & 0 \end{bmatrix}$$

then

| $k$ | $\text{dist}\{D_2(A), Q_k\}$ |
|---|---|
| 0 | $.2 \times 10^{-1}$ |
| 1 | $.5 \times 10^{-3}$ |
| 2 | $.1 \times 10^{-4}$ |
| 3 | $.3 \times 10^{-6}$ |
| 4 | $.8 \times 10^{-8}$ |

Clearly, convergence is taking place at the rate $(2/99)^k$.

### Problems

**P8.4.1** Suppose $\lambda$ is an eigenvalue of a symmetric tridiagonal matrix $T$. Show that if $\lambda$ has algebraic multiplicity $k$, then at least $k-1$ of $T$'s subdiagonal elements are zero.

**P8.4.2** Suppose $A$ is symmetric and has bandwidth $p$. Show that if we perform the shifted QR step $A - \mu I = QR$, $A = RQ + \mu I$, then $A$ has bandwidth $p$.

**P8.4.3** Suppose $B \in \mathbb{R}^{n \times n}$ is upper bidiagonal with diagonal entries $d(1{:}n)$ and superdiagonal entries $f(1{:}n-1)$. State and prove a singular value version of Theorem 8.4.1.

### Notes and References for Sec. 8.4

A bisection subroutine is in Eispack. The Algol program upon which it is based is described in

W. Barth, R.S. Martin, and J.H. Wilkinson (1967). "Calculation of the Eigenvalues of a Symmetric Tridiagonal Matrix by the Method of Bisection," *Numer. Math. 9*, 386–93. See also *HACLA*, pp. 249–56.

Another way to compute a specified subset of eigenvalues is via the rational QR algorithm. In this method, the shift is determined using Newton's method. This makes it possible to "steer" the iteration towards desired eigenvalues. See

C. Reinsch and F.L. Bauer (1968). "Rational QR Transformation with Newton's Shift for Symmetric Tridiagonal Matrices," *Numer. Math. 11*, 264–72. See also *HACLA*, pp. 257–65.

Papers concerned with the symmetric QR algorithm for banded matrices include

R.S. Martin and J.H. Wilkinson (1967). "Solution of Symmetric and Unsymmetric Band Equations and the Calculation of Eigenvectors of Band Matrices," *Numer. Math. 9*, 279–301. See also *HACLA*, pp. 70–92.

R.S. Martin, C. Reinsch, and J.H. Wilkinson (1970). "The QR Algorithm for Band Symmetric Matrices," *Numer. Math. 16*, 85–92. See also *HACLA*, pp. 266–72.

The following references are concerned with the method of simultaneous iteration for sparse symmetric matrices

M. Clint and A. Jennings (1970). "The Evaluation of Eigenvalues and Eigenvectors of Real Symmetric Matrices by Simultaneous Iteration," *Comp. J. 13*, 76–80.

H. Rutishauser (1970). "Simultaneous Iteration Method for Symmetric Matrices," *Numer. Math. 16*, 205–23. See also *HACLA*, pp. 284–302.

G.W. Stewart (1969). "Accelerating The Orthogonal Iteration for the Eigenvalues of a Hermitian Matrix," *Numer. Math. 13*, 362–76.

The literature on special methods for the symmetric eigenproblem is vast. Some representative papers include

K.J. Bathe and E.L. Wilson (1973). "Solution Methods for Eigenvalue Problems in Structural Mechanics," *Int. J. Numer. Meth. Eng. 6,* 213-26.

C.F. Bender and I. Shavitt (1970). "An Iterative Procedure for the Calculation of the Lowest Real Eigenvalue and Eigenvector of a Non-Symmetric Matrix," *J. Comp. Physics 6,* 146–49.

K.K. Gupta (1972). "Solution of Eigenvalue Problems by Sturm Sequence Method," *Int. J. Numer. Meth. Eng. 4,* 379–404.

P.S. Jenson (1972). "The Solution of Large Symmetric Eigenproblems by Sectioning," *SIAM J. Num. Anal. 9,* 534–45.

S.F. McCormick (1972). "A General Approach to One-Step Iterative Methods with Application to Eigenvalue Problems," *J. Comput. Sys. Sci. 6,* 354–72.

B.N. Parlett (1974b). "The Rayleigh Quotient Iteration and Some Generalizations for Nonnormal Matrices," *Math. Comp. 28,* 679-93.

A. Ruhe (1974). "SOR Methods for the Eigenvalue Problem with Large Sparse Matrices," *Math. Comp. 28,* 695–710.

A. Sameh, J. Lermit and K. Noh (1975). "On the Intermediate Eigenvalues of Symmetric Sparse Matrices," *BIT 12,* 543–54.

D.S. Scott (1984). "Computing a Few Eigenvalues and Eigenvectors of a Symmetric Band Matrix," *SIAM J. Sci. and Stat. Comp. 5,* 658–666.

G.W. Stewart (1974). "The Numerical Treatment of Large Eigenvalue Problems," *Proc. IFIP Congress 74,* North-Holland, pp. 666–72.

J. Vandergraft (1971). "Generalized Rayleigh Methods with Applications to Finding Eigenvalues of Large Matrices," *Lin. Alg. and Its Applic. 4,* 353–68.

Specially structured dense eigenproblems also submit to special techniques. See

G. Cybenko and C. Van Loan (1986). "Computing the Minimum Eigenvalue of a Symmetric Positive Definite Toeplitz Matrix," *SIAM J. Sci. and Stat. Comp. 7,* 123–131.

## 8.5 Jacobi Methods

Jacobi methods for the symmetric eigenvalue problem attract current attention because they are inherently parallel. They work by performing a sequence of orthogonal similarity updates $A \leftarrow Q^T AQ$ with the property that each new $A$, although full, is "more diagonal" than its predecessor. Eventually, the off-diagonal entries are small enough to be declared zero.

After surveying the basic ideas behind the Jacobi approach we develop a parallel Jacobi procedure that is suitable for a ring multiprocessor. All of the symmetric eigenvalue Jacobi procedures in this section have SVD analogs.

### 8.5.1 The Jacobi Idea

The idea behind Jacobi's method is to systematically reduce the quantity

$$\text{off}(A) = \sqrt{\sum_{i=1}^{n}\sum_{\substack{j=1 \\ j\neq i}}^{n} a_{ij}^2} \,,$$

i.e., the"norm" of the off-diagonal elements. The tools for doing this are rotations of the form

$$J(p,q,\theta) = \begin{bmatrix} 1 & \cdots & 0 & \cdots & 0 & \cdots & 0 \\ \vdots & \ddots & \vdots & & \vdots & & \vdots \\ 0 & \cdots & c & \cdots & s & \cdots & 0 \\ \vdots & & \vdots & \ddots & \vdots & & \vdots \\ 0 & \cdots & -s & \cdots & c & \cdots & 0 \\ \vdots & & \vdots & & \vdots & \ddots & \vdots \\ 0 & \cdots & 0 & \cdots & 0 & \cdots & 1 \end{bmatrix} \begin{matrix} \\ \\ p \\ \\ q \\ \\ \\ \end{matrix}$$

$$\qquad\qquad\qquad\qquad p \qquad\quad q$$

which we call *Jacobi rotations.* Jacobi rotations are no different from Givens rotations, c.f §5.1.8. We submit to the name change in this section to honor the inventor.

The basic step in a Jacobi eigenvalue procedure involves (1) choosing an index pair $(p,q)$ that satisfies $1 \le p < q \le n$, (2) computing a cosine-sine pair $(c,s)$ such that

$$\begin{bmatrix} b_{pp} & b_{pq} \\ b_{qp} & b_{qq} \end{bmatrix} = \begin{bmatrix} c & s \\ -s & c \end{bmatrix}^T \begin{bmatrix} a_{pp} & a_{pq} \\ a_{qp} & a_{qq} \end{bmatrix} \begin{bmatrix} c & s \\ -s & c \end{bmatrix} \tag{8.5.1}$$

is diagonal, and (3) overwriting $A$ with $B = J^TAJ$ where $J = J(p,q,\theta)$. Observe that the matrix $B$ agrees with $A$ except in rows and columns $p$ and $q$. Moreover, since the Frobenius norm is preserved by orthogonal transformations we find that

$$a_{pp}^2 + a_{qq}^2 + 2a_{pq}^2 = b_{pp}^2 + b_{qq}^2 + 2b_{pq}^2 = b_{pp}^2 + b_{qq}^2$$

and so

$$\begin{aligned} \text{off}(B)^2 &= \| B \|_F^2 - \sum_{i=1}^{n} b_{ii}^2 \\ &= \| A \|_F^2 - \sum_{i=1}^{n} a_{ii}^2 \;+\; (a_{pp}^2 + a_{qq}^2 - b_{pp}^2 - b_{qq}^2) \\ &= \text{off}(A)^2 - 2a_{pq}^2 \,. \end{aligned}$$

It is in this sense that $A$ moves closer to diagonal form with each Jacobi step.

Before we discuss how the index pair $(p, q)$ can be chosen, let us look at the actual computations associated with the $(p, q)$ subproblem.

### 8.5.2 The 2-by-2 Symmetric Schur Decomposition

To say that we diagonalize in (8.5.1) is to say that

$$0 = b_{pq} = a_{pq}(c^2 - s^2) + (a_{pp} - a_{qq})cs. \tag{8.5.2}$$

If $a_{pq} = 0$ then we just set $(c, s) = (1,0)$ . Otherwise define

$$\tau = \frac{a_{qq} - a_{pp}}{2a_{pq}} \quad \text{and} \quad t = s/c$$

and conclude from (8.5.2) that $t = \tan(\theta)$ solves the quadratic

$$t^2 + 2\tau t - 1 = 0 .$$

It turns out to be important to select the smaller of the two roots,

$$t = \frac{\text{sign}(\tau)}{|\tau| + \sqrt{1+\tau^2}} \tag{8.5.3}$$

whereupon $c$ and $s$ can be resolved from the formulae

$$c = 1/\sqrt{1+t^2} \qquad s = tc . \tag{8.5.4}$$

Choosing $t$ to be the smaller of the two roots ensures that $|\theta| \leq \pi/4$ and has the effect of minimizing the difference between $B$ and $A$ because

$$\| B - A \|_F^2 = 4(1-c) \sum_{\substack{i=1 \\ i \neq p,q}}^{n} (a_{ip}^2 + a_{iq}^2) + 2a_{pq}^2/c^2$$

We summarize the 2-by-2 computations as follows:

**Algorithm 8.5.1** Given an $n$-by-$n$ symmetric $A$ and integers $p$ and $q$ that satisfy $1 \leq p < q \leq n$, this algorithm computes a cosine-sine pair $(c\ ,\ s)$ such that if $B = J(p,q,\theta)^T AJ(p,q,\theta)$ then $b_{pq} = b_{qp} = 0$.

```
function: (c , s) = sym.schur2(A, p, q)
    if A(p,q) ≠ 0
        τ = (A(q,q) − A(p,p))/(2A(p,q))
        t = sign(τ)/(|τ| + √(1 + τ²)); c = 1/√(1 + t²); s = tc
    else
        c = 1; s = 0
    end
end sym.schur2
```

### 8.5.3 Jacobi Updates

As we mentioned above, only rows and columns $p$ and $q$ are altered when the $(p,q)$ subproblem is solved. Once **sym.schur2** determines the 2-by-2 rotation, then the update $A \leftarrow J(p,q,\theta)^T AJ(p,q,\theta)$ can be implemented in $6n$ flops if we exploit symmetry.

### 8.5.4 The Classical Jacobi Algorithm

We now turn our attention to subproblem determination, i.e., the choice of $(p,q)$. From the standpoint of maximizing the reduction of off$(A)$ in (8.5.2), it makes sense to choose $(p,q)$ so that $a_{pq}^2$ is maximal. This is the basis of the classical Jacobi algorithm.

**Algorithm 8.5.2 (Classical Jacobi)**
Given a symmetric $A \in \mathbb{R}^{n\times n}$ and a tolerance $tol > 0$, this algorithm overwrites $A$ with $V^TAV$ where $V$ is orthogonal and $\text{off}(V^TAV) \le tol\| A \|_F$.

$V = I_n$; $eps = tol\| A \|_F$
**while** off$(A) > eps$
  Choose indices $p$ and $q$ with $1 \le p < q \le n$ so
    $|a_{pq}| = \max_{i\neq j} |a_{ij}|$.
  $(c\ ,\ s) =$ **sym.schur2**$(A,p,q)$
  $A = J(p,q,\theta)^T AJ(p,q,\theta)$
  $V = VJ(p,q,\theta)$
**end**

Since $|a_{pq}|$ is the largest off-diagonal entry it follows that

$$\text{off}(A)^2 \;\le\; N(a_{pq}^2 + a_{qp}^2)$$

where $N = n(n-1)/2$. From (8.5.1) we therefore obtain

$$\text{off}(B)^2 \;\le\; \left(1 - \frac{1}{N}\right)\text{off}(A)^2 .$$

By induction, if $A^{(k)}$ denotes the matrix $A$ after $k$ Jacobi updates, then

$$\text{off}(A^{(k)})^2 \;\le\; \left(1 - \frac{1}{N}\right)^k \text{off}(A^{(0)})^2.$$

This implies that the classical Jacobi procedure converges at a linear rate.

However, the asymptotic convergence rate of the method is considerably better than linear. Schonhage (1964) and van Kempen (1966) show that for $k$ large enough, there is a constant $c$ such that

$$\text{off}(A^{(k+N)}) \;\le\; c\cdot \text{off}(A^{(k)})^2$$

i.e., quadratic convergence. An earlier paper by Henrici (1958) established the same result for the special case when $A$ has distinct eigenvalues. In the convergence theory for the Jacobi iteration, it is critical that $|\theta| \leq \pi/4$. Among other things this precludes the possibility of "interchanging" nearly converged diagonal entries. This follows from the formulae $b_{pp} = a_{pp} - ta_{pq}$ and $b_{qq} = a_{qq} + ta_{pq}$, which can be derived from equations (8.5.1) and (8.5.3).

It is customary to refer to $N$ Jacobi updates as a *sweep*. Thus, after a sufficient number of iterations, quadratic convergence is observed when examining off$(A)$ after every sweep.

**Example 8.5.1** Applying the classical Jacobi iteration to

$$A = \begin{bmatrix} 1 & 1 & 1 & 1 \\ 1 & 2 & 3 & 4 \\ 1 & 3 & 6 & 10 \\ 1 & 4 & 10 & 20 \end{bmatrix}$$

we find

| *sweep* | $O(\text{off}(A))$ |
|---|---|
| 0 | $10^2$ |
| 1 | $10^1$ |
| 2 | $10^{-2}$ |
| 3 | $10^{-11}$ |
| 4 | $10^{-17}$ |

There is no rigorous theory that enables one to predict the number of sweeps that are required to achieve a specified reduction in off$(A)$. However, Brent and Luk(1985) have argued heuristically that the number of sweeps is proportional to $\log(n)$ and this seems to be the case in practice.

### 8.5.5 The Cyclic-by-Row Algorithm

The trouble with the classical Jacobi method is that the updates involve $O(n)$ flops while the search for the optimal $(p, q)$ is $O(n^2)$. One way to address this imbalance is to fix the sequence of subproblems to be solved in advance. A reasonable possibility is to step through all the subproblems in row-by-row fashion. For example, if $n = 4$ we cycle as follows:

$$(p, q) = (1,2), (1,3), (1,4), (2,3), (2,4), (3,4), (1,2), \ldots$$

This ordering scheme is referred to as *cyclic-by-row* and it results in the following procedure:

**Algorithm 8.5.3 (Cyclic Jacobi)** Given a symmetric $A \in \mathbb{R}^{n \times n}$ and a tolerance $tol > 0$, this algorithm overwrites $A$ with $V^TAV$ where $V$ is orthogonal and $\text{off}(V^TAV) \leq tol\| A \|_F$.

$V = I_n$
$eps = tol\| A \|_F$
**while** $\text{off}(A) > eps$
  **for** $p = 1{:}n-1$
    **for** $q = p+1{:}n$
      $(c\,,\,s) = \textbf{sym.schur2}(A, p, q)$
      $A = J(p,q,\theta)^T A J(p,q,\theta)$
      $V = VJ(p,q,\theta)$
    **end**
  **end**
**end**

Cyclic Jacobi converges also quadratically. (See Wilkinson (1962) and van Kempen (1966).) However, since it does not require off-diagonal search, it is considerably faster than Jacobi's original algorithm.

**Example 8.5.2** If the cyclic Jacobi method is applied to the matrix in Example 8.5.1 we find

| Sweep | $O(\text{off}(A))$ |
|---|---|
| 0 | $10^2$ |
| 1 | $10^x$ |
| 2 | $10^{-1}$ |
| 3 | $10^{-6}$ |
| 4 | $10^{-16}$ |

## 8.5.6 Threshold Jacobi

When implementing cyclic Jacobi, it is sensible to skip the annihilation of $a_{pq}$ if its modulus is less than some small, sweep-dependent parameter, because the net reduction in $\text{off}(A)$ is not worth the cost. This leads to what is called the *threshold Jacobi method.* Details concerning this variant of Jacobi's algorithm may be found in Wilkinson (AEP, p.277 ff.). Appropriate thresholding can guarantee convergence.

## 8.5.7 Error Analysis

Using Wilkinson's error analysis it is possible to show that if $r$ sweeps are needed in Algorithm 8.5.3 then the computed $d_i$ satisfy

$$\sum_{i=1}^{n}(d_i - \lambda_i)^2 \;\leq\; (\delta + k_r)\| A \|_F \mathbf{u}$$

for some ordering of $A$'s eigenvalues $\lambda_i$. The parameter $k_r$ depends mildly on $r$.

### 8.5.8 Comparison with the Symmetric QR Algorithm

Although the cyclic Jacobi method converges quadratically, it is not generally competitive with the symmetric QR algorithm. For example, if we just count flops, then 2 sweeps of Jacobi is roughly equivalent to a complete QR reduction to diagonal form with accumulation of transformations. Thus, the Jacobi algorithm is generally inferior. However, the program is very simple compared to QR and so for small $n$ it may be attractive. Moreover, if an approximate eigenvector matrix $V$ is known then $V^TAV$ is almost diagonal, a situation that Jacobi can exploit but not QR.

### 8.5.9 A Parallel Ordering

Perhaps the most interesting distinction between the QR and Jacobi approaches to the symmetric eigenvalue problem is the rich inherent parallelism of the latter algorithm. To illustrate this, suppose $n = 4$ and group the six subproblems into three *rotation sets* as follows:

$$\begin{aligned} rot.set(1) &= \{(1,2),(3,4)\} \\ rot.set(2) &= \{(1,3),(2,4)\} \\ rot.set(3) &= \{(1,4),(2,3)\} \end{aligned}$$

Note that all the rotations within each of the three rotation sets are "nonconflicting." That is, subproblems (1,2) and (3,4) can be carried out in parallel. Likewise the (1,3) and (2,4) subproblems can be executed in parallel as can subproblems (1,4) and (2,3). In general, we say that

$$(i_1, j_1), (i_2, j_2), \ldots, (i_N, j_N) \qquad N = (n-1)n/2$$

is a *parallel ordering* of the set $\{(i,j)\,|\,1 \le i < j \le n\}$ if for $s = 1{:}n-1$ the rotation set

$$rot.set(s) \;=\; \{\, (i_r, j_r) : r = 1 + n(s-1)/2{:}ns/2 \,\}$$

consists of nonconflicting rotations. This requires $n$ to be even, which we assume throughout this section. (The odd $n$ case can be handled by bordering $A$ with a row and column of zeros and being careful when solving the subproblems that involve these augmented zeros.)

A good way to generate a parallel ordering is to visualize a chess tournament with $n$ players in which everybody must play everbody else exactly once. In the $n = 8$ case this entails 7 "rounds." During round one we have the following four games:

| 1 | 3 | 5 | 7 |
|---|---|---|---|
| 2 | 4 | 6 | 8 |

,

i.e., 1 plays 2, 3 plays 4, etc. This corresponds to

$$rot.set(1) = \{ (1,2), (3,4), (5,6), (7,8) \} .$$

To set up rounds 2 through 7, player 1 stays put and players 2 through 8 embark on a merry-go-round:

| 1 | 2 | 3 | 5 |
|---|---|---|---|
| 4 | 6 | 8 | 7 |

$$rot.set(2) = \{(1,4), (2,6), (3,8), (5,7)\}$$

| 1 | 4 | 2 | 3 |
|---|---|---|---|
| 6 | 8 | 7 | 5 |

$$rot.set(3) = \{(1,6), (4,8), (2,7), (3,5)\}$$

| 1 | 6 | 4 | 2 |
|---|---|---|---|
| 8 | 7 | 5 | 3 |

$$rot.set(4) = \{(1,8), (6,7), (4,5), (2,3)\}$$

| 1 | 8 | 6 | 4 |
|---|---|---|---|
| 7 | 5 | 3 | 2 |

$$rot.set(5) = \{(1,7), (5,8), (3,6), (2,4)\}$$

| 1 | 7 | 8 | 6 |
|---|---|---|---|
| 5 | 3 | 2 | 4 |

$$rot.set(6) = \{(1,5), (3,7), (2,8), (4,6)\}$$

| 1 | 5 | 7 | 8 |
|---|---|---|---|
| 3 | 2 | 4 | 6 |

$$rot.set(7) = \{(1,3), (2,5), (4,7), (6,8)\}$$

We can encode these operations in a pair of integer vectors $top(1{:}n/2)$ and $bot(1{:}n/2)$. During a given round $top(k)$ plays $bot(k)$ , $k = 1{:}n/2$. The pairings for the next round is obtained by updating $top$ and $bot$ as follows:

```
function: [top, bot] = music(top, bot)
    m = length(top)
    for k = 1:m
        if k = 2
            new.top(k) = bot(1)
        elseif k > 2
            new.top(k) = top(k - 1)
        end
        if k = m
            new.bot(k) = top(k)
        else
            new.bot(k) = bot(k + 1)
        end
    end
    top = new.top; bot = new.bot
end music
```

Using **music** we obtain the following parallel order Jacobi procedure.

**Algorithm 8.5.4 (Parallel Order Jacobi )** Given a symmetric $A \in \mathbb{R}^{n\times n}$ and a tolerance $tol > 0$, this algorithm overwrites $A$ with $V^TAV$ where $V$ is orthogonal and $\text{off}(V^TAV) \le tol\| A \|_F$ . It is assumed that $n$ is even.

$V = I_n$
$eps = tol\| A \|_F$
$top = 1{:}2{:}n; bot = 2{:}2{:}n$
**while** $\text{off}(A) > eps$
  **for** $set = 1{:}n-1$
    **for** $k = 1{:}n/2$
      $p = \min(top(k), bot(k))$
      $q = \max(top(k), bot(k))$
      $(c\,,\,s) = \textbf{sym.schur2}(A, p, q)$
      $A = J(p,q,\theta)^T AJ(p,q,\theta)$
      $V = VJ(p,q,\theta)$
    **end**
    $[top, bot] = \textbf{music}(top, bot)$
  **end**
**end**

Notice that the $k$-loop steps through $n/2$ independent, nonconflicting subproblems.

### 8.5.10 A Ring Procedure

We now show how to implement Algorithm 8.5.4 on a ring of $p$ processors. We assume that $p = n/2$ for the time being. At any instant, Proc($\mu$) houses two columns of $A$ and the corresponding $V$ columns. For example, if $n = 8$ then here is how the column distribution of $A$ proceeds from step to step:

| | Proc(1) | Proc(2) | Proc(3) | Proc(4) |
|---|---|---|---|---|
| Step 1: | [1 2] | [3 4] | [5 6] | [7 8] |
| Step 2: | [1 4] | [2 6] | [3 8] | [5 7] |
| Step 3: | [1 6] | [4 8] | [2 7] | [3 5] |
| | | etc. | | |

The ordered pairs denote the indices of the housed columns. The first index names the *left* column and the second index names the *right* column. Thus, the *left* and *right* columns in Proc(3) during step 3 are 2 and 7 respectively. This terminology is used in the parallel procedure below.

Note that in between steps, the columns are shuffled according to the permutation implicit in **music** and that nearest neighbor communication is maintained. At each step, each processor oversees a single subproblem.

This involves (a) computing an orthogonal $V_{small} \in \mathbb{R}^{2\times 2}$ that solves a local 2-by-2 Schur problem, (b) using the 2-by-2 $V_{small}$ to update the two housed columns of $A$ and $V$, (c) sending the 2-by-2 $V_{small}$ to all the other processors, and (d) receiving the $V_{small}$ matrices from the other processors and updating the local portions of $A$ and $V$ accordingly. Since $A$ is stored by column, communication is necessary to carry out the $V_{small}$ updates because they effect rows of $A$. For example, in the second step of the $n = 8$ problem, Proc(2) must receive the 2-by-2 rotations associated with subproblems (1,4), (3,8), and (5,7). These come from Proc(1), Proc(3), and Proc(4) respectively. In general, the sharing of the rotation matrices can be conveniently implemented by circulating the 2-by-2 $V_{small}$ matrices in "merry go round" fashion around the ring. Each processor copies a passing 2-by-2 $V_{small}$ into its local memory and then appropriately updates the locally housed portions of $A$ and $V$.

The termination criteria in Algorithm 8.5.4 poses something of a problem in a distributed memory environment in that the value of off($\cdot$) and $\| A \|_F$ require access to all of $A$. However, these global quantities can be computed during the $V$ matrix merry-go-round phase. Before the circulation of the $V$'s begins, each processor can compute its contribution to $\| A \|_F$ and off($\cdot$) . These quantities can then be summed by each processor if they are placed on the merry-go-round and read at each stop. By the end of one revolution each processor has its own copy of $\| A \|_F$ and off($\cdot$). However, for simplicity in the following algorithm we merely terminate after a fixed number of sweeps $M$.

**Algorithm 8.5.5** Suppose $A$ is an $n$-by-$n$ symmetric matrix with $n = 2p$. If each processor in a $p$-processor ring executes the following node program, then upon completion Proc($\mu$) houses $B(:, 2\mu - 1{:}2\mu)$ in $A_{loc}$ and $V(:, 2\mu - 1{:}2\mu)$ in $V_{loc}$ where $B = V^T AV$ is the matrix that results after $M$ sweeps in Algorithm 8.5.4. It is assumed that $p \geq 2$.

> **local.init**[ $p, \mu = my.id, left, right, n, A_{loc} = A(:, 2\mu - 1{:}2\mu), M$ ]
> $V_{loc} = I_n(:, 2\mu - 1{:}2\mu)$
> $top = 1{:}2{:}n$; $bot = 2{:}2{:}n$
> { Establish communication partners for the column shuffles.
>     $Lsend, Rsend$ = where the left (right) column is sent.
>     $Lrecv, Rrecv$ = where the new left (right) column originates.}
> **if** $\mu = 1$
>     $Lsend = 1; Rsend = 2; Lrecv = 1; Rrecv = 2$
> **elseif** $\mu = p$
>     $Lsend = p; Rsend = p - 1; Lrecv = p - 1; Rrecv = p;$
> **else**
>     $Lsend = \mu + 1; Rsend = \mu - 1; Lrecv = \mu - 1; Rrecv = \mu + 1$
> **end**

```
for sweep = 1:M
    for set = 1:n − 1
        p = min(top(μ), bot(μ)); q = max(top(μ), bot(μ))
        (c, s) = sym.schur2(A_loc([p, q], 1:2), 1, 2); V_small = [c s; −s c]
        A_loc = A_loc V_small; V_loc = V_loc V_small
        { Circulate the 2-by-2 V_small matrices around the ring. }
        λ = μ
        for k = 1:p
            send(V_small, right); recv(V_small, left)
            if λ = 1
                λ = p
            else
                λ = λ − 1
            end
            { Have the V_small from Proc(λ)}
            p_0 = min(top(λ), bot(λ)); q_0 = max(top(λ), bot(λ) )
            A_loc([p_0, q_0], :) = V_small^T A_loc([p_0, q_0], :);
        end
        { Column shuffling to prepare for next rotation set. }
        send(A_loc(:, 1), Lsend); send(A_loc(:, 2), Rsend)
        recv(A_loc(:, 1), Lrecv); recv(A_loc(:, 2), Rrecv)
        send(V_loc(:, 1), Lsend); send(V_loc(:, 2), Rsend)
        recv(V_loc(:, 1), Lrecv); recv(V_loc(:, 2), Rrecv)
        [top, bot] = music(top, bot)
    end
end
```

The order of the **send**'s and **recv**'s associated with the column shuffling is critical. The reader should appreciate why both the left and the right columns must both be sent before their replacements are requested.

### 8.5.11 Block Jacobi Procedures

It is usually the case when solving the symmetric eigenvalue problem on a $p$-processor machine that $n \gg p$. In this case a block version of the Jacobi algorithm may be appropriate. Block versions of the above procedures are straightforward. Suppose that $n = rN$ and that we partition the $n$-by-$n$ matrix $A$ as follows:

$$A = \begin{bmatrix} A_{11} & \cdots & A_{1N} \\ \vdots & & \vdots \\ A_{N1} & \cdots & A_{NN} \end{bmatrix}$$

Here, each $A_{ij}$ is $r$-by-$r$. In block Jacobi the $(p,q)$ subproblem involves computing the $2r$-by-$2r$ Schur decomposition

$$\begin{bmatrix} V_{pp} & V_{pq} \\ V_{qp} & V_{qq} \end{bmatrix}^T \begin{bmatrix} A_{pp} & A_{pq} \\ A_{qp} & A_{qq} \end{bmatrix} \begin{bmatrix} V_{pp} & V_{pq} \\ V_{qp} & V_{qq} \end{bmatrix} = \begin{bmatrix} D_{pp} & O \\ O & D_{qq} \end{bmatrix}$$

and then applying to $A$ the block Jacobi rotation made up of the $V_{ij}$ . If we call this block rotation $V$ then it is easy to show that

$$\text{off}(V^TAV)^2 = \text{off}(A)^2 - \left(2\| A_{pq} \|_F^2 + \text{off}(A_{pp})^2 + \text{off}(A_{qq})^2\right).$$

Block Jacobi procedures have many interesting computational aspects. For example, there are many ways to solve the subproblems and the choice appears to be critical. See Bischof (1987).

### 8.5.12 Jacobi SVD Procedures

Finally, we mention that for each of the above symmetric eigenvalue Jacobi procedures there corresponds an analogous algorithm for the SVD. The subproblems now involve 2-by-2 SVD computations, e.g.,

$$\begin{bmatrix} c_1 & s_1 \\ -s_1 & c_1 \end{bmatrix}^T \begin{bmatrix} a_{pp} & a_{pq} \\ a_{qp} & a_{qq} \end{bmatrix} \begin{bmatrix} c_2 & s_2 \\ -s_2 & c_2 \end{bmatrix} = \begin{bmatrix} d_p & 0 \\ 0 & d_q \end{bmatrix}.$$

For more details, see problems P8.5.1 and P8.5.2.

#### Problems

**P8.5.1** (a) Let

$$C = \begin{bmatrix} w & x \\ y & y \end{bmatrix}$$

be real. Give a stable algorithm for computing c and s with $c^2 + s^2 = 1$ such that

$$B = \begin{bmatrix} c & s \\ -s & c \end{bmatrix} C$$

is symmetric. (b) Combine (a) with the Jacobi trigonometric calculations in the text to obtain a stable algorithm for computing the SVD of $C$. (c) Part (b) can be used to develop a Jacobi-like algorithm for computing the SVD of $A \in \mathbb{R}^{n\times n}$. For a given $(p,q)$ with $p < q$, Jacobi transformations $J(p,q,\theta_1)$ and $J(p,q,\theta_2)$ are determined such that if

$$B = J(p,q,\theta_1)^T AJ(p,q,\theta_2),$$

then $b_{pq} = b_{qp} = 0$. Show

$$\text{off}(B)^2 = \text{off}(A)^2 - b_{pq}^2 - b_{qp}^2.$$

How might $p$ and $q$ be determined? How could the algorithm be adapted to handle the case when $A \in \mathbb{R}^{m\times n}$ with $m > n$?

**P8.5.2** Let $x$ and $y$ be in $\mathbb{R}^n$ and define the orthogonal matrix $Q$ by

$$Q = \begin{bmatrix} c & s \\ -s & c \end{bmatrix}.$$

Give a stable algorithm for computing $c$ and $s$ such that the columns of $[x, y]Q$ are orthogonal to one another. This is the basis of "one-sided" Jacobi-like SVD algorithms that find an orthogonal $V$ such that the columns of $B = AV$ are mutually orthogonal. The "$U$" matrix in the SVD can then be found through a column normalization.

**P8.5.3** Let the scalar $\gamma$ be given along with the matrix

$$A = \begin{bmatrix} w & x \\ x & z \end{bmatrix}.$$

It is desired to compute an orthogonal matrix

$$J = \begin{bmatrix} c & s \\ -s & c \end{bmatrix}$$

such that the (1, 1) entry of $J^TAJ$ equals $\gamma$. Show that this requirement leads to the equation

$$(w-\gamma)\tau^2 - 2x\tau + (z-\gamma) = 0,$$

where $\tau = c/s$. Verify that this quadratic has real roots if $\gamma$ satisfies $\lambda_2 \le \gamma \le \lambda_1$, where $\lambda_1$ and $\lambda_2$ are the eigenvalues of $A$.

**P8.5.4** Let $A \in \mathbb{R}^{n\times n}$ be symmetric. Using P8.5.3, give an algorithm that computes the factorization

$$Q^TAQ = \gamma I + F$$

where $Q$ is a product of Jacobi rotations, $\gamma = \text{trace}(A)/n$, and $F$ has zero diagonal entries. Discuss the uniqueness of $Q$.

**P8.5.5** Formulate Jacobi procedures for (a) skew symmetric matrices and (b) complex Hermitian matrices.

**P8.5.6** Partition the $n$-by-$n$ real symmetric matrix $A$ as follows:

$$A = \begin{bmatrix} a & v^T \\ v & A_1 \end{bmatrix} \begin{matrix} 1 \\ n-1 \end{matrix}$$
$$\qquad 1 \quad n-1$$

Let $Q$ be a Householder matrix such that if $B = Q^TAQ$, then $B(3{:}n, 1) = 0$. Let $J = J(1,2,\theta)$ be determined such that if $C = J^TBJ$, then $c_{12} = 0$ and $c_{11} \ge c_{22}$. Show $c_{11} \ge a + \| v \|_2$. La Budde (1964) formulated an algorithm for the symmetric eigenvalue probem based upon repetition of this Householder-Jacobi computation.

**P8.5.7** Organize function **music** so that it involves minimum workspace.

**P8.5.8** Modify Algorithm 8.5.5 so that it terminates when $\text{off}(A) \le tol\| A \|_F$

**P8.5.9** Adapt Algorithm 8.5.5 so that it computes the SVD.

### Notes and References for Sec. 8.5

Jacobi's original paper is one of the earliest references found in the numerical analysis literature

C.G.J. Jacobi (1846). "Uber ein Leichtes Verfahren Die in der Theorie der Sacularstroungen Vorkommendern Gleichungen Numerisch Aufzulosen," *Crelle's J. 30*, 51–94.

Prior to the QR algorithm, the Jacobi technique was the standard method for solving dense symmetric eigenvalue problems. Early attempts to improve upon it include

C.D. La Budde (1964). "Two Classes of Algorithms for Finding the Eigenvalues and Eigenvectors of Real Symmetric Matrices," *J. ACM 11*, 53–58.
M. Lotkin (1956). "Characteristic Values of Arbitrary Matrices," *Quart. Appl. Math. 14*, 267–75.

D.A. Pope and C. Tompkins (1957). "Maximizing Functions of Rotations: Experiments Concerning Speed of Diagonalization of Symmetric Matrices Using Jacobi's Method," *J. ACM 4,* 459–66.

The computational aspects of Jacobi method are described in Wilkinson (AEP, pp. 265 ff.), while an Algol procedure can be found in

H. Rutishauser (1966). "The Jacobi Method for Real Symmetric Matrices," *Numer. Math. 9,* 1–10. See also *HACLA*, pp. 202–11.

Jacobi methods are also of interest in minicomputing environments, where their compactness makes them attractive. See

J.C. Nash (1975). "A One-Sided Tranformation Method for the Singular Value Decomposition and Algebraic Eigenproblem," *Comp. J. 18,* 74–76.

They are also useful when a nearly diagonal matrix must be diagonalized. See

J.H. Wilkinson (1968). "Almost Diagonal Matrices with Multiple or Close Eigenvalues," *Lin. Alg. and Its Applic. I,* 1–12.

Establishing the quadratic convergence of the classical and cyclic Jacobi iterations has engrossed several authors:

K.W. Brodlie and M.J.D. Powell (1975). "On the Convergence of Cyclic Jacobi Methods," *J. Inst. Math. Applic. 15,* 279–87.
E.R. Hansen (1962). "On Quasicyclic Jacobi Methods," *ACM J. 9,* 118–35.
E.R. Hansen (1963). "On Cyclic Jacobi Methods," *SIAM J. Appl. Math. 11,* 448–59.
P. Henrici (1958). "On the Speed of Convergence of Cyclic and Quasicyclic Jacobi Methods for Computing the Eigenvalues of Hermitian Matrices," *SIAM J. Appl. Math. 6,* 144–62.
P. Henrici and K. Zimmermann (1968). "An Estimate for the Norms of Certain Cyclic Jacobi Operators," *Lin. Alg. and Its Applic. 1,* 489–501.
A. Schonhage (1964). "On the Quadratic Convergence of the Jacobi Process," *Numer. Math. 6,* 410–12.
H.P.M. van Kempen (1966). "On Quadratic Convergence of the Special Cyclic Jacobi Method," *Numer. Math. 9,* 19–22.
J.H. Wilkinson (1962). "Note on the Quadratic Convergence of the Cyclic Jacobi Process," *Numer. Math. 6,* 296–300.

Attempts have been made to extend the Jacobi iteration to other classes of matrices and to push through corresponding convergence results. The case of normal matrices is discussed in

H.H. Goldstine and L.P. Horowitz (1959). "A Procedure for the Diagonalization of Normal Matrices," *J. Assoc. Comp. Mach. 6,* 176–95.
G. Loizou (1972). "On the Quadratic Convergence of the Jacobi Method for Normal Matrices," *Comp. J. 15,* 274–76.
A. Ruhe (1972). "On the Quadratic Convergence of the Jacobi Method for Normal Matrices," *BIT 7,* 305–13.

See also

M.H.C. Paardekooper (1971). "An Eigenvalue Algorithm for Skew Symmetric Matrices," *Numer. Math. 17,* 189–202.

Essentially, the analysis and algorithmic developments presented in the text carry over to the normal case with minor modification. The same comment generally applies to

Jacobi-like methods for the SVD. See

Z. Bai (1988). "Note on the Quadratic Convergence of Kogbetliantz's Algorithm for Computing the Singular Value Decomposition," *Lin. Alg. and Its Applic. 104,* 131–140.

J.P. Charlier, M. Vanbegin, P. Van Dooren (1988). "On Efficient Implementaion of Kogbetliantz's Algorithm for Computing the Singular Value Decomposition," *Numer. Math. 52,* 279–300.

J.P. Charlier and P. Van Dooren (1987). "On Kogbetliantz's SVD Algorithm in the Presence of Clusters," *Lin. Alg. and Its Applic. 95,* 135–160.

G.E. Forsythe and P. Henrici (1960). "The Cyclic Jacobi Method for Computing the Principal Values of a Complex Matrix," *Trans. Amer. Math. Soc. 94,* 1–23.

P.C. Hansen (1988). "Reducing the Number of Sweeps in Hestenes Method," in *Singular Value Decomposition and Signal Processing,* ed. E.F. Deprettere, North Holland.

E.G. Kogbetliantz (1955). "Solution of Linear Equations by Diagonalization of Coefficient Matrix," *Quart. Appl. Math. 13,* 123–132.

C.C. Paige (1986). "On the Quadratic Convergence of Kogbetliantz's Algorithm for Computing the Singular Value Decomposition," *Lin. Alg. and Its Applic. 77,* 301–314.

P. Van Dooren and C.C. Paige (1986). "On the Quadratic Convergence of Kogbetliantz's Algorithm for Computing the Singular Value Decomposition," *Lin. Alg. and Its Applic. 77,* 301–313.

The Kogbetliantz algorithm is essentially the Jacobi SVD method sketched in §8.5.12.

For non-normal matrices, the situation is considerably more difficult. Consult

J. Boothroyd and P.J. Eberlein (1968). "Solution to the Eigenproblem by a Norm-Reducing Jacobi-Type Method (Handbook)," *Numer. Math. 11,* 1–12. See also *HACLA*, pp. 327–38.

P.J. Eberlein (1970). "Solution to the Complex Eigenprobelm by a Norm-Reducing Jacobi-type Method," *Numer. Math. 14,* 232–45. See also *HACLA*, pp. 404–17.

C.E. Froberg (1965). "On Triangularization of Complex Matrices by Two Dimensional Unitary Tranformations," *BIT 5,* 230–34.

J. Greenstadt (1955). "A Method for Finding Roots of Arbitrary Matrices," *Math. Tables and Other Aids to Comp. 9,* 47–52.

V. Hari (1982). "On the Global Convergence of the Eberlein Method for Real Matrices," *Numer. Math. 39,* 361–370.

C.P. Huang (1975). "A Jacobi-Type Method for Triangularizing an Arbitrary Matrix," *SIAM J. Num. Anal. 12,* 566–70.

A. Ruhe (1968). On the Quadratic Convergence of a Generalization of the Jacobi Method to Arbitrary Matrices," *BIT 8,* 210–31.

A. Ruhe (1969). "The Norm of a Matrix After a Similarity Transformation," *BIT 9,* 53–58.

G.W. Stewart (1985). "A Jacobi-Like Algorithm for Computing the Schur Decomposition of a Nonhermitian Matrix," *SIAM J. Sci. and Stat. Comp. 6,* 853–862.

Jacobi methods for complex symmetric matrices have also been developed. See

P. Anderson and G. Loizou (1973). "On the Quadratic Convergence of an Algorithm Which Diagonalizes a Complex Symmetric Matrix," *J. Inst. Math. Applic. 12,* 261–71.

P. Anderson and G. Loizou (1976). "A Jacobi-Type Method for Complex Symmetric Matrices (Handbook)," *Numer. Math. 25,* 347–63.

P.J. Eberlein (1971). "On the Diagonalization of Complex Symmetric Matrices," *J. Inst. Math. Applic. 7,* 377–83.

J.J. Seaton (1969). "Diagonalization of Complex Symmetric Matrices Using a Modified Jacobi Method," *Comp. J. 12,* 156–57.

Although the symmetric QR algorithm is generally much faster than the Jacobi method, there are special settings where the latter technique is of interest. As we illustrated, on a parallel-computer it is possible to perform several rotations concurrently, thereby accelerating the reduction of the off-diagonal elements. See

M. Berry and A. Sameh (1986). "Multiprocessor Jacobi Algorithms for Dense Symmetric Eigenvalue and Singular Value Decompositions," in *Proc. International Conference on Parallel Processing*, 433–440.
C.H. Bischof (1987). "The Two-Sided Block Jacobi Method on Hypercube Architectures," in *Hypercube Multiprocessors*, ed. M.T. Heath, SIAM Press, Philadelphia.
C.H. Bischof (1988). "Computing the Singular Value Decomposition on a Distributed System of Vector Processors," Cornell Computer Science Report 87-869, Ithaca, NY.
C.H. Bischof and C. Van Loan (1986). "Computing the SVD on a Ring of Array Processors," in *Large Scale Eigenvalue Problems,* eds. J. Cullum and R. Willoughby, North Holland, 51-66.
R.P. Brent and F.T. Luk (1985). "The Solution of Singular Value and Symmetric Eigenvalue Problems on Multiprocessor Arrays," *SIAM J. Sci. and Stat. Comp. 6,* 69–84.
R.P. Brent, F.T. Luk, and C. Van Loan (1985). "Computation of the Singular Value Decomposition Using Mesh Connected Processors," *J. VLSI Computer Systems 1,* 242–270.
P.J. Eberlein (1987). "On Using the Jacobi Method on a Hypercube," in *Hypercube Multiprocessors*, ed. M.T. Heath, SIAM Publications, Philadelphia.
F.T. Luk (1980). "Computing the Singular Value Decomposition on the ILLIAC IV," *ACM Trans. Math. Soft. 6,* 524–39.
F.T. Luk (1986). "A Triangular Processor Array for Computing Singular Values," *Lin. Alg. and Its Applic. 77,* 259–274.
J.J. Modi and J.D. Pryce (1985). "Efficient Implementation of Jacobi's Diagonalization Method on the DAP," *Numer. Math. 46,* 443–454.
A. Sameh (1971). "On Jacobi and Jacobi-like Algorithms for a Parallel Computer," *Math. Comp. 25,* 579–90.
R. Schreiber (1986). "Solving Eigenvalue and Singular Value Problems on an Undersized Systolic Array," *SIAM J. Sci. and Stat. Comp. 7,* 441–451.
D.S. Scott, M.T. Heath, and R.C. Ward (1986). "Parallel Block Jacobi Eigenvalue Algorithms Using Systolic Arrays," *Lin. Alg. and Its Applic. 77,* 345–356.
G. Shroff and R. Schreiber (1987). "Convergence of Block Jacobi Methods," Report 87-25, Comp. Sci. Dept., RPI, Troy , NY.

## 8.6 A Divide and Conquer Method

In this section we describe a "divide-and-conquer" algorithm for computing the eigenvalues and eigenvectors of the symmetric tridiagonal matrix

$$T = \begin{bmatrix} a_1 & b_1 & & \cdots & 0 \\ b_1 & a_2 & \ddots & & \vdots \\ & \ddots & \ddots & \ddots & \\ \vdots & & \ddots & \ddots & b_{n-1} \\ 0 & \cdots & & b_{n-1} & a_n \end{bmatrix}.$$

It computes the Schur decomposition

$$Q^T T Q \;=\; \Lambda \;=\; \mathrm{diag}(\lambda_1, \ldots, \lambda_n) \qquad Q^T Q = I \tag{8.6.1}$$

by "gluing" together the Schur decompositions of two special half-sized tridiagonal problems. Each of these reductions can in turn be specified by a pair of quarter-sized Schur decompositions and so on. The overall procedure, due to Dongarra and Sorensen (1987), is suitable for parallel computation.

## 8.6.1 Tearing

We first show how $T$ can be "torn" in half with a rank-one modification. For simplicity, assume $n = 2m$. Define $v \in \mathbb{R}^n$ as follows

$$v \;=\; \begin{bmatrix} e_m^{(m)} \\ \theta e_1^{(m)} \end{bmatrix}. \tag{8.6.2}$$

Note that for all $\rho \in \mathbb{R}$ the matrix $\tilde{T} = T - \rho v v^T$ is identical to $T$ except in its "middle four" entries:

$$\tilde{T}(m{:}m+1, m{:}m+1) \;=\; \begin{bmatrix} a_m - \rho & b_m - \rho\theta \\ b_m - \rho\theta & a_{m+1} - \rho\theta^2 \end{bmatrix}.$$

If we set $\rho\theta = b_m$ then

$$T \;=\; \begin{bmatrix} T_1 & 0 \\ 0 & T_2 \end{bmatrix} + \rho v v^T$$

where

$$T_1 \;=\; \begin{bmatrix} a_1 & b_1 & & \cdots & 0 \\ b_1 & a_2 & \ddots & & \vdots \\ & \ddots & \ddots & \ddots & \\ \vdots & & \ddots & \ddots & b_{m-1} \\ 0 & \cdots & & b_{m-1} & \tilde{a}_m \end{bmatrix},$$

$$T_2 \;=\; \begin{bmatrix} \tilde{a}_{m+1} & b_{m+1} & & \cdots & 0 \\ b_{m+1} & a_{m+2} & \ddots & & \vdots \\ & \ddots & \ddots & \ddots & \\ \vdots & & \ddots & \ddots & b_{n-1} \\ 0 & \cdots & & b_{n-1} & a_n \end{bmatrix},$$

and $\tilde{a}_m = a_m - \rho$ and $\tilde{a}_{m+1} = a_{m+1} - \rho\theta^2$.

### 8.6.2 Combining Schur Decompositions

Suppose that we have $m$-by-$m$ orthogonal matrices $Q_1$ and $Q_2$ such that $Q_1^T T_1 Q_1 = D_1$ and $Q_2^T T_2 Q_2 = D_2$ are each diagonal. If we set

$$U = \begin{bmatrix} Q_1 & 0 \\ 0 & Q_2 \end{bmatrix}$$

then

$$U^T T U = U^T \left( \begin{bmatrix} T_1 & 0 \\ 0 & T_2 \end{bmatrix} + \rho v v^T \right) U = D + \rho z z^T \tag{8.6.3}$$

where

$$D = \begin{bmatrix} D_1 & 0 \\ 0 & D_2 \end{bmatrix}$$

is diagonal and

$$z = U^T v = \begin{bmatrix} Q_1^T e_m \\ \theta Q_2^T e_1 \end{bmatrix}. \tag{8.6.4}$$

Comparing (8.6.1) and (8.6.3) we see the effective synthesis of the two half-sized Schur decompositions requires the quick and stable computation of an orthogonal $V$ such that

$$V^T(D + \rho z z^T)V = \Lambda = \text{diag}(\lambda_1, \ldots, \lambda_n). \tag{8.6.5}$$

After $V$ is computed we then set $Q = UV$ to obtain (8.6.1).

### 8.6.3 The Eigensystem of $D + \rho z z^T$

Fortunately, it is possible to diagonalize very rapidly a matrix of the form diagonal + rank-one. The technique rests upon some basic theoretical results which we now establish.

**Lemma 8.6.1** *Suppose* $D = \text{diag}(d_1, \ldots, d_n) \in \mathbb{R}^{n \times n}$ *has the property that* $d_1 > \cdots > d_n$ . *Assume that* $\rho \neq 0$ *and that* $z \in \mathbb{R}^n$ *has no zero components. If*

$$(D + \rho z z^T)v = \lambda v \qquad v \neq 0$$

*then* $z^T v \neq 0$ *and* $D - \lambda I$ *is nonsingular.*

**Proof.** If $\lambda \in \lambda(D)$ , then $\lambda = d_i$ for some $i$ and thus

$$0 = e_i^T[(D - \lambda I)v + \rho(z^T v)z] = \rho(z^T v)z_i.$$

Since $\rho$ and $z_i$ are nonzero we must have $0 = z^T v$ and so $Dv = \lambda v$. However, $D$ has distinct eigenvalues and therefore, $v \in \text{span}\{e_i\}$. But then $0 = z^T v = z_i$, a contradiction. Thus, $D$ and $D + \rho z z^T$ do not have any common eigenvalues and $z^T v \neq 0$. □

**Theorem 8.6.2** *Suppose* $D = \text{diag}(d_1, \ldots, d_n) \in \mathbb{R}^{n\times n}$ *and that the diagonal entries satisfy* $d_1 > \cdots > d_n$. *Assume that* $\rho \neq 0$ *and that* $z \in \mathbb{R}^n$ *has no zero components. If* $V \in \mathbb{R}^{m\times m}$ *is orthogonal such that*

$$V^T(D + \rho zz^T)V \;=\; \text{diag}(\lambda_1, \ldots, \lambda_n)$$

*with* $\lambda_1 \geq \cdots \geq \lambda_n$ *and* $V = [\, v_1, \ldots, v_n \,]$ *then*

**(a)** *The* $\lambda_i$ *are the* $m$ *zeros of* $f(\lambda) = 1 + \rho z^T(D - \lambda I)^{-1}z$.

**(b)** *If* $\rho > 0$ *then* $\lambda_1 > d_1 > \lambda_2 > \cdots > \lambda_m > d_n$.
*If* $\rho < 0$ *then* $d_1 > \lambda_1 > d_2 > \cdots > d_n > \lambda_n$.

**(c)** *The eigenvector* $v_i$ *is a multiple of* $(D - \lambda_i I)^{-1}z$.

**Proof.** If $(D + \rho zz^T)v = \lambda v$, then

$$(D - \lambda I)v + \rho(z^T v)z = 0. \tag{8.6.6}$$

We know from Lemma 8.6.1 that $D - \lambda I$ is nonsingular. Thus,

$$v \in \text{span}\{(D - \lambda I)^{-1}z\}$$

thereby establishing (c). Moreover, if we apply $z^T(D - \lambda I)^{-1}$ to both sides of equation (8.6.6) we obtain

$$z^T v \left(1 + \rho z^T(D - \lambda I)^{-1}z\right) \;=\; 0\,.$$

By Lemma 8.6.1, $z^T v \neq 0$ and so this shows that if $\lambda \in \lambda(D + \rho zz^T)$ then $f(\lambda) = 0$. We must show that all the zeros of $f$ are eigenvalues of $D + \rho zz^T$ and that the interlacing relations (b) hold.

To do this we look more carefully at the equations

$$\begin{aligned} f(\lambda) &= 1 + \rho\left(\frac{z_1^2}{d_1 - \lambda} + \cdots + \frac{z_n^2}{d_n - \lambda}\right) \\ f'(\lambda) &= -\rho\left(\frac{z_1^2}{(d_1 - \lambda)^2} + \cdots + \frac{z_n^2}{(d_n - \lambda)^2}\right) \end{aligned}$$

Note that $f$ is monotone in between its poles. This allows us to conclude that if $\rho > 0$ then $f$ has precisely $n$ roots, one in each of the intervals

$$(d_n, d_{n-1}), \ldots, (d_2, d_1), (d_1, \infty).$$

If $\rho < 0$ then $f$ has exactly $n$ roots, one in each of the intervals

$$(-\infty, d_n), (d_n, d_{n-1}), \ldots, (d_2, d_1).$$

In either case, it follows that the zeros of $f$ are precisely the eigenvalues of $D + \rho vv^T$. $\square$

The theorem suggests that to compute $V$ we (a) find the roots $\lambda_1, \ldots, \lambda_n$ of $f$ and then (b) compute the columns of $V$ by normalizing the vectors $(D - \lambda_i I)^{-1}z$ for $i = 1{:}n$. Before we look into the practical aspects of these calculations we show that the same plan of attack can be followed even if there are repeated $d_i$ and zero $z_i$.

**Theorem 8.6.3** *If $D = \text{diag}(d_1, ..., d_n)$ and $z \in \mathbb{R}^n$, then there exists an orthogonal matrix $V_1$ such that if $V_1^T D V_1 = \text{diag}(\mu_1, \ldots, \mu_n)$ and $w = V_1^T z$ then*

$$\mu_1 > \mu_2 > \cdots > \mu_r \geq \mu_{r+1} \geq \cdots \geq \mu_n ,$$

*$w_i \neq 0$ for $i = 1{:}r$, and $w_i = 0$ for $i = r + 1{:}n$.*

**Proof.** We give a constructive proof based upon two elementary operations. (a) Suppose $d_i = d_j$ for some $i < j$ . Let $J(i, j, \theta)$ be a Jacobi rotation in the $(i, j)$ plane with the property that the $j$th component of $J(i, j, \theta)^T z$ is zero. It is not hard to show that $J(i, j, \theta)^T D J(i, j, \theta) = D$. Thus, we can zero a component of $z$ if there is a repeated $d_i$. (b) If $z_i = 0$, $z_j \neq 0$ , and $i < j$, then let $P$ be the identity with columns $i$ and $j$ interchanged. It follows that $P^T D P$ is diagonal, $(P^T z)_i \neq 0$, and $(P^T z)_j = 0$. Thus, we can permute all the zero $z_i$ to the "bottom." Clearly, repetition of (a) and (b) eventually renders the desired canonical structure. $V_1$ is the product of the rotations. □

### 8.6.4 Practical Synthesis

The practical determination of the decomposition (8.6.5) involves the judicious application of Theorems 8.6.2 and 8.6.3. The sought after $V$ is a product $V = V_1 V_2$. The orthogonal matrix $V_1$ is obtained by performing the Theorem 8.6.3 reductions. Of course, in practice we need a criteria for deciding when two diagonal entries are distinct and when an element of $z$ can be regarded as zero. Suppose *tol* is a small multiple of the machine precision, e.g., $tol = \mathbf{u}(\| D \|_2 + |\rho| \| z \|_2))$. We then determine an orthogonal matrix $V_1$ and an integer $\hat{r}$ ( $1 \leq \hat{r} \leq n$) such that $V_1^T D V_1 = \text{diag}(\mu_1, \ldots, \mu_n)$ (modulo *tol*) and $w = V_1^T z$ with (a) $\mu_i - \mu_{i+1} > tol$ for $i = 1{:}\hat{r} - 1$, (b) $|w_i| \geq tol$ for $i = 1{:}\hat{r}$, and (c) $|w_i| < tol$ for $i = \hat{r} + 1{:}n$. Overwriting $D \leftarrow V_1^T D V_1$ and $z \leftarrow V_1^T z$, we then proceed with the problem of finding $\tilde{V}_2 \in \mathbb{R}^{\hat{r} \times \hat{r}}$ so

$$\tilde{V}_2^T (D(1{:}\hat{r}, 1{:}\hat{r}) + \rho z(1{:}\hat{r}) z(1{:}\hat{r})^T) \tilde{V}_2 = \text{diag}(\lambda_1, \ldots, \lambda_{\hat{r}}).$$

Noting that Theorem 8.6.2 is applicable to this subproblem, our first task is to find the $\hat{r}$ zeros of

$$f(\lambda) = 1 + \rho \sum_{i=1}^{\hat{r}} \frac{z_i^2}{d_i - \lambda} .$$

Assume $\rho > 0$. (If not replace $\rho$ by $-\rho$ and $D$ with $-D$.) Let $\lambda_i$ be the unique zero of $f$ in the open interval $(d_i, d_{i-1})$ where we assume $d_0 = \infty$. The nonlinear equation $f(\lambda) = 0$ in $(d_i, d_{i-1})$ can be solved by using a very fast Newton-like iteration that is based upon a simple rational model of $f$ at the current iterate. Because the secular equation is so nicely structured, it is possible to start the Newton process with a good initial guess. If $\hat{\lambda}_i$ is a computed zero of the secular equation then $i$th column of $\tilde{V}_2$ is found by normalizing the vector $(D(1:\hat{r}, 1:\hat{r}) - \lambda_i I_{\hat{r}})^{-1} z(1:\hat{r})$. Note that $V = V_1 V_2$ where $V_2 = \text{diag}(\tilde{V}_2, I_{n-\hat{r}})$. Details may be found in Dongarra and Sorensen (1987) along with an error analysis that addresses the orthonormality of the computed $V$.

### 8.6.5 The Overall Process with Parallelism

Having stepped through the tearing and synthesis operations, we are ready to illustrate the overall process and how it can be implemented on a multiprocessor. For clarity, assume that $n = 8N$ for some positive integer $N$ and that three levels of tearing are performed. We can depict this with a binary tree as follows:

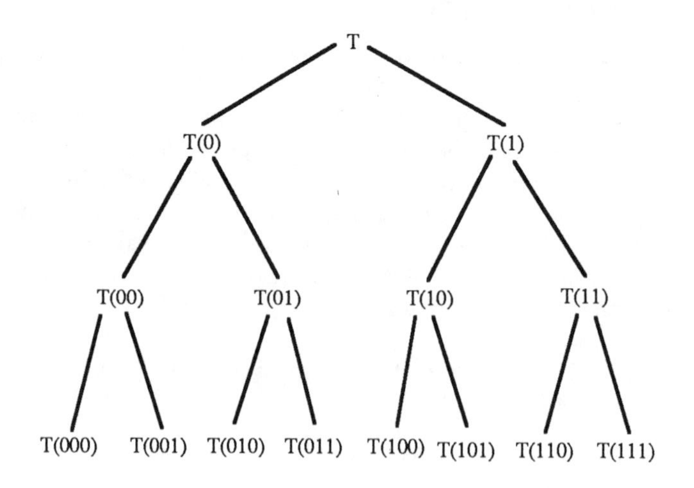


Figure 8.6.1 Computation Tree

Here, the indices are specified in binary and the schemata

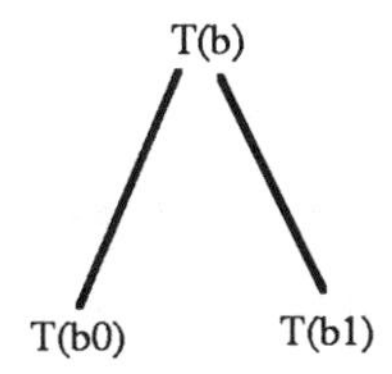


Figure 8.6.2. A Node

means that the eigensystem for the tridiagonal $T(b)$ is obtained from the eigensystems of the tridiagonals $T(b0)$ and $T(b1)$. For example, the eigensystems for the $N$-by-$N$ matrices $T(110)$ and $T(111)$ are combined to produce the eigensystem for the $2N$-by-$2N$ tridiagonal matrix $T(11)$.

With tree-structured algorithms there is always the danger that parallelism is lost as the tree is "climbed" towards the root, but this is not the case in our problem. To see this suppose we have 8 processors and that the first task of Proc(b) is to compute the Schur decomposition of $T(b)$ where $b = 000, 001, 010, 011, 100, 101, 110, 111$. This portion of the computation is perfectly load balanced and does not involve interprocessor communication. (We are ignoring the Theorem 8.6.3 deflations, which are unlikely to cause significant load imbalance.)

At the next level there are four gluing operations to perform: $T(00)$, $T(01)$, $T(10)$, $T(11)$. However, each of these computations neatly subdivides and we can assign two processors to each task. For example, once the secular equation that underlies the $T(00)$ synthesis is known to both Proc(000) and Proc(001), then they each can go about getting half of the eigenvalues and corresponding eigenvectors. Likewise, 4 processors can each be assigned to the $T(0)$ and $T(1)$ problem. All 8 processors can participate in computing the eigensystem of $T$. Thus, at every level full parallelism can be maintained because the eigenvalue/eigenvector computations are independent of one another.

### Problems

**P8.6.1** Give an algorithm for determining $\rho$ and $\theta$ in (8.6.3) with the property that $\theta \in \{-1, 1\}$ and $\min\{\, |a_r - \rho|, |a_{r+1} - \rho| \,\}$ is maximized.

**P8.6.2** Show that if $D = \text{diag}(d_1, \ldots, d_n)$, $z_i^2 + z_j^2 \neq 0$, $i < j$ , and $(G(i,j,\theta)^T z)_j = 0$ , then the off-diagonal portion of $J(i,j,\theta)^T D J(i,j,\theta)$ has Frobenius norm

$$2|d_i - d_j||z_i z_j|/(z_i^2 + z_j^2)$$

**P8.6.3** What communication is necessary between the processors assigned to a particular $T_b$? Is it possible to share the work associated with the processing of repeated $d_i$ and zero $z_i$ ?

**P8.6.4** If $T$ is positive definite, does it follow that the matrices $T_1$ and $T_2$ in §8.6.1 are positive definite?

**Notes and References for Sec. 8.6**

The algorithm discussed in this section is detailed in

J.J. Dongarra and D.C. Sorensen (1987). "A Fully Parallel Algorithm for the Symmetric Eigenvalue Problem," *SIAM J. Sci. and Stat. Comp. 8,* S139–S154.

Critical aspects of the algorithm have their origins in the following papers:

J.R. Bunch, C.P. Nielsen, and D.C. Sorensen (1978). "Rank-One Modification of the Symmetric Eigenproblem," *Numer. Math. 31,* 31–48.
J.J.M. Cuppen (1981). "A Divide and Conquer Method for the Symmetric Eigenproblem," *Numer. Math. 36,* 177–95.
G.H. Golub (1973). "Some Modified Matrix Eigenvalue Problems," *SIAM Review 15,* 318–44.

SVD versions and how to proceed for the general band case are discussed in

P. Arbenz, W. Gander, and G.H. Golub (1988). "Restricted Rank Modification of the Symmetric Eigenvalue Problem: Theoretical Considerations," *Lin. Alg. and Its Applic. 104,* 75–95.
P. Arbenz and G.H. Golub (1988). "On the Spectral Decomposition of Hermitian Matrices Subject to Indefinite Low Rank Perturbations with Applications," *SIAM J. Matrix Anal. Appl. 9,* 40–58.
E.R. Jessup and D.C. Sorensen (1988). "A Parallel Algorithm for Computing the Singular Value Decomposition," Report 102, Mathematics and Computer Science Divison, Argonne National Laboratory, Argonne, IL.

# 8.7 More Generalized Eigenvalue Problems

In the generalized eigenvalue problem $Ax = \lambda Bx$ it is frequently the case that $A$ is symmetric and $B$ is symmetric and positive definite. Pencils of this variety are referred to as *symmetric-definite pencils.* This property is preserved under congruence transformations:

$$\begin{array}{ccc} A - \lambda B & & (X^T AX) - \lambda(X^T BX) \\ & \Leftrightarrow & \\ \text{symmetric-definite} & & \text{symmetric-definite} \end{array}$$

In this section we present various structure-preserving procedures that solve such eigenproblems. The related generalized singular value decomposition problem is also discussed.

## 8.7.1 Mathematical Background

Although the $QZ$ algorithm of §7.7 can be used to solve the symmetric-definite problem, it destroys both symmetry and definiteness. What we seek is a stable, efficient algorithm that computes $X$ such that $X^TAX$ and $X^TBX$ are both in "canonical form." The obvious form to aim for is diagonal form.

**Theorem 8.7.1** *Suppose $A$ and $B$ are n-by-n symmetric matrices, and define $C(\mu)$ by*

$$C(\mu) = \mu A + (1-\mu)B \qquad \mu \in \mathbb{R}. \tag{8.7.1}$$

*If there exists a $\mu \in [0,1]$ such that $C(\mu)$ is non-negative definite and*

$$\text{null}(C(\mu)) = \text{null}(A) \cap \text{null}(B)$$

*then there exists a nonsingular $X$ such that both $X^TAX$ and $X^TBX$ are diagonal.*

**Proof.** Let $\mu \in [0,1]$ be chosen so that $C(\mu)$ is non-negative definite with the property that $\text{null}(C(\mu)) = \text{null}(A) \cap \text{null}(B)$. Let

$$Q_1^T C(\mu) Q_1 = \begin{bmatrix} D & 0 \\ 0 & 0 \end{bmatrix} \qquad D = \text{diag}(d_1, \ldots, d_k),\ d_i > 0$$

be the Schur decomposition of $C(\mu)$ and define $X_1 = Q_1 \text{diag}(D^{-1/2}, I_{n-k})$. If $A_1 = X_1^T A X_1$, $B_1 = X_1^T B X_1$, and $C_1 = X_1^T C(\mu) X_1$, then

$$C_1 = \begin{bmatrix} I_k & 0 \\ 0 & 0 \end{bmatrix} = \mu A_1 + (1-\mu) B_1.$$

Since $\text{span}\{e_{k+1}, \ldots, e_n\} = \text{null}(C_1) = \text{null}(A_1) \cap \text{null}(B_1)$ it follows that $A_1$ and $B_1$ have the following block structure:

$$A_1 = \begin{array}{c} \begin{bmatrix} A_{11} & 0 \\ 0 & 0 \end{bmatrix} \begin{array}{l} k \\ n-k \end{array} \\ \quad k \quad n-k \end{array} \qquad B_1 = \begin{array}{c} \begin{bmatrix} B_{11} & 0 \\ 0 & 0 \end{bmatrix} \begin{array}{l} k \\ n-k \end{array} \\ \quad k \quad n-k \end{array}.$$

Moreover $I_k = \mu A_{11} + (1-\mu) B_{11}$.

Suppose $\mu \neq 0$. It then follows that if $Z^T B_{11} Z = \text{diag}(b_1, \ldots, b_k)$ is the Schur decomposition of $B_{11}$ and we set $X = X_1 \text{diag}(Z, I_{n-k})$ then

$$X^T B X = \text{diag}(b_1, \ldots, b_k, 0, \ldots, 0) \equiv D_B$$

and

$$\begin{aligned} X^T A X &= \frac{1}{\mu} X^T \left( C(\mu) - (1-\mu) B \right) X \\ &= \frac{1}{\mu} \left( \begin{bmatrix} I_k & 0 \\ 0 & 0 \end{bmatrix} - (1-\mu) D_B \right) \equiv D_A. \end{aligned}$$

On the other hand, if $\mu = 0$, then let $Z^T A_{11} Z = \mathrm{diag}(a_1, \ldots, a_k)$ be the Schur decomposition of $A_{11}$ and set $X = X_1 \mathrm{diag}(Z, I_{n-k})$. It is easy to verify that in this case as well, both $X^T AX$ and $X^T BX$ are diagonal. □

Frequently, the conditions in Theorem 8.7.1 are satisfied because either $A$ or $B$ is positive definite.

**Corollary 8.7.2** *If $A - \lambda B \in \mathbb{R}^{n \times n}$ is symmetric-definite, then there exists a nonsingular $X = [\, x_1, \ldots, x_n \,]$ such that*

$$X^T AX = \mathrm{diag}(a_1, \ldots, a_n) \quad \textit{and} \quad X^T BX = \mathrm{diag}(b_1, \ldots, b_n).$$

*Moreover, $Ax_i = \lambda_i Bx_i$ for $i = 1{:}n$ where $\lambda_i = a_i/b_i$.*

**Proof.** By setting $\mu = 0$ in Theorem 8.7.1 we see that symmetric-definite pencils can be simultaneously diagonalized. The rest of the corollary is easily verified. □

**Example 8.7.1** If

$$A = \begin{bmatrix} 229 & 163 \\ 163 & 116 \end{bmatrix} \quad \text{and} \quad B = \begin{bmatrix} 81 & 59 \\ 59 & 43 \end{bmatrix}$$

then $A - \lambda B$ is symmetric-definite and $\lambda(A, B) = \{5, -1/2\}$. If

$$X = \begin{bmatrix} 3 & -5 \\ -4 & 7 \end{bmatrix}$$

then $X^T AX = \mathrm{diag}(5, -1)$ and $X^T BX = \mathrm{diag}(1, 2)$.

Stewart (1979b) has worked out a perturbation theory for symmetric pencils $A - \lambda B$ that satisfy

$$c(A, B) = \min_{\|x\|_2 = 1} (x^T Ax)^2 + (x^T Bx)^2 > 0 \tag{8.7.2}$$

The scalar $c(A, B)$ is called the *Crawford number* of the pencil $A - \lambda B$.

**Theorem 8.7.3** *Suppose $A - \lambda B$ is an n-by-n symmetric-definite pencil with eigenvalues*

$$\lambda_1 \geq \lambda_2 \geq \cdots \geq \lambda_n.$$

*Suppose $E_A$ and $E_B$ are symmetric n-by-n matrices that satisfy*

$$\epsilon^2 = \| E_A \|_2^2 + \| E_B \|_2^2 < c(A, B).$$

*Then $(A + E_A) - \lambda(B + E_B)$ is symmetric-definite with eigenvalues $\mu_1 \geq \cdots \geq \mu_n$ that satisfy*

$$|\arctan(\lambda_i) - \arctan(\mu_i)| \leq \arctan(\epsilon / c(A, B))$$

*for $i = 1{:}n$.*

**Proof.** See Stewart (1979b). □

## 8.7.2 Methods for the Symmetric-Definite Problem

Turning to algorithmic matters, we first present a method for solving the symmetric-definite problem that utilizes both the Cholesky factorization and the symmetric QR algorithm.

**Algorithm 8.7.1** Given $A = A^T \in \mathbb{R}^{n\times n}$ and $B = B^T \in \mathbb{R}^{n\times n}$ with $B$ positive definite, the following algorithm computes a nonsingular $X$ such that $X^T BX= I_n$ and $X^T AX = \text{diag}(a_1, \ldots, a_n)$.

Compute the Cholesky factorization $B = GG^T$
using Algorithm 4.2.2.
Compute $C = G^{-1}AG^{-T}$.
Use the symmetric QR algorithm to compute the Schur
decomposition $Q^T CQ = \text{diag}(a_1, \ldots, a_n)$.
Set $X = G^{-T}Q$.

This algorithm requires about $14n^3$ flops. In a practical implementation, $A$ can be overwritten by the matrix $C$. See Martin and Wilkinson (1968c) for details. Note that

$$\lambda(A, B) = \lambda(A, GG^T) = \lambda(G^{-1}AG^{-T}, I) = \lambda(C) = \{a_1, \ldots, a_n\}.$$

If $\hat{a}_i$ is a computed eigenvalue obtained by Algorithm 8.7.1, then it can be shown that $\hat{a}_i \in \lambda(G^{-1}AG^{-T} + E_i)$, where $\| E_i \|_2 \approx \mathbf{u}\| A \|_2\| B^{-1} \|_2$. Thus, if $B$ is ill-conditioned, then $\hat{a}_i$ may be severely contaminated with roundoff error even if $a_i$ is a well-conditioned generalized eigenvalue. The problem, of course, is that in this case, the matrix $C = G^{-1}AG^{-T}$ can have some very large entries if $B$, and hence $G$, is ill-conditioned. This difficulty can sometimes be overcome by replacing the matrix $G$ in Algorithm 8.7.1 with $VD^{-1/2}$ where $V^T BV = D$ is the Schur decomposition of $B$. If the diagonal entries of $D$ are ordered from smallest to largest, then the large entries in $C$ are concentrated in the upper left-hand corner. The small eigenvalues of $C$ can then be computed without excessive roundoff error contamination (or so the heuristic goes). For further discussion, consult Wilkinson (AEP, pp.337–38).

**Example 8.7.2** If

$$A = \begin{bmatrix} 1 & 2 & 3 \\ 2 & 4 & 5 \\ 3 & 5 & 6 \end{bmatrix} \quad \text{and} \quad G = \begin{bmatrix} .001 & 0 & 0 \\ 1 & .001 & 0 \\ 2 & 1 & .001 \end{bmatrix}$$

and $B = GG^T$, then the two smallest eigenvalues of $A - \lambda B$ are

$$a_1 = -0.619402940600584 \qquad a_2 = 1.627440079051887.$$

If 17-digit floating point arithmetic is used, then these eigenvalues are computed to full machine precision when the symmetric QR algorithm is applied to $fl(D^{-1/2}V^T AVD^{-1/2})$, where $B = VDV^T$ is the Schur decomposition of $B$. On the other hand, if Algorithm 8.7.1 is applied, then

$$\hat{a}_1 = -0.619373517376444 \qquad \hat{a}_2 = 1.627516601905228.$$

The reason for obtaining only four correct significant digits is that $\kappa_2(B) \approx 10^{18}$.

The condition of the matrix $X$ in Algorithm 8.7.1 can sometimes be improved by replacing $B$ with a suitable convex combination of $A$ and $B$. The connection between the eigenvalues of the modified pencil and those of the original are detailed in the proof of Theorem 8.7.1.

Other difficulties concerning Algorithm 8.7.1 revolve around the fact that $G^{-1}AG^{-T}$ is generally full even when $A$ and $B$ are sparse. This is a serious problem, since many of the symmetric-definite problems arising in practice are large and sparse.

Crawford (1973) has shown how to implement Algorithm 8.7.1 effectively when $A$ and $B$ are banded. Aside from this case, however, the simultaneous diagonalization approach is impractical for the large, sparse symmetric-definite problem.

An alternative idea is to extend the Rayleigh quotient iteration (8.4.4) as follows:

$$\begin{array}{l} x_0 \text{ given with } \| x_0 \|_2 = 1 \\ \textbf{for } k = 0, 1, \ldots \\ \quad \mu_k = x_k^T A x_k / x_k^T B x_k \\ \quad \text{Solve } (A - \mu_k B) z_{k+1} = B x_k \text{ for } z_{k+1}. \\ \quad x_{k+1} = z_{k+1} / \| z_{k+1} \|_2 \\ \textbf{end} \end{array} \tag{8.7.3}$$

The mathematical basis for this iteration is that

$$\lambda = \frac{x^T A x}{x^T B x} \tag{8.7.4}$$

minimizes

$$f(\lambda) = \| Ax - \lambda Bx \|_B \tag{8.7.5}$$

where $\| \cdot \|_B$ is defined by $\|z\|_B^2 = z^T B^{-1} z$. The mathematical properties of

(8.7.3) are similar to those of (8.4.4). Its applicability depends on whether or not systems of the form $(A - \mu B)z = x$ can be readily solved. A similar comment pertains to the following generalized orthogonal iteration:

$$\begin{array}{l} Q_0 \in \mathbb{R}^{n \times p} \text{ given with } Q_0^T Q_0 = I_p \\ \textbf{for } k = 1, 2, \ldots \\ \quad \text{Solve } BZ_k = AQ_{k-1} \text{ for } Z_k. \\ \quad Z_k = Q_k R_k \quad \text{(QR factorization)} \\ \textbf{end} \end{array} \tag{8.7.6}$$

This is mathematically equivalent to (7.3.4) with $A$ replaced by $B^{-1}A$. Its practicality depends on how easy it is to solve linear systems of the form $Bz = y$.

Sometimes $A$ and $B$ are so large that neither (8.7.3) nor (8.7.6) can be invoked. In this situation, one can resort to any of a number of gradient and coordinate relaxation algorithms. See Stewart (1976b) for an extensive guide to the literature.

### 8.7.3 The Generalized Singular Value Problem

We conclude with some remarks about symmetric pencils that have the form $A^TA - \lambda B^TB$ where $A \in \mathbb{R}^{m\times n}$ and $B \in \mathbb{R}^{p\times n}$. This pencil underlies the *generalized singular value decomposition* (GSVD), a decomposition that is useful in several constrained least squares problems. (Cf. §12.1.) Note that by Theorem 8.7.1 there exists a nonsingular $X \in \mathbb{R}^{n\times n}$ such that $X^T(A^TA)X$ and $X^T(B^TB)X$ are both diagonal. The value of the GSVD is that these diagonalizations can be achieved without forming $A^TA$ and $B^TB$.

**Theorem 8.7.4 (Generalized Singular Value Decomposition)** *If we have $A \in \mathbb{R}^{m\times n}$ with $m \geq n$ and $B \in \mathbb{R}^{p\times n}$, then there exist orthogonal $U \in \mathbb{R}^{m\times m}$ and $V \in \mathbb{R}^{p\times p}$ and an invertible $X \in \mathbb{R}^{n\times n}$ such that*

$$U^TAX = C = \text{diag}(c_1,\ldots,c_n) \qquad c_i \geq 0$$

*and*

$$V^TBX = S = \text{diag}(s_1,\ldots,s_q) \qquad s_i \geq 0$$

*where $q = \min(p,n)$.*

**Proof.** The proof of this decomposition appears in Van Loan (1976). We present a more constructive proof along the lines of Paige and Saunders (1981). For clarity we assume that $\text{null}(A) \cap \text{null}(B) = \{0\}$ and $p \geq n$. We leave it to the reader to extend the proof so that it covers theses cases.

Let

$$\begin{bmatrix} A \\ B \end{bmatrix} = \begin{bmatrix} Q_1 \\ Q_2 \end{bmatrix} R \qquad (8.7.6)$$

be a QR factorization with $Q_1 \in \mathbb{R}^{m\times n}$, $Q_2 \in \mathbb{R}^{p\times n}$, and $R \in \mathbb{R}^{n\times n}$. Paige and Saunders show that the SVD's of $Q_1$ and $Q_2$ are related in the sense that

$$Q_1 = UCW^T \qquad Q_2 = VSW^T \qquad (8.7.7)$$

Here, $U$,$V$, and $W$ are orthogonal, $C = \text{diag}(c_i)$ with $0 \leq c_1 \leq \cdots \leq c_n$, $S = \text{diag}(s_i)$ with $s_1 \geq \cdots \geq s_n$, and $C^TC + S^TS = I_n$. The decomposition (8.7.7) is a variant of the CS decomposition in §2.6 and from it we conclude that $A = Q_1R = UC(W^TR)$ and $B = Q_2R = VS(W^TR)$. The theorem follows by setting $X = (W^TR)^{-1}$, $D_A = C$, and $D_B = S$. The invertibility of $R$ follows from our assumption that $\text{null}(A) \cap \text{null}(B) = \{0\}$. We leave it to the reader to work out the details when $A$ and $B$ have common null vectors. □

The elements of the set $\sigma(A, B) \equiv \{ c_1/s_1, \ldots, c_n/s_q \}$ are referred to as the *generalized singular values* of $A$ and $B$. Note that $\sigma \in \sigma(A, B)$ implies that $\sigma^2 \in \lambda(A^TA, B^TB)$. The theorem is a generalization of the SVD in that if $B = I_n$, then $\sigma(A, B) = \sigma(A)$.

Our proof of the GSVD is of practical importance since Stewart (1982) and Van Loan (1985) have shown how to stably compute the CS decomposition. The only tricky part is the inversion of $W^TR$ to get $X$. Note that the columns of $X = [\, x_1, \ldots, x_n \,]$ satisfy

$$s_i^2 A^TAx_i = c_i^2 B^TBx_i \qquad i = 1{:}n$$

and so if $s_i \neq 0$ then $A^TAx_i = \sigma_i^2 B^TBx_i$ where $\sigma_i = c_i/s_i$. Thus, the $x_i$ are aptly termed the *generalized singular vectors* of the pair $(A, B)$.

In several applications an orthonormal basis for some designated generalized singular vector subspace space $\text{span}\{x_{i_1}, \ldots, x_{i_k}\}$ is required. We show how this can be accomplished without any matrix inversions or cross products:

- Compute the QR factorization

$$\begin{bmatrix} A \\ B \end{bmatrix} = \begin{bmatrix} Q_1 \\ Q_2 \end{bmatrix} R.$$

- Compute the CS decomposition

$$Q_1 = UCW^T \qquad Q_2 = VSW^T$$

  and order the diagonals of $C$ and $S$ so that

$$\{ c_1/s_1, \ldots, c_k/s_k \} = \{ c_{i_1}/s_{i_1}, \ldots, c_{i_k}/s_{i_k} \}.$$

- Compute orthogonal $Z$ and upper triangular $T$ so $TZ = W^TR$. (See P8.7.5.) Note that if $X^{-1} = W^TR = TZ$, then $X = Z^TT^{-1}$ and so the first $k$ rows of $Z$ are an orthonormal basis for $\text{span}\{x_1, \ldots, x_k\}$.

**Problems**

**P8.7.1** Suppose $A \in \mathbb{R}^{n\times n}$ is symmetric and $G \in \mathbb{R}^{n\times n}$ is lower triangular and nonsingular. Give an efficient algorithm for computing $C = G^{-1}AG^{-T}$.

**P8.7.2** Suppose $A \in \mathbb{R}^{n\times n}$ is symmetric and $B \in \mathbb{R}^{n\times n}$ is symmetric positive definite. Given an algorithm for computing the eigenvalues of $AB$ that uses the Cholesky factorization and the symmetric $QR$ algorithm.

**P8.7.3** Show that if $C$ is real and diagonalizable, then there exist symmetric matrices $A$ and $B$, $B$ nonsingular, such that $C = AB^{-1}$. This shows that symmetric pencils $A - \lambda B$ are essentially general.

**P8.7.4** Show how to convert an $Ax = \lambda Bx$ problem into a generalized singular value problem if $A$ and $B$ are both symmetric and non-negative definite.

**P8.7.5** Given $Y \in \mathbb{R}^{n\times n}$ show how to compute Householder matrices $H_2, \ldots, H_n$ so that $YH_n \cdots H_2 = T$ is upper triangular. Hint:$H_k$ zeros out the $k$th row.

**P8.7.6** Suppose

$$\begin{bmatrix} 0 & A \\ A^T & 0 \end{bmatrix}\begin{bmatrix} y \\ z \end{bmatrix} = \lambda\begin{bmatrix} B_1 & 0 \\ 0 & B_2 \end{bmatrix}\begin{bmatrix} y \\ z \end{bmatrix}$$

where $A \in \mathbb{R}^{m\times n}$, $B_1 \in \mathbb{R}^{m\times m}$, and $B_2 \in \mathbb{R}^{n\times n}$. Assume that $B_1$ and $B_2$ are positive definite with Cholesky triangles $G_1$ and $G_2$ respectively. Relate the generalized eigenvalues of this problem to the singular values of $G_1^{-1}AG_2^{-T}$

### Notes and References for Sec. 8.7

An excellent survey of computational methods for symmetric-definite pencils is given in

G.W. Stewart (1976b). "A Bibliographical Tour of the Large Sparse Generalized Eigenvalue Problem," in *Sparse Matrix Computations* , ed., J.R. Bunch and D.J. Rose, Academic Press, New York.

Some papers of particular interest include

C.R. Crawford (1973). "Reduction of a Band Symmetric Generalized Eigenvalue Problem," *Comm. ACM 16,* 41–44.

C.R. Crawford (1976). "A Stable Generalized Eigenvalue Problem," *SIAM J. Num. Anal. 13,* 854–60.

G. Fix and R. Heiberger (1972). "An Algorithm for the Ill-Conditioned Generalized Eigenvalue Problem," *SIAM J. Num. Anal. 9,* 78–88.

R.S. Martin and J.H. Wilkinson (1968c). "Reduction of a Symmetric Eigenproblem $Ax = \lambda Bx$ and Related Problems to Standard Form," *Numer. Math. 11,* 99–110.

G. Peters and J.H. Wilkinson (1969). "Eigenvalues of $Ax = \lambda Bx$ with Band Symmetric A and B," *Comp. J. 12,* 398-404.

A. Ruhe (1974). "SOR Methods for the Eigenvalue Problem with Large Sparse Matrices," *Math. Comp.* 28, 695–710.

See also

A. Bunse-Gerstner (1984). "An Algorithm for the Symmetric Generalized Eigenvalue Problem," *Lin. Alg. and Its Applic. 58,* 43–68.

C.R. Crawford (1986). "Algorithm 646 PDFIND: A Routine to Find a Positive Definite Linear Combination of Two Real Symmetric Matrices," *ACM Trans. Math. Soft. 12,* 278–282.

C.R. Crawford and Y.S. Moon (1983). "Finding a Positive Definite Linear Combination of Two Hermitian Matrices," *Lin. Alg. and Its Applic. 51,* 37–48.

M.T. Heath, A.J. Laub, C.C. Paige, and R.C. Ward (1986). "Computing the SVD of a Product of Two Matrices," *SIAM J. Sci. and Stat. Comp. 7,* 1147-1159.

The simultaneous reduction of two symmetric matrices to diagonal form is discussed in

A. Berman and A. Ben-Israel (1971). "A Note on Pencils of Hermitian or Symmetric Matrices," *SIAM J. Applic. Math. 21*, 51–54.
K.N. Majinder (1979). "Linear Combinations of Hermitian and Real Symmetric Matrices," *Lin. Alg. and Its Applic. 25*, 95–105.
F. Uhlig (1973). "Simultaneous Block Diagonalization of Two Real Symmetric Matrices," *Lin. Alg. and Its Applic. 7*, 281–89.
F. Uhlig (1976). "A Canonical Form for a Pair of Real Symmetric Matrices That Generate a Nonsingular Pencil," *Lin. Alg. and Its Applic. 14*, 189–210.

The perturbation theory that we presented for the symmetric-definite problem was taken from

G.W. Stewart (1979b). "Perturbation Bounds for the Definite Generalized Eigenvalue Problem," *Lin. Alg. and Its Applic. 23*, 69–86.

See also

L. Elsner and J. Guang Sun (1982). "Perturbation Theorems for the Generalized Eigenvalue Problem,; *Lin. Alg. and its Applic. 48*, 341-357.
C.C. Paige (1984). "A Note on a Result of Sun J.-Guang: Sensitivity of the CS and GSV Decompositions," *SIAM J. Numer. Anal. 21*, 186–191.
J. Guang Sun (1982). "A Note on Stewart's Theorem for Definite Matrix Pairs," *Lin. Alg. and Its Applic. 48*, 331–339.
J. Guang Sun (1983). "Perturbation Analysis for the Generalized Singular Value Problem," *SIAM J. Numer. Anal. 20*, 611–625.

The generalized SVD and some of its applications are discussed in

B. Kågström (1985). "The Generalized Singular Value Decomposition and the General $A - \lambda B$ Problem," *BIT 24*, 568-583.
C.F. Van Loan (1976). "Generalizing the Singular Value Decomposition," *SIAM J. Num. Anal. 13*, 76–83.
C.C. Paige and M. Saunders (1981). "Towards A Generalized Singular Value Decomposition," *SIAM J. Num. Anal. 18*, 398–405.

Stable methods for computing the CS and generalized singular value decompositions are described in

C.C. Paige (1986). "Computing the Generalized Singular Value Decomposition," *SIAM J. Sci. and Stat. Comp. 7*, 1126–1146.
G.W. Stewart (1983). "A Method for Computing the Generalized Singular Value Decomposition," in *Matrix Pencils* , ed. B. Kågström and A. Ruhe, Springer-Verlag, New York, pp. 207–20.
C. Van Loan (1985). "Computing the CS and Generalized Singular Value Decomposition," *Numer. Math. 46*, 479–492.

# Chapter 9

# Lanczos Methods



In this chapter we develop the Lanczos method, a technique that can be used to solve certain large, sparse, symmetric eigenproblems $Ax = \lambda x$. The method involves partial tridiagonalizations of the given matrix $A$. However, unlike the Householder approach, no intermediate, full submatrices are generated. Equally important, information about $A$'s extremal eigenvalues tends to emerge long before the tridiagonalization is complete. This makes the Lanczos algorithm particularly useful in situations where a few of $A$'s largest or smallest eigenvalues are desired.

The derivation and exact arithmetic attributes of the method are presented in §9.1. The key aspects of the Kaniel-Paige theory are detailed. This theory explains the extraordinary convergence properties of the Lanczos process.

Unfortunately, roundoff errors make the Lanczos method somewhat difficult to use in practice. The cental problem is a loss of orthogonality among the Lanczos vectors that the iteration produces. There are several ways to cope with this as we discuss §9.2.

In the final section we show how the "Lanczos idea" can be applied to solve an assortment of singular value, least squares, and linear equations problems. The unsymmetric Lanczos process is also discussed.

Of particular interest in §9.3 is the development of the conjugate gradient method for symmetric positive definite linear systems. The Lanczos-conjugate gradient connection is explored further in the next chapter.

## 9.1 Derivation and Convergence Properties

Suppose $A \in \mathbb{R}^{n \times n}$ is large, sparse, and symmetric and assume that a few of its largest and/or smallest eigenvalues are desired. This problem can be solved by a method attributed to Lanczos (1950). The method generates a sequence of tridiagonal matrices $T_j$ with the property that the extremal eigenvalues of $T_j \in \mathbb{R}^{j \times j}$ are progressively better estimates of $A$'s extremal eigenvalues. In this section, we derive the technique and investigate its exact arithmetic properties.

### 9.1.1 Krylov Subspaces

The derivation of the Lanczos algorithm can proceed in several ways. So that its remarkable convergence properties do not come as a complete surprise, we prefer to lead into the technique by considering the optimization of the Rayleigh quotient

$$r(x) = \frac{x^T A x}{x^T x} \qquad x \neq 0 .$$

Recall from Theorem 8.1.2 that the maximum and minimum values of $r(x)$ are $\lambda_1(A)$ and $\lambda_n(A)$, respectively. Suppose $\{q_i\} \subseteq \mathbb{R}^n$ is a sequence of orthonormal vectors and define the scalars $M_j$ and $m_j$ by

$$M_j = \lambda_1(Q_j^T A Q_j) = \max_{y \neq 0} \frac{y^T (Q_j^T A Q_j) y}{y^T y} = \max_{\|y\|_2 = 1} r(Q_j y) \leq \lambda_1(A)$$

$$m_j = \lambda_j(Q_j^T A Q_j) = \min_{y \neq 0} \frac{y^T (Q_j^T A Q_j) y}{y^T y} = \min_{\|y\|_2 = 1} r(Q_j y) \geq \lambda_n(A)$$

where $Q_j = [\, q_1, \ldots, q_j \,]$. The Lanczos algorithm can be derived by considering how to generate the $q_j$ so that $M_j$ and $m_j$ are increasingly better estimates of $\lambda_1(A)$ and $\lambda_n(A)$.

Suppose $u_j \in \text{span}\{q_1, \ldots, q_j\}$ is such that $M_j = r(u_j)$. Since $r(x)$ increases most rapidly in the direction of the gradient

$$\nabla r(x) = \frac{2}{x^T x}(Ax - r(x)x),$$

we can ensure that $M_{j+1} > M_j$ if $q_{j+1}$ is determined so

$$\nabla r(u_j) \in \text{span}\{q_1, \ldots, q_{j+1}\}. \tag{9.1.1}$$

(This assumes $\nabla r(u_j) \neq 0$.) Likewise, if $v_j \in \text{span}\{q_1, \ldots, q_j\}$ satisfies $r(v_j) = m_j$, then it makes sense to require

$$\nabla r(v_j) \in \text{span}\{q_1, \ldots, q_{j+1}\} \tag{9.1.2}$$

since $r(x)$ decreases most rapidly in the direction of $-\nabla r(x)$.

At first glance, the task of finding a single $q_{j+1}$ that satisfies these two requirements appears impossible. However, since $\nabla r(x) \in \text{span}\{x, Ax\}$, it is clear that (9.1.1) and (9.1.2) can be simultaneously satisfied provided

$$\text{span}\{q_1, \ldots, q_j\} = \text{span}\{q_1, Aq_1, \ldots, A^{j-1}q_1\} \equiv \mathcal{K}(A, q_1, j)$$

and we choose $q_{j+1}$ so $\mathcal{K}(A, q_1, j+1) = \text{span}\{q_1, \ldots, q_{j+1}\}$. Thus, we are led to the problem of computing orthonormal bases for the *Krylov subspaces* $\mathcal{K}(A, q_1, j)$.

### 9.1.2 Tridiagonalization

In order to find this basis efficiently we exploit the connection between the tridiagonalization of $A$ and the QR factorization of the *Krylov matrix*

$$K(A, q_1, n) = \left[\, q_1,\; Aq_1,\; A^2q_1, \ldots,\; A^{n-1}q_1 \,\right].$$

(See §7.4.4.) Specifically, if $Q^TAQ = T$ is tridiagonal with $Qe_1 = q_1$, then

$$K(A, q_1, n) = Q\left[\, e_1,\; Te_1,\; T^2e_1, \ldots,\; T^{n-1}e_1 \,\right]$$

is the QR factorization of $K(A, q_1, n)$. Thus, the $q_j$ can effectively be generated by tridiagonalizing $A$ with an orthogonal matrix whose first column is $q_1$.

Householder tridiagonalization, discussed in §8.2.1, can be adapted for this purpose. However, this approach is impractical if $A$ is large and sparse because Householder similarity transformations tend to destroy sparsity. As a result, unacceptably large, dense matrices arise during the reduction.

Loss of sparsity can sometimes be controlled by using Givens rather than Householder transformations. See Duff and Reid (1976). However, any method that computes $T$ by successively updating $A$ is not useful in the majority of cases when $A$ is sparse.

This suggests that we try to compute the elements of the tridiagonal matrix $T = Q^TAQ$ directly. Setting $Q = [\, q_1, \ldots, q_n \,]$ and

$$T = \begin{bmatrix} \alpha_1 & \beta_1 & & \cdots & 0 \\ \beta_1 & \alpha_2 & \ddots & & \vdots \\ & \ddots & \ddots & \ddots & \\ \vdots & & \ddots & \ddots & \beta_{n-1} \\ 0 & \cdots & & \beta_{n-1} & \alpha_n \end{bmatrix}$$

and equating columns in $AQ = QT$, we find

$$Aq_j = \beta_{j-1}q_{j-1} + \alpha_j q_j + \beta_j q_{j+1} \qquad \beta_0 q_0 \equiv 0$$

for $j = 1{:}n-1$. The orthonormality of the $q_i$ implies $\alpha_j = q_j^T A q_j$. Moreover, if $r_j = (A - \alpha_j I)q_j - \beta_{j-1}q_{j-1}$ is nonzero, then $q_{j+1} = r_j/\beta_j$ where $\beta_j = \pm\| r_j \|_2$. If $r_j = 0$ then the iteration breaks down but (as we shall see) not without the acquisition of valuable invariant subspace information. So by properly sequencing the above formulae we obtain the *Lanczos iteration*:

$$
\begin{array}{l}
r_0 = q_1;\ \beta_0 = 1;\ q_0 = 0;\ j = 0 \\
\textbf{while}\ (\beta_j \neq 0) \\
\quad q_{j+1} = r_j/\beta_j;\ j = j+1;\ \alpha_j = q_j^T A q_j \\
\quad r_j = (A - \alpha_j I)q_j - \beta_{j-1}q_{j-1};\ \beta_j = \| r_j \|_2 \\
\textbf{end}
\end{array}
\qquad (9.1.3)
$$

There is no loss of generality in choosing the $\beta_j$ to be positive. The $q_j$ are called *Lanczos vectors.*

### 9.1.3 Termination and Error Bounds

The iteration halts before complete tridiagonalization if $q_1$ is contained in a proper invariant subspace. This is one of several mathematical properties of the method that we summarize in the following theorem.

**Theorem 9.1.1** *Let $A \in \mathbb{R}^{n\times n}$ be symmetric and assume $q_1 \in \mathbb{R}^n$ has unit 2-norm. Then the Lanczos iteration (9.1.3) runs until $j = m$, where $m =$* rank$(K(A, q_1, n))$ *Moreover, for $j = 1{:}m$ we have*

$$AQ_j = Q_j T_j + r_j e_j^T \qquad (9.1.4)$$

*where*

$$
T_j = \begin{bmatrix}
\alpha_1 & \beta_1 & & \cdots & 0 \\
\beta_1 & \alpha_2 & \ddots & & \vdots \\
 & \ddots & \ddots & \ddots & \\
\vdots & & \ddots & \ddots & \beta_{j-1} \\
0 & \cdots & & \beta_{j-1} & \alpha_j
\end{bmatrix}
$$

*and $Q_j = [\, q_1, \ldots, q_j \,]$ has orthonormal columns that span $\mathcal{K}(A, q_1, j)$.*

**Proof.** The proof is by induction on $j$. Suppose the iteration has produced $Q_j = [\, q_1, \ldots, q_j \,]$ such that range$(Q_j) = \mathcal{K}(A, q_1, j)$ and $Q_j^T Q_j = I_j$. It is easy to see from (9.1.3) that (9.1.4) holds. Thus, $Q_j^T A Q_j = T_j + Q_j^T r_j e_j^T$. Since $\alpha_i = q_i^T A q_i$ for $i = 1{:}j$ and

$$q_{i+1}^T A q_i = q_{i+1}^T (A q_i - \alpha_i q_i - \beta_{i-1} q_{i-1}) = q_{i+1}^T (\beta_i q_{i+1}) = \beta_i$$

for $i = 1{:}j-1$, we have $Q_j^T A Q_j = T_j$. Consequently, $Q_j^T r_j = 0$.

If $r_j \neq 0$, then $q_{j+1} = r_j/\| r_j \|_2$ is orthogonal to $q_1, \ldots, q_j$ and

$$q_{j+1} \in \text{span}\{Aq_j, q_j, q_{j-1}\} \subseteq \mathcal{K}(A, q_1, j+1).$$

Thus, $Q_{j+1}^T Q_{j+1} = I_{j+1}$ and range$(Q_{j+1}) = \mathcal{K}(A, q_1, j+1)$. On the other hand, if $r_j = 0$, then $AQ_j = Q_j T_j$. This says that range$(Q_j) = \mathcal{K}(A, q_1, j)$ is invariant. From this we conclude that $j = m = \text{rank}(K(A, q_1, n))$. □

Encountering a zero $\beta_j$ in the Lanczos iteration is a welcome event in that it signals the computation of an exact invariant subspace. However, an exact zero or even a small $\beta_j$ is a rarity in practice. Nevertheless, the extremal eigenvalues of $T_j$ turn out to be surprisingly good approximations to $A$'s extremal eigenvalues. Consequently, other explanations for the convergence of $T_j$'s eigenvalues must be sought. The following result is a step in this direction.

**Theorem 9.1.2** *Suppose that $j$ steps of the Lanczos algorithm have been performed and that $S_j^T T_j S_j = \text{diag}(\theta_1, \ldots, \theta_j)$ is the Schur decomposition of the tridiagonal matrix $T_j$. If $Y_j = [\, y_1, \ldots, y_j \,] = Q_j S_j \in \mathbb{R}^{j \times j}$ then for $i = 1{:}j$ we have $\| Ay_i - \theta_i y_i \|_2 = |\beta_j|\,|s_{ji}|$ where $S_j = (s_{pq})$.*

**Proof.** Post-multiplying (9.1.4) by $S_j$ gives

$$AY_j = Y_j \text{diag}(\theta_1, \ldots, \theta_j) + r_j e_j^T S_j,$$

and so $Ay_i = \theta_i y_i + r_j(e_j^T S e_i)$. The proof is complete by taking norms and recalling that $\| r_j \|_2 = |\beta_j|$. □

The theorem provides computable error bounds for $T_j$'s eigenvalues:

$$\min_{\mu \in \lambda(A)} |\theta_i - \mu| \leq |\beta_j|\,|s_{ji}| \qquad i = 1{:}j$$

Note that in the terminology of Theorem 8.1.10, the $(\theta_i, y_i)$ are Ritz pairs for the subspace range$(Q_j)$.

Another way that $T_j$ can be used to provide estimates of $A$'s eigenvalues is described in Golub (1974) and involves the judicious construction of a rank-one matrix $E$ such that range$(Q_j)$ is invariant for $A+E$. In particular, if we use the Lanczos method to compute $AQ_j = Q_j T_j + r_j e_j^T$ and set $E = \tau w w^T$, where $\tau = \pm 1$ and $w = aq_j + br_j$, then it can be shown that

$$(A+E)Q_j = Q_j(T_j + \tau a^2 e_j e_j^T) + (1 + \tau ab) r_j e_j^T.$$

If $0 = 1 + \tau ab$, then the eigenvalues of $\tilde{T}_j = T_j + \tau a^2 e_j e_j^T$, a tridiagonal matrix, are also eigenvalues of $A+E$. We may then conclude from Corollary

8.1.5 that the intervals $[\lambda_i(T_j), \lambda_{i-1}(T_j)]$ each contain an eigenvalue of $A$ for $i = 2{:}j$.

These bracketing intervals depend on the choice of $\tau a^2$. Suppose we have an approximate eigenvalue of $\lambda$ of $A$. One possibility is to choose $\tau a^2$ so that $\det(T_j - \lambda I_j) = (\alpha_2 + \tau a^2 - \lambda)p_{j-1}(\lambda) - \beta_{j-1}^2 p_{j-2}(\lambda) = 0$ where the polynomials $p_i(x) = \det(T_i - xI_i)$ can be evaluated at $\lambda$ using the three-term recurrence (8.5.2). (This assumes that $p_{j-1}(\lambda) \neq 0$.) Eigenvalue estimation in this spirit is discussed in Lehmann (1963) and Householder (1968).

### 9.1.4 The Kaniel-Paige Convergence Theory

The preceding discussion indicates how eigenvalue estimates can be obtained via the Lanczos algorithm, but it reveals nothing about rate of convergence. Results of this variety constitute what is known as the *Kaniel-Paige theory*, a sample of which follows.

**Theorem 9.1.3** *Let $A$ be an n-by-n symmetric matrix with eigenvalues $\lambda_1 \geq \cdots \geq \lambda_n$ and corresponding orthonormal eigenvectors $z_1, \ldots, z_n$. If $\theta_1 \geq \cdots \geq \theta_j$ are the eigenvalues of the matrix $T_j$ obtained after $j$ steps of the Lanczos iteration, then*

$$\lambda_1 \geq \theta_1 \geq \lambda_1 - \frac{(\lambda_1 - \lambda_n)\tan(\phi_1)^2}{(c_{j-1}(1+2\rho_1))^2}$$

*where $\cos(\phi_1) = |q_1^T z_1|$, $\rho_1 = (\lambda_1 - \lambda_2)/(\lambda_2 - \lambda_n)$, and $c_{j-1}(x)$ is the Chebyshev polynomial of degree $j-1$.*

**Proof.** From Theorem 8.1.2, we have

$$\theta_1 = \max_{y \neq 0} \frac{y^T T_j y}{y^T y} = \max_{y \neq 0} \frac{(Q_j y)^T A (Q_j y)}{(Q_j y)^T (Q_j y)} = \max_{0 \neq w \in \mathcal{K}(A, q_1, j)} \frac{w^T A w}{w^T w}$$

Since $\lambda_1$ is the maximum of $w^T A w / w^T w$ over all nonzero $w$, it follows that $\lambda_1 \geq \theta_1$. To obtain the lower bound for $\theta_1$, note that

$$\theta_1 = \max_{p \in \mathcal{P}_{j-1}} \frac{q_1^T p(A) A p(A) q_1}{q_1^T p(A)^2 q_1}$$

where $\mathcal{P}_{j-1}$ is the set of $j-1$ degree polynomials. If $q_1 = \sum_{i=1}^{n} d_i z_i$ then

$$\frac{q_1^T p(A) A p(A) q_1}{q_1^T p(A)^2 q_1} = \frac{\sum_{i=1}^{n} d_i^2 p(\lambda_i)^2 \lambda_i}{\sum_{i=1}^{n} d_i^2 p(\lambda_i)^2}$$

$$\geq \lambda_1 - (\lambda_1 - \lambda_n) \frac{\sum_{i=2}^{n} d_i^2 p(\lambda_i)^2}{d_1^2 p(\lambda_1)^2 + \sum_{i=2}^{n} d_i^2 p(\lambda_i)^2} .$$

We can make the lower bound tight by selecting a polynomial $p(x)$ that is large at $x = \lambda_1$ in comparison to its value at the remaining eigenvalues. One way of doing this is to set

$$p(x) = c_{j-1}\left(-1 + 2\frac{x - \lambda_n}{\lambda_2 - \lambda_n}\right)$$

where $c_{j-1}(z)$ is the $(j-1)$-st Chebyshev polynomial generated via the recursion

$$c_j(z) = 2zc_{j-1}(z) - c_{j-2}(z) \qquad c_0 = 1,\ c_1 = z.$$

These polynomials are bounded by unity on $[-1, 1]$, but grow very rapidly outside this interval. By defining $p(x)$ this way, it follows that $|p(\lambda_i)|$ is bounded by unity for $i = 2{:}n$, while $p(\lambda_1) = c_{j-1}(1 + 2\rho_1)$. Thus,

$$\theta_1 \geq \lambda_1 - (\lambda_1 - \lambda_n) \frac{1 - d_1^2}{d_1^2} \frac{1}{c_{j-1}(1 + 2\rho_1)^2} .$$

The desired lower bound is obtained by noting that $\tan(\phi_1)^2 = (1 - d_1^2)/d_1^2$. □

An analogous result pertaining to $\theta_j$ follows immediately from this theorem:

**Corollary 9.1.4** *Using the same notation as the theorem,*

$$\lambda_n \leq \theta_j \leq \lambda_n + \frac{(\lambda_1 - \lambda_n)\tan(\phi_n)^2}{c_{j-1}(1 + 2\rho_n)^2}$$

*where* $\rho_n = (\lambda_{n-1} - \lambda_n)/(\lambda_1 - \lambda_{n-1})$ *and* $\cos(\theta_n) = q_n^T z_n$.

**Proof.** Apply Theorem 9.1.3 with $A$ replaced by $-A$. □

### 9.1.5 The Power Method Versus the Lanczos Method

It is worthwhile to compare $\theta_1$ with the corresponding power method estimate of $\lambda_1$. (See §7.3.1.) For clarity, assume $\lambda_1 \geq \cdots \geq \lambda_n \geq 0$. After $j-1$ power method steps applied to $q_1$, a vector is obtained in the direction of

$$v = A^{j-1}q_1 = \sum_{i=1}^{n} c_i \lambda_i^{j-1} z_i$$

along with an eigenvalue estimate

$$\gamma_1 = \frac{v^T A v}{v^T v}.$$

Using the proof and notation of Theorem 9.1.3, it is easy to show that

$$\lambda_1 \geq \gamma_1 \geq \lambda_1 - (\lambda_1 - \lambda_n)\tan(\phi_1)^2 \left(\frac{\lambda_2}{\lambda_1}\right)^{2j-1}. \qquad (9.1.5)$$

(Hint: Set $p(x) = x^{j-1}$ in the proof.) Thus, we can compare the tightness of the lower bounds for $\theta_1$ and $\gamma_1$ by comparing

$$L_{j-1} \equiv 1 \Big/ \left[c_{j-1}\left(2\frac{\lambda_1}{\lambda_2} - 1\right)\right]^2 \geq 1 \Big/ [c_{j-1}(1 + 2\rho_1)]^2$$

and

$$R_{j-1} = \left(\frac{\lambda_2}{\lambda_1}\right)^{2(j-1)}.$$

This is done in following table for representative values of $j$ and $\lambda_2/\lambda_1$.

| $\lambda_1/\lambda_2$ | $j=5$ | $j=10$ | $j=15$ | $j=20$ | $j=25$ |
|---|---|---|---|---|---|
| 1.50 | $\frac{1.1\times10^{-4}}{3.9\times10^{-2}}$ | $\frac{2.0\times10^{-10}}{6.8\times10^{-4}}$ | $\frac{3.9\times10^{-16}}{1.2\times10^{-5}}$ | $\frac{7.4\times10^{-22}}{2.0\times10^{-7}}$ | $\frac{1.4\times10^{-27}}{3.5\times10^{-9}}$ |
| 1.10 | $\frac{2.7\times10^{-2}}{4.7\times10^{-1}}$ | $\frac{5.5\times10^{-5}}{1.8\times10^{-1}}$ | $\frac{1.1\times10^{-7}}{6.9\times10^{-2}}$ | $\frac{2.1\times10^{-10}}{2.7\times10^{-2}}$ | $\frac{4.2\times10^{-13}}{1.0\times10^{-2}}$ |
| 1.01 | $\frac{5.6\times10^{-1}}{9.2\times10^{-1}}$ | $\frac{1.0\times10^{-1}}{8.4\times10^{-1}}$ | $\frac{1.5\times10^{-2}}{7.6\times10^{-1}}$ | $\frac{2.0\times10^{-3}}{6.9\times10^{-1}}$ | $\frac{2.8\times10^{-4}}{6.2\times10^{-1}}$ |

Table 9.1.1 $L_{j-1}/R_{j-1}$

The superiority of the Lanczos estimate is self-evident. This should be no surprise, since $\theta_1$ is the maximum of $r(x) = x^T Ax/x^T x$ over *all* of $\mathcal{K}(A, q_1, j)$, while $\gamma_1 = r(v)$ for a particular $v$ in $\mathcal{K}(A, q_1, j)$, namely $v = A^{j-1}q_1$.

## 9.1.6 Convergence of Interior Eigenvalues

We conclude with some remarks about error bounds for $T_j$'s interior eigenvalues. The key idea in the proof of Theorem 9.1.3 is the use of the translated Chebyshev polynomial. With this polynomial we amplified the component of $q_1$ in the direction $z_1$. A similar idea can be used to obtain bounds for an interior Ritz value $\theta_1$. However, the bounds are not as satisfactory because the "amplifying polynomial" has the form $q(x)\Pi_{k=1}^{j-1}(x-\lambda_k)$ , where $q(x)$ is the $(j-1)$ degree Chebyshev polynomial on the interval $[\lambda_{i+1}, \lambda_n]$. For details, see Kaniel (1966), Paige (1971), or Scott (1978).

### Problems

**P9.1.1** Suppose $A \in \mathbf{R}^{n\times n}$ is skew-symmetric. Derive a Lanczos-like algorithm for computing a skew-symmetric tridiagonal matrix $T_m$ such that $AQ_m = Q_mT_m$, where $Q_m^TQ_m = I_m$.

**P9.1.2** Let $A \in \mathbf{R}^{n\times n}$ be symmetric and define $r(x) = x^TAx/x^Tx$. Suppose $S \subseteq \mathbf{R}^n$ is a subspace with the property that $x \in S$ implies $\nabla r(x) \in S$. Show that $S$ is invariant for $A$.

**P9.1.3** Show that if a symmetric matrix $A \in \mathbf{R}^{n\times n}$ has a multiple eigenvalue, then the Lanczos iteration terminates prematurely.

**P9.1.4** Show that the index $m$ in Theorem 9.1.1 is the dimension of the smallest invariant subspace for $A$ that contains $q_1$.

**P9.1.5** Let $A \in \mathbf{R}^{n\times n}$ be symmetric and consider the problem of determining an orthonormal sequence $q_1, q_2, \ldots$ with the property that once $Q_j = [\, q_1, \ldots, q_j\,]$ is known, $q_{j+1}$ is chosen so as to minimize $\mu_j = \| (I - Q_{j+1}Q_{j+1}^T)AQ_j \|_F$. Show that if $\text{span}\{q_1, \ldots, q_j\} = \mathcal{K}(A, q_1, j)$, then it is possible to choose $q_{j+1}$ so $\mu_j = 0$. Explain how this optimization problem leads to the Lanczos iteration.

### Notes and References for Sec. 9.1

Good "global" references for the Lanczos algorithm include Parlett (SEP) and

J.K. Cullum and R.A. Willoughby (1985a). *Lanczos Algorithms for Large Symmetric Eigenvalue Computations, Vol. I Theory,* Birkhaüser, Boston.

The classic reference for the Lanczos method is

C. Lanczos (1950). "An Iteration Method for the Solution of the Eigenvalue Problem of Linear Differential and Integral Operators," *J. Res. Nat. Bur. Stand. 45,* 255–82.

Although the convergence of the Ritz values is alluded to this paper, for more details we refer the reader to

S. Kaniel (1966). "Estimates for Some Computational Techniques in Linear Algebra," *Math. Comp. 20,* 369–78.

C.C. Paige (1971). "The Computation of Eigenvalues and Eigenvectors of Very Large Sparse Matrices," Ph.D. thesis, London University.

The Kaniel-Paige theory set forth in these papers is nicely summarized in

D. Scott (1978). "Analysis of the Symmetric Lanczos Process," Electronic Research Laboratory Technical Report UCB/ERL M78/40, University of California, Berkeley.

See also Wilkinson (AEP, pp. 270 ff.), Parlett (SEP, chapter 13), and

Y. Saad (1980). "On the Rates of Convergence of the Lanczos and the Block Lanczos Methods," *SIAM J. Num. Anal.17,* 687–706.

The connections between the Lanczos algorithm, orthogonal polynomials, and the theory of moments are discussed in

G.H. Golub (1974). "Some Uses of the Lanczos Algorithm in Numerical Linear Algebra," *in Topics in Numerical Analysis*, ed., J.J.H. Miller, Academic Press, New York.
A.S. Householder (1968). "Moments and characteristic Roots II," *Numer. Math. 11*, 126–28.
N.J. Lehmann (1963). "Optimale Eigenwerteinschliessungen," *Numer. Math. 5*, 246–72.

We motivated our discussion of the Lanczos algorithm by discussing the inevitability of fill-in when Householder or Givens transformations are used to tridiagonalize. Actually, fill-in can sometimes be kept to an acceptable level if care is exercised. See

I.S. Duff (1974). "Pivot Selection and Row Ordering in Givens Reduction on Sparse Matrices," *Computing 13*, 239–48.
I.S. Duff and J.K. Reid (1976). "A Comparison of Some Methods for the Solution of Sparse Over-Determined Systems of Linear Equations," *J. Inst. Maths. Applic. 17*, 267–80.
L. Kaufman (1979). "Application of Dense Householder Transformations to a Sparse Matrix," *ACM Trans. Math. Soft. 5*, 442–50.

## 9.2 Practical Lanczos Procedures

Rounding errors greatly affect the behavior of the Lanczos iteration. The basic difficulty is caused by loss of orthogonality among the Lanczos vectors, a phenomenon that muddies the issue of termination and complicates the relationship between $A$'s eigenvalues and those of the tridiagonal matrices $T_j$. This troublesome feature, coupled with the advent of Householder's perfectly stable method of tridiagonalization, explains why the Lanczos algorithm was disregarded by numerical analysts during the 1950's and 1960's. However, interest in the method was rejuvenated with the development of the Kaniel-Paige theory and because the pressure to solve large, sparse eigenproblems increased with increased computer power. With many fewer than $n$ iterations typically required to get good approximate extremal eigenvalues, the Lanczos method became attractive as a sparse matrix technique rather than as a competitor of the Householder approach.

Successful implementations of the Lanczos iteration involve much more than a simple encoding of (9.1.3). In this section we outline some of the practical ideas that have been proposed to make Lanczos procedure viable in practice.

### 9.2.1 Exact Arithmetic Implementation

With careful overwriting in (9.1.3) and exploitation of the formula

$$\alpha_j = q_j^T(Aq_j - \beta_{j-1}q_{j-1}),$$

the whole Lanczos process can be implemented with just two $n$-vectors of storage.

**Algorithm 9.2.1. (The Lanczos Algorithm)** Given a symmetric $A \in \mathbb{R}^{n\times n}$ and $w \in \mathbb{R}^n$ having unit 2-norm, the following algorithm computes a $j$-by-$j$ symmetric tridiagonal matrix $T_j$ with the property that $\lambda(T_j) \subset \lambda(A)$. It assumes the existence of a function **A.mult**($w$) that returns the matrix-vector product $Aw$. The diagonal and subdiagonal elements of $T_j$ are stored in $\alpha(1:j)$ and $\beta(1:j-1)$ respectively.

```
v(1:n) = 0; β_0 = 1; j = 0
while β_j ≠ 0
    if j ≠ 0
        for i = 1:n
            t = w_i; w_i = v_i/β_j; v_i = −β_j t
        end
    end
    v = v + A.mult(w)
    j = j + 1; α_j = w^T v; v = v − α_j w; β_j = || v ||_2
end
```

Note that $A$ is not altered during the entire process. Only a procedure **A.mult**($\cdot$) for computing matrix-vector products involving $A$ need be supplied. If $A$ has an average of about $k$ nonzeros per row then approximately $(2k+8)n$ flops are involved in a single Lanczos step.

Upon termination the eigenvalues of $T_j$ can be found using the symmetric tridiagonal QR algorithm or any of the special methods of §8.4, such as bisection.

The Lanczos vectors are generated in the $n$-vector $w$. If they are desired for later use, then special arrangements must be made for their storage. In the typical sparse matrix setting they could be stored on a disk or some other secondary storage device until required.

## 9.2.2 Roundoff Properties

The development of a practical, easy-to-use Lanczos procedure requires an appreciation of the fundamental error analyses of Paige (1971, 1976a, 1980). An examination of his results is the best way to motivate the several modified Lanczos procedures of this section.

After $j$ steps of the algorithm we obtain the matrix of computed Lanczos vectors $\hat{Q}_j = [\hat{q}_1, \ldots, \hat{q}_j]$ and the associated tridiagonal matrix

$$\hat{T}_j = \begin{bmatrix} \hat{\alpha}_1 & \hat{\beta}_1 & & \cdots & 0 \\ \hat{\beta}_1 & \hat{\alpha}_2 & \ddots & & \vdots \\ & \ddots & \ddots & \ddots & \\ \vdots & & \ddots & \ddots & \hat{\beta}_{j-1} \\ 0 & \cdots & & \hat{\beta}_{j-1} & \hat{\alpha}_j \end{bmatrix}.$$

Paige (1971, 1976a) shows that if $\hat{r}_j$ is the computed analog of $r_j$, then

$$A\hat{Q}_j = \hat{Q}_j\hat{T}_j + \hat{r}_j e_j^T + E_j \tag{9.2.1}$$

where

$$\| E_j \|_2 \approx \mathbf{u}\| A \|_2 . \tag{9.2.2}$$

This indicates that the important equation $AQ_j = Q_jT_j + r_je_j^T$ is satisfied to working precision.

Unfortunately, the picture is much less rosy with respect to the orthogonality among the $\hat{q}_i$ . (Normality is not an issue. The computed Lanczos vectors essentially have unit length.) If $\hat{\beta}_j = fl(\| \hat{r}_j \|_2)$ and we compute $\hat{q}_{j+1} = fl(\hat{r}_j/\hat{\beta}_j)$, then a simple analysis shows that $\hat{\beta}_j\hat{q}_{j+1} \approx \hat{r}_j + w_j$ where $\| w_j \|_2 \approx \mathbf{u}\| \hat{r}_j \|_2 \approx \mathbf{u}\| A \|_2$. Thus, we may conclude that

$$|\hat{q}_{j+1}^T\hat{q}_i| \approx \frac{|\hat{r}_j^T\hat{q}_i| + \mathbf{u}\| A \|_2}{|\hat{\beta}_j|}$$

for $i = 1{:}j$. In other words, significant departures from orthogonality can be expected when $\hat{\beta}_j$ is small, *even* in the ideal situation where $\hat{r}_j^T\hat{Q}_j$ is zero. A small $\hat{\beta}_j$ implies cancellation in the computation of $\hat{r}_j$. We stress that loss of orthogonality is due to this cancellation and is not the result of the gradual accumulation of roundoff error.

**Example 9.2.1** The matrix

$$A = \begin{bmatrix} 2.64 & -.48 \\ -.48 & 2.36 \end{bmatrix}$$

has eigenvalues $\lambda_1 = 3$ and $\lambda_2 = 2$. If the Lanczos algorithm is applied to this matrix with $q_1 = [\,.810 \; -.586\,]^T$ and three-digit floating point arithmetic is performed, then $\hat{q}_2 = [\,.707 \; .707\,]^T$. Loss of orthogonality occurs because span$\{q_1\}$ is almost invariant for $A$. (The vector $x = (.8 \; -.6)^T$ is the eigenvector affiliated with $\lambda_1$.)

Further details of the Paige analysis are given shortly. Suffice it to say now that loss of orthogonality always occurs in practice and with it, an apparent deterioration in the quality of $\hat{T}_j$'s eigenvalues. This can be quantified by combining (9.2.1) with Theorem 8.1.11. In particular, if in that theorem we set $F_1 = \hat{r}_j e_j^T + E_j$, $X_1 = \hat{Q}_j$, $S = \hat{T}_j$, and $\tau = \| \hat{Q}_j^T \hat{Q}_j - I_j \|_2$, then we may conclude that there exist eigenvalues $\mu_1, \ldots, \mu_j \in \lambda(A)$ such that

$$|\mu_i - \lambda_i(T_j)| \leq \sqrt{2}\, \frac{\| \hat{r}_j \|_2 + \| E_j \|_2}{\sigma_j(Q_j)} + \tau \| A \|_2$$

for $i = 1{:}j$. An obvious way to prevent the denominator in the upper bound from going to zero is to orthogonalize each newly computed Lanczos vector against its predecessors. This leads directly to our first "practical" Lanczos procedure.

### 9.2.3 Lanczos with Complete Reorthogonalization

Let $r_0, \ldots, r_{j-1} \in \mathbb{R}^n$ be given and suppose that Householder matrices $H_0, \ldots, H_{j-1}$ have been computed such that $(H_0 \cdots H_{j-1})^T [\, r_0, \ldots, r_{j-1} \,]$ is upper triangular. Let $[\, q_1, \ldots, q_j \,]$ denote the first $j$ columns of the Householder product $(H_0 \cdots H_{j-1})$. Now suppose that we are given a vector $r_j \in \mathbb{R}^n$ and wish to compute a unit vector $q_{j+1}$ in the direction of

$$w = r_j - \sum_{i=1}^{j} (q_i^T r_j) q_i \in \text{span}\{q_1, \ldots, q_j\}^{\perp}.$$

If a Householder matrix $H_j$ is determined so $(H_0 \cdots H_j)^T [\, r_0, \ldots, r_j \,]$ is upper triangular, then it follows that the $(j+1)$-st column of $H_0 \cdots H_j$ is the desired unit vector.

If we incorporate these Householder computations into the Lanczos process, then we can produce Lanczos vectors that are orthogonal to working accuracy:

$r_0 = q_1$ (given unit vector)
Determine Householder $H_0$ so $H_0 r_0 = e_1$.
$\alpha_1 = q_1^T A q_1$
**for** $j = 1{:}n-1$
  $r_j = (A - \alpha_j I) q_j - \beta_{j-1} q_{j-1} \qquad (\beta_0 q_0 \equiv 0)$ (9.2.3)
  $w = (H_{j-1} \cdots H_0) r_j$
  Determine Householder $H_j$ so $H_j w = (w_1, \ldots, w_j, \beta_j, 0, \ldots, 0)^T$.
  $q_{j+1} = H_0 \cdots H_j e_{j+1}$; $\alpha_{j+1} = q_{j+1}^T A q_{j+1}$
**end**

This is an example of a *complete reorthorgonalization* Lanczos scheme. A thorough analysis may be found in Paige (1970). The idea of using Householder matrices to enforce orthogonality appears in Golub, Underwood, and Wilkinson (1972).

That the computed $\hat{q}_i$ in (9.2.3) are orthogonal to working precision follows from the roundoff properties of Householder matrices. Note that by virtue of the definition of $q_{j+1}$ , it makes no difference if $\beta_j = 0$. For this reason, the algorithm may safely run until $j = n-1$. (However, in practice one would terminate for a much smaller value of $j$.)

Of course, in any implementation of (9.2.3), one stores the Householder vectors $v_j$ and never explicitly forms the corresponding $P_j$. Since we have $H_j(1{:}j, 1{:}j) = I_j$ there is no need to compute the first $j$ components of $w = (H_{j-1} \cdots H_0) r_j$, for in exact arithmetic these components would be zero.

Unfortunately, these economies make but a small dent in the computational overhead associated with complete reorthogonalization. The Householder calculations increase the work in the $j$th Lanczos step by $O(jn)$ flops. Moreover, to compute $q_{j+1}$, the Householder vectors associated with $H_0, \ldots, H_j$ must be accessed. For large $n$ and $j$, this usually implies a prohibitive amount of data transfer.

Thus, there is a high price associated with complete reorthogonalization. Fortunately, there are more effective courses of action to take, but these demand that we look more closely at how orthogonality is lost.

### 9.2.4 Selective Orthogonalization

A remarkable, ironic consequence of the Paige (1971) error analysis is that loss of orthogonality goes hand in hand with convergence of a Ritz pair. To be precise, suppose the symmetric QR algorithm is applied to $\hat{T}_j$ and renders computed Ritz values $\hat{\theta}_1, \ldots, \hat{\theta}_j$ and a nearly orthogonal matrix of eigenvectors $\hat{S}_j = (\hat{s}_{pq})$. If $\hat{Y}_j = [\, \hat{y}_1, \ldots, \hat{y}_j \,] = fl(\hat{Q}_j \hat{S}_j)$, then it can be shown that for $i = 1{:}j$ we have

$$|\hat{q}_{j+1}^T \hat{y}_i| \approx \frac{\mathbf{u} \| A \|_2}{|\hat{\beta}_j| \, |\hat{s}_{ji}|} \tag{9.2.4}$$

and

$$\| A\hat{y}_i - \hat{\theta}_i \hat{y}_i \|_2 \approx |\hat{\beta}_j| \, |\hat{s}_{ji}| \,. \tag{9.2.5}$$

That is, the most recently computed Lanczos vector $\hat{q}_{j+1}$ tends to have a nontrivial and unwanted component in the direction of any converged Ritz vector. Consequently, instead of orthogonalizing $\hat{q}_{j+1}$ against all of the previously computed Lanczos vectors, we can achieve the same effect by orthogonalizing it against the much smaller set of converged Rtiz vectors.

The practical aspects of enforcing orthogonality in this way are discussed in Parlett and Scott (1979). In their scheme, known as *selective orthogonalization,* a computed Ritz pair $(\hat{\theta}, \hat{y})$ is called "good" if it satisfies

$$\| A\hat{y} - \hat{\theta}\hat{y} \|_2 \approx \sqrt{\mathbf{u}} \| A \|_2 \,.$$

As soon as $\hat{q}_{j+1}$ is computed, it is orthogonalized against each good Ritz vector. This is much less costly than complete reorthogonalization, since there are usually many fewer good Ritz vectors than Lanczos vectors.

One way to implement selective orthogonalization is to diagonalize $\hat{T}_j$ at each step and then examine the $\hat{s}_{ji}$ in light of (9.2.4) and (9.2.5). A much more effcient approach is to estimate the loss-of-orthogonality measure $\| I_j - \hat{Q}_j^T \hat{Q}_j \|_2$ using the following result:

**Lemma 9.2.1** *Suppose* $S_+ = [\, S \; d\,]$ *where* $S \in \mathbb{R}^{n \times j}$ *and* $d \in \mathbb{R}^n$. *If* $S$ *satisfies* $\| I_j - S^T S \|_2 \le \mu$ *and* $|1 - d^T d| \le \delta$ *then* $\| I_{j+1} - S_+^T S_+ \|_2 \le \mu_+$ *where*

$$\mu_+ = \frac{1}{2}\left( \mu + \delta + \sqrt{(\mu - \delta)^2 + 4\| S^T d \|_2^2} \right)$$

**Proof.** See Kahan and Parlett (1974) or Parlett and Scott (1979). □

Thus, if we have a bound for $\| I_j - \hat{Q}_j^T \hat{Q}_j \|_2$ we can generate a bound for $\| I_{j+1} - \hat{Q}_{j+1}^T \hat{Q}_{j+1} \|_2$ by applying the lemma with $S = \hat{Q}_j$ and $d = \hat{q}_{j+1}$. (In this case $\delta \approx \mathbf{u}$ and we assume that $\hat{q}_{j+1}$ has been orthogonalized against the set of currently good Ritz vectors.) It is possible to estimate the norm of $\hat{Q}_j^T \hat{q}_{j+1}$ from a simple recurrence that spares one the need for accessing $\hat{q}_1, \ldots, \hat{q}_j$. See Kahan and Parlett (1974) or Parlett and Scott (1979). The overhead is minimal, and when the bounds signal loss of orthogonality, it is time to contemplate the enlargement of the set of good Ritz vectors. Then and only then is $\hat{T}_j$ diagonalized.

### 9.2.5 The Ghost Eigenvalue Problem

Considerable effort has been spent in trying to develop a workable Lanczos procedure that does not involve any kind of orthogonality enforcement. Research in this direction focusses on the problem of "ghost" or "spurious" eigenvalues. These are multiple eigenvalues of $\hat{T}_j$ that correspond to simple eigenvalues of $A$. They arise because the iteration essentially restarts itself when orthogonality to a converged Ritz vector is lost. (By way of analogy, consider what would happen during orthogonal iteration §7.3.2 if we "forgot" to orthogonalize.)

The problem of identifying ghost eigenvalues and coping with their presence is discussed in Cullum and Willoughby (1979) and Parlett and Reid (1981). It is a particularly pressing problem in those applications where all of $A$'s eigenvalues are desired, for then the above orthogonalization procedures are too expensive to implement.

Difficulties with the Lanczos iteration can be expected even if $A$ has a genuinely multiple eigenvalue. This follows because the $\hat{T}_j$ are unreduced, and unreduced tridiagonal matrices cannot have multiple eigenvalues, see Theorem 7.4.3. Our next practical Lanczos procedure attempts to circumvent this difficulty.

### 9.2.6 Block Lanczos

Just as the simple power method has a block analog in simultaneous iteration, so does the Lanczos algorithm have a block version. Suppose $n = rp$ and consider the decomposition

$$Q^TAQ = \bar{T} = \begin{bmatrix} M_1 & B_1^T & & \cdots & 0 \\ B_1 & M_2 & \ddots & & \vdots \\ & \ddots & \ddots & \ddots & \\ \vdots & & \ddots & \ddots & B_{r-1}^T \\ 0 & \cdots & & B_{r-1} & M_r \end{bmatrix} \qquad (9.2.6)$$

where

$$Q = [\, X_1, \ldots, X_r \,] \qquad X_i \in \mathbb{R}^{n\times p}$$

is orthogonal, each $M_i \in \mathbb{R}^{p\times p}$, and each $B_i \in \mathbb{R}^{p\times p}$ is upper triangular. Comparing blocks in $AQ = Q\bar{T}$ shows that

$$AX_j = X_{j-1}B_{j-1}^T + X_jM_j + X_{j+1}B_j \qquad X_0B_0 \equiv 0$$

for $j = 1{:}r-1$. From the orthogonality of $Q$ we have $M_j = X_j^TAX_j$ for $j = 1{:}r$. Moreover, if we define

$$R_j = AX_j - X_jM_j - X_{j-1}B_{j-1}^T \in \mathbb{R}^{n\times p}$$

then $X_{j+1}B_j = R_j$ is a QR factorization of $R_j$. These observations suggest that the block tridiagonal matrix $\bar{T}$ in (9.2.6) can be generated as follows:

$X_1 \in \mathbb{R}^{p\times p}$ given with $X_1^TX_1 = I_p$.
$M_1 = X_1^TAX_1$
**for** $j = 1{:}r-1$ (9.2.7)
  $R_j = AX_j - X_jM_j - X_{j-1}B_{j-1}^T \qquad (X_0B_0^T \equiv 0)$
  $X_{j+1}B_j = R_j \qquad$ (QR factorization of $R_j$)
  $M_{j+1} = X_{j+1}^TAX_{j+1}$
**end**

At the beginning of the $j$-th pass through the loop we have

$$A[\, X_1, \ldots, X_j \,] = [\, X_1, \ldots, X_j \,]\bar{T}_j + R_j[\, 0, \ldots, 0, I_p \,] \qquad (9.2.8)$$

where

$$\bar{T}_j = \begin{bmatrix} M_1 & B_1^T & & \cdots & 0 \\ B_1 & M_2 & \ddots & & \vdots \\ & \ddots & \ddots & \ddots & \\ \vdots & & \ddots & \ddots & B_{j-1}^T \\ 0 & \cdots & & B_{j-1} & M_j \end{bmatrix}.$$

Using an argument similar to the one used in the proof of Theorem 9.1.1, we can show that the $X_j$ are mutually orthogonal provided none of the $R_j$ are rank-deficient. However if $\text{rank}(R_j) < p$ for some $j$, then it is possible to choose the columns of $X_{j+1}$ such that $X_{j+1}^T X_i = 0$, for $i = 1{:}j$. See Golub and Underwood (1977).

Because $\bar{T}_j$ has bandwidth $p$, it can be efficiently reduced to tridiagonal form using an algorithm of Schwartz (1968). Once tridiagonal form is achieved, the Ritz values can be obtained via the symmetric QR algorithm.

In order to intelligently decide when to use block Lanczos, it is necessary to understand how the block dimension affects convergence of the Ritz values. The following generalization of Theorem 9.1.3 sheds light on this issue.

**Theorem 9.2.2** *Let $A$ by an $n$-by-$n$ symmetric matrix with eigenvalues $\lambda_1 \geq \cdots \geq \lambda_n$ and corresponding orthonormal eigenvectors $z_1, \ldots, z_n$. Let $\mu_1 \geq \cdots \geq \mu_p$ be the $p$ largest eigenvalues of the matrix $\bar{T}_j$ obtained after $j$ steps of the block Lanczos iteration (9.2.7). If $Z_1 = [\, z_1, \ldots, z_p \,]$ and $\cos(\theta_p) = \sigma_p(Z_1^T X_1) > 0$, then for $k = 1{:}p$, $\lambda_k \;\geq\; \mu_k \;\geq\; \lambda_k - \epsilon_k^2$ where*

$$\epsilon_k^2 = \frac{(\lambda_1 - \lambda_k)\tan^2(\theta_p)}{\left[c_{j-1}\left(\frac{1+\gamma_k}{1-\gamma_k}\right)\right]^2} \qquad \gamma_k = \frac{\lambda_k - \lambda_{p+1}}{\lambda_k - \lambda_n}$$

*and $c_{j-1}(z)$ is the $(j-1)$-st Chebyshev polynomial.*

**Proof.** See Underwood (1975). □

Analogous inequalities can be obtained for $\bar{T}_j$'s smallest eigenvalues by applying the theorem with $A$ replaced by $-A$.

Based on Theorem 9.2.2 and scrutiny of the block Lanczos iteration (9.2.7) we may conclude that:

- the error bound for the Ritz values improve with increased $p$.
- the amount of work required to compute $\bar{T}_j$'s eigenvalues is proportional to $p^2$.
- the block dimension should be at least as large as the largest multiplicity of any sought-after eigenvalue.

How to determine block dimension in the face of these trade-offs is discussed in detail by Scott (1979).

Loss of orthogonality also plagues the block Lanczos algorithm. However, all of the orthogonality enforcement schemes described above can be applied in the block setting. See Scott (1979), Lewis (1977), and Ruhe (1979) for details.

### 9.2.7 s-Step Lanczos

The block Lanczos algorithm (9.2.7) can be used in an iterative fashion to calculate selected eigenalues of $A$. To fix ideas, suppose we wish to calculate the $p$ largest eigenvalues. If $X_1 \in \mathbb{R}^{n \times p}$ is a given matrix having orthonormal columns, we may proceed as follows:

**until** $\| AX_1 - X_1\bar{T}_s \|_F$ is small enough
  Generate $X_2, \ldots, X_s \in \mathbb{R}^{n \times p}$ via the block Lanczos algorithm.
  Form $\bar{T}_s = [\, X_1, \ldots, X_s \,]^T A [\, X_1, \ldots, X_s \,]$, an $sp$-by-$sp$,
    $p$-diagonal matrix.
  Compute an orthogonal matrix $U = [\, u_1, \ldots, u_{sp} \,]$ such that
    $U^T\bar{T}_sU = \text{diag}(\theta_1, \ldots, \theta_{sp})$ with $\theta_1 \geq \cdots \geq \theta_{sp}$.
  Set $X_1 = [\, X_1, \ldots, X_s \,][\, u_1, \ldots, u_p \,]$.
**end**

This is the block analog of the *s-step Lanczos algorithm* , which has been extensively analyzed by Cullum and Donath (1974) and Underwood (1975).

The same idea can also be used to compute several of $A$'s smallest eigenvalues or a mixture of both large and small eigenvalues. See Cullum (1978). The choice of the parameters $s$ and $p$ depends upon storage constraints as well as upon the factors we mentioned above in our discussion of block dimension. The block dimension $p$ may be diminished as the good Ritz vectors emerge. However this demands that orthogonality to the converged vectors be enforced. See Cullum and Donath (1974).

#### Problems

**P9.2.1** Prove Lemma 9.2.1.

**P9.2.2** If $\text{rank}(R_j) < p$ in (9.2.7), does it follow that $\text{range}([\, X_1, \ldots, X_j \,])$ contains an eigenvector of $A$?

#### Notes and References for Sec. 9.2

Good general references for the practical Lanczos method include Parlett (SEP) and

J.K. Cullum and R.A. Willoughby (1985b). *Lanczos Algorithms for Large Symmetric Eigenvalue Computations, Vol. II Programs,* Birkhaüser, Boston.

Of the several computational variants of the Lanczos Method, Algorithm 9.2.1 is the most stable. For details, see

C.C. Paige (1972). "Computational Variants of the Lanczos Method for the Eigenproblem," *J. Inst. Math. Applic. 10,* 373–81.

Other practical details associated with the implementation of the Lanczos procedure are discussed in

B.N. Parlett and B. Nour-Omid (1985). "The Use of a Refined Error Bound When Updating Eigenvalues of Tridiagonals," *Lin. Alg. and Its Applic. 68,* 179–220.

B.N. Parlett, H. Simon, and L.M. Stringer (1982). "On Estimating the Largest Eigenvalue with the Lanczos Algorithm," *Math. Comp. 38,* 153–166.
D.S. Scott (1979). "How to Make the Lanczos Algorithm Converge Slowly," *Math. Comp. 33,* 239–47.

The behavior of the Lanczos method in the presence of roundoff error was originally reported in

C.C. Paige (1971). "The Computation of Eigenvalues and Eigenvectors of Very Large Sparse Matrices," Ph.D. thesis, University of London.

Important follow-up papers include

C.C. Paige (1976a). "Error Analysis of the Lanczos Algorithm for Tridiagonalizing Symmetric Matrix," *J. Inst. Math. Applic. 18,* 341–49.
C.C. Paige (1980). "Accuracy and Effectiveness of the Lanczos Algorithm for the Symmetric Eigenproblem," *Lin. Alg. and Its Applic. 34,* 235–58.

Details pertaining to complete reorthogonalization may be found in

G.H. Golub, R. Underwood, and J.H. Wilkinson (1972). "The Lanczos Algorithm for the Symmetric $Ax = \lambda Bx$ Problem," Report STAN-CS-72-270, Department of Computer Science, Stanford University, Stanford, California.
C.C. Paige (1970). "Practical Use of the Symmetric Lanczos Process with Reorthogonalization," *BIT 10,* 183–95.
H. Simon (1984). "Analysis of the Symmetric Lanczos Algorithm with Reorthogonalization Methods," *Lin. Alg. and Its Applic. 61,* 101–132.

The more efficient approach of maintaining orthogonality to the recently converged Ritz vectors is explored in

D.S. Scott (1978). "Analysis of the Symmetric Lanczos Process," UCB-ERL Technical Report M78/40, University of California, Berkeley.
B.N. Parlett and D.S. Scott (1979). "The Lanczos Algorithm with Selective Orthogonalization," *Math. Comp. 33,* 217–38.

Without any reorthogonalization it is necessary either to monitor the loss of orthogonality and quit at the appropriate instant or else to devise some scheme that will aid in the distinction between the ghost eigenvalues and the actual eigenvalues. See

J. Cullum and R.A. Willoughby (1979). "Lanczos and the Computation in Specified Intervals of the Spectrum of Large, Sparse Real Symmetric Matrices, in *Sparse Matrix Proc.* , 1978, ed. I.S. Duff and G.W. Stewart, SIAM Publications, Philadelphia, PA.
W. Kahan and B.N. Parlett (1974). "An Analysis of Lanczos Algorithms for Symmetric Matrices," ERL-M467, University of California, Berkeley.
W. Kahan and B.N. Parlett (1976). "How Far Should You Go with the Lanczos Process?" in *Sparse Matrix Computations*, ed. J. Bunch and D. Rose, Academic Press, New York, pp. 131-44.
J. Lewis (1977). "Algorithms for Sparse Matrix Eigenvlaue Problems," Report STAN-CS-77-595, Department of Computer Science, Stanford University, Stanford, California.
B.N. Parlett and J.K. Reid (1981). "Tracking the Progress of the Lanczos Algorithm for Large Symmetric Eigenproblems," *IMA J. Num. Anal. 1,* 135–55.

The block Lanczos algorithm is discussed in

J. Cullum (1978). "The Simultaneous Computation of a Few of the Algebraically Largest and Smallest Eigenvalues of a Large Sparse Symmetric Matrix," *BIT 18,* 265–75.

J. Cullum and W.E. Donath (1974). "A Block Lanczos Algorithm for Computing the $q$ Algebraically Largest Eigenvalues and a Corresponding Eigenspace of Large Sparse Real Symmetric Matrices," *Proc. of the 1974 IEEE Conf. on Decision and Control*, Phoenix, Arizona, pp. 505–9.

G.H. Golub and R. Underwood (1977). "The Block Lanczos Method for Computing Eigenvalues," in *Mathematical Software III*, ed. J. Rice, Academic Press, New York, pp. 364–77.

A. Ruhe (1979). "Implementation Aspects of Band Lanczos Algorithms for Computation of Eigenvalues of Large Sparse Symmetric Matrices," *Math. Comp. 33,* 680-87.

D.S. Scott (1979). "Block Lanczos Software for Symmetric Eigenvalue Problems," Report ORNL/CSD-48, Oak Ridge National Laboratory, Union Carbide Corporation, Oak Ridge, Tennessee.

R. Underwood (1975). "An Iterative Block Lanczos Method for the Solution of Large Sparse Symmetric Eigenproblems," Report STAN-CS-75-495, Department of Computer Science, Stanford University, Stanford, California.

The block Lanczos algorithm generates a symmetric band matrix whose eigenvalues can be computed in any of several ways. One approach is described in

H.R. Schwartz (1968). "Tridiagonalization of a Symmetric Band Matrix," *Numer. Math. 12,* 231–41. See also *HACLA*, pp. 273–83.

In some applications it is necessary to obtain estimates of interior eigenvalues. The Lanczos algorithm, however, tends to find the extreme eigenvalues first. The following papers deal with this issue:

A.K. Cline, G.H. Golub, and G.W. Platzman (1976). "Calculation of Normal Modes of Oceans Using a Lanczos Method," in *Sparse Matrix Computations*, ed. J.R. Bunch and D.J. Rose, Academic Press, New York, pp. 409–26.

T. Ericsson and A. Ruhe (1980). "The Spectral Transformation Lanczos Method for the Numerical Solution of Large Sparse Generalized Symmetric Eigenvalue Problems," *Math. Comp. 35,* 1251–68.

## 9.3 Applications and Extensions

In this section, we briefly show how the Lanczos iteration can be embellished to solve large sparse linear equation and least squares problems. We also discuss the Arnoldi and the unsymmetric Lanczos processes.

### 9.3.1 Symmetric Positive Definite Systems

Suppose $A \in \mathbb{R}^{n \times n}$ is symmetric and positive definite and consider the functional $\phi(x)$ defined by

$$\phi(x) = \frac{1}{2}x^T Ax - x^T b$$

where $b \in \mathbb{R}^n$. Since $\nabla\phi(x) = Ax - b$, it follows that $x = A^{-1}b$ is the unique minimizer of $\phi$. Hence, an approximate minimizer of $\phi$ can be regarded as an approximate solution to $Ax = b$.

One way to produce a sequence $\{x_j\}$ that converges to $x$ is to generate a sequence of orthonormal vectors $\{q_j\}$ and to let $x_j$ minimize $\phi$ over $\text{span}\{q_1,\ldots,q_j\}$ where $j = 1{:}n$. Let $Q_j = [\, q_1,\ldots,q_j \,]$. Since

$$x \in \text{span}\{q_1,\ldots,q_j\} \quad \Rightarrow \quad \phi(x) = \frac{1}{2}y^T(Q_j^T A Q_j)y - y^T(Q_j^T b)$$

for some $y \in \mathbb{R}^j$, it follows that

$$x_j \;=\; Q_j y_j \tag{9.3.1}$$

where

$$(Q_j^T A Q_j)y_j \;=\; Q_j^T b. \tag{9.3.2}$$

Note that $Ax_n = b$.

We now consider how this approach to solving $Ax = b$ can be made effective when $A$ is large and sparse. There are two hurdles to overcome:

- the linear system (9.3.2) must be "easily" solved.
- we must be able to compute $x_j$ *without* having to refer to $q_1,\ldots,q_j$ explicitly as (9.3.1) suggests. Otherwise there would be an excessive amount of data movement.

We show that both of these requirements are met if the $q_j$ are Lanczos vectors.

After $j$ steps of the Lanczos algorithm we obtain the factorization

$$AQ_j \;=\; Q_j T_j + r_j e_j^T \tag{9.3.3}$$

where

$$T_j \;=\; Q_j^T A Q_j \;=\; \begin{bmatrix} \alpha_1 & \beta_1 & & \cdots & 0 \\ \beta_1 & \alpha_2 & \ddots & & \vdots \\ & \ddots & \ddots & \ddots & \\ \vdots & & \ddots & \ddots & \beta_{j-1} \\ 0 & \cdots & & \beta_{j-1} & \alpha_j \end{bmatrix}. \tag{9.3.4}$$

With this approach (9.3.2) becomes a symmetric positive definite tridiagonal system which may be solved via the $LDL^T$ factorization. (See Algorithm 4.3.6.) In particular, by setting

$$L_j \;=\; \begin{bmatrix} 1 & 0 & 0 & 0 \\ \mu_1 & 1 & 0 & 0 \\ \vdots & \ddots & \ddots & \vdots \\ 0 & \cdots & \mu_{j-1} & 1 \end{bmatrix} \qquad D_j \;=\; \begin{bmatrix} d_1 & 0 & \cdots & 0 \\ 0 & d_2 & & \vdots \\ \vdots & & \ddots & 0 \\ 0 & \cdots & 0 & d_j \end{bmatrix}$$

we find by comparing entries in

$$T_j = L_j D_j L_j^T \tag{9.3.5}$$

that

$$\begin{array}{l} d_1 = \alpha_1 \\ \textbf{for } i = 2{:}j \\ \qquad \mu_{i-1} = \beta_{i-1}/d_{i-1};\ d_i = \alpha_i - \beta_{i-1}\mu_{i-1} \\ \textbf{end} \end{array}$$

Note that we need only calculate the quantities

$$\mu_{j-1} = \beta_{j-1}/d_{j-1} \qquad\qquad d_j = \alpha_j - \beta_{j-1}\mu_{j-1} \tag{9.3.6}$$

in order to obtain $L_j$ and $D_j$ from $L_{j-1}$ and $D_{j-1}$.

As we mentioned, it is critical to be able to compute $x_j$ in (9.3.1) efficiently. To this end we define $C_j \in \mathbb{R}^{n\times j}$ and $p_j \in \mathbb{R}^j$ by the equations

$$C_j L_j^T = Q_j \qquad\qquad L_j D_j p_j = Q_j^T b \tag{9.3.7}$$

and observe that

$$x_j \;=\; Q_j T_j^{-1} Q_j^T b \;=\; Q_j (L_j D_j L_j^T)^{-1} Q_j^T b \;=\; C_j p_j .$$

Let $C_j = [\, c_1, \ldots, c_j \,]$ be a column partitioning. It follows from (9.3.7) that

$$[\, c_1,\ \mu_1 c_1 + c_2,\ \cdots,\ \mu_{j-1}c_{j-1} + c_j \,] \;=\; [\, q_1, \ldots, q_j \,]$$

and therefore

$$C_j \;=\; [\; C_{j-1} \quad c_j \;] \qquad c_j = q_j - \mu_{j-1} c_{j-1} .$$

Also observe that if we set $p_j = [\, \rho_1, \ldots, \rho_j \,]^T$ in $L_j D_j p_j = Q_j^T b$, then that equation becomes

$$\left[\begin{array}{c|c} L_{j-1}D_{j-1} & 0 \\ \hline 0 \cdots 0 \ \ \mu_{j-1}d_{j-1} & d_j \end{array}\right] \left[\begin{array}{c} \rho_1 \\ \rho_2 \\ \vdots \\ \rho_{j-1} \\ \rho_j \end{array}\right] = \left[\begin{array}{c} q_1^T b \\ q_2^T b \\ \vdots \\ q_{j-1}^T b \\ q_j^T b \end{array}\right].$$

Since $L_{j-1} D_{j-1} p_{j-1} = Q_{j-1}^T b$, it follows that

$$p_j \;=\; \left[\begin{array}{c} p_{j-1} \\ \rho_j \end{array}\right] \qquad \rho_j = \left(q_j^T b - \mu_{j-1} d_{j-1} \rho_{j-1}\right)/d_j$$

and thus, $x_j = C_j p_j = C_{j-1}p_{j-1} + \rho_j c_j = x_{j-1} + \rho_j c_j$. This is precisely the kind of recursive formula for $x_j$ that we need. Together with (9.3.6) and (9.3.7) it enables us to make the transition from $(q_{j-1}, c_{j-1}, x_{j-1})$ to $(q_j, c_j, x_j)$ with a minimal amount of work and storage.

A further simplification results if we set $q_1 = b/\beta_0$, where $\beta_0 = \| b \|_2$. For this choice of a Lanczos starting vector we see that $q_i^T b = 0$ for $i \geq 2$. It follows from (9.3.3) that

$$Ax_j = AQ_j y_j = Q_j T_j y_j + r_j e_j^T y_j = Q_j Q_j^T b + r_j e_j^T y_j = b + r_j e_j^T y_j .$$

Thus, if $\beta_j = \| r_j \|_2 = 0$ in the Lanczos iteration, then $Ax_j = b$. Moreover, since $\| Ax_j - b \|_2 = \beta_j |e_j^T y_j|$, the iteration provides estimates of the current residual. Overall, we have the following procedure.

**Algorithm 9.3.1** If $b \in \mathbb{R}^n$ and $A \in \mathbb{R}^{n \times n}$ is symmetric positive definite, then this algorithm computes $x \in \mathbb{R}^n$ such that $Ax = b$.

$r_0 = b$; $\beta_0 = \| b \|_2$; $q_0 = 0$; $j = 0$; $x_0 = 0$
**while** $\beta_j \neq 0$
  $q_{j+1} = r_j/\beta_j$; $j = j + 1$; $\alpha_j = q_j^T A q_j$
  $r_j = (A - \alpha_j I)q_j - \beta_{j-1}q_{j-1}$; $\beta_j = \| r_j \|_2$
  **if** $j = 1$
    $d_1 = \alpha_1; c_1 = q_1; \rho_1 = \beta_0/\alpha_1; x_1 = \rho_1 q_1$
  **else**
    $\mu_{j-1} = \beta_{j-1}/d_{j-1}$; $d_j = \alpha_j - \beta_{j-1}\mu_{j-1}$
    $c_j = q_j - \mu_{j-1}c_{j-1}$; $\rho_j = -\mu_{j-1}d_{j-1}\rho_{j-1}/d_j$
    $x_j = x_{j-1} + \rho_j c_j$
  **end**
**end**
$x = x_j$

This algorithm requires one matrix-vector multiplication and a couple of saxpy operations per iteration. The numerical behavior of Algorithm 9.3.1 is discussed in the next chapter, where it is rederived and identified as the widely known method of *conjugate gradients.*

### 9.3.2 Symmetric Indefinite Systems

A key feature in the above development is the idea of computing the $LDL^T$ factorization of the tridiagonal matrices $T_j$. Unfortunately, this is potentially unstable if $A$, and consequently $T_j$, is not positive definite. A way around this difficulty proposed by Paige and Saunders (1975) is to develop the recursion for $x_j$ via an "LQ" factorization of $T_j$. In particular, at the

$j$th step of the iteration, we have Givens rotations $J_1, \ldots, J_{j-1}$ such that

$$T_j J_1 \cdots J_{j-1} = L_j = \begin{bmatrix} d_1 & 0 & 0 & \cdots & \cdots & \cdots & 0 \\ e_1 & d_2 & 0 & \cdots & \cdots & \cdots & 0 \\ f_1 & e_2 & d_3 & \cdots & \cdots & \cdots & 0 \\ \vdots & \ddots & \ddots & \ddots & & & \\ & & & \ddots & \ddots & \ddots & 0 \\ 0 & 0 & 0 & \cdots & f_{j-2} & e_{j-1} & \bar{d}_j \end{bmatrix}.$$

Note that with this factorization $x_j$ is given by

$$x_j = Q_j y_j = Q_j T_j^{-1} Q_j^T b = W_j s_j$$

where $W_j \in \mathbb{R}^{n \times j}$ and $s_j \in \mathbb{R}^j$ are defined by

$$W_j = Q_j J_1 \cdots J_{j-1} \qquad\qquad L_j s_j = Q_j^T b.$$

Scrutiny of these equations enables one to develop a formula for computing $x_j$ from $x_{j-1}$ and an easily computed multiple of $w_j$, the last column of $W_j$. For details, see Paige and Saunders (1975).

### 9.3.3 Bidiagonalization and the SVD

Suppose $U^T A V = B$ represents the bidiagonalization of $A \in \mathbb{R}^{m \times n}$ with

$$\begin{aligned} U &= [\, u_1, \ldots, u_m \,] \qquad U^T U = I_m \\ V &= [\, v_1, \ldots, v_n \,] \qquad V^T V = I_n \end{aligned}$$

and

$$B = \begin{bmatrix} \alpha_1 & \beta_1 & & \cdots & 0 \\ 0 & \alpha_2 & \ddots & & \vdots \\ & \ddots & \ddots & \ddots & \\ \vdots & & \ddots & \ddots & \beta_{n-1} \\ 0 & \cdots & & 0 & \alpha_n \end{bmatrix}. \tag{9.3.8}$$

Recall (§5.4.3) that this factorization may be computed using Householder transformations and that it serves as a front end for the SVD algorithm.

Unfortunately, if $A$ is large and sparse, then we can expect large, dense submatrices to arise during the Householder bidiagonalization. Consequently, it would be nice to develop a means for computing $B$ directly without any orthogonal updates of the matrix $A$.

Proceeding just as we did in §9.1.2 we compare columns in the equations $AV = UB$ and $A^TU = VB^T$ for $j = 1{:}n$ and obtain

$$\begin{aligned} Av_j &= \alpha_j u_j + \beta_{j-1}u_{j-1} \qquad \beta_0 u_0 \equiv 0 \\ A^T u_j &= \alpha_j v_j + \beta_j v_{j+1} \qquad \beta_n v_{n+1} \equiv 0 \end{aligned} \tag{9.3.9}$$

Defining $r_j = Av_j - \beta_{j-1}u_{j-1}$ and $p_j = A^Tu_j - \alpha_j v_j$, we may conclude from orthonormality that $\alpha_j = \pm\| r_j \|_2$, $u_j = r_j/\alpha_j$, $\beta_j = \pm\| p_j \|_2$, and $v_{j+1} = p_j/\beta_j$. Properly sequenced, these equations define the Lanczos method for bidiagonalizing a rectangular matrix:

$v_1$ = given unit 2-norm $n$-vector
$p_0 = v_1$; $\beta_0 = 1$; $j = 0$; $u_0 = 0$
**while** $\beta_j \neq 0$
  $v_{j+1} = p_j/\beta_j$; $j = j+1$
  $r_j = Av_j - \beta_{j-1}u_{j-1}$; $\alpha_j = \| r_j \|_2$; $u_j = r_j/\alpha_j$     (9.3.10)
  $p_j = A^Tu_j - \alpha_j v_j$; $\beta_j = \| p_j \|_2$
**end**

If rank$(A) = n$ then we can guarantee that no zero $\alpha_j$ arise. Indeed, if $\alpha_j = 0$ then span$\{Av_1,\ldots,Av_j\} \subset$ span$\{u_1,\ldots,u_{j-1}\}$ which imples rank deficiency.

If $\beta_j = 0$ then it is not hard to verify that

$$\begin{aligned} A\,[\,v_1,\ldots,v_j\,] &= [\,u_1,\ldots,u_j\,]\,B_j \\ A^T\,[\,u_1,\ldots,u_j\,] &= [\,v_1,\ldots,v_j\,]\,B_j^T \end{aligned}$$

where $B_j = B(1{:}j, 1{:}j)$ and $B$ is prescribed by (9.3.8). Thus, the $v$ vectors and the $u$ vectors are singular vectors and $\sigma(B_j) \subset \sigma(A)$. Lanczos bidiagonalization is discussed in Paige (1974). See also Cullum and Willoughby (1985a, 1985b). It is essentially equivalent to applying the Lanczos tridiagonalization scheme to the symmetric matrix

$$C = \begin{bmatrix} 0 & A \\ A^T & 0 \end{bmatrix}.$$

We showed that $\lambda_i(C) = \sigma_i(A) = -\lambda_{n+m-i+1}(C)$ for $i = 1{:}n$ at the beginning of §8.3. Because of this, it is not surprising that the large singular values of the bidiagonal matrix tend to be very good approximations to the large singular values of $A$. The small singular values of $A$ correspond to the interior eigenvalues of $C$ and are not so well approximated. The equivalent of the Kaniel-Paige theory for the Lanczos bidiagonalization may be found in Luk (1978) as well as in Golub, Luk, and Overton (1981). The analytic, algorithmic, and numerical developments of the previous two sections all carry over naturally to the bidiagonalization.

### 9.3.4 Least Squares

The full-rank LS problem min $\| Ax - b \|_2$ can be solved via the bidiagonalization. In particular,

$$x_{LS} = Vy_{LS} = \sum_{i=1}^{n} y_i v_i$$

where $y = (y_1, \ldots, y_n)^T$ solves the system $By = (u_1^T b, \ldots, u_n^T b)^T$. Note that because $B$ is *upper* bidiagonal, we cannot solve for $y$ until the bidiagonalization is complete. Moreover, we are required to save the vectors $v_1, \ldots, v_n$, an unhappy circumstance if $n$ is large.

The development of a sparse least squares algorithm based on the bidiagonalization can be accomplished more favorably if $A$ is reduced to lower bidiagonal form

$$U^T AV = B = \begin{bmatrix} \alpha_1 & 0 & \cdots & & 0 \\ \beta_1 & \alpha_2 & \cdots & & \vdots \\ & \ddots & \ddots & & \\ \vdots & & \ddots & \ddots & 0 \\ 0 & \cdots & & \ddots & \alpha_n \\ 0 & \cdots & & 0 & \beta_n \end{bmatrix}$$

where $V = [\, v_1, \ldots, v_n \,]$ and $U = [\, u_1, \ldots, u_{n+1} \,]$. It is straightforward to develop a Lanczos procedure for doing this, and the resulting algorithm is very similar to (9.3.10).

Let $V_j = [\, v_1, \ldots, v_j \,]$ and $U_j = [\, u_1, \ldots, u_j \,]$, and $B_j = B(1{:}j+1, 1{:}j)$ and consider minimizing $\| Ax - b \|_2$ over all vectors of the form $x = V_j y$, where $y \in \mathbb{R}^j$. Since

$$\| AV_j y - b \|_2 = \| U^T AV_j y - U^T b \|_2 = \| B_j y - U_{j+1}^T b \|_2 + \sum_{i=j+2}^{m} (u_i^T b)^2$$

it follows that $x_j = V_j y_j$ is the minimizing $x$ that we are after where $y_j$ minimizes the $(j+1)$-by-$j$ LS problem min$\| B_j y - U_{j+1}^T b \|_2$.

Since $B_j$ is lower bidiagonal, it is easy to compute Givens rotations $J_1, \ldots, J_j$ such that

$$J_j \cdots J_1 B_j = \begin{bmatrix} R_j \\ 0 \end{bmatrix} \begin{matrix} j \\ 1 \end{matrix}$$

is upper bidiagonal. If

$$J_j \cdots J_1 U_{j+1}^T b = \begin{bmatrix} d_j \\ u \end{bmatrix} \begin{matrix} j \\ 1 \end{matrix}$$

then it follows that $x_j = V_j y_j = W_j d_j$ where $W_j = V_j R_j^{-1}$. Paige and Saunders (1975) show how $x_j$ can be obtained from $x_{j-1}$ via a simple recursion that involves the last column of $W_j$. The net result is a sparse LS algorithm that requires only a few $n$-vectors of storage to implement.

### 9.3.5 The Arnoldi Idea

One way to extend the Lanczos process to unsymmetric matrices is due to Arnoldi (1951) and revolves around the Hessenberg reduction $Q^TAQ = H$. In particular, if $Q = [\, q_1, \ldots, q_n \,]$ and we compare columns in $AQ = QH$ then

$$Aq_j = \sum_{i=1}^{j+1} h_{ij} q_i \qquad 1 \le j \le n-1 .$$

Thus, if we know $q_1, \ldots, q_j$ and compute $h_{ij} = q_i^T A q_j$ for $i = 1{:}j$, then $q_{j+1} = r_{j+1}/\| r_{j+1} \|_2$ where

$$r_{j+1} = Aq_j - \sum_{i=1}^{j} h_{ij} q_i .$$

If $r_{j+1} = 0$ then it is not hard to show that $\text{span}\{q_1, \ldots, q_j\}$ is an invariant subspace for $A$. Overall we have

$r = q_1;\ \beta = 1;\ j = 0$
**while** $\beta \neq 0$
  $h_{j+1,j} = \beta;\ q_{j+1} = r_j/\beta;\ j = j+1$
  $w = Aq_j;\ r = w$
  **for** $i = 1{:}j$
    $h_{ij} = q_i^T w;\ r = r - h_{ij} q_i$
  **end**
  $\beta = \| r \|_2$
  **if** $j < n$
    $h_{j+1,j} = \beta$
  **end**
**end**

This is referred to as the *Arnoldi iteration*. Some of its numerical properties are discussed in Wilkinson (AEP, 382ff). As in the symmetric Lanczos iteration, loss of orthogonality among the $q_j$ is an issue. Also note that each computed $q$ vector is referenced each step.

It is possible to build sparse linear equation solvers upon the Arnoldi process just as we developed symmetric $Ax = b$ solvers from the Lanczos iteration. The central idea is to compute a sequence of approximate solutions

$\{x_j\}$ with the property that $x_j$ solves the problem

$$\min_{x \in \mathcal{K}(A,q_1,j)} \| Ax - b \|_2$$

where $\mathcal{K}(A, q_1, j) = \text{span}\{q_1, Aq_1, \ldots, A^{j-1}q_1\} = \text{span}\{q_1, \ldots, q_j\}$. References to these Krylov subspace methods are given at the end of the section.

### 9.3.6 Unsymmetric Lanczos Tridiagonalization

Another way to extend the unsymmetric Lanczos process is to reduce $A$ to tridiagonal form using a general similarity transformation. However, as is pointed out in Wilkinson (AEP, pp. 388–405), the similarity reduction of a general matrix to tridiagonal form is inadvisable for reasons of stability. Despite this we feel that it is of interest to look at the unsymmetric Lanczos process, if only to highlight the added difficulties associated with the unsymmetric eigenvalue problem.

Suppose $A \in \mathbb{R}^{n \times n}$ and that a nonsingular matrix $X$ exists so

$$X^{-1}AX = T = \begin{bmatrix} \alpha_1 & \gamma_1 & & \cdots & 0 \\ \beta_1 & \alpha_2 & \ddots & & \vdots \\ & \ddots & \ddots & \ddots & \\ \vdots & & \ddots & \ddots & \gamma_{n-1} \\ 0 & \cdots & & \beta_{n-1} & \alpha_n \end{bmatrix}.$$

With the column partitionings

$$\begin{aligned} X &= [\, x_1, \ldots, x_n \,] \\ X^{-T} = Y &= [\, y_1, \ldots, y_n \,] \end{aligned}$$

we find upon comparing columns in $AX = XT$ and $A^TY = YT^T$ that

$$Ax_j = \gamma_{j-1}x_{j-1} + \alpha_j x_j + \beta_j x_{j+1} \qquad \gamma_0 x_0 \equiv 0$$

and

$$A^Ty_j = \beta_{j-1}y_{j-1} + \alpha_j y_j + \gamma_j y_{j+1} \qquad \beta_0 y_0 \equiv 0$$

for $j = 1{:}n-1$. These equations together with $Y^TX = I_n$ imply

$$\alpha_j = y_j^T A x_j$$

and

$$\begin{aligned} \beta_j x_{j+1} = r_j &= (A - \alpha_j I)x_j - \gamma_{j-1}x_{j-1} \\ \gamma_j y_{j+1} = p_j &= (A - \alpha_j I)^T y_j - \beta_{j-1}y_{j-1}. \end{aligned}$$

There is some flexibility in choosing the scale factors $\beta_j$ and $\gamma_j$. A canonical choice is to set $\beta_j = \| r_j \|_2$ and $\gamma_j = x_{j+1}^T p_j$ giving

$$
\begin{array}{l}
x_1,\ y_1 \text{ given nonzero vectors.} \\
j = 0;\ \beta_0 = 1;\ x_0 = 0;\ r_0 = x_1;\ y_0 = 0;\ p_0 = y_1 \\
\textbf{while } \beta_j \neq 0 \wedge r_j^T p_j \neq 0 \\
\quad \gamma_j = r_j^T p_j / \beta_j \\
\quad x_{j+1} = r_j / \beta_j;\ y_{j+1} = p_j / \gamma_j \\
\quad j = j + 1 \\
\quad \alpha_j = y_j^T A x_j;\ r_j = (A - \alpha_j I)x_j - \gamma_{j-1} x_{j-1};\ \beta_j = \| r_j \|_2 \\
\quad p_j = (A - \alpha_j I)^T y_j - \beta_{j-1} y_{j-1} \\
\textbf{end}
\end{array}
\qquad (9.3.11)
$$

Defining $X_j = [\, x_1, \ldots, x_j \,]$, $Y_j = [\, y_1, \ldots, y_j \,]$ and $T_j$ to be the leading $j$-by-$j$ principal submatrix of $T$, it is easy to verify that

$$AX_j = X_j T_j + r_j e_j^T \quad \text{and} \quad A^T Y_j = Y_j T_j^T + p_j e_j^T .$$

Thus, whenever $\beta_j = \| r_j \|_2 = 0$, the columns of $X_j$ define an invariant subspace for $A$. Termination in this regard is therefore a welcome event. Unfortunately, if $\gamma_j = x_{j+1}^T p_j = 0$, then the iteration terminates without any invariant subspace information for either $A$ or $A^T$.

This problem is one of the several difficulties associated with (9.3.3). Other problems include a lack of convergence of $T_j$'s eigenvalues and near dependence among the $x_i$ and among the $y_i$. We are unaware of any successful implementations of the unsymmetric Lanczos algorithm.

### Problems

**P9.3.1** Show that the vector $x_j$ in Algorithm 9.3.1 satisfies $\| Ax_j - b \|_2 = \beta_j |e_j^T y_j|$ with $T_j y_j = Q_j^T b$.

**P9.3.2** Give an example of a starting vector for which the unsymmetric Lanczos iteration (9.3.12) breaks down without rendering any invariant subspace information. Use

$$A = \begin{bmatrix} 1 & 6 & 2 \\ 3 & 0 & 2 \\ 1 & 3 & 5 \end{bmatrix}.$$

**P9.3.3** Modify Algorithm 9.3.1 so that it implements the indefinite symmetric solver outlined in §9.3.2.

**P9.3.4** How many vector workspaces are required to implement efficiently (9.3.10)?

**P9.3.5** Supose $A$ is rank deficient and $\alpha_j = 0$ in (9.3.10). How could $u_j$ be obtained so that the iteration could continue?

**P9.3.6** Work out the lower bidiagonal version of (9.3.10) and detail the least square solver sketched in §9.3.4.

### Notes and References for Sec. 9.3

Much of the material in this section has been distilled from the following papers:

C.C. Paige (1974a). "Bidiagonalization of Matrices and Solution of Linear Equations," *SIAM J. Num. Anal. 11,* 197–209.
C.C. Paige and M.A. Saunders (1978). "Solution of Sparse Indefinite Systems of Linear Equations," *SIAM J. Num. Anal. 12,* 617–29.
C.C. Paige and M.A. Saunders (1975). "A Bidiagonalization Algorithm for Sparse Linear Equations and Least Squares Problems," Report SOL 78-19, Department of Operations Research, Stanford University, Stanford, California.
C.C. Paige and M.A. Saunders (1982a). "LSQR: An Algorithm for Sparse Linear Equations and Sparse Least Squares," *ACM Trans. Math. Soft. 8,* 43–71.
C.C. Paige and M.A. Saunders (1982b). "Algorithm 583 LSQR: Sparse Linear Equations and Least Squares Problems," *ACM Trans. Math. Soft. 8,* 195–209.

See also

J. Cullum, R.A. Willoughby, and M. Lake (1983). "A Lanczos Algorithm for Computing Singular Values and Vectors of Large Matrices," *SIAM J. Sci. and Stat. Comp. 4,* 197–215.
J. Cullum and R.A. Willoughby (1985a). *Lanczos Algorithms for Large Symmetric Eigenvalue Computations, Vol. I Theory,* Birkhaüser, Boston.
J. Cullum and R.A. Willoughby (1985b). *Lanczos Algorithms for Large Symmetric Eigenvalue Computations, Vol. II Programs,* Birkhaüser, Boston.
G.H. Golub, F.T. Luk, and M. Overton (1981). "A Block Lanczos Method for Computing the Singular Values and Corresponding Singular Vectors of a Matrix," *ACM Trans. Math. Soft. 7,* 149–69.
F.T. Luk (1978). "Sparse and Parallel Matrix Computations," Report STAN-CS-78-685, Department of Computer Science, Stanford University, Stanford, California.
B.N. Parlett (1980a). "A New Look at the Lanczos Algorithm for Solving Symmetric Systems of Linear Equations," *Lin. Alg. and Its Applic. 29,* 323–46.
Y. Saad (1987). "On the Lanczos Method for Solving Symmetric Systems with Several Right Hand Sides," *Math. Comp. 48,* 651–662.
O. Widlund (1978). "A Lanczos Method for a Class of Nonsymmetric Systems of Linear Equations," *SIAM J. Numer. Anal. 15,* 801–12.

Sparse unsymmetric $Ax = b$ solvers based on the Lanczos/Arnoldi ideas are discussed in

W.E. Arnoldi (1951). "The Principle of Minimized Iterations in the Solution of the Matrix Eigenvalue Problem," *Quart. Appl. Math. 9,* 17–29.
Y. Saad (1982). "The Lanczos Biorthogonalization Algorithm and Other Oblique Projection Metods for Solving Large Unsymmetric Systems," *SIAM J. Numer. Anal. 19,* 485–506.
Y. Saad (1984). "Practical Use of Some Krylov Subspace Methods for Solving Indefinite and Nonsymmetric Linear Systems," *SIAM J. Sci. and Stat. Comp. 5,* 203–228.
Y. Saad and M. Schultz (1986). "GMRES: A Generalized Minimal Residual Algorithm for Solving Nonsymmetric Linear Systems," *SIAM J. Scientific and Stat. Comp. 7,* 856–869.
H.F. Walker (1988). "Implementation of the GMRES Method Using Householder Transformations," *SIAM J. Sci. Stat. Comp. 9,* 152–163.

Extensions of the Lanczos process to other types of eigenproblems are discussed in

G.H. Golub, R. Underwood, and J.H. Wilkinson (1972). "The Lanczos Algorithm for the Symmetric $Ax = \lambda Bx$ Problem," Report STAN-CS-72-270, Department of Computer Science, Stanford University, Stanford, California.
H.A. Van der Vorst (1982). "A Generalized Lanczos Scheme," *Math. Comp. 39,* 559–562.

# Chapter 10

# Iterative Methods for Linear Systems



We concluded the previous chapter by showing how the Lanczos iteration could be used to solve various linear equation and least squares problems. The methods developed were suitable for large sparse problems because they did not require the factorization of the underlying matrix. In this section, we continue the discussion of linear equation solvers that have this property.

The first section is a brisk review of the classical iterations: Jacobi, Gauss-Seidel, SOR, Chebyshev semi-iterative, and so on. Our treatment of these methods is brief because our principal aim in this chapter is to highlight the method of conjugate gradients. In §10.2, we carefully develop this important technique in a natural way from the method of steepest descent. Recall that the conjugate gradient method has already been introduced via the Lanczos iteration in §9.3. The reason for deriving the method again is to motivate some of its practical variants, which are the subject of §10.3.

We warn the reader of an inconsistency in the notation of this chapter In §10.1, methods are developed at the "$(i, j)$ level" necessitating the use of superscripts: $x_i^{(k)}$ denotes the $i$-th component of a vector $x^{(k)}$. In the other sections, however, algorithmic developments can proceed without explicit mention of vector/matrix entries. Hence, in §10.2 and §10.3 we can dispense with superscripts and denote vector sequences by $\{x_k\}$.

## 10.1 The Standard Iterations

The linear equation solvers in Chapters 3 and 4 involve the factorization of the coefficient matrix $A$. Methods of this type are called *direct methods.* Direct methods can be impractical if $A$ is large and sparse, because the sought-after factors can be dense. An exception to this occurs when $A$ is banded (cf. §4.3). Yet in many band matrix problems even the band itself is sparse making algorithms such as band Cholesky difficult to implement.

One reason for the great interest in sparse linear equation solvers is the importance of being able to obtain numerical solutions to partial differential equations. Indeed, researchers in computational PDE's have been responsible for many of the sparse matrix techniques that are presently in general use.

Roughly speaking, there are two approaches to the sparse $Ax = b$ problem. One is to pick an appropriate direct method and adapt it to exploit $A$'s sparsity. Typical adaptation strategies involve the intelligent use of data structures and special pivoting strategies that minimize fill-in. The literature in this area is vast, and the interested reader should consult the books by George and Liu (1981) and Duff, Erisman, and Reid (1986).

In contrast to the direct methods are the *iterative methods.* These methods generate a sequence of approximate solutions $\{x^{(k)}\}$ and essentially involve the matrix $A$ only in the context of matrix-vector multiplication. The evaluation of an iterative method invariably focusses on how quickly the iterates $x^{(k)}$ converge. In this section, we present some basic iterative methods, discuss their practical implementation, and prove a few representative theorems concerned with their behavior.

### 10.1.1 The Jacobi and Gauss-Seidel Iterations

Perhaps the simplest iterative scheme is the *Jacobi iteration.* It is defined for matrices that have nonzero diagonal elements. The method can be motivated by rewriting the 3-by-3 system $Ax = b$ as follows:

$$\begin{aligned} x_1 &= (b_1 - a_{12}x_2 - a_{13}x_3)/a_{11} \\ x_2 &= (b_2 - a_{21}x_1 - a_{23}x_3)/a_{22} \\ x_3 &= (b_3 - a_{31}x_1 - a_{32}x_2)/a_{33} \end{aligned}$$

Suppose $x^{(k)}$ is an approximation to $x = A^{-1}b$. A natural way to generate a new approximation $x^{(k+1)}$ is to compute

$$\begin{aligned} x_1^{(k+1)} &= (b_1 - a_{12}x_2^{(k)} - a_{13}x_3^{(k)})/a_{11} \\ x_2^{(k+1)} &= (b_2 - a_{21}x_1^{(k)} - a_{23}x_3^{(k)})/a_{22} \\ x_3^{(k+1)} &= (b_3 - a_{31}x_1^{(k)} - a_{32}x_2^{(k)})/a_{33} \end{aligned} \qquad (10.1.1)$$

This defines the Jacobi iteration for the case $n = 3$. For general $n$ we have

**for** $i = 1{:}n$

$$x_i^{(k+1)} = \left( b_i - \sum_{j=1}^{i-1} a_{ij} x_j^{(k)} - \sum_{j=i+1}^{n} a_{ij} x_j^{(k)} \right) \Big/ a_{ii} \qquad (10.1.2)$$

**end**

Note that in the Jacobi iteration one does not use the most recently available information when computing $x_i^{(k+1)}$. For example, $x_1^{(k)}$ is used in the calcuation of $x_2^{(k+1)}$ even though component $x_1^{(k+1)}$ is known. If we revise the Jacobi iteration so that we always use the most current estimate of the exact $x_i$ then we obtain

**for** $i = 1{:}n$

$$x_i^{(k+1)} = \left( b_i - \sum_{j=1}^{i-1} a_{ij} x_j^{(k+1)} - \sum_{j=i+1}^{n} a_{ij} x_j^{(k)} \right) \Big/ a_{ii} \qquad (10.1.3)$$

**end**

This defines what is called the *Gauss-Seidel iteration.*

For both the Jacobi and Gauss-Seidel iterations, the transition from $x^{(k)}$ to $x^{(k+1)}$ can be succinctly described in terms of the matrices $L$, $D$, and $U$ defined by:

$$L = \begin{bmatrix} 0 & 0 & \cdots & \cdots & 0 \\ a_{21} & 0 & \cdots & & \vdots \\ a_{31} & a_{32} & \ddots & & 0 \\ \vdots & & & 0 & 0 \\ a_{n1} & a_{n2} & \cdots & a_{n,n-1} & 0 \end{bmatrix}$$

$$D = \operatorname{diag}(a_{11}, \ldots, a_{nn}) \qquad (10.1.4)$$

$$U = \begin{bmatrix} 0 & a_{12} & \cdots & \cdots & a_{1n} \\ 0 & 0 & \cdots & & \vdots \\ 0 & 0 & \ddots & & a_{n-2,n} \\ \vdots & & & & a_{n-1,n} \\ 0 & 0 & \cdots & 0 & 0 \end{bmatrix}$$

In particular, the Jacobi step has the form $M_J x^{(k+1)} = N_J x^{(k)} + b$ where $M_J = D$ and $N_J = -(L+U)$. On the other hand, Gauss-Seidel is defined by $M_G x^{(k+1)} = N_G x^{(k)} + b$ with $M_G = (D+L)$ and $N_G = -U$.

### 10.1.2 Splittings and Convergence

The Jacobi and Gauss-Seidel procedures are typical members of a large family of iterations that have the form

$$Mx^{(k+1)} = Nx^{(k)} + b \tag{10.1.5}$$

where $A = M - N$ is a *splitting* of the matrix $A$. For the iteration (10.1.5) to be practical, it must be "easy" to solve a linear system with $M$ as the matrix. Note that for Jacobi and Gauss-Seidel, $M$ is diagonal and lower triangular respectively.

Whether or not (10.1.5) converges to $x = A^{-1}b$ depends upon the eigenvalues of $M^{-1}N$. In particular, if the *spectral radius* of an $n$-by-$n$ matrix $G$ is defined by

$$\rho(G) = \max\{ |\lambda| : \lambda \in \lambda(G) \},$$

then it is the size of $\rho(M^{-1}N)$ is critical to the success of (10.1.5).

**Theorem 10.1.1** *Suppose $b \in \mathbb{R}^n$ and $A = M + N \in \mathbb{R}^{n\times n}$ is nonsingular. If $M$ is nonsingular and the spectral radius of $M^{-1}N$ satisfies the inequality $\rho(M^{-1}N) < 1$, then the iterates $x^{(k)}$ defined by $Mx^{(k+1)} = Nx^{(k)} + b$ converge to $x = A^{-1}b$ for any starting vector $x^{(0)}$.*

**Proof.** Let $e^{(k)} = x^{(k)} - x$ denote the error in the $k$th iterate. Since $Mx = Nx + b$ it follows that $M(x^{(k+1)} - x) = N(x^{(k)} - x)$, and thus, the error in $x^{(k+1)}$ is given by $e^{(k+1)} = M^{-1}Ne^{(k)} = (M^{-1}N)^k e^{(0)}$. From Lemma 7.3.2 we know that $(M^{-1}N)^k \to 0$ iff $\rho(M^{-1}N) < 1$. □

This result is central to the study of iterative methods where algorithmic development typically proceeds along the following lines:

- A splitting $A = M - N$ is proposed where linear systems of the form $Mz = d$ are "easy" to solve.
- Classes of matrices are identified for which the iteration matrix $G = M^{-1}N$ satisfies $\rho(G) < 1$.
- Further results about $\rho(G)$ are established to gain intuition about how the error $e^{(k)}$ tends to zero.

For example, consider the Jacobi iteration, $Dx^{(k+1)} = -(L + U)x^{(k)} + b$. One condition that guarantees $\rho(M_J^{-1}N_J) < 1$ is strict diagonal dominance. Indeed, if $A$ has that property (defined in §3.4.10), then

$$\rho(M_J^{-1}N_j) \le \| D^{-1}(L + U) \|_\infty = \max_{1\le i\le n} \sum_{\substack{j=1 \\ j\ne i}}^{n} \left| \frac{a_{ij}}{a_{ii}} \right| < 1$$

Usually, the "more dominant" the diagonal the more rapid the convergence but there are counterexamples. See P10.1.8.

A more complicated spectral radius argument is needed to show that Gauss-Seidel converges for symmetric positive definite $A$.

**Theorem 10.1.2** *If $A \in \mathbb{R}^{n\times n}$ is symmetric and positive definite, then the Gauss-Seidel iteration (10.1.3) converges for any $x^{(0)}$.*

**Proof.** Write $A = L + D + L^T$ where $D = \text{diag}(a_{ii})$ and $L$ is strictly lower triangular. In light of Theorem 10.1.1 our task is to show that the matrix $G = -(D+L)^{-1}L^T$ has eigenvalues that are inside the unit circle. Since $D$ is positive definite we have $G_1 \equiv D^{1/2}GD^{-1/2} = -(I+L_1)^{-1}L_1^T$, where $L_1 = D^{-1/2}LD^{-1/2}$. Since $G$ and $G_1$ have the same eigenvalues, we must verify that $\rho(G_1) < 1$. If $G_1x = \lambda x$ with $x^Hx = 1$, then we have $-L_1^Tx = \lambda(I + L_1)x$ and thus, $-x^HL_1^Tx = \lambda(1 + x^HL_1x)$. Letting $a + bi = x^HL_1x$ we have

$$|\lambda|^2 = \left|\frac{-a+bi}{1+a+bi}\right|^2 = \frac{a^2+b^2}{1+2a+a^2+b^2}.$$

However, since $D^{-1/2}AD^{-1/2} = I + L_1 + L_1^T$ is positive definite, it is not hard to show that $0 < 1 + x^HL_1x + x^HL_1^Tx = 1 + 2a$ implying $|\lambda| < 1$. $\square$

This result is frequently applicable because many of the matrices that arise from discretized elliptic PDE's are symmetric positive definite. Numerous other results of this flavor appear in the literature. See Varga (1962), Young (1971) and Hageman and Young (1981).

### 10.1.3 Practical Implementation of Gauss-Seidel

We now focus on some practical details associated with the Gauss-Seidel iteration. With overwriting the Gauss-Seidel step (10.1.3) is particularly simple to implement:

**for** $i = 1{:}n$

$$x_i = \left(b_i - \sum_{j=1}^{i-1} a_{ij}x_j - \sum_{j=i+1}^{n} a_{ij}x_j\right) \Big/ a_{ii}$$

**end**

This computation requires about twice as many flops as there are nonzero entries in the matrix $A$. It makes no sense to be more precise about the work involved because the actual implementation depends greatly upon the structure of the problem at hand.

In order to stress this point we consider the application of Algorithm 10.1.1 to the $NM$-by-$NM$ block tridiagonal system

$$\begin{bmatrix} T & -I_N & & \cdots & 0 \\ -I_N & T & \ddots & & \vdots \\ & \ddots & \ddots & \ddots & \\ \vdots & & \ddots & \ddots & -I_N \\ 0 & \cdots & & -I_N & T \end{bmatrix} \begin{bmatrix} g_1 \\ g_2 \\ \vdots \\ \vdots \\ g_M \end{bmatrix} = \begin{bmatrix} f_1 \\ f_2 \\ \vdots \\ \vdots \\ f_M \end{bmatrix} \qquad (10.1.6)$$

where

$$T = \begin{bmatrix} 4 & -1 & & \cdots & 0 \\ -1 & 4 & \ddots & & \vdots \\ & \ddots & \ddots & \ddots & \\ \vdots & & \ddots & \ddots & -1 \\ 0 & \cdots & & -1 & 4 \end{bmatrix}, \quad g_j = \begin{bmatrix} G(1,j) \\ G(2,j) \\ \vdots \\ \vdots \\ G(N,j) \end{bmatrix}, \quad f_j = \begin{bmatrix} F(1,j) \\ F(2,j) \\ \vdots \\ \vdots \\ F(N,j) \end{bmatrix}.$$

This problem arises when the Poisson equation is discretized on a rectangle. It is easy to show that the matrix $A$ is positive definite.

With the convention that $G(i,j) = 0$ whenever $i \in \{0, N+1\}$ or $j \in \{0, M+1\}$ we see that with overwriting the Gauss-Seidel step takes on the form:

```
for j = 1:M
    for i = 1:N
        G(i,j) = (F(i,j) + G(i-1,j) + G(i+1,j)+
                     G(i,j-1) + G(i,j+1))/4
    end
end
```

Note that in this problem no storage is required for the matrix $A$.

### 10.1.4 Successive Over-Relaxation

The Gauss-Seidel iteration is very attractive because of its simplicity. Unfortunately, if the spectral radius of $M_G^{-1}N_G$ is close to unity, then it may be prohibitively slow because the error tends to zero like $\rho(M_G^{-1}N_G)^k$. To rectify this, let $\omega \in \mathbb{R}$ and consider the following modification of the Gauss-Seidel step:

**for** $i = 1{:}n$

$$x_i^{(k+1)} = \omega \left( b_i - \sum_{j=1}^{i-1} a_{ij} x_j^{(k+1)} - \sum_{j=i+1}^{n} a_{ij} x_j^{(k)} \right) \Big/ a_{ii} + (1-\omega) x_i^{(k)} \qquad (10.1.7)$$

**end**

This defines the method of *successive over-relaxation* (SOR). Using (10.1.4) we see that in matrix terms, the SOR step is given by

$$M_\omega x^{(k+1)} = N_\omega x^{(k)} + \omega b \qquad (10.1.8)$$

where $M_\omega = D + \omega L$ and $N_\omega = (1-\omega)D - \omega U$. For a few structured (but important) problems such as (10.1.6), the value of the *relaxation parameter* $\omega$ that minimizes $\rho(M_\omega^{-1} N_\omega)$ is known. Moreover, a significant reduction in $\rho(M_1^{-1} N_1) = \rho(M_G^{-1} N_G)$ can result. In more complicated problems, however, it may be necessary to perform a fairly sophisticated eigenvalue analysis in order to determine an appropriate $\omega$. A complete survey of "SOR theory" appears in Young (1971). Some practical schemes for estimating the optimum $\omega$ are discussed in O'Leary (1976) and Wachpress (1966).

## 10.1.5 The Chebyshev Semi-Iterative Method

Another way to accelerate the convergence of an iterative method makes use of Chebyshev polynomials. Suppose $x^{(1)}, \ldots, x^{(k)}$ have been generated via the iteration $Mx^{(j+1)} = Nx^{(j)} + b$ and that we wish to determine coefficients $\nu_j(k)$, $j = 0{:}k$ such that

$$y^{(k)} = \sum_{j=0}^{k} \nu_j(k) x^{(j)} \qquad (10.1.9)$$

represents an improvement over $x^{(k)}$. If $x^{(0)} = \cdots = x^{(k)} = x$, then it is reasonable to insist that $y^{(k)} = x$. Hence, we require

$$\sum_{j=0}^{k} \nu_j(k) = 1 . \qquad (10.1.10)$$

Subject to this constraint, we consider how to choose the $\nu_j(k)$ so that the error in $y^{(k)}$ is minimized.

Recalling from the proof of Theorem 10.1.1 that $x^{(k)} - x = (M^{-1}N)^k e^{(0)}$ where $e^{(0)} = x^{(0)} - x$, we see that

$$y^{(k)} - x = \sum_{j=0}^{k} \nu_j(k)(x^{(j)} - x) = \sum_{j=0}^{k} \nu_j(k)(M^{-1}N)^j e^{(0)} .$$

Working in the 2-norm we therefore obtain

$$\| y^{(k)} - x \|_2 \le \| p_k(G) \|_2 \| e^{(0)} \|_2 \qquad (10.1.11)$$

where $G = M^{-1}N$ and

$$p_k(z) = \sum_{j=0}^{k} \nu_j(k) z^j .$$

Note that the condition (10.1.10) implies $p_k(1) = 1$.

At this point we assume that $G$ is symmetric with eigenvalues $\lambda_i$ that satisfy $-1 < \alpha \le \lambda_n \le \cdots \le \lambda_1 \le \beta < 1$. It follows that

$$\| p_k(G) \|_2 = \max_{\lambda_i \in \lambda(A)} |p_k(\lambda_i)| \le \max_{\alpha \le \lambda \le \beta} |p_k(\lambda)| .$$

Thus, to make the norm of $p_k(G)$ small, we need a polynomial $p_k(z)$ that is small on $[\alpha, \beta]$ subject to the constraint that $p_k(1) = 1$.

Consider the Chebyshev polynomials $c_j(z)$ generated by the recursion $c_j(z) = 2zc_{j-1}(z) - c_{j-2}(z)$ where $c_0(z) = 1$ and $c_1(z) = z$. These polynomials satisfy $|c_j(z)| \le 1$ on [-1, 1] but grow rapidly off this interval. As a consequence, the polynomial

$$p_k(z) = \frac{c_k\left(-1 + 2\dfrac{z-\alpha}{\beta-\alpha}\right)}{c_k(\mu)}$$

where

$$\mu = -1 + 2\frac{1-\alpha}{\beta-\alpha} = 1 + 2\frac{1-\beta}{\beta-\alpha}$$

satisfies $p_k(1) = 1$ and tends to be small on $[\alpha, \beta]$. From the definition of $p_k(z)$ and equation (10.1.11) we see

$$\| y^{(k)} - x \|_2 \le \frac{\| x - x^{(0)} \|_2}{|c_k(\mu)|} .$$

Thus, the larger $\mu$ is, the greater the acceleration of convergence.

In order for the above to be a practical acceleration procedure, we need a more efficient method for calculating $y^{(k)}$ than (10.1.9). We have been tacitly assuming that $n$ is large and thus the retrieval of $x^{(0)}, \ldots, x^{(k)}$ for large $k$ would be inconvenient or even impossible.

Fortunately, it is possible to derive a three-term recurrence among the $y^{(k)}$ by exploiting the three-term recurrence among the Chebyshev polynomials. In particular, it can be shown that if

$$\omega_{k+1} = 2\frac{2-\beta-\alpha}{\beta-\alpha}\frac{c_k(\mu)}{c_{k+1}(\mu)}$$

then

$$\begin{aligned} y^{(k+1)} &= \omega_{k+1}(y^{(k)} - y^{(k-1)} + \gamma z^{(k)}) + y^{(k-1)} \\ Mz^{(k)} &= b - Ay^{(k)} \\ \gamma &= 2/(2-\alpha-\beta), \end{aligned} \tag{10.1.12}$$

where $y^{(0)} = x^{(0)}$ and $y^{(1)} = x^{(1)}$. We refer to this scheme as the Chebyshev semi-iterative method associated with $My^{(k+1)} = Ny^{(k)} + b$. For the acceleration to be effective we need good lower and upper bounds $\alpha$ and $\beta$. As in SOR, these parameters may be difficult to ascertain except in a few structured problems.

Chebyshev semi-iterative methods are extensively analyzed in Varga (1962, chapter 5), as well as in Golub and Varga (1961).

### 10.1.6 Symmetric SOR

In deriving the Chebyshev acceleration we assumed that the iteration matrix $G = M^{-1}N$ was symmetric. Thus, our simple analysis does not apply to the unsymmetric SOR iteration matrix $M_\omega^{-1}N_\omega$. However, it is possible to symmetrize the SOR method making it amenable to Chebyshev acceleration. The idea is to couple SOR with the *backward SOR* scheme

**for** $i = n{:}-1{:}1$

$$x_i^{(k+1)} = \omega\left(b_i - \sum_{j=1}^{i-1} a_{ij}x_j^{(k+1)} - \sum_{j=i+1}^{n} a_{ij}x_j^{(k)}\right)\Big/ a_{ii} + (1-\omega)x_i^{(k)} \tag{10.1.13}$$

**end**

This iteration is obtained by updating the unknowns in reverse order in (10.1.7). Backward SOR can be described in matrix terms using (10.1.4). In particular, we have $\tilde{M}_\omega x^{(k+1)} = \tilde{N}_\omega x^{(k)} + \omega b$ where

$$\tilde{M}_\omega = D + \omega U \qquad \text{and} \qquad \tilde{N}_\omega = (1-\omega)D - \omega L. \tag{10.1.14}$$

If $A$ is symmetric ($U = L^T$), then $\tilde{M}_\omega = M_\omega^T$ and $\bar{N}_\omega = N_\omega^T$, and we have the iteration

$$\begin{aligned} M_\omega x^{(k+1/2)} &= N_\omega x^{(k)} + \omega b \\ M_\omega^T x^{(k+1)} &= N_\omega^T x^{(k+1/2)} + \omega b. \end{aligned} \qquad (10.1.15)$$

It is clear that $G = M_\omega^{-T} N_\omega^T M_\omega^{-1} N_\omega$ is the iteration matrix for this method. From the definitions of $M_\omega$ and $N_\omega$ it follows that

$$G = M^{-1}N \equiv (M_\omega D^{-1} M_\omega^T)^{-1}(N_\omega^T D^{-1} N_\omega). \qquad (10.1.16)$$

If $D$ has positive diagonal entries and $KK^T = (N_\omega^T D^{-1} N_\omega)$ is the Cholesky factorization, then $K^T G K^{-T} = K^T(M_\omega D^{-1} M_\omega^T)^{-1} K$. Thus, $G$ is similar to a symmetric matrix and has real eigenvalues.

The iteration (10.1.15) is called the *symmetric successive over-relaxation* (SSOR) method. It is frequently used in conjunction with the Chebyshev semi-iterative acceleration.

### Problems

**P10.1.1** Show that the Jacobi iteration can be written in the form $x^{(k+1)} = x^{(k)} + Hr^{(k)}$ where $r^{(k)} = b - Ax^{(k)}$. Repeat for the Gauss-Seidel iteration.

**P10.1.2** Show that if $A$ is strictly diagonally dominant, then the Gauss-Siedel iteration converges.

**P10.1.3** Show that the Jacobi iteration converges for 2-by-2 symmetric positive definite systems.

**P10.1.4** Show that if $A = M - N$ is singular, then we can never have $\rho(M^{-1}N) < 1$ even if $M$ is nonsingular.

**P10.1.5** Prove (10.1.16).

**P10.1.6** Prove the converse of Theorem 10.1.1. In other words, show that if the iteration $Mx^{(k+1)} = Nx^{(k)} + b$ always converges, then $\rho(M^{-1}N) < 1$ .

**P10.1.7** (Supplied by R.S. Varga) Suppose that $A_1 = \begin{bmatrix} 1 & -1/2 \\ -1/2 & 1 \end{bmatrix}$ and $A_2 = \begin{bmatrix} 1 & -3/4 \\ -1/12 & 1 \end{bmatrix}$. Let $J_1$ and $J_2$ be the associated Jacobi iteration matrices. Show that $\rho(J_1) > \rho(J_2)$ thereby refuting the claim that greater diagonal dominance implies more rapid Jacobi convergence.

### Notes and References for Sec. 10.1

Three comprehensive volumes survey the area of iterative methods in far greater depth than our own treatment:

L.A. Hageman and D.M. Young (1981). *Applied Iterative Methods*, Academic Press, New York.

R.S. Varga (1962). *Matrix Iterative Analysis*, Prentice-Hall, Englewood Cliffs, New Jersey.

D.M. Young (1971). *Iterative Solution of Large Linear Systems*, Academic Press, New York.

We also highly recommend chapter 7 of

J. Ortega (1972). *Numerical Analysis: A Second Course*, Academic Press, New York.

The direct (non-iterative) solution of large sparse systems is often a preferred course of action. See

I.S. Duff, A.M. Erisman, and J.K. Reid (1986). *Direct Methods for Sparse Matrices,* Oxford University Press.
J.A. George and J.W. Liu (1981). *Computer Solution of Large Sparse Positive Definite Systems*, Prentice-Hall, Englewood Cliffs, NJ.

We have seen that the condition $\kappa(A)$ is an important issue when direct methods are applied to $Ax = b$. However, the condition of the system also has a bearing on iterative method performance. See

M. Arioli and F. Romani (1985). "Relations Between Condition Numbers and the Convergence of the Jacobi Method for Real Positive Definite Matrices," *Numer. Math. 46,* 31–42.

As we mentioned, Young (1971) has the most comprehensive treatment of the SOR method. The object of "SOR theory" is to guide the user in choosing the relaxation parameter $\omega$. In this setting, the ordering of equations and unknowns is critical. See

M.J.M. Bernal and J.H. Verner (1968). "On Generalizing of the Theory of Consistent Orderings for Successive Over-Relaxation Methods," *Numer. Math. 12,* 215–22.
R.A. Nicolaides (1974). "On a Geometrical Aspect of SOR and the Theory of Consistent Ordering for Positive Definite Matrices," *Numer. Math. 12,* 99–104.
D.M. Young (1970). "Convergence Properties of the Symmetric and Unsymmetric Over-Relaxation Methods," *Math. Comp. 24,* 793–807.
D.M. Young (1972). "Generalization of Property A and Consistent Ordering," *SIAM J. Num. Anal. 9,* 454–63.

Heuristic methods for estimating $\omega$ are also of interest. See

D.P. O'Leary (1976). "Hybrid Conjugate Gradient Algorithms," Report STAN-CS-76-548, Department of Computer Science, Stanford University, (Ph.D. thesis).
E.L. Wachpress (1966). *Iterative Solution of Elliptic Systems*, Prentice-Hall, Englewood Cliffs, New Jersey.

An analysis of the Chebyshev semi-iterative method appears in

G.H. Golub and R.S. Varga (1961). "Chebychev Semi-Iterative Methods, Successive Over-Relaxation Iterative Methods, and Second-Order Richardson Iterative Methods, Parts I and II," *Numer. Math. 3,* 147–56, 157–68.

This work is premised on the assumption that the underlying iteration matrix has real eigenvalues. How to proceed when this is not the case is discussed in

M. Eiermann and W. Niethammer (1983). "On the Construction of Semi-iterative Methods," *SIAM J. Numer. Anal. 20,* 1153–1160.
G.H. Golub and M. Overton (1988). "The Convergence of Inexact Chebychev and Richardson Iterative Methods for Solving Linear Systems," *Numer. Math. 53,* 571–594.

T.A. Manteuffel (1977). "The Tchebychev Iteration for Nonsymmetric Linear Systems," *Numer. Math. 28*, 307–27.
W. Niethammer and R.S. Varga (1983). "The Analysis of k-step Iterative Methods for Linear Systems from Summability Theory," *Numer. Math. 41*, 177–206.

Sometimes it is possible to "symmetrize" an iterative method, thereby simplifying the acceleration process, since all the relevant eigenvalues are real. This is the case for the SSOR method discussed in

J.W. Sheldon (1955). "On the Numerical Solution of Elliptic Difference Equations," *Math. Tables Aids Comp. 9*, 101–12.

The parallel implementation of the classical iterations has received some attention. See

D.J. Evans (1984). "Parallel SOR Iterative Methods," *Parallel Computing 1*, 3–18.
S.L. Johnsson and C.T. Ho (1987). "Multiple Tridiagonal Systems, the Alternating Direction Methods, and Boolean Cube Configured Multiprocessors," Report YALEU DCS RR-532, Dept. of Computer Science, Yale University, New Haven, CT.
N. Patel and H. Jordan (1984). "A Parallelized Point Rowwise Successive Over-Relaxation Method on a Multiprocessor," *Parallel Computing 1*, 207–222.
R.J. Plemmons (1986). "A Parallel Block Iterative Scheme Applied to Computations in Structural Analysis," *SIAM J. Alg. and Disc. Methods 7*, 337–347.

## 10.2 The Conjugate Gradient Method

A difficulty associated with the SOR, Chebyshev semi-iterative, and related methods is that they depend upon parameters that are sometimes hard to choose properly. For example, for Chebyshev acceleration to be successful we need good estimates of the largest and smallest eigenvalue of the underlying iteration matrix $M^{-1}N$. Unless this matrix is sufficiently structured, this may be analytically impossible and/or computationally expensive.

In this section, we present a method without this difficulty, the well-known Hestenes-Stiefel conjugate gradient method. We derived this method in §9.3.1 where we discussed how to use the Lanczos algorithm to solve symmetric positive definite linear systems. The reason for rederiving the conjugate gradient method is to set the stage for §10.3 where we show its connection with other iterative $Ax = b$ schemes and describe some of its useful generalizations.

### 10.2.1 Steepest Descent

The starting point in the derivation is to consider how we might go about minimizing the functional $\phi(x)$, defined by

$$\phi(x) = \frac{1}{2}x^T Ax - x^T b$$

where $b \in \mathbb{R}^n$ and $A \in \mathbb{R}^{n\times n}$ is assumed to be positive definite and symmetric. The minimum value of $\phi$ is $-b^T A^{-1}b/2$, achieved by setting $x = A^{-1}b$. Thus, minimizing $\phi$ and solving $Ax = b$ are equivalent problems.

One of the simplest strategies for minimizing $\phi$ is the *method of steepest descent.* At a current point $x_c$ the function $\phi$ decreases most rapidly in the direction of the negative gradient $-\nabla\phi(x_c) = b - Ax_c$. We call

$$r_c = b - Ax_c$$

the *residual* of $x_c$. If the residual is nonzero, then there exists a positive $\alpha$ such that $\phi(x_c + \alpha r_c) < \phi(x_c)$. In the method of steepest descent (with exact line search) we set $\alpha = r_c^T r_c / r_c^T A r_c$ thereby minimizing $\phi(x_c + \alpha r_c)$. This gives

$$\begin{array}{l} k = 0;\ x_0 = 0;\ r_0 = b \\ \textbf{while } r_k \neq 0 \\ \quad k = k+1 \\ \quad \alpha_k = r_{k-1}^T r_{k-1} / r_{k-1}^T A r_{k-1} \\ \quad x_k = x_{k-1} + \alpha_k r_{k-1} \\ \quad r_k = b - Ax_k \\ \textbf{end} \end{array} \qquad (10.2.1)$$

The starting vector $x_0 = 0$ is not restrictive. If $\tilde{x}$ is a more appropriate initial guess, then we merely apply (10.2.1) with $b$ replaced by $b - A\tilde{x}$. The vectors $x_k + \tilde{x}$ are then the desired approximate solutions.

The global convergence of the steepest descent method follows from the easily established inequality

$$\phi(x_k) + \frac{1}{2}b^T A^{-1} b \;\le\; \left(1 - \frac{1}{\kappa_2(A)}\right)\left(\phi(x_{k-1}) + \frac{1}{2}b^T A^{-1} b\right). \qquad (10.2.2)$$

Unfortunately, the speed of convergence may be prohibitively slow if $\kappa_2(A) = \lambda_1(A)/\lambda_n(A)$ is large. In this situation, the level curves of $\phi$ are very elongated hyperellipsoids and minimization corresponds to finding the lowest point on a relatively flat, steep-sided valley. In steepest descent, we are forced to traverse back and forth *across* the valley rather than *down* the valley. The gradient directions that arise during the iteration are too similar thus slowing progress towards the minimum point.

### 10.2.2 General Search Directions

To avoid the pitfalls of steepest descent, we consider the successive minimization of $\phi$ along a set of directions $\{p_1, p_2, \ldots\}$ that do not necessarily correspond to the residuals $\{r_0, r_1, \ldots\}$. It is easy to show that to minimize $\phi(x_{k-1} + \alpha p_k)$ with respect to $\alpha$, we merely set

$$\alpha \;=\; \alpha_k \;=\; p_k^T r_{k-1} / p_k^T A p_k.$$

With this choice, it can be shown that

$$\phi(x_{k-1} + \alpha_k p_k) \;=\; \phi(x_{k-1}) - \frac{1}{2}(p_k^T r_{k-1})^2 p_k^T A p_k. \qquad (10.2.3)$$

Thus, to ensure a reduction in the size of $\phi$ we must insist that $p_k$ not be orthogonal to $r_{k-1}$. This leads to the following minimization strategy:

$$
\begin{array}{l}
k = 0;\ x_0 = 0;\ r_0 = b \\
\textbf{while } r_k \neq 0 \\
\quad k = k + 1 \\
\quad \text{Choose a direction } p_k \text{ such that } p_k^T r_{k-1} \neq 0. \\
\quad \alpha_k = p_k^T r_{k-1} / p_k^T A p_k \\
\quad x_k = x_{k-1} + \alpha_k p_k \\
\quad r_k = b - A x_k \\
\textbf{end}
\end{array}
\qquad (10.2.4)
$$

Notice that $x_k \in \text{span}\{p_1, \ldots, p_k\}$ . The problem, of course, is how to choose these vectors so as to guarantee global convergence and at the same time avoid the pitfalls of steepest descent.

### 10.2.3 A-Conjugate Search Directions

A seemingly ideal approach would be to choose linearly independent $p_i$ with the property that each $x_k$ in (10.2.4) solves

$$\min_{x \in \text{span}\{p_1, \ldots, p_k\}} \phi(x) \,. \qquad (10.2.5)$$

This would guarantee not only global convergence but finite termination as well because we must have $Ax_n = b$. Note that what we are seeking is a vector $p_k$ such that when we solve the one-dimensional minimization problem

$$\min_{\alpha} \ \phi(x_{k-1} + \alpha p_k)$$

we also solve the $k$-dimensional minimization (10.2.5). It turns out that this stringent requirement can be met.

Let $P_k = [\, p_1, \ldots, p_k \,] \in \mathbb{R}^{n \times k}$ be the matrix of search directions. If $x \in \text{range}(P_k)$ then $x = P_{k-1}y + \alpha p_k$ for some $y \in \mathbb{R}^{k-1}$ and $\alpha \in \mathbb{R}$. If $x$ has this form then it is easy to show that

$$\phi(x) \ = \ \phi(P_{k-1}y) \ + \ \alpha y^T P_{k-1}^T A p_k \ + \ \frac{\alpha^2}{2} p_k^T A p_k \ - \ \alpha p_k^T b \,.$$

The presence of the "cross term" $\alpha y^T P_{k-1}^T A p_k$ complicates the minimization. Without it the minimization of $\phi$ over range($P_k$) would decouple into a minimization over range($P_{k-1}$), whose solution $x_{k-1}$ is assumed known, and a simple minimization involving the scalar $\alpha$. One way to effect this decoupling is to insist that $p_k$ is *A-conjugate* to $p_1, \ldots, p_{k-1}$ meaning that

$$P_{k-1}^T A p_k \;=\; 0. \tag{10.2.6}$$

If we do this and define $x_{k-1} \in \text{range}(P_{k-1})$ and $\alpha_k \in \mathbb{R}$ by

$$\phi(x_{k-1}) = \min_y \; \phi(P_{k-1}y)$$

and $\alpha_k \;=\; p_k^T b / p_k^T A p_k$, then

$$\min_{y,\alpha} \; \phi(P_{k-1}y + \alpha_k p_k) = \min_y \; \phi(P_{k-1}y) + \min_\alpha \left\{ \frac{\alpha^2}{2} p_k^T A p_k - \alpha p_k^T b \right\}$$

is solved by setting $P_{k-1}y = x_{k-1}$ and $\alpha = \alpha_k$. Since $x_{k-1}$ is linear combination of $p_1, \ldots, p_{k-1}$ it follows that $p_k^T A x_{k-1} = 0$ and so

$$\alpha_k = p_k^T r_{k-1} / p_k^T A p_k,$$

precisely the recipe for $\alpha_k$ appearing in (10.2.4). Thus, we obtain

$k = 0$; $x_0 = 0$; $r_0 = b$
**while** $r_k \neq 0$
  $k = k + 1$
  Choose $p_k \in \text{span}\{Ap_1, \ldots, Ap_{k-1}\}^\perp$ so $p_k^T r_{k-1} \neq 0$.  (10.2.7)
  $\alpha_k = p_k^T r_{k-1} / r_{k-1}^T A r_{k-1}$
  $x_k = x_{k-1} + \alpha_k p_k$
  $r_k = b - Ax_k$
**end**

Here, the vector $x_k = x_{k-1} + \alpha_k p_k$ minimizes $\phi$ over the span of the search directions, i.e., the subspace $\text{span}\{p_1, \ldots, p_k\}$ .

### 10.2.4 The Conjugate Gradient Method

In order to carry (10.2.7) we must be sure that it is possible to choose $p_k$ such that $p_k^T A p_k \;=\; 0$ and $p_k^T r_{k-1} \neq 0$. Suppose $p_1, \ldots, p_{k-1}$ are nonzero and satisfy $p_i^T A p_j \;=\; 0$ for all $i \neq j$. Since $x_{k-1} \in \text{span}\{p_1, \ldots, p_{k-1}\}$ , it follows that for any $p \in \mathbb{R}^n$ we have $p^T r_{k-1} \;=\; p^T b - p^T A P_{k-1} z$ for some $z \in \mathbb{R}^{k-1}$. If $p^T r_{k-1} = 0$ for every $p$ that is A-conjugate to $p_1, \ldots, p_{k-1}$, then we must have $b \in \text{span}\{Ap_1, \ldots, Ap_{k-1}\}$ . But this implies $x = A^{-1}b \in \text{span}\{p_1, \ldots, p_{k-1}\}$ and thus, $x_{k-1} = x$, i.e., $r_{k-1} = 0$. Hence, if $r_{k-1} \neq 0$ then we can find a nonzero $p_k$ that is A-conjugate to $p_1, \ldots, p_{k-1}$ and satisfies $p_k^T r_{k-1} \neq 0$.

Since our aim is to bring about the swift reduction in the size of the residuals, it is natural to choose $p_k$ to be the closest vector to $r_{k-1}$ that is $A$-conjugate to $p_1, \ldots, p_{k-1}$. This defines "version zero" of the *method of conjugate gradients*:

$k = 0$; $x_0 = 0$ ; $r_0 = b$
**while** $r_k \neq 0$
  $k = k + 1$
  **if** $k = 1$
    $p_1 = r_0$
  **else** (10.2.8)
    Let $p_k$ minimize $\| p - r_{k-1} \|_2$ over all vectors
      $p \in \text{span}\{Ap_1, \ldots, Ap_{k-1}\}^{\perp}$
  **end**
  $\alpha_k = p_k^T r_{k-1} / p_k^T A p_k$
  $x_k = x_{k-1} + \alpha_k p_k$
  $r_k = b - Ax_k$
**end**
$x = x_k$

The finite termination of this algorithm is guaranteed since the $p_k$ are nonzero and nonzero $A$-conjugate vectors are independent. In particular, either $r_{k-1} = 0$ for some $k \leq n$ or we compute $x_n$ which minimizes $\phi$ over $\text{span}\{p_1, \ldots, p_n\} = \mathbb{R}^n$.

### 10.2.5 Some Critical Observations

To make (10.2.8) an effective sparse $Ax = b$ solver, we need an efficient method for computing $p_k$. As a first step in this direction we characterize $p_k$ as the solution to a certain LS problem.

**Lemma 10.2.1** *For $k \geq 2$ the vectors $p_k$ generated by (10.2.8) satisfy*

$$p_k = r_{k-1} - AP_{k-1} z_{k-1},$$

*where $z_{k-1}$ solves* $\min_z \| r_{k-1} - AP_{k-1} z \|_2$.

**Proof.** Suppose $z_{k-1}$ solves the above LS problem and let $p = r_{k-1} - AP_{k-1} z_{k-1}$ be the minimum residual. It follows that $p^T AP_{k-1} = 0$. Moreover, since $p = [I - (AP_{k-1})(AP_{k-1})^+] r_{k-1}$ is the orthogonal projection of $r_{k-1}$ into $\text{range}(AP_{k-1}^{\perp})$, $p$ is the closest vector in $\text{range}(AP_{k-1}^{\perp})$ to $r_{k-1}$. Thus, $p = p_{k-1}$. □

With this result we can establish a number of important relationships between the residuals $r_i$ and the search directions $p_i$.

**Theorem 10.2.2** *After k iterations of (10.2.8), the conjugate gradient method, we have*

$$\begin{aligned} r_j &= r_{j-1} - \alpha_j A p_j && (10.2.9) \\ P_j^T r_j &= 0 && (10.2.10) \\ \text{span}\{p_1, \ldots, p_j\} &= \text{span}\{r_0, \ldots, r_{j-1}\} && (10.2.11) \\ &= \text{span}\{b, Ab, \ldots, A^{j-1}b\}\ . \end{aligned}$$

**Proof.** Equation (10.2.9) follows by applying $A$ to both sides of $x_j = x_{j-1} + \alpha_j p_j$ and using the definition of the residual.

To establish (10.2.10), observe that if $(P_j^T A P_j)y_j = P_j^T b$, then $y_j$ minimizes

$$\phi(P_j y) = \frac{1}{2} y^T (P_j^T A P_j) y - y^T P_j^T b.$$

Thus, $P_j^T A P_j y_j = P_j^T b$ and so $x_j = P_j y_j$ and $P_j^T r_j = P_j^T (b - A P_j y_j) = 0$.

Induction is necessary to prove (10.2.11). It is clearly true for $j = 1$. Suppose it holds for some $j$ satisfying $1 \leq j < k$. From (10.2.9) it follows that $r_j \in \text{span}\{b, Ab, \ldots, A^j b\}$ and thus $p_{j+1} = r_j - AP_j z_j \in \text{span}\{b, Ab, \ldots, A^j b\}$. Since $\text{span}\{r_0, \ldots, r_j\}$ and $\text{span}\{p_1, \ldots, p_{j+1}\}$ each have dimension $j+1$, they must equal $\text{span}\{b, Ab, \ldots, A^j b\}$ . □

We next establish that the residuals $r_i$ are mutually orthogonal.

**Theorem 10.2.3** *After $k-1$ steps of the conjugate gradient algorithm (10.2.8) the residuals $r_0, \ldots, r_{k-1}$ are mutually orthogonal.*

**Proof.** From the equation $x_i = x_{i-1} + \alpha_i p_i$ it follows that $r_i = r_{i-1} - \alpha_i A p_i$. Thus, from Lemma 10.2.1 we have

$$p_j = r_{j-1} - [\, Ap_1, \ldots, Ap_{j-1}\,]\, z_{j-1} \in \text{span}\{r_0, \ldots, r_{j-1}\}$$

for $j = 1{:}k$. Hence, $[\, p_1, \ldots, p_k\,] = [\, r_0, \ldots, r_{k-1}\,] T$ for some upper triangular matrix $T \in \mathbb{R}^{k \times k}$. Since the $p_i$ are independent, $T$ is nonsingular and so

$$[\, r_0, \ldots, r_{k-1}\,] = [\, p_1, \ldots, p_k\,] T^{-1}.$$

This implies $r_{j-1} \in \text{span}\{p_1, \ldots, p_j\}$ for $j = 1{:}k$. But from (10.2.10) we know that $P_{k-1}^T r_{k-1} = 0$. Since $r_0, \ldots, r_{k-2} \in \text{span}\{p_1, \ldots, p_{k-1}\}$ we must have $r_j^T r_{k-1} = 0$ for $j = 0{:}k-2$. □

We now show that $p_k$ can be expressed as a linear combination of $p_{k-1}$ and $r_{k-1}$. Partitioning the vector $z_{k-1}$ of Lemma 10.2.1 as

$$z_{k-1} = \begin{bmatrix} w \\ \mu \end{bmatrix} \begin{matrix} k-2 \\ 1 \end{matrix}$$

and using the identity $r_{k-1} = r_{k-2} - \alpha_{k-1}Ap_{k-1}$, we see that

$$\begin{aligned} p_k &= r_{k-1} - AP_{k-1}z_{k-1} = r_{k-1} - AP_{k-2}w - \mu Ap_k \\ &= \left(1 + \frac{\mu}{\alpha_{k-1}}\right) r_{k-1} + s_{k-1} \end{aligned}$$

where

$$s_{k-1} = -\frac{\mu}{\alpha_{k-1}} r_{k-2} - AP_{k-2}w.$$

Because the $r_i$ are mutually orthogonal, it follows that $s_{k-1}$ and $r_{k-1}$ are orthogonal to each other. Thus, the LS problem of Lemma 10.2.1 boils down to choosing $\mu$ and $w$ such that

$$\| p_k \|_2^2 = \left(1 + \frac{\mu}{\alpha_{k-1}}\right)^2 \| r_{k-1} \|_2^2 + \| s_{k-1} \|_2^2$$

is minimum. Since the 2-norm of $r_{k-2} - AP_{k-2}z$ is minimized by $z_{k-2}$ giving residual $p_{k-1}$, it follows that $s_{k-1}$ is a multiple of $p_{k-1}$. Consequently, $p_k \in \text{span}\{r_{k-1}, p_{k-1}\}$. Without loss of generality we may assume $p_k = r_{k-1} + \beta_k p_{k-1}$ and since $p_{k-1}^T A p_k = 0$ and $p_{k-1}^T r_{k-1} = 0$, it follows that

$$\beta_k = -\frac{p_{k-1}^T A r_{k-1}}{p_{k-1}^T A p_{k-1}} \qquad \alpha_k = \frac{r_{k-1}^T r_{k-1}}{p_k^T A p_k}.$$

This leads us to "version 1" of the conjugate gradient method:

$$\begin{array}{l} k = 0;\ x_0 = 0;\ r_0 = b \\ \textbf{while } r_k \neq 0 \\ \quad k = k + 1 \\ \quad \textbf{if } k = 1 \\ \quad\quad p_1 = r_0 \\ \quad \textbf{else} \\ \quad\quad \beta_k = -p_{k-1}^T A r_{k-1} / p_{k-1}^T A p_{k-1} \\ \quad\quad p_k = r_{k-1} + \beta_k p_{k-1} \\ \quad \textbf{end} \\ \quad \alpha_k = r_{k-1}^T r_{k-1} / p_k^T A p_k \\ \quad x_k = x_{k-1} + \alpha_k p_k \\ \quad r_k = b - A x_k \\ \textbf{end} \\ x = x_k \end{array} \tag{10.2.12}$$

As it stands, this iteration seems to require three separate matrix-vector multiplications each time through the loop. However, by computing residuals recursively via $r_k = r_{k-1} - \alpha_k A p_k$ and substituting

$$r_{k-1}^T r_{k-1} = -\alpha_{k-1} r_{k-1}^T A p_{k-1} \tag{10.2.13}$$

$$r_{k-2}^T r_{k-2} = \alpha_{k-1} p_{k-1}^T A p_{k-1} \tag{10.2.14}$$

into the formula for $\beta_k$, we obtain the following more efficient implementation.

**Algorithm 10.2.1 [Conjugate Gradients]** If $A \in \mathbb{R}^{n\times n}$ is symmetric positive definite and $b \in \mathbb{R}^n$ then the following algorithm computes $x \in \mathbb{R}^n$ so $Ax = b$.

$k = 0$; $x_0 = 0$; $r_0 = b$
**while** $r_k \neq 0$
    $k = k + 1$
    **if** $k = 1$
        $p_1 = r_0$
    **else**
        $\beta_k = r_{k-1}^T r_{k-1} / r_{k-2}^T r_{k-2}$
        $p_k = r_{k-1} + \beta_k p_{k-1}$
    **end**
    $\alpha_k = r_{k-1}^T r_{k-1} / p_k^T A p_k$
    $x_k = x_{k-1} + \alpha_k p_k$
    $r_k = r_{k-1} - \alpha_k A p_k$
**end**
$x = x_k$

This procedure is essentially the form of the conjugate gradient algorithm that appears in the original paper by Hestenes and Stiefel (1952). The details associated with its practical implementation are discussed in §10.3. However, we point out here that only one matrix-vector multiplication is required per iteration.

## 10.2.6 The Lanczos Connection

In §9.3.1 we derived the conjugate gradient method from the Lanczos algorithm. Now let us look at the connections between these two algorithms in the reverse direction by "deriving" the Lanczos process from conjugate gradients. Define the matrix of residuals $R_k \in \mathbb{R}^{n\times k}$ by $R_k = [\, r_0, \ldots, r_{k-1}\,]$ and the upper bidiagonal matrix $B_k \in \mathbb{R}^{k\times k}$ by

$$B_k = \begin{bmatrix} 1 & -\beta_2 & & \cdots & 0 \\ & 1 & -\beta_3 & & \vdots \\ & \ddots & \ddots & \ddots & \\ \vdots & & \ddots & \ddots & -\beta_k \\ 0 & \cdots & & & 1 \end{bmatrix}.$$

From the equations $p_j = r_{j-1} + \beta_j p_{j-1}$, $j = 2{:}k$, and $p_1 = r_0$ it follows that $R_k = P_k B_k$. Since the columns of $P_k = [\, p_1, \ldots, p_k \,]$ are $A$-conjugate, we see that $R_k^T A R_k = B_k^T \mathrm{diag}(p_1^T A p_1, \ldots, p_k^T A p_k) B_k$ is tridiagonal. From (10.2.11) and Theorem 10.2.3 it follows that if

$$\Delta = \mathrm{diag}(\rho_0, \ldots, \rho_{k-1}) \qquad \rho_i = \| \, r_i \, \|_2$$

then the columns of $R_k \Delta^{-1}$ form an orthonormal basis for the subspace $\mathrm{span}\{b, Ab, \ldots, A^{k-1}b\}$. Consequently, the columns of this matrix are essentially the Lanczos vectors of Algorithm 9.3.1, i.e., $q_i = \pm r_{i-1}/\rho_{i-1}$, where $i = 1{:}k$. Moreover, the tridiagonal matrix associated with these Lanczos vectors is given by

$$T_k = \Delta^{-1} B_k^T \mathrm{diag}(p_i^T p_i) B_k \Delta^{-1}. \tag{10.2.16}$$

The diagonal and subdiagonal of this matrix involve quantities that are readily available during the conjugate gradient iteration. Thus, we can obtain good estimates of $A$'s extremal eigenvalues (and condition number) as we generate the $x_k$ in Algorithm 10.2.1.

### 10.2.7 Some Practical Details

The termination criteria in Algorithm 10.2.1 is unrealistic. Rounding errors lead to a loss of orthogonality among the residuals and finite termination is not mathematically guaranteed. Moreover, when the conjugate gradient method is applied, $n$ is usually so big that $O(n)$ iterations represents an unacceptable amount of work. As a consequence of these observations, it is customary to regard the method as a genuinely iterative technique with termination based upon an iteration maximum $k_{max}$ and the residual norm. This leads to the following practical version of Algorithm 10.2.1:

$k = 0$; $x = 0$; $r = b$ $\rho_0 = \| \, r \, \|_2^2$
**while** $\sqrt{\rho_k} > \epsilon \| \, b \, \|_2 \wedge k < k_{max}$
    $k = k + 1$
    **if** $k = 1$
        $p = r$
    **else**
        $\beta_k = \rho_{k-1}/\rho_{k-2}$
        $p = r + \beta_k p$
    **end**
    $w = Ap$
    $\alpha_k = \rho_{k-1}/p^T w$
    $x = x + \alpha_k p$
    $r = r - \alpha_k w$
    $\rho_k = \| \, r \, \|_2^2$
**end**

This algorithm requires one matrix-vector multiplication and 10n flops per iteration. Notice that just four $n$-vectors of storage are essential: $x$, $r$, $p$, and $w$. The subscripting of the scalars is not necessary and is only done here to facilitate comparison with Algorithm 10.2.1.

It is also possible to base the termination criteria on heuristic estimates of the error $A^{-1}r_k$ by approximating $\| A^{-1} \|_2$ with the reciprocal of the smallest eigenvalue of the tridiagonal matrix $T_k$ given in (10.2.16).

The idea of regarding conjugate gradients as an iterative method began with Reid (1971b). The iterative point of view is useful but then the *rate* of convergence is central to the method's success.

### 10.2.8 Convergence Properties

We conclude this section by examining the convergence of the conjugate gradient iterates $\{x_k\}$. Two results are given and they both say that the method performs well when $A$ is near the identity either in the sense of a low rank perturbation or in the sense of norm.

**Theorem 10.2.4** *If $A = I + B$ is an n-by-n symmetric positive definite matrix and rank$(B) = r$ then Algorithm 10.2.1 converges in at most $r + 1$ steps.*

**Proof.** If $S_k = \text{span}\{b, Ab, \ldots, A^{k-1}b\}$ then it follows from rank$(A - I) = r$ that $\dim(S_k) \leq r+1$ for all $k$. Since span$\{p_1, \ldots, p_k\} = S_k$ and the $p_i$ are independent, it follows that the iteration cannot progress beyond $r+1$ steps. □

An important metatheorem follows from Theorem 10.2.4:

- If $A$ is close to a rank $r$ correction to the identity, then Algorithm 10.2.1 almost converges after $r + 1$ steps.

We show how this heuristic can be exploited in the next section.

An error bound of a different flavor can be obtained in terms of the $A$-norm which we define as follows:

$$\| w \|_A = \sqrt{w^T A w}\,.$$

**Theorem 10.2.5** *Suppose $A \in \mathbb{R}^{n \times n}$ is symmetric positive definite and $b \in \mathbb{R}^n$. If Algorithm 10.2.1 produces iterates $\{x_k\}$ and $\kappa = \kappa_2(A)$ then*

$$\| x - x_k \|_A \leq 2\| x - x_0 \|_A \left(\frac{\sqrt{\kappa} - 1}{\sqrt{\kappa} + 1}\right)^k .$$

**Proof.** See Luenberger (1973, p.187). □

The accuracy of the $\{x_k\}$ is often much better than this theorem predicts. However, a heuristic version of Theorem 10.2.5 turns out to be very useful:

- The conjugate gradient method converges very fast in the $A$-norm if $\kappa_2(A) \approx 1$.

In the next section we show how we can sometimes convert a given $Ax = b$ problem into a related $\tilde{A}\tilde{x} = \tilde{b}$ problem with $\tilde{A}$ being close to the identity.

### Problems

**P10.2.1** Verify that the residuals in (10.2.1) satisfy $r_i^T r_j = 0$ whenever $j = i + 1$.

**P10.2.2** Verify (10.2.2).

**P10.2.3** Verify (10.2.3).

**P10.2.4** Verify (10.2.13) and (10.2.14).

**P10.2.5** Give formula for the entries of the tridiagonal matrix $T_k$ in (10.2.16).

**P10.2.6** Compare the work and storage requirements associated with the practical implementation of Algorithms 9.3.1 and 10.2.1.

### Notes and References for Sec. 10.2

The conjugate gradient method is a member of a larger class of methods that are referred to as *conjugate direction* algorithms. In a conjugate direction algorithm the search directions are all $B$-conjugate for some suitably chosen matrix $B$. A discussion of these methods appears in

J.E. Dennis Jr. and K. Turner (1987). "Generalized Conjugate Directions," *Lin. Alg. and Its Applic. 88/89,* 187–209.

G.W. Stewart (1973). "Conjugate Direction Methods for Solving Systems of Linear Equations," *Numer. Math. 21,* 284–97.

The classic reference for the conjugate gradient method is

M.R. Hestenes and E. Stiefel (1952). "Methods of Conjugate Gradients for Solving Linear Systems," *J. Res. Nat. Bur. Stand. 49,* 409–36.

A exact arithmetic analysis of the method may be found in chapter 2 of

M.R. Hestenes (1980). *Conjugate Direction Methods in Optimization*, Springer-Verlag, Berlin.

See also

O. Axelsson (1977). "Solution of Linear Systems of Equations: Iterative Methods," in *Sparse Matrix Techniques: Copenhagen*, 1976, ed. V.A. Barker, Springer-Verlag, Berlin.

D.K. Faddeev and V.N. Faddeeva (1963). *Computational Methods of Linear Algebra*, W.H. Freeman and Co., San Francisco, California.

For a discussion of conjugate gradient convergence behavior, see

D. G. Luenberger (1973). *Introduction to Linear and Nonlinear Programming*, Addison-Wesley, New York.

A. van der Sluis and H.A. Van Der Vorst (1986). "The Rate of Convergence of Conjugate Gradients," *Numer. Math. 48,* 543–560.

The idea of using the conjugate gradient method as an iterative method was first discussed in

J.K. Reid (1971b). " On the Method of Conjugate Gradients for the Solution of Large Sparse Systems of Linear Equations," *in Large Sparse Sets of Linear Equations* , ed. J.K. Reid, Academic Press, New York, pp. 231–54.

Several authors have attempted to explain the algorithm's behavior in finite precision arithmetic. See

J. Cullum and R.A. Willoughby (1977). "The Equivalence of the Lanczos and the Conjugate Gradient Algorithms," IBM Research Report RE-6903.
A. Greenbaum (1981). " Behavior of the Conjugate Gradient Algorithm in Finite Precision Arithmetic," Report UCRL 85752, Lawrence Livermore Laboratory, Livermore, California.
H. Wozniakowski (1980). "Roundoff Error Analysis of a New Class of Conjugate Gradient Algorithms," *Lin. Alg. and Its Applic. 29,* 507–29.

See also the analysis in

J. Cullum and R. Willoughby (1980). "The Lanczos Phenomena: An Interpretation Based on Conjugate Gradient Optimization," *Lin. Alg. and Its Applic. 29,* 63–90.
A. Jennings (1977a). "Influence of the Eigenvalue Spectrum on the Convergence Rate of the Conjugate Gradient Method," *J. Inst. Math. Applic. 20,* 61–72.
G.W. Stewart (1975a). "The Convergence of the Method of Conjugate Gradients at Isolated Extreme Points in the Spectrum," *Numer. Math. 24,* 85–93.

Finally, we mention that the method can be used to compute an eigenvector of a large sparse symmetric matrix. See

A. Ruhe and T. Wiberg (1972). "The The Method of Conjugate Gradients Used in Inverse Iteration," *BIT 12,* 543–54.

## 10.3 Preconditioned Conjugate Gradients

We concluded the previous section by observing that the method of conjugate gradients works well on matrices that are either well conditioned or have just a few distinct eigenvalues. (The latter being the case when $A$ is a lowe rank perturbation of the identity.) In this section we show how to *precondition* a linear system so that the matrix of coefficients assumes one of these nice forms. Our treatment is quite brief and informal. Golub and Meurant (1983) and Axelssson (1985) have more comprehensive expositions.

### 10.3.1 Derivation

Consider the $n$-by-$n$ symmetric positive definite linear system $Ax = b$. The idea behind preconditioned conjugate gradients is to apply the "regular" conjugate gradient method to the transformed system

$$\tilde{A}\tilde{x} = \tilde{b}, \tag{10.3.1}$$

where $\tilde{A} = C^{-1}AC^{-1}$, $\tilde{x} = Cx$, $\tilde{b} = C^{-1}b$, and $C$ is symmetric positive definite. In view of our remarks in §10.2.8, we should try to choose $C$ so that $\tilde{A}$ is well conditioned or a matrix with clustered eigenvalues. For reasons that will soon emerge, the matrix $C^2$ must also be "simple."

If we apply Algorithm 10.2.1 to (10.3.1) then we obtain the iteration

$$
\begin{array}{l}
k = 0;\ \tilde{x}_0 = 0;\ \tilde{r}_0 = \tilde{b} \\
\textbf{while } \tilde{r}_k \neq 0 \\
\quad k = k+1 \\
\quad \textbf{if } k = 1 \\
\quad\quad \tilde{p}_1 = \tilde{r}_0 \\
\quad \textbf{else} \\
\quad\quad \beta_k = \tilde{r}_{k-1}^T \tilde{r}_{k-1} / \tilde{r}_{k-2}^T \tilde{r}_{k-2} \\
\quad\quad \tilde{p}_k = \tilde{r}_{k-1} + \beta_k \tilde{p}_{k-1} \\
\quad \textbf{end} \\
\quad \alpha_k = \tilde{r}_{k-1}^T \tilde{r}_{k-1} / \tilde{p}_k^T C^{-1}AC^{-1}\tilde{p}_k \\
\quad \tilde{x}_k = \tilde{x}_{k-1} + \alpha_k \tilde{p}_k \\
\quad \tilde{r}_k = \tilde{r}_{k-1} - \alpha_k C^{-1}AC^{-1}\tilde{p}_k \\
\textbf{end} \\
\tilde{x} = \tilde{x}_k
\end{array}
\qquad (10.3.2)
$$

Here, $\tilde{x}_k$ should be regarded as an approximation to $\tilde{x}$ and $\tilde{r}_k$ is the residual in the transformed coordinates, i.e., $\tilde{r}_k = \tilde{b} - \tilde{A}\tilde{x}_k$. Of course, once we have $\tilde{x}$ then we can obtain $x$ via the equation $x = C^{-1}\tilde{x}$. However, it is possible to avoid explicit reference to the matrix $C^{-1}$ by defining $\tilde{p}_k = Cp_k$, $\tilde{x}_k = Cx_k$, and $\tilde{r}_k = C^{-1}r_k$. Indeed, if we substitute these definitions into (10.3.2) and recall that $\tilde{b} = C^{-1}b$ and $\tilde{x} = Cx$, then we obtain

$$
\begin{array}{l}
k = 0;\ Cx_0 = 0;\ C^{-1}r_0 = C^{-1}b \\
\textbf{while } C^{-1}r_k \neq 0 \\
\quad k = k+1 \\
\quad \textbf{if } k = 1 \\
\quad\quad Cp_1 = C^{-1}r_0 \\
\quad \textbf{else} \\
\quad\quad \beta_k = (C^{-1}r_{k-1})^T(C^{-1}r_{k-1})/(C^{-1}r_{k-2})^T(C^{-1}r_{k-2}) \\
\quad\quad Cp_k = C^{-1}r_{k-1} + \beta_k Cp_{k-1} \\
\quad \textbf{end} \\
\quad \alpha_k = (C^{-1}r_{k-1})^T(C^{-1}r_{k-1})/(Cp_k)^T(C^{-1}AC^{-1})(Cp_k) \\
\quad Cx_k = Cx_{k-1} + \alpha_k Cp_k \\
\quad C^{-1}r_k = C^{-1}r_{k-1} - \alpha_k(C^{-1}AC^{-1})Cp_k \\
\textbf{end} \\
Cx = Cx_k
\end{array}
\qquad (10.3.3)
$$

If we define the *preconditioner* $M$ by $M = C^2$ (also positive definite) and let $z_k$ be the solution of the system $Mz_k = r_k$ then (10.3.3) simplifies to

**Algorithm 10.3.1 [Preconditioned Conjugate Gradients]** Given a symmetric positive definite $A \in \mathbb{R}^{n\times n}$ and $b \in \mathbb{R}^n$, the following algorithm solves the linear system $Ax = b$ using the method of conjugate gradients with preconditioner $M \in \mathbb{R}^{n\times n}$.

$k = 0$; $x_0 = 0$; $r_0 = b$
**while** ($r_k \neq 0$)
  Solve $Mz_k = r_k$.
  $k = k + 1$
  **if** $k = 1$
    $p_1 = z_0$
  **else**
    $\beta_k = r_{k-1}^T z_{k-1} / r_{k-2}^T z_{k-2}$
    $p_k = z_{k-1} + \beta_k p_{k-1}$
  **end**
  $\alpha_k = r_{k-1}^T z_{k-1} / p_k^T A p_k$
  $x_k = x_{k-1} + \alpha_k p_k$
  $r_k = r_{k-1} - \alpha_k A p_k$
**end**
$x = x_k$

A number of important observations should be made about this procedure:

- It can be shown that the residuals and search directions satisfy

$$r_j^T M^{-1} r_i = 0 \qquad i \neq j \tag{10.3.4}$$

$$p_j^T (C^{-1}AC^{-1}) p_i = 0 \qquad i \neq j \tag{10.3.5}$$

- The denominators $r_{k-2}^T z_{k-2} = z_{k-2}^T M z_{k-2}$ never vanish because $M$ is positive definite.

- Although the transformation $C$ figured heavily in the derivation of the algorithm , its action is only felt through the preconditioner $M = C^2$.

- For Algorithm 10.3.1 to be an effective sparse matrix technique, linear systems of the form $Mz = r$ must be easily solved *and* convergence must be rapid.

The choice of a good preconditoner can have a dramatic effect upon the rate of convergence. Some of the possibilities are now discussed.

### 10.3.2 Incomplete Cholesky Preconditioners

One of the most important preconditioning strategies involves computing an *incomplete Cholesky factorization* of $A$. The idea behind this approach is to calculate a lower triangular matrix $H$ with the property that $H$ has some tractable sparsity structure and is somehow "close" to $A$'s exact Cholesky factor $G$. The preconditioner is then taken to be $M = HH^T$. To appreciate this choice consider the following facts:

- There exists a unique symmetric positive definite matrix $C$ such that $M = C^2$.
- There exists an orthogonal $Q$ such that $C = QH^T$, i.e., $H^T$ is the upper triangular factor of a QR factorization of $C$.

We therefore obtain the heuristic

$$\begin{aligned} \tilde{A} &= C^{-1}AC^{-1} = C^{-T}AC^{-1} \\ &= (HQ^T)^{-1}A(QH^T)^{-1} = Q(H^{-1}GG^TH^{-T})Q^T \approx I \end{aligned} \tag{10.3.6}$$

Thus, the better $H$ approximates $G$ the smaller the condition of $\tilde{A}$, and the better the performance of Algorithm 10.3.1.

An easy but effective way to determine such a simple $H$ that approximates $G$ is to step through the Cholesky reduction setting $h_{ij}$ to zero if the corresponding $a_{ij}$ is zero. Pursuing this with the outer product version of Cholesky we obtain

```
for k = 1:n
    A(k,k) = sqrt(A(k,k))
    for i = k+1:n
        if A(i,k) ≠ 0
            A(i,k) = A(i,k)/A(k,k)
        end
    end                                                  (10.3.7)
    for j = k+1:n
        for i = j:n
            if A(i,j) ≠ 0
                A(i,j) = A(i,j) - A(i,k)A(j,k)
            end
        end
    end
end
```

In practice, the matrix $A$ and its incomplete Cholesky factor $H$ would be stored in an appropriate data structure and the looping in the above algorithm would take on a very special appearance.

Unfortunately, (10.3.7) is not always stable. Classes of positive definite matrices for which incomplete Cholesky is stable are identified in Manteuffel (1979). See also Elman (1986).

### 10.3.3 Incomplete Block Preconditioners

As with just about everything else in this book, the incomplete factorization ideas outlined in the previous subsection have a block analog. We illustrate this by looking at the *incomplete block Cholesky factorization* of the symmetric, positive definite, block tridiagonal matrix

$$A = \begin{bmatrix} A_1 & E_1^T & 0 \\ E_1 & A_2 & E_2^T \\ 0 & E_2 & A_3 \end{bmatrix}.$$

For purposes of illustration, we assume that the $A_i$ are tridiagonal and the $E_i$ are diagonal. Matrices with this structure arise from the standard 5-point discretization of self-adjoint elliptic partial differential equations over a two-dimensional domain.

The 3-by-3 case is sufficiently general. Our discussion is based upon Concus, Golub, and Meurant (1985). Let

$$G = \begin{bmatrix} G_1 & 0 & 0 \\ F_1 & G_2 & 0 \\ 0 & F_2 & G_3 \end{bmatrix}$$

be the exact block Cholesky factor of $A$. Although $G$ is sparse as a block matrix, the individual blocks are dense with the exception of $G_1$. This can be seen from the required computations:

$$\begin{aligned} G_1G_1^T &= B_1 \equiv A_1 \\ F_1 &= E_1G_1^{-1} \\ G_2G_2^T &= B_2 \equiv A_2 - F_1F_1^T = A_2 - E_1B_1^{-1}E_1^T \\ F_2 &= E_2G_2^{-1} \\ G_3G_3^T &= B_3 \equiv A_3 - F_2F_2^T = A_3 - E_2B_2^{-1}E_2^T \end{aligned}$$

We therefore seek an approximate block Cholesky factor of the form

$$\tilde{G} = \begin{bmatrix} \tilde{G}_1 & 0 & 0 \\ \tilde{F}_1 & \tilde{G}_2 & 0 \\ 0 & \tilde{F}_2 & \tilde{G}_3 \end{bmatrix}$$

so that we can easily solve systems that involve the preconditioner $M = \tilde{G}\tilde{G}^T$. This involves the imposition of sparsity on $\tilde{G}$'s blocks and here is a reasonable approach given that the $A_i$ are tridiagonal and the $E_i$ are diagonal:

$$\begin{aligned}
\tilde{G}_1\tilde{G}_1^T &= \tilde{B}_1 \equiv A_1 \\
\tilde{F}_1 &= E_1\tilde{G}_1^{-1} \\
\tilde{G}_2\tilde{G}_2^T &= \tilde{B}_2 \equiv A_2 - E_1\Lambda_1E_1^T, \quad \Lambda_1 \text{ (tridiagonal)} \approx \tilde{B}_1^{-1} \\
\tilde{F}_2 &= E_2\tilde{G}_2^{-1} \\
\tilde{G}_3\tilde{G}_3^T &= \tilde{B}_3 \equiv A_3 - E_2\Lambda_2E_2^T, \quad \Lambda_2 \text{ (tridiagonal)} \approx \tilde{B}_2^{-1}
\end{aligned}$$

Note that all the $\tilde{B}_i$ are tridiagonal. Clearly, the $\Lambda_i$ must be carefully chosen to ensure that the $\tilde{B}_i$ are also symmetric and positive definite. It then follows that the $\tilde{G}_i$ are lower bidiagonal. The $\tilde{F}_i$ are full, but they need not be explicitly formed. For example, in the course of solving the system $Mz = r$ we must solve a system of the form

$$\begin{bmatrix} \tilde{G}_1 & 0 & 0 \\ \tilde{F}_1 & \tilde{G}_2 & 0 \\ 0 & \tilde{F}_2 & \tilde{G}_3 \end{bmatrix} \begin{bmatrix} w_1 \\ w_2 \\ w_3 \end{bmatrix} = \begin{bmatrix} r_1 \\ r_2 \\ r_3 \end{bmatrix}.$$

Forward elimination can be used to carry out matrix-vector products that involve the $\tilde{F}_i = \tilde{E}_i\tilde{G}_i^{-1}$:

$$\begin{aligned}
\tilde{G}_1w_1 &= r_1 \\
\tilde{G}_2w_2 &= r_2 - \tilde{F}_1w_1 = r_2 - E_1\tilde{G}_1^{-1}w_1 \\
\tilde{G}_3w_3 &= r_3 - \tilde{F}_2w_2 = r_3 - E_2\tilde{G}_2^{-1}w_2
\end{aligned}$$

The choice of $\Lambda_i$ is delicate as the resulting $\tilde{B}_i$ must be positive definite. As we have organized the computation, the central issue is how to approximate the inverse of an $m$-by-$m$ symmetric, positive definite, tridiagonal matrix $T = (t_{ij})$ with a symmetric tridiadiagonal matrix $\Lambda$. There are several reasonable approaches:

- Set $\Lambda = \text{diag}(1/t_{11}, \ldots, 1/t_{nn})$.

- Take $\Lambda$ to be the tridiagonal part of $T^{-1}$. This can be efficiently computed since there exist $u, v \in \mathbb{R}^m$ such that the lower triangular part of $T^{-1}$ is the lower triangular part of $uv^T$. See Asplund(1959).

- Set $\Lambda = U^TU$ where $U$ is the lower bidiagonal portion of $G^{-1}$ where $T = GG^T$ is the Cholesky factorization. This can be found in $O(m)$ flops.

For a discussion of these approximations and what they imply about the associated preconditioners, see Concus, Golub, and Meurant (1985).

### 10.3.4 Domain Decomposition Ideas

The numerical solution of elliptic partial differential equations often leads to linear systems of the form

$$\begin{bmatrix} A_1 & \cdots & & \cdots & B_1 \\ \vdots & A_2 & & & B_2 \\ & & \ddots & & \vdots \\ \vdots & & & A_p & B_p \\ B_1^T & B_2^T & \cdots & B_p^T & Q \end{bmatrix} \begin{bmatrix} x_1 \\ x_2 \\ \vdots \\ x_p \\ z \end{bmatrix} = \begin{bmatrix} d_1 \\ d_2 \\ \vdots \\ d_p \\ f \end{bmatrix} \qquad (10.3.8)$$

if the unknowns are properly sequenced. See Meurant (1984). Here, the $A_i$ are symmetric positive definite, the $B_i$ are sparse, and the last block column is generally much narrower than the others.

An example with $p = 2$ serves to connect (10.3.8) and its block structure with the underlying problem geometry and the chosen *domain decomposition*. Suppose we are to solve an elliptic problem with a domain shaped as follows:

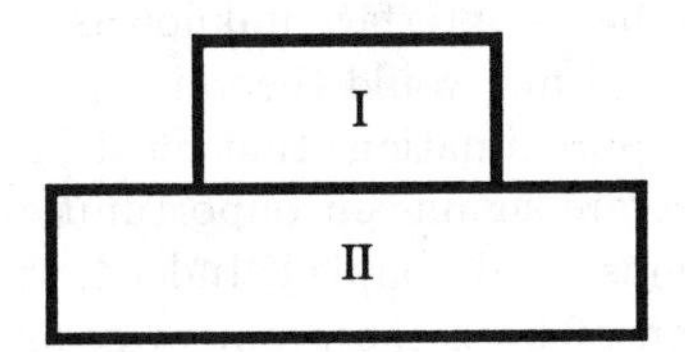


Figure 10.3.1. Domain Decomposition

There are three "types" of variables: those interior to subdomain I (aggregated in the subvector $x_1$), those interior to subdomain II (aggregated in the subvector $x_2$), and those on the interface between the two subdomains (aggregated in the subvector $z$). Note that the interior unknowns of one subdomain are not coupled to the interior unknowns of another subdomain, which accounts for the zero blocks in (10.3.8). Also observe that the number of interface unknowns is typically small compared to the overall number of unknowns.

Now let us explore the preconditioning possibilities associated with (10.3.8). We continue with the $p = 2$ case for simplicity. If we set

$$M = L \begin{bmatrix} M_1^{-1} & 0 & 0 \\ 0 & M_2^{-1} & 0 \\ 0 & 0 & S^{-1} \end{bmatrix} L^T$$

where

$$L = \begin{bmatrix} M_1 & 0 & 0 \\ 0 & M_2 & 0 \\ B_1^T & B_2^T & S \end{bmatrix}$$

then

$$M = \begin{bmatrix} M_1 & 0 & B_1 \\ 0 & M_2 & B_2 \\ B_1^T & B_2^T & S_* \end{bmatrix} \qquad (10.3.9)$$

with $S_* = S + B_1^T M_1^{-1} B_1 + B_2^T M_2^{-1} B_2$. Let us consider how we might choose the block parameters $M_1$, $M_2$, and $S$ so as to produce an effective preconditioner.

If we compare (10.3.9) with the $p = 2$ version of (10.3.8) we see that it makes sense for $M_i$ to approximate $A_i$ and for $S_*$ to approximate $Q$. The latter is achieved if $S \approx Q - B_1^T M_1^{-1} B_1 - B_2 M_2^{-1} B_2$ . There are several approaches to selecting $S$ and they all address the fact that we cannot form the dense matrices $B_i M_i^{-1} B_i^T$. For example, as discussed in the previous subsection, tridiagonal approximations of the $M_i^{-1}$ could be used. See Meurant (1989).

If the subdomains are sufficiently regular and it is feasible to solve linear systems that involve the $A_i$ exactly (say by using a fast Poisson solver), then we can set $M_i = A_i$. It follows that $M = A + E$ where the rank$(E) = m$ with $m$ being the number of interface unknowns. Thus, the preconditioned conjugate gradient algorithm would theoretically converge in $m + 1$ steps.

Regardless of the approximations that must be incorporated in the process, we see that there are significant opportunities for parallelism because the subdomain problems are decoupled. Indeed, the number of subdomains $p$ is usually a function of both the problem geometry and the number of processors that are available for the computation.

### 10.3.5 Polynomial Preconditioners

The vector $z$ defined by the preconditioner system $Mz = r$ should be thought of as an approximate solution to $Az = r$ insofar as $M$ is an approximation of $A$. One way to obtain such an approximate solution is to apply $p$ steps of a stationary method $M_1 z^{(k+1)} = N_1 z^{(k)} + r$, $z^{(0)} = 0$. It follows that if $G = M_1^{-1} N_1$ then

$$z \equiv z^{(p)} = (I + G + \cdots G^{p-1}) M_1^{-1} r.$$

Thus, if $M^{-1} = (I + G + \cdots G^{p-1}) M_1^{-1}$ then $Mz = r$ and we can think of $M$ as a preconditioner. Of course, it is important that $M$ be symmetric positive definite and this constrains the choice of $M_1$, $N_1$, and $p$. Because $M$ is a polynomial in $G$ it is referred to as a *polynomial preconditioner.* This type of preconditioner is attractive from the vector/parallel point of view and has therefore attracted considerable attention.

### 10.3.6 A Concluding Perspective

The polynomial preconditioner discussion points to an important connection between the classical iterations and the preconditioned conjugate gradient algorithm. Many iterative methods have as their basic step

$$x_k = x_{k-2} + \omega_k(\gamma_k z_{k-1} + x_{k-1} - x_{k-2}) \qquad (10.3.10)$$

where $Mz_{k-1} = r_{k-1} = b - Ax_{k-1}$. For example, if we set $\omega_k = 1$, and $\gamma_k = 1$, then

$$x_k = M^{-1}(b - Ax_{k-1}) + x_{k-1},$$

i.e., $Mx_k = Nx_{k-1} + b$, where $A = M - N$. Thus, the Jacobi, Gauss-Seidel, SOR, and SSOR methods of §10.1 have the form (10.3.10). So also does the Chebyshev semi-iterative method (10.1.12).

Following Concus, Golub, and O'Leary 1976), it is also possible to organize Algorithm 10.3.1 with a central step of the form (10.3.10):

$x_{-1} = 0;\ x_0 = 0;\ k = 0;\ r_0 = b$
**while** $r_k \neq 0$
  $k = k + 1$
  Solve $Mz_{k-1} = r_{k-1}$ for $z_{k-1}$.
  $\gamma_{k-1} = z_{k-1}^T M z_{k-1} / z_{k-1}^T A z_{k-1}$
  **if** $k = 1$ (10.3.11)
    $\omega_1 = 1$
  **else**

$$\omega_k = \left(1 - \frac{\gamma_{k-1}}{\gamma_{k-2}} \frac{z_{k-1}^T M z_{k-1}}{z_{k-2}^T M z_{k-2}} \frac{1}{\omega_{k-1}}\right)^{-1}$$

  **end**
  $x_k = x_{k-2} + \omega_k(\gamma_{k-1} z_{k-1} + x_{k-1} - x_{k-2})$
  $r_k = b - Ax_k$
**end**
$x = x_n$

Thus, we can think of the scalars $\omega_k$ and $\gamma_k$ in (10.3.11) as acceleration parameters that can be chosen to speed the convergence of the iteration $Mx_k = Nx_{k-1} + b$. Hence, any iterative method based on the splitting $A = M - N$ can be accelerated by the conjugate gradient algorithm so long as $M$ (the preconditioner) is symmetric and positive definite.

**Problems**

**P10.3.1** Detail an incomplete factorization procedure that is based on gaxpy Cholesky, i.e., Algorithm 4.2.1.

**P10.3.2** How many $n$-vectors of storage is required by a practical implementation of

Algorithm 10.3.1? Ignore workspaces that may be required when $Mz = r$ is solved.

**Notes and References for Sec. 10.3**

Our discussion of the preconditioned conjugate gradient method is patterned after

O. Axelsson (1985). "A Survey of Preconditioned Iterative Methods for Linear Systems of Equations," *BIT 25,* 166–187.
P. Concus, G.H. Golub, and G. Meurant (1985). "Block Preconditioning for the Conjugate Gradient Method," *SIAM J. Sci. and Stat. Comp. 6,* 220–252.
P. Concus, G.H. Golub, and D.P. O'Leary (1976). " A Generalized Conjugate Gradient Method for the Numerical Solution of Elliptic Partial Differential Equations," in *Sparse Matrix Computations* , ed. J.R. Bunch and D.J. Rose, Academic Press, New York.
G.H. Golub and G. Meurant (1983). *Résolution Numérique des Grandes Systèmes Linéaires,* Collection de la Direction des Etudes et Recherches de l'Electricité de France, vol. 49, Eyolles, Paris.

Other references concerned with extensions of the basic cg method include

J.H. Bramble, J.E. Pasciak, and A.H. Schatz (1986). "The construction of Preconditioners for Elliptic Problems by Substructuring I," *Math. Comp. 47,* 103–134.
J.H. Bramble, J.E. Pasciak, and A.H. Schatz (1986). "The construction of Preconditioners for Elliptic Problems by Substructuring II," *Math. Comp. 49,* 1–17.
R.C. Chin, T.A. Manteuffel, and J. de Pillis (1984). "ADI as a Preconditioning for Solving the Convection-Diffusion Equation," *SIAM J. Sci. and Stat. Comp. 5,* 281–299.
D.P. O'Leary (1980a). "The Block Conjugate Gradient Algorithm and Related Methods," *Lin. Alg. and Its Applic. 29,* 293–322.
J.K. Reid (1972). "The Use of Conjugate Gradients for Systems of Linear Equations Possessing Property A," *SIAM J. Num. Anal. 9,* 325–32.

Incomplete factorization ideas are detailed in

T.F. Chan, K.R. Jackson, and B. Zhu (1983). "Alternating Direction Incomplete Factorizations," *SIAM J. Numer. Anal. 20,* 239–257.
H. Elman (1986). "A Stability Analysis of Incomplete LU Factorization," *Math. Comp. 47,* 191–218.
T.A. Mantueffel (1979). " Shifted Incomplete Cholesky Factorization," in *Sparse Matrix Proceedings* , 1978, ed. I.S. Duff and G.W. Stewart, SIAM Publications, Philadelphia, PA.
J.A. Meijerink and H.A. Van der vorst (1977). "An Iterative Solution Method for Linear Equation Systems of Which the Coefficient Matrix is a Symmetric $M$-Matrix," *Math. Comp. 31,* 148–62.
G. Roderigue and D. Wolitzer (1984). "Preconditioning by Incomplete Block Cyclic Reduction," *Math. Comp. 42,* 549–566.

An excellent overview of domain decomposition in matrix terms is given in

G. Meurant (1989). "Domain Decomposition Methods for Partial Differential Equations on Parallel Computers," to appear *Int'l J. Supercomputing Applications.*

The conjugate gradient algorithm is based on the functional $\phi(x)$. However, the whole development goes through if the minimization of $\| Ax - b \|_2$ is considered instead. The resulting technique, called the *minimum residual variant,* is analyzed in

A.K. Cline (1976b). " Several Observations on the Use of Conjugate Gradient Methods," ICASE Report 76, NASA Langley Research Center, Hampton, Virginia.

It can be "derived" by replacing all the inner products in Algorithm 10.2.1 with $A$-inner products, i.e., $u^T v$ becomes $u^T Av$. The minimum residual variant sometimes brings about a slightly faster reduction in the norm of the residual. Similar modifications of the basic cg algorithm are classified in

S. Ashby, T.A. Manteuffel, and P.E. Saylor (1988). "A Taxonomy for Conjugate Gradient Methods," Report UCRL-98508, Lawrence Livermore National Laboratory, Livermore, CA.

while

J.E. Dennis Jr. and K. Turner (1987). "Generalized Conjugate Directions," *Lin. Alg. and Its Applic. 88/89,* 187–209.

offers a unifying perspective for the family of conjugate direction procedures.

Some representative papers concerned with the development of nonsymmetric conjugate gradient procedures include

D.M. Young and K.C. Jea (1980). "Generalized Conjugate Gradient Acceleration of Nonsymmetrizable Iterative Methods," *Lin. Alg. and Its Applic. 34,* 159–94.
O. Axelsson (1980). "Conjugate Gradient Type Methods for Unsymmetric and Inconsistent Systems of Linear Equations," *Lin. Alg. and Its Applic. 29,* 1–16.
Y. Saad (1981). "Krylov Subspace Methods for Solving Large Unsymmetric Linear Systems," *Math. Comp. 37,* 105–126.
K.C. Jea and D.M. Young (1983). "On the Simplification of Generalized Conjugate Gradient Methods for Nonsymmetrizable Linear Systems," *Lin. Alg. and Its Applic. 52/53,* 399–417.
V. Faber and T. Manteuffel (1984). "Necessary and Sufficient Conditions for the Existence of a Conjugate Gradient Method," *SIAM J. Numer. Anal. 21* 352–362.
H.A. Van der Vorst (1986). "An Iterative Solution Method for Solving $f(A)x = b$ Using Krylov Subspace Information Obtained for the Symmetric Positive Definite Matrix $A$," *J. Comp. and App. Math. 18,* 249–263.

Polynomial preconditioners are discussed in

O.G. Johnson, C.A. Micchelli, and G. Paul (1983). "Polynomial Preconditioners for Conjugate Gradient Calculations," *SIAM J. Numer. Anal. 20,* 362–376.
S.C. Eisenstat (1984). "Efficient Implementation of a Class of Preconditioned Conjugate Gradient Methods," *SIAM J. Sci. and Stat. Computing 2,* 1–4.
L. Adams (1985). "m-step Preconditioned Congugate Gradient Methods," *SIAM J. Sci. and Stat. Comp. 6,* 452–463.
S.F. Ashby (1987). "Polynomial Preconditioning for Conjugate Gradient Methods," Ph.D. Thesis, Dept. of Computer Science, University of Illinois.

Various vector/parallel implementations are discussed in

C.C. Ashcraft and R. Grimes (1988). "On Vectorizing Incomplete Factorization and SSOR Preconditioners," *SIAM J. Sci. and Stat. Comp. 9,* 122–151.
O. Axelsson and B. Polman (1986). "On Approximate Factorization Methods for Block Matrices Suitable for Vector and Parallel Processors," *Lin. Alg. and Its Applic. 77,* 3–26.
P.F. Dubois, A. Greenbaum, and G.H. Rodrigue (1979). "Approximating the Inverse of a Matrix for Use on Iterative Algorithms on Vector Processors," *Computing 22,* 257–268.

W.D. Gropp and D.E. Keyes (1988). "Complexity of Parallel Implementation of Domain Decomposition Techniques for Elliptic Partial Differential Equations," *SIAM J. Sci. and Stat. Comp. 9*, 312–326.

T. Jordan (1984). "Conjugate Gradient Preconditioners for Vector and Parallel Processors," in G. Birkoff and A. Schoenstadt (eds), *Proceedings of the Conference on Elliptic Problem Solvers*, Academic Press, NY.

R. Melhem(1987). "Toward Efficient Implementation of Preconditioned Conjugate Gradient Methods on Vector Supercomputers," *Int'l J. Supercomputing Applications 1*, 70–98.

G. Meurant (1984). "The Block Preconditioned Conjugate Gradient Method on Vector Computers," *BIT 24*, 623–633.

E.L. Poole and J.M. Ortega (1987). "Multicolor ICCG Methods for Vector Computers," *SIAM J. Numer. Anal. 24*, 1394–1418.

M.K. Seager (1986). "Parallelizing Conjugate Gradient for the Cray X-MP," *Parallel Computing 3*, 35–47.

H.A. Van der Vorst (1982). "A Vectorizable Variant of Some ICCG Methods," *SIAM J. Sci. and Stat. Comp. 3*, 350–356.

H.A. Van der Vorst (1986). "The Performance of Fortran Implementations for Preconditioned Conjugate Gradients on Vector Computers," *Parallel Computing 3*, 49–58.

An interesting application of preconditioned cg to Toeplitz systems is discussed in

T.F. Chan (1988). "An Optimal Circulant Preconditioner for Toeplitz Systems," *SIAM. J. Sci. Stat. Comp. 9*, 766-771.

# Chapter 11

# Functions of Matrices



Computing a function $f(A)$ of an $n$-by-$n$ matrix $A$ is a frequently occurring problem in control theory and other application areas. Roughly speaking, if the scalar function $f(z)$ is defined on $\lambda(A)$, then $f(A)$ is defined by substituting "$A$" for "$z$" in the "formula" for $f(z)$. For example, if $f(z) = (1+z)/(1-z)$ and $1 \notin \lambda(A)$, then $f(A) = (I+A)(I-A)^{-1}$.

The computations get particularly interesting when the function $f$ is transcendental. One approach in this more complicated situation is to compute an eigenvalue decomposition $A = YBY^{-1}$ and use the formula $f(A) = Yf(B)Y^{-1}$. If $B$ is sufficiently simple then it is often possible to calculate $f(B)$ directly. This is illustrated in §11.1 for the Jordan and Schur decompositions. Not surprisingly, reliance on the latter decomposition results in a more stable $f(A)$ procedure.

Another class of methods for the matrix function problem is to approximate the desired function $f(A)$ with an easy-to-calculate function $g(A)$. For example, $g$ might be a truncated Taylor series approximate to $f$. Error bounds associated with the approximation of matrix functions are given in §11.2.

In the last section we discuss the special and very important problem of computing the matrix exponential $e^A$.

## 11.1 Eigenvalue Methods

Given an $n$-by-$n$ matrix $A$ and a scalar function $f(z)$, there are several ways to define the *matrix function* $f(A)$. A very informal definition might be to substitute "$A$" for "$z$" in the formula for $f(z)$. For example, if $p(z) = 1 + z$ and $r(z) = (1 - (z/2))^{-1}(1 + (z/2))$ for $z \neq 2$, then it is certainly reasonable to define $p(A)$ and $r(A)$ by

$$p(A) = I + A$$

and

$$r(A) = \left(I - \frac{A}{2}\right)^{-1}\left(I + \frac{A}{2}\right) \qquad 2 \notin \lambda(A).$$

"$A$-for-$z$" substitution also works for transcendental functions, i.e.,

$$e^A = \sum_{k=0}^{\infty} \frac{A^k}{k!}.$$

To make subsequent algorithmic developments precise, however, we need a more precise definition of $f(A)$.

### 11.1.1 A Definition

There are many ways to establish rigorously the notion of a matrix function. See Rinehart (1955). Perhaps the most elegant approach is in terms of a line integral. Suppose $f(z)$ is analytic inside on a closed contour $\Gamma$ which encircles $\lambda(A)$. We define $f(A)$ to be the matrix

$$f(A) = \frac{1}{2\pi i}\oint_\Gamma f(z)(zI - A)^{-1}dz. \qquad (11.1.1)$$

This definition is immediately recognized as a matrix version of the Cauchy integral theorem. The integral is defined on an element-by-element basis:

$$f(A) = (f_{kj}) \implies f_{kj} = \frac{1}{2\pi i}\oint_\Gamma f(z)e_k^T(zI - A)^{-1}e_j dz.$$

Notice that the entries of $(zI - A)^{-1}$ are analytic on $\Gamma$ and that $f(A)$ is defined whenever $f(z)$ is analytic in a neighborhood of $\lambda(A)$.

### 11.1.2 The Jordan Characterization

Although fairly useless from the computational point of view, the definition (11.1.1) can be used to derive more practical characterizations of $f(A)$. For example, if $f(A)$ is defined and

$$A = XBX^{-1} = X\text{diag}(B_1, \ldots, B_p)X^{-1}, \qquad B_i \in \mathbb{C}^{n_i \times n_i}$$

then it is easy to verify that

$$f(A) = Xf(B)X^{-1} = X\text{diag}(f(B_1),\ldots,f(B_p))X^{-1}. \qquad (11.1.2)$$

For the case when the $B_i$ are Jordan blocks we obtain the following:

**Theorem 11.1.1** *Let $X^{-1}AX = \text{diag}(J_1,\ldots,J_p)$ be the Jordan canonical form (JCF) of $A \in \mathbb{C}^{n\times n}$ with*

$$J_i = \begin{bmatrix} \lambda_i & 1 & & \cdots & 0 \\ 0 & \lambda_i & 1 & & \vdots \\ & \ddots & \ddots & \ddots & \\ \vdots & & \ddots & \ddots & 1 \\ 0 & \cdots & & 0 & \lambda_i \end{bmatrix}$$

*being an $m_i$-by-$m_i$ Jordan block. If $f(z)$ is analytic on an open set containing $\lambda(A)$, then*

$$f(A) = X\text{diag}(f(J_1),\ldots,f(J_p))X^{-1}$$

*where*

$$f(J_i) = \begin{bmatrix} f(\lambda_i) & f^{(1)}(\lambda_i) & & \cdots & \dfrac{f^{(m_i-1)}(\lambda_i)}{(m_i-1)!} \\ 0 & f(\lambda_i) & \ddots & & \vdots \\ & & \ddots & \ddots & \\ \vdots & & & \ddots & f^{(1)}(\lambda_i) \\ 0 & \cdots & & & f(\lambda_i) \end{bmatrix}.$$

**Proof.** In view of the remarks preceding the statement of the theorem, it suffices to examine $f(G)$ where

$$G = \lambda I + E \qquad E = (\delta_{i,j-1})$$

is a $q$-by-$q$ Jordan block. Suppose $(zI - G)$ is nonsingular. Since

$$(zI - G)^{-1} = \sum_{k=0}^{q-1} \frac{E^k}{(z-\lambda)^{k+1}}$$

it follows from Cauchy's integral theorem that

$$f(G) = \sum_{k=0}^{q-1} \left[\frac{1}{2\pi i}\oint_{\Gamma} \frac{f(z)}{(z-\lambda)^{k+1}}dz\right] E^k = \sum_{k=0}^{q-1} \frac{f^{(k)}(\lambda)}{k!}E^k .$$

The theorem follows from the observation that $E^k = (\delta_{i,j-k})$. □

**Corollary 11.1.2** *If $A \in \mathbb{C}^{n\times n}$, $A = X\text{diag}(\lambda_1, \ldots, \lambda_n)X^{-1}$, and $f(A)$ is defined, then $f(A) = X\text{diag}(f(\lambda_1), \ldots, f(\lambda_n))X^{-1}$.*

**Proof.** The Jordan blocks are all 1-by-1. □

These results illustrate the close connection between $f(A)$ and the eigensystem of $A$. Unfortunately, the JCF approach to the matrix function problem has dubious computational merit unless $A$ is diagonalizable with a well-conditioned matrix of eigenvectors. Indeed, rounding errors of order $\mathbf{u}\kappa_2(X)$ can be expected to contaminate the computed result, since a linear system involving the matrix $X$ must be solved. The following example suggests that ill-conditioned similarity transformations should be avoided when computing a function of a matrix.

**Example 11.1.1** If

$$A = \begin{bmatrix} 1+10^{-5} & 1 \\ 0 & 1-10^{-5} \end{bmatrix}$$

then any matrix of eigenvectors is a column scaled version of

$$X = \begin{bmatrix} 1 & -1 \\ 0 & 2(1-10^{-5}) \end{bmatrix}$$

and has a 2-norm condition number of order $10^5$. Using a computer with machine precision $\mathbf{u} \approx 10^{-7}$ we find

$$fl[X^{-1}\text{diag}(\exp(1+10^{-5}), \exp(1-10^{-5}))X] = \begin{bmatrix} 2.718307 & 2.750000 \\ 0.000000 & 2.718254 \end{bmatrix}$$

while

$$e^A = \begin{bmatrix} 2.718309 & 2.718282 \\ 0.000000 & 2.718255 \end{bmatrix}.$$

### 11.1.3 A Schur Decomposition Approach

Some of the difficulties associated with the Jordan approach to the matrix function problem can be circumvented by relying upon the Schur decomposition. If $A = QTQ^H$ is the Schur decomposition of $A$ then

$$f(A) = Qf(T)Q^H.$$

For this to be effective, we need an algorithm for computing functions of upper triangular matrices. Unfortunately, an explicit expression for $f(T)$ is very complicated as the following theorem shows.

**Theorem 11.1.3** *Let $T = (t_{ij})$ be an n-by-n upper triangular matrix with $\lambda_i = t_{ii}$ and assume $f(T)$ is defined. If $f(T) = (f_{ij})$, then $f_{ij} = 0$ if $i > j$, $f_{ij} = f(\lambda_i)$ for $i = j$, and for all $i < j$ we have*

$$f_{ij} = \sum_{(s_0,\ldots,s_k)\in S_{ij}} t_{s_0,s_1}t_{s_1,s_2}\cdots t_{s_{k-1},s_k}f[\lambda_{s_0},\ldots,\lambda_{s_k}],$$

*where $S_{ij}$ is the set of all strictly increasing sequences of integers that start at $i$ and end at $j$ and $f[\lambda_{s_0},\ldots,\lambda_{s_k}]$ is the $k$th order divided difference of $f$ at $\{\lambda_{s_0},\ldots,\lambda_{s_k}\}$.*

**Proof.** See Descloux (1963), Davis (1973), or Van Loan (1975b). □

Computing $f(T)$ via Theorem 11.1.3 would require $O(2^n)$ flops. Fortunately, Parlett (1974a) has derived an elegant recursive method for determining the strictly upper triangular portion of the matrix $F = f(T)$. It requires only $2n^3/3$ flops and can be derived from the following commutivity result:

$$FT = TF. \qquad (11.1.3)$$

Indeed, by comparing $(i,j)$ entries in this equation, we find

$$\sum_{k=i}^{j} f_{ik}t_{kj} = \sum_{k=i}^{j} t_{ik}f_{kj} \qquad j > i$$

and thus, if $t_{ii}$ and $t_{jj}$ are distinct,

$$f_{ij} = t_{ij}\frac{f_{jj}-f_{ii}}{t_{jj}-t_{ii}} + \sum_{k=i+1}^{j-1} \frac{t_{ik}f_{kj}-f_{ik}t_{kj}}{t_{jj}-t_{ii}}. \qquad (11.1.4)$$

From this we conclude that $f_{ij}$ is a linear combination of its neighbors to its left and below in the matrix $F$. For example, the entry $f_{25}$ depends upon $f_{22}$, $f_{23}$, $f_{24}$, $f_{55}$, $f_{45}$, and $f_{35}$. Because of this, the entire upper triangular portion of $F$ can be computed one superdiagonal at a time beginning with the diagonal, $f(t_{11}),\ldots, f(t_{nn})$. The complete procedure is as follows:

**Algorithm 11.1.1** This algorithm computes the matrix function $F = f(T)$ where $T$ is upper triangular with distinct eigenvalues and $f$ is defined on $\lambda(T)$.

```
for i = 1:n
    f_ii = f(t_ii)
end
for p = 1:n-1
    for i = 1:n-p
        j = i+p
        s = t_ij(f_jj - f_ii)
        for k = i+1:j-1
            s = s + t_ik f_kj - f_ik t_kj
        end
        f_ij = s/(t_jj - t_ii)
    end
end
```

This algorithm requires $2n^3/3$ flops. Assuming that $T = QAQ^H$ is the Schur form of $A$, $f(A) = QFQ^H$ where $F = f(T)$. Clearly, most of the work in computing $f(A)$ by this approach is in the computation of the Schur decomposition.

**Example 11.1.2** If

$$T = \begin{bmatrix} 1 & 2 & 3 \\ 0 & 3 & 4 \\ 0 & 0 & 5 \end{bmatrix}$$

and $f(z) = (1+z)/z$ then $F = (f_{ij}) = f(T)$ is defined by

$$\begin{aligned}
f_{11} &= (1+1)/1 = 2 \\
f_{22} &= (1+3)/3 = 4/3 \\
f_{33} &= (1+5)/5 = 6/5 \\
f_{12} &= t_{12}(f_{22} - f_{11})/(t_{22} - t_{11}) = -2/3 \\
f_{23} &= t_{23}(f_{33} - f_{22})/(t_{33} - t_{22}) = -4/15 \\
f_{13} &= [t_{13}(f_{33} - f_{11}) + (t_{12}f_{23} - f_{12}t_{23})]/(t_{33} - t_{11}) = -1/15.
\end{aligned}$$

### 11.1.4 A Block Schur Approach

If $A$ has close or multiple eigenvalues, then Algorithm 11.1.1 leads to poor results. In this case, it is advisable to use a block version of Algorithm 11.1.1. We outline such a procedure due to Parlett (1974a). The first step is to choose $Q$ in the Schur decomposition such that close or multiple eigenvalues are clustered in blocks $T_{11}, \ldots, T_{pp}$ along the diagonal of $T$. In particular, we must compute a partitioning

$$T = \begin{bmatrix} T_{11} & T_{12} & \cdots & T_{1p} \\ 0 & T_{22} & \cdots & T_{2p} \\ \vdots & \vdots & \ddots & \vdots \\ 0 & 0 & \cdots & T_{pp} \end{bmatrix} \qquad F = \begin{bmatrix} F_{11} & F_{12} & \cdots & F_{1p} \\ 0 & F_{22} & \cdots & F_{2p} \\ \vdots & \vdots & \ddots & \vdots \\ 0 & 0 & \cdots & F_{pp} \end{bmatrix}$$

where $\lambda(T_{ii}) \cap \lambda(T_{jj}) \neq \emptyset$, $i \neq j$. The actual determination of the block sizes can be done using the methods of §7.6.

Next, we compute the submatrices $F_{ii} = f(T_{ii})$ for $i = 1{:}p$. Since the eigenvalues of $T_{ii}$ are presumably close, these calculations require special methods. (Some possibilities are discussed in the next two sections.) Once the diagonal blocks of $F$ are known, the blocks in the strict upper triangle of $F$ can be found recursively, as in the scalar case. To derive the governing equations, we equate $(i,j)$ blocks in $FT = TF$ for $i < j$ and obtain the following generalization of (11.1.4):

$$F_{ij}T_{jj} - T_{ii}F_{ij} = T_{ij}F_{jj} - F_{ii}T_{ij} + \sum_{k=i+1}^{j-1} (T_{ik}F_{kj} - F_{ik}T_{kj}). \quad (11.1.5)$$

This is a linear system whose unknowns are the elements of the block $F_{ij}$ and whose right-hand side is "known" if we compute the $F_{ij}$ one block super-diagonal at a time. We can solve (11.1.5) using the Bartels-Stewart algorithm (Algorithm 7.6.2).

The block Schur approach described here is useful when computing real functions of real matrices. After computing the real Schur form $A = QTQ^T$, the block algorithm can be invoked in order to handle the 2-by-2 bumps along the diagonal of $T$.

### Problems

**P11.1.1** Using the definition (11.1.1) show that (a) $Af(A) = f(A)A$, (b) $f(A)$ is upper triangular if $A$ is upper triangular, and (c) $f(A)$ is Hermitian if $A$ is Hermitian.

**P11.1.2** Rewrite Algorithm 11.1.1 so that $f(T)$ is computed column by column.

**P11.1.3** Suppose $A = X\text{diag}(\lambda_i)X^{-1}$ where $X = [\,x_1, \ldots, x_n\,]$ and $X^{-1} = [\,y_1, \ldots, y_n\,]^H$. Show that if $f(A)$ is defined, then

$$f(A) = \sum_{k=1}^{n} f(\lambda_i)x_i y_i^H .$$

**P11.1.4** Show that

$$T = \begin{bmatrix} T_{11} & T_{12} \\ 0 & T_{22} \end{bmatrix} \begin{matrix} p \\ q \end{matrix} \quad \Rightarrow \quad f(T) = \begin{bmatrix} F_{11} & F_{12} \\ 0 & F_{22} \end{bmatrix} \begin{matrix} p \\ q \end{matrix}$$
$$\qquad\; p \quad\; q \qquad\qquad\qquad\qquad\qquad\quad p \quad\; q$$

where $F_{11} = f(T_{11})$ and $F_{22} = f(T_{22})$. Assume $f(T)$ is defined.

### Notes and References for Sec. 11.1

The following texts have a better-than-average treatment of the matrix function problem:

R. Bellman (1969). *Introduction to Matrix Analysis*, McGraw–Hill, New York.
F. Gantmacher (1959). *The Theory of Matrices*, vols. 1–2, Chelsea Publishing Co., New York.
L. Mirsky (1955). *An Introduction to Linear Algebra*, Oxford University Press, London.

The contour integral representation of $f(A)$ given in the text is useful in functional analysis because of its generality. See

N. Dunford and J. Schwartz (1958). *Linear Operators, Part I*, Interscience, New York.

As we discussed, other definitions of $f(A)$ are possible. However, for the matrix functions typically encountered in practice, all these definitions are equivalent. See

R.F. Rinehart (1955). "The Equivalence of Definitions of a Matric Function," *Amer. Math. Monthly 62*, 395–414.

Various aspects of the Jordan representation are detailed in

J.S. Frame (1964a). "Matrix Functions and Applications, Part II," *IEEE Spectrum 1 (April)*, 102–8.

J.S. Frame (1964b). "Matrix Functions and Applications, Part IV," *IEEE Spectrum 1 (June)*, 123–31.

The following are concerned with the Schur decomposition and its relationship to the $f(A)$ problem:

D. Davis (1973). "Explicit Functional Calculus," *Lin. Alg. and Its Applic. 6*, 193–99.

J. Descloux (1963). "Bounds for the Spectral Norm of Functions of Matrices," *Numer. Math. 5*, 185–90.

C.F. Van Loan (1975b). "A Study of the Matrix Exponential," Numerical Analysis Report No. 10, Dept. of Maths., University of Manchester, England.

Algorithm 11.1.1 and the various computational difficulties that arise when it is applied to a matrix having close or repeated eigenvalues are discussed in

B.N. Parlett (1974a). "Computation of Functions of Triangular Matrices," Memorandum No. ERL–M481, Electronics Research Laboratory, College of Engineering, University of California, Berkeley.

B.N. Parlett (1976). "A Recurrence Among the Elements of Functions of Triangular Matrices," *Lin. Alg. and Its Applic. 14*, 117–21.

A compromise between the Jordan and Schur approaches to the $f(A)$ problem results if $A$ is reduced to block diagonal form as described in §7.6.3. See

B. Kågström (1977). "Numerical Computation of Matrix Functions," Department of Information Processing Report UMINF-58.77, University of Ümea, Sweden.

## 11.2 Approximation Methods

We now consider a class of methods for computing matrix functions which prima facie do not involve eigenvalues. These techniques are based on the idea that if $g(z)$ approximates $f(z)$ on $\lambda(A)$, then $f(A)$ approximates $g(A)$, e.g.,

$$e^A \approx I + A + \frac{A^2}{2!} + \cdots + \frac{A^q}{q!}.$$

We begin by bounding $\| f(A) - g(A) \|$ using the Jordan and Schur matrix function representations. We follow this discussion with some comments on the evaluation of matrix polynomials.

### 11.2.1 A Jordan Analysis

The Jordan representation of matrix functions (Theorem 11.1.1) can be used to bound the error in an approximant $g(A)$ of $f(A)$.

**Theorem 11.2.1** *Let $X^{-1}AX = \text{diag}(J_1, \ldots, J_p)$ be the JCF of $A \in \mathbb{C}^{n\times n}$ with*

$$J_i = \begin{bmatrix} \lambda_i & 1 & & \cdots & 0 \\ 0 & \lambda_i & 1 & & \vdots \\ & & \ddots & \ddots & \\ \vdots & & & \ddots & 1 \\ 0 & \cdots & & & \lambda_i \end{bmatrix}$$

*being an $m_i$-by-$m_i$ Jordan block. If $f(z)$ and $g(z)$ are analytic on an open set containing $\lambda(A)$, then*

$$\| f(A) - g(A) \|_2 \le \kappa_2(X) \max_{\substack{1\le i\le p \\ 0\le r\le m_i-1}} m_i \frac{\left|f^{(r)}(\lambda_i) - g^{(r)}(\lambda_i)\right|}{r!}.$$

**Proof.** Defining $h(z) = f(z) - g(z)$ we have

$$\begin{aligned} \| f(A) - g(A) \|_2 &= \| X \text{diag}(h(J_1), \ldots, h(J_p)) X^{-1} \|_2 \\ &\le \kappa_2(X) \max_{1\le i\le p} \| h(J_i) \|_2 . \end{aligned}$$

Using Theorem 11.1.1 and equation (2.3.8) we conclude that

$$\| h(J_i) \|_2 \le m_i \max_{0\le r\le m_i-1} \frac{\left|h^{(r)}(\lambda_i)\right|}{r!}$$

thereby proving the theorem. □

### 11.2.2 A Schur Analysis

If we rely on the Schur instead of the Jordan decomposition we obtain an alternative bound.

**Theorem 11.2.2** *Let $Q^H AQ = T = \text{diag}(\lambda_i) + N$ be the Schur decomposition of $A \in \mathbb{C}^{n\times n}$, with $N$ being the strictly upper triangular portion of $T$. If $f(z)$ and $g(z)$ are analytic on a closed convex set $\Omega$ whose interior contains $\lambda(A)$, then*

$$\| f(A) - g(A) \|_F \le \sum_{r=0}^{n-1} \delta_r \frac{\| \, |N|^r \, \|_F}{r!}$$

*where*

$$\delta_r = \sup_{z\in\Omega} \left|f^{(r)}(z) - g^{(r)}(z)\right| .$$

**Proof.** Let $h(z) = f(z) - g(z)$ and set $H = (h_{ij}) = h(A)$. Let $S_{ij}^{(r)}$ denote the set of strictly increasing integer sequences $(s_0, \ldots, s_r)$ with the property that $s_0 = i$ and $s_r = j$. Notice that

$$S_{ij} = \bigcup_{r=1}^{j-i} S_{ij}^{(r)}$$

and so from Theorem 11.1.3, we obtain the following for all $i < j$:

$$h_{ij} = \sum_{r=1}^{j-1} \sum_{s \in S_{ij}^{(r)}} n_{s_0,s_1} n_{s_1,s_2} \cdots n_{s_{r-1},s_r} h\left[\lambda_{s_0}, \ldots, \lambda_{s_r}\right].$$

Now since $\Omega$ is convex and $h$ analytic, we have

$$\left|h\left[\lambda_{s_0}, \ldots, \lambda_{s_r}\right]\right| \leq \sup_{z \in \Omega} \frac{\left|h^{(r)}(z)\right|}{r!} = \frac{\delta_r}{r!}. \qquad (11.2.1)$$

Furthermore if $|N|^r = (n_{ij}^{(r)})$ for $r \geq 1$, then it can be shown that

$$n_{ij}^{(r)} = \begin{cases} 0 & j < i + r \\ \displaystyle\sum_{s \in S_{ij}^{(r)}} \left|n_{s_0,s_1} n_{s_1,s_2} \cdots n_{s_{r-1},s_r}\right| & j \geq i + r \end{cases} \qquad (11.2.2)$$

The theorem now follows by taking absolute values in the expression for $h_{ij}$ and then using (11.2.1) and (11.2.2). □

The bounds in the above theorems suggest that there is more to approximating $f(A)$ than just approximating $f(z)$ on the spectrum of $A$. In particular, we see that if the eigensystem of $A$ is ill-conditioned and/or $A$'s departure from normality is large, then the discrepancy between $f(A)$ and $g(A)$ may be considerably larger than the maximum of $|f(z) - g(z)|$ on $\lambda(A)$. Thus, even though approximation methods avoid eigenvalue computations, they appear to be influenced by the structure of $A$'s eigensystem, a point that we pursue further in the next section.

**Example 11.2.1** Suppose

$$A = \begin{bmatrix} -.01 & 1 & 1 \\ 0 & 0 & 1 \\ 0 & 0 & .01 \end{bmatrix}$$

If $f(z) = e^z$ and $g(z) = 1 + z + z^2/2$ then $\| f(A) - g(A) \| \approx 10^{-5}$ in either the Frobenius norm or the 2-norm. Since $\kappa_2(X) \approx 10^7$, the error predicted by Theorem 11.2.1 is $O(1)$, rather pessimistic. On the other hand, the error predicted by the Schur decomposition approach is $O(10^{-2})$.

### 11.2.3 Taylor Approximants

A popular way of approximating a matrix function such as $e^A$ is through the truncation of its Taylor series. The conditions under which a matrix function $f(A)$ has a Taylor series representation are easily established.

**Theorem 11.2.3** *If $f(z)$ has a power series representation*

$$f(z) = \sum_{k=0}^{\infty} c_k z^k$$

*on an open disk containing $\lambda(A)$, then*

$$f(A) = \sum_{k=0}^{\infty} c_k A^k .$$

**Proof.** We prove the theorem for the case when $A$ is diagonalizable. In P11.2.1, we give a hint as to how to proceed without this assumption. Suppose $X^{-1}AX = D = \text{diag}(\lambda_1, \ldots, \lambda_n)$. Using Corollary 11.1.2, we have

$$\begin{aligned} f(A) &= X \text{diag}(f(\lambda_1), \ldots, f(\lambda_n)) X^{-1} \\ &= X \text{diag}\left( \sum_{k=0}^{\infty} c_k \lambda_1^k, \ldots, \sum_{k=0}^{\infty} c_k \lambda_n^k \right) X^{-1} \\ &= X \left( \sum_{k=0}^{\infty} c_k D^k \right) X^{-1} = \sum_{k=0}^{\infty} c_k (XDX^{-1})^k = \sum_{k=0}^{\infty} c_k A^k . \ \square \end{aligned}$$

Several important transcendental matrix functions have particularly simple series representations:

$$\log(I - A) = \sum_{k=1}^{\infty} \frac{A^k}{k} \qquad |\lambda| < 1, \ \lambda \in \lambda(A)$$

$$\sin(A) = \sum_{k=0}^{\infty} (-1)^k \frac{A^{2k+1}}{(2k+1)!}$$

$$\cos(A) = \sum_{k=0}^{\infty} (-1)^k \frac{A^{2k}}{(2k)!} .$$

The following theorem bounds the errors that arise when matrix functions such as these are approximated via truncated Taylor series.

**Theorem 11.2.4** *If $f(z)$ has the Taylor series*

$$f(z) = \sum_{k=0}^{\infty} \alpha_k z^k$$

*on an open disk containing the eigenvalues of $A \in \mathbb{C}^{n\times n}$, then*

$$\| f(A) - \sum_{k=0}^{q} \alpha_k A^k \|_2 \le \frac{n}{(q+1)!} \max_{0\le s\le 1} \| A^{q+1} f^{(q+1)}(As) \|_2 .$$

**Proof.** Define the matrix $E(s)$ by

$$f(As) = \sum_{k=0}^{q} \alpha_k (As)^k + E(s) \qquad 0 \le s \le 1 . \tag{11.2.3}$$

If $f_{ij}(s)$ is the $(i,j)$ entry of $f(As)$, then it is necessarily analytic and so

$$f_{ij}(s) = \left( \sum_{k=0}^{q} \frac{f_{ij}^{(k)}(0)}{k!} s^k \right) + \frac{f_{ij}^{(q+1)}(\varepsilon_{ij})}{(q+1)!} s^{q+1} \tag{11.2.4}$$

where $\varepsilon_{ij}$ satisfies $0 \le \varepsilon_{ij} \le s \le 1$.

By comparing powers of $s$ in (11.2.3) and (11.2.4) we conclude that $e_{ij}(s)$, the $(i,j)$ entry of $E(s)$, has the form

$$e_{ij}(s) = \frac{f_{ij}^{(q+1)}(\varepsilon_{ij})}{(q+1)!} s^{q+1} .$$

Now $f_{ij}^{(q-1)}(s)$ is the $(i,j)$ entry of $A^{q+1} f^{(q+1)}(As)$ and therefore

$$|e_{ij}(s)| \le \max_{0\le s\le 1} \frac{f_{ij}^{(q+1)}(s)}{(q+1)!} \le \max_{0\le s\le 1} \frac{\| A^{q+1} f^{(q+1)}(As) \|_2}{(q+1)!} .$$

The theorem now follows by applying (2.3.8). □

**Example 11.2.2** If

$$A = \begin{bmatrix} -49 & 24 \\ -64 & 31 \end{bmatrix}$$

then

$$e^A = \begin{bmatrix} -0.735759 & .0551819 \\ -1.471518 & 1.103638 \end{bmatrix} .$$

For $q = 59$, Theorem 11.2.4 predicts that

$$\| e^A - \sum_{k=0}^{q} \frac{A^k}{k!} \|_2 \le \frac{n}{(q+1)!} \max_{0\le s\le 1} \| A^{q+1} e^{As} \|_2 \le 10^{-60} .$$

However, if $\beta = 16$, $t = 6$, chopped arithmetic is used we find

$$fl\left(\sum_{k=0}^{59} \frac{A^k}{k!}\right) = \begin{bmatrix} -22.25880 & -1.4322766 \\ -61.49931 & -3.474280 \end{bmatrix}.$$

The problem is that some of the partial sums have large elements. For example, $I + \cdots + A^{17}/17!$ has entries of order $10^7$. Since the machine precision is approximately $10^{-7}$, rounding errors larger than the norm of the solution are sustained.

Example 11.2.2 highlights a shortcoming of truncated Taylor series approximation: It tends to be worthwhile only near the origin. The problem can sometimes be circumvented through a change of scale. For example, by repeated application of the *double angle* formulae:

$$\cos(2A) = 2\cos(A)^2 - I \qquad \sin(2A) = 2\sin(A)\cos(A)$$

it is possible to "build up" the sine and cosine of a matrix from suitably truncated Taylor series approximates:

$S_0$ = Taylor approximate to $\sin(A/2^k)$
$C_0$ = Taylor approximate to $\cos(A/2^k)$
**for** $j = 1{:}k$
  $S_j = 2S_{j-1}C_{j-1}$
  $C_j = 2C_{j-1}^2 - I$
**end**

Here $k$ is a positive integer chosen so that, say, $\| A \|_\infty \approx 2^k$. See Serbin and Blalock (1979).

### 11.2.4 Evaluating Matrix Polynomials

Since the approximation of transcendental matrix functions so often involves the evaluation of polynomials, it is worthwhile to look at the details of computing

$$p(A) = b_0 I + b_1 A + \cdots + b_q A^q$$

where the scalars $b_0, \ldots, b_q \in \mathbb{R}$ are given. The most obvious approach is to invoke Horner's scheme:

**Algorithm 11.2.1** Given a matrix $A$ and $b(0{:}q)$, the following algorithm computes $F = b_q A^q + \cdots + b_1 A + b_0 I$.

$F = b_q A + b_{q-1} I$
**for** $k = q-2{:}-1{:}0$
  $F = AF + b_k I$
**end**

This requires $q-1$ matrix multiplications. However, unlike the scalar case, this summation process is not optimal. To see why, suppose $q = 9$ and observe that

$$\begin{aligned} p(A) &= A^3(A^3(b_9A^3 + (b_8A^2 + b_7A + b_6I)) \\ &\quad +(b_5A^2 + b_4A + b_3I)) + b_2A^2 + b_1A + b_0I. \end{aligned}$$

Thus, $F = p(A)$ can be evaluated with only four matrix multiplies:

$$\begin{aligned} A_2 &= A^2 \\ A_3 &= AA_2 \\ F_1 &= b_9A_3 + b_8A_2 + b_7A + b_6I \\ F_2 &= A_3F_1 + b_5A_2 + b_4A + b_3I \\ F &= A_3F_2 + b_2A_2 + b_1A + b_0I. \end{aligned}$$

In general, if $s$ is any integer satisfying $1 \le s \le \sqrt{q}$ then

$$p(A) = \sum_{k=0}^{r} B_k(A^s)^k \qquad r = \text{floor}(q/s) \tag{11.2.5}$$

where

$$B_k = \begin{cases} b_{sk+s-1}A^{s-1} + \cdots + b_{sk+1}A + b_{sk}I & k = 0{:}r-1 \\ \\ b_qA^{q-sr} + \cdots + b_{sr+1}A + b_kI & k = r\,. \end{cases}$$

Once $A^2, \ldots, A^s$ are computed, Horner's rule can be applied to (11.2.5) and the net result is that $p(A)$ can be computed with $s + r - 1$ matrix multiplies. By choosing $s = \text{floor}(\sqrt{q})$, the number of matrix multiplies is approximately minimized. This technique is discussed in Paterson and Stockmeyer (1973). Van Loan (1978b) shows how the procedure can be implemented without storage arrays for $A^2, \ldots, A^s$.

## 11.2.5 Computing Powers of a Matrix

The problem of raising a matrix to a given power deserves special mention. Suppose it is required to compute $A^{13}$. Noting that $A^4 = (A^2)^2$, $A^8 = (A^4)^2$ and $A^{13} = A^8A^4A$, we see that this can be accomplished with just 5 matrix multiplies. In general we have

**Algorithm 11.2.2 (Binary Powering)** Given a positive integer $s$ and $A \in \mathbb{R}^{n\times n}$, the following algorithm computes $F = A^s$ where $s$ is a positive integer and $A \in \mathbb{R}^{n\times n}$.

Let $s = \sum_{k=0}^{t} \beta_k 2^k$ be the binary expansion of $s$ with $\beta_t \neq 0$.

```
Z = A; q = 0
while β_q = 0
    Z = Z^2; q = q + 1
end
F = Z
for k = q + 1:t
    Z = Z^2
    if β_k ≠ 0
        F = FZ
    end
end
```

This algorithm requires at most 2 floor[$\log_2(s)$] matrix multiplies. If $s$ is a power of 2, then only $\log_2(s)$ matrix multiplies are needed.

### 11.2.6 Integrating Matrix Functions

We conclude this section with some remarks on the integration of matrix functions. Suppose $f(At)$ is defined for all $t \in [a, b]$ and that we wish to compute

$$F = \int_a^b f(At)dt.$$

As in (11.1.1) the integration is on an element-by-element basis.

Ordinary quadrature rules can be applied to $F$. For example, with Simpson's rule, we have

$$F \approx \tilde{F} = \frac{h}{3}\sum_{k=0}^{m} w_k f(A(a+kh)) \tag{11.2.6}$$

where $m$ is even, $h = (b-a)/m$ and

$$w_k = \begin{cases} 1 & k = 0, m \\ 4 & k \text{ odd} \\ 2 & k \text{ even}, k \neq 0, m\,. \end{cases}$$

If $(d^4/dz^4)f(zt) = f^{(4)}(zt)$ is continuous for $t \in [a, b]$ and if $f^{(4)}(At)$ is defined on this same interval, then it can be shown that $\tilde{F} = F + E$ where

$$\| E \|_2 \le \frac{nh^4(b-a)}{180} \max_{a \le t \le b} \| f^{(4)}(At) \|_2\,. \tag{11.2.7}$$

Let $f_{ij}$ and $e_{ij}$ denote the $(i,j)$ entries of $F$ and $E$, respectively. Under the above assumptions we can apply the standard error bounds for Simpson's rule and obtain

$$|e_{ij}| \le \frac{h^4(b-a)}{180} \max_{a \le t \le b} |e_i^T f^{(4)}(At)e_j| .$$

The inequality (11.2.7) now follows since $\| E \|_2 \le n \max |e_{ij}|$ and

$$\max_{a \le t \le b} |e_i^T f^{(4)}(At)e_j| \le \max_{a \le t \le b} \| f^{(4)}(At) \|_2 .$$

Of course, in the practical application of (11.2.6), the function evaluations $f(A(a+kh))$ normally have to be approximated. Thus, the overall error involves the error in approximating $f(A(a+kh)$ as well as the Simpson rule error.

### Problems

**P11.2.1** (a) Suppose $G = \lambda I + E$ is a $p$-by-$p$ Jordan block, where $E = (\delta_{i,j-1})$. Show that

$$(\lambda I + E)^k = \sum_{j=0}^{\min\{p-1,k\}} \binom{k}{j} \lambda^{k-j} E^j .$$

(b) Use (a) and Theorem 11.1.1 to prove Theorem 11.2.3.

**P11.2.2** Verify (11.2.2).

**P11.2.3** Show that if $\| A \|_2 < 1$, then $\log(I+A)$ exists and satisfies the bound $\| \log(I+A) \|_2 \le \| A \|_2 / (1 - \| A \|_2)$.

**P11.2.4** Let $A$ by an $n$-by-$n$ symmetric positive definite matrix. (a) Show that there exists a unique symmetric positive definite $X$ such that $A = X^2$. (b) Show that if $X_0 = I$ and $X_{k+1} = (X_k + AX_k^{-1})/2$ then $X_k \rightarrow \sqrt{A}$ quadratically where $\sqrt{A}$ denotes the matrix $X$ in part (a).

**P11.2.5** Specialize Algorithm 11.2.1 to the case when $A$ is symmetric. Repeat for the case when $A$ is upper triangular. In both instances, give the associated flop counts.

**P11.2.6** Show that $X(t) = C_1 \cos(t\sqrt{A}) + C_2\sqrt{A^{-1}}\sin(t\sqrt{A})$ solves the initial value problem $\ddot{X}(t) = -AX(t)$, $X(0) = C_1$, $\dot{X}(0) = C_2$. Assume that $A$ is symmetric positive definite.

**P11.2.7** Using Theorem 11.2.4, bound the error in the approximations:

$$\sin(A) \approx \sum_{k=0}^{q} (-1)^k \frac{A^{2k+1}}{(2k+1)!} \qquad \cos(A) \approx \sum_{k=0}^{q} (-1)^k \frac{A^{2k}}{(2k)!} .$$

### Notes and References for Sec. 11.2

The optimality of Horner's rule for polynomial evaluation is discussed in

D. Knuth (1981). *The Art of Computer Programming , vol. 2. Seminumerical Algorithms* , 2nd ed., Addison-Wesley, Reading, Massachusetts.
M.S. Paterson and L.J. Stockmeyer (1973). "On the Number of Nonscalar Multiplications Necessary to Evaluate Polynomials," *SIAM J. Comp. 2,* 60–66.

The Horner evaluation of matrix polynomials is analyzed in

C.F. Van Loan (1978b). "A Note on the Evaluation of Matrix Polynomials," *IEEE Trans. Auto. Cont. AC-24,* 320–21.

The Newton and Language representations for $f(A)$ and their relationship to other matrix function definitions is discussed in

R.F. Rinehart (1955). "The Equivalence of Definitions of a Matric Function," *Amer. Math. Monthly 62,* 395–414.

The "double angle" method for computing the cosine of matrix is analyzed in

S. Serbin and S. Blalock (1979). "An Algorithm for Computing the Matrix Cosine," *SIAM J. Sci. Stat. Comp. 1,* 198–204.

The square root is a particularly important matrix function and several approaches are possible. See

Å. Björck and S. Hammarling (1983). "A Schur Method for the Square Root of a Matrix," *Lin. Alg. and Its Applic. 52/53,* 127–140.
N.J. Higham (1986a). "Newton's Method for the Matrix Square Root," *Math. Comp. 46,* 537–550.
N.J. Higham (1987c). "Computing Real Square Roots of a Real Matrix," *Lin. Alg. and Its Applic. 88/89,* 405–430.

## 11.3 The Matrix Exponential

One of the most frequently computed matrix functions is the exponential

$$e^{At} = \sum_{k=0}^{\infty} \frac{(At)^k}{k!} .$$

Numerous algorithms for computing $e^{At}$ have been proposed, but most of them are of dubious numerical quality, as is pointed out in the survey article by Moler and Van Loan (1978). In order to illustrate what the computational difficulties are, we present a brief perturbation analysis of the matrix exponential problem and then use it to assess one of the better $e^{At}$ algorithms.

### 11.3.1 Perturbation Theory

The starting point in the discussion is the initial value problem

$$\dot{X}(t) = AX(t) \qquad X(0) = I$$

where $A, X(t) \in \mathbb{R}^{n\times n}$. This has the unique solution $X(t) = e^{At}$, a characterization of the matrix exponential that can be used to establish the identity

$$e^{(A+E)t} - e^{At} = \int_0^t e^{A(t-s)} E e^{(A+E)s} ds .$$

From this it follows that

$$\frac{\| e^{(A+E)t} - e^{At} \|_2}{\| e^{At} \|_2} \leq \frac{\| E \|_2}{\| e^{At} \|_2} \int_0^t \| e^{A(t-s)} \|_2 \| e^{(A+E)s} \|_2 ds .$$

Further simplifications result if we bound the norms of the exponentials that appear in the integrand. One way of doing this is through the Schur decomposition. If $Q^H AQ = \text{diag}(\lambda_i) + N$ is the Schur decomposition of $A \in \mathbb{C}^{n\times n}$ then it can be shown that

$$\| e^{At} \|_2 \leq e^{\alpha(A)t M_S(t)},$$

where

$$\alpha(A) = \max \{\text{Re}(\lambda) : \lambda \in \lambda(A) \}$$

and

$$M_S(t) = \sum_{k=0}^{n-1} \frac{\| Nt \|_2^k}{k!} .$$

With a little manipulation, these results can be used to establish the inequality

$$\frac{\| e^{(A+E)t} - e^{At} \|_2}{\| e^{At} \|_2} \leq t\| E \|_2 M_S(t)^2 \exp(t M_S(t) \| E \|_2) .$$

Notice that $M_S(t) \equiv 1$ if and only if $A$ is normal, suggesting that the matrix exponential problem is "well behaved" if $A$ is normal. This observation is confirmed by the behavior of the *matrix exponential condition number* $\nu(A,t)$, defined by

$$\nu(A,t) = \max_{\| E \|_2 \leq 1} \left\| \int_0^t e^{A(t-s)} E e^{As} ds \right\|_2 \frac{\| A \|_2}{\| e^{At} \|_2} .$$

This quantity, discussed in Van Loan (1977a), measures the sensitivity of the map $A \rightarrow e^{At}$ in that for a given $t$, there is a matrix $E$ for which

$$\frac{\| e^{(A+E)t} - e^{At} \|_2}{\| e^{At} \|_2} \approx \nu(A,t) \frac{\| E \|_2}{\| A \|_2} .$$

Thus, if $\nu(A,t)$ is large, small changes in $A$ can induce relatively large changes in $e^{At}$. Unfortunately, it is difficult to characterize precisely those

$A$ for which $\nu(A,t)$ is large. (This is in contrast to the linear equation problem $Ax = b$, where the ill-conditioned $A$ are neatly described in terms of SVD.) One thing we can say, however, is that $\nu(A,t) \geq tA$, with equality holding for all non-negative $t$ if and only if $A$ is normal.

### 11.3.2 A Padé Approximation Method

Following the discussion in §11.2, if $g(z) \approx e^z$ then $g(A) \approx e^A$. A very useful class of approximants for this purpose are the Padé functions defined by

$$R_{pq}(z) \;=\; D_{pq}(z)^{-1}N_{pq}(z),$$

where

$$N_{pq}(z) \;=\; \sum_{k=0}^{p} \frac{(p+q-k)!p!}{(p+q)!k!(p-k)!}z^k$$

and

$$D_{pq}(z) \;=\; \sum_{k=0}^{q} \frac{(p+q-k)!q!}{(p+q)!k!(q-k)!}(-z)^k\,.$$

Notice that $R_{po}(z) = 1+z+\cdots+z^p/p!$ is the $p$th order Taylor polynomial.

Unfortunately, the Padé approximants are good only near the origin, as the following identity reveals:

$$e^A = R_{pq}(A)+\frac{(-1)^q}{(p+q)!}A^{p+q+1}D_{pq}(A)^{-1}\int_0^1 u^p(1-u)^q e^{A(1-u)}du. \quad (11.3.1)$$

However, this problem can be overcome by exploiting the fact that $e^A = (e^{A/m})^m$. In particular, we can scale $A$ by $m$ such that $F_{pq}= R_{pq}(A/m)$ is a suitably accurate approximation to $e^{A/m}$. We then compute $F_{pq}^m$ using Algorithm 11.2.2. If $m$ is a power of two then this amounts to repeated squaring and so is very efficient. The success of the overall procedure depends on the accuracy of the approximant

$$F_{pq} \;=\; \left(R_{pq}\left(\frac{A}{2^j}\right)\right)^{2^j}.$$

In Moler and Van Loan (1978) it is shown that if

$$\frac{\|A\|_\infty}{2^j} \;\leq\; \frac{1}{2}$$

then there exists an $E \in \mathbb{R}^{n\times n}$ such that

$$\begin{aligned}
F_{pq} &= e^{A+E}\\
AE &= EA\\
\|E\|_\infty &\leq \varepsilon(p,q)\|A\|_\infty\\
\varepsilon(p,q) &= 2^{3-(p+q)}\frac{p!q!}{(p+q)!(p+q+1)!}\,.
\end{aligned}$$

These results form the basis of an effective $e^A$ procedure with error control. Using the above formulae it is easy to establish the inequality:

$$\frac{\| e^A - F_{pq} \|_\infty}{\| e^A \|_\infty} \le \epsilon(p,q)\| A \|_\infty e^{\epsilon(p,q)\| A \|_\infty} .$$

The parameters $p$ and $q$ can be determined according to some relative error tolerance. Note that since $F_{pq}$ requires about $j + \max(p,q)$ matrix multiplies it makes sense to set $p = q$ as this choice minimizes $\epsilon(p,q)$ for a given amount of work. Encapsulating these ideas we obtain

**Algorithm 11.3.1** Given $\delta > 0$ and $A \in \mathbb{R}^{n\times n}$, the following algorithm computes $F = e^{A+E}$ where $\| E \|_\infty \le \delta \| A \|_\infty$

$j = \max(0, 1 + \text{floor}(\log_2(\| A \|_\infty)))$
$A = A/2^j$
Let $q$ be the smallest non-negative integer such that $\epsilon(q,q) \le \delta$.
$D = I;\ N = I;\ X = I;\ c = 1$
**for** $k = 1{:}q$
  $c = c(q-k+1)/[(2q-k+1)k]$
  $X = AX;\ N = N + cX;\ D = D + (-1)^k cX$
**end**
Solve $DF = N$ for $F$ using Gaussian elimination.
**for** $k = 1{:}j$
  $F = F^2$
**end**

This algorithm requires about $2(q + j + 1/3)n^3$flops.

The special Horner techniques of §11.2 can be applied to quicken the computation of $D = D_{qq}(A)$ and $N = N_{qq}(A)$. For example, if $q = 8$ we have $N_{qq}(A) = U + AV$ and $D_{qq}(A) = U - AV$ where

$$U = c_0 I + c_2 A^2 + (c_4 I + c_6 A^2 + c_8 A^4)A^4$$

and

$$V = c_1 I + c_3 A^2 + (c_5 I + c_7 A^2)A^4 .$$

Clearly, $N$ and $D$ can be found in 5 matrix multiplies rather than the 7 required by Algorithm 11.3.1.

The roundoff error properties of Algorithm 11.3.1 have essentially been analyzed by Ward (1977). The bounds he derives are computable and incorporated in a Fortran subroutine.

### 11.3.3 Some Stability Issues

There are some interesting questions pertaining to the stability of Algorithm 11.3.1. A potential difficulty arises during the squaring process if $A$ is a matrix whose exponential grows before it decays. If

$$G = R_{qq}\left(\frac{A}{2^j}\right) \approx e^{A/2^j}$$

then it can be shown that rounding errors of order

$$\gamma = \mathbf{u}\| G^2 \|_2 \| G^4 \|_2 \| G^8 \|_2 \cdots \| G^{2^{j-1}} \|_2$$

can be expected to contaminate the computed $G^{2^j}$. If $\| e^{At} \|_2$ has a substantial initial growth, then it may be the case that

$$\gamma \gg \mathbf{u}\| G^{2^j} \|_2 \approx \mathbf{u}\| e^A \|_2$$

thus ruling out the possibility of small relative errors.

If $A$ is normal, then so is the matrix $G$ and therefore $\| G^m \|_2 = \| G \|_2^m$ for all positive integers. Thus, $\gamma \approx \mathbf{u}\| G^{2^j} \|_2 \approx \mathbf{u}\| e^A \|_2$ and so the initial growth problems disappear. The algorithm can essentially be guaranteed to produce small relative error when $A$ is normal. On the other hand, it is more difficult to draw conclusions about the method when $A$ is non-normal because the connection between $\nu(A,t)$ and the initial growth phenomena is unclear. Judging from limited experiments, however, it seems that Algorithm 11.3.1 fails to produce a relatively accurate $e^A$ only when $v(A,1)$ is correspondingly large.

### Problems

**P11.3.1** Show that $e^{(A+B)t} = e^{At}e^{Bt}$ for all $t$ if and only if $AB = BA$. (Hint: Express both sides as a power series in $t$ and compare the coefficient of $t$.)

**P11.3.2** Show that $\lim_{t\to\infty} e^{At} = 0$ if and only if $\alpha(A) < 0$, where $\alpha(A)$ is defined by $\alpha(A) = \max\{\text{Re}(\lambda) : \lambda \in \lambda(A)\}$.

**P11.3.3** Suppose that $A$ is skew-symmetric. Show that both $e^A$ and the (1,1) Padé approximate $R_{11}(A)$ are orthogonal. Are there any other values of $p$ and $q$ for which $R_{pq}(A)$ is orthogonal?

**P11.3.4** Show that if $A$ is nonsingular, then there exists a matrix $X$ such that $A = e^X$. Is $X$ unique?

**P11.3.5** Show that if

$$\exp\left(\begin{bmatrix} -A^T & P \\ 0 & A \end{bmatrix} z\right) = \begin{bmatrix} F_{11} & F_{12} \\ 0 & F_{22} \end{bmatrix} \begin{matrix} n \\ n \end{matrix}$$
$$\qquad\qquad\qquad\qquad\qquad\qquad n \quad\; n$$

then

$$F_{11}^T F_{12} = \int_0^z e^{A^T t} P e^{At} dt.$$

### Notes and References for Sec. 11.3

Much of what appears in this section and an extensive bibliography may be found in the following survey article:

C.B. Moler and C.F. Van Loan (1978). "Nineteen Dubious Ways to Compute the Exponential of a Matrix," *SIAM Review 20,* 801–36.

Scaling and squaring with Padé approximants (Algorithm 11.3.1) and a careful implementation of Parlett's Schur decomposition method (Algorithm 11.1.1) were found to be among the less dubious of the nineteen methods scrutinized. Various aspects of Padé approximation of the matrix exponential are discussed in

W. Fair and Y. Luke (1970). "Padé Approximations to the Operator Exponential," *Numer. Math. 14,* 379–82.
C.F. Van Loan (1977a). "On the Limitation and Application of Padé Approximation to the Matrix Exponential," in *Padé and Rational Approximation,* ed. E.B. Saff and R.S. Varga, Academic Press, New York.
R.C. Ward (1977). "Numerical Computation of the Matrix Exponential with Accuracy Estimate," *SIAM J. Num. Anal. 14,* 600–14.
A. Wragg (1973). "Computation of the Exponential of a Matrix I: Theoretical Considerations," *J. Inst. Math. Applic. 11,* 369–75.
A. Wragg (1975). "Computation of the Exponential of a Matrix II: Practical Considerations," *J. Inst. Math. Applic. 15,* 273–78.

A proof of equation (11.3.1) for the scalar case appears in

R.S. Varga (1961). "On Higher-Order Stable Implicit Methods for Solving Parabolic Partial Differential Equations," *J. Math. Phys. 40,* 220–31.

There are many applications in control theory calling for the computation of the matrix exponential. In the linear optimal regular problem, for example, various integrals involving the matrix exponential are required. See

E.S. Armstrong and A.K. Caglayan (1976). "An Algorithm for the Weighting Matrices in the Sample-Data Optimal Linear Regulator Problem," NASA Technical Note, TN D-8372.
J. Johnson and C.L. Phillips (1971). "An Algorithm for the Computation of the Integral of the State Transition Matrix," *IEEE Trans. Auto. Cont. AC-16,* 204–5.
C.F. Van Loan (1978a). "Computing Integrals Involving the Matrix Exponential," *IEEE Trans. Auto. Cont. AC-16,* 395–404.

An understanding of the map $A \rightarrow \exp(At)$ and its sensitivity is helpful when assessing the performance of algorithms for computing the matrix exponential. Work in this direction includes

B. Kågström (1977). "Bounds and Perturbation Bounds for the Matrix Exponential," *BIT 17,* 39–57.
C.F. Van Loan (1977b). "The Sensitivity of the Matrix Exponential," *SIAM J. Num. Anal. 14,* 971–81.

The computation of a logarithm of a matrix is an important area demanding much more work. These calculations arise in various "system identification" problems. See

B. Singer and S. Spilerman (1976). "The Representation of Social Processes by Markov Models," *Amer. J. Sociology 82,* 1–54.
B.W. Helton (1968). "Logarithms of Matrices," *Proc. Amer. Math. Soc. 19,* 733–36.

# Chapter 12

# Special Topics



In this final chapter we discuss an assortment of problems that represent important applications of the SVD and QR factorization. We first consider least squares minimization with constraints. Two types of constraints are considered in §12.1, quadratic inequality and linear equality. The following two sections are also concerned with variations on the standard LS problem. In §12.2 we consider how $b$ might be approximated by some subset of $A$'s columns, a course of action that is sometimes appropriate when $A$ is rank-deficient. In §12.3 we consider a variation of ordinary regression known as total least squares that has appeal when $A$ is contaminated with error. More applications of the SVD are considered in §12.4, where various subspace calculations are considered. Some variations of the symmetric eigenvalue problem are discussed in §12.5. In §12.6 we investigate the updating of the QR factorization when $A = QR$ it is altered by a rank-one matrix.

## 12.1 Constrained Least Squares

In the least squares setting it is sometimes natural to minimize $\| Ax - b \|_2$ over a proper subset of $\mathbb{R}^n$. For example, we may wish to predict $b$ as best we can with $Ax$ subject to the constraint that $x$ is a unit vector. Or, perhaps the solution defines a fitting function $f(t)$ which is to have prescribed values at a finite number of points. This can lead to an equality constrained least squares problem. In this section we show how these problems can be solved using the QR factorization and the SVD.

### 12.1.1 The Problem LSQI

Least squares minimization with a quadratic inequality constraint—the *LSQI problem*—is a technique that can be used whenever the solution to the ordinary LS problem needs to be *regularized.* A simple LSQI problem that arises when attempting to fit a function to noisy data is

$$\text{minimize } \| Ax - b \|_2 \qquad \text{subject to } \| Bx \|_2 \le \alpha \tag{12.1.1}$$

where $A \in \mathbb{R}^{m \times n}$, $b \in \mathbb{R}^m$, $B \in \mathbb{R}^{n \times n}$ (nonsingular), and $\alpha \ge 0$. The constraint defines a hyperellipsoid in $\mathbb{R}^n$ and is usually chosen to damp out excessive oscillation in the fitting function. This can be done, for example, if $B$ is a discretized second derivative operator.

More generally, we have the problem

$$\text{minimize } \| Ax - b \|_2 \qquad \text{subject to } \| Bx - d \|_2 \le \alpha \tag{12.1.2}$$

where $A \in \mathbb{R}^{m \times n}$ $(m \ge n)$, $b \in \mathbb{R}^m$, $B \in \mathbb{R}^{p \times n}$, $d \in \mathbb{R}^p$, and $\alpha \ge 0$. The generalized singular value decomposition of §8.7.3 sheds light on the solvability of (12.1.2). Indeed, if

$$\begin{aligned} U^T AX &= \operatorname{diag}(\alpha_1, \ldots, \alpha_n) \qquad U^T U = I_m \\ V^T BX &= \operatorname{diag}(\beta_1, \ldots, \beta_q) \qquad V^T V = I_p, \quad q = \min\{p, n\} \end{aligned} \tag{12.1.3}$$

is the generalized singular value decomposition of $A$ and $B$, then (12.1.2) transforms to

$$\text{minimize } \| D_A y - \tilde{b} \|_2 \qquad \text{subject to } \| D_B y - \tilde{d} \|_2 \le \alpha$$

where $\tilde{b} = U^T b$, $\tilde{d} = V^T d$, and $y = X^{-1}x$. The simple form of the objective function

$$\| D_A y - \tilde{b} \|_2^2 = \sum_{i=1}^{n} (\alpha_i y_i - \tilde{b}_i)^2 + \sum_{i=n+1}^{m} \tilde{b}_i^2 \tag{12.1.4}$$

and the constraint equation

$$\| D_B y - \tilde{d} \|_2^2 = \sum_{i=1}^{r} (\beta_i y_i - \tilde{d}_i)^2 + \sum_{i=r+1}^{p} \tilde{d}_i^2 \le \alpha^2 \tag{12.1.5}$$

facilitate the analysis of the LSQI problem. Here, $r = \text{rank}(B)$ and we assume that $\beta_{r+1} = \cdots = \beta_q = 0$.

To begin with, the problem has a solution if and only if

$$\sum_{i=r+1}^{p} \tilde{d}_i^2 \le \alpha^2 .$$

If we have equality in this expression then consideration of (12.1.4) and (12.1.5) shows that the vector defined by

$$y_i = \begin{cases} \tilde{d}_i/\beta_i & i = 1{:}r \\ \tilde{b}_i/\alpha_i & i = r+1{:}n, \alpha_i \neq 0 \\ 0 & i = r+1{:}n, \alpha_i = 0 \end{cases} \tag{12.1.6}$$

solves the LSQI problem. Otherwise

$$\sum_{i=r+1}^{p} \tilde{d}_i^2 < \alpha^2. \tag{12.1.7}$$

and we have more alternatives to pursue. Now the vector $y \in \mathbb{R}^n$, defined by

$$y_i = \begin{cases} \tilde{b}_i/\alpha_i & \alpha_i \neq 0 \\ \tilde{d}_i/\beta_i & \alpha_i = 0 \end{cases} \qquad i = 1{:}n$$

is a minimizer of $\| D_A y - \tilde{b} \|_2$. If this vector is also feasible, then we have a solution to (12.1.2). (This is not necessarily the solution of minimum 2-norm, however.) We therefore assume that

$$\sum_{\substack{i=1 \\ \alpha_i \neq 0}}^{q} \left( \beta_i \frac{\tilde{b}_i}{\alpha_i} - \tilde{d}_i \right)^2 + \sum_{i=q+1}^{p} \tilde{d}_i^2 > \alpha^2. \tag{12.1.8}$$

This implies that the solution to the LSQI problem occurs on the boundary of the feasible set. Thus, our remaining goal is to

$$\text{minimize } \| D_A y - \tilde{b} \|_2 \qquad \text{subject to } \| D_B y - \tilde{d} \|_2 = \alpha .$$

To solve this problem, we use the method of Lagrange multipliers. Defining

$$h(\lambda, y) = \| D_A y - \tilde{b} \|_2^2 + \lambda \left( \| D_B y - \tilde{d} \|_2^2 - \alpha^2 \right)$$

we see that the equations $0 = \partial h / \partial y_i$ , $i = 1{:}n$, lead to the linear system

$$(D_A^T D_A + \lambda D_B^T D_B) y = D_B^T \tilde{b} + \lambda D_B^T \tilde{d}.$$

Assuming that the matrix of coefficients is nonsingular, this has a solution $y(\lambda)$ where

$$y_i(\lambda) = \begin{cases} \dfrac{\alpha_i \tilde{b}_i + \lambda \beta_i \tilde{d}_i}{\alpha_i^2 + \lambda \beta_i^2} & i = 1{:}q \\ \tilde{b}_i / \alpha_i & i = q+1{:}n \end{cases}$$

To determine the Lagrange parameter we define

$$\phi(\lambda) \equiv \| D_B y(\lambda) - \tilde{d} \|_2^2 = \sum_{i=1}^{r} \alpha_i \frac{\beta_i \tilde{b}_i - \alpha_i \tilde{d}_i^2}{\alpha_i^2 + \lambda \beta_i^2} + \sum_{i=r+1}^{p} \tilde{d}_i^2$$

and then seek a solution to $\phi(\lambda) = \alpha^2$. Equations of this type are referred to as *secular equations.* (See §8.6.) From (12.1.8) we see that $\phi(0) > \alpha^2$. Now $\phi(\lambda)$ is monotone decreasing for $\lambda > 0$, and therefore, (12.1.8) implies the existence of a unique positive $\lambda^*$ for which $\phi(\lambda^*) = \alpha^2$. It is easy to show that this is the desired root. It can be found through the application of any standard root-finding technique, such as Newton's method. The solution of the original LSQI problem is then $x = Xy(\lambda^*)$.

### 12.1.2 LS Minimization Over a Sphere

For the important case of minimization over a sphere ($B = I_n$, $d = 0$), we have the following procedure:

**Algorithm 12.1.1** Given $A \in \mathbb{R}^{m \times n}$ with $m \geq n$, $b \in \mathbb{R}^m$, and $\alpha > 0$, the following algorithm computes a vector $x \in \mathbb{R}^n$ such that $\| Ax - b \|_2$ is minimum, subject to the constraint that $\| x \|_2 \leq \alpha$.

Compute the SVD $A = U\Sigma V^T$ and save $V = [\, v_1, \ldots, v_n \,]$.
$b = U^T b$; $r = \text{rank}(A)$
**if** $\displaystyle\sum_{i=1}^{r} \left( \frac{b_i}{\sigma_i} \right)^2 > \alpha^2$

Find $\lambda^*$ such that $\displaystyle\sum_{i=1}^{r} \left( \frac{\sigma_i b_i}{\sigma_i^2 + \lambda^*} \right)^2 = \alpha^2$.

$$x = \sum_{i=1}^{r} \left( \frac{\sigma_i b_i}{\sigma_i^2 + \lambda^*} \right) v_i$$

**else**

$$x = \sum_{i=1}^{r} \left( \frac{b_i}{\sigma_i} \right) v_i$$

**end**

The SVD is the dominant computation in this algorithm.

**Example 12.1.1** The secular equation for the problem

$$\min_{\| x \|_2 = 1} \left\| \begin{bmatrix} 2 & 0 \\ 0 & 1 \\ 0 & 0 \end{bmatrix} \begin{bmatrix} x_1 \\ x_2 \end{bmatrix} - \begin{bmatrix} 4 \\ 2 \\ 3 \end{bmatrix} \right\|_2$$

is given by

$$\left(\frac{8}{\lambda+4}\right)^2 + \left(\frac{2}{\lambda+1}\right)^2 = 1 .$$

For this problem we find $\lambda^* = 4.57132$ and $x = (.93334\ .35898)^T$.

### 12.1.3 Ridge Regression

The problem solved by Algorithm 12.1.1 is equivalent to the Lagrange multiplier problem of determining $\lambda > 0$ such that

$$(A^TA + \lambda I)x = A^Tb \qquad (12.1.9)$$

and $\| x \|_2 = \alpha$. Now (12.1.9) is precisely the normal equation formulation for the *ridge regression* problem

$$\min_x \left\| \begin{bmatrix} A \\ \sqrt{\lambda} I \end{bmatrix} x - \begin{bmatrix} b \\ 0 \end{bmatrix} \right\|_2^2 = \min_x \| Ax - b \|_2^2 + \lambda \| x \|_2^2 .$$

In the general ridge regression problem one has some criteria for selecting the ridge parameter $\lambda$, e.g., $\| x(\lambda) \|_2 = \alpha$ for some given $\alpha$. We describe a $\lambda$-selection procedure that is discussed in Golub, Heath, and Wahba (1979).

Set $D_k = I - e_k e_k^T = \text{diag}(1, \ldots, 1, 0, 1, \ldots, 1) \in \mathbb{R}^{m \times m}$ and let $x_k(\lambda)$ solve

$$\min_x \| D_k(Ax - b) \|_2^2 + \lambda \| x \|_2^2 . \qquad (12.1.10)$$

Thus, $x_k(\lambda)$ is the solution to the ridge regression problem with the $k$th row of $A$ and $k$th component of $b$ deleted, i.e., the $k$th experiment is ignored. Now consider choosing $\lambda$ so as to minimize the *cross-validation weighted square error* $C(\lambda)$ defined by

$$C(\lambda) = \frac{1}{m} \sum_{k=1}^{m} w_k (a_k^T x_k(\lambda) - b_k)^2 .$$

Here, $w_1, \ldots, w_m$ are non-negative weights and $a_k^T$ is the $k$th row of $A$. Noting that

$$\| Ax_k(\lambda) - b \|_2^2 = \| D_k(Ax_k(\lambda) - b) \|_2^2 + (a_k^T x_k(\lambda) - b_k)^2$$

we see that $[a_k^T x_k(\lambda) - b_k]^2$ is the increase in the sum of squares resulting when the $k$th row is "reinstated." Minimizing $C(\lambda)$ is tantamount to choosing $\lambda$ such that the final model is not overly dependent on any one experiment.

A more rigorous analysis can make this statement precise and also suggest a method for minimizing $C(\lambda)$. Assuming that $\lambda > 0$, an algebraic manipulation shows that

$$x_k(\lambda) = x(\lambda) + \frac{a_k^T x(\lambda) - b_k}{1 - z_k^T a_k} z_k \qquad (12.1.11)$$

where $z_k = (A^TA + \lambda I)^{-1} a_k$ and $x(\lambda) = (A^TA + \lambda I)^{-1} A^T b$. Applying $-a_k^T$ to (12.1.11) and then adding $b_k$ to each side of the resulting equation gives

$$b_k - a_k^T x_k(\lambda) = \frac{e_k^T (I - A(A^TA + \lambda I)^{-1} A^T) b}{e_k^T (I - A(A^TA + \lambda I)^{-1} A^T) e_k}. \qquad (12.1.12)$$

Noting that the residual $r = (r_1, \ldots, r_m)^T = b - Ax(\lambda)$ is given by the formula $r = [I - A(A^TA + \lambda I)^{-1} A^T] b$, we see that

$$C(\lambda) = \frac{1}{m} \sum_{k=1}^{m} w_k \left( \frac{r_k}{\partial r_k / \partial b_k} \right)^2 .$$

The quotient $r_k / (\partial r_k / \partial b_k)$ may be regarded as an inverse measure of the "impact" of the $k$th observation $b_k$ on the model. When $\partial r_k / \partial b_k$ is small, this says that the error in the model's prediction of $b_k$ is somewhat independent of $b_k$. The tendency for this to be true is lessened by basing the model on the $\lambda^*$ that minimizes $C(\lambda)$.

The actual determination of $\lambda^*$ is simplified by computing the SVD of $A$. Indeed, if $U^T A V = \mathrm{diag}(\sigma_1, \ldots, \sigma_n)$ with $\sigma_1 \geq \ldots \geq \sigma_n$ and $\tilde{b} = U^T b$, then it can be shown from (12.1.12) that

$$C(\lambda) = \frac{1}{m} \sum_{k=1}^{m} w_k \left[ \frac{\tilde{b}_k - \sum_{j=1}^{r} u_{kj} \tilde{b}_j \left( \frac{\sigma_j^2}{\sigma_j^2 + \lambda} \right)}{1 - \sum_{j=1}^{r} u_{kj}^2 \left( \frac{\sigma_j^2}{\sigma_j^2 + \lambda} \right)} \right]^2 .$$

The minimization of this expression is discussed in Golub, Heath, and Wahba (1979).

### 12.1.4 Equality Constrained Least Squares

We conclude the section by considering the least squares problem with linear equality constraints:

$$\min_{Bx=d} \| Ax - b \|_2 \tag{12.1.13}$$

Here $A \in \mathbb{R}^{m \times n}$, $B \in \mathbb{R}^{p \times n}$, $b \in \mathbb{R}^m$, $d \in \mathbb{R}^p$, and $\text{rank}(B) = p$. We refer to (12.1.13) as the LSE problem. By setting $\alpha = 0$ in (12.1.2) we see that the LSE problem is a special case of the LSQI problem. However, it is simpler to approach the LSE problem directly rather than through Lagrange multipliers.

Assume for clarity that both $A$ and $B$ have full rank. Let

$$Q^T B^T = \begin{bmatrix} R \\ 0 \end{bmatrix} \begin{matrix} p \\ n-p \end{matrix}$$

be the QR factorization of $B^T$ and set

$$AQ = [\, \underset{p}{A_1} \quad \underset{n-p}{A_2} \,] \qquad Q^T x = \begin{bmatrix} y \\ z \end{bmatrix} \begin{matrix} p \\ n-p \end{matrix} .$$

It is clear that with these transformations (12.1.13) becomes

$$\min_{R^T y = d} \| A_1 y + A_2 z - b \|_2 .$$

Thus, $y$ is determined from the constraint equation $R^T y = d$ and the vector $z$ is obtained by solving the unconstrained $LS$ problem

$$\min_{z} \| A_2 z + (b - A_1 y) \|_2 .$$

Combining the above, we see that $x = Q \begin{bmatrix} y \\ z \end{bmatrix}$ solves (12.1.13).

**Algorithm 12.1.2** Suppose $A \in \mathbb{R}^{m \times n}$, $B \in \mathbb{R}^{p \times n}$, $b \in \mathbb{R}^m$, and $d \in \mathbb{R}^p$. If $\text{rank}(A) = n$ and $\text{rank}(B) = p$, then the following algorithm minimizes $\| Ax - b \|_2$ subject to the constraint $Bx = d$ .

$B^T = QR$ ( QR factorization)
Solve $R(1{:}p, 1{:}p)^T y = d$ for $y$.
$A = AQ$
Find $z$ so $\| A(:, p+1{:}n)z - (b - A(:, 1{:}p)y) \|_2$ is minimized.
$x = Q(:, 1{:}p)y + Q(:, p+1{:}n)z$

Note that this approach to the LSE problem involves two factorizations and a matrix multiplication.

### 12.1.5 The Method of Weighting

An interesting way to obtain an approximate solution to (12.1.13) is to solve the unconstrained LS problem

$$\min_x \left\| \begin{bmatrix} A \\ \lambda B \end{bmatrix} x - \begin{bmatrix} b \\ \lambda d \end{bmatrix} \right\|_2 \tag{12.1.14}$$

for large $\lambda$. The generalized singular value decomposition of §8.7.3 sheds light on the quality of the approximation. Let

$$\begin{aligned} U^T A X &= \text{diag}(\alpha_1, \ldots, \alpha_n) = D_A \in \mathbb{R}^{m \times n} \\ V^T B X &= \text{diag}(\beta_1, \ldots, \beta_p) = D_B \in \mathbb{R}^{p \times n} \end{aligned}$$

be the GSVD of $(A, B)$ and assume that both matrices have full rank for clarity. If $U = [\, u_1, \ldots, u_m \,]$, $V = [\, v_1, \ldots, v_p \,]$, and $X = [\, x_1, \ldots, x_n \,]$, then it is easy to show that

$$x = \sum_{i=1}^{p} \frac{v_i^T d}{\beta_i} x_i + \sum_{i=p+1}^{n} \frac{u_i^T b}{\alpha_i} x_i \tag{12.1.15}$$

is the exact solution to (12.1.13), while

$$x(\lambda) = \sum_{i=1}^{p} \frac{\alpha_i u_i^T b + \lambda^2 \beta_i^2 v_i^T d}{\alpha_i^2 + \lambda^2 \beta_i^2} x_i + \sum_{i=p+1}^{n} \frac{u_i^T b}{\alpha_i} x_i \tag{12.1.16}$$

solves (12.1.14). Since

$$x(\lambda) - x = \sum_{i=1}^{p} \frac{\alpha_i (\beta_i u_i^T b - \alpha_i v_i^T d)}{\beta_i (\alpha_i^2 + \lambda^2 \beta_i^2)} x_i \tag{12.1.17}$$

it follows that $x(\lambda) \rightarrow x$ as $\lambda \rightarrow \infty$.

The appeal of this approach to the LSE problem is that no special subroutines are required: an ordinary LS solver will do. However, for large values of $\lambda$ numerical problems can arise and it is necessary to take precautions. See Powell and Reid (1968) and Van Loan (1982a).

**Example 12.1.2** The problem

$$\min_{x_1 = x_2} \left\| \begin{bmatrix} 1 & 2 \\ 3 & 4 \\ 5 & 6 \end{bmatrix} \begin{bmatrix} x_1 \\ x_2 \end{bmatrix} - \begin{bmatrix} 7 \\ 1 \\ 3 \end{bmatrix} \right\|_2$$

has solution $x = (.3407821,\ .3407821)^T$. This can be approximated by solving

$$\min \left\| \begin{bmatrix} 1 & 2 \\ 3 & 4 \\ 5 & 6 \\ 1000 & -1000 \end{bmatrix} \begin{bmatrix} x_1 \\ x_2 \end{bmatrix} - \begin{bmatrix} 7 \\ 1 \\ 3 \\ 0 \end{bmatrix} \right\|_2$$

which has solution $x = (.3407810\ .3407829)^T$.

**Problems**

**P 12.1.1** (a) Show that if null$(A) \cap$ null$(B) \neq \{0\}$, then (12.1.2) cannot have a unique solution. (b) Give an example which shows that the converse is not true. (Hint: $A^+b$ feasible.)

**P12.1.2** Let $p_0(x), \ldots, p_n(x)$ be given polynomials and $(x_0, y_0), \ldots, (x_m, y_m)$ a given set of coordinate pairs with $x_i \in [a, b]$. It is desired to find a polynomial $p(x) = \sum_{k=0}^{n} a_k p_k(x)$ such that $\sum_{i=0}^{m}(p(x_i) - y_i)^2$ is minimizes subject to the constraint that

$$\int_a^b [p''(x)]^2 dx \;\approx\; h \sum_{i=0}^{N} \left(\frac{p(z_{i-1}) - 2p(z_i) + p(z_{i+1})}{h^2}\right)^2 \;\leq\; \alpha^2$$

where $z_i = a + ih$ and $b = a + Nh$. Show that this leads to an LSQI problem of the form (12.1.1).

**P12.1.3** Suppose $Y = [\, y_1, \ldots, y_k \,] \in \mathbb{R}^{m \times k}$ has the property that

$$Y^T Y \;=\; \text{diag}(d_1^2, \ldots, d_k^2) \qquad d_1 \geq d_2 \geq \cdots \geq d_k > 0\,.$$

Show that if $Y = QR$ is the QR factorization of $Y$, then $R$ is diagonal with $|r_{ii}| = d_i$.

**P12.1.4** (a) Show that if $(A^TA + \lambda I)x = A^Tb$, $\lambda > 0$, and $\|\, x \,\|_2 = \alpha$, then $z = (Ax - b)/\lambda$ solves the *dual equations* $(AA^T + \lambda I)z = -b$ with $\|\, A^Tz \,\|_2 = \alpha$. (b) Show that if $(AA^T + \lambda I)z = -b$, $\|\, A^Tz \,\|_2 = \alpha$, then $x = -A^Tz$ satisfies $(A^TA + \lambda I)x = A^Tb$, $\|\, x \,\|_2 = \alpha$.

**P12.1.5** Suppose $A$ is the $m$-by-1 matrix of ones and let $b \in \mathbb{R}^m$. Show that the cross-validation technique with unit weights prescribes an optimal $\lambda$ given by

$$\lambda \;=\; \left(\left(\frac{\bar{b}}{s}\right)^2 - \frac{1}{m}\right)^{-1}$$

where $\bar{b}^T = (b_1 + \cdots + b_m)/m$ and $s = \sum_{i=1}^{m}(b_i - \bar{b})^2/(m-1)$.

**P12.1.6** Establish equations (12.1.15) through (12.1.17).

**P12.1.7** Develop an SVD version of Algorithm 12.1.2 that can handle rank deficiency in $A$ and $B$.

**P12.1.8** Suppose

$$A \;=\; \begin{bmatrix} A_1 \\ A_2 \end{bmatrix}$$

where $A_1 \in \mathbb{R}^{n \times n}$ is nonsingular and $A_2 \in \mathbb{R}^{(m-n) \times n}$. Show that

$$\sigma_{min}(A) \;\geq\; \sqrt{1 + \sigma_{min}(A_2A_1^{-1})^2}\;\sigma_{min}(A_1)\,.$$

**P12.1.9** Consider the problem

$$\min_{\substack{x^TBx = \beta^2 \\ x^TCx = \gamma^2}} \;\|\, Ax - b \,\|_2 \qquad A \in \mathbb{R}^{m \times n},\; b \in \mathbb{R}^m,\; B, C \in \mathbb{R}^{n \times n}$$

Assume that $B$ and $C$ are positive definite and that $Z \in \mathbb{R}^{n \times n}$ is a nonsingular matrix with the property that $Z^TBZ = \text{diag}(\lambda_1, \ldots, \lambda_n)$ and $Z^TCZ = I_n$. Assume that

$\lambda_1 \geq \cdots \geq \lambda_n$. (a) Show that the the set of feasible $x$ is empty unless $\lambda_n \leq \beta^2/\gamma^2 \leq \lambda_1$. (b) Using $Z$, show how the two constraint problem can be converted to a single constraint problem of the form

$$\min_{y^T W y = \beta^2 - \lambda_n \gamma^2} \| \tilde{A}x - b \|_2$$

where $W = \text{diag}(\lambda_1, \ldots, \lambda_n) - \lambda_n I$.

**Notes and References for Sec. 12.1**

Roughly speaking, regularization is a technique for transforming a poorly conditioned problem into a stable one. Quadratically constrained least squares is an important example. See

L. Eldèn (1977a). "Algorithms for the Regularization of Ill-Conditioned Least Squares Problems," *BIT 17,* 134–45.
L. Eldèn (1977b). "Numerical Analysis of Regularization and Constrained Least Square Methods," Ph.D. thesis, Linkoping Studies in Science and Technology, Dissertion no. 20, Linkoping, Sweden.

References for cross-validation include

L. Eldèn (1985). "A Note on the Computation of the Generalized Cross-Validation Function for Ill-Conditioned Least Squares Problems," *BIT 24,* 467–472.
G.H. Golub, M. Heath, and G. Wahba (1979). "Generalized Cross-Validation as a Method for Choosing a Good Ridge Parameter," *Technometrics 21,* 215–23.

The LSQI problem is discussed in

L. Eldèn (1980). "Perturbation Theory for the Least Squares Problem with Linear Equality Constraints," *SIAM J. Num. Anal. 17,* 338–50.
L. Eldèn (1983). "A Weighted Pseudoinverse, Generalized Singular Values, and Constrained Least Squares Problems," *BIT 22* , 487–502.
G.E. Forsythe and G.H. Golub (1965). "On the Stationary Values of a Second-Degree Polynomial on the Unit Sphere," *SIAM J. App. Math. 14,* 1050–68.
W. Gander (1981). "Least Squares with a Quadratic Constraint," *Numer. Math. 36,* 291–307.
G.W. Stewart (1984). "On the Asymptotic Behavior of Scaled Singular Value and QR Decompositions," *Math. Comp. 43,* 483–490.

Other computational aspects of the LSQI problem involve updating, the handling of banded and sparse problems. See

Å. Björck (1984). "A General Updating Algorithm for Constrained Linear Least Squares Problems," *SIAM J. Sci. and Stat. Comp. 5,* 394–402.
L. Eldèn (1984). "An Algorithm for the Regularization of Ill-Conditioned, Banded Least Squares Problems," *SIAM J. Sci. and Stat. Comp. 5,* 237–254.
D.P. O'Leary and J.A. Simmons (1981). "A Bidiagonalization-Regularization Procedure for Large Scale Discretizations of Ill-Posed Problems," *SIAM J. Sci. and Stat. Comp. 2,* 474–89.
K. Schittkowski and J. Stoer (1979). "A Factorization Method for the Solution of Constrained Linear Least Squares Problems Allowing for Subsequent Data changes," *Numer. Math. 31,* 431–63.

The LSE problem is discussed in Lawson and Hansen (SLE, chapter 22). The problems associated with the method of weights are analyzed in

J.L. Barlow, N.K. Nichols, and R.J. Plemmons (1988). "Iterative Methods for Equality Constrained Least Squares Problems," SIAM J. Sci. and Stat. Comp. 9, 892–906.
M.J.D. Powell and J.K. Reid (1968). "On Applying Householder's Method to Linear Least Squares Problems," *Proc. IFIP Congress*, pp. 122–26.
C. Van Loan (1985). "On the Method of Weighting for Equality Constrained Least Squares Problems," *SIAM J. Numer. Anal. 22*, 851–864.

GSVD is also useful in the study of generalized least squares problems. See

C.C. Paige (1985). "The General Limit Model and the Generalized Singular Value Decomposition," *Lin. Alg. and Its Applic. 70*, 269–284.

## 12.2 Subset Selection Using the SVD

As described in §5.5, the rank-deficient LS problem min $\| Ax - b \|_2$ can be approached by approximating the minimum norm solution

$$x_{LS} = \sum_{i=1}^{r} \frac{u_i^T b}{\sigma_i} v_i \qquad r = \text{rank}(A)$$

with

$$x_{\tilde{r}} = \sum_{i=1}^{\tilde{r}} \frac{u_i^T b}{\sigma_i} v_i \qquad \tilde{r} \le r$$

where

$$A = U\Sigma V^T = \sum_{i=1}^{r} \sigma_i u_i v_i^T \tag{12.2.1}$$

is the SVD of $A$ and $\tilde{r}$ is some numerically determined estimate of $r$. Note that $x_{\tilde{r}}$ minimizes $\| A_{\tilde{r}} x - b \|_2$ where

$$A_{\tilde{r}} = \sum_{i=1}^{\tilde{r}} \sigma_i u_i v_i^T$$

is the closest matrix to $A$ that has rank $\tilde{r}$. See Theorem 2.5.2.

Replacing $A$ by $A_{\tilde{r}}$ in the LS problem amounts to filtering the small singular values and can make a great deal of sense in those situations where $A$ is derived from noisy data. In other applications, however, rank deficiency implies redundancy among the factors that comprise the underlying model. In this case, the model-builder may not be interested in a predictor such as $A_{\tilde{r}} x_{\tilde{r}}$ that involves all $n$ redundant factors. Instead, a predictor $Ay$ may be sought where $y$ has at most $\tilde{r}$ nonzero components. The position of the nonzero entries determines which columns of $A$, i.e., which factors in the model, are to be used in approximating the observation vector $b$. How to pick these columns is the problem of *subset selection* and is the subject of this section.

### 12.2.1 QR with Column Pivoting

QR with column pivoting can be regarded as a method for selecting an independent subset of $A$'s columns from which $b$ might be predicted. Suppose we apply Algorithm 5.4.1 to $A \in \mathbb{R}^{m\times n}$ and compute an orthogonal $Q$ and a permutation $\Pi$ such that $R = Q^TA\Pi$ is upper triangular. If $R(1{:}\tilde{r}, 1{:}\tilde{r})z = \tilde{b}(1{:}\tilde{r})$ where $\tilde{b} = Q^Tb$ and we set

$$y = \Pi \begin{bmatrix} z \\ 0 \end{bmatrix},$$

then $Ay$ is an approximate LS predictor of $b$ that involves the first $\tilde{r}$ columns of $A\Pi$ .

### 12.2.2 Using the SVD

Although QR with column pivoting is a fairly reliable way to handle near rank deficiency, the SVD is sometimes preferable for reasons discussed in §5.5. We therefore describe an SVD-based subset selection procedure due to Golub, Klema, and Stewart (1976) that proceeds as follows:

- Compute the SVD $A = U\Sigma V^T$ and use it to determine a rank estimate $\tilde{r}$.
- Calculate a permutation matrix $P$ such that the columns of the matrix $B_1 \in \mathbb{R}^{m\times\tilde{r}}$ in $AP = [\, B_1 \; B_2 \,]$ are "sufficiently independent."
- Predict $b$ with the vector $Ay$ where $y = P\begin{bmatrix} z \\ 0 \end{bmatrix}$ and $z \in \mathbb{R}^{\tilde{r}}$ minimizes $\| B_1z - b \|_2$

The second step is key. Since

$$\min_{z \in \mathbb{R}^{\tilde{r}}} \| B_1z - b \|_2 = \| Ay - b \|_2 \geq \min_{x \in \mathbb{R}^n} \| Ax - b \|_2$$

it can be argued that the permutation $P$ should be chosen to make the residual $(I - B_1B_1^+)b$ as small as possible. Unfortunately, such a solution procedure can be unstable. For example, if

$$A = \begin{bmatrix} 1 & 1 & 0 \\ 1 & 1+\epsilon & 1 \\ 0 & 0 & 1 \end{bmatrix}, \qquad b = \begin{bmatrix} 1 \\ -1 \\ 0 \end{bmatrix},$$

$\tilde{r} = 2$, and $P = I$, then $\min \| B_1z - b \|_2 = 0$, but $\| B_1^+b \|_2 = O(1/\epsilon)$. On the other hand, any proper subset involving the third column of $A$ is strongly independent but renders a much worse residual.

This example shows that there can be a trade-off between the independence of the chosen columns and the norm of the residual that they render. How to proceed in the face of this trade-off requires additional mathematical machinery in the form of useful bounds on $\sigma_{\tilde{r}}(B_1)$, the smallest singular value of $B_1$.

**Theorem 12.2.1** *Let the SVD of $A \in \mathbb{R}^{m\times n}$ be given by (12.2.1), and define the matrix $B_1 \in \mathbb{R}^{m\times \tilde{r}}$, $\tilde{r} \leq \text{rank}(A)$, by*

$$AP = \begin{bmatrix} \underset{\tilde{r}}{B_1} & \underset{n-\tilde{r}}{B_2} \end{bmatrix}$$

*where $P \in \mathbb{R}^{n\times n}$ is a permutation. If*

$$P^T V = \begin{bmatrix} \tilde{V}_{11} & \tilde{V}_{12} \\ \tilde{V}_{21} & \tilde{V}_{22} \end{bmatrix} \begin{matrix} \tilde{r} \\ n-\tilde{r} \end{matrix} \qquad (12.2.2)$$
$$\quad \tilde{r} \quad n-\tilde{r}$$

*and $\tilde{V}_{11}$ is nonsingular, then*

$$\frac{\sigma_{\tilde{r}}(A)}{\| \tilde{V}_{11}^{-1} \|_2} \leq \sigma_{\tilde{r}}(B_1) \leq \sigma_{\tilde{r}}(A) .$$

**Proof.** The upper bound follows from the minimax characterization of singular values given in §8.3.1

To establish the lower bound, partition the diagonal matrix of singular values as follows:

$$\Sigma = \begin{bmatrix} \Sigma_1 & 0 \\ 0 & \Sigma_2 \end{bmatrix} \begin{matrix} \tilde{r} \\ m-\tilde{r} \end{matrix} .$$
$$\quad \tilde{r} \quad n-\tilde{r}$$

If $w \in \mathbb{R}^{\tilde{r}}$ is a unit vector with the property that $\| B_1 w \|_2 = \sigma_{\tilde{r}}(B_1)$, then

$$\begin{aligned} \sigma_{\tilde{r}}(B_1)^2 &= \| B_1 w \|_2^2 = \left\| U\Sigma V^T P \begin{bmatrix} w \\ 0 \end{bmatrix} \right\|_2^2 \\ &= \| \Sigma_1 \tilde{V}_{11}^T w \|_2^2 + \| \Sigma_2 \tilde{V}_{12}^T w \|_2^2 . \end{aligned}$$

The theorem now follows because $\| \Sigma_1 \tilde{V}_{11}^T w \|_2 \geq \sigma_{\tilde{r}}(A)/\| \tilde{V}_{11}^{-1} \|_2$. □

This result suggests that in the interest of obtaining a sufficiently independent subset of columns, we choose the permutation $P$ such that the resulting $\tilde{V}_{11}$ submatrix is as well-conditioned as possible. A heuristic solution to this problem can be obtained by computing the QR with column-pivoting factorization of the matrix $[\, V_{11}^T \; V_{21}^T \,]$, where

$$V = \begin{bmatrix} V_{11} & V_{12} \\ V_{21} & V_{22} \end{bmatrix} \begin{matrix} \tilde{r} \\ n-\tilde{r} \end{matrix} \\ \quad\quad\quad \begin{matrix} \tilde{r} & n-\tilde{r} \end{matrix}$$

is a partitioning of the matrix $V$ in (12.2.1). In particular, if we apply QR with column pivoting (Algorithm 5.4.1) to compute

$$Q^T[\, V_{11}^T \; V_{21}^T \,]P = [\, \underset{\tilde{r}}{R_{11}} \quad \underset{n-\tilde{r}}{R_{12}} \,]$$

where $Q$ is orthogonal, $P$ is a permutation matrix, and $R_{11}$ is upper triangular, then (12.2.2) implies:

$$\begin{bmatrix} \tilde{V}_{11} \\ \tilde{V}_{21} \end{bmatrix} = P^T \begin{bmatrix} V_{11} \\ V_{21} \end{bmatrix} = \begin{bmatrix} R_{11}^T Q^T \\ R_{12}^T Q^T \end{bmatrix}.$$

Note that $R_{11}$ is nonsingular and that $\| \tilde{V}_{11}^{-1} \|_2 = \| R_{11}^{-1} \|_2$. Heuristically, column pivoting tends to produce a well-conditioned $R_{11}$, and so the overall process tends to produce a well-conditoned $\tilde{V}_{11}$. Thus we obtain

**Algorithm 12.2.1** Given $A \in \mathbb{R}^{m \times n}$ and $b \in \mathbb{R}^m$ the following algorithm computes a permutation $P$, a rank estimate $\tilde{r}$, and a vector $z \in \mathbb{R}^{\tilde{r}}$ such that the first $\tilde{r}$ columns of $B = AP$ are independent and such that $\| B(:, 1{:}\tilde{r})z - b \|_2$ is minimized.

Compute the SVD $U^T AV = \text{diag}(\sigma_1, \ldots, \sigma_n)$ and save $V$.
Determine $\tilde{r} \le \text{rank}(A)$.
Apply QR with column pivoting: $Q^T V(:, 1{:}\tilde{r})^T P = [\, R_{11} \; R_{12} \,]$ and set
  $AP = [\, B_1 \; B_2 \,]$ with $B_1 \in \mathbb{R}^{m \times \tilde{r}}$ and $B_2 \in \mathbb{R}^{m \times (n-\tilde{r})}$.
Determine $z \in \mathbb{R}^{\tilde{r}}$ such that $\| b - B_1 z \|_2 = \min$.

**Example 12.2.1** Let

$$A = \begin{bmatrix} 3 & 4 & 1 \\ 7 & 4 & -3 \\ 2 & 5 & 3 \\ -1 & 4 & 5 \end{bmatrix}, \qquad b = \begin{bmatrix} 1 \\ 1 \\ 1 \\ 1 \end{bmatrix},$$

$\text{rank}(A) = 2$, and $x_{LS} = [\, .0815 \; .1545 \; .0730 \,]^T$. If Algorithm 12.2.1 is applied, we find $P = [\, e_2 \; e_1 \; e_3 \,]$ and solution $x = [\, .0845 \; .2275 \; .0000 \,]^T$. Note: $\| b - Ax_{LS} \|_2 \approx \| b - Ax \|_2 = .1966$.

### 12.2.3 More on Column Independence vs. Residual

We return to the discussion of the trade-off between column independence and norm of the residual. In particular, to assess the above method of subset selection, we need to examine the residual of the vector $y$ that it produces $r_y = b - Ay = b - B_1 z = (I - B_1 B_1^+)b$. Here, $B_1 = B(:, 1{:}\tilde{r})$ with $B = AP$. To this end, it is appropriate to compare $r_y$ with $r_{x_{\tilde{r}}} = b - Ax_{\tilde{r}}$ since we are regarding $A$ as a rank-$\tilde{r}$ matrix and since $x_{\tilde{r}}$ solves the nearest rank-$\tilde{r}$ LS problem, namely, min $\| A_{\tilde{r}}x - b \|_2$.

**Theorem 12.2.2** *If $r_y$ and $r_{x_{\tilde{r}}}$ are defined as above and if $\tilde{V}_{11}$ is the leading r-by-r principal submatrix of $P^T V$, then*

$$\| r_{x_{\tilde{r}}} - r_y \|_2 \;\leq\; \frac{\sigma_{\tilde{r}+1}(A)}{\sigma_{\tilde{r}}(A)} \| \tilde{V}_{11}^{-1} \|_2 \| b \|_2$$

**Proof.** Note that $r_{x_{\tilde{r}}} = (I - U_1 U_1^T)b$ and $r_y = (I - Q_1 Q_1^T)b$ where

$$U = [\; \underset{\tilde{r}}{U_1} \quad \underset{m-\tilde{r}}{U_2} \;]$$

is a partitioning of the matrix $U$ in (12.2.1) and where $Q_1 = B_1(B_1^T B_1)^{-1/2}$. From §2.6.x we obtain

$$\| r_{x_{\tilde{r}}} - r_y \|_2 \;\leq\; \| U_1 U_1^T - Q_1 Q_1^T \|_2 \, \| b \|_2 \;=\; \| U_2^T Q_1 \|_2 \, \| b \|_2$$

while Theorem 12.2.1 permits us to conclude that

$$\begin{aligned} \| U_2^T Q_1 \|_2 \;&\leq\; \| U_2^T B_1 \|_2 \| (B_1^T B_1)^{-1/2} \|_2 \;\leq\; \sigma_{\tilde{r}+1}(A)\frac{1}{\sigma_{\tilde{r}}(B_1)} \\ &\leq\; \frac{\sigma_{\tilde{r}+1}(A)}{\sigma_{\tilde{r}}(A)} \| \tilde{V}_{11}^{-1} \|_2 . \;\Box \end{aligned}$$

Noting that $\| r_{x_{\tilde{r}}} - r_y \|_2 \;=\; \| B_1 y - \sum_{i=1}^{r} (u_i^T b) u_i \|_2$ we see that Theorem 12.2.2 sheds light on how well $B_1 y$ can predict the "stable" component of $b$, i.e., $U_1^T b$. Any attempt to approximate $U_2^T b$ can lead to a large norm solution. Moreover, the theorem says that if $\sigma_{\tilde{r}+1}(A) \ll \sigma_{\tilde{r}}(A)$, then any reasonably independent subset of columns produces essentially the same-sized residual. On the other hand, if there is no well-defined gap in the singular values, then the determination of $\tilde{r}$ becomes difficult and the entire subset selection problem more complicated.

**Problems**

**P12.2.1** Suppose $A \in \mathbb{R}^{m \times n}$ and that $\| u^T A \|_2 = \sigma$ with $u^T u = 1$. Show that if

$u^T(Ax - b) = 0$ for $x \in \mathbb{R}^n$ and $b \in \mathbb{R}^m$, then $\| x \|_2 \geq |u^T b|/\sigma$.

**P12.2.2** Show that if $B_1 \in \mathbb{R}^{m\times k}$ is comprised of $k$ columns from $A \in \mathbb{R}^{m\times n}$ then $\sigma_k(B_1) \leq \sigma_k(A)$.

**P12.2.3** In equation (12.2.2) we know that the matrix

$$P^T V = \begin{bmatrix} \tilde{V}_{11} & \tilde{V}_{12} \\ \tilde{V}_{21} & \tilde{V}_{22} \end{bmatrix} \begin{matrix} \tilde{r} \\ n-\tilde{r} \end{matrix}$$
$$\qquad\qquad \tilde{r} \quad n-\tilde{r}$$

is orthogonal. Thus, $\| \tilde{V}_{11}^{-1} \|_2 = \| \tilde{V}_{22}^{-1} \|_2$ from the CS decomposition (Theorem 2.4.1). Show how to compute $P$ by applying the QR with column pivoting algorithm to $[\tilde{V}_{22}^T \; \tilde{V}_{12}^T]$. (For $\tilde{r} > n/2$, this procedure would be more economical than the technqiue discussed in the text.) Incorporate this observation in Algorithm 12.2.1.

**Notes and References for Sec. 12.2**

The material in this section is derived from

G.H. Golub, V. Klema and G.W. Stewart (1976). "Rank Degeneracy and Least Squares Problems," Technical Report TR-456, Department of Computer Science, University of Maryland, College Park, MD.

A subset selection procedure based upon the total least squares fitting technique of §12.3 is given in

S. Van Huffel and J. Vandewalle (1987). "Subset Selection Using the Total Least Squares Approach in Collinearity Problems with Errors in the Variables," *Lin. Alg. and Its Applic. 88/89*, 695–714.

The literature on subset selection is vast and we refer the reader to

H. Hotelling (1957). "The Relations of the Newer Multivariate Statistical Methods to Factor Analysis," *Brit. J. Stat. Psych. 10*, 69–79.

## 12.3 Total Least Squares

The problem of minimizing $\| D(Ax - b) \|_2$ where $A \in \mathbb{R}^{m\times n}$, and $D = \text{diag}(d_1, \ldots, d_m)$ is nonsingular can be recast as follows:

$$\min_{b+r \,\in\, \text{range}(A)} \| Dr \|_2 \qquad r \in \mathbb{R}^m . \tag{12.3.1}$$

In this problem, there is a tacit assumption that the errors are confined to the "observation" $b$. When error is also present in the "data" $A$, then it may be more natural to consider the problem

$$\min_{b+r \,\in\, \text{range}(A+E)} \| D[\, E \; r\,]T \|_F \qquad E \in \mathbb{R}^{m\times n},\; r \in \mathbb{R}^m \tag{12.3.2}$$

where $D = \text{diag}(d_1, \ldots, d_m)$ and $T = \text{diag}(t_1, \ldots, t_{n+1})$ are nonsingular. This problem, discussed in Golub and Van Loan (1980), is referred to as the *total least squares* (TLS) problem.

If a minimizing $[\,E_0\; r_0\,]$ can be found for (12.3.2) then any $x$ satisfying $(A+E_0)x = b+r_0$ is called a TLS solution. It should be realized however, that (12.3.2) may fail to have a solution altogether. For example, if

$$A = \begin{bmatrix} 1 & 0 \\ 0 & 0 \\ 0 & 0 \end{bmatrix}, \quad b = \begin{bmatrix} 1 \\ 1 \\ 1 \end{bmatrix}, \quad D = I_3,\; T = I_3, \text{ and } E_\epsilon = \begin{bmatrix} 0 & 0 \\ 0 & \epsilon \\ 0 & \epsilon \end{bmatrix}$$

then for all $\epsilon > 0$, $b \in \text{range}(A + E_\epsilon)$. However, there is no smallest value of $\|\,[\,E\; r\,]\,\|_F$ for which $b + r \in \text{range}(A + E)$.

A generalization of (12.3.2) results if we allow multiple right-hand sides. In particular, if $B \in \mathbb{R}^{m \times k}$ then we have the problem

$$\min_{\text{range}(B+R)\,\subseteq\,\text{range}(A+E)} \|\, D[\,E\; R\,]T\,\|_F \qquad (12.3.3)$$

where $E \in \mathbb{R}^{m\times n}$ and $R \in \mathbb{R}^{m\times k}$ and the matrices $D = \text{diag}(d_1,\ldots,d_m)$ and $T = \text{diag}(t_1,\ldots,t_{n+k})$ are nonsingular. If $[E_0\; R_0]$ solves (12.3.3), then any $X \in \mathbb{R}^{n\times k}$ that satisfies $(A+E_0)X = (B+R_0)$ is said to be a TLS solution to (12.3.3). In this section we discuss some of the mathematical properties of the total least squares problem and show how it can be solved using the SVD.

## 12.3.1 Mathematical Background

The following theorem gives conditions for the uniqueness and existence of a TLS solution to the multiple right-hand side problem.

**Theorem 12.3.1** *Let $A$, $B$, $D$, and $T$ be as above and assume $m \geq n+k$. Let*

$$C = D[\,A\; B\,]T = [\,\underset{n}{C_1}\quad \underset{k}{C_2}\,]$$

*have SVD $U^TCV = \text{diag}(\sigma_1,\ldots,\sigma_{n+k}) = \Sigma$ where $U$, $V$, and $\Sigma$ are partitioned as follows:*

$$U = [\,\underset{n}{U_1}\quad \underset{k}{U_2}\,] \qquad V = \begin{bmatrix} V_{11} & V_{12} \\ V_{21} & V_{22} \end{bmatrix}\begin{matrix} n \\ k \end{matrix} \quad (\text{columns } n,\ k)$$

$$\Sigma = \begin{bmatrix} \Sigma_1 & 0 \\ 0 & \Sigma_2 \end{bmatrix}\begin{matrix} n \\ k \end{matrix} \quad (\text{columns } n,\ k)$$

*If $\sigma_n(C_1) > \sigma_{n+1}(C)$ then the matrix $[\,E_0\; R_0\,]$ defined by*

$$D[\,E_0\; R_0\,]T = -U_2\Sigma_2[\,V_{12}^T\; V_{22}^T\,] \qquad (12.3.4)$$

*solves (12.3.3). If* $T_1 = \text{diag}(t_1, \ldots, t_n)$ *and* $T_2 = \text{diag}(t_{n+1}, \ldots, t_{n+k})$ *then the matrix*

$$X_{TLS} = -T_1 V_{12} V_{22}^{-1} T_2^{-1}$$

*exists and is the unique solution to* $(A + E_0)X = B + R_0$.

**Proof.** We first establish two results that follow from the assumption $\sigma_n(C_1) > \sigma_{n+1}(C)$. From the equation $CV = U\Sigma$ we have $C_1 V_{12} + C_2 V_{22} = U_2 \Sigma_2$. We wish to show that $V_{22}$ is nonsingular. Suppose $V_{22}x = 0$ for some unit 2-norm $x$. It follows from $V_{12}^T V_{12} + V_{22}^T V_{22} = I$ that $\| V_{12}x \|_2 = 1$. But then

$$\sigma_{n+1}(C) \geq \| U_2 \Sigma_2 x \|_2 = \| C_1 V_{12} x \|_2 \geq \sigma_n(C_1),$$

a contradiction. Thus, the submatix $V_{22}$ is nonsingular.

The other fact that follows from $\sigma_n(C_1) > \sigma_{n+1}(C)$ concerns the strict separation of $\sigma_n(C)$ and $\sigma_{n+1}(C)$. From Corollary 8.3.3, we have $\sigma_n(C) \geq \sigma_n(C_1)$ and so $\sigma_n(C) \geq \sigma_n(C_1) > \sigma_{n+1}(C)$.

Now we are set to prove the theorem. If $\text{range}(B + R) \subset \text{range}(A + E)$, then there is an $X$ ($n$-by-$k$) so $(A + E)X = B + R$, i.e.,

$$\{ D[\, A\ B\,]T + D[\, E\ R\,]T \} T^{-1} \begin{bmatrix} X \\ -I_k \end{bmatrix} = 0. \tag{12.3.5}$$

Thus, the matrix in curly brackets has, at most, rank $n$. By following the argument in Theorem 2.5.2, it can be shown that

$$\| D[\, E\ R\,]T \|_F \geq \sum_{i=n+1}^{n+k} \sigma_i(C)^2$$

and that the lower bound is realized by setting $[\, E\ R\,] = [\, E_0\ R_0\,]$. The inequality $\sigma_n(C) > \sigma_{n+1}(C)$ ensures that $[\, E_0\ R_0\,]$ is the unique minimizer. The null space of

$$\{D[\, A\ B\,]T + D[E_0 R_0]T\} = U_1 \Sigma_1 [\, V_{11}^T\ V_{21}^T\,]$$

is the range of $\begin{bmatrix} V_{12} \\ V_{22} \end{bmatrix}$. Thus, from (12.3.5)

$$T^{-1} \begin{bmatrix} X \\ -I_k \end{bmatrix} = \begin{bmatrix} V_{12} \\ V_{22} \end{bmatrix} S$$

for some $k$-by-$k$ matrix $S$. From the equations $T_1^{-1}X = V_{12}S$ and $-T_2^{-1} = V_{22}S$ we see that $S = -V_{22}^{-1}T_2^{-1}$. Thus, we must have

$$X = T_1 V_{12} S = -T_1 V_{12} V_{22}^{-1} T_2^{-1} = X_{TLS}. \square$$

If $\sigma_n(C) = \sigma_{n+1}(C)$ then the TLS problem may still have a solution, although it may not be unique. In this case, it may be desirable to single out a "minimal norm" solution. To this end, consider the $\tau$-norm defined on $\mathbb{R}^{n\times k}$ by $\| Z \|_\tau = \| T_1^{-1} Z T_2 \|_2$. If $X$ is given by (12.3.5), then from Theorem 2.4.1 we have

$$\| X \|_\tau^2 \;=\; \| V_{12}V_{22}^{-1} \|_2^2 \;=\; \frac{1-\sigma_k(V_{22})^2}{\sigma_k(V_{22})^2}$$

This suggests choosing $V$ in Theorem 12.3.1 so that $\sigma_k(V_{22})$ is maximized.

## 12.3.2 Computations for the k=1 Case

We show how to maximize $V_{22}$ in the important $k = 1$ case. Suppose the singular values of $C$ satisfy $\sigma_{n-p} > \sigma_{n-p+1} = \cdots = \sigma_{n+1}$ and let $V = [\, v_1, \ldots, v_{n+1} \,]$ be a column partitioning of $V$. If $\tilde{Q}$ is a Householder matrix such that

$$V(:, n+1-p{:}n+1)\tilde{Q} \;=\; \begin{array}{c} \left[\begin{array}{cc} W & z \\ 0 & \alpha \end{array}\right] \begin{array}{c} n \\ 1 \end{array} \\ \begin{array}{cc} p & 1 \end{array} \end{array}$$

then $\left[\begin{array}{c} z \\ \alpha \end{array}\right]$ has the largest $(n+1)$-st component of all the vectors in $\text{span}\{v_{n+1-p}, \ldots, v_{n+1}\}$. If $\alpha = 0$, the TLS problem has no solution. Otherwise $x_{TLS} = -T_1 z/(t_{n+1}\alpha)$. Moreover,

$$\left[\begin{array}{cc} I_{n-1} & 0 \\ 0 & Q \end{array}\right] U^T(D[\, A\; b \,]T)V \left[\begin{array}{cc} I_{n-p} & 0 \\ 0 & \tilde{Q} \end{array}\right] \;=\; \Sigma$$

and so

$$D[\, E_0 \; r_0 \,]T \;=\; -D[\, A \; b \,]T \left[\begin{array}{c} z \\ \alpha \end{array}\right] [\, z^T \; \alpha \,].$$

Overall, we have the following algorithm:

**Algorithm 12.3.1** Given $A \in \mathbb{R}^{m\times n}$ $(m > n)$, $b \in \mathbb{R}^m$, and nonsingular $D = \text{diag}(d_1, \ldots, d_m)$ and $T = \text{diag}(t_1, \ldots, t_{n+1})$, the following algorithm computes (if possible) a vector $x_{TLS} \in \mathbb{R}^n$ such that $(A + E_0)x = (b + r_0)$ and $\| D[\, E_0 \; r_0 \,]T \|_F$ is minimal.

> Compute the SVD $U^T(D[\, A\; b \,]T)V = \text{diag}(\sigma_1, \ldots, \sigma_{n+1})$. Save $V$.
> Determine $p$ such that $\sigma_1 \geq \cdots \geq \sigma_{n-p} > \sigma_{n-p+1} = \cdots = \sigma_{n+1}$.
> Compute a Householder matrix $P$ such that if $\tilde{V} = VP$ then
>     $\tilde{V}(n+1, n-p+1{:}n) = 0$
> **if** $\tilde{v}_{n+1,n+1} \neq 0$
>     **for** $i = 1{:}n$
>         $x_i = -t_i \tilde{v}_{i,n+1}/(t_{n+1}\tilde{v}_{n+1,n+1})$
>     **end**
> **end**

This algorithm requires about $2mn^2 + 12n^3$ flops and most of these are associated with the SVD computation.

**Example 12.3.1** The TLS problem

$$\min_{(a+e)x=b+r} \| [\, e \; r \,] \|_F$$

where $a = [\,1\;2\;3\;4\,]^T$ and $b = [\,2.01\;3.99\;5.80\;8.30\,]^T$ has solution $x_{TLS} = 2.0212$, $e = (-.0045\; -.0209\; -.1048\; .0855)^T$, and $r = (.0022\; .0103\; .0519\; -.4023)^T$. Note: for this data $x_{LS} = 2.0197$.

### 12.3.3 Geometric Interpretation

It can be shown that the TLS solution $x_{TLS}$ minimizes

$$\psi(x) \;=\; \sum_{i=1}^{m} d_i^2 \frac{|a_i^T x - b_i|^2}{x^T T_1^{-2} x + t_{n+1}^{-2}}$$

where $a_i^T$ is $i$th row of $A$ and $b_i$ is the $i$th component of $b$. A geometrical interpretation of the TLS problem is made possible by this observation. Indeed,

$$\frac{|a_i^T x - b_i|^2}{x^T T_1^{-2} x + t_{n+1}^{-2}}$$

is the square of the distance from $\begin{bmatrix} a_i \\ b_i \end{bmatrix} \in \mathbb{R}^{n+1}$ to the nearest point in the subspace

$$P_x \;=\; \left\{ \begin{bmatrix} a \\ b \end{bmatrix} : a \in \mathbb{R}^n,\; b \in \mathbb{R},\; b = x^T a \right\}$$

where distance in $\mathbb{R}^{n+1}$ is measured by the norm $\| z \| = \| Tz \|_2$. A great deal has been written about this kind of fitting. See Pearson (1901) and Madansky (1959).

#### Problems

**P12.3.1** Consider the TLS problem (12.3.2) with nonsingular $D$ and $T$. (a) Show that if $\text{rank}(A) < n$, then (12.3.2) has a solution if and only if $b \in \text{range}(A)$. (b) Show that if $\text{rank}(A) = n$, then (12.3.2) has no solution if $A^T D^2 b = 0$ and $|t_{n+1}| \| Db \|_2 \geq \sigma_n(DAT_1)$ where $T_1 = \text{diag}(t_1, \ldots, t_n)$.

**P12.3.2** Show that if $C = D[\, A \; b \,]T = [\, A_1 \; d \,]$ and $\sigma_n(C) > \sigma_{n+1}(C)$, then the TLS solution $x$ satisfies $(A_1^T A_1 - \sigma_{n+1}(C)^2 I)x = A_1^T d$.

**P12.3.3** Show how to solve (12.1.2) with the added constraint that the first $p$ columns of the minimizing $E$ are zero.

**Notes and References for Sec. 12.3**

This section is based upon

G.H. Golub and C.F. Van Loan (1980). "An Analysis of the Total Least Squares Problem," *SIAM J. Num. Anal. 17,* 883–93.

The most detailed study of the TLS problem is

S. Van Huffel (1987). "Analysis of the Total Least Squares Problem and Its Use in Parameter Estimation," Doctoral Thesis, Department of Electrical Engineering, K.U. Leuven.

See also

S. Van Huffel (1988). "Comments on the Solution of the Nongeneric Total Least Squares Problem," Report ESAT-KUL-88/3, Department of Electrical Engineering, K.U. Leuven.

If some of the columns of $A$ are known exactly then it is sensible to force the TLS perturbation matrix $E$ to be zero in the same columns. Aspects of this constrained TLS problem are discussed in

J.W. Demmel (1987). "The smallest perturbation of a submatrix which lowers the rank and constrained total least squares problems, *SIAM J. Numer. Anal.* 24, 199–206.
S. Van Huffel and J. Vandewalle (1988). "The Partial Total Least Squares Algorithm," *J. Comp. and App. Math. 21,* 333–342.

Other references concerned with least squares fitting when there are errors in the data matrix include

W.G. Cochrane (1968). "Errors of Measurement in Statistics," *Technometrics 10,* 637–66.
R.F. Gunst, J.T. Webster, and R.L. Mason (1976). "A Comparison of Least Squares and Latent Root Regression Estimators," *Technometrics 18,* 75–83.
I. Linnik (1961). *Method of Least Squares and Principles of the Theory of Observations,* Pergamon Press, New York.
A. Madansky (1959). "The Fitting of Straight Lines When Both Variables Are Subject to Error," *J. Amer. Stat. Assoc. 54,* 173–205.
K. Pearson (1901). "On Lines and Planes of Closest Fit to Points in Space," *Phil. Mag. 2,* 559–72.
G.W. Stewart (1977c). "Sensitivity Coefficients for the Effects of Errors in the Independent Variables in a Linear Regression," Technical Report TR-571, Department of Computer Science, University of Maryland, College Park, MD.
A. Van der Sluis and G.W. Veltkamp (1979). "Restoring Rank and Consistency by Orthogonal Projection," *Lin. Alg. and Its Applic. 28,* 257–78.

## 12.4 Comparing Subspaces Using the SVD

It is sometimes necessary to investigate the relationship between two given subspaces. How close are they? Do they intersect? Can one be "rotated" into the other? And so on. In this section we show how questions like these can be answered using the singular value decomposition.

### 12.4.1 Rotation of Subspaces

Suppose $A \in \mathbb{R}^{m \times p}$ is a data matrix obtained by performing a certain set of experiments. If the same set of experiments is performed again, then a different data matrix, $B \in \mathbb{R}^{m \times p}$, is obtained. In the *orthogonal Procrustes problem* the possibility that $B$ can be rotated into $A$ is explored by solving the following problem:

$$\text{minimize } \| A - BQ \|_F \qquad \text{subject to } Q^T Q = I_p \tag{12.4.1}$$

Note that if $Q \in \mathbb{R}^{p \times p}$ is orthogonal, then

$$\| A - BQ \|_F^2 = \text{trace}(A^T A) + \text{trace}(B^T B) - 2\,\text{trace}(Q^T B^T A)\,.$$

Thus, (12.4.1) is equivalent to the problem of maximizing $\text{trace}(Q^T B^T A)$.

The maximizing $Q$ can be found by calculating the SVD of $B^T A$. Indeed, if $U^T (B^T A) V = \Sigma = \text{diag}(\sigma_1, \ldots, \sigma_p)$ is the SVD of this matrix and we define the orthogonal matrix $Z$ by $Z = V^T Q^T U$, then

$$\text{trace}(Q^T B^T A) = \text{trace}(Q^T U \Sigma V^T) = \text{trace}(Z\Sigma) = \sum_{i=1}^{p} z_{ii}\sigma_i \le \sum_{i=1}^{p} \sigma_i$$

Clearly, the upper bound is attained by setting $Q = UV^T$ for then $Z = I_p$. This gives the following algorithm:

**Algorithm 12.4.1** Given $A$ and $B$ in $\mathbb{R}^{m \times p}$, the following algorithm finds an orthogonal $Q \in \mathbb{R}^{p \times p}$ such that $\| A - BQ \|_F$ is minimum.

$C = B^T A$
Compute the SVD $U^T C V = \Sigma$. Save $U$ and $V$.
$Q = UV^T$.

The solution matrix $Q$ is the orthogonal polar factor of $B^T A$. See P4.2.9.

**Example 12.4.1** If

$$A = \begin{bmatrix} 1 & 2 \\ 3 & 4 \\ 5 & 6 \\ 7 & 8 \end{bmatrix} \quad \text{and} \quad B = \begin{bmatrix} 1.2 & 2.1 \\ 2.9 & 4.3 \\ 5.2 & 6.1 \\ 6.8 & 8.1 \end{bmatrix}$$

then

$$Q = \begin{bmatrix} .9999 & -.0126 \\ .0126 & .9999 \end{bmatrix}$$

minimizes $\| A - BQ \|_F$ over all 2-by-2 orthogonal matrices $Q$ and the minimum value is .4661.

### 12.4.2 Intersection of Null Spaces

Let $A \in \mathbb{R}^{m \times n}$ and $B \in \mathbb{R}^{p \times n}$ be given, and consider the problem of finding an orthonormal basis for $\text{null}(A) \cap \text{null}(B)$. One approach is to compute the null space of the matrix

$$C = \begin{bmatrix} A \\ B \end{bmatrix}$$

since $Cx = 0 \Leftrightarrow x \in \text{null}(A) \cap \text{null}(B)$ However, a more economical procedure results if we exploit the following theorem.

**Theorem 12.4.1** *Suppose* $A \in \mathbb{R}^{m \times n}$ *and let* $\{z_1, \ldots, z_t\}$ *be an orthonormal basis for* $\text{null}(A)$. *Define* $Z = [\, z_1, \ldots, z_t \,]$ *and let* $\{w_1, \ldots, w_q\}$ *be an orthonormal basis for* $\text{null}(BZ)$ *where* $B \in \mathbb{R}^{p \times n}$. *If* $W = [\, w_1, \ldots, w_q \,]$, *then the columns of* $ZW$ *form an orthonormal basis for* $\text{null}(A) \cap \text{null}(B)$.

**Proof.** Since $AZ = 0$ and $(BZ)W = 0$, we clearly have $\text{range}(ZW) \subset \text{null}(A) \cap \text{null}(B)$. Now suppose $x$ is in both $\text{null}(A)$ and $\text{null}(B)$. It follows that $x = Za$ for some $0 \neq a \in \mathbb{R}^t$. But since $0 = Bx = BZa$, we must have $a = Wb$ for some $b \in \mathbb{R}^q$. Thus, $x = ZWb \in \text{range}(ZW)$. □

When the SVD is used to compute the orthonormal bases in this theorem we obtain the following procedure:

**Algorithm 12.4.2** Given $A \in \mathbb{R}^{m \times n}$ and $B \in \mathbb{R}^{p \times n}$, the following algorithm computes and integer $s$ and a matrix $Y = [\, y_1, \ldots, y_s \,]$ having orthonormal columns which span $\text{null}(A) \cap \text{null}(B)$. If the intersection is trivial then $s = 0$.

> Compute the SVD $U_A^T A V_A = \text{diag}(\sigma_i)$. Save $V_A$ and set
>     $r = \text{rank}(A)$.
> **if** $r < n$
>     $C = BV_A(:, r+1{:}n)$
>     Compute the SVD $U_C^T C V_C = \text{diag}(\gamma_i)$. Save $V_C$ and set
>         $q = \text{rank}(C)$.
>     **if** $q < n - r$
>         $s = n - r - q$
>         $Y = V_A(:, r+1{:}n)V_C(:, q+1{:}n-r)$
>     **else**
>         $s = 0$
>     **end**
> **else**
>     $s = 0$
> **end**

The amount of work required by this algorithm depends upon the relative sizes of $m$, $n$, $p$, and $r$.

We mention that a practical implementation of this algorithm requires a means for deciding when a computed singular value $\hat{\sigma}_i$ is negligible. The use of a tolerance $\delta$ for this purpose (e.g. $\hat{\sigma}_i < \delta \Rightarrow \hat{\sigma}_i = 0$) implies that the columns of the computed $\hat{Y}$ "almost" define a common null space of $A$ and $B$ in the sense that $\| A\hat{Y} \|_2 \approx \| B\hat{Y} \|_2 \approx \delta$.

**Example 12.4.2** If

$$A = \begin{bmatrix} 1 & -1 & 1 \\ 1 & -1 & 1 \\ 1 & -1 & 1 \end{bmatrix} \quad \text{and} \quad B = \begin{bmatrix} 4 & 2 & 0 \\ 2 & 1 & 0 \\ 6 & 3 & 0 \end{bmatrix}$$

then $\text{null}(A) \cap \text{null}(B) = \text{span}\{x\}$, where $x = (1\ -2\ -3)^T$. Applying Algorithm 12.4.2 we find

$$V_{2A}V_{2C} = \begin{bmatrix} -.8165 & .0000 \\ -.4082 & .7071 \\ .4082 & .7071 \end{bmatrix} \begin{bmatrix} -.3273 \\ -.9449 \end{bmatrix} \approx \begin{bmatrix} .2673 \\ -.5345 \\ -.8018 \end{bmatrix} \approx .2673 \begin{bmatrix} 1 \\ -2 \\ -3 \end{bmatrix}.$$

### 12.4.3 Angles Between Subspaces

Let $F$ and $G$ be subspaces in $\mathbb{R}^m$ whose dimensions satisfy

$$p = \dim(F) \geq \dim(G) = q \geq 1.$$

The *principal angles* $\theta_1, \ldots, \theta_q \in [0, \pi/2]$ between $F$ and $G$ are defined recursively by

$$\cos(\theta_k) = \max_{u \in F} \max_{v \in G} u^T v = u_k^T v_k$$

subject to:

$$\begin{array}{ll} \| u \| = \| v \| = 1 & \\ u^T u_i = 0 & i = 1{:}k-1 \\ v^T v_i = 0 & i = 1{:}k-1 \end{array}$$

Note that the principal angles satisfy $0 \leq \theta_1 \leq \cdots \leq \theta_q \leq \pi/2$. The vectors $\{u_1, \ldots, u_q\}$ and $\{v_1, \ldots, v_q\}$ are called the *principal vectors* between the subspaces $F$ and $G$.

Principal angles and vectors arise in many important statistical applications. The largest principal angle is related to the notion of distance between equidimensional subspaces that we discussed in §2.6.3 If $p = q$ then $\text{dist}(F, G) = \sqrt{1 - \cos(\theta_p)^2} = \sin(\theta_p)$.

If the columns of $Q_F \in \mathbb{R}^{m\times p}$ and $Q_G \in \mathbb{R}^{m\times q}$ define orthonormal bases for $F$ and $G$ respectively, then

$$\max_{\substack{u\in F\\ \|u\|_2=1}} \max_{\substack{v\in G\\ \|v\|_2=1}} u^T v = \max_{\substack{y\in \mathbf{R}^p\\ \|y\|_2=1}} \max_{\substack{z\in \mathbf{R}^q\\ \|z\|_2=1}} y^T(Q_F^T Q_G)z$$

From the minimax characterization of singular values given in §8.3.1 it follows that if

$$Y^T(Q_F^T Q_G)Z = \text{diag}(\sigma_1,\ldots,\sigma_q)$$

is the SVD of $Q_F^T Q_G$, then we may define the $u_k$, $v_k$, and $\theta_k$ by

$$\begin{aligned} [\,u_1,\ldots,u_p\,] &= Q_F Y \\ [\,v_1,\ldots,v_q\,] &= Q_G Z \\ \cos(\theta_k) &= \sigma_k \qquad k = 1{:}q \end{aligned}$$

Typically, the spaces $F$ and $G$ are defined as the ranges of given matrices $A \in \mathbb{R}^{m\times p}$ and $B \in \mathbb{R}^{m\times q}$. When this is the case, the desired orthonormal bases can be obtained by computing the QR factorizations of these two matrices.

**Algorithm 12.4.3** Given $A \in \mathbb{R}^{m\times p}$ and $B \in \mathbb{R}^{m\times q}$ $(p \ge q)$ each with linearly independent columns, the following algorithm computes the orthogonal matrices $U = [\,u_1,\ldots,u_q\,]$ and $V = [\,v_1,\ldots,v_q\,]$ and $\cos(\theta_1),\ldots\cos(\theta_q)$ such that the $\theta_k$ are the principal angles between range($A$) and range($B$) and $u_k$ and $v_k$ are the associated principal vectors.

Use Algorithm 5.2.1 to compute the QR factorizations

$$\begin{aligned} A = Q_A R_A \quad & Q_A^T Q_A = I_p \quad R_A \in \mathbb{R}^{p\times p} \\ B = Q_B R_B \quad & Q_B^T Q_B = I_q \quad R_B \in \mathbb{R}^{q\times q} \end{aligned}$$

$C = Q_A^T Q_B$
Compute the SVD $Y^T C Z = \text{diag}(\cos(\theta_k))$.
$Q_A Y(:,1{:}q) = [\,u_1,\ldots,u_q\,]$
$Q_B Z = [\,v_1,\ldots,v_q\,]$

This algorithm requires about $4m(q^2+2p^2)+2pq(m+q)+12q^3$ flops.

The idea of using the SVD to compute the principal angles and vectors is due to Björck and Golub (1973). The problem of rank deficiency in $A$ and $B$ is also treated in this paper.

### 12.4.4 Intersection of Subspaces

Algorithm 12.4.3 can also be used to compute an orthonormal basis for $\text{range}(A) \cap \text{range}(B)$ where $A \in \mathbb{R}^{m \times p}$ and $B \in \mathbb{R}^{m \times q}$

**Theorem 12.4.2** *Let $\{\cos(\theta_k), u_k, v_k\}_{k=1}^{q}$ be defined by Algorithm 12.4.3. If the index $s$ is defined by $1 = \cos(\theta_1) = \cdots = \cos(\theta_s) > \cos(\theta_{s+1})$, then we have*

$$\text{range}(A) \cap \text{range}(B) = \text{span}\{u_1, \ldots, u_s\} = \text{span}\{v_1, \ldots, v_s\} .$$

**Proof.** The proof follows from the observation that if $\cos(\theta_k) = 1$, then $u_k = v_k$. □

With inexact arithmetic, it is necessary to compute the approximate multiplicity of the unit cosines in Algorithm 12.4.3.

**Example 12.4.3** If

$$A = \begin{bmatrix} 1 & 2 \\ 3 & 4 \\ 5 & 6 \end{bmatrix} \quad \text{and} \quad B = \begin{bmatrix} 1 & 5 \\ 3 & 7 \\ 5 & -1 \end{bmatrix}$$

then the cosines of the principal angles between $\text{range}(A)$ and $\text{range}(B)$ are 1.000 and .856.

#### Problems

**P12.4.1** Show that if $A$ and $B$ are $m$-by-$p$ matrices, with $p \le m$, then

$$\min_{Q^TQ=I_p} \| A - BQ \|_F^2 = \sum_{i=1}^{p} (\sigma_i(A)^2 - 2\sigma_i(B^TA) + \sigma_i(B)^2).$$

**P12.4.2** Extend Algorithm 12.4.2 so that it can compute an orthonormal basis for $\text{null}(A_1) \cap \cdots \cap \text{null}(A_s)$.

**P12.4.3** Extend Algorithm 12.4.3 to handle the case when $A$ and $B$ are rank deficient.

**P12.4.4** Relate the principal angles and vectors between $\text{range}(A)$ and $\text{range}(B)$ to the eigenvalues and eigenvectors of the generalized eigenvalue problem

$$\begin{bmatrix} 0 & A^TB \\ B^TA & 0 \end{bmatrix} \begin{bmatrix} y \\ z \end{bmatrix} = \sigma \begin{bmatrix} A^TA & 0 \\ 0 & B^TB \end{bmatrix} \begin{bmatrix} y \\ z \end{bmatrix} .$$

#### Notes and References for Sec. 12.4

The problem of minimizing $\| A - BQ \|_F$ over all orthogonal matrices arises in psychometrics. See

I.Y. Bar-Itzhack (1975). "Iterative Optimal Orthogonalization of the Strapdown Matrix," *IEEE Trans. Aerospace and Electronic Systems 11,* 30–37.
B. Green (1952). "The Orthogonal Approximation of an Oblique Structure in Factor Analysis," *Psychometrika 17,* 429–40.
R.J. Hanson and M.J. Norris (1981). "Analysis of Measurements Based on the Singular Value Decomposition," *SIAM J. Sci. and Stat. Comp. 2,* 363–374.
P. Schonemann (1966). "A Generalized Solution of the Orthogonal Procrustes Problem," *Psychometrika 31,* 1–10.

When $B = I$, this problem amounts to finding the closest orthogonal matrix to $A$. This is equivalent to the polar decomposition problem (P4.2.9). See

A. Björck and C. Bowie (1971). "An Iterative Algorithm for Computing the Best Estimate of an Orthogonal Matrix," *SIAM J. Num. Anal. 8,* 358–64.
N.J. Higham (1986b). "Computing the Polar Decomposition—with Applications," *SIAM J. Sci. and Stat. Comp.* 7, 1160–1174.
N.J. Higham and R.S. Schreiber (1988). "Fast Polar Decomposition of an Arbitrary Matrix," Technical report 88-942, Department of Computer Science, Cornell University, Ithaca, New York.

If $A$ is reasonably close to being orthogonal itself, then Björck and Bowie's technique is more efficient than the SVD algorithm. It would be interesting to extend their iteration to the case when $B$ is not the identity.

The problem of minimizing $\| AX - B \|_F$ subject to the constraint that $X$ is symmetric is studied in

N.J. Higham (1988b). "The Symmetric Procrustes Problem," *BIT 28,* 133–43.

Using the SVD to solve the canonical correlation problem was originally proposed in

A. Björck and G.H. Golub (1973). "Numerical Methods for Computing Angles Between Linear Subspaces," *Math. Comp. 27,* 579–94.

# 12.5 Some Modified Eigenvalue Problems

In this section some variations of the standard eigenvalue problem are considered. The discussion underscores the wide applicability of the matrix methods that have been described throughout the book.

## 12.5.1 Stationary Values of a Constrained Quadratic Form

Let $A \in \mathbb{R}^{n\times n}$ be symmetric. The gradient of $r(x) = x^T Ax/x^T x$ is zero if and only if $x$ is an eigenvector of $A$. The stationary values of $r(x)$ are therefore the eigenvalues of $A$.

In certain applications it is necessary to find the stationary values of $r(x)$ subject to the constraint $C^T x = 0$ where $C \in \mathbb{R}^{n\times p}$ with $n \geq p$. Suppose

$$Q^T CZ = \begin{bmatrix} S & 0 \\ 0 & 0 \end{bmatrix} \begin{matrix} r \\ n-r \end{matrix} \qquad r = \text{rank}(C)$$
$$\qquad\qquad r \quad p-r$$

is a complete orthogonal decomposition of $C$. Define $B \in \mathbb{R}^{n \times n}$ by

$$Q^T A Q = B = \begin{bmatrix} B_{11} & B_{12} \\ B_{21} & B_{22} \end{bmatrix} \begin{matrix} r \\ n-r \end{matrix}$$
$$\qquad\qquad \begin{matrix} r & n-r \end{matrix}$$

and set

$$y = Q^T x = \begin{bmatrix} u \\ v \end{bmatrix} \begin{matrix} r \\ n-r \end{matrix} .$$

Since $C^T x = 0$ transforms to $S^T u = 0$, the original problem becomes one of finding the stationary values of $r(y) = y^T B y / y^T y$ subject to the constraint that $u = 0$. But this amounts merely to finding the stationary values (eigenvalues) of the $(n-r)$-by-$(n-r)$ symmetric matrix $B_{22}$.

### 12.5.2 An Inverse Eigenvlaue Problem

Consider the problem of finding the stationary values of $x^T Ax/x^T x$ subject to the constraint $c^T x = 0$ where $A \in \mathbb{R}^{n \times n}$ is symmetric and $c \in \mathbb{R}^n$ is nonzero. From the above it is easy to show that the desired stationary values $\tilde{\lambda}_1, \ldots, \tilde{\lambda}_{n-1}$ interlace the eigenvalues $\lambda_i$ of $A$:

$$\lambda_n \le \tilde{\lambda}_{n-1} \le \lambda_{n-1} \le \cdots \le \lambda_2 \le \tilde{\lambda}_1 \le \lambda_1$$

Now suppose that $A$ has distinct eigenvalues and that we are *given* the values $\tilde{\lambda}_1, \ldots, \tilde{\lambda}_{n-1}$ that satisfy

$$\lambda_n < \tilde{\lambda}_{n-1} < \lambda_{n-1} < \cdots < \lambda_2 < \tilde{\lambda}_1 < \lambda_1 .$$

We seek to determine a unit vector $c \in \mathbb{R}^n$ such that the $\tilde{\lambda}_i$ are the stationary values of $x^T Ax$ subject to $x^T x = 1$ and $c^T x = 0$.

In order to determine the properties that $c$ must have, we use the method of Lagrange multipliers. Equating the gradient of

$$\phi(x, \lambda, \mu) = x^T Ax - \lambda(x^T x - 1) + 2\mu x^T c$$

to zero we obtain the important equation $(A - \lambda I)x = -\mu c$. This implies that $\lambda = x^T Ax$, i.e., $\lambda$ is a stationary value. Thus, $A - \lambda I$ is nonsingular and so $x = -\mu(A - \lambda I)^{-1} c$. Applying $c^T$ to both sides and substituting the eigenvalue decomposition $Q^T AQ = \text{diag}(\lambda_i)$ we obtain

$$0 = \sum_{i=1}^{n} \frac{d_i^2}{\lambda_i - \lambda}$$

where $d = Q^T c$, i.e.,

$$p(\lambda) \equiv \sum_{i=1}^{n} d_i^2 \prod_{\substack{j=1 \\ j \ne i}}^{n} (\lambda_j - \lambda) = 0 .$$

Notice that $1 = \| c \|_2^2 = \| d \|_2 = d_1^2 + \cdots + d_n^2$ is the coefficient of $(-\lambda)^{n-1}$. Since $p(\lambda)$ is a polynomial having zeroes $\tilde{\lambda}_1, \ldots, \tilde{\lambda}_{n-1}$ we must have

$$p(\lambda) = \prod_{j=1}^{n-1} (\tilde{\lambda}_j - \lambda) .$$

It follows from these two formulas for $p(\lambda)$ that

$$d_k^2 = \frac{\displaystyle\prod_{j=1}^{n-1} (\tilde{\lambda}_j - \lambda_k)}{\displaystyle\prod_{\substack{j=1 \\ j \neq k}}^{n-1} (\lambda_j - \lambda_k)} .$$

Since there are two possible signs for $d_k$, there are $2^n$ different solutions to this problem.

### 12.5.3 Eigenvalues of a Matrix Modified by a Rank-One Matrix

Given $\sigma \in \mathbb{R}$, $u \in \mathbb{R}^n$, and $D = \text{diag}(d_1, \ldots, d_n)$ satisfying $d_1 \geq \cdots \geq d_n$, it is sometimes necessary to compute the eigenvalues $\lambda_1 \geq \cdots \geq \lambda_n$ of

$$C = D + \sigma u u^T .$$

We discussed problem in §8.6 and now make a few extra observations. Using the minimax characterization of eigenvalues, it can be shown that

(i) if $\sigma \geq 0$
$$d_i \leq \lambda_i \leq d_{i-1}, \qquad i = 2{:}n$$
$$d_2 \leq \lambda_1 \leq d_1 + \sigma u^T u$$
(ii) if $\sigma \leq 0$
$$d_{i+1} \leq \lambda_i \leq d_i, \qquad i = 1{:}n-1$$
$$d_n + \sigma u^T u \leq \lambda_n \leq d_n .$$

One approach to computing $\lambda(C)$ is to apply Newton's method to the function $p(\lambda) = \det(D + \sigma u u^T - \lambda I)$. It can be shown that if

$$\begin{aligned}
&r_1(\lambda) = 1 \\
&p_1(\lambda) = (d_1 - \lambda) + \sigma u_1^2 \\
&\textbf{for } k = 2{:}n \\
&\qquad r_k(\lambda) = (d_{k-1} - \lambda) r_{k-1}(\lambda) \\
&\qquad p_k(\lambda) = (d_k - \lambda) p_{k-1}(\lambda) + \sigma u_k^2 r_k(\lambda) \\
&\textbf{end}
\end{aligned} \qquad (12.5.2)$$

then $p_n(\lambda) = p(\lambda)$. Therefore, the evaluation of both $p(\lambda)$ and $p'(\lambda)$ is straightforward. The interlacing properties can be used to obtain good starting values.

An alternative method for finding $\lambda(C)$ leads to a generalized tridiagonal eigenvalue problem. Let $P$ be a permutation matrix such that if $v = Pu$ then

$$v_1 = \cdots = v_s = 0 \quad \text{and } 0 < |v_{s+1}| \leq \cdots \leq |v_n|.$$

Furthermore, let $K$ be the bidiagonal matrix

$$K = \begin{bmatrix} 1 & r_1 & & \cdots & 0 \\ & 1 & r_2 & & \vdots \\ & & \ddots & \ddots & \\ \vdots & & & \ddots & r_{n-1} \\ 0 & \cdots & & & 1 \end{bmatrix} \qquad r_i = \begin{cases} 0 & i = 1{:}s-1 \\ -v_i/v_{i+1} & i = s{:}n-1 \end{cases}$$

Notice that $Kv = v_n e_n$ and thus the equation

$$(D + \sigma uu^T)x = \lambda x$$

transforms to $\left(KPDP^TK^T + v_n^2 e_n e_n^T\right) y = \lambda KK^T y$ where $x = PK^T y$. The matrices $KPDP^TK^T + \sigma v_n^2 e_n e_n^T$ and $KK^T$ are both symmetric and tridiagonal. A straightforward generalization of the bisection scheme of §8.4.1 can be used to find the desired eigenvalues.

### 12.5.4 Another Inverse Eigenvalue Problem

Consider the problem of finding a tridiagonal matrix

$$T = \begin{bmatrix} \alpha_1 & \beta_1 & & \cdots & 0 \\ \beta_1 & \alpha_2 & \ddots & & \vdots \\ & \ddots & \ddots & \ddots & \\ \vdots & & \ddots & \ddots & \beta_{n-1} \\ 0 & \cdots & & \beta_{n-1} & \alpha_n \end{bmatrix}$$

such that $T$ and its leading $n-1$ order principal submatrix $\tilde{T}$ have prescribed eigenvalues. In other words, we are to find $T$ given $= \lambda(T) = \{\lambda_1, \ldots, \lambda_n\}$ and $\lambda(\tilde{T}) = \{\tilde{\lambda}_1, \ldots, \tilde{\lambda}_{n-1}\}$ satisfying

$$\lambda_1 > \tilde{\lambda}_1 > \lambda_2 > \cdots > \lambda_{n-1} > \tilde{\lambda}_{n-1} > \lambda_n \,.$$

For these relationships to hold, there must be an orthogonal $Q$ such that $Q^TTQ = \Lambda = \text{diag}(\lambda_1, \ldots, \lambda_n)$ and such that the stationary values of

$$\frac{x^TTx}{x^Tx} \qquad \text{subject to } e_1^Tx = 0$$

are given by $\tilde{\lambda}_1, \ldots, \tilde{\lambda}_{n-1}$. In other words, the $\tilde{\lambda}_i$ are stationary values of

$$\frac{y^T \Lambda y}{y^T y} \qquad \text{subject to } d^T y = 0$$

where $d = Q^T e_1$, the first column of $Q$. From the above we know that

$$d_k^2 = \frac{\prod_{j=1}^{n-1} (\tilde{\lambda}_j - \lambda_k)}{\prod_{\substack{j=1 \\ j \neq k}}^{n-1} (\lambda_j - \lambda_k)} \qquad k = 1{:}n$$

Consider applying the Lanczos iteration (9.1.3) with $A = \Lambda$ and $q_1 = d$. If the algorithm runs for $n$ iterations, we generate an orthogonal matrix $Q$ with the property that $Q^T \Lambda Q = T$ is tridiagonal. Since the first column of $Q$ is $d$, it follows that $T$ has the desired properties.

### Problems

**P12.5.1** Let $A \in \mathbb{R}^{m \times n}$ and consider the problem of finding the stationary values of

$$R(x, y) = \frac{y^T A x}{\| y \|_2 \| x \|_2} \qquad y \in \mathbb{R}^m, x \in \mathbb{R}^n$$

subject to the constraints

$$\begin{array}{lll} C^T x = 0 & C \in \mathbb{R}^{n \times p} & n \geq p \\ D^T y = 0 & D \in \mathbb{R}^{m \times q} & m \geq q \end{array}$$

Show how to solve this problem by first computing complete orthogonal decompositions of $C$ and $D$ and then computing the SVD of a certain submatrix of a transformed $A$.

**P12.5.2** Suppose $A \in \mathbb{R}^{m \times n}$ and $B \in \mathbb{R}^{p \times n}$. Assume that rank$(A) = n$ and rank$(B) = p$. Using the methods of this section, show how to solve

$$\min_{Bx=0} \frac{\| b - Ax \|_2^2}{\| x \|_2^2 + 1} = \min_{Bx=0} \frac{\left\| [\, A \; b \,] \begin{bmatrix} x \\ -1 \end{bmatrix} \right\|_2^2}{\left\| \begin{bmatrix} x \\ -1 \end{bmatrix} \right\|_2^2}$$

Show that this is a constrained TLS problem. Is there always a solution?

**P12.5.3** Suppose $A \in \mathbb{R}^{n \times n}$ is symmetric and that $B \in \mathbb{R}^{p \times n}$ has rank $p$. Let $d \in \mathbb{R}^p$. Show how to solve the problem of minimizing $x^T A x$ subject to the constraints $\| x \|_2 = 1$ and $Bx = d$. Indicate when a solution fails to exist.

### Notes and References for Sec. 12.5

Many of the problems discussed in this section appear in the following survey articles:

D. Boley and G.H. Golub (1987). "A Survey of Matrix Inverse Eigenvalue Problems," *Inverse Problems 3*, 595–622.

G.H. Golub (1973). "Some Modified Matrix Eigenvalue Problems," *SIAM Review 15*, 318–44.

References for the stationary value problem include

G.E. Forsythe and G.H. Golub (1965). "On the Stationary Values of a Second-Degree Polynomial on the Unit Sphere," *SIAM J. App. Math. 13*, 1050–68.
G.H. Golub and R. Underwood (1970). "Stationary Values of the Ratio of Quadratic Forms Subject to Linear Constraints," *Z. Angew. Math. Phys. 21*, 318–26.

Using the QR algorithm to determine the Gauss-type quadrature rules is discussed in

G.H. Golub and J.H. Welsch (1969). "Calculation of Gauss Quadrature Rules," *Math. Comp. 23*, 221–30.

Lanczos methods for various inverse eigenvalue problems are detailed in

D.L. Boley and G.H. Golub (1978). "The Matrix Inverse Eigenvalue Problem for Periodic Jacobi Matrices," in *Proc. Fourth Symposium on Basic Problems of Numerical Mathematics*, Prague, pp. 63–76.
D. Boley and G.H. Golub (1984). "A Modified Method for Restructuring Periodic Jacobi Matrices," *Math. Comp. 42*, 143–150.

Recent algorithmic developments include

S. Friedland, J. Nocedal, and M.L. Overton (1987). "The Formulation and Analysis of Numerical Methods for Inverse Eigenvalue Problems," *SIAM J. Numer. Anal. 24*, 634–667.
W.B. Gragg and W.J. Harrod (1984). "The Numerically Stable Reconstruction of Jacobi Matrices from Spectral Data," *Numer. Math. 44*, 317–336.
J. Kautsky and G.H. Golub (1983). "On the Calculation of Jacobi Matrices," *Lin. Alg. and Its Applic. 52/53*, 439–456.

Another important class of inverse eigenvalue problems involve finding a diagonal matrix $D$ so that $AD$ (or sometimes $A + D$) has prescribed eigenvalues. See

S. Friedland (1975). "On Inverse Multiplicative Eigenvalue Problems for Matrices," *Lin. Alg. and Its Applic. 12*, 127–38.

## 12.6 Updating the QR Factorization

In many applications it is necessary to re-factor a given matrix $A \in \mathbb{R}^{m \times n}$ after it has been altered in some minimal sense. For example, given that we have the QR factorization of $A$, we may need to calculate the QR factorization of a matrix $\tilde{A}$ that is obtained by (a) adding a general rank-one matrix to $A$, (b) appending a row (or column) to $A$, or (c) deleting a row (or column) from $A$. In this section we show that in situations like these, it is much more efficient to "update" $A$'s QR factorization than to generate it from scratch.

Before beginning, we mention that there are also techniques for updating the factorizations $PA = LU$, $A = GG^T$, and $A = LDL^T$. Updating these factorizations, however, can be quite delicate because of pivoting requirements and because when we tamper with a positive definite matrix the result may not be positive definite. See Gill, Golub, Murray, and Saunders (1974) and Stewart (1979c). Along these lines we briefly discuss hyperbolic transformations and their use in the Cholesky downdating problem.

### 12.6.1 Rank-One Changes

Suppose we have the QR factorization $QR = B \in \mathbb{R}^{m \times n}$ and that we need to compute the QR factorization $B + uv^T = Q_1 R_1$ where $u, v \in \mathbb{R}^n$ are given. Observe that

$$B + uv^T \;=\; Q(R + wv^T) \tag{12.6.1}$$

where $w = Q^T u$. Suppose that we compute rotations $J_{n-1}, \ldots, J_2, J_1$ such that

$$J_1^T \cdots J_{n-1}^T w \;=\; \pm\| w \|_2 e_1 \,.$$

Here, each $J_k$ is a rotation in planes $k$ and $k+1$. (For details, see Algorithm 5.3.1.) If these same Givens rotations are applied to $R$, it can be shown that

$$H = J_1^T \cdots J_{n-1}^T R \tag{12.6.2}$$

is upper Hessenberg. For example, in the $n = 4$ case we start with

$$R \;=\; \begin{bmatrix} \times & \times & \times & \times \\ 0 & \times & \times & \times \\ 0 & 0 & \times & \times \\ 0 & 0 & 0 & \times \end{bmatrix} \qquad w \;=\; \begin{bmatrix} \times \\ \times \\ \times \\ \times \end{bmatrix}$$

and then update as follows:

$$R \;=\; J_3^T R \;=\; \begin{bmatrix} \times & \times & \times & \times \\ 0 & \times & \times & \times \\ 0 & 0 & \times & \times \\ 0 & 0 & \times & \times \end{bmatrix} \qquad w \;=\; J_3^T w \;=\; \begin{bmatrix} \times \\ \times \\ \times \\ 0 \end{bmatrix}$$

$$R \;=\; J_2^T R \;=\; \begin{bmatrix} \times & \times & \times & \times \\ 0 & \times & \times & \times \\ 0 & \times & \times & \times \\ 0 & 0 & \times & \times \end{bmatrix} \qquad w \;=\; J_2^T w \;=\; \begin{bmatrix} \times \\ \times \\ 0 \\ 0 \end{bmatrix}$$

$$H \;=\; J_1^T R \;=\; \begin{bmatrix} \times & \times & \times & \times \\ \times & \times & \times & \times \\ 0 & \times & \times & \times \\ 0 & 0 & \times & \times \end{bmatrix} \qquad w \;=\; J_1^T w \;=\; \begin{bmatrix} \times \\ 0 \\ 0 \\ 0 \end{bmatrix}$$

Consequently,

$$(J_1^T \cdots J_{n-1}^T)(R + wv^T) \;=\; H \pm \| w \|_2 e_1 v^T = H_1 \tag{12.6.3}$$

is also upper Hessenberg.

In §5.2.3, we showed how to compute the QR factorization of an upper Hessenberg matrix in $O(n^2)$ flops. In particular, we can find Givens rotations $G_k$ , $k = 1{:}n-1$ such that

$$G_{n-1}^T \cdots G_1^T H_1 \;=\; R_1 \tag{12.6.4}$$

is upper triangular. Combining (12.6.1) through (12.6.4) we obtain the QR factorization $B + uv^T \;=\; Q_1 R_1$ where

$$Q_1 \;=\; QJ_{n-1} \cdots J_1 G_1 \cdots G_{n-1}.$$

A careful assessment of the work reveals that about $26n^2$ flops are required. The vector $w = Q^T u$ requires $2n^2$ flops. Computing $H$ and accumulating the $J_k$ into $Q$. involves $12n^2$ flops. Finally, computing $R_1$ and multiplying the $G_k$ into $Q$ involves $12n^2$ flops.

The technique readily extends to the case when $B$ is rectangular. It can also be generalized to compute the QR factorization of $B + UV^T$ where rank $(UV^T) = p > 1$.

### 12.6.2 Appending or Deleting a Column

Assume that we have the QR factorization

$$QR \;=\; A \;=\; [\, a_1, \ldots, a_n \,] \qquad a_i \in \mathbb{R}^m \tag{12.6.5}$$

and partition the upper triangular matrix $R \in \mathbb{R}^{m \times n}$ as follows:

$$R \;=\; \begin{bmatrix} R_{11} & v & R_{13} \\ 0 & r_{kk} & w^T \\ 0 & 0 & R_{33} \end{bmatrix} \begin{matrix} k-1 \\ 1 \\ m-k \end{matrix} \quad . \qquad \begin{matrix} k-1 & 1 & n-k \end{matrix}$$

Now suppose that we want to compute the QR factorization of

$$\tilde{A} \;=\; [\; a_1 \;\cdots\; a_{k-1} \;\; a_{k+1} \;\cdots\; a_n \;] \in \mathbb{R}^{m \times (n-1)} .$$

Note that $\tilde{A}$ is just $A$ with its $k$th column deleted and that

$$Q^T \tilde{A} \;=\; \begin{bmatrix} R_{11} & R_{13} \\ 0 & w^T \\ 0 & R_{33} \end{bmatrix} \;=\; H$$

is upper Hessenberg, e.g.,

$$H \;=\; \begin{bmatrix} \times & \times & \times & \times & \times \\ 0 & \times & \times & \times & \times \\ 0 & 0 & \times & \times & \times \\ 0 & 0 & \times & \times & \times \\ 0 & 0 & 0 & \times & \times \\ 0 & 0 & 0 & 0 & \times \\ 0 & 0 & 0 & 0 & 0 \end{bmatrix} \qquad m = 7,\; n = 6,\; k = 3$$

Clearly, the unwanted subdiagonal elements $h_{k+1,k}, \ldots, h_{n,n-1}$ can be zeroed by a sequence of Givens rotations: $G_{n-1}^T \cdots G_k^T H = R_1$. Here, $G_i$ is a rotation in planes $i$ and $i+1$ for $i = k{:}n-1$. Thus, if $Q_1 = QG_k \cdots G_{n-1}$ then $\tilde{A} = Q_1 R_1$ is the QR factorization of $\tilde{A}$.

The above update procedure can be executed in $O(n^2)$ flops and is very useful in certain least squares problems. For example, one may wish to examine the significance of the $k$th factor in the underlying model by deleting the $k$th column of the corresponding data matrix and solving the resulting LS problem.

In a similar vein, it is useful to be able to compute efficiently the solution to the LS problem after a column has been appended to $A$. Suppose we have the QR factorization (12.6.5) and now wish to compute the QR factorization of

$$\tilde{A} = \left[\, a_1 \ \cdots \ a_k \quad z \quad a_{k+1} \ \cdots \ a_n \,\right]$$

where $z \in \mathbb{R}^m$ is given. Note that if $w = Q^T z$ then

$$Q^T A = \left[\, Q^T a_1 \ \cdots \ Q^T a_k \quad w \quad Q^T a_{k+1} \ \cdots \ Q^T a_n \,\right] = \tilde{A}$$

is upper triangular except for the presence of a "spike" in its $k+1$-st column, e.g.,

$$\tilde{A} = \begin{bmatrix} \times & \times & \times & \times & \times & \times \\ 0 & \times & \times & \times & \times & \times \\ 0 & 0 & \times & \times & \times & \times \\ 0 & 0 & 0 & \times & \times & \times \\ 0 & 0 & 0 & \times & 0 & \times \\ 0 & 0 & 0 & \times & 0 & 0 \\ 0 & 0 & 0 & \times & 0 & 0 \end{bmatrix} \qquad m = 7,\ n = 5,\ k = 3$$

It is possible to determine Givens rotations $J_{m-1}, \ldots, J_{k+1}$ so that

$$J_{k+1}^T \cdots J_{m-1}^T w = \begin{bmatrix} w_1 \\ \vdots \\ w_{k+1} \\ 0 \\ \vdots \\ 0 \end{bmatrix}$$

with $J_{k+1}^T \cdots J_{m-1}^T \tilde{A} = \tilde{R}$ upper triangular. We illustrate this by continuing with the above example:

$$H = J_6^T \tilde{A} = \begin{bmatrix} \times & \times & \times & \times & \times & \times \\ 0 & \times & \times & \times & \times & \times \\ 0 & 0 & \times & \times & \times & \times \\ 0 & 0 & 0 & \times & \times & \times \\ 0 & 0 & 0 & \times & 0 & \times \\ 0 & 0 & 0 & \times & 0 & 0 \\ 0 & 0 & 0 & 0 & 0 & 0 \end{bmatrix}$$

$$H = J_5^T H = \begin{bmatrix} \times & \times & \times & \times & \times & \times \\ 0 & \times & \times & \times & \times & \times \\ 0 & 0 & \times & \times & \times & \times \\ 0 & 0 & 0 & \times & \times & \times \\ 0 & 0 & 0 & \times & 0 & \times \\ 0 & 0 & 0 & 0 & 0 & \times \\ 0 & 0 & 0 & 0 & 0 & 0 \end{bmatrix}$$

$$H = J_4^T H = \begin{bmatrix} \times & \times & \times & \times & \times & \times \\ 0 & \times & \times & \times & \times & \times \\ 0 & 0 & \times & \times & \times & \times \\ 0 & 0 & 0 & \times & \times & \times \\ 0 & 0 & 0 & 0 & \times & \times \\ 0 & 0 & 0 & 0 & 0 & \times \\ 0 & 0 & 0 & 0 & 0 & 0 \end{bmatrix}$$

This update requires $O(mn)$ flops.

### 12.6.3 Appending or Deleting a Row

Suppose we have the QR factorization $QR = A \in \mathbb{R}^{m \times n}$ and now wish to obtain the QR factorization of

$$\tilde{A} = \begin{bmatrix} w^T \\ A \end{bmatrix}$$

where $w \in \mathbb{R}^n$. Note that

$$\text{diag}(1, Q^T)\tilde{A} = \begin{bmatrix} w^T \\ R \end{bmatrix} = H$$

is upper Hessenberg. Thus, Givens rotations $J_1, \ldots, J_n$ could be determined so $J_n^T \cdots J_1^T H = R_1$ is upper triangular. It follows that

$$\tilde{A} = Q_1 R_1$$

is the desired QR factorization, where $Q_1 = \text{diag}(1, Q) J_1 \cdots J_n$.

No essential complications result if the new row is added between rows $k$ and $k+1$ of $A$. We merely apply the above with $A$ replaced by $PA$ and $Q$ replaced by $PQ$ where

$$P = \begin{bmatrix} 0 & I_{m-k} \\ I_k & 0 \end{bmatrix}.$$

Upon completion $\text{diag}(1, P^T)Q_1$ is the desired orthogonal factor.

Lastly, we consider how to update the QR factorization $QR = A \in \mathbb{R}^{m \times n}$ when the first row of $A$ is deleted. In particular, we wish to compute the QR factorization of the submatrix $A_1$ in

$$A = \begin{bmatrix} z^T \\ A_1 \end{bmatrix} \begin{matrix} 1 \\ m-1 \end{matrix}$$

(The procedure is similar when an arbitrary row is deleted.) Let $q^T$ be the first row of $Q$ and compute Givens rotations $G_1, \ldots, G_{m-1}$ such that

$$G_1^T \cdots G_{m-1}^T q = \alpha e_1,$$

where $\alpha = \pm 1$. Note that

$$H = G_1^T \cdots G_{m-1}^T R = \begin{bmatrix} v^T \\ R_1 \end{bmatrix} \begin{matrix} 1 \\ m-1 \end{matrix}$$

is upper Hessenberg and that

$$QG_{m-1} \cdots G_1 = \begin{bmatrix} \alpha & 0 \\ 0 & Q_1 \end{bmatrix}$$

where $Q_1 \in \mathbb{R}^{(m-1)\times(m-1)}$ is orthogonal. Thus,

$$A = \begin{bmatrix} z^T \\ A_1 \end{bmatrix} = (QG_{m-1} \cdots G_1)(G_1^T \cdots G_{m-1}^T R) = \begin{bmatrix} \alpha & 0 \\ 0 & Q_1 \end{bmatrix} \begin{bmatrix} v^T \\ R_1 \end{bmatrix}$$

from which we conclude that $A_1 = Q_1 R_1$ is the desired QR factorization.

### 12.6.4 Hyperbolic Transformation Methods

Recall that the "$R$" in $A = QR$ is the transposed Cholesky factor in $A^T A = GG^T$. Thus, there is a close connection between the QR modifications just discussed and analogous modifications of the Cholesky factorization. We illustrate this with the *Cholesky downdating problem* which corresponds to the removal of an $A$-row in QR. In the Cholesky downdating problem we have the Cholesky factorization

$$GG^T = A^T A = \begin{bmatrix} z^T \\ A_1 \end{bmatrix}^T \begin{bmatrix} z^T \\ A_1 \end{bmatrix} \qquad (12.6.6)$$

where $A \in \mathbb{R}^{m \times n}$ with $m > n$ and $z \in \mathbb{R}^n$. Our task is to find a lower triangular $G_1$ such that $G_1 G_1^T = A_1^T A_1$. There are several approaches to this interesting and important problem. Simply because it is an oppportunity to introduce some new ideas, we present a downdating procedure that relies on *hyperbolic transformations.*

We start with a definition. $H \in \mathbb{R}^{m \times m}$ is *pseudo-orthogonal* with respect to the *signature matrix* $S = \text{diag}(\pm 1) \in \mathbb{R}^{m \times m}$ if $H^T S H = S$. Now from (12.6.6) we have $A^T A = A_1^T A_1 + zz^T = GG^T$ and so

$$A_1^T A_1 \;=\; A^T A - zz^T \;=\; GG^T - zz^T \;=\; [\,G\, z\,]\begin{bmatrix} I_n & 0 \\ 0 & -1 \end{bmatrix}\begin{bmatrix} G^T \\ z^T \end{bmatrix}.$$

Define the signature matrix

$$S \;=\; \begin{bmatrix} I_n & 0 \\ 0 & -1 \end{bmatrix} \qquad (12.6.7)$$

and suppose that we can find $H \in \mathbb{R}^{(n+1)\times(n+1)}$ such that $H^T S H = S$ with the property that

$$H\begin{bmatrix} G^T \\ z^T \end{bmatrix} = \begin{bmatrix} G_1^T \\ 0 \end{bmatrix} \qquad (12.6.8)$$

is upper triangular. It follows that

$$A_1^T A_1 \;=\; [\,G\, z\,]\, H^T S H \begin{bmatrix} G^T \\ z^T \end{bmatrix} \;=\; [G_1\, 0\,]\, S \begin{bmatrix} G_1 \\ 0 \end{bmatrix} \;=\; G_1 G_1^T$$

is the sought after Cholesky factorization.

We now show how to construct the hyperbolic transformation $H$ in (12.6.8) using *hyperbolic rotations.* A 2-by-2 hyperbolic rotation has the form

$$H \;=\; \begin{bmatrix} \cosh(\theta) & -\sinh(\theta) \\ -\sinh(\theta) & \cosh(\theta) \end{bmatrix} \;=\; \begin{bmatrix} c & -s \\ -s & c \end{bmatrix}.$$

Note that if $H \in \mathbb{R}^{2\times 2}$ is a hyperbolic rotation then $H^T S H = S$ where $S = \text{diag(-1,1)}$. Paralleling our Givens rotations developments, let us see how hyperbolic rotations can be used for zeroing. From

$$\begin{bmatrix} c & -s \\ -s & c \end{bmatrix}\begin{bmatrix} x_1 \\ x_2 \end{bmatrix} = \begin{bmatrix} r \\ 0 \end{bmatrix} \qquad c^2 - s^2 = 1$$

we obtain the equation $cx_2 = sx_1$. Note that there is no solution to this equation if $x_1 = x_2 \neq 0$, a clue that hyperbolic rotations are not as numerically solid as their Givens rotation counterparts. If $x_1 \neq x_2$ then it is possible to compute the cosh-sinh pair:

$$
\begin{array}{l}
\textbf{if } x_2 = 0 \\
\quad s = 0;\ c = 1 \\
\textbf{else} \\
\quad \textbf{if } |x_2| < |x_1| \\
\quad\quad \tau = x_2/x_1;\ c = 1/\sqrt{1-\tau^2};\ s = c\tau \\
\quad \textbf{elseif } |x_1| < |x_2| \\
\quad\quad \tau = x_1/x_2;\ s = 1/\sqrt{1-\tau^2};\ c = s\tau \\
\quad \textbf{end} \\
\textbf{end}
\end{array}
\tag{12.6.9}
$$

Observe that the norm of the hyperbolic rotation produced by this algorithm gets large as $x_1$ gets close to $x_2$.

Now any matrix $H = H(p, n+1, \theta) \in \mathbb{R}^{(n+1)\times(n+1)}$ that is the identity everywhere except $h_{p,p} = h_{n+1,n+1} = \cosh(\theta)$ and $h_{p,n+1} = h_{n+1,p} = -\sinh(\theta)$ satisfies $H^TSH = S$ where $S$ is prescribed in (12.6.7). Using (12.6.9), we attempt to generate hyperbolic rotations $H_k = H(1, k, \theta_k)$ for $k = 2{:}n+1$ so that

$$
H_n \cdots H_1 \begin{bmatrix} G^T \\ z^T \end{bmatrix} = \begin{bmatrix} \tilde{G}^T \\ 0 \end{bmatrix}.
$$

This turns out to be possible if $A$ has full column rank. Hyperbolic rotation $H_k$ zeros entry $(k+1, k)$. In other words, if $A$ has full column rank, then it can be shown that each call to (12.6.9) results in a cosh-sinh pair. See Alexander, Pan, and Plemmons (1988).

### Problems

**P12.6.1** Suppose we have the QR factorization for $A \in \mathbf{R}^{m\times n}$ and now wish to minimize $\| A + uv^T)x - b \|_2$ where $u, b \in \mathbf{R}^m$ and $v \in \mathbf{R}^n$ are given. Give an algorithm for solving this problem that requires $O(mn)$ flops. Assume that $Q$ must be updated.

**P12.6.2** Suppose we have the QR factorization $QR = A \in \mathbf{R}^{m\times n}$. Give an algorithm for computing the QR factorization of the matrix $A$ obtained by deleting the $k$th row of $A$. Your algorithm should require $O(mn)$ flops.

**P12.6.3** Suppose $T \in \mathbf{R}^{n\times n}$ is tridiagonal and symmetric and that $v \in \mathbf{R}^n$. Give an $O(n^2)$ method for computing an orthogonal $Q \in \mathbf{R}^{n\times n}$ such that $Q^T(T + vv^T)Q = \tilde{T}$ is also tridiagonal.

**P12.6.4** Suppose

$$
A = \begin{bmatrix} c^T \\ B \end{bmatrix} \qquad c \in \mathbf{R}^n,\ B \in \mathbf{R}^{(m-1)\times n}
$$

with $m > n$. Using the Sherman-Morrison-Woodbury formula show that

$$
\frac{1}{\sigma_{min}(B)} \le \frac{1}{\sigma_{min}(A)} + \frac{\| (A^TA)^{-1}c \|_2^2}{1 - c^T(A^TA)^{-1}c}.
$$

**P12.6.5** As a function of $x_1$ and $x_2$, what is the 2-norm of the hyperbolic rotation produced by (12.6.9)?

**P12.6.6** Show that the hyperbolic reduction in §12.6.4 does not breakdown if $A$ has full column rank.

### Notes and References for Sec. 12.6

Numerous aspects of the updating problem are presented in

P.E. Gill, G.H. Golub, W. Murray, and M.A. Saunders (1974). "Methods for Modifying Matrix Factorizations," *Math. Comp. 28,* 505-35.

Further references include

R.H. Bartels (1971). "A Stabilization of the Simplex Method," *Numer. Math. 16,* 414-434

J. Daniel, W.B. Gragg, L. Kaufman, and G.W. Stewart (1976). "Reorthogonaization and Stable Algorithms for Updating the Gram-Schmidt QR Factorization," *Math. Comp. 30,* 772-95.

P.E. Gill, W. Murray, and M.A. Saunders (1975). "Methods for Computing and Modifying the LDV Factors of a Matrix," *Math. Comp. 29,* 1051–77.

D. Goldfarb (1976). "Factored Variable Metric Methods for Unconstrained Optimization," *Math. Comp. 30,* 796–811.

Fortran programs for updating the QR and Cholesky factorizations are included in Linpack (chapter 10). The stability of downdating the Cholesky factorization is analyzed in

G.W. Stewart (1979c). "The Effects of Rounding Error on an Algorithm for Downdating a Cholesky Factorization," *J. Inst. Math. Applic. 23,* 203–13.

Hyperbolic transformations are like Givens transformations but are based upon cosh and sinh instead of cos and sin. The use of hyperbolic transformations in updating factorizations is described in

S.T. Alexander, C.T. Pan, and R.J. Plemmons (1988). "Analysis of a Recursive Least Squares Hyperbolic Rotation Algorithm for Signal Processing," *Lin. Alg. and Its Applic. 98,* 3–40.

A.W. Bojanczyk, R.P. Brent, P. Van Dooren, and F.R. de Hoog (1987). "A Note on Downdating the Cholesky Factorization," *SIAM J. Sci. and Stat. Comp. 8,* 210–221.

G.H. Golub (1969). "Matrix Decompositions and Statistical Computation," in *Statistical Computation*, ed., R.C. Milton and J.A. Nelder, Academic Press, New York, pp. 365–97.

Work on parallel updating and downdating procedures has just begun. See

C.S. Henkel, M.T. Heath, and R.J. Plemmons (1988). "Cholesky Downdating on a Hypercube," in G. Fox (1988), 1592–1598.

The role of updating techniques in the area of nonlinear equations and optimization is surveyed in

J.E. Dennis and R.B. Schnabel (1983). *Numerical Methods for Unconstrained Optimization and Nonlinear Equations,* Prentice-Hall, Englewood Cliffs, NJ.

# Bibliography

J.O. Aasen (1971). "On the Reduction of a Symmetric Matrix to Tridiagonal Form", *BIT 11*, 233–42.

N.N. Abdelmalck (1971). "Roundoff Error Analysis for Gram-Schmidt Method and Solution of Linear Least Squares Problems," *BIT 11,* 1345–68.

L. Adams (1985). "m-step Preconditioned Congugate Gradient Methods," *SIAM J. Sci. and Stat. Comp.,* 6, 452–463.

L. Adams and T. Crockett (1984). "Modeling Algorithm Execution Time on Processor Arrays," *Computer 17,* 38–43.

S.T. Alexander, C.T. Pan, and R.J. Plemmons (1988). "Analysis of a Recursive Least Squares Hyperbolic Rotation Algorithm for Signal Processing," *Lin. Alg. and Its Applic. 98,* 3–40.

E.L. Allgower (1973). "Exact Inverses of Certain Band Matrices," *Numer. Math. 21,* 279–84.

A.R. Amir-Moez (1965). *Extremal Properties of Linear Transformations and Geometry of Unitary Spaces,* Texas Tech University Mathematics Series, no. 243, Lubbock, TX.

N. Anderson and I. Karasalo (1975). "On Computing Bounds for the Least Singular Value of a Triangular Matrix,"*BIT 15,* 1–4.

P. Anderson and G. Loizou (1973). "On the Quadratic Convergence of an Algorithm that Diagonalizes a Complex Symmetric Matrix," *J. Inst. Math. Applic. 12,* 261–71.

P. Anderson and G. Loizou (1976). "A Jacobi-Type Method for Complex Symmetric Matrices (Handbook)," *Numer. Math. 25,* 347–63.

T.W. Anderson, I. Olkin, and L.G. Underhill (1987). "Generation of Random Orthogonal Matrices," *SIAM J. Sci. and Stat. Comp. 8,* 625–629.

G. Andrews and F.B. Schneider (1983). "Concepts and Notations for Concurrent Programming," *Computing Surveys 15,* 1–43.

W.E. Arnoldi (1951). "The Principle of Minimized Iterations in the Solution of the Matrix Eigenvalue Problem," *Quart. Appl. Math. 9,* 17-29.

P. Arbenz and G.H. Golub (1987). "On the Spectral Decomposition of Hermitian Matrices Subject to Indefinite Low Rank Perturbations with Applications," Stanford Numerical Analysis Report NA 87-07, Dept. of Comp. Sci., Stanford University, Stanford, CA 94305.

P. Arbenz, W. Gander, and G.H. Golub (1988). "Restricted Rank Modification of the Symmetric Eigenvalue Problem: Theoretical Considerations," *Lin. Alg. and Its Applic. 104,* 75–95.

P. Arbenz and G.H. Golub (1988). "On the Spectral Decomposition of Hermitian Matrices Subject to Indefinite Low Rank Perturbations with Applications," *SIAM J. Matrix Anal. Appl. 9,* 40-58.

M. Arioli, J.W. Demmel, and I.S. Duff (1988). "Solving Sparse Linear Systems with Sparse Backward Error," Report CSS 214, Computer Science and Systems Division, AERE Harwell, Didcot, England.

M. Arioli and F. Romani (1985). "Relations Between Condition Numbers and the Convergence of the Jacobi Method for Real Positive Definite Matrices," *Numer. Math.*

*46,* 31–42.

M. Arioli and A. Laratta (1985). "Error Analysis of an Algorithm for Solving an Underdetermined System," *Numer. Math. 46,* 255–268.

E.S. Armstrong and A.K. Caglayan (1976). "An Algorithm for the Weighting Matrices in the Sample-Data Optimal Linear Regulator Problem," NASA Technical Note, TN D-8372.

W.F. Arnold and A.J. Laub (1984). "Generalized Eigenproblem Algorithms and Software for Algebraic Riccati Equations," *Proc. IEEE 72,* 1746–1754.

S.F. Ashby (1987). "Polynomial Preconditioning for Conjugate Gradient Methods," Ph.D. Thesis, Dept. of Computer Science, University of Illinois.

S. Ashby, T.A. Manteuffel, and P.E. Saylor (1988). "A Taxonomy for Conjugate Gradient Methods," Report UCRL-98508, Lawrence Livermore National Laboratory, Livermore, CA.

E. Asplund (1959). "Inverse of Matrices $\{a_{ij}\}$ Which Satisfy $a_{ij} = 0$, $j > i + p$," *Math. Scand. 7,* 57–60.

O. Axelsson (1977). "Solution of Linear Systems of Equations: Iterative Methods," in *Sparse Matrix Techniques: Copenhagen, 1976*, ed. V.A. Barker, Springer-Verlag, Berlin.

O. Axelsson (1980). "Conjugate Gradient Type Methods for Unsymmetric and Inconsistent Systems of Linear Equations," *Lin. Alg. and Its Applic. 29,* 1–66.

O. Axelsson (1985). "A Survey of Preconditioned Iterative Methods for Linear Systems of Equations," *BIT 25,* 166–187.

O. Axelsson and B. Polman (1986). "On Approximate Factorization Methods for Block Matrices Suitable for Vector and Parallel Processors," *Lin. Alg. and Its Applic. 77,* 3–26.

Z. Bai (1988). "Note on the Quadratic Convergence of Kogbetliantz's Algorithm for Computing the Singular Value Decomposition," *Lin. Alg. and Its Applic. 104,* 131–140.

D. Bailey (1988). "Extra High Speed Matrix Multiplication on the Cray-2," *SIAM J. Sci. and Stat. Comp. 9,* 603–607.

I.Y. Bar-Itzhack (1975). "Iterative Optimal Orthogonalization of the Strapdown Matrix," *IEEE Trans. Aerospace and Electronic Systems 11,* 30–37.

J. L. Barlow (1987). "On the Smallest Positive Singular Value of an $M$-Matrix with Applications to Ergodic Markov Chains," *SIAM J. Alg. and Disc. Struct. 7,* 414–424.

J.L. Barlow, N.K. Nichols, and R.J. Plemmons (1988). "Iterative Methods for Equality Constrained Least Squares Problems," *SIAM J. Sci. and Stat. Comp. 9,* 892–906.

S. Barnett and C. Storey (1968). "Some Applications of the Lyapunov Matrix Equation," *J. Inst. Math. Applic. 4,* 33–42.

I. Barrodale and C. Phillips (1975). "Algorithm 495: Solution of an Overdetermined System of Linear Equations in the Chebychev Norm," *ACM Trans. Math. Soft. 1,* 264–70.

I. Barrodale and F.D.K. Roberts (1973). "An Improved Algorithm for Discrete $L_1$ Linear Approximation," *SIAM J. Num. Anal. 10,* 839–48.

R.H. Bartels (1971). "A Stabilization of the Simplex Method," *Numer. Math. 16,* 414-434

R.H. Bartels, A.R. Conn, and C. Charalambous (1978). "On Cline's Direct Method for Solving Overdetermined Linear Systems in the $L_\infty$ Sense,"*SIAM J. Num. Anal. 15,* 255–70.

R.H. Bartels, A.R. Conn, and J.W. Sinclair (1978). "Minimization Techniquees for Piecewise Differentiable Functions: The $L_1$ Solution to an Overdetermined Linear System," *SIAM J. Num. Anal. 15,* 224–41.

R.H. Bartels and G.W. Stewart (1972). "Solution of the Equation $AX + XB = C$," *Comm. ACM 15,* 820–26.

W. Barth, R.S. Martin and J.H. Wilkinson (1967). "Calculation of the Eigenvalues of a Symmetric Tridiagonal Matrix by the Method of Bisection," *Num. Math. 9,* 386-93. See also *HACLA*, pp. 249–56.

V. Barwell and J.A. George (1976). "A Comparison of Algorithms for Solving Symmetric Indefinite Systems of Linear Equations," *ACM Trans. Math. Soft. 2,* 242–51.

K.J. Bathe and E.L. Wilson (1973). "Solution Methods for Eigenvalue Problems in Structural Mechanics," *Int. J. Numer. Meth. Eng. 6,* 213–26.

F.L. Bauer and C.T. Fike (1960). "Norms and Exclusion Theorems," *Numer. Math. 2,* 137–44.

F.L. Bauer (1963). "Optimally Scaled Matrices," *Numer. Math. 5,* 73–87.

F.L. Bauer (1965). "Elimination with Weighted Row Combinations for Solving Linear Equations and Least Squares Problems," *Numer. Math. 7,* 338–52. See also *HACLA*, pp. 119-33.

F.L. Bauer and C.T. Fike (1960). "Norms and Exclusion Theorems," *Numer. Math. 2,* 137–44.

F.L. Bauer and C. Reinsch (1968). "Rational QR Transformation with Newton Shift for Symmetric Tridiagonal Matrices," *Numer. Math. 11,* 264-72. See also *HACLA*, pp. 257–65.

F.L. Bauer and C. Reinsch (1970). "Inversion of Positive Definite Matrices by the Gauss-Jordan Methods," in *HACLA*, pp. 45–49.

C. Bavely and G.W. Stewart (1979). "An Algorithm for Computing Reducing Subspaces by Block Diagonalization," *SIAM J. Num. Anal. 16,* 359–67.

R. Bellman (1970). *Introduction to Matrix Analysis*, 2nd ed., McGraw-Hill, New York.

E. Beltrami (1873). "Sulle Funzioni Bilineari," *Giornale di Mathematiche 11,* 98–106.

C.F. Bender and I. Shavitt (1970). "An Iterative Procedure for the Calculation of the Lowest Real Eigenvalue and Eigenvector of a Non-Symmetric Matrix," *J. Comp. Physics 6,* 146–49.

A. Berman and A. Ben-Israel (1971). "A Note on Pencils of Hermitian of Symmetric Matrices," *SIAM J. Appl. Math. 21,* 51–54.

M.J.M. Bernal and J.H. Verner (1968). "On Generalizations of the Theory of Consistent Orderings for Successive Over-Relaxation Methods," *Numer. Math. 12,* 215–22.

M. Berry and A. Sameh (1986). "Multiprocessor Jacobi Algorithms for Dense Symmetric Eigenvalue and Singular Value Decompositions," in *Proc. International Conference on Parallel Processing*, 433–440.

C.H. Bischof (1987). "The Two-Sided Block Jacobi Method on Hypercube Architectures," in *Hypercube Multiprocessors*, ed. M.T. Heath, SIAM Press, Philadelphia.

C.H. Bischof (1988). "Computing the Singular Value Decomposition on a Distributed System of Vector Processors," Cornell Computer Science Report 87 869, Ithaca, NY.

C.H. Bischof (1988). "QR Factorization Algorithms for Coarse Grain Distributed Systems," PhD Thesis, Dept. of Computer Science, Cornell University, Ithaca, NY.

C.H. Bischof and C. Van Loan (1986). "Computing the SVD on a Ring of Array Processors," in *Large Scale Eigenvalue Problems,* eds. J. Cullum and R. Willoughby, North Holland, p.51-66.

C.H. Bischof and C. Van Loan (1987). "The WY Representation for Products of Householder Matrices," *SIAM J. Sci. and Stat. Comp. 8,* s2–s13.

Å. Björck (1967a). "Iterative Refinement of Linear Least Squares Solution I," *BIT 7,* 257–78.

Å. Björck (1967b). "Solving Linear Least Squares Problems by Gram-Schmidt Orthogonalization," *BIT 7,* 1–21.

Å. Björck (1968). "Iterative Refinement of Linear Least Squares Solution II," *BIT 8,* 8–30.

Å. Björck (1984). "A General Updating Algorithm for Constrained Linear Least Squares Problems," *SIAM J. Sci. and Stat. Comp. 5,* 394–402.

Å. Björck (1987). "Stability Analysis of the Method of Seminormal Equations," *Lin. Alg. and Its Applic. 88/89,* 31–48.

Å. Björck (1988). *Least Squares Methods: Handbook of Numerical Analysis Vol. 1 Solution of Equations in* $R^N$, Elsevier North Holland.

Å. Björck and C. Bowie (1971). "An Iterative Algorithm for Computing the Best Estimate of an Orthogonal Matrix," *SIAM J. Num. Anal. 8,* 358–64.

Å. Björck and T. Elfving (1973). "Algorithms for Confluent Vandermonde Systems," *Numer. Math. 21,* 130–37.

Å. Björck and G.H. Golub (1967). "Iterative Refinement of Linear Least Squares Solutions by Householder Transformation," *BIT 7,* 322–37.

Å. Björck and G.H. Golub (1973). "Numerical Methods for Computing Angles Between Linear Subspaces," *Math. Comp. 27,* 579–94.

Å. Björck and V. Pereyra (1970). "Solution of Vandermonde Systems of Equations," *Math. Comp. 24,* 893–903.

Å. Björck, R.J. Plemmons, and H. Schneider (1981). *Large-Scale Matrix Problems.* North-Holland, New York.

Å. Björck and S. Hammarling (1983). "A Schur Method for the Square Root of a Matrix," *Lin. Alg. and Its Applic. 52/53,* 127–140.

J.M. Blue (1978). "A Portable FORTRAN Program to Find the Euclidean Norm of a Vector," *ACM Trans. Math. Soft. 4,* 15–23.

Z. Bohte (1975). "Bounds for Rounding Errors in the Gaussian Elimination for Band Systems," *J. Inst. Math. Applic. 16,* 133–42.

A.W. Bojanczyk, R.P. Brent, and F.R. de Hoog (1986). "QR Factorization of Toeplitz Matrices," *Numer. Math. 49,* 81-94.

A.W. Bojanczyk, R.P. Brent, P. Van Dooren, and F.R. de Hoog (1987). "A Note on Downdating the Cholesky Factorization," *SIAM J. Sci. and Stat. Comp. 8,* 210–221.

D.L. Boley and G.H. Golub (1978). "The Matrix Inverse Eigenvalue Problem for Periodic Jacobi Matrices," in *Proc. Fourth Symposium on Basic Problems of Numerical Mathematics*, Prague, pp. 63–76.

D. Boley and G.H. Golub (1984). "A Modified Method for Restructuring Periodic Jacobi Matrices," *Math. Comp. 42,* 143–150.

J. Boothroyd and P.J. Eberlein (1968). "Solution to the Eigenproblem by a Norm-Reducing Jacobi-Type Method (Handbook)," *Numer. Math. 11,* 1-12. See also *HACLA,* pp. 327–38.

H.J. Bowdler, R.S. Martin, G. Peters, and J.H. Wilkinson (1966). "Solution of Real and Complex Systems of Linear Equations,"*Numer. Math. 8,* 217-34. See also *HACLA,* pp. 93–110.

H. Bowdler, R.S. Martin, C. Reinsch, and J.H. Wilkinson (1968). "The QR and QL Algorithms for Symmetric Matrices," *Numer. Math. 11,* 293-306. See also *HACLA,* pp. 227–40.

J. Boyle, R. Butler, T. Disz, B. Glickfield, E. Lusk, R. Overbeek, J. Patterson, and R. Stevens (1987). *Portable Programs for Parallel Processors,* Holt, Rinehart and Winston.

J.H. Bramble, J.E. Pasciak, and A.H. Schatz (1986a). "The construction of Preconditioners for Elliptic Problems by Substructuring I," *Math. Comp. 47,* 103–134.

J.H. Bramble, J.E. Pasciak, and A.H. Schatz (1986b). "The construction of Preconditioners for Elliptic Problems by Substructuring II," *Math. Comp. 49,* 1–17.

R.P. Brent (1970). "Error Analysis of Algorithms for Matrix Multiplication and Triangular Decomposition Using Winograd's Identity," *Numer. Math. 16,* 145–156.

R.P. Brent and F.T. Luk (1985). "The Solution of Singular Value and Symmetric Eigenvalue Problems on Multiprocessor Arrays," *SIAM J. Sci. and Stat. Comp. 6,* 69–84.

R.P. Brent, F.T. Luk, and C. Van Loan (1985). "Computation of the Singular Value Decomposition Using Mesh Connected Processors," *J. VLSI Computer Systems 1,* 242–270.

K.W. Brodlie and M.J. D. Powell (1975). "On the Convergence of Cyclic Jacobi Methods," *J. Inst. Math. Applic. 15,* 279–87.

C.G. Broyden (1973). "Some Condition Number Bounds for the Gaussian Elimiantion Process," *J. Inst. Math. Applic. 12,* 273–86.

A. Buckley (1974). "A Note on Matrices $A = 1 + H$, $H$ Skew-Symmetric," *Z. Angew. Math. Mech. 54,* 125–26.

A. Buckley (1977). "On the Solution of Certain Skew-Symmetric Linear Systems," *SIAM J. Num. Anal. 14,* 566–70.

J.R. Bunch (1971a). "Analysis of the Diagonal Pivoting Method," *SIAM J. Num. Anal. 8,* 656–80.

J.R. Bunch (1971b). "Equilibration of Symmetric Matrices in the Max–Norm," *J. ACM 18,* 566–72.

J.R. Bunch (1974). "Partial Pivoting Strategies for Symmetric Matrices," *SIAM J. Num. Anal. 11,* 521–28.

J.R. Bunch (1976). "Block Methods for Solving Sparse Linear Systems," in *Sparse Matrix Computations,* eds. J.R. Bunch and D.J. Rose, Academic Press, New York.

J.R. Bunch and K. Kaufman (1977). "Some Stable Methods for Calculating Inertia and Solving Symmetric Linear Systems," *Math. Comp. 31,* 162–79.

J.R. Bunch, K. Kaufman, and B.N. Parlett (1976). "Decomposition of a Symmetric Matrix," *Numer. Math. 27,* 95–109.

J.R. Bunch, C.P. Nielsen, and D.C. Sorensen (1978). "Rank-One Modification of the Symmetric Eigenproblem," *Numer. Math. 31,* 31-48.

J.R. Bunch and B.N. Parlett (1971). "Direct Methods for Solving Symmetric Indefinite Systems of Linear Equations," *SIAM J. Num. Anal. 8,* 639–55.

J.R. Bunch and D.J. Rose, eds. (1976). *Sparse Matrix Computations,* Academic Press, New York.

J.R. Bunch (1982). "A Note on the Stable Decomposition of Skew Symmetric Matrices," *Math. Comp. 158,* 475–480.

J.R. Bunch (1985). "Stability of Methods for Solving Toeplitz Systems of Equations," *SIAM J. Sci. Stat. Comp. 6,* 349–364.

J.R. Bunch (1987). 'The Weak and Strong Stability of Algorithms in Numerical Linear Algebra," *Lin. Alg. and Its Applic. 88/89,* 49–66.

O. Buneman (1969). "A Compact Non–Interative Poisson Solver," Report 294, Stanford University Institute for Plasma Research, Stanford, CA.

A. Bunse-Gerstner (1984). "An Algorithm for the Symmetric Generalized Eigenvalue Problem," *Lin. Alg. and Its Applic. 58,* 43–68.

A. Bunse-Gerstner and W.B. Gragg (1988). "Singular Value Decompositions of Complex Symmetric Matrices," *J. Comp. Applic. Math. 21,* 41–54.

P.A. Businger (1968). "Matrices Which Can be Optimally Scaled," *Numer. Math. 12,* 346–48.

P.A. Businger (1969). "Reducing a Matrix to Hessenberg Form," *Math. Comp. 23,* 819–21.

P.A. Businger (1971a). "Monitoring the Numerical Stability of Gaussian Elimination," *Numer. Math. 16,* 360–61.

P.A. Businger (1971b). "Numerically Stable Deflation of Hessenberg and Symmetric Tridiagonal Matrices," *BIT 11,* 262–70.

P.A. Businger and G.H. Golub (1965). "Linear Least Squares Solutions by Householder Transformations," *Numer. Math. 7,* 269–76. See also *HACLA,* pp. 111–18.

P.A. Businger and G.H. Golub (1969). "Algorithm 358: Singular Value Decomposition of a Complex Matrix," *Comm. ACM 12,* 564–65.

B.L. Buzbee and F.W. Dorr (1974). "The Direct Solution of the Biharmonic Equation on Rectangular Regions and the Poisson Equation on Irregular Regions," *SIAM J. Num. Anal. 11,* 753–63.

B.L. Buzbee, F.W. Dorr, J.A. George, and G.H. Golub (1971). "The Direct Solution of the Discrete Poisson Equation on Irregular Regions," *SIAM J. Num. Anal. 8,* 722–36.

B.L. Buzbee, G.H. Golub, and C.W. Nielson (1970). "On Direct Methods for Solving Poisson's Equations," *SIAM J. Num. Anal. 7,* 627–56.

B.L. Buzbee (1986) "A Strategy for Vectorization," *Parallel Computing 3,* 187–192.

R. Byers (1983). "Hamiltonian and Symplectic Algorithms for the Algebraic Riccati Equation," PhD Thesis, Center for Applied Mathematics, Cornell University.

R. Byers (1984). "A Linpack–Style Condition Estimator for the Equation $AX - XB^T = C$," *IEEE Trans. Auto. Cont. AC-29,* 926–928.

R. Byers (1986) "A Hamiltonian QR Algorithm," *SIAM J. Sci. and Stat. Comp. 7,* 212–229.

R. Byers and S.G. Nash (1987). "On the Singular Vectors of the Lyapunov Operator," *SIAM J. Alg. and Disc. Methods 8,* 59–66.

D.A. Calihan (1986). "Block-Oriented, Local-Memory-Based Linear Equation Solution on the Cray-2: Uniprocessor Algorithms," *Proceedings of the 1986 Conference on Parallel Processing,* pp. 375–378.

S.P. Chan and B.N. Parlett (1977). "Algorithm 517: A Program for Computing the Condition Numbers of Matrix Eigenvalues without Computing Eigenvectors," *ACM. Trans. Math. Soft. 3,* 186–203.

T.F. Chan (1982a). "An Improved Algorithm for Computing the Singular Value Decomposition," *ACM Trans. Math. Soft. 8,* 72–83.

T.F. Chan (1982b). "Algorithm 581: An Improved Algorithm for Computing the Singular Value Decompositon," *ACM Trans. Math. Soft. 8,* 84–88.

T.F. Chan (1984). "Deflated Decomposition Solutions of Nearly Singular Systems," *SIAM J. Numer. Anal. 21,* 738–754.

T.F. Chan (1985). "On the Existence and Computation of LU Factorizations with small pivots," *Math. Comp. 42,* 535–548.

T.F. Chan (1987). "Rank-Revealing QR Factorizations," *Lin. Alg. and Its Applic. 88/89,* 67–82.

T.F. Chan (1988). "An Optimal Circulant Preconditioner for Toeplitz Systems," *SIAM J. Sci. and Stat. Comp. 9,* 766-771.

T.F. Chan, K.R. Jackson, and B. Zhu (1983). "Alternating Direction Incomplete Factorizations," *SIAM J. Numer. Anal. 20,* 239–257.

H.Y. Chang and M.Salama (1988). "A Parallel Householder Tridiagonalization Strategm Using Scattered Square Decomposition," *Parallel Computing 6,* 297–312.

J.P. Charlier, M. Vanbegin, P. Van Dooren (1988). "On Efficient Implementaion of Kogbetliantz's Algorithm for Computing the Singular Value Decomposition," *Numer. Math. 52,* 279–300.

J.P. Charlier and P. Van Dooren (1987). "On Kogbetliantz's SVD Algorithm in the Presence of Clusters," *Lin. Alg. and Its Applic. 95,* 135–160.

S. Chen, J. Dongarra, and C. Hsuing (X). "Multiprocessing Linear Algebra Algorithms on the Cray X–MP–2: Experiences with Small Granularity," *J. Parallel and Distributed Computing 1,* 22–31.

S. Chen, D. Kuck, and A. Sameh (1978). "Practical Parallel Band Triangular Systems Solvers," *ACM Trans. Math. Soft. 4,* 270–277.

K.H. Cheng and S. Sahni (1987). "VLSI Systems for Band Matrix Multiplication," *Parallel Computing 4,* 239–258.

R.C. Chin, T.A. Manteuffel, and J. de Pillis (1984). "ADI as a Preconditioning for Solving the Convection-Diffusion Equation," *SIAM J. Sci. and Stat. Comp. 5,* 281–299.

A.K. Cline (1973). "An Elimination Method for the Solution of Linear Least Squares Problems," *SIAM J. Num. Anal. 10,* 283–89.

A.K. Cline (1976a). "A Descent Method for the Uniform Solution to Overdetermined Systems of Equations", *SIAM J. Num. Anal. 13,* 293–309.

A.K. Cline (1976b). "Several observations on the Use of Conjugate Gradient Methods," ICASE Report 76-22. NASA Langley Research Center, Hampton, VA.

A.K. Cline and R.K. Rew (1983). "A Set of Counter examples to Three Condition Number Estimators," *SIAM J. Sci. and Stat. Comp. 4*, 602–611.

A.K. Cline, A.R. Conn, and C. Van Loan (1982). "Generalizing the LINPACK Condition Estimator," in Numerical Analysis, ed. J.P. Hennart, Lecture Notes in Mathematics, no. 909, Springer-Verlag, NY.

A.K. Cline, G.H. Golub, and G.W. Platzman (1976). "Calculation of Normal Modes of Oceans Using a Lanczos Method," in *Sparse Matrix Computations*, ed. J.R. Bunch and D.J. Rose, Academic Press, NY, pp. 409–26.

A.K. Cline, C.B. Moler, G.W. Stewart, and J.H. Wilkinson (1979). "An Estimate for the Condition Number of a Matrix," *SIAM J. Num. Anal. 16*, 368–75.

R.E. Cline and R.J. Plemmons (1976). "$L_1$-Solutions to Underdetermined Linear Systems," *SIAM Review 18*, 92–106.

M. Clint and A. Jennings (1970). "The Evaluation of Eigenvalues and Eigenvectors of Real Symmetric Matrix by Simultaneous Iteration," *Comp. J. 13*, 76–80.

M. Clint and A. Jennings (1971). "A Simultaneous Iteration Method for the Unsymmetric Eigenvalue Problem," *J. Inst. Math. Applic. 8*, 111–21.

W.G. Cochrane (1968). "Errors of Measurement in Statistics," *Technometrics 10*, 637–66.

A.M. Cohen (1974). "A Note on Pivot Size in Gaussian Elimination," *Lin. Alg. and Its Applic. 8*, 361–68.

T. Coleman and C. Van Loan (1988). *Handbook for Matrix Computations*, SIAM Publications, Phil.

P. Concus and G.H. Golub (1973). "Use of Fast Direct Methods for the Efficient Numerical Solution of Nonseparable Elliptic Equations," *SIAM J. Num. Anal. 10*, 1103–20.

P. Concus, G.H. Golub, and G. Meurant (1985). "Block Preconditioning for the Conjugate Gradient Method," *SIAM J. Sci. and Stat. Comp. 6*, 220–252.

P. Concus, G.H. Golub, and D.P. O'Leary (1976). "A Generalized Conjugate Gradient Method for the Numerical Solution of Elliptic Partial Differential Equations," in *Sparse Matrix Computations*, ed. J.R. Bunch and D.J. Rose, Academic Press, NY.

S.D. Conte and C. de Boor (1980). *Elementary Numerical Analysis: An Algorithmic Approach*, 3rd. ed., McGraw-Hill, NY.

J.E. Cope and B.W. Rust (1979). "Bounds on Solutions of Systems with Inaccurate Data," *SIAM J. Num. Anal. 16*, 950–63.

M. Costnard, J.M. Muller, and Y. Robert (1986). "Parallel QR Decomposition of a Rectangular Matrix," *Numer. Math. 48*, 239–250.

M. Costnard, M. Marrakchi, and Y. Robert (1988). "Parallel Gaussian Elimination on an MIMD Computer," *Parallel Computing 6*, 275–296.

R.W. Cottle (1974). "Manifestations of the Schur Complement," *Lin. Alg. and Applic. 8*, 189–211.

M.G. Cox (1981). "The Least Squares Solution of Overdetermined Linear Equations having Band or Augmented Band Structure," *IMA J. Numer. Anal. 1*, 3–22.

C.R. Crawford (1973). "Reduction of a Band Symmetric Generalized Eigenvalue Problem," *Comm. ACM 16*, 41–44.

C.R. Crawford (1976). "A Stable Generalized Eigenvalue Problem," *SIAM J. Num. Anal. 13*, 854–60.

C.R. Crawford (1986). "Algorithm 646 PDFIND: A Routine to Find a Positive Definite Linear Combination of Two Real Symmetric Matrices," *ACM Trans. Math. Soft. 12*, 278–282.

C.R. Crawford and Y.S. Moon (1983). "Finding a Positive Definite Linear Combination of Two Hermitian Matrices," *Lin. Alg. and Its Applic. 51*, 37–48.

C.W. Cryer (1968). "Pivot Size in Gaussian Elimination," *Numer. Math.12*, 335–45.

J. Cullum (1978). "The Simultaneous Computation of a Few of the Algebraically Largest and Smallest Eigenvalues of a Large Sparse Symmetric Matrix," *BIT 18*, 265–75.

J. Cullum and W.E. Donath (1974). "A Block Lanczos Algorithm for Computing the Q Algebraically Largest Eigenvalues and a Corresponding Eigenspace of Large, Sparse Real Symmetric Matrices," *Proc. of the 1974 IEEE Conf. on Decision and Control,* Phoenix, AZ, pp. 505–9.

J. Cullum and R.A. Willoughby (1977). "The Equivalence of the Lanczos and the Conjugate Gradient Algorithms," *IBM Research Report* RC-6903.

J. Cullum and R.A. Willoughby (1979). "Lanczos and the Computation in Specified Intervals of the Spectrum of Large, Sparse Real Symmetric Matrices," in *Sparse Matrix Proc., 1978* , eds. I.S. Duff and G.W. Stewart, SIAM Publications, Philadelphia, PA.

J. Cullum and R.A. Willoughby (1980). "The Lanczos Phenomena: An Interpretation Based on Conjugate Gradient Optimization," *Lin. Alg. and Its Applic. 29,* 63–90.

J. Cullum and R.A. Willoughby (1985a). *Lanczos Algorithms for Large Symmetric Eigenvalue Computations, Vol. I Theory,* Birkhaüser, Boston.

J. Cullum and R.A. Willoughby (1985b). *Lanczos Algorithms for Large Symmetric Eigenvalue Computations, Vol. II Programs,* Birkhaüser, Boston.

J. Cullum and R.A. Willoughby (eds) (1986). *Large Scale Eigenvalue Problems,* North Holland.

J. Cullum, R.A. Willoughby, and M. Lake (1983). "A Lanczos Algorithm for Computing Singular Values and Vectors of Large Matrices," *SIAM J. Sci. and Stat. Comp. 4,* 197–215.

J.J.M. Cuppen (1981). "A Divide and Conquer Method for the Symmetric Eigenproblem," *Numer. Math. 36,* 177–95.

J.J.M. Cuppen (1983). "The Singular Value Decomposition in Product Form," *SIAM J. Sci. and Stat. Comp. 4,* 216–222.

J.J.M. Cuppen (1984). "On Updating Triangular Products of Householder Matrices," *Numer. Math. 45,* 403–410.

E. Cuthill (1972). "Several Strategies for Reducing the Bandwidth of Matrices," in *Sparse Matrices and Their Applications,* ed., D.J. Rose and R.A. Willoughby, Plenum Press, NY.

G. Cybenko (1978). "Error Analysis of Some Signal Processing Algorithms," Ph.D. thesis, Princeton University, Princeton, NJ.

G. Cybenko (1980). "The Numerical Stability of the Levinson-Durbin Algorithm for Toeplitz Systems of Equations," *SIAM J. Sci. and Stat. Comp. 1,* 303–10.

G. Cybenko (1984). "The Numerical Stability of the Lattice Algorithm for Least Squares Linear Prediction Problems," *BIT 24,* 441–455.

G. Cybenko and C. Van Loan (1986). "Computing the Minimum Eigenvalue of a Symmetric Positive Definite Toeplitz Matrix," *SIAM J. Sci. and Stat. Comp. 7,* 123–131.

J. Daniel, W.B. Gragg, L. Kaufman, and G.W. Stewart (1976). "Reorthogonalization and Stable Algorithms for Updating the Gram-Schmidt QR Factorization," *Math. Comp. 30,* 772–95.

C. Davis (1973). "Explicit Functional Calculus," *Lin. Alg. and Its Applic 6,.* 193–99.

B.N. Datta, C.R. Johnson. M.A. Kaashoek, R. Plemmons, and E.D. Sontag (1988), *Linear Algebra in Signals, Systems, and Control,* SIAM Publications, Philadelphia.

G.J. Davis (1986). "Column LU Pivoting on a Hypercube Multiprocessor," *SIAM J. Alg. and Disc. Methods* 7, 538–550.

C. Davis and W.M. Kahan (1970). "The Rotation of Eigenvectors by a Perturbation III," *SIAM J. Num. Anal. 7,* 1–46.

A. Dax and S. Kaniel (1977). "Pivoting Techniques for Symmetric Gaussian Elimination," *Numer. Math. 28,* 221–42.

J. Day and B. Peterson (1988). "Growth in Gaussian Elimination," *Amer. Math. Monthly 95,* 489-513.

M.J. Dayde and I.S. Duff (1988). "Use of Level-3 BLAS in LU Factorization on the Cray-2, the ETA-10P, and the IBM 3090-200/VF," Report CSS-229, Computer Science and Systems Division, Harwell Laboratory, Oxon OX11 ORA, England.

C. de Boor and A. Pinkus (1977). "A Backward Error Analysis for Totally Positive Linear Systems," *Numer. Math. 27,* 485–90.

P. Deift, T. Nande, and C. Tome (1983). "Ordinary Differential Equations and the Symmetric Eigenvalue Problem," SIAM J. Numer. Anal. 20, 1–22.

T.J. Dekker and J.F. Traub (1971). "The Shifted QR Algorithm for Hermitian Matrices," *Lin. Alg. and Its Applic. 4,* 137–54.

J.M. Delosme and I.C.F. Ipsen (1986). "Parallel Solution of Symmetric Positive Definite Systems with Hyperbolic Rotations," *Lin. Alg. and Its Applic. 77,* 75–112.

J.W. Demmel (1983a). "A Numerical Analyst's Jordan Canonical Form," Ph.D. Thesis, Univ. of California, Berkeley.

J.W. Demmel (1983b). "The Condition Number of Equivalence Transformations that Block Diagonalize Matrix Pencils," *SIAM J. Numer. Anal. 20,* 599–610.

J.W. Demmel (1984). "Underflow and the Reliability of Numerical Software," *SIAM J. Sci. and Stat. Comp. 5,* 887–919.

J.W. Demmel (1987a). "On the Distance to the Nearest Ill-Posed Problem," *Numer. Math. 51,* 251–289.

J.W. Demmel (1987b). "A Counterexample for two Conjectures About Stability," *IEEE Trans. Auto. Cont. AC-32,* 340–342.

J.W. Demmel (1987c). "Three Methods for Refining Estimates of Invariant Subspaces," *Computing 38,* 43–57.

J.W. Demmel (1987d). "The smallest perturbation of a submatrix which lowers the rank and constrained total least squares problems, *SIAM J. Numer. Anal.* 24, 199–206.

J.W. Demmel (1988). "The Probability that a Numerical Analysis Problem is Difficult," *Math. Comp. 50,* 449–480.

J.W. Demmel and B. Kågström (1987). "Computing Stable Eigendecompositions of Matrix Pencils," *Linear Alg. and Its Applic. 88/89,* 139–186.

J.W. Demmel and B. Kågström (1988). "Accurate Solutions of Ill-Posed Problems in Control Theory," *SIAM J. Matrix Anal. Appl.* 126–145.

J.E. Dennis and R. Schnabel (1983). *Numerical Methods for Unconstrained Optimization and Nonlinear Equations*, Prentice-Hall, Englewood Cliffs, NJ.

J.E. Dennis Jr and K. Turner (1987). "Generalized Conjugate Directions," *Lin. Alg. and Its Applic 88/89,* 187–209.

J. Descloux (1963). "Bounds for the Spectral Norm of Functions of Matrices," *Numer. Math. 5,* 185–90.

M.A. Diamond and D.L.V. Ferreira (1976). "On a Cyclic Reduction Method for the Solution of Poisson's Equation, *SIAM J. Num. Anal. 13,* 54–70.

J.J. Dongarra (1983). "Improving the Accuracy of Computed Singular Values," *SIAM J. Sci. and Stat. Comp. 4,* 712–719.

J.J. Dongarra, J.R. Bunch, C.B. Moler, and G.W. Stewart (1978). *LINPACK Users Guide,* SIAM Publications, Philadelphia, PA.

J.J. Dongarra, J. Du Croz, I.S. Duff, and S. Hammarling (1988). "A Set of Level 3 Basic Linear Algebra Subprograms," Argonne National Laboratory Report, ANL-MCS-TM-88.

J.J. Dongarra, J. Du Croz, S. Hammarling, and R.J. Hanson (1988a). "An Extended Set of Fortran Basic Linear Algebra Subprograms," *ACM Trans. Math. Soft. 14,* 1–17.

J.J. Dongarra, J. Du Croz, S. Hammarling, and R.J. Hanson (1988b). "Algorithm 656 An Extended Set of Fortran Basic Linear Algebra Subprograms: Model Implementation and Test Programs," *ACM Trans. Math. Soft. 14,* 18–32.

J.J. Dongarra and S. Eisenstat (1984). "Squeezing the Most Out of an Algorithm in Cray Fortran," *ACM Trans. Math. Soft. 10,* 221–230.

J.J. Dongarra, F.G. Gustavson, and A. Karp (1984). "Implementing Linear Algebra Algorithms for Dense Matrices on a Vector Pipeline Machine," *SIAM Review 26,* 91–112.

J.J. Dongarra, S. Hammarling, and D.C. Sorensen (1987). "Block Reduction of Matrices to Condensed form for Eigenvalue Computations," ANL-MCS- TM 99, Argonne National Laboratory, Argonne, Illinois.

J. Dongarra and T. Hewitt (1986). "Implementing Dense Linear Algebra Algorithms Using Multitasking on the Cray X-MP-4 (or Approaching the Gigaflop)," *SIAM J. Sci. and Stat. Comp. 7,* 347–350.

J. Dongarra and A. Hinds (1979). "Unrolling Loops in Fortran," *Software Practice and Experience 9,* 219–229.

J.J. Dongarra and R.E. Hiromoto (1984). "A Collection of Parallel Linear Equation Routines for the Denelcor HEP," *Parallel Computing 1,* 133–142.

J.J. Dongarra, L. Kaufman, and S. Hammarling (1986). "Squeezing the Most Out of Eigenvalue Solvers on High Performance Computers," *Lin. Alg. and Its Applic. 77,* 113–136.

J.J. Dongarra, C.B. Moler, and J.H. Wilkinson (1983). "Improving the Accuracy of Computed Eigenvalues and Eigenvectors," *SIAM J. Numer. Anal. 20,* 23–46.

J.J. Dongarra and A.H. Sameh (1984). "On Some Parallel Banded System Solvers," *Parallel Computing 1,* 223–235.

J.J. Dongarra, A. Sameh, and D. Sorensen (1986). "Implementation of Some Concurrent Algorithms for Matrix Factorization," *Parallel Computing 3,* 25–34.

J.J. Dongarra and D.C. Sorensen (1986). "Linear Algebra on High Performance Computers," *Appl. Math. and Comp. 20,* 57–88.

J.J. Dongarra and D.C. Sorensen (1987). "A Fully Parallel Algorithm for the Symmetric Eigenvalue Problem," *SIAM J. Sci. and Stat. Comp. 8,* S139–S154.

J.J. Dongarra and D.C. Sorensen (1987). "A Portable Environment for Developing Parallel Programs," *Parallel Computing 5,* 175–186.

F.W. Dorr (1970). "The Direct Solution of the Discrete Poisson Equation on a Rectangle," *SIAM Review 12,* 248–63.

F.W. Dorr (1973). "The Direct Solution of the Discrete Poisson Equation in $O(n^2)$ Operations," *SIAM Review 15,* 412–15.

P.F. Dubois, A. Greenbaum, and G.H. Rodrigue (1979). "Approximating the Inverse of a Matrix for Use on Iterative Algorithms on Vector Processors," *Computing 22,* 257–268.

A. Dubrulle (1970). "A Short Note on the Implicit QL Algorithm for Symmetric Tridiagonal Matrices," *Numer. Math.15,* 450.

A. Dubrulle, R.S. Martin, and J.H. Wilkinson (1968). "The Implicit QL Algorithm," *Numer. Math. 12,* 377-83. See also*HACLA*, pp. 241–48.

I.S. Duff (1974). "Pivot Selection and Row Ordering in Givens Reduction on Sparse Matrices," *Computing 13,* 239–48.

I.S. Duff (1977). "A Survey of Sparse Matrix Research," *Proc. IEEE 65,* 500–35.

I.S. Duff and J.K. Reid (1975). "On the Reduction of Sparse Matrices to Condensed Forms by Similarity Transformations," *J. Inst. Math. Applic. 15,* 217–24.

I.S. Duff and J.K. Reid (1976). "A Comparison of Some Methods for the Solution of Sparse Over-Determined Systems of Linear Equations," *J. Inst. Math. Applic. 17,* 267–80.

I.S. Duff and G.W. Stewart, eds. (1979). *Sparse Matrix Proceedings, 1978,* SIAM Publications, Philadelphia, PA.

I.S. Duff, A.M. Erisman, and J.K. Reid (1986). *Direct Methods for Sparse Matrices,* Oxford University Press.

N. Dunford and J. Schwartz (1958). *Linear Operators, Part I,* Interscience, New York.

J. Durbin (1960). "The Fitting of Time Series Models," *Rev. Inst. Int. Stat. 28,* 233–43.

P.J. Eberlein (1965). "On Measures of Non-normality for Matrices," *Amer. Math. Soc. Monthly 72,* 995–96.

P.J. Eberlein (1970). "Solution to the Complex Eigenproblem by a Norm-Reducing Jacobi-Type Method," *Numer. Math. 14,* 232–45. See also *HACLA*, pp. 404–17.

P.J. Eberlein (1971). "On the Diagonalization of Complex Symmetric Matrices," *J. Inst. Math. Applic. 7,* 377–83.

P.J. Eberlein (1987). "On Using the Jacobi Method on a Hypercube," in *Hypercube Multiprocessors*, ed. M.T. Heath, SIAM Publications, Philadelphia.

C. Eckart and G. Young (1939). "A Principal Axis Transformation for Non-Hermitian Matrices," *Bull. Amer. Math. Soc. 45,* 118–21.

M. Eiermann and W. Niethammer (1983). "On the Construction of Semi-iterative Methods," *SIAM J. Numer. Anal. 20*, 1153–1160.

S.C. Eisenstat (1984). "Efficient Implementation of a Class of Preconditioned Conjugate Gradient Methods," *SIAM J. Sci. and Stat. Computing 2,* 1–4.

S.C Eisenstat, M.T. Heath, C.S. Henkel, and C.H. Romine (1988). "Modified Cyclic Algorithms for Solving Triangular Systems on Distributed Memory Multiprocessors," *SIAM J. Sci. and Stat. Comp. 9,* 589–600.

L. Eldèn (1977a). "Algorithms for the Regularization of Ill-Conditioned Least Squares Problems," *BIT 17,* 134–45.

L. Eldèn (1977b). "Numerical Analysis of Regularization and Constrained Least Square Methods," Ph.D. thesis, Linkoping Studies in Science and Technology Dissertation, no. 20, Linkoping, Sweden.

L. Eldèn (1980). "Perturbation Theory for the Least Squares Problem with Linear Equality Constraints," *SIAM J. Num. Anal. 17,* 338–50.

L. Eldèn (1983). "A Weighted Pseudoinverse, Generalized Singular Values, and Constrained Least Squares Problems," *BIT 22* , 487–502.

L. Eldèn (1984). "An Algorithm for the Regularization of Ill-Conditioned, Banded Least Squares Problems," *SIAM J. Sci. and Stat. Comp. 5*, 237–254.

L. Eldèn (1985). "A Note on the Computation of the Generalized Cross-Validation Function for Ill-Conditioned Least Squares Problems," *BIT 24,* 467–472.

L. Eldèn (1988) "A Parallel QR Decomposition Algorithm," Report LiTh Mat R 1988-02, Dept. of Math., Linkoping University, Sweden.

L. Eldèn and R. Schreiber (1986). "An Application of Systolic Arrays to Linear Discrete Ill-Posed Problems," *SIAM J. Sci. and Stat. Comp. 7,* 892–903.

H. Elman (1986). "A Stability Analysis of Incomplete LU Factorization," *Math. Comp. 47*, 191–218.

L. Elsner and J. Guang Sun (1982). "Perturbation Theorems for the Generalized Eigenvalue Problem," *Lin. Alg. and its Applic. 48,* 341–357.

A. Elster and A.P. Reeves (1988). "Block Matrix Operations Using Orthogonal Trees," in G. Fox (1988), 1554–1561.

W. Enright (1979). "On the Efficient and Reliable Numerical Solution of Large Linear Systems of O.D.E.'s," *IEEE Trans. Auto. Cont. AC-24,* 905–8.

I. Erdelyi (1967). "On the Matrix Equation $Ax = \lambda Bx$," *J. Math. Anal. and Applic. 17,* 119–32.

T. Ericsson and A. Ruhe (1980). "The Spectral Transformation Lanczos Method for the Numerical Solution of Large Sparse Generalized Symmetric Eigenvalue Problems," *Math. Comp. 35,* 1251–68.

A.M. Erisman and J.K. Reid (1974). "Monitoring the Stability of the Triangular Factorization of a Sparse Matrix," *Numer. Math. 22,* 183–86.

D.J. Evans (1984). "Parallel SOR Iterative Methods," *Parallel Computing 1,* 3–18.

D.J. Evans and R. Dunbar (1983). "The Parallel Solution of Triangular Systems of Equations," *IEEE Trans. Comp. C-32,* 201–204.

V. Faber and T. Manteuffel (1984). "Necessary and Sufficient Conditions for the Existence of a Conjugate Gradient Method," *SIAM J. Numer. Anal. 21* 352–362.

D.K. Faddeev and V.N. Faddeva (1963). *Computational Methods of Linear Algebra*, W.H. Freeman and Co., San Francisco, CA.

V. Fadeeva and D. Fadeev (1977). "Parallel Computations in Linear Algebra," *Kibernetica 6,* 28–40.

W. Fair and Y. Luke (1970). "Padé Approximations to the Operator Exponential," *Numer. Math. 14,* 379–82.

D.G. Feingold and R.S. Varga (1962). "Block Diagonally Dominant Matrices and Generalizations of the Gershgorin Circle Theorem," *Pacific J. Math. 12,* 1241–50.

T. Fenner and G. Loizou (1974). "Some New Bounds on the Condition Numbers of Optimally Scaled Matrices", *J. ACM 1,* 514–24.

C. Fischer and R.A. Usmani (1969). "Properties of Some Tridiagonal Matrices and Their Application to Boundary Value Problems," *SIAM J. Num. Anal. 6,* 127–42.

G. Fix and R. Heiberger (1972). "An Algorithm for the Ill-Conditioned Generalized Eigenvalue Problem," *SIAM J. Num. Anal. 9,* 78–88.

R. Fletcher (1976). "Factorizing Symmetric Indefinite Matrices", *Lin. Alg. and Its Applic. 14,* 257–72.

G.E. Forsythe (1960). "Crout with Pivoting," *Comm. ACM 3,* 507–8.

G.E. Forsythe and G.H. Golub (1965). "On the Stationary Values of a Second-Degree Polynomial on the Unit Sphere," *SIAM J. App. Math.13,* 1050–68.

G.E. Forsythe and P. Henrici (1960). "The Cyclic Jacobi Method for Computing the Principal Values of a Complex Matrix," *Trans. Amer. Math. Soc. 94,* 1–23.

G.E. Forsythe, M.A. Malcolm, and C.B. Moler (1977). *Computer Methods for Mathematical Computations*, Prentice-Hall, Englewood Cliffs, NJ.

G.E. Forsythe and C.B. Moler (1967). *Computer Solution of Linear Algebraic Systems*, Prentice-Hall, Englewood Cliffs, NJ.

L.V. Foster (1986). "Rank and Null Space Calculations Using Matrix Decomposition without Column Interchanges," *Lin. Alg. and Its Applic. 74,* 47–71.

R. Fourer (1984). "Staircase Matrices and Systems," *SIAM Review 26,* 1–71.

L. Fox (1964). *An Introduction to Numerical Linear Algebra*, Oxford University Press, Oxford, England.

G. Fox (ed) (1988). *The Third Conference on Hypercube Concurrent Computers and Applications, Vol. II – Applications,* ACM Press, New York.

G. Fox, S.W. Otto, and A.J. Hey (1987). "Matrix Algorithms on a Hypercube I: Matrix Multiplication," *Parallel Computing 4,* 17-31.

G. Fox, M. Johnson, G. Lyzenga, S. Otto, J. Salmon, and D. Walker (1988). *On Concurrent Processors Vol I: General Techiques and Regular Problems*, Prentice-Hall, Englewood Cliffs, NJ.

J.S. Frame (1964a). "Matrix Functions and Applications, Part II," *IEEE Spectrum* I (April), 102–8.

J.S. Frame (1964b). "Matrix Functions and Applications, Part IV," *IEEE Spectrum* 1 (June), 123–31.

J.G.F. Francis (1961). "The QR Transformation: A Unitary Analogue to the LR Transformation, Parts I and II," *Comp. J. 4,* 265-72, 332–45.

S. Friedland (1975). "On Inverse Multiplicative Eigenvalue Problems for Matrices," *Lin. Alg. and Its Applic. 12,* 127–38.

S. Friedland (1977). "Inverse Eigenvalue Problems," *Lin. Alg. and Its Applic. 17,* 15–52.

S. Friedland, J. Nocedal, and M.L. Overton (1987). "The Formulation and Analysis of Numerical Methods for Inverse Eigenvalue Problems," *SIAM J. Numer. Anal. 24,* 634–667.

C.E. Froberg (1965). "On Triangularization of Complex Matrices by Two-Dimensional Unitary Transformations," *BIT 5,* 230–34.

R.E. Funderlic and A. Geist (1986). "Torus Data Flow for Parallel Computation of Missized Matrix Problems," *Lin. Alg. and Its Applic. 77,* 149–164.

R.E. Funderlic, M. Neuman, and R.J. Plemmons (1982). "Generalized Diagonally Dominant Matrices," *Numer. Math. 40,* 57–70.

G. Galimberti and V. Pereyra (1970). "Numerical Differentian and the Solution of Multidimensional Vandermonde Systems," *Math. Comp. 24,* 357–64.

G. Galimberti and V. Pereyra (1971). "Solving Cofluent Vandermonde Systems of Hermite Type," *Numer. Math. 18,* 44–60.

K. Gallivan, W. Jalby, and U. Meier (1987). "The Use of BLAS3 in Linear Algebra on a Parallel Processor with a Hierarchical Memory," *SIAM J. Sci. and Stat. Comp. 8,* 1079–1084.

K. Gallivan, W. Jalby, U. Meier, and A.H. Sameh (1988). "Impact of Hierarchical Memory Systems on Linear Algebra Algorithm Design," *Int'l J. Supercomputer Applic. 2,* 12–48.

W. Gander (1981). "Least Squares with a Quadratic Constraint," *Numer. Math. 36,* 291–307.

D. Gannon and J. Van Rosendale (1984). "On the Impact of Communication Complexity on the Design of Parallel Numerical Algorithms," *IEEE Trans. Comp. C-33,* 1180–1194.

F.R. Gantmacher (1959). *The Theory of Matrices*, vols. 1-2, Chelsea, New York.

B.S. Garbow, J.M. Boyle, J.J. Dongarra, and C.B. Moler (1972). *Matrix Eigensystem Routines: EISPACK Guide Extension*, Springer-Verlag, New York.

W. Gautschi (1975a). "Norm Estimates for Inverses of Vandermonde Matrices," *Numer. Math. 23,* 337–47.

W. Gautschi (1975b). "Optimally Conditioned Vandermonde Matrices," *Numer. Math. 24,* 1–12.

G.A. Geist and M.T. Heath (1985). "Parallel Cholesky Factorization on a Hypercube Multiprocessor," Report ORNL 6190, Oak Ridge Laboratory, Oak Ridge, TN.

G.A. Geist and M.T. Heath (1986). "Matrix Factorization on a Hypercube," in *Hypercube Multiprocessors* , ed. M.T. Heath, SIAM Press, 161–180.

G.A. Geist, R.C. Ward, G.J. Davis, and R.E. Funderlic (1988). "Finding Eigenvalues and Eigenvectors of Unsymmetric Matrices Using a Hypercube Multiprocessor," in G. Fox (1988), 1577-1582.

W.M. Gentleman (1973a). "Error Analysis of QR Decompositions by Givens Transformations," *Lin. Alg. and Its Applic. 10,* 189–97.

W.M. Gentleman (1973b). "Least Squares Computations by Givens Transformations Without Square Roots," *J. Inst. Math. Appl. 12,* 329–36.

W.M. Gentleman and H.T. Kung (1982). "Matrix Triangularization by Systolic Arrays," SPIE Proceedings, Vol. 298, 19–26.

J.A. George (1973). "Nested Dissection of a Regular Finite Element Mesh," *SIAM J. Num. Anal. 10,* 345–63.

J.A. George (1974). "On Block Elimination for Sparse Linear Systems," *SIAM J. Num. Anal. 11,* 585–603.

J.A. George and M.T. Heath (1980). "Solution of Sparse Linear Least Squares Problems Using Givens Rotations," *Lin. Alg. and Its Applic. 34,* 69–83.

J.A. George, M.T. Heath, and J. Liu (1986). "Parallel Cholesky Factorization on a Shared Memory Multiprocessor," *Lin. Alg. and Its Applic. 77,* 165–187.

J.A. George and J.W. Liu (1981). *Computer Solution of Large Sparse Positive Definite Systems*, Prentice-Hall, Englewood Cliffs, NJ.

N.E. Gibbs and W.G. Poole, Jr. (1974). "Tridiagonalization by Permutations," *Comm. ACM 17,* 20–24.

N.E. Gibbs, W.G. Poole, and P.K. Stockmeyer (1976a). "A Comparison of Several Bandwidth and Profile Reduction Algorithms," *ACM Trans. Math. Soft. 2,* 322–30.

N.E. Gibbs, W.G. Poole, and P.K. Stockmeyer (1976b). "An Algorithm for Reducing the Bandwidth and Profile of a Sparse Matrix," *SIAM J. Num. Anal.* 13, 236–50.

P.E. Gill, G.H. Golub, W. Murray, and M.A. Saunders (1974). "Methods for Modifying Matrix Factorizations," *Math. Comp. 28,* 505–35.

P.E. Gill and W. Murray (1976). "The Orthogonal Factorization of a Large Sparse Matrix," in *Sparse Matrix Computations*, ed. J.R. Bunch and D.J. Rose, Academic Press, New York, pp. 177–200.

P.E. Gill, W. Murray and M.A. Saunders (1975). "Methods for Computing and Modifying the LDV Factors of a Matrix," *Math. Comp. 29,* 1051–77.

T. Ginsburg (1971). "The Conjugate Gradient Method," in *HACLA*, pp. 57–69.

W. Givens (1958). "Computation of Plane Unitary Rotations Transforming a General Matrix to Triangular Form," *SIAM J. App. Math. 6,* 26–50.

I.C. Gohberg and M.G. Krein (1969). *Introduction to the Theory of Linear Non-Self-Adjoint Operators,* Amer. Math. Soc., Providence, RI.

I.C. Gohberg, P. Lancaster, and L. Rodman (1986). *Invariant Subspaces of Matrices With Applications,* John Wiley and Sons, New York.

H.H. Goldstine and L.P. Horowitz (1959). "A Procedure for the Diagonalization of Normal Matrices," *J. ACM 6,* 176–95.

D. Goldfarb (1976). "Factorized Variable Metric Methods for Unconstrained Optimization," *Math. Comp. 30,* 796–811.

G.H. Golub (1965). "Numerical Methods for Solving Linear Least Squares Problems," *Numer. Math. 7,* 206–16.

G.H. Golub (1969). "Matrix Decompositions and Statistical Computation," in *Statistical Computation*, ed., R.C. Milton and J.A. Nelder, Academic Press, New York, pp. 365–97.

G.H. Golub (1973). "Some Modified Matrix Eigenvalue Problems," *SIAM Review 15,* 318–44.

G.H. Golub (1974). "Some Uses of the Lanczos Algorithm in Numerical Linear Algebra," in *Topics in Numerical Analysis*, ed., J.J.H. Miller, Academic Press, New York.

G.H. Golub, M. Heath, and G. Wahba (1979). "Generalized Cross-Validation as a Method for Choosing a Good Ridge Parameter," *Technometrics 21,* 215–23.

G.H. Golub, A. Hoffman, and G.W. Stewart (1988). "A Generalization of the Eckart-Young-Mirsky Approximation Theorem." *Lin. Alg. and Its Applic. 88/89,* 317–328

G.H. Golub and W. Kahan (1965). "Calculating the Singular Values and Pseudo-Inverse of a Matrix," *SIAM J. Num. Anal. Ser. B 2,* 205–24.

G.H. Golub, V. Klema, and G.W. Stewart (1976). "Rank Degeneracy and Least Squares Problems," Technical Report TR-456, Department of Computer Science, University of Maryland, College Park, MD.

G.H. Golub, F.T. Luk, and M. Overton (1981). "A Block Lanczos Mesthod for Computing the Singular Values and Corresponding Singular Vectors of a Matrix," *ACM Trans. Math. Soft. 7,* 149–69.

G.H. Golub and G. Meurant (1983). *Résolution Numérique des Grandes Systèmes Linéaires,* Collection de la Direction des Etudes et Recherches de l'Electricité de France, vol. 49, Eyolles, Paris.

G.H. Golub and C.D. Meyer (1986). "Using the QR Factorization and Group Inversion to Compute, Differentiate, and estimate the Sensitivity of Stationary Probabilities for Markov Chains," *SIAM J. Alg. and Dis. Methods,* 7, 273–281.

G.H. Golub, S. Nash, and C. Van Loan (1979). "A Hessenberg-Schur Method for the Matrix Problem $AX + XB = C$," *IEEE Trans. Auto. Cont. AC-24,* 909-913.

G.H. Golub and M. Overton (1988). "The Convergence of Inexact Chebychev and Richardson Iterative Methods for Solving Linear Systems," *Numer. Math. 53,* 571–594.

G.H. Golub and V. Pereyra (1973). "The Differentiation of Pseudo-Inverses and Nonlinear Least Squares Problems Whose Variables Separate," *SIAM J. Num. Anal. 10,* 413–32.

G.H. Golub and V. Pereyra (1976). "Differentiation of Pseudo-Inverses, Separable Nonlinear Least Squares Problems and Other Tales," in *Generalized Inverses and Applications*, ed. M.Z. Nashed, Academic Press, New York, pp. 303–24.

G.H. Golub and C. Reinsch (1970). "Singular Value Decomposition and Least Squares Solutions," *Numer. Math. 14,* 403-20. See also *HACLA*, pp. 134–51.

G.H. Golub and W.P. Tang (1981). "The Block Decomposition of a Vandemronde Matrix and Its Applications," *BIT 21,* 505–17.

G.H. Golub and R. Underwood (1970). "Stationary Values of the Ratio of Quadratic Forms Subject to Linear Constraints," *Z. Angew. Math. Phys. 21,* 318–26.

G.H. Golub and R. Underwood (1977). "The Block Lanczos Method for Computing Eigenvalues," in *Mathematical Software III*, ed. J. Rice, Academic Press, New York, pp. 364–77.

G.H. Golub, R. Underwood, and J.H. Wilkinson (1972). "The Lanczos Algorithm for the Symmetric $Ax = \lambda Bx$ Problem," Report STAN-CS-72-270, Department of Computer Science, Stanford University, Stanford, CA.

G.H. Golub and C.F. Van Loan (1979). "Unsymmetric Positive Definite Linear Systems," *Lin. Alg. and Its Applic. 28,* 85–98.

G.H. Golub and C.F. Van Loan (1980). "An Analysis of the Total Least Squares Problem," *SIAM J. Num. Anal. 17,* 883–93.

G.H. Golub and J.M. Varah (1974). "On a Characterization of the Best $1_2$-Scaling of a Matrix," *SIAM J. Num. Anal. 11,* 472–79.

G.H. Golub and R.S. Varga (1961). "Chebychev Semi-Iterative Methods, Successive Over-Relaxation Iterative Methods, and Second-Order Richardson Iterative Methods," Parts I and II, *Numer. Math. 3,* 147–56, 157–68.

G.H. Golub and J.H. Welsch (1969). "Calculation of Gauss Quadrature Rules," *Math. Comp. 23,* 221–30.

G.H. Golub and J.H. Wilkinson (1966). "Note on the Iterative Refinement of Least Squares Solution," *Numer. Math.* 9, 139–48.

G.H. Golub and J.H. Wilkinson (1976). "Ill-Conditioned Eigensystems and the Computation of the Jordan Canonical Form," *SIAM Review 18,* 578–619.

A.R. Gourlay (1970). "Generalization of Elementary Hermitian Matrices," *Comp. J. 13,* 411–12.

W. B. Gragg (1986). "The QR Algorithm for Unitary Hessenberg Matrices," *J. Comp. Appl. Math.* 16, 1–8.

W.B. Gragg and W.J. Harrod (1984). "The Numerically Stable Reconstruction of Jacobi Matrices from Spectral Data," *Numer. Math. 44,* 317–336.

B. Green (1952). "The Orthogonal Approxmation of an Oblique Structure in Factor Analysis," *Psychometrika 17,* 429–40.

A. Greenbaum (1981). "Behavior of the Conjugate Gradient Algorithm in Finite Precision Arithmetic," Report UCRL 85752, Lawrence Livermore Laboratory, Livermore, CA.

R.G. Grimes and J.G. Lewis (1981). "Condition Number Estimation for Sparse Matrices," *SIAM J. Sci. and Stat. Comp. 2,* 384–88.

R.F. Gunst, J.T. Webster, and R.L. Mason (1976). "A Comparison of Least Squares and Latent Root Regression Estimators," *Technometrics 18,* 75–83.

K.K. Gupta (1972). "Solution of Eigenvalue Problems by Sturm Sequence Method," *Int. J. Num. Meth. Eng. 4,* 379–404.

L.A. Hageman and D.M. Young (1981). *Applied Iterative Methods,* Academic Press, New York.

W. Hager (1984). "Condition Estimates," *SIAM J. Sci. and Stat. Comp. 5,* 311–316.

W. Hager (1988). *Applied Numerical Linear Algebra,* Prentice-Hall, Englewood Cliffs, NJ.

*HACLA.* See Wilkinson and Reinsch (1971).

P. Halmos (1958). *Finite Dimensional Vector Spaces,* Van Nostrand, New York.

S. Hammarling (1974). "A Note on Modifications to the Givens Plane Rotation," *J. Inst. Math. Appl.13,* 215–18.

S.J. Hammarling (1985). "The Singular Value Decomposition in Multivariate Statistics," *ACM SIGNUM Newsletter 20,* 2–25.

E.R. Hansen (1962). "On Quasicyclic Jacobi Methods," *J. ACM 9,* 118–35.

E.R. Hanson (1963). "On Cyclic Jacobi Methods," *SIAM J. Aplied Math. 11,* 448–59.

P.C. Hansen (1987). "The Truncated SVD as a Method for Regularization," *BIT* 27, 534–553.

P.C. Hansen (1988). "Reducing the Number of Sweeps in Hestenes Method," in *Singular Value Decomposition and Signal Processing*, in (ed.) E.F. Deprettere, North Holland.

R.J. Hanson and C.L. Lawson (1969). "Extensions and Applications of the Householder Algorithm for Solving Linear Least Squares Problems," *Math. Comp. 23,* 787–812.

R.J. Hanson and M.J. Norris (1981). "Analysis of Measurements Based on the Singular Value Decomposition," *SIAM J. Sci. and Stat. Comp. 2,* 363–374.

V. Hari (1982). "On the Global Convergence of the Eberlein Method for Real Matrices," *Numer. Math. 39,* 361–370.

M.T. Heath (1978). "Numerical Algorithms for Nonlinearly Constrained Optimization," Report STAN-CS-78-656, Department of Computer Science, Stanford University (Ph.D. thesis).

M.T. Heath (ed) (1986). *Proceedings of First SIAM Conference on Hypercube Multiprocessors*, SIAM Publications, Philadelphia, Pa.

M.T. Heath (ed) (1987). *Hypercube Multiprocessors*, SIAM Publications, Philadelphia, Pa.

M.T. Heath, A.J. Laub, C.C. Paige, and R.C. Ward (1986). "Computing the SVD of a Product of Two Matrices," *SIAM J. Sci. and Stat. Comp. 7,* 1147–1159.

M.T. Heath and C.H. Romine (1988). "Parallel Solution of Triangular Systems on Distributed Memory Multiprocessors," *SIAM J. Sci. and Stat. Comp. 9,* 558–588.

M.T. Heath and D.C. Sorensen (1986). "A Pipelined Method for Computing the QR Factorization of a Sparse Matrix," *Lin Alg. and Its App. 77,* 189–203.

D. Heller (1976). "Some Aspects of the Cyclic Reduction Algorithm for Block Tridiagonal Linear Systems," *SIAM J. Num. Anal. 13,* 484–96.

D. Heller (1978). "A Survey of Parallel Algorithms in Numerical Linear Algebra," *SIAM Review 20,* 740–777.

D.E. Heller and I.C.F. Ipsen (1983). "Systolic Networks for Orthogonal Decompositions," *SIAM J. Sci. and Stat. Comp. 4,* 261–269.

B.W. Helton (1968). "Logarithms of Matrices," *Proc. Amer. Math. Soc. 19,* 733–36.

C.S. Henkel, M.T. Heath, and R.J. Plemmons (1988). "Cholesky Downdating on a Hypercube," in G. Fox (1988), 1592–1598.

P. Henrici (1958)."On the Speed of Convergence of Cyclic and Quasicyclic Jacobi Mesthods for Computing the Eigenvalues of of Hermitian Matrices," *SIAM J. App. Math. 6,* 144–62.

P. Henrici (1962). "Bounds for Iterates, Inverses, Spectral Variation, and Fields of Values of Non-Normal Matrices," *Numer. Math. 4,* 24–40.

P. Henrici and K. Zimmermann (1968). "An Estimate for the Norms of Certain cyclic Jacobi Operators," *Lin. Alg. and Its Applic.* 1, 489–501.

M.R. Hestenes (1980). *Conjugate Direction Methods in Optimization,* Springer-Verlag, Berlin.

M.R. Hestenes and E. Stiefel (1952). "Methods of Conjugate Gradients for Solving Linear Systems," *J. Res. Nat. Bur. Stand. 49,* 409–36.

N.J. Higham (1985). "Nearness Problems in Numerical Linear Algebra," PhD Thesis, University of Manchester, England.

N.J. Higham (1986a). "Newton's Method for the Matrix Square Root," *Math. Comp. 46*, 537–550.

N.J. Higham (1986b). "Computing the Polar Decomposition with Applications," *SIAM J. Sci. and Stat. Comp. 7,* 1160–1174.

N.J. Higham (1986c). "Efficient Algorithms for computing the condition number of a tridiagonal matrix," *SIAM J. Sci. and Stat. Comp. 7,* 150–165.

N.J. Higham (1987a). "A Survey of Condition Number Estimation for Triangular Matrices," *SIAM Review 29,* 575–596.

N.J. Higham (1987b). "Error Analysis of the Björck-Pereyra Algorithms for Solving Vandermonde Systems," *Numer. Math. 50,* 613–632.

N.J. Higham (1987c). "Computing Real Square Roots of a Real Matrix," *Lin. Alg. and Its Applic. 88/89,* 405–430.

N.J. Higham (1988a). "Fast Solution of Vandermonde-like Systems Involving Orthogonal Polynomials," *IMA J. Numer. Anal. 8*, 473-486.

N.J. Higham (1988b). "Computing a Nearest Symmetric Positive Semidefinite Matrix," *Lin. Alg. and Its Applic. 103,* 103–118.

N.J, Higham (1988c). "The Symmetric Procrustes Problem," *BIT 28,* 133–143.

N.J. Higham (1988d). "The Accuracy of Solutions to Triangular Systems," Report 158, University of Manchester, Department of Mathematics.

N.J. Higham (1988e). "Matrix Nearness Problems and Applications," Report 161, University of Manchester, Department of Mathematics. To appear in S. Barnett and M.J.C. Gover (eds), *Proceedings of the IMA Conference on Applications of Matrix Theory.*

N.J. Higham (1988f). "Fortran Codes for Estimating the One-Norm of a Real or Complex Matrix, with Applications to Condition Estimation," *ACM Trans. Math. Soft. 14,* 381–396.

N.J. Higham(1989) "Analysis of the Cholesky Decomposition of a Semi-definite Matrix," in *Reliable Numerical Computation*, eds. M.G. Cox and S.J. Hammarling, Oxford University Press.

N.J. Higham and D.J. Higham (1988). "Large Growth Factors in Gaussian Elimination with Pivoting," Report 152, Department of Mathematics, University of Manchester, M13 9PL, England, to appear *SIAM J. Matrix Analysis and Applications.*

N.J. Higham and R.S. Schreiber (1988). "Fast Polar Decomposition of an Arbitrary Matrix," Report 88-942, Dept. of Computer Science, Cornell University, Ithaca, NY 14853.

D. Hoaglin (1977). "Mathematical Software and Exploratory Data Analysis," in *Mathematical Software III,* ed., John Rice, Academic Press, New York, pp. 139–59.

R.W. Hockney (1965). "A Fast Direct Solution of Poisson's Equation Using Fourier Analysis," *J. ACM 12,* 95–113.

R. Hockney(1983). "Characterizing Computers and Optimizing the FACR($\ell$) Poisson Solver on Parallel Unicomputers," *IEEE Trans. Comp. C-32,* 933–941.

R.W. Hockney and C.R. Jesshope (1988). *Parallel Computers 2*, Adam Hilger, Bristol and Philadelphia.

W. Hoffmann and B.N. Parlett (1978). "A New Proof of Global Convergence for the Tridiagonal QL Algorithm," *SIAM J. Num. Anal. 15,* 929–37.

H. Hotelling (1957). "The Relations of the Newer Multivariate Statistical Mesthods to Factor Analysis," *Brit. J. Stat. Psych. 10,* 69–79.

A.S. Householder (1958). "Unitary Triangularization of a Nonsymmetric Matrix," *J. ACM 5,* 339–42.

A.S. Householder (1968). "Moments and Characteristic Roots II," *Numer. Math. 11,* 126–28.

A.S. Householder (1974). *The Theory of Matrices in Numerical Analysis,* Dover Publications, New York.

C.P. Huang (1975). "A Jacobi-Type Method for Triangularizing an Arbitrary Matrix," *SIAM J. Num. Anal. 12,* 566–70.

C.P. Huang (1981). "On the Convergence of the QR Algorithm with Origin Shifts for Normal Matrices," *IMA J. Num. Anal. 1,* 127–33.

T.E. Hull and J.R. Swenson (1966). "Tests of Probabilistic Models for Propagation of Roundoff Errors," *Comm. ACM 9,* 108–13.

Y. Ikebe (1979). "On Inverses of Hessenberg Matrices," *Lin. Alg. and Its Applic. 24,* 93–97.

I.C.F. Ipsen, Y. Saad, and M. Schultz (1986). "Dense Linear Systems on a Ring of Processors," *Lin. Alg. and Its Applic. 77,* 205–239.

A. Iserles and M.J.D. Powell (eds) (1987). *The State of the Art in Numerical Analysis,* Oxford University Press.

C.G.J. Jacobi (1846). "Uber ein Leichtes Verfahren Die in der Theorie der Sacularstorungen Vorkommendern Gleichungen Numerisch Aufzulosen," *Crelle's J. 30,* 51–94.

M. Jankowski and M. Wozniakowski (1977). "Iterative Refinement Implies Numerical Stability," *BIT 17*, 303–311.

K.C. Jea and D.M. Young (1983). "On the Simplification of Generalized Conjugate Gradient Methods for Nonsymmetrizable Linear Systems," *Lin. Alg. and Its Applic. 52/53*, 399–417.

A. Jennings (1977a). "Influence of the Eigenvalue Spectrum on the Convergence Rate of the Conjugate Gradient Method," *J. Inst. Math. Applic. 20*, 61–72.

A. Jennings (1977b). *Matrix Computation for Engineers and Scientists*, John Wiley and Sons, New York.

A. Jennings and D.R.L. Orr (1971). "Application of the Simultaneous Iteration Method to Undamped Vibration Problems," *Inst. J. Numer. Meth. Eng. 3*, 13–24.

A. Jennings and M.R. Osborne (1977). "Generalized Eigenvalue Problems for Certain Unsymmetric Band Matrices," *Lin. Alg. and Its Applic. 29*, 139–50.

A. Jennings and W.J. Stewart (1975). "Simultaneous Iteration for the Partial Eigensolution of Real Matrices," *J. Inst. Math. Applic. 15*, 351–62.

L.S. Jennings and M.R. Osborne (1974). "A Direct Error Analysis for Least Squares," *Numer. Math. 22*, 322–32.

P.S. Jenson (1972). "The Solution of Large Symmetric Eigenproblems by Sectioning," *SIAM J. Num. Anal. 9*, 534–45.

J. Johnson and C.L. Phillips (1971). "An Algorithm for the Computation of the Integral of the State Transition Matrix," *IEEE Trans. Auto. Cont. AC-16*, 204–5.

O.G. Johnson, C.A. Micchelli, and G. Paul (1983). "Polynomial Preconditioners for Conjugate Gradient Calculations," *SIAM J. Numer. Anal. 20*, 362–376.

S.L. Johnsson (1984). "Odd-Even Cyclic Reduction on Ensemble Architectures and the Solution of Tridiagonal Systems of Equations," Report YALEU/CSD/RR-339, Dept. of Comp. Sci., Yale University, New Haven, CT.

S.L. Johnsson (1985). "Solving Narrow Banded Systems on Ensemble Architectures," *ACM Trans. Math. Soft. 11*, 271–x.

S.L. Johnsson (1986). "Band Matrix System Solvers on Ensemble Architectures," in *Supercomputers: Algorithms, Architectures, and Scientific Computation*, eds. F.A. Matsen and T. Tajima, University of Texas Press, Austin TX., 196–216.

S.L. Johnsson (1987a). "Solving Tridiagonal Systems on Ensemble Architectures," *SIAM J. Sci. and Stat. Comp., 8*, 354–392.

S.L. Johnsson (1987b). "Communication Efficient Basic Linear Algebra Computations on Hypercube Multiprocessors," *J. Parallel and Distributed Computing 4*, 133–172.

S.L. Johnsson and C.T. Ho (1987a). "Multiple Tridiagonal Systems, the Alternatine Direction Methods, and Boolean Cube Configured Multiprocessors," Report YALEU DCS RR-532, Dept. of Computer Science, Yale University, New Haven, CT.

S.L. Johnsson and C.T. Ho (1987b). "Algorithms for Multiplying Matrices of Arbitrary Shapes Using Shared Memory Primatives on a Boolean Cube," Report YALEU DCS RR-569, Dept. of Computer Science, Yale University, New Haven, CT.

S.L. Johnsson and C.T. Ho (1987c). "Matrix Transposition on Boolean n-cube Configured Ensemble Architectures," Report YALEU/DCS/RR-574, Dept. of Computer Science, Yale University, New Haven, CT., to appear in *SIAM J. Alg. and Discrete Methods.*

R.L. Johnston (1971). "Gershgorin Theorems for Partitioned Matrices," *Lin. Alg. and Its Applic. 4*, 205–20.

H. Jordan (1987). "Interpreting Parallel Processor Performance Measurements," *SIAM J. Sci. and Stat. Comp. 8*, s220–s226.

T. Jordan (1984). "Conjugate Gradient Preconditioners for Vector and Parallel Processors," in G. Birkoff and A. Schoenstadt (eds), *Proceedings of the Conference on Elliptic Problem Solvers*, Academic Press, NY.

B. Kågström (1977a). "Bounds and Perturbation Bounds for the Matrix Exponential," *BIT 17*, 39–57.

B. Kågström (1977b). "Numerical Computation of Matrix Functions," Department of Information Processing Report UMINF-58.77, University of Umea, Umea, Sweden.

B. Kågström and A. Ruhe (1980a). "An Algorithm for Numerical Computation of the Jordan Normal Form of a Complex Matrix," *ACM Trans. Math. Soft. 6,* 398–419.

B. Kågström and A. Ruhe (1980b). "Algorithm 560 JNF: An Algorithm for Numerical Computation of the Jordan Normal Form of a Complex Matrix," *ACM Trans. Math. Soft. 6,* 437–43.

B. Kågström (1985). "The Generalized Singular Value Decomposition and the General $A - \lambda B$ Problem," *BIT 24,* 568–583.

B. Kågström (1986). "RGSVD: An Algorithm for Computing the Kronecker Structure and Reducing Subspaces of Singular $A - \lambda B$ Pencils," *SIAM J. Sci. and Stat. Comp.* 7,185–211.

B. Kågström and P. Ling (1988). "Level 2 and 3 BLAS Routines for the IBM 3090 VF/400: Implementation and Experiences," Report UMINF–154.88, Inst. of Inf. Proc., University of Umeå, S-901 87 Umeå, Sweden.

B. Kågström, L. Nyström, and P. Poromaa (1987). "Parallel Algorithms for Solving the Triangular Sylvester Equation on a Hypercube Multiprocessor," Report UMINF–136.87, Inst. of Inf. Proc., University of Umeå, S-901 87 Umeå, Sweden.

B. Kågström, L. Nyström, and P. Poromaa (1988). "Parallel Shared Memory Algorithms for Solving the Triangular Sylvester Equation," Report UMINF–155.88, Inst. of Inf. Proc., University of Umeå, S-901 87 Umeå, Sweden.

B. Kågström and A. Ruhe (1983). *Matrix Pencils,* Proc. Pite Havsbad, 1982, Lecture Notes in Mathemnatics 973, Springer-Verlag, New York and Berlin.

B. Kågström and L. Westin (1987). "GSYLV- Fortran Routines for the Generalized Schur Method with $\text{dif}^{-1}$ estimators for Solving the Generalized Sylvester Equation," Report UMINF–132.86, Inst. of Inf. Proc., University of Umeå, S-901 87 Umeå, Sweden.

W. Kahan (1966). "Numerical Linear Algebra," *Canadian Math. Bull. 9,* 757–801.

W. Kahan (1967). Inclusion Theorems for Clusters of Eigenvlaues of Hermitian Matrices," Computer Science Report, University of Toronto, Toronto, Canada.

W. Kahan (1975). "Spectra of Nearly Hermitian Matrices," *Proc. Amer. Math. Soc. 48,* 11–17.

W. Kahan and B.N. Parlett (1974). "An Analysis of Lanczos Algorithms for Symmetric Matrices," ERL-M467, University of California, Berkeley, CA.

W. Kahan and B.N. Parlett (1976). "How Far Should You Go with the Lanczos Process?," in *Sparse Matrix Computations,* ed., J. Bunch and D. Rose, Academic Press, New York, pp. 131–44.

W. Kahan, B.N. Parlett, and E. Jiang (1982). "Residual Bounds on Approximate Eigensystems of Nonnormal Matrices," *SIAM J. Numer. Anal. 19,* 470–484.

D. Kahaner, C.B. Moler, and S. Nash (1988). *Numerical Methods and Software,* Prentice-Hall, Englewood Cliffs, NJ.

S. Kaniel (1966). "Estimates for Some Computational Techniques in Linear Algebra," *Math. Comp. 20,* 369–78.

R.N. Kapur and J.C. Browne (1984). "Techniques for Solving Block Tridiagonal Systems on Reconfigurable Array Computers," *SIAM J. Sci. and Stat. Comp. 5,* 701–719.

I. Karasalo (1974). "A Criterion for Truncation of the QR Decomposition Algorithm for the Singular Linear Least Squares Problem," *BIT 14,* 156–66.

T. Kato (1966). *Perturbation Theory for Linear Operators,* Springer-Verlag, New York, NY.

L. Kaufman (1974). "The LZ Algorithm to Solve the Generalized Eigenvalue Problem," *SIAM J. Num. Anal. 11,* 997–1024.

L. Kaufman (1977). "Some Thoughts on the QZ Algorithm for Solving the Generalized Eigenvalue Problem," *ACM Trans. Math. Soft. 3,* 65–75.

L. Kaufman (1979). "Application of Dense Householder Transformations to a Sparse Matrix," *ACM Trans. Math. Soft. 5,* 442–50.

L. Kaufman (1983). "Matrix Methods for Queueing Problems," *SIAM J. Sci. and Stat. Comp. 4,* 525–552.

L. Kaufman (1987). "The Generalized Householder Transformation and Sparse Matrices," *Lin. Alg. and Its Applic. 90,* 221–234.

J. Kautsky and G.H. Golub (1983). "On the Calculation of Jacobi Matrices," *Lin. Alg. and Its Applic. 52/53,* 439–456.

D. Kershaw(1982). "Solution of Single Tridiagonal Linear Systems and Vectorization of the ICCG Algorithm on the Cray-1," in *Parallel Computation*, ed. G. Roderigue, Academic Press, NY, 1982.

A. Kielbasinski (1987). "A Note on Rounding Error Analysis of Cholesky Factorization," *Lin. Alg. and Its Applic. 88/89,* 487–494.

D. Knuth (1981). "The Art of Computer Programming, Vol. 2, *Seminumerical-Algorithms,* Second Edition, Additon-Wesley, Reading, MA.

E.G.Kogbetliantz (1955). "Solution of Linear Equations by Diagonalization of Coefficient Matrix," *Quart. Appl. Math. 13,* 123–132.

S. Kourouklis and C.C. Paige (1981). "A Constrained Least Squares Approach to the General Gauss-Markov Linear Model," *J. Amer. Stat. Assoc. 76,* 620–25.

A.S. Krishnakuma and M. Morf (1986). "Eigenvalues of a Symmetric Tridiagonal Matrix: A Divide and Conquer Approach," *Numer. Math. 48,* 349–368.

V.N. Kublanovskaya (1961). "On Some Algorithms for the Solution of the Complete Eigenvalue Problem," *USSR Comp. Math. Phys. 3,* 637–57.

V.N. Kublanovskaya and V.N. Fadeeva (1964). "Computational Methods for the Solution of a Generalized Eigenvalue Problem," *Amer. Math. Soc. Transl. 2,* 271–90.

U.W. Kulisch and W.L. Miranker (1986). "The Arithmetic of the Digital Computer," *SIAM Review 28,* 1–40.

H.T. Kung (1982). "Why Systolic Architectures?," *Computer 15,* 37–46.

C.D. La Budde (1964). "Two Classes of Algorithms for Finding the Eigenvalues and Eigenvectors of Real Symmetric Matrices," *J. ACM 11,* 53–58.

J. Lambiotte and R.G. Voigt (1975)."The Solution of Tridiagonal Linear Systems on the CDC-STAR 100 Computer," *ACM Trans. Math. Soft.1,* 308–29.

P. Lancaster (1970). "Explicit Solution of Linear Matrix Equations," *SIAM Review 12,* 544–66.

P. Lancaster and M. Tismenetsky (1985). *The Theory of Matrices,* 2nd Edition, Academic Press, New York.

C. Lanczos (1950). "An Iteration Method for the Solution of the Eigenvalue Problem of Linear Differential and Integral Operators," *J. Res. Nat. Bur. Stand. 45,* 255–82.

J. Larson and A. Sameh (1978). "Efficient Calculation of the Effects of Roundoff Errors," *ACM Trans. Math. Soft. 4,* 228–36.

A. Laub (1981). "Efficient Multivariable Frequency Response Computations," *IEEE Trans. Auto. Cont. AC-26,* 407–8.

A. Laub (1985). "Numerical Linear Algebra Aspects of Control Design Computations," *IEEE Trans. Auto. Cont. AC-30,* 97–108.

C.L. Lawson and R.J. Hanson (1969). "Extensions and Applications of the Householder Algorithm for Solving Linear Least Squares Problems," *Math. Comp. 23,* 787–812.

C.L. Lawson and R.J. Hanson (1974). *Solving Least Squares Problems,* Prentice-Hall, Englewood Cliffs, NJ.

C.L. Lawson, R.J. Hanson, D.R. Kincaid, and F.T. Krogh (1979). "Basic Linear Algebra Subprograms for FORTRAN Usage," *ACM Trans. Math. Soft. 5,* 308–23.

C.L. Lawson, R.J. Hanson, D.R. Kincaid, and F.T. Krogh (1979). "Algorithm 539, Basic Linear Algebra Subprograms for FORTRAN Usage," *ACM Trans. Math. Soft. 5,* 324–25.

N.J. Lehmann (1963). "Optimale Eigenwerteinschliessungen," *Numer. Math. 5,* 246–72.

F. Lemeire (1973). "Bounds for Condition Numbers of Triangular and Trapezoid Matrices," *BIT 15,* 58–64.

S.J. Leon (1980). *Linear Algebra with Applications.* Macmillan, New York, NY.

N. Levinson (1947). "The Weiner RMS Error Criterion in Filter Design and Prediction," *J. Math. Phys. 25,* 261–78.

J. Lewis (1977). "Algorithms for Sparse Matrix Eigenvalue Problems," Technical Report STAN-CS-77-595, Department of Computer Science, Stanford University, Stanford, CA.

G. Li and T. Coleman (1988). "A Parallel Triangular Solver for a Distributed-Memory Multiprocessor," *SIAM J. Sci. and Stat. Comp. 9,* 485-502.

I. Linnik (1961). *Method of Least Squares and Principles of the Theory of Observation,* Pergamon Press, New York, NY.

S. Lo, B. Philippe, and A. Sameh (1987). "A Multiprocessor Algorithm for the Symmetric Tridiagonal Eigenvalue Problem," *SIAM J. Sci. and Stat. Comp. 8,* s155–s165.

G. Loizou (1969). "Nonnormality and Jordan Condition Numbers of Matrices," *J. Assoc. Comp. Mach. 16,* 580–84.

G. Loizou (1972). "On the Quadratic Convergence of the Jacobi Method for Normal Matrices," *Comp. J. 15,* 274–76.

M. Lotkin (1956). "Characteristic Values of Arbitrary Matrices," *Quart. Appl. Math. 14,* 267–75.

D. G. Luenberger (1973). *Introduction to Linear and Nonlinear Programming,* Addison-Wesley, New York.

F.T. Luk (1978). "Sparse and Parallel Matrix Computations," PhD Thesis, Report STANS-CS-78-685, Department of Computer Science, Stanford University, Stanford, CA.

F.T. Luk (1980). "Computing the Singular Value Decomposition on the ILLIAC IV," *ACM Trans. Math. Soft. 6,* 524–39.

F.T. Luk (1986a). "A Rotation Method for Computing the QR Factorization," *SIAM J. Sci. and Stat. Comp. 7,* 452–459.

F.T. Luk (1986b). "A Triangular Processor Array for Computing Singular Values," *Lin. Alg. and Its Applic. 77,* 259–274.

E. Lusk and R. Overbeek (1983). "Implementation of Monitors with Macros: A Programming Aid for the HEP and other Parallel Processors," Argonne Report 83–97.

C. McCarthy and G. Strang (1973). "Optimal Conditioning of Matrices," *SIAM J. Num. Anal. 10,* 370–88.

S.F. McCormick (1972). "A General Approach to One-Step Iterative Methods with Application to Eigenvalue Problems," *J. Comput. Sys. Sci. 6,* 354–72.

W.M. McKeeman (1962). "Crout with Equilibration and Iteration," *Comm. ACM 5,* 553–55.

A. Madansky (1959). "The Fitting of Straight Lines When Both Variables Are Subject to Error," *J. Amer. Stat. Assoc. 54,* 173–205.

N. Madsen, G. Roderigue, and J. Karush (1976). "Matrix Multiplication by Diagonals of a Vector Parallel Processor," *Information Processing Letters,* 41–45.

K.N. Mahindar (1979). "Linear Combinations of Hermitian and Real Symmetric Matrices," *Lin. Alg. and Its Applic. 25,* 95–105.

J. Makhoul (1975). "Linear Prediction: A Tutorial Review," *Proc. IEEE 63(4),* 561–80.

M.A. Malcolm and J. Palmer (1974). "A Fast Method For Solving a Class of Tridiagonal Systems of Linear Equations," *Comm. ACM 17,* 14–17.

T.A. Manteuffel (1977). "The Tchebychev Iteration for Nonsymmetric Linear Systems," *Numer. Math. 28,* 307–27.

T.A. Manteuffel (1979). "Shifted Incomplete Cholesky Factorization," in *Sparse Matrix Proceedings, 1978,* ed. I.S. Duff and G.W. Stewart, SIAM Publications, Philadelphia, PA.

M. Marcus and H. Minc (1964). *A Survey of Matrix Theory and Matrix Inequalities,* Allyn and Bacon, Boston, MA.

J. Markel and A. Gray (1976). *Linear Prediction of Speech,* Springer-Verlag, Berlin and New York.

R.S. Martin, G. Peters, and J.H. Wilkinson (1965). "Symmetric Decomposition of a Positive Definite Matrix," *Numer. Math. 7,* 362–83. See also *HACLA,* pp. 9–30.

R.S. Martin, G. Peters, and J.H. Wilkinson (1966). "Iterative Refinement of the Solution of a Positive Definite System of Equations," *Numer. Math. 8,* 203–16. See also *HACLA,* pp. 31–44.

R.S. Martin, G. Peters, and J.H. Wilkinson (1970). "The QR Algorithm for Real Hessenberg Matrices," *Numer. Math. 14,* 219–31. See also *HACLA,* pp. 359–71.

R.S. Martin, C. Reinsch, and J.H. Wilkinson (1970). "The QR Algorithm for Band Symmetric Matrices," *Numer. Math. 16,* 85–92. See also *HACLA,* pp. 266–272.

R.S. Martin and J.H. Wilkinson (1965). "Symmetric Decomposition of Positive Definite Band Matrices," *Numer. Math. 7,* 355–61. See also *HACLA,* pp. 50–56.

R.S. Martin and J.H. Wilkinson (1967). "Solution of Symmetric and Unsymmetric Band Equations and the Calculation of Eigenvalues of Band Matrices," *Numer. Math. 9,* 279–301. See also *HACLA,* pp. 70–92.

R.S. Martin and J.H. Wilkinson (1968a). "Householder's Tridiagonalization of a Symmetric Matrix," *Numer. Math. 11,* 181–95. See also *HACLA,* pp. 212–26.

R.S. Martin and J.H. Wilkinson (1968b). "The Modified LR Algorithm for Complex Hessenberg Matrices," *Numer. Math. 12,* 369–76. See also *HACLA,* pp. 396-403.

R.S. Martin and J.H. Wilkinson (1968c). "Reduction of the Symmetric Eigenproblem $Ax = \lambda Bx$ and Related Problems to Standard Form," *Numer. Math. 11,* 99–110.

R.S. Martin and J.H. Wilkinson (1968d). "Similarity Reduction of a General Matrix to Hessenberg Form," *Numer. Math. 12,* 349–68. See also *HACLA,* pp. 339–58.

O. McBryan and E.F. van de Velde (1987). "Hypercube Algorithms and Implementations," *SIAM J. Sci. and Stat. Comp. 8,* s227–s287.

J.A. Meijerink and H.A. Van der Vorst (1977). "An Iterative Solution Method for Linear Equations Systems of Which the Coefficient Matrix is a Symmetric $M$-Matrix," *Math. Comp. 31,* 148–62.

J. Meinguet (1983). "Refined Error Analyses of Cholesky Factorization," *SIAM J. Numer. Anal. 20,* 1243–1250.

R. Melhem(1987). "Toward efficient Implementation of Preconditioned Conjugate Gradient Methods on Vector Supercomputers," *Int'l J. Supercomputing Applications 1,* 70–98.

M.L. Merriam (1985). "On the Factorization of Block Tridiagonals With Storage Constraints," *SIAM J. Sci. and Stat. Comp. 6,* 182–192.

G. Meurant (1984). "The Block Preconditioned Conjugate Gradient Method on Vector Computers," *BIT 24,* 623–633.

G. Meurant (1989). "Domain Decomposition Methods for Partial Differential Equations on Parallel Computers," to appear *Int'l J. Supercomputing Applications.*

W. Miller (1975). "Computational Complexity and Numerical Stability," *SIAM J. Computing 4,* 97–107.

W. Miller and D. Spooner (1978). "Software for Roundoff Analysis, II," *ACM Trans. Math. Soft. 4,* 369–90.

G. Miminis and C.C. Paige (1982). "An Algorithm for Pole Assignment of Time Invariant Linear Systems," *International J. of Control 35,* 341–354.

L. Mirsky (1955). *An Introduction to Linear Algebra,* Oxford University Press, London, England.

L. Mirsky (1960). "Symmetric Gauge Functions and Unitarily Invariant Norms," *Quart. J. Math. 11,* 50–59.

J.J. Modi (1988).*Parallel Algorithms and Matrix Computation,* Oxford University Press, Oxford.

J.J. Modi and M.R.B. Clarke (1986). "An Alternative Givens Ordering," *Numer. Math. 43,* 83–90.

J.J. Modi and J.D. Pryce (1985). "Efficient Implementation of Jacobi's Diagonalization Method on the DAP," *Numer. Math. 46,* 443–454.

C.B. Moler (1967). "Iterative Refinement in Floating Point," *J. ACM 14,* 316–71.

C.B. Moler (1980). "MATLAB User's Guide," Technical Report CS81-1, Department of Computer Science, University of New New Mexico, Albuquerque, NM.

C.B. Moler (1986). "Matrix Computations on Distributed Memory Multiprocessors," in *Hypercube Multiprocessors*, ed. M.T. Heath, SIAM Press, Philadelphia, Pa., 1986.

C.B. Moler, J.N. Little, and S. Bangert (1987). *PC-Matlab Users Guide*, The Math Works Inc., 20 N. Main St., Sherborn, Mass.

C.B. Moler and D. Morrison (1983). "Singular Value Analysis of Cryptograms," *Amer. Math. Monthly 90,* 78–87.

C.B. Moler and G.W. Stewart (1973). "An Algorithm for Generalized Matrix Eigenvalue Problems," *SIAM J. Num. Anal. 10,* 241–56.

C.B. Moler and C.F. Van Loan (1978). "Nineteen Dubious Ways to Compute the Exponential of a Matrix," *SIAM Review 20,* 801–36.

R. Montoye and D. Laurie (1982). "A Practical Algorithm for the Solution of Triangular Systems on a Parallel Processing System," *IEEE Trans. Comp. C-31,* 1076–1082.

D. Mueller (1966). "Householder's Method for Complex Matrices and Hermitian Matrices," *Numer. Math. 8,* 72-92.

F.D. Murnaghan and A. Wintner (1931). "A Canonical Form for Real Matrices Under Orthogonal Transformations," *Proc. Nat. Acad. Sci. 17,* 417–20.

T. Nanda (1985). "Differential Equations and the QR Algorithm," *SIAM J. Numer. Anal. 22,* 310–321.

J.C. Nash (1975)."A One-Sided Transformation Method for the Singular Value Decomposition and Algebraic Eigenproblem," *Comp. J. 18,* 74–76.

M.Z. Nashed (1976). *Generalized Inverses and Applications,* Academic Press, New York, NY.

R.A. Nicolaides (1974). "On a Geometrical Aspect of SOR and the Theory of Consistent Ordering for Positive Definite Matrices," *Numer. Math. 23,* 99–104.

W. Niethammer and R.S. Varga (1983). "The Analysis of $k$-step Iterative Methods for Linear Systems from Summability Theory," *Numer. Math. 41,* 177–206.

B. Noble and J.W. Daniel (1977). *Applied Linear Algebra,* Prentice-Hall, Englewood Cliffs, NJ.

A. Noor and R. Voigt (1975). "Hypermatrix Scheme for the STAR-100 Computer," *Computers and Structures 5,* 287–296.

W. Oettli and W. Prager (1964). "Compatibility of Approximate Solutions of Linear Equations with Given Error Bounds for Coefficients and Right Hand Sides," *Numer. Math. 6,* 405–09.

D.P. O'Leary (1976). "Hybrid Conjugate Gradient Algorithms," Report STAN-CS-76-548, Department of Computer Science, Stanford University (Ph.D. thesis).

D.P. O'Leary (1980a). "The Block Conjugate Gradient Algorithm and Related Methods," *Lin. Alg. and Its Applic. 29,* 293–322.

D.P. O'Leary (1980b). "Estimating Matrix Condition Numbers," *SIAM J. Sci. and Stat. Comp. 1,* 205–9.

D.P. O'Leary and J.A. Simmons (1981). "A Bidiagonalization - Regularization Procedure for Large Scale Discretizations of Ill-Posed Problems," *SIAM J. Sci. and Stat. Comp. 2,* 474–89.

D.P. O'Leary and G.W. Stewart (1985). "Data Flow Algorithms for Parallel Matrix Computations," *Comm. of the ACM 28,* 841–853.

D.P. O'Leary and G.W. Stewart (1986). "Assignment and Scheduling in Parallel Matrix Factorization," *Lin. Alg. and Its Applic. 77,* 275–300.

A.V. Oppenheim (1978). *Applications of Digital Signal Processing,* Prentice-Hall, Englewood Cliffs, NJ.

J.M. Ortega (1972). *Numerical Analysis: A Second Course,* Academic Press, New York, NY.

J.M. Ortega (1988). *Matrix Theory: A Second Course*, Plenum Press, New York.

J.M. Ortega and C.H. Romine (1988). "The $ijk$ Forms of Factorization Methods II: Parallel Systems," *Parallel Computing 7,* 149–162.

J.M. Ortega and R.G. Voigt (1985). "Solution of Partial Differential Equations on Vector and Parallel Computers," *SIAM Review 27,* 149–240.

E.E. Osborne (1960). "On Preconditioning of Matrices," *J. ACM 7,* 338–45.

M.H.C. Paardekooper (1971). "An Eigenvalue Algorithm for Skew Symmetric Matrices," *Numer. Math. 17,* 189–202.

C.C. Paige (1970). "Practical Use of the Symmetric Lanczos Process with Reorthogonalization," *BIT 10,* 183–95.

C.C. Paige (1971). "The Computation of Eigenvalues and Eigenvectors of Very Large Sparse Matrices," Ph.D. thesis, London University, London, England.

C.C. Paige (1973). "An Error Analysis of a Method for Solving Matrix Equations," *Math. Comp. 27,* 355–59.

C.C. Paige (1974a). "Bidiagonalization of Matrices and Solution of Linear Equations," *SIAM J. Num. Anal. 11,* 197–209.

C.C. Paige (1974b). "Eigenvalues of Perturbed Hermitian Matrices," *Lin. Alg. and Its Applic. 8,* 1–10.

C.C. Paige (1976). "Error Analysis of the Lanczos Algorithm for Tridiagonalizing a Symmetric Matrix," *J. Inst. Math. Applic. 18,* 341–49.

C.C. Paige (1979a). "Computer Solution and Perturbation Analysis of Generalized Least Squares Problems," *Math. Comp. 33,* 171–84.

C.C. Paige (1979b). "Fast Numerically Stable Computations for Generalized Linear Least Squares Problems," *SIAM J. Num. Anal. 16,* 165–71.

C.C. Paige (1980). "Accuracy and Effectivenss of the Lanczos Algorithm for the Symmetric Eigenproblem," *Lin. Alg. and Its Applic. 34,* 235–58.

C.C. Paige (1981). "Properties of Numerical Algorithms Related to Computing Controllability," *IEEE Trans. Auto. Cont. AC-26,* 130–38.

C.C. Paige (1984). "A Note on a Result of Sun J.-Guang: Sensitivity of the CS and GSV Decompositions," *SIAM J. Numer. Anal. 21,* 186–191.

C.C. Paige (1985). "The General Linear Model and the Generalized Singular Value Decomposition," *Lin. Alg. and Its Applic. 70,* 269–284.

C.C. Paige (1986). "Computing the Generalized Singular Value Decomposition," *SIAM J. Sci. and Stat. Comp. 7,* 1126–1146.

C.C. Paige (1986). "On the Quadratic Convergence of Kogbetliantz's Algorithm for Computing the Singular Value Decomposition," *Lin. Alg. and Its Applic. 77,* 301–314.

C.C. Paige and M.A. Saunders (1975). "Solution of Sparse Indefinite Systems of Linear Equations," *SIAM J. Num. Anal. 12,* 617–29.

C.C. Paige and M.A. Saunders (1978). "A Bidiagonalization Algorithm for Sparse Linear Equations and Least Squares Problems," Report SOL 78-19, Department of Operations Research, Stanford University, Stanford, CA.

C.C. Paige and M. Saunders (1981). "Towards A Generalized Singular Value Decomposition," *SIAM J. Num. Anal. 18,* 398–405.

C.C. Paige and M.A. Saunders (1982a). "LSQR: An Algorithm for Sparse Linear Equations and Sparse Least Squares," *ACM Trans. Math. Soft. 8,* 43–71.

C.C. Paige and M.A. Saunders (1982b). "Algorithm 583 LSQR: Sparse Linear Equations and Least Squares Problems," *ACM Trans. Math. Soft. 8,* 195–209.

C.C. Paige and C. Van Loan (1981). "A Schur Decomposition for Hamiltonian Matrices," *Lin. Alg. and Its Applic. 41,* 11–32.

V. Pan (1984). "How Can We Speed Up Matrix Multiplication," *SIAM Review 26,* 393–416.

B.N. Parlett (1965), "Convergence of the Q-R Algorithm," *Numer. Math. 7,* 187–93. (Correction in *Numer. Math. 10,* 163–64).

B.N. Parlett (1966). "Singular and Invariant Matrices Under the QR Algorithm," *Math. Comp.* 20, 611–15.

B.N. Parlett (1967). "Canonical Decomposition of Hessenberg Matrices," *Math. Comp. 21,* 223–27.

B.N. Parlett (1968). "Global Convergence of the Basic QR Algorithm on Hessenberg Matrices," *Math. Comp. 22,* 803–17.

B.N. Parlett (1971). "Analysis of Algorithms for Reflections in Bisectors," *SIAM Review 13,* 197–208.

B.N. Parlett (1974a). "Computation of Functions of Triangular Matrices," Memorandum no. ERL-M481, Electronics Research Laboratory, College of Engineering, University of California, Berkeley, CA.

B.N. Parlett (1974b). "The Rayleigh Quotient Iteration and Some Generalizations for Nonnormal Matrices," *Math. Comp. 28,* 679–93.

B.N. Parlett (1976). "A Recurrence Among the Elements of Functions of Triangular Matrices," *Lin. Alg. and Its Applic. 14,* 117–21.

B.N. Parlett (1980a). "A New Look at the Lanczos Algorithm for Solving Symmetric Systems and Linear Equations," *Lin. Alg. and Its Applic. 29,* 323–46.

B.N. Parlett (1980b). *The Symmetric Eigenvalue Problem,* Prentice-Hall, Englewood Cliffs, NJ.

B.N. Parlett and B. Nour-Omid (1985). "The Use of a Refined Error Bound When Updating Eigenvalues of Tridiagonals," *Lin. Alg. and Its Applic. 68,* 179–220.

B.N. Parlett and W.G. Poole (1973). "A Geometric Theory for the QR, LU, and Power Iterations," *SIAM J. Num. Anal. 10,* 389–412.

B.N. Parlett and J.K. Reid (1970). "On the Solution of a System of Linear Equations Whose Matrix is Symmetric But Not Definite," *BIT 10,* 386–97.

B.N. Parlett and J.K. Reid (1981). "Tracking the Progress of the Lanczos Algorithm for Large Symmetric Eigenproblems," *IMA J. Num. Anal. 1,* 135–55.

B.N. Parlett and C. Reinsch (1969). "Balancing a Matrix for Calculation fo Eigenvalues and Eigenvectors," *Numer. Math. 13,* 292-304. See also *HACLA*, pp. 315–26.

B.N. Parlett and D.S. Scott (1979). "The Lanczos Algorithm with Selective Orthogonalization," *Math. Comp. 33,* 217–38.

B.N. Parlett, H. Simon, and L.M. Stringer (1982). "On Estimating the Largest Eigenvalue with the Lanczos Algorithm," *Math. Comp. 38,* 153–166.

N. Patel and H. Jordan (1984). "A Parallelized Point Rowwise Successive Over-Relaxation Method on a Multiprocessor," *Parallel Computing 1,* 207–222.

M.S. Paterson and L.J. Stockmeyer (1973). "On the Number of Nonscalar Multiplications Necessary to Evaluate Polynomials," *SIAM J. Comp. 2,* 60–66.

K. Pearson (1901). "On Lines and Planes of Closest Fit to Points in Space," *Phil. Mag. 2,* 559–72.

G. Peters and J.H. Wilkinson (1969). "Eigenvalue of $Ax = \lambda Bx$ with Band Symmetric $A$ and $B$," *Comp. J. 12,* 398–404.

G. Peters and J.H. Wilkinson (1970a). "$Ax = \lambda Bx$ and the Generalized Eigenproblem," *SIAM J. Num. Anal. 7,* 479–92.

G. Peters and J.H. Wilkinson (1970b). "The Least Squares Problem and Pseudo-Inverses," *Comp. J. 13,* 309–16.

G. Peters and J.H. Wilkinson (1971). "The Calculation of Specified Eigenvectors by Inverse Iteration," in *HACLA*, pp. 418–39.

G. Peters and J.H. Wilkinson (1979). "Inverse Iteration, Ill-Conditoned Equations, and Newton's Method," *SIAM Review 21,* 339–60.

J.L. Phillips (1971). "The Triangular Decomposition of Hankel Matrices," *Math. Comp. 25,* 599–602.

R.J. Plemmons (1974). "Linear Least Squares by Elimination and MGS," *J. ACM 21,* 581–85.

R.J. Plemmons (1986). "A Parallel Block Iterative Scheme Applied to Computations in Structural Analysis," *SIAM J. Alg. and Disc. Methods 7,* 337–347.

E.L. Poole and J.M. Ortega (1987). "Multicolor ICCG Methods for Vector Computers," *SIAM J. Numer. Anal. 24,* 1394–1418.

D.A. Pope and C. Tompkins (1957). "Maximizing Functions of Rotations: Experiments Concerning Speed of Diagonalization of Symmetric Matrices Using Jacobi's Method," *J. ACM 4,* 459–66.

A. Pothen, S. Jha, and U. Vemapulati (1987). "Orthogonal Factoprization on a Distributed Memory Multiprocessor," in *Hypercube Multiprocessors,* ed. M.T. Heath, SIAM Press, 1987.

M.J.D. Powell and J.K. Reid (1968). "On Applying Householder's Method to Linear Least Squares Problems," *Proc. IFIP Congress,* pp. 122–26.

J.D. Pryce (1984). "A New Measure of Relative Error for Vectors," *SIAM J. Numer. Anal. 21,* 202–21.

J.D. Pryce (1985). "Multiplicative Error Analysis of Matrix Transformation Algorithms," *IMA J. Numer. Anal. 5,* 437–445.

W. Rath (1982). "Fast Givens Rotations for Orthogonal Similarity," *Numer. Math.* 40, 47–56.

J.K. Reid (1967). "A Note on the Least Squares Solution of a Band System of Linear Equations by Householder Reductions," *Comp. J. 10,* 188–89.

J.K. Reid (1971a). "A Note on the Stability of Gaussian Elimination," *J. Inst. Math. Applic. 8,* 374–75.

J.K. Reid (1971b). "On the Method of Conjugate Gradients for the Solution of Large Sparse Linear Equations," in *Large Sparse Sets of Linear Equations,* ed. J.K. Reid, Academic Press, New York, pp. 231–54.

J.K. Reid (1972). "The Ue of Conjugate Gradients for Systems of Linear Equations Possessing Property A," *SIAM J. Num. Anal. 9,* 325–32.

C. Reinsch and F.L. Bauer (1968). "Rational QR Transformations with Newton's Shift for Symmetric Tridiagonal Matrices," *Numer. Math. 11,* 264–72. See also *HACLA*, pp. 257–65.

J.R. Rice (1966a). "Experiments on Gram-Schmidt Orthogonalization," *Math. Comp. 20,* 325–28.

J. R. Rice (1966b). "A Theory of Condition," *SIAM J. Num. Anal. 3,* 287–310.

J.R. Rice (1981). *Matrix Computations and Mathematical Software*, Academic Press, New York.

R.F. Rinehart (1955). "The Equivalence of Definitions of a Matrix Function," *Amer. Math. Monthly 62,* 395–414.

J. Rissanen (1973). "Algorithms for Triangular Decomposition of block Hankel and Toeplitz Matrices with Application to Factoring Positive Matrix Polynomials," *Math. Comp. 27,* 147–54.

H.H. Robertson (1977). "The Accuracy of Error Estimates for Systems of Linear Algebraic Equations," *J. Inst. Math. Applic. 20,* 409–14.

G. Rodrigue (1973). "A Gradient Method for the Matrix Eigenvalue Problem $Ax = \lambda Bx$," *Numer. Math. 22,* 1–16.

G. Roderigue and D. Wolitzer (1984). "Preconditioning by Incomplete Block Cyclic Reduction," *Math. Comp. 42,* 549–566.

G. Roderigue (ed) (1982). *Parallel Computations,* Academic Press, New York.

D.J. Rose (1969). "An Algorithm for Solving a Special Class of Tridiagonal Systems of Linear Equations," *Comm. ACM 12,* 234–36.

D.J. Rose and R.A. Willoughby, eds. (1972). *Sparse Matrices and Their Applications,* Plenum Press, New York, NY.

G. Roderigue and D. Wolitzer (1984). "Preconditioning by Incomplete Block Cyclic Reduction," *Math. Comp. 42,* 549–566.

G. Roderigue (ed) (1982). *Parallel Computations,* Academic Press, New York.

C.H. Romine and J.M. Ortega (1988). "Parallel Solution of Triangular Systems of Equations," *Parallel Computing 6,* 109–114.

A. Ruhe (1967). "On the Quadratic Convergence of the Jacobi Method for Normal Matrices," *BIT 7,* 305–13.

A. Ruhe (1968). "On the Quadratic Convergence of a Generalization of the Jacobi Method to Arbitrary Matrices," *BIT 8,* 210–31.

A. Ruhe (1969). "The Norm of a Matrix after a Similarity Transformation," *BIT 9,* 53–58.

A. Ruhe (1969). "An Algorithm for Numerical Determination of the Structure of a General Matrix," *BIT 10,* 196–216.

A. Ruhe (1970b). "Perturbation Bounds for Means of Eigenvalues and Invariant Subspaces," *BIT 10,* 343–54.

A. Ruhe (1970c). "Properties of a Matrix with a Very Ill-Conditioned Eigenproblem," *Numer. Math. 15,* 57–60.

A. Ruhe (1974). "SOR Methods for the Eigenvalue Problem with Large Sparse Matrices," *Math. Comp. 28,* 695–710.

A. Ruhe (1975). "On the Closeness of Eigenvalues and Singular Values for Almost Normal Matrices," *Lin. Alg. and Its Applic. 11,* 87–94.

A. Ruhe (1978). "A Note on the Efficient Solution of Matrix Pencil Systems," *BIT 18,* 276–81.

A. Ruhe (1979). "Implementation Aspects of Band Lanczos Algorithms for Computation of Eigenvalues of Large Sparse Symmetric Matrices," *Math. Comp. 33,* 680–87.

A. Ruhe (1983). "Numerical Aspects of Gram-Schmidt Orthogonalization of Vectors," *Lin. Alg. and Its Applic. 52/53,* 591–602.

A. Ruhe (1987). "Closest Normal Matrix Found!," *BIT 27,* 585–598.

A. Ruhe and T. Wiberg (1972). "The Method of Conjugate Gradients Used in Inverse Iteration," *BIT 12,* 543–54.

H. Rutishauser (1958). "Solution of Eigenvalue Problems with the WR Transformation," *Nat. Bur. Stand. App. Math. Ser. 49,* 47–81.

H. Rutishauser (1966). "The Jacobi Method for Real Symmetric Matrices," *Numer. Math. 9,* 1-10. See also *HACLA,* pp. 202–11.

H. Rutishauser (1969). "Computation Aspects of F.L. Bauer's Simultaneous Iteration Method," *Numer. Math. 13,* 4–13.

H. Rutishauser (1970). "Simultaneous Iteration Method for Symmetric Matrices," *Numer. Math. 16,* 205–23. See also *HACLA,* pp. 284–302.

Y. Saad (1980). "On the Rates of Convergence of the Lanczos and the Block Lanczos Methods," *SIAM J. Num. Anal.17,* 687–706.

Y. Saad (1984). "Practical Use of Some Krylov Subspace Methods for Solving Indefinite and Nonsymmetric Linear Systems," *SIAM J. Sci. and Stat. Comp. 5,* 203–228.

Y. Saad (1987). "On the Lanczos Method for Solving Symmetric Systems with Several Right Hand Sides," *Math. Comp. 48,* 651–662.

Y. Saad (1981). "Krylov Subspace Methods for Solving Large Unsymmetric Linear Systems," *Math. Comp. 37,* 105–126.

Y. Saad (1982). "The Lanczos Biorthogonalization Algorithm and Other Oblique Projection Methods for Solving Large Unsymmetric Systems," *SIAM J. Numer. Anal. 19,* 485–506.

Y. Saad (1986). "On the Condition Number of Some Gram Matrices Arising from Least Squares Approximation in the Complex Plane," *Numer. Math. 48,* 337–348.

Y. Saad and M.H. Schultz (1985). "Data Communication in Hypercubes," Report YALEU/DCS/RR-428, Dept. of Computer Science, Yale University, New Haven, CT.

Y. Saad and M.H. Schultz (1985). "Topological Properties of Hypercubes," Report YALEU/DCS/RR-389, Dept. of Computer Science, Yale University, New Haven, CT.

Y. Saad and M. Schultz (1986). "GMRES: A Generalized Minimal Residual Algorithm for Solving Nonsymmetric Linear Systems," *SIAM J. Scientific and Stat. Comp. 7,* 856–869.

A. Sameh (1971). "On Jacobi and Jacobi-Like Algorithsm for a Parallel Computer," *Math. Comp. 25,* 579–90.

A. Sameh and D. Kuck (1978). "On Stable Parallel Linear System Solvers," *J. ACM 25,* 81–91.

A. Sameh, J. Lermit, and K. Noh (1975). "On the Intermediate Eigenvalues of Symmetric Sparse Matrices," *BIT 12,* 543–54.

K. Schittkowski and J. Stoer (1979). "A Factorization Method for the Solution of Constrained Linear Least Squares Problems Allowing for Subsequent Data Changes," *Numer. Math. 31,* 431–63.

P. Schoenemann (1966). "A Generalized Solution of the Orthogonal Procrustes Problem," *Psychometrika 31,* 1–10.

A. Schonage (1964). "On the Quadratic Convergence of the Jacobi Process," *Numer. Math. 6,* 410–12.

A. Schonage (1979). "Arbitrary Perturbations of Hermitian Matrices," *Lin. Alg. and Its Applic. 24,* 143–49.

W. Schönauer (1987). *Scientific Computing on Vector Computers,* North Holland, Amsterdam.

R. Schreiber (1986). "Solving Eigenvalue and Singular Value Problems on an Undersized Systolic Array," *SIAM J. Sci. and Stat. Comp. 7,* 441–451.

R. Schreiber and W.P. Tang (1986). "On Systolic Arrays for Updating the Cholesky Factorization," *BIT 26*, 451–466.

R. Schreiber and B.N. Parlett (1987). "Block Reflectors: Theory and Computation," *SIAM J. Numer. Anal. 25,* 189-205.

R. Schreiber and C. Van Loan (1989). "A Storage Efficient WY Representation for Products of Householder Transformations," *SIAM J. Sci. and Stat. Comp.*, to appear.

I. Schur (1909). "On the Characteristic Roots of a Linear Substitution with an Application to the Theory of Integral Equations," *Math. Ann. 66,* 488–510 (German).

H.R. Schwartz (1968). "Tridiagonalization of a Symmetric Band Matrix," *Numer. Math. 12,* 231–41. See also *HACLA*, pp. 273–83.

H.R. Schwartz (1974). "The Method of Coordinate Relaxation for $(A - \lambda B)x = 0$," *Numer. Math. 23,* 135–52.

D.S. Scott (1978). "Analysis of the Symmetric Lanczos Process," UCB-ERL Technical Report M78/40, University of California, Berkeley, CA.

D.S. Scott (1979a). "Block Lanczos Software for Symmetric Eigenvalue Problems," Report ORNL/CSD-48, Oak Ridge National Laboratory, Union Carbide Corporation, Oak Ridge, TN.

D.S. Scott (1979b). "How to Make the Lanczos Algorithm Converge Slowly," *Math. Comp. 33,* 239–47.

D.S. Scott (1984). "Computing a Few Eigenvalues and Eigenvectors of a Symmetric Band Matrix," *SIAM J. Sci. and Stat. Comp. 5,* 658–666.

D.S. Scott (1985). "On the Accuracy of the Gershgorin Circle Theorem for Bounding the Spread of a Real Symmetric Matrix," *Lin. Alg. and Its Applic. 65,* 147–155

D.S. Scott, M.T. Heath, and R.C. Ward (1986). "Parallel Block Jacobi Eigenvalue Algorithms Using Systolic Arrays," *Lin. Alg. and Its Applic. 77,* 345–356.

M.K. Seager (1986). "Parallelizing Conjugate Gradient for the Cray X-MP," *Parallel Computing 3,* 35–47.

J.J. Seaton (1969). "Diagonalization of Complex Symmetric Matrices Using a Modified Jacobi Method," *Comp. J. 12,* 156–57.

S. Serbin (1980). "On Factoring a Class of Complex Symmetric Matrices Without Pivoting," *Math. Comp. 35,* 1231–1234.

S. Serbin and S. Blalock (1979). "An Algorithm for Computing the Matrix Cosine," *SIAM J. Sci. and Stat. Comp. 1,* 198–204.

J.W. Sheldon (1955). "On the Numerical Solution of Elliptic Difference Equations," *Math. Tables Aids Comp. 9,* 101–12.

G. Shroff and R. Schreiber (1987). "Convergence of Block Jacobi Methods," Report 87–25, Comp. Sci. Dept., RPI, Troy , NY.

H. Simon (1984). "Analysis of the Symmetric Lanczos Algorithm with Reorthogonalization Methods," *Lin. Alg. and Its Applic. 61,* 101–132.

B. Singer and S. Spilerman (1976). "The Representation of Social Processes by Markov Models," *Amer. J. Sociology 82,* 1–54.

R.D. Skeel (1979). "Scaling for Numerical Stability in Gaussian Elimination," *J. ACM 26,* 494–526.

R.D. Skeel (1980). "Iterative Refinement Implies Numerical Stability for Gaussian Elimination," *Math. Comp. 35,* 817–832.

R.D. Skeel (1981). "Effect of Equilibration on Residual Size for Partial Pivoting," *SIAM J. Num. Anal. 18,* 449–55.

B.T. Smith, J.M. Boyle, Y. Ikebe, V.C. Klema, and C.B. Moler (1970). *Matrix Eigensystem Routines: EISPACK Guide*, 2nd ed., Springer-Verlag, New York, NY.

R.A. Smith (1967). "The Condition Numbers of the Matrix Eigenvalue Problem," *Numer. Math. 10,* 232–40.

F. Smithies (1970). *Integral Equations*, Cambridge University Press, Cambridge, England.

D. Sorensen (1985). "Analysis of Pairwise Pivoting in Gaussian Elimination," *IEEE Trans. on Computers C-34,* 274–278.

D. Stevenson (1981). "A Proposed Standard for Binary Floating Point Arithmetic," *Computer 14 (March),* 51–62.

G.W. Stewart (1969). "Accelerating the Orthogonal Iteration for the Eigenvalue sof a Hermitian Matrix," *Numer. Math. 13,* 362–6.

G.W. Stewart (1970). "Incorporating Origin Shifts into the QR Algorithm for Symmetric Tridiagonal Matrices," *Comm. ACM 13,* 365–67.

G.W. Stewart (1971). "Error Bounds for Approximate Invariant Subspaces of Closed Linear Operators," *SIAM J. Num. Anal. 8,* 796–808.

G.W. Stewart (1972). "On the Sensitivity of the Eigenvalue Problem $Ax = \lambda Bx$," *SIAM J. Num. Anal. 9,* 669–86.

G.W. Stewart (1973a). "Conjugate Direction Methods for Solving Systems of Linear Equations," *Numer. Math.- 21,* 284–97.

G.W. Stewart (1973b). "Error and Perturbation Bounds for Subspaces Associated with Certain Eigenvalue Problems," *SIAM Review 15,* 727–64.

G.W. Stewart (1973c). *Introduction to Matrix Computations.* Academic Press, New York, NY.

G.W. Stewart (1974). "The Numerical Treatment of Large Eigenvalue Problems," *Proc. IFIP Congress 74*, North-Holland, pp. 666–72.

G.W. Stewart (1975a). "The Convergenc eof the Method of Conjugate Gradients at Isolated Extreme Points in the Spectrum," *Numer. Math. 24,* 85–93.

G.W. Stewart (1975b). "Gershgorin Theory for the Generalized Eigenvalue Problem $Ax = \lambda Bx$," *Math. Comp. 29,* 600–606.

G.W. Stewart (1975c). "Methods of Simultaneous Iteration for Calculating Eigenvectors of Matrices," in *Topics in Numerical Analysis II*, ed. J.H. Miller, Academic Press, New York, pp. 185–96.

G.W. Stewart (1976a). "Algorithm 406: HQR3 and EXCHNG: FORTRAN Subroutines for Calculating and Ordering and Eigenvalues of a Real Upper Hessenberg Matrix," *ACM Trans. Math. Soft. 2,* 275–80.

G.W. Stewart (1976b). "A Bibliographical Tour of the Large Sparse Generalized Eigenvalue Problem," in *Sparse Matrix Computations,* ed. J.R. Bunch and D.J. Rose, Academic Press, New York, NY.

G.W. Stewart (1976c). "The Economical Storage of Plane Rotations," *Numer. Math. 25,* 137–38.

G.W. Stewart (1976d). "Simultaneous Iteration for Computing Invariant Subspaces of Non-Hermitian Matrices," *Numer. Math. 25,* 12–36.

G.W. Stewart (1977a). "On the Perturbation of Pseudo-Inverses, Projections, and Linear Least Squares Problems," *SIAM Review 19,* 634–62.

G.W. Stewart (1977b). "Perturbation Bounds for the QR Factorization of a Matrix," *SIAM J. Num. Anal. 14,* 509–18.

G.W. Stewart (1977c). "Sensitivity Coefficients for the Effects of Errors in the Independent Variables in a Linear Regression," Technical Report TR-571, Department of Computer Science, University of Maryland, College Park, MD.

G.W. Stewart (1978). "Perturbation Theory for the Generalized Eigenvalue Problem," in *Recent Advances in Numerical Analysis*, ed. C. de Boor and G.H. Golub, Academic Press, New York, NY.

G.W. Stewart (1979a). "A Note on the Perturbation of Singular Values," *Lin. Alg. and Its Applic. 28,* 213–16.

G.W. Stewart (1979b). "Perturbation Bounds for the Definite Generalized Eigenvalue Problem," *Lin. Alg. and Its Applic. 23,* 69–86.

G.W. Stewart (1979c). "The Effects of Rounding Error on an Algorithm for Downdating a Cholesky Factorization," *J. Inst. Math. Applic. 23,* 203–13.

G.W. Stewart (1980). "The Efficient Generation of Random Orthogonal Matrices with an Application to Condition Estimators," *SIAM J. Num. Anal. 17,* 403–9.

G.W. Stewart (1981). "On the Implicit Deflation of Nearly Singular Systems of Linear Equations," *SIAM J. Sci. and Stat. Comp. 2,* 136–140.

G.W. Stewart (1983). "A Method for Computing the Generalized Singular Value Decomposition," in *Matrix Pencils*, ed. B. Kågström and A. Ruhe, Springer-Verlag, New York, pp. 207–20.

G.W. Stewart (1984a). "Rank Degeneracy," *SIAM J. Sci. and Stat. Comp. 5,* 403–413.

G.W. Stewart (1984b). "On the Asymptotic Behavior of Scaled Singular Value and QR Decompositions," *Math. Comp. 43,* 483–490.

G.W. Stewart (1984c). "A Second Order Perturbation Expansion for Small Singular Values," *Lin. Alg. and Its Applic. 56,* 231–236.

G.W. Stewart (1984d). "On the Invariance of Perturbed Null Vectors Under Column Scaling," *Numer. Math. 33,34* 61–66.

G.W. Stewart (1985). "A Jacobi-Like Algorithm for Computing the Schur Decomposition of a Nonhermitian Matrix," *SIAM J. Sci. and Stat. Comp. 6,* 853–862.

G.W. Stewart (1987). "Collinearity and Least Squares Regression," *Statistical Science 2,* 68–100.

H.S. Stone (1973). "An Efficient Parallel Algorithm for the Solution of a Tridiagonal Linear System of Equations," *J. ACM 20,* 27–38.

H.S. Stone (1975). "Parallel Tridiagonal Equation Solvers," *ACM Trans. Math. Soft. 1,* 289–307.

G. Strang (1988). *Linear Algebra and Its Applications*, 3rd ed., Academic Press, New York.

H. Stone (1975). "Parallel Tridiagonal Equation Solvers," *ACM Trans. Math. Soft. 1,* 289–307.

V. Strassen (1969). "Gaussian Elimination is Not Optimal," *Numer. Math. 13,* 354–356.

J. Guang Sun (1982). "A Note on Stewart's Theorem for Definite Matrix Pairs," *Lin. Alg. and Its Applic. 48,* 331–339.

J. Guang Sun (1983). "Perturbation Analysis for the Generalized Singular Value Problem," *SIAM J. Numer. Anal. 20,* 611–625.

P.N. Swarztrauber (1979). "A Parallel Algorithm for Solving General Tridiagonal Equations," *Math. Comp. 33,* 185–199.

P.N. Swarztrauber and R.A. Sweet (1973). "The Direct Solution of the Discrete Poisson Equation on a Disk," *SIAM J. Num. Anal. 10,* 900–907.

R.A. Sweet (1974). "A Generalized Cyclic Reduction Algorithm," *SIAM J. Num. Anal. 11,* 506–20.

R.A. Sweet (1977). "A Cyclic Reduction Algorithm for Solving Block Tridiagonal Systems of Arbitrary Dimension," *SIAM J. Num. Anal. 14,* 706–20.

H.J. Symm and J.H. Wilkinson (1980). "Realistic Error Bounds for a Simple Eigenvalue and Its Associated Eigenvector," *Numer. Math. 35,* 113–26.

W.P. Tang and G.H. Golub (1981). "The Block Decomposition of a Vandermonde Matrix and Its Applications," *BIT 21,* 505–517.

G.L. Thompson and R.L. Weil (1970). "Reducing the Rank of $A - \lambda B$," *Proc. Amer. Math. Soc. 26,* 548-54.

G.L. Thompson and R.L. Weil (1972). "Roots of Matrix Pencils $Ay = \lambda By$: Existence, Calculations, and Relations to Game Theory," *Lin. Alg. and Its Applic. 5,* 207–26.

L.N. Trefethen and R.S. Schreiber (1987). "Average Case Stability of Gaussian Elimination," Numer. Anal. Report 88-3, Department of Mathematics, MIT.

W.F. Trench (1964). "An Algorithm for the Inversion of Finite Toeplitz Matrices," *J. SIAM 12,* 515–22.

W.F. Trench (1974). "Inversion of Toeplitz Band Matrices," *Math. Comp. 28,* 1089–95.

N.K. Tsao (1975). "A Note on Implementing the Householder Transformation," *SIAM J. Num. Anal. 12,* 53–58.

H.W. Turnbull and A.C. Aitken (1961). *An Introduction to the Theory of Canonical Matrices,* Dover, NY.

F. Uhlig (1973). "Simultaneous Block Diagonalization of Two Real Symmetric Matrices," *Lin. Alg. and Its Applic. 7,* 281–89.

F. Uhlig (1976). "A Canonical Form for a Pair of Real Symmetric Matrices that Generate a Nonsingular Pencil," *Lin. Alg. and Its Applic. 14,* 189–210.

R. Underwood (1975). "An Iterative Block Lanczos Method for the Solution of Large Sparse Symmetric Eigenproblems," Report STAN-CS-75-496, Department of Computer Science, Stanford University, Stanford, CA.

J. Vandergraft (1971). "Generalized Rayleigh Methods with Applications to Finding Eigenvalues of Large Matrices," *Lin. Alg. and Its Applic. 4,* 353–68.

A. van der Sluis (1969). "Condition Numbers and Equilibration Matrices," *Numer. Math. 14,* 14–23.

A. van der Sluis (1970). "Condition, Equilibration, and Pivoting in Linear Algebraic Systems," *Numer. Math. 15,* 74–86.

A. van der Sluis (1975a). "Perturbations of Eigenvalues of Nonnormal Matrices," *Comm. ACM 18,* 30–36.

A. van der Sluis (1975b). "Stability of the Solutions of Linear Least Squares Problem," *Numer. Math. 23,* 241–54.

A. van der Sluis and G.W. Veltkamp (1979). "Restoring Rank and Consistency by Orthogonal Projection," *Lin. Alg. and Its Applic. 28,* 257–78.

A. van der Sluis and H.A. Van Der Vorst (1986). "The Rate of Convergence of Conjugate Gradients," *Numer. Math. 48,* 543–560.

H. Van de Vel (1977). "Numerical Treatment of a Generalized Vandermonde System of Equations," *Lin. Alg. and Its Applic. 17,* 149–74.

H.A. Van der Vorst (1982). "A Vectorizable Variant of Some ICCG Methods," *SIAM J. Sci. and Stat. Comp. 3,* 350–356.

H.A. Van der Vorst (1982). "A Generalized Lanczos Scheme," *Math. Comp. 39,* 559–562.

P. Van Dooren (1979). "The Computation of Kronecker's Canonical Form of a Singular Pencil," *Lin. Alg. and Its Applic. 27,* 103–40.

P. Van Dooren (1981). "A Generalized Eigenvalue Approach for Solving Riccati Equations," *SIAM J. Sci. and Stat. Comp. 2,* 121–135.

P. Van Dooren (1981). "The Generalized Eigenstructure Problem in Linear System Theory," *IEEE Trans. Auto. Cont. AC-26,* 111-128.

P. Van Dooren (1983). "Reducing Subspaces: definitions, properties and Algorithms," in *Matrix Pencils,* eds. B. Kågström and A. Ruhe, Springer-Verlag, New York, 1983, 58-73.

P. Van Dooren and C.C. Paige (1986). "On the Quadratic Convergence of Kogbetliantz's Algorithm for Computing the Singular Value Decomposition," *Lin. Alg. and Its Applic. 77,* 301–313.

S. Van Huffel (1987). "Analysis of the Total Least Squares Problem and Its Use in Parameter Estimation," Doctoral Thesis, Department of Electrical Engineering, K.U. Leuven.

S. Van Huffel (1988). "Comments on the Solution of the Nongeneric Total Least Squares Problem," Report ESAT-KUL-88/3, Department of Electrical Engineering, K.U. Leuven.

S. Van Huffel and J. Vandewalle (1988). "The Partial Total Least Squares Algorithm," *J. Comp. and App. Math. 21,* 333–342.

S. Van Huffel and J. Vandewalle (1987). "Subset Selection Using the Total Least Squares Approach in Collinearity Problems with Errors in the Variables," *Lin. Alg. and Its Applic. 88/89,* 695–714.

S. Van Huffel, J. Vandewalle, and A. Haegemans (1987). "An Efficient and Reliable Algorithm for Computing the Singular Subspace of a Matrix Associated with its Smallest Singular Values," *J. Comp. and Appl. Math. 19,* 313–330.

S. Van Huffel and J. Vandewalle (1988). "The Partial Total Least Squares Algorithm," *J. Comp. and App. Math. 21,* 333–342.

J.M. Van Kats and H.A. Van der Vorst (1977). "Automatic Monitoring of Lanczos Schemes for Symmetric or Skew-Symmetric Generalized Eigenvalue Problems," Report TR 7, Academisch Computer Centru, Utrecht, The Netherlands.

H.P.M. van Kempen (1966). "On Quadratic Convergence of the Special Cyclic Jacobi Method," *Numer. Math. 9,* 19–22.

C.F. Van Loan (1973). "Generalized Singular Values with Algorithms and Applications," Ph.D. thesis, University of Michigan, Ann Arbor, MI.

C.F. Van Loan (1975a). "A General Matrix Eigenvalue Algorithm," *SIAM J. Num. Anal. 12,* 819–34.

C.F. Van Loan (1975b). "A Study of the Matrix Exponential," Numerical Analysis Report no. 10, University of Manchester, England.

C.F. Van Loan (1976). "Generalizing the Singular Value Decomposition," *SIAM J. Num. Anal. 13,* 76–83.

C.F. Van Loan (1977a). "On the Limitation and Application of the Pade Approximation to the Matrix Exponential" in *Pade and Rational Approximation,* ed. E.B. Saff and R.S. Varga, Academic Press, New York, NY.

C.F. Van Loan (1977b). "The Sensitivity of the Matrix Exponential," *SIAM J. Num. Anal. 14,* 971–81.

C.F. Van Loan (1978a). "Computing Integrals Involving the Matrix Exponential," *IEEE Trans. Auto. Cont. AC-23,* 395–404.

C.F. Van Loan (1978b). "A Note on the Evaluation of Matrix Polynomials," *IEEE Trans. Auto. Cont. AC-24,* 320–21.

C.F. Van Loan (1982a). "A Generalized SVD Analysis of Some Weighting Methods for Equality-Constrained Least Squares," in *Proceedings of the Conference on Matrix Pencils,* ed. B. Kagstrom and A. Ruhe, Springer-Verlag, New York, NY.

C.F. Van Loan (1982b). "Using the Hessenberg Decomposition in Control Theory," in *Algorithms and Theory in Filtering and Control,* ed. D.C. Sorensen and R.J. Wets, Mathematical Programming Study no. 18, North Holland, Amsterdam, pp. 103.

C.F. Van Loan (1984). "A Symplectic Method for Approximating All the Eigenvalues of a Hamiltonian Matrix," *Lin. Alg. and Its Applic. 61,* 233–252.

C.F. Van Loan (1985). "Computing the CS and Generalized Singular Value Decomposition," *Numer. Math. 46,* 479–492.

C.F. Van Loan (1985). "On the Method of Weighting for Equality Constrained Least Squares Problems," *SIAM J. Numer. Anal. 22,* 851–864.

C.F. Van Loan (1985). "How Near is a Stable Matrix to an Unstable Matrix?," Contemporary Mathematics, Vol. 47.,465–477.

C.F. Van Loan (1987). "On Estimating the Condition of Eigenvalues and Eigenvectors," *Lin. Alg. and Its Applic.* 88/89, 715–732.

J.M. Varah (1968a). "The Calculation of the Eigenvectors of a General Complex Matrix by Inverse Iteration," *Math. Comp. 22,* 785–91.

J.M. Varah (1968b). "Rigorous Machine Bounds for the Eigensystem of a General Complex Matrix," *Math. Comp. 22,* 793–801.

J.M. Varah (1970). "Computing Invariant Subspaces of a General Matrix when the Eigensystem is Poorly Determined," *Math. Comp. 24,* 137–49.

J.M. Varah (1979). "On the Separation of Two Matrices," *SIAM J. Num. Anal. 16,* 216–22.

J.M. Varah (1972). "On the Solution of Block-Tridiagonal Systems Arising from Certain Finite-Difference Equations," *Math. Comp. 26,* 859–68.

J.M. Varah (1973). "On the Numerical Solution of Ill-Conditioned Linear Systems with Applications to Ill-Posed Problems," *SIAM J. Num. Anal. 10,* 257–67.

J.M. Varah (1975). "A Lower Bound for the Smallest Singular Value of a Matrix," *Lin. Alg. and Its Applic. 11,* 1–2.

R.S. Varga (1961). "On Higher-Order Stable Implicit Methods for Solving Parabolic Partial Differential Equations," *J. Math. Phys. 40,* 220–31.

R.S. Varga (1962). *Matrix Iterative Analysis*, Prentice-Hall, Englewood Cliffs, NJ.

R.S. Varga (1970). "Minimal Gershgorin Sets for Partitioned Matrices," *SIAM J. Num. Anal. 7,* 493–507.

R.S. Varga (1976). "On Diagonal Dominance Arguments for Bounding $\|A^{-1}\|$," *Lin. Alg. and Its Applic. 14,* 211–17.

W.J. Vetter (1975). "Vector Structures and Solutions of Linear Matrix Equations," *Lin. Alg. and Its Applic. 10,* 181–88.

E.L. Wachpress (1966). *Iterative Solution of Elliptic Systems*, Prentice-Hall, Englewood Cliffs, NJ.

D.W. Walker, T. Aldcroft, A. Cisneros, G. Fox, and W. Furmanski (1988). "LU Decomposition of Banded Matrices and the Solution of Linear Systems on Hypercubes," in G. Fox (ed.) (1988), 1635–1655.

H.F. Walker (1988). "Implementation of the GMRES Method Using Householder Transformations," *SIAM J. Sci. Stat. Comp. 9,* 152–163.

R.C. Ward (1975). "The Combination Shift QZ Algorithm," *SIAM J. Num. Anal. 12,* 835–53.

R.C. Ward (1977). "Numerical Computation of the Matrix Exponential with Accuracy Estimate," *SIAM J. Num. Anal. 14,* 600–614.

R.C. Ward (1981). "Balancing the Generalized Eigenvalue Problem," *SIAM J. Sci and Stat. Comp. 2,* 141–152.

R.C. Ward and L.J. Gray (1978). "Eigensystem Computation for Skew-Symmetric and a Class of Symmetric Matrices," *ACM Trans. Math. Soft. 4,* 278–85.

D.S. Watkins (1982). "Understanding the QR Algorithm," *SIAM Review 24,* 427–440.

G.A. Watson (1973). "An Algorithm for the Inversion of Block Matrices of Toeplitz Form," *J. ACM 20,* 409–15.

G.A. Watson (1988). "The Smallest Perturbation of a Submatrix which Lowers the Rank of the Matrix," *IMA J. Numer. Anal. 8,* 295–304.

P.Å. Wedin (1972). "Perturbation Bounds in Connection with the Singular Value Decomposition," *BIT 12,* 99–111.

P.Å. Wedin (1973a). "On the Almost Rank-Deficient Case of the Least Squares Problem," *BIT 13,* 344–54.

P.Å. Wedin (1973b). "Perturbation Theory for Pseudo-Inverses," *BIT 13,* 217–32.

O. Widlund (1978). "A Lanczos Method for a Class of Nonsymmetric Systems of Linear Equarions," *SIAM J. Numer. Anal. 15,* 801–12.

J.H. Wilkinson (1961). "Error Analysis of Direct Methods of Matrix Inversion," *J. ACM 10,* 281–330.

J.H. Wilkinson (1963). *Rounding Errors in Algebraic Processes*, Prentice-Hall, Englewood Cliffs, NJ.

J.H. Wilkinson (1965a). *The Algebraic Eigenvalue Problem*, Claredon Press, Oxford, England.

J.H. Wilkinson (1965b). "Convergence of the LR, QR, and Related Algorithms," *Comp. J. 8*, 77–84.

J.H. Wilkinson (1968a). "Almost Diagonal Matrices with Multiple or Close Eigenvalues," *Lin. Alg. and Its Applic.* 1, 1–12.

J.H.Wilkinson (1968b). "Global Convergence of Tridiagonal QR Algorithm with Origin Shifts," *Lin. Alg. and Its Applic. 1*, 409–20.

J.H. Wilkinson (1968c). "A Priori Error Analysis of Algebraic Processes," *Proc. International Congress Math.* (Moscow: Izdat. Mir, 1968), pp. 629–39.

J.H. Wilkinson (1971). "Modern Error Analysis," *SIAM Review 14*, 548–68.

J.H.Wilkinson (1972). "Note on Matrices with a Very Ill-Conditioned Eigenproblem," *Numer. Math.* 19, 176–78.

J.H. Wilkinson (1977). "Some Recent Advances in Numerical Linear Algebra," in *The State of the Art in Numerical Analysis*, ed. D.A. H. Jacobs, Academic Press, New York, pp. 1–53.

J.H. Wilkinson (1978). "Linear Differential Equations and Kronecker's Canonical Form," in *Recent Advances in Numerical Analysis*, ed. C. deBoor and G.H. Golub, Academic Press, New York, pp. 231–65.

J.H. Wilkinson (1979). "Kronecker's Canonical Form and the QZ Algorithm," *Lin. Alg. and Its Applic. 28*, 285–303.

J.H. Wilkinson and C. Reinsch, eds. (1971). *Handbook for Automatic Computation.* Vol. 2, Linear Algebra, Springer-Verlag, New York. (Here abbreviated as *HACLA*).

J.H. Wilkinson (1984). "On Neighboring Matrices with Quadratic Elementary Divisors," *Numer. Math.* 44, 1–21.

H. Wimmer and A.D. Ziebur (1972). "Solving the Matrix Equation $\sum f_p(A)Xg_p(A)$," *SIAM Review 14*, 318–23.

S. Winograd (1968). "A New Algorithm for Inner Product," *IEEE Trans. Comp. C-17*, 693–694.

H. Wozniakowski (1980). "Roundoff Error Analysis of a New Class of Conjugate Gradient Algorithms," *Lin. Alg. and Its Applic. 29*, 507–29.

A. Wragg (1973). "Computation of the Exponential of a Matrix I: Theoretical Considerations," *J. Inst. Math. Applic. 11*, 369–75.

A. Wragg (1975). "Computation of the Exponential of a Matrix II: Practical Considerations," *J. Inst. Math. Applic. 15*, 273–78.

J.M. Yohe (1979). "Software for Interval Arithmetic: A Reasonable Portable Package," *ACM Trans. Math. Soft. 5*, 50–63.

D.M. Young (1970). "Convergence Properties of the Symmetric and Unsymmetric Over-Relaxation Mesthods," *Math. Comp. 24*, 793–807.

D.M. Young (1971). *Iterative Solution of Large Linear Systems*, Academic Press, New York.

D.M. Young (1972). "Generalization of Property *A* and Consistent Ordering," *SIAM J. Num. Anal. 9*, 454–63.

D.M. Young and K.C. Jea (1980) "Generalized Conjugate Gradient Acceleration of Non-symmetrizable Iterative Methods," *Lin. Alg. and Its Applic. 34*, 159–94.

S. Zohar (1969)."Toeplitz Matrix Inversion: The Algorithm of W.F. Trench," *J. ACM 16*, 592–601.

# Index